NASA SP-289

APOLLO 15

Preliminary Science Report

PREPARED BY
NASA MANNED SPACECRAFT CENTER

Scientific and Technical Information Office   1972
NATIONAL AERONAUTICS AND SPACE ADMINISTRATION
Washington, D.C.

EDITORIAL BOARD

The material submitted for the "Apollo 15 Preliminary Science Report" was reviewed by a NASA Editorial Review Board consisting of the following members: Joseph P. Allen (Chairman), Keith F. Anderson, Richard R. Baldwin, Roy L. Cox, Helen N. Foley, Robert L. Giesecke, Richard H. Koos, Robert Mercer, William C. Phinney, Floyd I. Roberson, and Scott H. Simpkinson.

Foreword

In richness of scientific return, the Apollo 15 voyage to the plains at Hadley compares with voyages of Darwin's H.M.S. *Beagle*, and those of the *Endeavour* and *Resolution*. Just as those epic ocean voyages set the stage for a revolution in the biological sciences and exploration generally, so also the flight of *Falcon* and *Endeavor* did the same in planetary and Earth sciences and will guide the course of future explorations.

The boundary achievements of Apollo 15 cannot now be established. As an author of a following paper points out, the mission was not finished at splashdown in the Pacific, nor later with painstaking analysis in scores of laboratories of the samples and cores brought back, nor with careful study of the photographic imagery and instrument traces returned home. For the distinctive fact is that the mission is not yet over. Data still flow in daily from the isotope-powered station emplaced on the plain at Hadley, and from the Moon-encircling scientific satellite left in orbit. This data flow is of exceptional value because it now affords, for the first time, a triangulation of lunar events perceived by the three physically separated scientific stations that man has left on the Moon.

This volume is the first, though assuredly not the final, effort to assemble a comprehensive accounting of the scientific knowledge so far acquired through this remarkable mission.

                                Dr. James C. Fletcher
                                Administrator
                                National Aeronautics and Space Administration

December 8, 1971

Contents

	Page
INTRODUCTION *A.J. Calio*	xi
1. MISSION DESCRIPTION *Richard R. Baldwin*	1-1
2. SUMMARY OF SCIENTIFIC RESULTS *Joseph P. Allen*	2-1
3. PHOTOGRAPHIC SUMMARY *John W. Dietrich and Uel S. Clanton*	3-1
4. CREW OBSERVATIONS *David R. Scott, Alfred M. Worden, and James B. Irwin*	4-1
5. PRELIMINARY GEOLOGIC INVESTIGATION OF THE APOLLO 15 LANDING SITE *G.A. Swann, N.G. Bailey, R.M. Batson, V.L. Freeman, M.H. Hait, J.W. Head, H.E. Holt, K.A. Howard, J.B. Irwin, K.B. Larson, W.R. Muehlberger, V.S. Reed, J.J. Rennilson, G.G. Schaber, D.R. Scott, L.T. Silver, R.L. Sutton, G.E. Ulrich, H.G. Wilshire, and E.W. Wolfe*	5-1
6. PRELIMINARY EXAMINATION OF LUNAR SAMPLES *The Lunar Sample Preliminary Examination Team*	6-1
7. SOIL-MECHANICS EXPERIMENT *J.K. Mitchell, L.G. Bromwell, W.D. Carrier, III, N.C. Costes, W.N. Houston, and R.F. Scott*	7-1
8. PASSIVE SEISMIC EXPERIMENT *Gary V. Latham, Maurice Ewing, Frank Press, George Sutton, James Dorman, Yosio Nakamura, Nafi Toksoz, David Lammlein, and Fred Duennebier*	8-1
9. LUNAR-SURFACE MAGNETOMETER EXPERIMENT *P. Dyal, C.W. Parkin, and C.P. Sonett*	9-1
10. SOLAR-WIND SPECTROMETER EXPERIMENT *Douglas R. Clay, Bruce E. Goldstein, Marcia Neugebauer, and Conway W. Snyder*	10-1

11. HEAT-FLOW EXPERIMENT　　　　　　　　　　　　　　　　　　　　　　11-1
    Marcus G. Langseth, Jr., Sydney P. Clark, Jr., John L. Chute, Jr., Stephen J. Keihm, and Alfred E. Wechsler

12. SUPRATHERMAL ION DETECTOR EXPERIMENT (LUNAR-IONOSPHERE DETECTOR)　　　　　　　　　　　　　　　　　　　　　　　　　　　　12-1
    H. Kent Hills, Jurg C. Meister, Richard R. Vondrak, and John W. Freeman, Jr.

13. COLD CATHODE GAGE EXPERIMENT (LUNAR-ATMOSPHERE DETECTOR)　　　　　　　　　　　　　　　　　　　　　　　　　　　　　　13-1
    F.S. Johnson, D.E. Evans, and J.M. Carroll

14. LASER RANGING RETROREFLECTOR　　　　　　　　　　　　　　　　　14-1
    J.E. Faller, C.O. Alley, P.L. Bender, D.G. Currie, R.H. Dicke, W.M. Kaula, G.J.F. MacDonald, J.D. Mulholland, H.H. Plotkin, E.C. Silverberg, and D.T. Wilkinson

15. SOLAR-WIND COMPOSITION EXPERIMENT　　　　　　　　　　　　　　15-1
    J. Geiss, F. Buehler, H. Cerutti, and P. Eberhardt

16. GAMMA-RAY SPECTROMETER EXPERIMENT　　　　　　　　　　　　　16-1
    James R. Arnold, Laurence E. Peterson, Albert E. Metzger, and Jack I. Trombka

17. X-RAY FLUORESCENCE EXPERIMENT　　　　　　　　　　　　　　　　17-1
    I. Adler, J. Trombka, J. Gerard, R. Schmadebeck, P. Lowman, H. Blodgett, L. Yin, E. Eller, R. Lamothe, P. Gorenstein, P. Bjorkholm, B. Harris, and H. Gursky

18. ALPHA-PARTICLE SPECTROMETER EXPERIMENT　　　　　　　　　　　18-1
    Paul Gorenstein and P. Bjorkholm

19. LUNAR ORBITAL MASS SPECTROMETER EXPERIMENT　　　　　　　　19-1
    J.H. Hoffman, R.R. Hodges, and D.E. Evans

20. S-BAND TRANSPONDER EXPERIMENT　　　　　　　　　　　　　　　　20-1
    W.L. Sjogren, P. Gottlieb, P.M. Muller, and W.R. Wollenhaupt

21. SUBSATELLITE MEASUREMENTS OF PLASMAS AND SOLAR PARTI-PARTICLES　　　　　　　　　　　　　　　　　　　　　　　　　　　　　21-1
    K.A. Anderson, L.M. Chase, R.P. Lin, J.E. McCoy, and R.E. McGuire

22. THE PARTICLES AND FIELDS SUBSATELLITE MAGNETOMETER EXPERIMENT　　　　　　　　　　　　　　　　　　　　　　　　　　　22-1
    Paul J. Coleman, Jr., G. Schubert, C.T. Russell, and L.R. Sharp

23. BISTATIC-RADAR INVESTIGATION　　　　　　　　　　　　　　　　　　23-1
    H.T. Howard and G.L. Tyler

24. APOLLO WINDOW METEOROID EXPERIMENT　　　　　　　　　　　　24-1
    Burton G. Cour-Palais, Robert E. Flaherty, and Milton L. Brown

25. ORBITAL-SCIENCE PHOTOGRAPHY   25-1

    PART A. VISUAL OBSERVATIONS FROM LUNAR ORBIT   25-1
    Farouk El-Baz and Alfred M. Worden

    PART B. PHOTOGRAMMETRIC ANALYSIS OF APOLLO 15 RECORDS 25-27
    Frederick J. Doyle

    PART C. PHOTOGRAMMETRY OF APOLLO 15 PHOTOGRAPHY   25-36
    Sherman S.C. Wu, Francis J. Schafer, Raymond Jordan, Gary M. Nakata, and James L. Derick

    PART D. APOLLO 15 LASER ALTIMETER   25-48
    F.I. Roberson and W.M. Kaula

    PART E. SURFACE DISTURBANCES AT THE APOLLO 15 LANDING SITE   25-50
    N.W. Hinners and Farouk El-Baz

    PART F. REGIONAL GEOLOGY OF HADLEY RILLE   25-53
    Keith A. Howard and James W. Head

    PART G. LINEAMENTS THAT ARE ARTIFACTS OF LIGHTING   25-58
    Keith A. Howard and Bradley R. Larsen

    PART H. SKETCH MAP OF THE REGION AROUND CANDIDATE LITTROW APOLLO LANDING SITES   25-63
    M.H. Carr

    PART I. THE CINDER FIELD OF THE TAURUS MOUNTAINS   25-66
    Farouk El-Baz

    PART J. PRELIMINARY GEOLOGIC MAP OF THE REGION AROUND THE CANDIDATE PROCLUS APOLLO LANDING SITE   25-72
    Don E. Wilhelms

    PART K. GEOLOGIC SKETCH MAP OF THE CANDIDATE PROCLUS APOLLO LANDING SITE   25-76
    Baerbel Koesters Lucchitta

    PART L. SELECTED VOLCANIC FEATURES   25-81
    Mareta N. West

    PART M. MARE IMBRIUM LAVA FLOWS AND THEIR RELATIONSHIP TO COLOR BOUNDARIES   25-83
    Ewen A. Whitaker

    PART N. AN UNUSUAL MARE FEATURE   25-84
    Ewen A. Whitaker

    PART O. REGIONAL VARIATIONS IN THE MAGNITUDE OF HEILIGENSCHEIN AND CAUSAL CONNECTIONS   25-86
    Robert L. Wildey

PART P. THE PROCESS OF CRATER REMOVAL IN THE LUNAR
    MARIA   25-87
    L.A. Soderblom

PART Q. CRATER-SHADOWING EFFECTS AT LOW SUN ANGLES   25-92
    H.J. Moore

PART R. NEAR-TERMINATOR PHOTOGRAPHY   25-95
    J.W. Head and D.D. Lloyd

PART S. FIRST EARTHSHINE PHOTOGRAPHY FROM LUNAR ORBIT   25-101
    D.D. Lloyd and J.W. Head

PART T. ASTRONOMICAL PHOTOGRAPHY   25-108
    L. Dunkelman, R.D. Mercer, C.L. Ross, and A. Worden

APPENDIX A—Glossary   A-1

APPENDIX B—Acronyms   B-1

APPENDIX C—Units and Unit-Conversion Factors   C-1

APPENDIX D—Lunar-Surface Panoramic Views   D-1

Introduction

The Apollo 15 mission was the first of the Apollo missions to utilize the full capability of a complex set of spacecraft and launch vehicles, the design, development, and construction of which have occupied the major efforts of the U.S. space program for the last decade. The reliability and capability of the Apollo spacecraft, launch vehicles, and ancillary equipment such as space suits were extensively tested and demonstrated in the preceding Apollo missions. These missions also provided the necessary experience in orbital maneuvering and extravehicular activity that enabled the Apollo 15 crew and Mission Control Center personnel to undertake a mission that was defined almost entirely in terms of its exploratory and scientific objectives.

The scope of the Apollo 15 mission differed from that of previous missions in three distinct ways: (1) the command-service module carried a diverse set of experiments aimed at the study of the lunar surface from orbit, (2) the lunar module carried to the surface an electrically powered vehicle that extended the exploration range on the lunar surface by more than a factor of 5 over that of previous missions, and (3) the stay time on the lunar surface was extended to twice that of previous landings. The full utilization of this enhanced capability provided results that furnish many new insights into lunar history and structure. Perhaps most important of all, this mission provided results that give a meaningful overall picture of the Moon.

The scientific endeavors of the Apollo 15 mission can be divided into three distinct kinds of activities: (1) the orbital experiments, (2) the package of lunar-surface experiments, and (3) the surface sampling and observation.

The orbital experiments were aimed toward a determination and understanding of regional variations in the chemical composition of the lunar surface, further study of the gravitational field of the Moon, determination of the induced and permanent magnetic field of the Moon, and detailed study of the morphology and albedo of the lunar surface. These experiments were carried out by the use of X-ray and gamma-ray sensors deployed from the scientific instrument module bay of the command-service module, by high-resolution and metric cameras that photographed more than 12.5 percent of the surface of the Moon, and by a subsatellite with magnetic sensors and S-band transponders that will remain in lunar orbit for more than 1 yr. The preliminary results of these experiments summarized in this report indicate that the Moon has some remarkable characteristics. The highland and mare regions appear to be made up of rocks that have very different aluminum concentrations. The abundance of aluminum varies by more than a factor of 3 over very wide areas. Large areas of the highlands appear to be underlain by rocks that contain more than 25 percent, by weight, of aluminum oxide. The existence of aluminum-rich rocks—in particular, plagioclase-rich rocks—was, in fact, anticipated from the samples returned from previous missions. Their areal extent, however, was only a matter of conjecture.

Regional differences in the abundance of radioactive elements were also observed. The radioactivity of typical highland areas was much lower than had been inferred from nonmare samples that were returned from previous missions. The observed variations imply that there is a wide range of radioactive-element content in different highland materials and also suggest that most mare basalts have a higher radioactivity than average highland materials.

Local magnetic anomalies associated with craters were observed for the first time by the subsatellite magnetometer. The anomaly patterns, along with the direct-current magnetic fields and remanent magnetism of returned samples, suggest that the observed magnetic field of the Moon is a result of the magnetization of surface rocks. Preliminary data from detailed studies of three mascon basins show that these basins are, indeed, topographic lows several kilometers below the surrounding regions. The gravitational anomalies associated with these basins suggest that they are filled to a depth of more than 10 km by materials that exceed the density of the average crust by 0.3 g/cm^3. Examination of the wealth of photographic information returned from the lunar surface has only begun. The detail revealed in these photographs extends to features as small as the lunar module landing craft, which can be readily observed on the photographs of the Hadley region.

The Apollo lunar surface experiments package included the first attempt to determine the temperature gradient of the upper few meters of the lunar surface. This gradient and the measured conductivity of the lunar regolith provided the first estimate of the heat, or energy, flux coming from the lunar interior. Although difficulty was encountered in deploying this experiment, it was possible to measure this flux with remarkable accuracy. The measured value, if it is representative of that of the whole Moon, implies a radioactive element content for the whole Moon that is substantially higher than that usually accepted for the Earth.

The impact of the Saturn IVB stage was recorded by both the Apollo 12 and Apollo 14 seismometers. This seismic refraction experiment provides the first substantial evidence for very marked sound-velocity contrasts in the lunar subsurface. The velocity contrasts are so great that significant chemical-composition variations or phase-composition variations must be inferred in the upper 70 km of the Moon.

The samples and photographs returned from both the Apennine Front and the Hadley Rille region reveal an immense variety of materials and structure at this site. A core tube that penetrated more than 2-1/2 m into the surface was obtained from the mare region. Preliminary studies of this core tube reveal a remarkably detailed stratigraphic record. The length of time represented by this sample of the lunar regolith may be a significant fraction of the total history of the Moon, and the sample could easily provide data on variations in solar activity that extend hundreds of millions of years into the past.

Among the documented samples returned from this mission are rocks that are very rich in aluminum and plagioclase, which probably can be related to the aluminum-rich highland areas observed by the X-ray experiment and, thus, may provide a more detailed chemical picture of these regions. Other rocks, such as the large block that was ejected from Dune Crater and the shock-melted mare basalt (sample 15256), will provide unprecedented opportunities to determine the age of major postmare craters such as Aristillus and Autolycus.

The study of both the Hadley Rille wall and the samples from the rille and mare regions provides clear evidence that the Hadley plain is underlain by a series of lava flows similar to those found in terrestrial lava fields. Age determinations made on one sample from these flows suggest the very intriguing possibility that the time of the volcanic activity at this site coincides almost exactly with that found at the Apollo 12 site.

It must be emphasized that most of the observations and all of the conclusions mentioned here are very preliminary. Many of the experiments require long-term observations. Others require complex data-reduction procedures. In addition, the return

INTRODUCTION

of detailed data from the Moon by means of magnetic tapes from distant receiving stations often involves several weeks between the time when an event occurs and the time when it becomes known to a principal investigator. The description of samples given in this report is clearly preliminary. The detailed study of these samples is, in fact, just beginning as this report goes to press. Thus, even though this report is the most detailed and diverse Apollo science report compiled to date, it cannot do more than anticipate most of the scientific results and the understanding that we feel confident will ultimately become the hallmark of the Apollo 15 mission.

A. J. CALIO
NASA Manned Spacecraft Center

1. Mission Description

Richard R. Baldwin[a]

The successful Apollo 15 manned lunar-landing mission was the first in a series of three missions of this type planned for the Apollo Program. As compared with previous Apollo manned lunar-landing missions, these missions are characterized by increased hardware capability, a larger scientific payload, and a battery-powered lunar roving vehicle (Rover). Benefits resulting from these additions to Apollo 15 were a mission duration of 12-1/3 days, a lunar stay time of nearly 67 hr, a lunar-surface traverse distance of 27.9 km traveled at an average speed of 9.6 km/hr, and a scientific instrument module (SIM) containing equipment for orbital experiments and photographic tasks not performed on previous missions.

The primary scientific objectives of the mission were to perform selenological inspection, survey, and sampling of materials and surface features in a preselected area of the Hadley-Apennine region; to emplace and activate surface experiments; and to conduct inflight experiments and photographic tasks from lunar orbit. Apollo 15 scientific activities scheduled to satisfy these objectives involved collecting a lunar-surface contingency sample and conducting 11 lunar-surface experiments, 11 lunar-orbital experiments, service module orbital tasks, and command module photographic tasks. Specific experiments and photographic tasks were as follows.

1. Lunar-surface activities
    a. Emplaced experiments
        (1) Apollo lunar surface experiments package
            (a) Heat flow
            (b) Lunar-surface magnetometer
            (c) Passive seismometer
            (d) Cold cathode gage
            (e) Solar-wind spectrometer
            (f) Suprathermal ion detector
            (g) Lunar dust detector
        (2) Laser ranging retroreflector
        (3) Solar-wind composition
    b. Inspection, survey, and sampling
        (1) Contingency-sample collection
        (2) Lunar geological investigation
            (a) Geologic soil and rock samples
            (b) Core-tube samples
            (c) Trench-soil samples
            (d) Drill-core sample
            (e) Descent-engine-exhaust contamination sample
        (3) Soil-mechanics experiment
2. Lunar-orbital activities
    a. Orbital experiments
        (1) Gamma-ray spectrometer
        (2) X-ray fluorescence
        (3) Alpha-particle spectrometer
        (4) Mass spectrometer
        (5) Subsatellite
            (a) Particle shadows/boundary layer
            (b) Magnetometer
            (c) S-band transponder
        (6) Bistatic radar
        (7) S-band transponder
        (8) Apollo window meteoroid
    b. Photographic tasks
        (1) Ultraviolet photography of Earth and Moon
        (2) Photography of gegenschein from lunar orbit
        (3) Service module orbital photographic tasks
        (4) Command module photographic tasks

During the mission, several relatively minor anomalies occurred that were associated with spacecraft

---

[a] NASA Manned Spacecraft Center.

operational equipment and experiment instruments. Despite the problems caused by these anomalies, scientific objectives were satisfied by the experimental and photographic activities accomplished in Earth orbit, during translunar coast, in lunar orbit, on the lunar surface, and during transearth coast.

MISSION OPERATIONAL DESCRIPTION

The space vehicle, manned by David R. Scott, commander; Alfred J. Worden, command module pilot; and James B. Irwin, lunar module (LM) pilot, was launched on schedule from the NASA Kennedy Space Center, Florida, at 9:34:00 a.m. e.d.t. (13:34:00 G.m.t.) on July 26, 1971. The combined command-service module (CSM), LM, and SIVB booster stage were inserted 11 min 44 sec later into an Earth orbit of 91.5 by 92.5 n. mi.

After normal systems checkout in Earth orbit, a nominal translunar injection was accomplished 2 hr 50 min after launch. Shortly after CSM separation from the SIVB stage, the color-television camera was activated to view the separated SIVB stage and to monitor CSM/LM docking. The auxiliary service propulsion system of the SIVB stage was then fired to impact the stage at a preselected target point on the Moon for generation of a seismic source signal. Impact occurred at 20:58:42 G.m.t. on July 29 at selenographic coordinates of latitude 1°21' S and longitude 11°48' W, a point approximately 146 km from the target point and 185 km east-northeast of the Apollo 14 landing site. Strong signals resulting from the impact were recorded by the passive seismic experiments emplaced during the previous Apollo 12 and 14 missions.

During the separation and docking maneuvers, the service propulsion system thrust light on the entry monitor system panel illuminated erroneously. The first midcourse correction, performed to test fire the service propulsion system, verified that both primary and secondary systems were operative and that a short circuit existed in the control circuitry. Contingency procedures, developed to compensate for the short, were used in all subsequent firings.

Only two of the four planned midcourse corrections were performed during translunar coast to place the spacecraft in the trajectory required for lunar-orbit insertion. The first midcourse correction occurred at 18:14:22 G.m.t. on July 27, and the second occurred at 15:05:15 G.m.t. on July 29. The SIM door was jettisoned approximately 4-1/2 hr before the lunar-orbit-insertion maneuver at 20:05:47 G.m.t. on July 29, which placed the spacecraft into an elliptical lunar orbit of 170.1 by 57.7 n. mi.

Descent-orbit insertion, executed at 00:13:49 G.m.t. on July 30, changed the spacecraft orbit to 58.5 by 9.6 n. mi.; this maneuver was followed approximately 13-1/2 hr later by a trim maneuver that increased the decaying perilune altitude from 7.2 to 9.6 n. mi. After CSM/LM separation and two revolutions before the lunar landing, a circularization maneuver at 19:12:59 G.m.t. placed the CSM into a 65.2- by 54.8-n. mi. orbit.

At 22:04:09 G.m.t., the LM descent propulsion system was fired for powered-descent initiation. The LM landed approximately 12 min later in the Hadley-Apennine region of the Moon; sufficient propellant remained to provide an additional hover time of 103 sec. had it been required. The best estimate of the landing location is latitude.26°06'04" N and longitude 3°39'10" E (ref. 1-1). The Apollo 15 landing site and its relationship to the Apollo 11, 12, and 14 landing sites are shown in figure 1-1.

A standup extravehicular activity (EVA) and three periods of surface EVA were performed on the lunar surface. During this time, the CSM circled the Moon 34 times, functioning as a manned scientific satellite. At 10:45:33 G.m.t. on August 2, a plane change maneuvered the vehicle into a 64.5- by 53.6-n. mi. orbit and to the inclination required for rendezvous with the LM ascent stage.

While observed by the ground-command television assembly mounted on the Rover, the LM ascent stage lifted off the Moon at 17:11:23 G.m.t. on August 2 into a lunar orbit of 42.5 by 9.0 n. mi. A nominal LM-active rendezvous was achieved from this orbit, and docking with the CSM was completed approximately 2 hr after lift-off. Crew predocking activities included photographic and television coverage of the docking maneuvers and visual inspection of the LM ascent stage, the CSM, and the SIM bay of the service module.

Jettison of the LM ascent stage was delayed one revolution because of tunnel-venting problems noted by the crew during the hatch-integrity check. When no evidence of contamination was found after the hatches were removed and inspected, the hatches were reinstalled and a successful hatch-integrity test was performed. The LM ascent stage was jettisoned at 01:04:01 G.m.t. on August 3; approximately 34 min

FIGURE 1-1.—Landing sites of Apollo lunar-landing missions. Apollo 11 landed in Mare Tranquillitatis on July 16, 1969; Apollo 12 near the unmanned Surveyor III spacecraft on November 14, 1969; Apollo 14 in the Fra Mauro highlands on January 31, 1971; and Apollo 15 in the Hadley-Apennine region on July 30, 1971. Apollo 13 was aborted during translunar coast because of spacecraft-equipment malfunctions.

later, the deorbit maneuver was performed. The discarded ascent stage impacted the lunar surface at 03:03:36 G.m.t. at latitude 26°21′ N and longitude 0°15′ E, approximately 22 km from the targeted point and approximately 93 km west of the Apollo 15 landing site. Impact signals were recorded by seismic stations emplaced during this mission and during the Apollo 12 and 14 missions.

Approximately 2-1/2 hr before the transearth injection maneuver, which placed the CSM in a return-to-Earth trajectory, an orbit-shaping burn was performed; an hour later, the subsatellite was launched from the SIM into a lunar orbit of approximately 76.3 by 55.1 n. mi. Photographs were taken of the subsatellite separating from the service module. The firing for transearth injection occurred at 21:22:46 G.m.t. on August 4 to initiate transearth coast. During this period, the command module pilot conducted an EVA to retrieve film cassettes from the panoramic and mapping cameras located in the SIM bay and to inspect the condition of the instruments. This activity required approximately 38 min of the scheduled 58 min and was highlighted by television coverage.

Only one of the three planned midcourse-correction maneuvers was performed during transearth coast, reducing the flight-path angle to −6.49°, which was acceptable for entry into the atmosphere of the Earth. The service module was jettisoned at 20:17:56 G.m.t. on August 7, and entry interface occurred 15 min later at an altitude of 122 km and a distance of approximately 2150 km from the landing point. Although entry was nominal and all three main parachutes deployed initially, one parachute collapsed before splashdown. However, the command module was landed safely at 20:45:53 G.m.t., 2 km from the target point and 9.8 km from the prime recovery ship, the U.S.S. *Okinawa*. The landing point, as determined by personnel of the U.S.S. *Okinawa*, was latitude 26°07'30" N and longitude 158°09'00" W. These coordinates differed slightly from the spacecraft-onboard-computer coordinates of latitude 26°07'48" N and longitude 158°07'12" W.

LUNAR-SURFACE ACTIVITIES

The LM touched down at the Hadley-Apennine landing site at 22:16:29 G.m.t. on July 30. During a lunar stay of 66 hr 54 min 53 sec, a 33-min standup EVA and three periods of surface EVA totaling approximately 18-1/2 hr were performed. Lunar-surface activities involved collecting a contingency surface sample; emplacing seven experiments composing the Apollo lunar surface experiments package; deploying the laser ranging retroreflector and solar-wind composition experiments; accomplishing the lunar geological investigation (which involved collecting approximately 76 kg of lunar material including soil, rock, core-tube, and deep-core samples); and performing the soil-mechanics experiment (which required penetration and plate-load tests on the lunar surface and near an excavated trench to aid in defining the mechanical characteristics of the lunar soil).

Traverses during the three EVA periods were enhanced by use of the Rover. An average speed of 9.6 km/hr was achieved, and speeds up to 12 km/hr were attained over level lunar terrain. The total distance traveled, as measured by the Rover odometer, was 27.9 km, corresponding to a map distance of approximately 25.3 km.

Commander Scott, his upper body extending through the top hatch of the LM, performed the 33-min standup EVA. From this elevated vantage point, he described the lunar terrain, used the Sun compass to establish the location of the LM with respect to recognizable selenographic features, and obtained (1) panoramic photographs of lunar terrain with the Hasselblad electric data camera with a 60-mm lens and (2) photographs of interesting distant lunar features with the long-focal-length camera (Hasselblad electric data camera with a 500-mm lens).

Initial activities during the first EVA included collecting the lunar-surface contingency sample, unstowing the Rover and scientific equipment from the LM, and checking out the Rover. Because the front wheels of the vehicle did not respond to steering commands during the checkout, rear-wheel steering was employed during the first EVA. When steering functions returned to normal before the second EVA, dual-Ackerman steering was used on both the second and third traverses.

The first EVA began at 13:13:17 G.m.t. on July 31 and lasted 6 hr 33 min. Activities included a 10.3-km geological traverse and deployment of lunar-surface experiments. Starting at the LM, the crew traversed southward across the mare to the edge of Hadley Rille, south along the rille edge to Elbow Crater and to an area near St. George Crater, and then north past Elbow Crater and across the mare back to the LM. The lunar geological investigation and the soil-mechanics experiment were conducted during this traverse.

After returning to the LM, the crew deployed the Apollo lunar surface experiments package, which contained the heat-flow, lunar-surface magnetometer, passive seismic, cold cathode gage, solar-wind spectrometer, suprathermal ion detector, and lunar dust detector experiments and associated central station and radioisotope thermoelectric generator; the laser ranging retroreflector experiment; and the solar-wind composition experiment. The central station (which relays control and telemetry data between the Moon and the Earth) and the radioisotope thermoelectric generator were deployed approximately 110 m west of the LM, the laser ranging retroreflector experiment about 43 m southwest of the central station, and the solar-wind composition experiment approximately 15 m west of the LM. Telemetry data from the central station indicate that all experiments, the central station, and the radioisotope thermoelectric generator functioned as planned. Signals from the 300-reflector laser ranging retroreflector were acquired initially on August 3.

The second EVA began at 11:48:48 G.m.t. on August 1 and lasted 7 hr 12 min. It featured a 12.5-km traverse southeast across the mare and near Index, Arbeit, Crescent, Dune, and Spur Craters. The return traverse closely followed the outbound route.

Soil and rock samples, with photographic and television documentation, were obtained at several sampling stops during the traverse. Crew attempts to obtain a deep-core sample at the deployment site of the Apollo lunar surface experiments package were suspended because of operational problems with the lunar-surface drill. However, the two holes for the heat-flow experiment were drilled successfully, and the heat probes were inserted. The EVA was terminated after the U.S. flag was deployed near the LM.

The third EVA began at 08:52:14 G.m.t. on August 2, approximately 1-1/2 hr later than planned to allow additional rest for the crew. This EVA lasted 4 hr 50 min, shortened to meet the lift-off time line, and involved a 5.1-km traverse west to Scarp Crater, northwest along the edge of Hadley Rille, and back east across the mare to the LM. As on the other EVA periods, samples were obtained and documented; these samples included a deep-core sample retrieved at the lunar-surface-drill site. Near the end of the EVA, the crew retrieved the aluminum foil of the solar-wind composition experiment deployed on the first EVA. The foil had to be rolled manually when it failed to roll mechanically.

Lift-off of the LM ascent stage occurred at 17:11:23 G.m.t. on August 2 and was monitored by the ground-commanded television assembly mounted on the Rover. Commanded from Earth, the television assembly was planned to provide coverage after lift-off of the lunar surface and of the lunar eclipse on August 6. Although the television assembly operated successfully during all three EVA periods, the elevation clutch began to slip during the second EVA, and operation deteriorated during the rest of the mission. When activated about 40 hr after LM lift-off, the unit operated satisfactorily for 13 min. Signals were then lost, and subsequent activation attempts failed.

INFLIGHT EXPERIMENTS AND PHOTOGRAPHIC TASKS

Required inflight experiments and photographic tasks were performed in Earth orbit and lunar orbit and during translunar coast and transearth coast. The camera equipment needed to satisfy the requirements of ultraviolet photography of the Earth and Moon, of photography of gegenschein from lunar orbit, and of the command module photographic tasks was stowed in the command module. All other scientific equipment was housed in the SIM bay. As illustrated in figure 1-2, this equipment included the gamma-ray spectrometer, X-ray fluorescence experiment, alpha-particle spectrometer, mass spectrometer, subsatellite (including the S-band transponder, magnetometer, and particle shadows/boundary layer experiments), and instruments for the service module orbital photographic tasks involving the panoramic camera, mapping camera, and laser altimeter. Existing command module S-band very-high-frequency (VHF) communications systems were used for the bistatic-radar and S-band transponder (CSM/LM) experiments. Operational periods for these experiments and tasks are shown in figure 1-3, and the groundpath envelope of the orbiting spacecraft is shown in figure 1-4. The Apollo window meteoroid experiment was a passive experiment and had no influence on crew and spacecraft requirements during the mission.

Inflight scientific activities were initiated when the SIM door was jettisoned at 15:40:47 G.m.t. on July 29 and terminated approximately 7 hr before splashdown. During this 214-hr interval, the following tasks were accomplished: photography of most of the lunar area overflown in sunlight; mapping of the bulk chemical composition of the lunar surface overflown; determination of the geometric shape of the Moon along groundtracks; a visual geological survey of various lunar regions in sunlight and other significant geological features; investigation of lunar-atmosphere composition; and astronomical surveys of gamma-ray and X-ray galactic sources, including detailed observations in seven different galactic directions.

Objectives of the gamma-ray spectrometer experiment were to measure, in lunar orbit, the gamma-ray tracted boom), and 10.2 hr (LM attached and extended boom) were dominated by radiation from the plutonium-fuel capsule in the radioisotope thermoelectric generator mounted on the LM.

Objectives of the X-ray fluorescence experiment were to measure, in lunar orbit, fluorescent X-ray flux from the lunar surface, and, during transearth coast, X-ray flux from galactic objects. Data were collected for a total of 116.6 hr. Of these data, 26.5 hr were collected during transearth coast and 90.1 hr in lunar orbit (79.9 hr of prime data and 10.2 hr with the LM attached to the CSM). Collection of fluorescent X-ray flux was limited to the sunlit portion of the Moon because the Sun was the primary excitation source.

The objective of the alpha-particle spectrometer experiment was to locate craters or fissures in the lunar surface through which radon gas had recently escaped. This locating was done by detecting alpha particles emitted by two radon isotopes, products of uranium and thorium. This experiment collected prime data in lunar orbit for 79.4 hr, data with the CSM docked to the LM for 10.2 hr, and background data during transearth coast for 55.3 hr.

The objective of the mass spectrometer experiment was to measure the composition and density of molecules in the lunar atmosphere. This experiment collected 33 hr of prime data in the 60-n. mi. circular orbit with the CSM flying in the $-X$ direction, 7 hr in the 60- by 8-n. mi. orbit with the docked CSM/LM flying in the $+X$ direction, and 48.5 hr during transearth coast.

At 20:13:29 G.m.t. on August 4, the subsatellite was launched into a 76.3- by 55.1-n. mi. orbit at an inclination of 28.7°. Photographs of the instrument as it separated from the spacecraft were obtained with the 70-mm Hasselblad electric camera and with the 16-mm data-acquisition camera. When activated, all three experiments operated as designed. However, because the battery was charging at less than the nominal rate, the S-band transponder experiment was operated every 12th instead of every third revolution as initially scheduled. All subsatellite experiments were turned off while the battery was being recharged after each tracking revolution. Both the magnetometer and particle shadows/boundary layer experiments were acquiring data on all revolutions except those when the battery was being charged.

The objective of the bistatic-radar experiment was to obtain S-band and VHF signals reflected from the

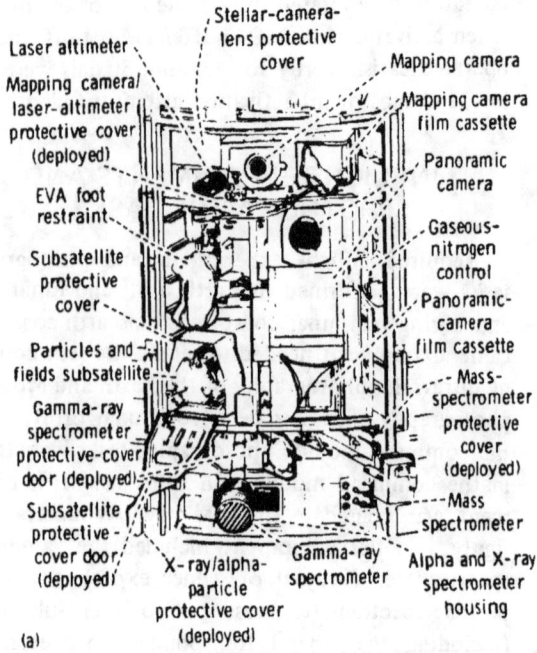

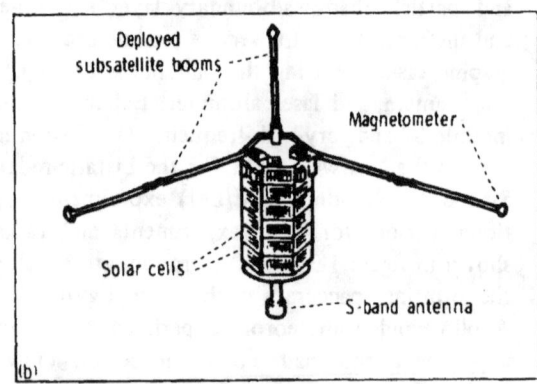

FIGURE 1-2.—Scientific equipment located in the SIM of the service module. Included are orbital experiment instruments and photographic equipment flown for the first time on Apollo 15. After deployment, the subsatellite remained in orbit with a life expectancy of at least 1 yr. (a) Drawing of SIM bay. (b) Deployed-subsatellite configuration.

flux radiated from the lunar surface and, during transearth coast, the background flux of galactic sources, of the CSM, and of the SIM. Data were obtained for a total of 148.7 hr. Of the 94.2 hr of data collected in lunar orbit, 61.8 hr were prime data (extended boom and open mapping-camera/laser-altimeter cover), 5.8 hr were slightly degraded (extended boom and open mapping-camera/laser-altimeter cover), 16.4 hr were seriously degraded (re-

MISSION DESCRIPTION

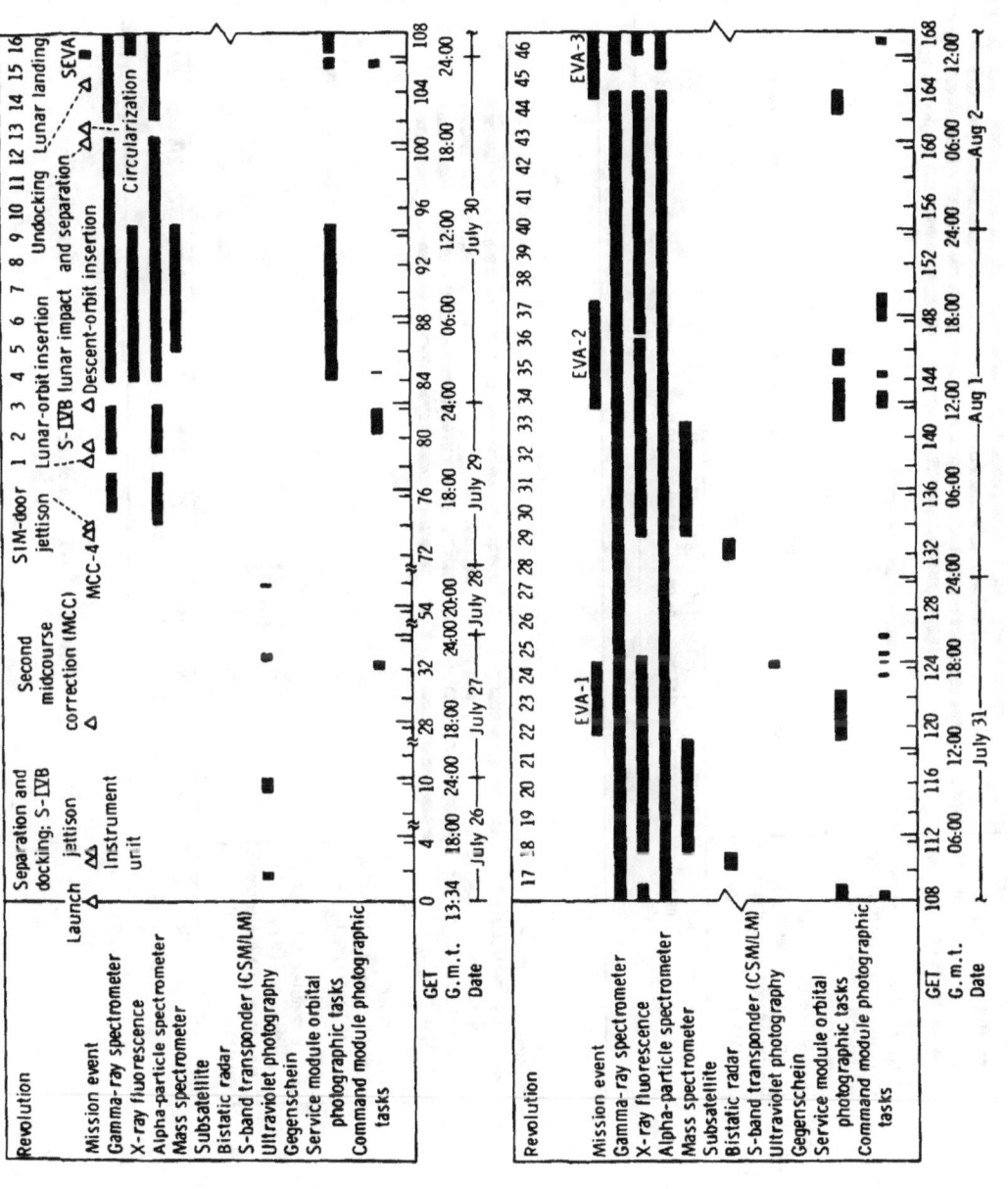

FIGURE 1-3.—Major mission events and experimental-data-collection periods correlated to G.m.t. and ground elapsed time (GET).

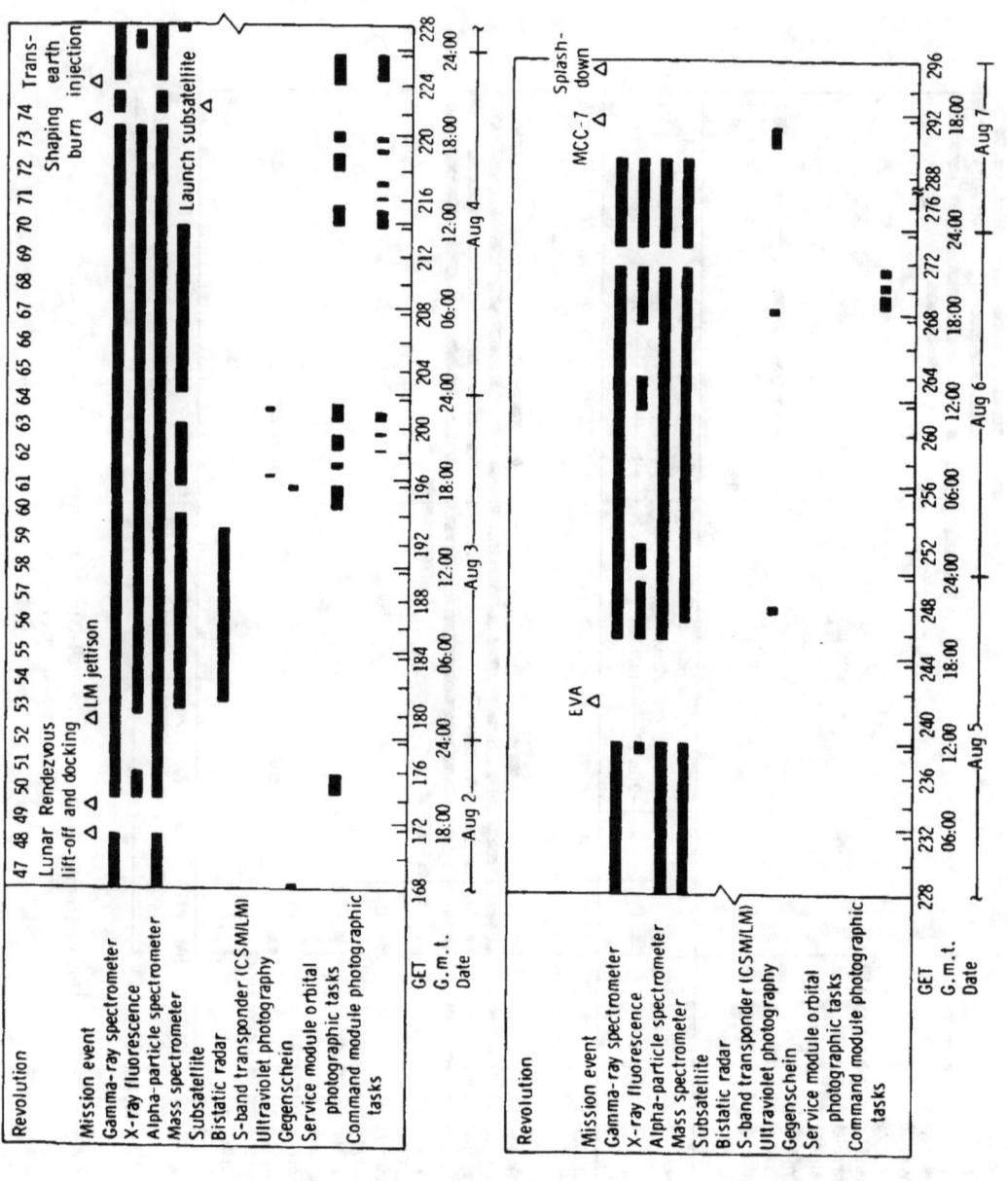

FIGURE 1-3.—Concluded.

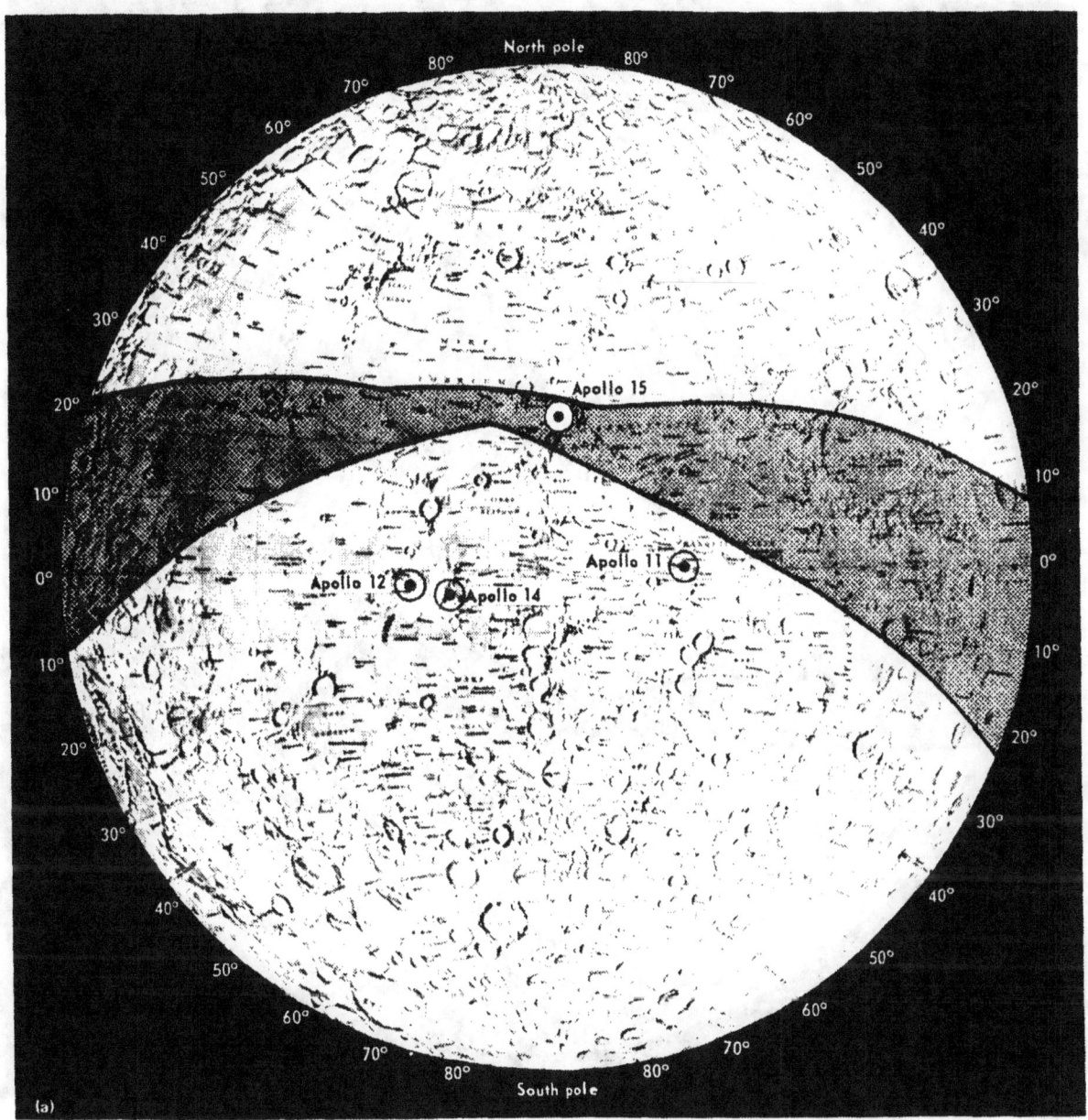

FIGURE 1-4.—Lunar-surface groundtrack envelope of the Apollo 15 orbiting spacecraft for revolutions 1 to 74. Areas of additional data coverage outside the envelope are determined by the fields of view of experiment instruments and photographic cameras. (a) Near side.

Moon for determining the geologic structure and electrical characteristics of the lunar crust. The experiment was performed as scheduled. The S-band and VHF signals were transmitted simultaneously during near-side passes on revolutions 17 and 28, and VHF-only signals were transmitted during near-side passes on revolutions 53 to 57, while the crew slept. The S-band signals were received by the 64-m-diameter antenna at Goldstone, California, and VHF signals were received by the 46-m-diameter antenna at Stanford University, California.

The objective of the S-band transponder (CSM/LM) experiment was to obtain data on S-band doppler tracking of the CSM and LM to determine the distribution of mass along the lunar-surface groundtracks. In this experiment, the spacecraft

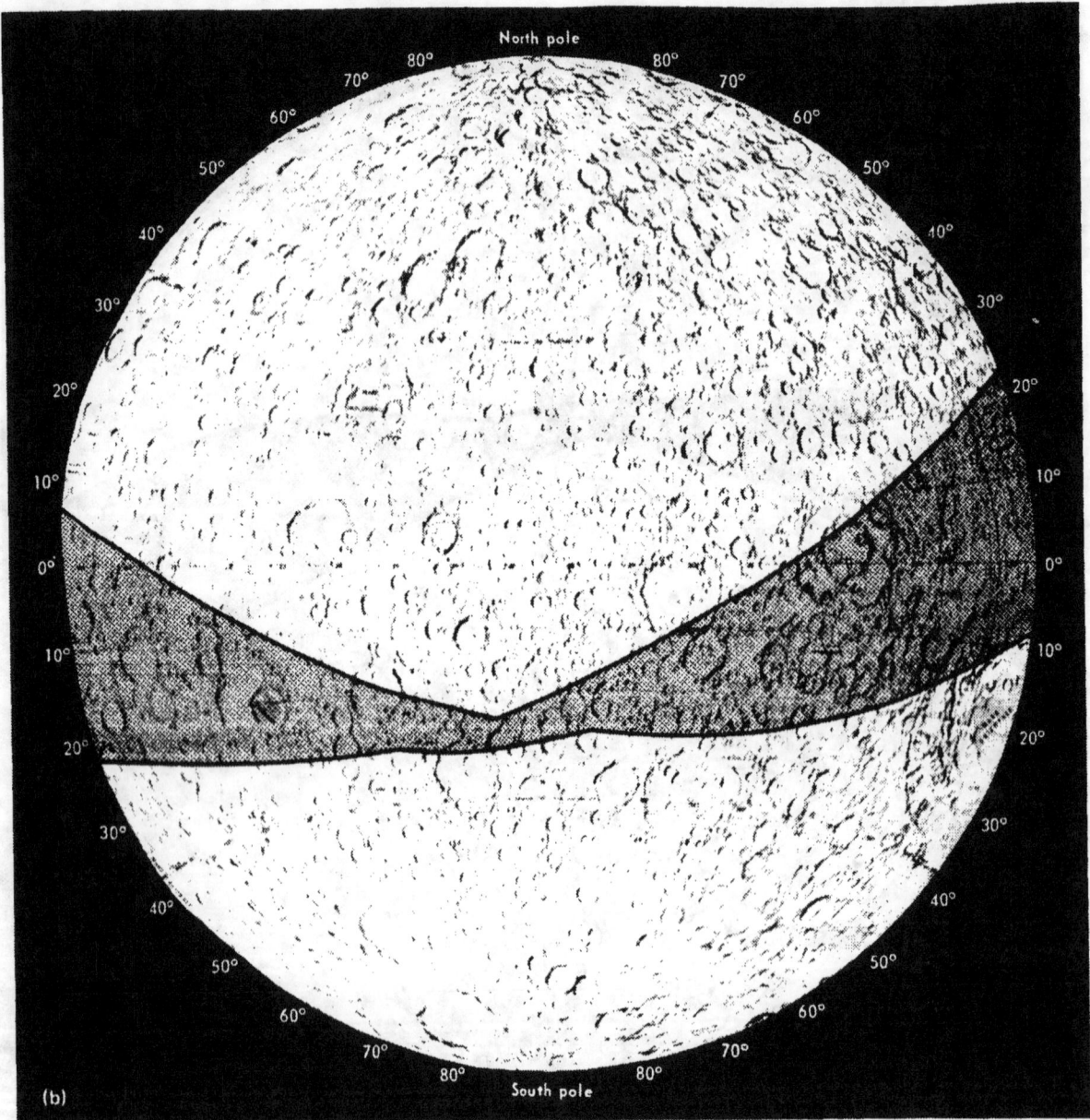

FIGURE 1-4.—(b) Far side.

S-band communications system was used in conjunction with S-band doppler tracking performed by the Manned Space Flight Network. Data were obtained of the docked CSM/LM in lunar orbit, the undocked CSM in lunar orbit, the undocked LM during descent, and the LM ascent stage after deorbit for impact on the lunar surface.

The objectives of the service module orbital photographic tasks were to obtain high-resolution panoramic and high-quality metric lunar-surface photographs and altitude data from lunar orbit. These tasks involved operating the panoramic camera, the mapping camera, and the laser altimeter. During the first and subsequent passes of the panoramic camera on revolution 4, down-link telemetry indicated that the velocity/height sensor was improperly resetting to a 60-n. mi. altitude even with the CSM at altitudes within the normal 40- to 80-n. mi. instrument range;

however, good-quality photographs were obtained of all areas necessary to satisfy the goals established before the mission.

Mapping-camera operation was nominal except for problems in the deployment mechanism (i.e., deployment and retraction times during the mission varied 3 to 5 min, as compared with the 2 min expected from premission analyses). In addition, in an unparalleled example of good fortune with regard to the timing of failures, the mechanism failed to retract after the last scheduled deployment during transearth coast.

Laser altimeter operation, normal through revolution 24, then started to fail and became progressively worse until revolution 38, when it failed completely. An attempt to revive the laser altimeter on revolution 63 was unsuccessful.

Objectives of the command module photographic tasks were to obtain photographs of lunar-surface features of scientific interest from lunar orbit and during transearth coast and of low-brightness astronomical and terrestrial sources. The photographic equipment used included the 70-mm Hasselblad electric camera with 80- and 250-mm lenses and both black-and-white and color film; the 16-mm data-acquisition camera with 18- and 75-mm lenses; and the 35-mm camera with a 55-mm lens. Objectives were satisfied by photographs obtained of 21 of 23 preplanned lunar-surface targets and of numbers of unplanned targets, solar corona, zodiacal light, the Moon during lunar eclipse as it entered and exited the umbra of the Earth, star fields through the command module sextant, and specific areas of the lunar surface in earthshine and in low light levels near the terminator.

REFERENCE

1-1. Rima Hadley Lunar Photomap (Lunar Orbiter V, Site 26.1). First ed., U.S. Army Topographic Command, Apr. 1970.

2. Summary of Scientific Results

Joseph P. Allen[a]

The major scientific objectives of the Apollo 15 mission were to carry out extensive geological exploration, comprehensive sampling, and photographic documentation of the Apennine Front at Hadley Delta, Hadley Rille, and the mare plain; to emplace the Apollo lunar surface experiments package (ALSEP) near the landing site; and to perform a series of survey experiments with the scientific instrument module (SIM) equipment from lunar orbit and during transearth coast. The main scientific phase of the mission began when the Apollo 15 lunar module (LM) landed as planned on the mare plain at the eastern margin of the multiringed Imbrian basin just inside the arcuate Apennine mountain range. The scientific adventure by no means ended with command module (CM) splashdown in the Pacific Ocean, however; rather, the adventure continues as the returned samples and photographs are studied and as the data transmitted daily from the ALSEP and the orbiting subsatellite are analyzed. Only the initial results of these scientific investigations are contained in the Apollo 15 Preliminary Science Report. Whenever possible, data trends from each of the experiments are indicated in this summary, and tentative interpretations based on these trends are pointed out. It should be emphasized that, because the results are preliminary, the interpretations based on the results possibly will change as more data become available and the analyses continue.

GEOLOGIC INVESTIGATION

Because of the extended capability of the life-support equipment and the new mobility provided by the lunar roving vehicle (Rover), the Apollo 15 astronauts explored a much larger area than had been possible on previous missions. The three major geological objectives investigated during the traverses were the Apennine Front along Hadley Delta, Hadley Rille at locations west and southwest of the landing site, and the mare plain at various locations. Extensive information also was obtained about the secondary crater cluster near the Hadley Delta scarp and, although not visited, about the North Complex by photographing the south-facing exposures of this positive feature.

The Apennine Mountains, which rise above the Imbrian plain to heights of nearly 5 km, are thought to be fault blocks uplifted and segmented by the Imbrian impact. The frontal scarp of Hadley Delta, consequently, is interpreted as an exposed section of the pre-Imbrian lunar crust. For this reason, the frontal scarp of Hadley Delta was of highest priority for exploration during the mission. The mountain front was visited on both the first and second traverses; and it was sampled, photographed, and described extensively during this time. In general, the Apennine Mountains show gentle to moderate slopes and are sparsely cratered, with very subdued, rounded outlines. Large blocks are extremely scarce on the mountain flanks, which suggests a gravitationally transported, thick regolith cover on the lower portions of the mountain with a thinner cover of debris on the upper slopes. Sets of stark, sharply etched, parallel linear patterns, completely unexpected before the mission, appear on many of the mountain faces. These major lineaments may represent the expressions of sets of compositional layers or regional fractures showing through the regolith. However, the ambiguities introduced by the oblique lighting of the vertical exposures make difficult an unequivocal interpretation of these linear patterns. For example, the linear ribs clearly present in photographs of Silver Spur and vividly described in real time by the crew may be the expression of gently dipping massive rock layers, or they may reflect near-vertical geologic

---

[a]NASA Manned Spacecraft Center.

structure. The dark band observed by the crew near the base of Mt. Hadley is intermittently visible in both the surface photographs and the panoramic-camera photographs of the landing site. This feature is quite possibly the remnant of a high-lava mark left after the subsidence of the mare basalts following a partial lava drain-back or a cooling shrinkage during one episode of basin filling.

The rocks collected from the mountain front are mainly breccias; many are glass coated. The absence of clasts of older breccias within them distinguish the Apollo 15 rocks from the Apollo 14 samples. The samples from the mountain front are of three types: (1) friable breccias with clasts of nonmare-type basalt, of mare-type basalt, and of glass fragments; (2) coherent breccias with a vitreous matrix that contains clasts of nonmare-type basalt and granulated olivines and pyroxenes; and (3) well-lithified breccias with abundant granulated feldspathic clasts.

Hadley Rille, interpreted as one of the freshest sinuous rilles found on the Moon, was visited during the first and third traverses. The exposed rille walls, on both the near and far sides, were photographed in detail, and the rille rim and several massive outcrops on the near side were extensively sampled. The exposed bedrock strata visible in the photographs have thicknesses as great as 60 m, are distinctly layered, and exhibit varying surface textures and albedos. These characteristics are indicative of a number of individual flow units. All the layers are nearly horizontal. The talus deposits over the lower sections of the rille walls contain enormous blocks shed from the poorly jointed outcrops above. Unbroken blocks of the sizes seen (approaching 20 m in dimension) are uncommon on Earth. The detailed shape of the rille, the regolith cover of the rims, the lithologies of the outcrops and talus deposits, and the stratigraphy displayed in the rille walls are discussed in greater detail in sections 5 and 25.

The dark plain of the mare surface is generally smooth to gently undulating and hummocky. Rocks cover approximately 1 percent of the surface, except for the rougher ejecta blankets around the numerous subdued craters. The morphology of the craters in the surface indicates the mare age to be late Imbrian to early Eratosthenian, and the specific sampling sites visited by the crew span this age range. For example, a 15-m-diameter crater with a widespread glassy ejecta blanket probably represents the youngest surface feature (station 9) yet sampled on the Moon.

Initial study of the panoramic-camera photography of the landing site indicates a possible subdivision of the mare into four geological units characterized by differences in crater population and surface texture. The rocks collected from the mare and from the exposed outcrops at the rille edge consist mainly of basalts with abundant, coarse, yellow-green to brown pyroxene and olivine phyric basalts. In addition to the characteristics of the major geologic features, the optical properties of the surface materials, as recorded in the many photographs; a number of smaller scale features, such as the craters, fillets, and lineaments that are typical of the Hadley region; and the individual samples themselves are discussed in detail in section 5.

PRELIMINARY EXAMINATION OF LUNAR SAMPLES

A total of 77 kg of samples was returned from the Hadley area. These samples consist of rocks that weigh from 1 g to 9.5 kg, three core tubes, a deep-drill corestem, and a variety of soil samples taken from the two distinct selenologic regions at the Hadley site (the mare plain and the base of Hadley Delta). The Lunar Sample Preliminary Examination Team has made a macroscopic study of the more than 350 individual rock samples and additional petrographic and chemical studies of a selected few of these samples. The rocks from the mare plain fit into two categories: (1) extrusive and hypabyssal basalts and (2) glass-covered breccias. The rocks from the base of Hadley Delta exhibit a variety that ranges from breccias to possible metaigneous rocks.

The mare basalts appear to be fresh igneous rocks with textures that range from dense to scoriaceous. The chemical composition of these rocks is very similar to the compositions of those basalts returned from the Apollo 11 and 12 and Luna 16 mare sites. In particular, the mare basalts are high in iron, with a correspondingly high iron-oxide-to-magnesium-oxide ratio, and low in sodium oxide, in contrast to terrestrial basalts. Thin-section examinations of 13 of these basalts reveal four different textural types: (1) porphyritic-clinopyroxene basalt with 3- to 9-mm-long prisms, (2) porphyritic-clinopyroxene-basalt vitrophyre with 1- to 7-mm-long skeletal prisms, (3) porphyritic-olivine basalt, and (4) highly vesicular basalt.

A number of rock fragments from Spur Crater and

many of the clasts in breccias from the mountain front are basalts that are distinctly different from the mare basalts. Specific differences are: (1) The plagioclase-to-mafic-mineral ratio is 1, compared with a ratio of 0.5 for the mare basalts; (2) the pyroxene is light brown to tan with no zoning, compared with the cinnamon-brown, zoned pyroxene of the mare basalts; (3) the grain size of all mineral phases is less than 1 mm, compared with some grains as large as 1 cm in mare basalts; and (4) no vugs or vesicles are found in the nonmare basalts, in sharp contrast to the often highly vesiculated mare-basalt samples.

Many types of clastic and metamorphosed rocks were found along the traverse routes at the Hadley site, including several unique specimens that are discussed individually in section 6. The greatest variety of these rocks was concentrated along the base of Hadley Delta, where all the samples have undergone shock metamorphism and brecciation.

The soils returned from the Hadley-Apennine area are similar in most respects to soil samples returned from previous missions. The soil is composed primarily of the following particle types: (1) agglutinates plus brown-glass droplets, (2) basalt fragments of different textures, (3) mineral fragments, (4) microbreccias, and (5) glasses of varying color and angularity, including a particular component of green-glass spheres never before observed in lunar soils. The chemical composition of the soil samples, particularly from the mare regions, is distinctly different from the composition of the rocks from presumably the same locales. A linear correlation involving the iron oxide and aluminum oxide constituents of the soil and rocks from the Apollo sites suggests that the soil may be derived from a range of rock material, with the two end members being the iron-rich mare basalt and the aluminum-rich, iron-poor nonmare basalt.

A total of 4.6 kg of material was returned from the Hadley site in the form of core samples. A deep-drill corestem of six sections was driven to a depth of 2.4 m into the regolith near the landing site. All except part of the lowest section was returned completely full. Stereoscopic X-radiographs of this deep-drill core reveal significant variations in pebble concentration and in the density of the material along the core. These variations indicate the presence of more than 50 individual layers with thicknesses from 0.5 to 21 cm. In addition, three drive tubes with a maximum penetration as great as 70 cm were returned. As with the deep-drill core, X-radiographs of the lunar material within these tubes reveal distinct layering and a spectrum of soil textures and fragment sizes. The deep-drill core and the three drive cores all show that the lunar regolith has a substantial stratigraphic history.

The gamma-ray spectra of 19 samples have been measured to determine the concentrations of primordial radioactivity of potassium-40, uranium-238, and thorium-232 and of the cosmic-ray-induced radioactivity of aluminum-27 and sodium-22. In general, the radionuclide abundance is similar to that seen at previous sites. The potassium-to-uranium ratio of both the mare basalts and soils at the Hadley site is strikingly different from the ratio measured in terrestrial samples, which is further evidence in support of an earlier hypothesis concerning differences between Earth and Moon materials.

Concentrations of noble-gas isotopes measured in the samples from the Hadley-Apennine area are similar to the abundances previously measured in lunar materials. Variations in the argon-40 component of the soils are found at this site, as was the case for samples from previous sites, which suggests, for example, differing concentrations of argon-40 in the lunar atmosphere or differing argon-40 retention efficiencies of the soils. Concentrations of the spallation-produced isotopes, neon-21, krypton-80, and xenon-126, result in rock exposure ages in the range of 50 to 500 $\times$ 10^6 yr, which is similar to the range of exposure ages measured in the Apollo 11 and 12 samples.

The total carbon content for several samples has been determined. As found for the materials from previous lunar sites, the total carbon content for the soils and breccias is higher than for the igneous samples. This continued systematic difference in carbon content seems to confirm the idea that much of the carbon in the lunar soil may originate in the solar wind.

SOIL-MECHANICS EXPERIMENT

The objectives of the soil-mechanics investigation are to examine the physical characteristics (such as particle sizes, shapes, and distributions) and the mechanical properties (such as particle density, strength, and compressibility) of the in situ lunar soil and to examine the variation of these parameters laterally over the areas traversed at the Hadley site. The longer duration of the extravehicular activities and the correspondingly larger distances covered with the Rover, the variety of the geological units found at the Hadley-Apennine site, and the quantitative measurements provided by the self-recording penetrometer (a device carried for the first time on this mission) have resulted in a number of conclusions.

The lunar surface at the Hadley site is similar in color and texture to the surfaces at the previous landing sites. Although the variability of grain-size distribution of samples from the Apollo 15 site appears to be less than the variability found at the Apollo 12 and 14 sites, considerable variety exists, both with depth and laterally, in the soil properties of strength and compressibility. For example, the compressibility ranges from soft along the mountain front to much firmer near the rim of Hadley Rille. Evidence exists of downslope movement of surficial material on the walls of Hadley Rille; however, no evidence of deep-seated slope failures along the mountain front was found.

Soil densities derived from both the core-tube and the deep-drill corestem samples exhibit considerable variability that ranges from approximately 1.3 to 2.2 g/cm^3. The self-recording-penetrometer data indicate an in situ density of approximately 2.0 g/cm^3, a high soil strength, and a low soil compressibility. When coupled with additional data from the soil-mechanics trench dug near the ALSEP site, the penetrometer information can be used to estimate the cohesion and friction angle of the lunar soil. The values for both these parameters are higher than the values that resulted from experiments conducted during previous missions.

PASSIVE SEISMIC EXPERIMENT

The purpose of the passive seismic experiment is to study the lunar-surface vibrations, from which interpretations of the internal structure and physical state of the Moon can be determined. Sources of seismic energy may be internal (from moonquakes) or external (from impacts of both meteoroids and spent space hardware). In either case, a straightforward determination of the unambiguous source locations requires at least three vibration-sensing instruments monitoring the event of interest. The Apollo 15 passive-seismometer station represents the third of a network of seismometers now operating on the lunar surface; thus, the successful deployment of this instrument marked a vitally important step in the investigation of the Moon.

Seismic data accumulated over the first 45 days of operation have been analyzed, and the preliminary results are summarized as follows. Seismic evidence for a lunar crust and mantle has been found. The thickness of the crust is between 25 and 70 km in the region of the Apollo 12 and 14 landing sites. The velocity of compressional waves in the crustal material is between 6.0 and 7.5 km/sec, which is a range that spans the velocities expected for the feldspar-rich rocks found on the lunar surface. The transition from the crustal material to the mantle material may be gradual, starting at a depth of approximately 25 km, or rapid, with a sharp discontinuity at a depth of 45 to 70 km. In either case, the compressional-wave velocity reaches 9 km/sec in the subcrustal mantle material, and the contrast in elastic properties of the rocks comprising these two major layers is at least as great as the contrast that exists between the materials comprising the crust and mantle units of the Earth.

The major part of the natural lunar seismic energy detected by the network is in the form of periodic moonquakes that occur near times of perigee and apogee and that originate from at least 10 separate locations. However, a single focal zone at a depth of approximately 800 km, with a dimension less than 10 km and with an epicenter approximately 600 km south-southwest of the Apollo 12 and 14 sites, accounts for 80 percent of the seismic energy detected. The release of seismic energy at these depths (which are slightly greater than any known earthquake sources) suggests that the lunar interior at these depths must be rigid enough to support appreciable stress. This fact, in turn, places strong constraints on realistic thermal models of the lunar interior.

In addition to the periodic moonquakes, episodes of frequent small moonquakes have been discovered. Individual events may occur as frequently as every 2 hr for periods lasting up to several days. The source of the moonquake swarms is at present unknown, but they may well result from continuing minor adjustments to stresses in the outer shell of the Moon.

The average rate of seismic-energy release within the Moon is far below that of the Earth. Thus, the outer crust and mantle of the Moon appear to be relatively cold and stable compared with that of the Earth, and significant internal convection currents causing lunar tectonism seem to be absent; however, the discovery of moonquakes at great depths suggests the possibility of some very deep convective motion.

Seismic energy deposited at the lunar surface by an impacting meteoroid or manmade object is confined for a surprisingly long time in the near-source area by efficient scattering near the surface. Nevertheless, the energy slowly dissipates through interior propagation to more distant parts of the Moon, and, for this reason, all but the smallest of impact signals from all parts of the Moon are probably detected at the operating seismic stations.

LUNAR-SURFACE MAGNETOMETER EXPERIMENT

The Apollo 15 magnetometer, which is the third and all-important member of the magnetometer network now on the lunar surface, was deployed to study intrinsic remanent magnetic fields and to observe the global magnetic response of the Moon to large-scale solar and terrestrial magnetic fields imposed on it. Fundamental properties of the lunar interior, such as electrical conductivity, magnetic permeability, and temperature profile, can be calculated from these magnetic measurements. The measuring and understanding of these properties are obvious requirements for meaningful theoretical descriptions of the origin and evolution of the Moon.

The three fluxgate sensors of the Apollo 15 instrument show a steady magnetic field of approximately 5 γ at the Hadley site, which is considerably smaller than the 38-γ field measured at the Apollo 12 site and the 103- and 43-γ fields measured at the two locations at the Apollo 14 site. The bulk relative permeability of the Moon is calculated from the magnetometer data to be near unity. The electrical conductivity of the lunar interior is obtained from measurements of the response of the Moon to externally imposed, variable magnetic fields; and these measurements can be interpreted in terms of a spherically symmetric, three-layer model that has a thin outer crust (extending from the radius of the Moon to 0.95 the radius of the Moon) of very low conductivity, an intermediate layer (extending from 0.95 the radius of the Moon to 0.6 the radius of the Moon) with a conductivity of approximately 10^{-4} mho/m, and an inner core (of a diameter approximately 0.6 the radius of the Moon) of conductivity greater than 10^{-2} mho/m. In the case of an olivine Moon, these values correspond to a temperature profile of 440 K for the crustal layer, approximately 800 K for the intermediate layer, and greater than 1240 K for the central core.

SOLAR-WIND SPECTROMETER EXPERIMENT

Two identical solar-wind spectrometer experiments now operate 1100 km apart at the Apollo 12 and the Apollo 15 sites. Solar-wind plasma, magnetosphere plasma, and magnetopause crossings have been observed by both instruments, which show good internal agreement of observations; for example, simultaneous (within 15 sec) changes in proton densities and velocities are detected at both sites. As first measured with the Apollo 12 instrument, the solar plasma at the lunar surface is indistinguishable from the solar plasma some distance out from the surface (monitored by orbiting instruments), when the Moon is both ahead of and behind the magnetic bow shock of the Earth.

HEAT-FLOW EXPERIMENT

The heat-flow experiment is designed to make temperature and thermal-property measurements within the lunar subsurface in order to determine the rate at which heat is flowing out of the interior of the Moon. This heat loss is directly related to the rate of internal heat production and to the internal temperature profile; hence, the measurements result in information about the abundances of long-lived radioisotopes within the Moon and, in turn, result in an increased understanding of the thermal evolution of the body.

Emplacement of the first heat-flow experiment into the lunar surface was completed during the second period of extravehicular activity at the Hadley site. Initial measurements with this instrument show a subsurface temperature at 1.0 m below the surface of approximately 252.4 K at one probe site and 250.7 K at the other, which are temperatures that are approximately 35 K above the mean surface temperature. From 1.0 to 1.5 m below the surface, the temperature increases at the rate of 1.75 K/m (±2 percent). In situ conductivity measurements result in values between 1.4×10^{-4} and 2.5×10^{-4} W/cm-K at depth and are found to be greater than the conductivity values of the surface regolith by a factor of 7 to 10, which indicates that conductivity increases with depth as well.

Preliminary analysis of these results indicates that the heat flow from below the Hadley-Apennine site is 3.3×10^{-6} W/cm^2 (±15 percent). This value is approximately one-half the average heat flow of the Earth. By assuming that this value is an accurate representation of the heat flow at the Hadley site (while realizing that data accumulation over a number of lunations will be required to establish this accuracy) and by further assuming that this value is representative of the moonwide heat-flow value, then consideration of the Moon as a sphere with uniform internal heat generation results in a picture of the Moon as a far more radioactive body than had been previously suspected, and a far more radioactive body than suggested by the ordinary chondrites and the type 1 carbonaceous chondrites that have been used to construct the standard models of the Earth and the Moon to date.

SUPRATHERMAL ION DETECTOR EXPERIMENT (LUNAR-IONOSPHERE DETECTOR)

The suprathermal ion detector experiment deployed at the Apollo 15 site is identical, except for the ion mass ranges covered, to the instruments operating at the Apollo 12 and 14 sites, and the Apollo 15 experiment is the third member of this ion-monitoring network. In the first days of operation, a number of energy and mass spectra of positive ions were measured, primarily from the gas clouds vented by the spacecraft and other mission-associated equipment. At lunar lift-off, for example, a marked decrease in the magnetosheath-ion fluxes was observed. This decrease lasted approximately 8 min and is attributed to either a change in the ion flow direction (because of the exhaust-gas cloud) or to energy loss of the ions passing through the exhaust-gas cloud. Some hours later, the ascent stage impacted the lunar surface nearly 100 km west-northwest of the Hadley site, and the ions that resulted from the impact-generated cloud were monitored.

Multiple-site observations of ion events that possibly correlate with seismic events of an impact character (recorded at the seismic stations) have resulted in information about the apparent motions of the ion clouds. Typical travel velocities have been calculated to be approximately 80 m/sec. Numbers of single-site ion events have been detected, some with ions in the mass range of 16 to 20 amu/Q, which corresponds quite possibly to the release of water vapor from deep below the lunar surface. The 500- to 1000-eV ions streaming along the magnetosheath have been observed simultaneously by all three suprathermal ion detectors. This ion flux is strongly peaked in the down-Sun direction, which is a fact established by the different look directions of the individual instruments.

COLD CATHODE GAGE EXPERIMENT (LUNAR-ATMOSPHERE DETECTOR)

The cold cathode gage experiment that was deployed at the Hadley site is similar to the instruments at the Apollo 12 and 14 sites and is intended to measure the density of the tenuous lunar atmosphere at the lunar surface. This anticipated thin concentration of gases is a result of the solar wind, the possible continued release of molecules from the lunar interior through the lunar crust, and certain venting and outgassing from the LM descent stage and other gear left on the Moon. The contamination, however, should decrease with time in a recognizable way.

As might be expected from cold cathode gage experiment results from the earlier missions, the gas concentrations observed during the lunar days appear to be overwhelmingly a result of contaminants released by the LM and associated equipment. However, during the lunar nights, the observed concentrations, which are typically less than 2×10^5 particles/cm^3, are lower even than the concentrations expected from the neon component of the solar wind alone. This fact suggests that the contaminant gases from spacecraft equipment remain adsorbed at the low nighttime temperatures and that the lunar surface itself is not saturated with neon, but rather absorbs this gas much more readily than releases it. Except for mission-associated phenomena, no easily recognizable correlations have been found between transient gas events, as seen on this instrument, and the response of the suprathermal ion detector or of the solar-wind spectrometer.

LASER RANGING RETROREFLECTOR

The third and largest U.S. laser ranging retroreflector was delivered to the lunar surface and deployed at the Hadley site approximately 40 m from the ALSEP site. Successful range measurements to this 300-corner-cube array were made shortly after the LM lifted off several days later, and subsequent measurements indicate that no degradation of the reflective properties of the unit resulted from dust being kicked up during the extravehicular activities or by ascent-engine residue. The better signal-to-noise ratio available with this larger retroreflector will enable more frequent ranging measurements to be made and will enable measurements to be carried out by telescopes of smaller aperture than heretofore possible. Additionally, this third array now provides the important long north-south base-line separation with the Apollo 11 and the Apollo 14 retroreflectors. For the accumulation of data important to the planned astronomical, geophysical, and general relativity experiments, range measurements will be required over a period of years.

SOLAR-WIND COMPOSITION EXPERIMENT

The solar-wind composition experiment, which is similar to the experiments conducted during the Apollo 11, 12, and 14 missions, was deployed at the end of the first extravehicular activity and exposed to the solar wind for a period of 41 hr, nearly twice the exposure time obtained during the previous mission. Initial samples of the aluminum foil have been analyzed, and isotopes of helium and neon have been detected. The helium flux during the Apollo 15 exposure is nearly four times that detected during the Apollo 14 exposure; yet, interestingly enough, the relative abundance of helium and neon and the relative isotopic abundances of these elements are very similar to the earlier abundances. A positive correlation has been found between the general level of solar activity and the helium-4 to helium-3 ratio, a correlation first suggested perhaps by the helium-to-hydrogen ratio measurements of the Explorer 34 spacecraft.

GAMMA-RAY SPECTROMETER EXPERIMENT

The gamma-ray spectrometer experiment is designed to measure, from lunar orbit, the gamma-ray activity of the lunar-surface materials. The gamma-ray flux from the lunar surface is expected to contain two components, one resulting from naturally occurring radioisotopes (primarily of potassium, uranium, and thorium) and the other resulting from cosmic-ray-induced interactions at the lunar surface. The gamma-ray intensity from the naturally occurring radionuclides is a sensitive function of the degree of chemical differentiation undergone by the Moon; and, thus, the measured intensity relates directly to the origin and evolution of the planet.

Analysis of the very preliminary data printout (unfortunately the only data available for analysis before the preparation of this document) shows a strong contrast in gamma-ray count rates over different regions of the Moon. Specifically, the regions of highest activity are the western maria, followed by Mare Tranquillitatis and Mare Serenitatis. Considerably lower activity is found in the highlands of the far side, with the eastern portion containing nearly an order of magnitude less gamma-ray activity than that found in Oceanus Procellarum and Mare Imbrium. The preliminary data show intensity peaks that correspond to the characteristic energies of the isotopes of iron, aluminum, uranium, potassium, and thorium; however, more data than available in the preliminary printout are required to verify this identification unambiguously.

X-RAY FLUORESCENCE EXPERIMENT

The purpose of this experiment is to map the principal elemental constitutents of the upper layer of the lunar surface by measuring the fluorescent X-rays produced by the interaction of solar X-rays with the surface material. Secondarily, the experiment is used to observe X-radiation from astronomical objects during the transearth-coast phase of the mission. The X-ray detector assembly consists of three proportional counters, two X-ray filters (and associated collimators), temperature monitors, and the necessary support electronics. Inflight energy, resolution, and efficiency calibrations are made with X-ray sources carried with the instrument.

Preliminary analysis of the initial data from this experiment is very exciting. In general, the suspected major compositional differences between the two fundamental lunar features, the maria and the highlands, are confirmed by the X-ray data, and more subtle compositional differences within both the maria and the highlands are strongly suggested. For example, the aluminum-to-silicon intensity ratio is highest over the terrae, lowest over the maria, and intermediate over the rim areas of the maria. The extremes for this ratio vary from 0.58 to 1.37, with a tendency for the value to increase from the western mare to the highlands of the eastern limb. Furthermore, a striking correlation exists between the aluminum-to-silicon intensity ratio and the values of surface albedo along the groundtrack surveyed by the X-ray experiment.

Although the data from the X-ray fluorescence experiment are still in the initial stages of analysis, a number of tentative conclusions may be drawn about the fundamental properties of the lunar surface. The sharply varying aluminum-to-silicon ratio confirms that the maria and the highlands are indeed chemically different, and the distinguishing albedo differences between these major features must be, in part, the signature of this difference. The anorthositic component of the returned lunar samples is certainly related to the high aluminum content measured in the highland regions, and the correspondingly low aluminum content of the returned mare basalts is consistent with the low measurement values over the maria. The experimental X-ray data, thus, further support the theory that the Moon, shortly after formation, developed a differentiated, aluminum-rich crust. The sharp change in the aluminum-to-silicon intensity ratio between the highland and mare areas places stringent limitations on the amount of horizontal displacement of the aluminum-rich material after the mare flooding. Indications definitely exist in the more gradual data trends that the circular maria have a lower aluminum content than the irregular maria; and within particular maria (for example, Crisium and Serenitatis), the centers have a lower aluminum content than the edges. Finally, the large ejecta blankets, such as the Fra Mauro formation, seem to be chemically different from the unmantled highlands.

During the transearth coast, X-ray data were obtained from three discrete X-ray sources and from four locations dominated by the diffuse X-ray flux. The count rate from two of the sources, Sco X-1 and Cyg X-1, did show significant changes in intensity of approximately 10 percent over time periods of several minutes; however, a final analysis of Apollo data is required to rule out completely the possibility that changes in spacecraft attitude during the counting periods might account for the counting-rate variations.

ALPHA-PARTICLE SPECTROMETER EXPERIMENT

The alpha-particle spectrometer experiment consists of 10 totally depleted silicon surface-barrier solid-state detectors of 3 cm^2 area each that are particularly sensitive to alpha particles in the energy range between 5 and 12 MeV. The experiment is designed to map, from lunar orbit, possible uranium and thorium concentration differences across the surface by measuring the alpha-particle emissions from the two gaseous daughter products of uranium and thorium, radon-222 and radon-220, respectively. Because trace quantities of these radioactive gases would be included in any outgassing of, for example, water or carbon dioxide from the lunar interior, detection of radon also would provide a sensitive probe of remanent volcanic activity or of local release of common volatiles.

Preliminary analysis of the alpha-particle data indicates that the alpha-particle activity of the Moon is at most equal to the observed count rate of 0.004 count/cm^2-sec-sr (± 1 percent) in the energy band from 4.7 to 9.1 MeV. No significant difference in this count rate was observed between the dark and sunlit sides of the Moon. This measured alpha-particle activity is considerably less than was anticipated before the mission. For example, if the uranium and thorium concentrations measured in the samples returned from the Apollo 11 and 12 sites are typical moonwide values, then the alpha-particle counting rate that results from radon emission is at least a factor of 60 lower than the rate predicted by radon-diffusion models. Complete analysis of all the Apollo 15 alpha-particle data should result in another order of magnitude more sensitivity than reported here for the detection of alpha-particle emission from lunar sources.

LUNAR ORBITAL MASS SPECTROMETER EXPERIMENT

The lunar orbital mass spectrometer is designed to measure the composition and density of neutral gas molecules along the flight path of the command-service module (CSM) in order to better understand the origin of the lunar atmosphere and the related transport processes in planetary exospheres in general. The instrument was mounted in the SIM on a bistem boom and was operated from the CM. The preliminary results indicate that a large number of gas molecules of many species exist near the spacecraft in lunar orbit; none, however, have an obvious lunar origin or a sufficient intensity to be detected above the background of molecules from all sources. The gas cloud apparently moves with the vehicle because the measured density is essentially independent of the angle of attack of the spectrometer entrance plenum. The spacecraft is thought to be the source of most of this cloud, even though the intensity of the apparent contamination is a strong function of the orbital parameters. For example, during the transearth coast, the detected amplitudes of all species of molecules were reduced by a factor of 5 to 10 from the amplitudes measured in lunar orbit.

S-BAND TRANSPONDER EXPERIMENT

During the last near-side pass before transearth injection, a small scientific spacecraft was launched into lunar orbit from the SIM bay of the service module. This subsatellite is instrumented to measure plasma and energetic-particle fluxes and vector magnetic fields and is equipped with an S-band transponder to enable precision tracking of the spacecraft. The S-band transponder experiment uses the precise doppler-tracking data of this currently orbiting satellite and the tracking data of the CSM and of the LM taken during the mission to provide detailed information about the gravitational field of the near side of the Moon. The data consist of the minute changes in the spacecraft speed, as measured by the Earth-based radio tracking system (which has a resolution of 0.65 mm/sec). The initial data indicate that the subsatellite is operating normally and, because the periapsis altitudes are following closely the predicted altitudes, suggest that the spacecraft will have an orbital lifetime of at least the planned one year. The subsatellite transponder experiment should provide data for a detailed gravity map for the area between $\pm 95°$ longitude and $\pm 30°$ latitude.

Analyses of the low-altitude CSM data have resulted in new gravity profiles of the Serenitatis and Crisium mascons; these results are in good agreement with the Apollo 14 data analysis and strongly suggest that the mascons are near-surface features with a mass distribution per unit area of approximately 500 kg/cm^2. The Apennine Mountains show a local gravity high of 85 mgal but have undergone partial isostatic compensation, and the Marius Hills likewise have a gravity high of 62 mgal.

SUBSATELLITE MEASUREMENTS OF PLASMAS AND SOLAR PARTICLES

The main objectives of the subsatellite plasma and particles experiment are to monitor the various plasma regimes in which the Moon moves, to determine how the Moon interacts with the fields and plasmas within these regimes, and to investigate certain features of the structure and dynamics of the Earth magnetosphere. The experiment consists of two solid-state particle-detector telescopes that are sensitive to electrons with energies as large as approximately 320 keV and to protons with energies as large as approximately 4 MeV, and of five electrostatic-analyzer assemblies that are sensitive to electrons in the energy range from 0.5 to 15 keV.

Detailed observations have been made of particle fluxes around the Moon as the Moon moves through interplanetary space, through the magnetosphere, and through the bow shock of the Earth. Analysis of these preliminary data leads to several tentative conclusions. Solar electrons were measured at the subsatellite following a large solar flare on September 1, 1971. In the energy range of 6 to 300 keV, the electron spectrum is reproduced by the power-law equation $dJ/dE = (3 \times 10^3) E^{-1.5}$ electrons/cm^2-sr-sec-keV. Additionally, an electron flux of 20 electrons/cm^2-sr-sec of energy from 25 to 30 keV is found to move predominantly in a sunward direction for several days while the Moon is upstream from the Earth. It is not known as yet whether this flux is of solar or terrestrial origin. Finally, a distinct shadow in the fast-electron component of the solar wind is formed by the Moon. For the case when the interplanetary magnetic field is nearly alined along the solar-wind flow, this electron shadow corresponds closely to the optical shadow behind the Moon. However, for the case when the interplanetary magnetic field alines more perpendicular to the solar-wind flow, the fast-electron shadow region broadens to a diameter much greater than the lunar diameter and becomes extremely complex.

SUBSATELLITE MAGNETOMETER EXPERIMENT

The major objectives of the subsatellite magnetometer experiment are to extend the measurements of the permanent and induced components of the lunar magnetic field by systematically mapping the remanent magnetic field of the Moon and by measuring the magnetic effects of the interactions between cislunar plasmas and the lunar field. Initial data from the two subsatellite fluxgate sensors indicate that detailed mapping of the remanent magnetization, although complex, is entirely feasible with the present experiment. For example, preliminary analysis shows a fine structure in the magnetic field associated with the large craters Hertzsprung, Korolev, Gagarin, Milne, Mare Smythii I, and, in particular, Van de Graaff, which produces a 1-γ variation in the field measured by the subsatellite passing overhead. Furthermore, magnetic fields induced within the Moon by externally imposed interplanetary magnetic fields are detectable at the subsatellite orbit. Estimated variations of lunar conductivity as a function of latitude and longitude will be possible from magnetometer data. Finally, the data show that the plasma void that forms behind the Moon when it is in the solar wind extends probably to the lunar surface, and the flow of the solar wind is itself rather strongly disturbed near the limbs of the Moon.

BISTATIC-RADAR INVESTIGATION

The bistatic-radar experiment uses the S-band and the very high frequency (VHF) communication systems in the CSM to transmit toward the portion of the Moon that scatters the strongest echoes to Earth-based receivers. The echoes are received from an area approximately 10 km in diameter that moves across the lunar surface near the orbiting CSM. The characteristics of these echoes are compared to those of the directly transmitted signals in order to derive information about such lunar crustal properties as the dielectric constant, density, surface roughness, and average slope.

The VHF data obtained during this mission have approximately one order of magnitude higher signal-to-noise ratio than previously obtained, and the effects of the bulk electrical properties and slope statistics of the surface are clearly present in the data. The S-band data show the areas surveyed during the mission to be similar to those regions sampled at latitudes farther south during the Apollo 14 mission. Distinct variations in the slopes of the lunar terrain in the centimeter-to-meter range exist, and some areas contain an unusually heavy population of centimeter-size rock fragments. The bistatic-radar data are currently being combined with the CSM ephemeris data to correlate these results with orbital photography and corresponding geological interpretations in order to better distinguish between adjacent and subjacent geological units.

APOLLO WINDOW METEOROID EXPERIMENT

The Apollo window meteoroid experiment involves a careful study of the CM heat-shield window surfaces for pits caused by meteoroid impacts. These tiny craters, when identified, are further examined to obtain information on crater morphology and possible meteoroid residence. So far, 10 possible impacts of 50 μm diameter and larger have been identified in the windows of the Apollo 7, 8, 9, 10, 12, 13, and 14 spacecraft. These findings correspond to a meteoroid flux below that expected from theoretical calculations but are in good agreement with the flux value derived by an examination of the Surveyor shroud returned by the Apollo 12 mission.

ANCILLARY EXPERIMENTS

The numerous orbital-science and orbital-photography experiments conducted during the mission are discussed in a separate section (sec. 25) of this report. The individual experiments will not be discussed in this summary of scientific results.

Several other experiments and tests were conducted during the Apollo 15 mission that will not be discussed in detail in this report. The reader is particularly referred to the documents of the NASA Medical Research and Operations Directorate for the biomedical evaluation of the mission-related medical experiments (i.e., bone-mineral measurement, total-body gamma-ray spectrometry, and visual light-flash-phenomena experiment) and to the Apollo 15 Mission Report for a discussion of the many engineering tests conducted during the mission. Four additional experiments not reported elsewhere are discussed in the following paragraphs.

An ultraviolet (uv) photography experiment[1] was conducted primarily to obtain imagery of the Earth and Moon for comparison with similar photographs of Mars and Venus. Both of these latter planets show mysterious behavior in the uv-wavelength region; in particular, Mars exhibits a peculiar lack of detail in

---

[1] Private communication with T. Owen, Earth and Space Sciences Dept., State University of New York, October 1971.

SUMMARY OF SCIENTIFIC RESULTS

this radiation region, and, in contrast, Venus exhibits major detail only in this part of the photographically accessible spectrum. Comparison of uv data for the Earth with corresponding data from the less well understood planets should aid in the understanding of these planets. The experiment equipment consisted of a Hasselblad camera fitted with a 105-mm uv lens, two filters with pass bands centered at 3400 and 3750×10^{-10} m, and a uv cutoff filter to obtain comparison photographs in visible light. A special uv-transmitting window is used in the CM. Photographs were taken of both the Earth and the Moon from a variety of distances throughout the mission. Preliminary examination of these photographs indicates that the surface of the Earth is still clearly visible down to 3400×10^{-10} m, and no large-scale changes in the detection of aerosols occur between 3750 and 3400×10^{-10} m. This experiment will be extended on subsequent missions in wavelength coverage and in imagery of the Earth over more land mass.

A lunar dust detector experiment[2] has been deployed as a part of the ALSEP central station on all of the manned lunar landings to date. The experiment has three purposes: (1) to measure the accumulation of dust from the LM ascent or from slow accretion processes, (2) to measure the lunar-surface brightness temperature from reflected infrared radiation, and (3) to measure long-term high-energy-proton radiation damage to solar cells in the lunar-surface environment.

The Apollo 15 dust detector experiment is mounted on top of the ALSEP central station sunshield, with the vertically mounted infrared temperature sensor facing west. Three solar cells (2 ohm-cm, N-on-P, 1- by 2-cm corner dart cells) are mounted on a horizontal Kovar metal mounting plate. For the purpose of determining radiation damage, one cell is bare; the other two cells each have 6-mil cover glasses attached. Solar-cell temperature is monitored by a thermistor on the cell mounting plate.

Results from the dust detectors deployed so far have shown (1) no measurable dust accumulation as a result of the LM ascent during the Apollo 11, 12, 14, and 15 missions; (2) a rapid surface-temperature drop of approximately 185 K during the August 6, 1971,

lunar eclipse; and (3) no long-term radiation damage thus far to any of the dust detectors. However, a difference of approximately 8.5 percent in the amount of solar-radiation energy received at the lunar surface has been measured by the Apollo 12 dust detector; this variation is a result of the change of the Moon from lunar aphelion to perihelion.

During the time between spacecraft touchdown and the powerdown of the primary guidance, navigation, and control system of the LM, nearly 19 min of data[3] were obtained from the pulsed integrating pendulous accelerometers, which are instruments normally used with the inertial measurement unit for operational guidance and navigation of the LM. In this case, however, these data provide a direct measurement of the acceleration of gravity (g) at the Hadley site. The mean of the accelerometer data results in a value for g of 162 706 mgal, with a standard deviation of 12 mgal. If a spherical mass distribution is assumed for the Moon, this same quantity can be calculated from the familiar equation $g = GM/R^2$, where GM is the product of the universal gravitational constant and the lunar mass, and R is the lunar radius at the Hadley site, as determined from a combination of doppler-shift tracking of the spacecraft around the lunar center of mass and optical tracking of the landing site from the orbiting spacecraft. If the values $GM = 4902.78$ km^3/sec^2 and $R = 1735.64$ km are used in the equation, then a value for g of 162 752 mgal is obtained, which is in good agreement with the directly measured value of acceleration at the Hadley site.

During the final minutes of the third extravehicular activity, a short demonstration experiment was conducted. A heavy object (a 1.32-kg aluminum geological hammer) and a light object (a 0.03-kg falcon feather) were released simultaneously from approximately the same height (approximately 1.6 m) and were allowed to fall to the surface. Within the accuracy of the simultaneous release, the objects were observed to undergo the same acceleration and strike the lunar surface simultaneously, which was a result predicted by well-established theory, but a result nonetheless reassuring considering both the number of viewers that witnessed the experiment and the fact that the homeward journey was based critically on the validity of the particular theory being tested.

---

[2]Private communication with James R. Bates, NASA Manned Spacecraft Center, October 1971.

[3]Private communication with R.L. Nance, NASA Manned Spacecraft Center, October 1971.

3. Photographic Summary

John W. Dietrich[a] and Uel S. Clanton[a]

The photographic objectives of the Apollo 15 mission were designed to support a wide variety of scientific and operational experiments, to provide high-resolution panoramic photographs and precisely oriented metric photographs of the lunar surface, and to document operational tasks on the lunar surface and in flight. Detailed premission planning integrated the photographic tasks with the other mission objectives to produce a balanced mission that has returned more data than any previous space voyage. The lift-off of the Apollo 15 vehicle was photographed using ground-based cameras at Kennedy Space Center, Florida (fig. 3-1).

The return of photographic data was enhanced by new equipment, the high latitude of the landing site, and greater time in lunar orbit. New camera systems that were mounted in the scientific instrument module (SIM) bay of the service module provided a major photographic capability that was not available on any previous lunar mission, manned or unmanned. Additional camera equipment available for use within the command module (CM) and on the lunar surface increased the photographic potential of the Apollo 15 mission over previous manned flights. The orbital inclination that was required for a landing at the Hadley-Apennine site carried the Apollo 15 crew over terrain far north and south of the equatorial band observed during earlier Apollo missions. During the 6 days that the Apollo 15 command-service module (CSM) remained in lunar orbit, the Moon rotated more than 75°. This longer stay time increased the total surface area illuminated during the mission and provided opportunities to photograph specific features in a wide range of illumination.

The Apollo 15 crew returned an unprecedented number of photographs. The 61-cm-focal-length panoramic camera exposed 1570 high-resolution photographs. Each frame is 11.4 cm wide and 114.8 cm long; assuming the nominal spacecraft altitude of 110 km, each frame includes a lunar-surface area of 21 by 330 km. Mapping-camera coverage consists of 3375 11.4- by 11.4-cm frames from the 7.6-cm-focal-length camera; a companion 35-mm frame exposed in the stellar camera permits precise reconstruction of the camera-system orientation for each photograph. Approximately 375 photographs were exposed between transearth injection and the deep-space extravehicular activity (EVA). Some of the mapping-camera frames exposed in lunar orbit contain no usable surface imagery because the camera system was operated during selected dark-side passes to support the laser altimeter with stellar-camera orientation data. The Apollo 15 crew also returned approximately 2350 frames of 70-mm photography, 148 frames of 35-mm photography, and 11 magazines of exposed 16-mm film. At the time this report was prepared, the photography had been screened and indexed; lunar-surface footprints of most orbital photography had been plotted on lunar charts; and index data had been transmitted to the appropriate agency for the printing of index maps.

Photographic activity began in Earth orbit when the crew exposed the first of several sets of ultraviolet (uv) photographs scheduled for the Apollo 15 mission. A special spacecraft window designed to transmit energy at uv wavelengths, a uv-transmitting lens for the electric Hasselblad (EL) camera, a spectroscopic film sensitive to the shorter wavelengths, and a set of four filters were required to record spectral data for the experiment. Earth and its atmospheric envelope were photographed from various distances during the lunar mission to provide calibration data to support the study of planetary atmospheres by telescopic observations in the uv spectrum. The uv photographs of the Moon were scheduled for use in the investigation of short-wavelength radiation from the lunar surface.

The special CM window installed for the uv-

---

[a]NASA Manned Spacecraft Center.

photography experiment had to be covered with a filter most of the time to block the uv radiation. The high levels of uv radiation in direct sunlight or in light reflected from the lunar surface to an orbiting spacecraft limited the safe exposure time of the crew to a few minutes per 24 hr. Although the shield appeared transparent at visual wavelengths, preflight tests predicted a moderate degradation in the resolution of photographs exposed through the filter. This potential degradation influenced the selection of orbital-science photographic targets.

A nominal transposition, docking, and lunar module (LM) extraction maneuver was telecast in real time and recorded by data-acquisition-camera (DAC) and Hasselblad photography (fig. 3-2). After CSM/LM extraction, the spent SIVB stage was photographed (fig. 3-3) as it began a planned trajectory toward an impact on the lunar surface near the Apollo 12 and 14 sites of the Apollo lunar surface experiments packages.

Photographic activity was at a low level through most of the 75-hr translunar-coast phase of the mission. Three sets of uv photographs recorded the spectral signature of the Earth from distances of 50 000, 125 000, and 175 000 n. mi. (fig. 3-4). A fourth set recorded the uv signature of the Moon from a point approximately 50 000 n. mi. from Earth. The initial inspection of the LM was telecast from deep space. A 16-mm DAC sequence recorded the sextant-photography test performed near the halfway point in the trip from Earth to the Moon. During translunar coast, the SIM camera systems were cycled twice to advance the film and reduce the danger of film set.

Approximately 4.5 hr before lunar orbit insertion, the SIM door was jettisoned. The DAC photographs recorded the movements of the slowly tumbling door, which was jettisoned so as to pose no danger to the crew. Removal of the protective door increased the housekeeping requirements of the SIM camera systems because temperatures had to be maintained within operational limits.

The Apollo 15 crew provided extensive and detailed descriptions of lunar-surface features after insertion into lunar orbit. Because of the length of the mission and the amount of film budgeted for specific targets, not all features described on revolution 1 were photographed. The science-support-team requests for photographs to be taken on later revolutions were forwarded by the Mission Control Center.

A busy schedule of orbital-science Hasselblad photography on revolution 2 was followed by the first of several sets of terminator photographs.

After the successful descent-orbit-insertion burn, the crew completed some of the real-time requests for photography of features they had described during the initial pass over the Seas of Crises and Serenity (fig. 3-5). Far-side terminator photographs taken with the Hasselblad camera at the end of revolution 3 were followed almost immediately by a 12-min operating period for the SIM camera systems (figs. 3-6 to 3-10). In this pass, the mapping camera and the panoramic camera photographed terrain that would become dark as the sunset terminator moved during the first lunar rest period, which occurred during revolutions 5 to 8. Representative photographs that document orbital and lunar-surface activities are shown in figures 3-11 to 3-52.

Operational tasks preparatory to the LM undocking were interrupted during revolution 9 for the telecast of the landing site and surrounding terrain. Operating difficulties delayed the undocking for approximately 25 min on revolution 12. Scheduled photography of the maneuver was not affected, but the delay canceled a low-altitude tracking pass on a landmark within the landing site. After a far-side engine firing to circularize the CSM orbit, the command module pilot (CMP) successfully tracked the landing-site landmark on revolution 13. The DAC photographs through the sextant documented the high-altitude tracking operation.

During the period that the CSM and LM were operated separately, the CMP followed a busy schedule of operational, experimental, and housekeeping tasks in orbit. The experimental tasks included photographic assignments covering a wide range of targets and requiring the use of various combinations of cameras, lenses, and films (table 3-I) or operation of the complex SIM camera systems (table 3-II). The dominant photographic task performed by the CMP, measured in terms of time and budgeted film, was lunar-surface photography. Other tasks included the documentation of operations with the LM and the photography of Earth and deep-space targets in support of specific experiments.

The LM crew used an enlarged inventory of photographic equipment (table 3-III) to document the descent, surface-operations, and ascent phases of the mission. After photographing the delayed CSM/LM separation on revolution 12 and the landing site and other lunar-surface targets during the low-altitude

TABLE 3-I.—*Photographic Equipment Used In Command Module*

Camera	Features	Film size and type	Remarks
Hasselblad EL	Electric; interchangeable lenses of 80-, 105-, and 250-mm focal length. The 105-mm lens will transmit uv wavelengths	70-mm; SO-368 Ektachrome MS color-reversal film, ASA 64; 3414 high-definition aerial film, aerial exposure index (AEI) 6; 2485 black-and-white film, ASA 6000; IIa-O spectroscopic film (uv sensitive)	Used with 80-mm lens and color film to document operations and maneuvers involving more than one vehicle. Used with appropriate lens-film combinations to photograph preselected orbital-science lunar targets, different types of terrain at the lunar terminator, astronomical phenomena, views of the Moon after transearth injection, Earth from various distances, and special uv spectral photographs of Earth and Moon
Nikon	Mechanically operated; through-the-lens viewing and metering; 55-mm lens	35-mm; 2485 black-and-white film, ASA 6000	Used for dim-light photography of astronomical phenomena and photography of lunar-surface targets illuminated by earthshine
DAC	Electric; interchangeable lenses of 10-, 18-, and 75-mm focal length; variable frame rates of 1, 6, 12, and 24 frames/sec	16-mm; SO-368 Ektachrome MS color-reversal film, ASA 64; SO-368 Ektachrome EF color-reversal film, exposed and developed at ASA 1000; 2485 black-and-white film, ASA 6000; AEI 16 black-and-white film	Bracket-mounted in CSM rendezvous window to document maneuvers with the LM and CM entry; hand-held to document nearby objects such as SIM door after jettison and subsatellite after launch, and to photograph general targets inside and outside the CSM; bracket-mounted on sextant to document landmark tracking

pass on revolution 13, the LM crew completed the preparations for the lunar landing. After powered descent initiation on revolution 14, the lunar module pilot (LMP) actuated the 16-mm DAC mounted in the right-hand LM window to record the view ahead and to the right from time of pitchover through touchdown.

The 67-hr stay time of the LM on the lunar surface accommodated three EVA periods for a total of almost 38 man-hr of lunar-surface activity. While on the surface, the crew took 1151 photographs with the Hasselblad cameras.

This mission differed significantly from previous missions not only because of the different priorities placed on the experiments and because of new equipment (such as the lunar roving vehicle (Rover) for extended mobility), but also because of changes in the schedule of crew activities after the landing. The first crew activity after powering down the LM was the standup EVA. This activity consisted of the commander's removing the docking hatch of the LM and standing on the ascent-engine cover. From that vantage point, he photographed and described the surrounding area. The value of this early photography under the low-Sun-angle lighting is easily demonstrated by comparing the standup EVA photography with the surface photography.

The first EVA began with the Rover deployment and the drive to Hadley Rille. Throughout the stay on the lunar surface, a television camera mounted on the Rover provided real-time viewing of much of the surface activities. After the crew oriented the high-gain antenna at each stop, the camera was remotely controlled from Earth. These transmissions permitted observers to evaluate the operational capabilities of the crew and to observe the collection of samples to be returned to Earth. In addition to the usual photography at station 1, the commander (CDR) used the

TABLE 3-II.—*Photographic Equipment in the Scientific Instrument Module*

Camera	Features	Film size and type	Remarks
Mapping	Electric; controls in CSM; 7.6-cm-focal-length lens; 74° by 74° field of view; a square array of 121 reseau crosses, 8 fiducial marks, and the camera serial number recorded on each frame with auxiliary data of time, altitude, shutter speed, and forward-motion control setting	457.2 m of 127-mm film type 3400	The 11.4- by 11.4-cm frames with 78-percent forward overlap provide the first Apollo photographs of mapping quality. Data recorded on the film and telemetered to Earth will permit reconstruction of lunar-surface geometry with an accuracy not available with earlier systems.
Stellar	Part of mapping-camera subsystem; 7.6-cm lens; viewing angle at 96° to mapping-camera view; a square array of 25 reseau crosses, 4 edge fiducial marks, and the lens serial number recorded on each frame with binary-coded time and altitude	155.4 m of 35-mm film type 3401	A 3.2-cm circular image with 2.4-cm flats records the star field at a fixed point in space relative to the mapping-camera axis. Reduction of the stellar data permits accurate determination of camera orientation for each mapping-camera frame.
Panoramic	Electric; controls in CSM; 61-cm lens; 10°46' by 108° field of view; fiducial marks printed along both edges; IRIG B time code printed along forward edge; data block includes frame number, time, mission data, V/h, and camera-pointing altitude	1981.2 m of 127-mm film type EK 3414	The 11.4- by 114.8-cm images are tilted alternately forward and backward 12.5° in stereo mode. Consecutive frames of similar tilt have 10-percent overlap; stereopairs, 100-percent overlap. Panoramic photographs provide high-resolution stereoscopic coverage of a strip approximately 330 km wide, centered on the groundtrack.

Hasselblad camera with the 500-mm telephoto lens to record details of the far side of Hadley Rille. The two panoramas at station 2 on the side of St. George Crater provide excellent detail along the length and bottom of Hadley Rille. The Apollo lunar surface experiments package (ALSEP) was deployed near the end of EVA-1; the only major difficulty occurred during the emplacement of the heat-flow experiment.

On EVA-2, the crew again went south to the base of Hadley Delta to sample the material of the Apennine Front. The telephotographs from station 6A provide excellent detail of lineations in Mt. Hadley. These lineations appear to dip approximately 30° to the northwest and can be traced from the summit to the base of the mountain, a distance of more than 3000 m. The "Genesis" rock was collected at Spur Crater, station 7. A study of the surface photography indicates that this anorthositic fragment was not in situ but occurred as a clast within a breccia fragment. A vesicular basalt with vesicles as large as 9 cm in diameter was documented at Dune Crater, station 4.

The crew then returned to the LM area to finish the ALSEP tasks that had not been completed during EVA-1. The CDR continued the heat-flow-experiment installation and coring, while the LMP completed the three panoramas around the LM and began the trenching and soil-mechanics measurements near the ALSEP location. The second EVA ended with the coring completed but with the corestem still in the hole.

The duration of the third EVA was shortened

TABLE 3-III.—*Photographic Equipment Used in LM and on Lunar Surface*

Camera	Features	Film size and type	Remarks
Hasselblad data camera (DC), 2	Electric; 60-mm focal-length lens; reseau plate	70-mm; SO-168 Ektachrome EF color-reversal film exposed and developed at ASA 160; 3401 Plus-XX black-and-white film, AEI 64	Handheld within the LM; bracket-mounted on the remote-control unit for EVA photography; used for photography during standup EVA and through the LM window and for documentation of surface activities, sample sites, and experiment installation
Hasselblad DC	Electric; 500-mm lens; reseau plate	70-mm; 3401 Plus-XX black-and-white film, AEI 64	Handheld; used to photograph distant objects during standup EVA and from selected points during the three EVA periods
DAC	Electric; 10-mm lens	16-mm; SO-368 Ektachrome MS color-reversal film, ASA 64	Mounted in the LM right-hand window to record low-altitude views of the landing site one revolution before landing, to record the LMP view of the lunar scene during descent and ascent, and to document maneuvers with the CSM
Lunar DAC	Electric; 10-mm lens; battery pack and handle	16-mm; SO-368 Ektachrome MS color-reversal film, ASA 64	Handheld or mounted on the Rover to document lunar-surface operations; photography from this camera seriously degraded by intermittent malfunction of film feed

from the planned 6 hr to 4.5 hr. After considerable time and effort, both crewmembers recovered the core from the drill hole and continued the traverse to Hadley Rille. At stations 9A and 10, the CDR again obtained telephotographs of the west wall of Hadley Rille. These stations are approximately 300 m apart, and the combined telephotography from these two locations provides exceptional stereocoverage of Hadley Rille features, such as outcrops with massive and thin bedding, columnar jointing, boulder trails, and faulting.

The time used in recovering the core at the start of EVA-3 prevented the traverse to the North Complex from continuing as planned. The crew returned to the LM for closeout. The Rover was parked east of the LM so that the lift-off could be transmitted by the television system on the Rover. Some additional photography, both normal and telephotographic, during the closeout period provided additional coverage of the LM area and of the Apennine Front. The telephotographs obtained before lift-off provide additional detail of the Hadley Delta and the Mt. Hadley areas.

This mission, more than any other, demonstrated the detailed photographic information that is gained and lost as a function of changing Sun angle.

The lift-off of the LM ascent stage from the lunar surface was telecast for the first time by the remotely controlled television camera on the Rover. The DAC photographs of the lunar surface from the right-hand window of the LM during ascent were particularly interesting. The ALSEP site, boot and tire tracks, and surface features along and within Hadley Rille are clearly visible in these photographs. As the LM ascended and moved westward, the DAC field of view moved northward to cover Hadley Rille from the landing site to the point where the sinuous depression swings northward beyond the end of Hill 305.

After the LM was cleared to remain on the lunar surface, the CMP began his full schedule of orbital tasks. On revolutions 15 and 16, the Hasselblad cameras photographed four orbital-science targets and extra targets of opportunity within the Sea of Serenity. Two periods of mapping-camera operation totaled 2 hr 25 min and included a complete revolution of continuous mapping-camera, stellar-camera, and laser-altimeter operation. Two periods of panoramic-camera operation totaled 55 min. Telemetered data indicated an intermittent anomaly in the panoramic-camera velocity-over-height (V/h) sensor during all periods of operation. Real-time analysis suggested that as much as 30 percent of the panoramic frames exposed on revolution 16 were affected by the anomaly. Postmission tests have demonstrated that image smear, though present, generally does not seriously degrade the usefulness of the imagery. The second lunar rest period occurred during revolutions 18 to 21.

A broad spectrum of photographic tasks was scheduled during revolutions 22 to 28. Mapping-camera operations during three revolutions totaled almost 4 hr. Panoramic-camera operation totaled only slightly more than 1 min, but that brief period of activity provided high-resolution photographs of the lunar surface after the completion of EVA-1. A long strip of Hasselblad photographs across the Lick-Littrow orbital-science target, two periods of solar-corona photography, uv photographs of the Earth above the lunar horizon, and photographs supporting studies of dim-light phenomena completed the photographic activities during this period. The mapping camera was operated continuously during revolution 22 and the lighted half of revolution 23. A CSM maneuver near the far-side terminator on revolution 23 tilted the camera axis from the local vertical to provide forward-oblique photographs. The mapping camera was again activated for vertical photography across the lighted half of revolution 27 before the third lunar-orbit rest period, which occurred during revolutions 29 to 32.

Photographic tasks completed during revolutions 33 to 39 included operation of the mapping and panoramic cameras, Hasselblad photography of the orbital-science targets, 35-mm photography of the lunar surface illuminated by earthshine, and 35-mm photography of the gegenschein and zodiacal light in support of dim-light studies. The mapping camera was operated during revolution 33 and the lighted half of revolution 38 with the camera axis alined along the local vertical. Terminator-to-terminator passes of oblique photographs with the camera axis pointed backward and then northward were completed on revolutions 34 and 35, respectively. Three periods of panoramic-camera operation, one on revolution 33 and two on revolution 38, totaled slightly more than 40 min. Revolutions 40 to 43 were reserved for the fourth CMP rest period during lunar orbit.

Preparation for the return of the LM dominated activities in the CSM for the remainder of the solo phase of the mission. The mapping camera was operated through the lighted half of revolution 44, and gegenschein photography was performed after spacecraft earthset on revolution 46. The sextant-mounted DAC recorded the lunar scene during landmark tracking of the LM in preparation for lift-off, which occurred during CSM revolution 48.

The rendezvous between the LM and CSM, which occurred during revolution 49, was documented by DAC and Hasselblad photographs from both spacecraft (fig. 3-53). Before the nominal docking maneuver, the LM crew inspected the SIM bay from close range to determine whether a foreign object was intermittently blocking the V/h sensor port and thus causing the erratic signals from the panoramic camera. All SIM bay equipment appeared to be in satisfactory condition. Mapping-camera photography scheduled for the lighted half of revolution 50 was terminated on instructions from the Mission Control Center after only 42 min of operation. Panoramic-camera photographs obtained during the 2.5-min period of operation document the Hadley Rille site after the LM ascent. After documentation of the delayed LM jettison on revolution 52, photographic tasks were canceled to facilitate CSM cleanup and to complete essential operational tasks. The fifth rest period in lunar orbit occurred during revolutions 55 to 59.

Photographic schedules for revolutions 60 to 64 were revised to permit the accomplishment of most tasks delayed by the jettison problems and the long rest period. The mapping camera operated a total of 3 hr and provided terminator-to-terminator coverage on revolutions 60, 62, and 63. The panoramic camera was operated a total of approximately 40 min; a 10-min operation on revolution 60 was followed by two test sequences near the landing site, a brief period across the target point for LM impact on revolution 61, and continuous operation for 26 min on revolution 63. Hasselblad photography of orbital-science targets was supplemented by unscheduled telephotographs of selected lunar-surface features. The lunar-surface Hasselblad camera was returned to the CSM and, equipped with a 500-mm lens, was used to document targets of opportunity. Hasselblad photography supporting special experiments during this period included near- and far-side terminator photographs and two series of uv photographs recording the spectral data for lunar maria and highlands. The final rest period in lunar orbit occurred during revolutions 65 to 68.

The mapping camera was operated a total of 3 hr to provide terminator-to-terminator vertical photography during revolutions 70 and 72 as well as a sequence of unusually worthwhile oblique photographs that cover terrain out to the southern horizon on revolution 71. Film for the panoramic camera was exhausted near the end of a scheduled 24-min operating period during revolution 72. Other photographic tasks completed before the orbit-shaping engine firing on revolution 73 included orbital-science photography, unscheduled telephotography of targets of opportunity, solar-corona photography near the far-side terminator, and four sets of terminator photography.

The subsatellite launch on revolution 74 was documented with 16-mm DAC and 70-mm Hasselblad photographs (fig. 3-54). A sequence of DAC photographs recorded the initial spin rate and orientation of the small instrumented platform. The DAC sequence and Hasselblad photographs taken at random intervals documented the condition of the surfaces of the subsatellite and confirmed proper deployment of the booms.

Near the end of revolution 74, the successful firing of the service propulsion system engine injected the CSM into a transearth trajectory. Hasselblad and mapping-camera photography of the lunar surface recorded the changing aspects of the visible disk as the distance between the Moon and the CSM increased (figs. 3-55 and 3-56). Operation of the mapping camera during the transearth-coast phase totaled 3 hr 35 min. After a rest period of approximately 7 hr, the crew completed the solar-corona window-calibration photography and prepared to recover the film in the SIM bay. Video, DAC, and Hasselblad cameras were used to document the EVA (fig. 3-57). A sequence of uv photographs of Earth, originally scheduled to be taken before the EVA was performed, was completed shortly after the CMP returned to the CM.

The photographic workload was comparatively light for the remainder of the mission. During the final 2 days, the crew took two more sets of uv photographs of the Earth, performed a sextant-photography test, and exposed a sequence of photographs documenting the lunar eclipse. Documentation of entry phenomena visible from the CM window was the final photographic task assigned the crew in this most scientifically important mission to date. Carrier-based helicopters photographed the crew's somewhat rapid, but successful, splashdown (fig. 3-58).

FIGURE 3-1.—Apollo 15 lift-off from pad A, launch complex 39 (S-71-41411).

FIGURE 3-2.—During the translunar-coast phase of the mission, the CSM completed a transposition and docking maneuver to extract the LM from the SIVB stage. The top hatch and docking target on the LM are clearly visible in this predocking photograph (AS15-91-12333).

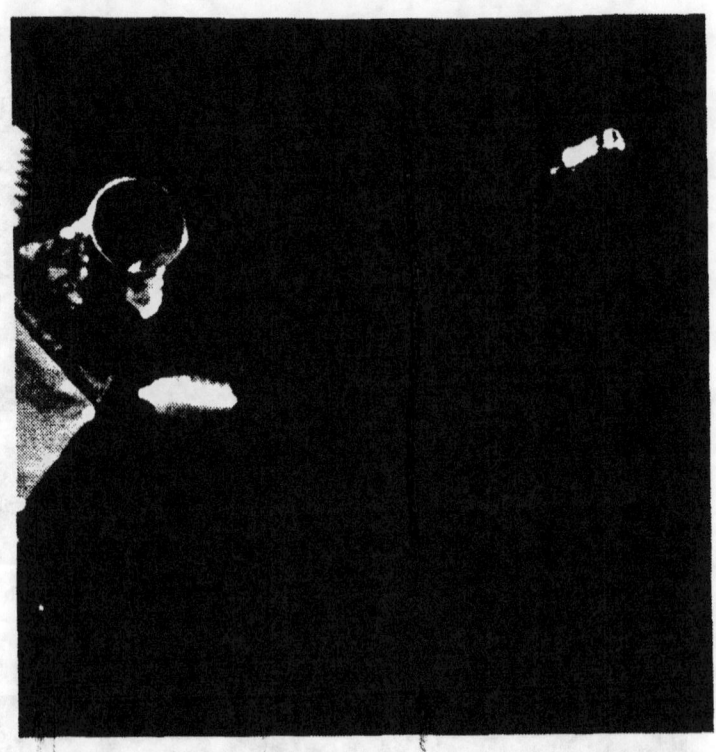

FIGURE 3-3.—The SIVB stage is visible to the right of the LM thrusters. The compartment formerly occupied by the LM is to the right; the rocket skirt is to the left. After the docking maneuver and the extraction of the LM, the SIVB stage was targeted to impact the Moon. The impact on the lunar surface provided an energy source of known magnitude that was used to calibrate the seismometers emplaced by the Apollo 12 and 14 crews (AS15-91-12341).

FIGURE 3-4.—The Apollo 15 crew saw a gibbous Earth after the transposition, docking, and LM extraction maneuver. South America, nearly free of clouds, is in the bottom center; Central America and North America are upper left; and the western coasts of Africa and Europe are at the top right (AS15-91-12343).

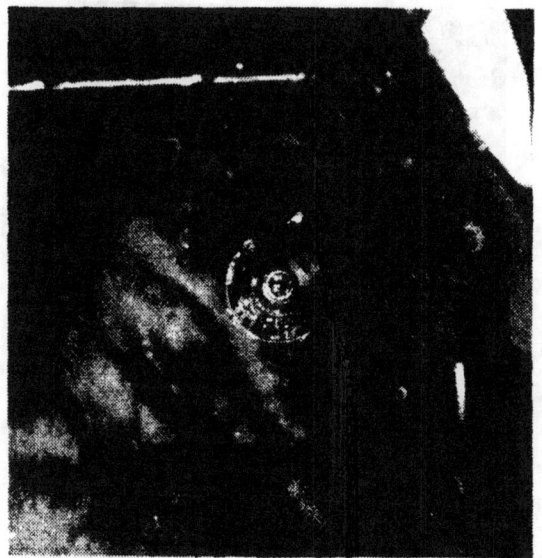

FIGURE 3-5.—The CSM is photographed through the LM window during stationkeeping and just before powered descent initiation. The eastern edge of the Sea of Serenity, south of the crater le Monnier, is below the spacecraft; north is to the right (AS15-87-11696).

FIGURE 3-6.—Apollo 15 landing site and vicinity. A broad area near the Hadley-Apennine landing site is documented by the mapping camera in this oblique view northward to the horizon. The Apennine Mountains, including the peaks of Mt. Hadley and Hadley Delta; Hadley Rille and the Hadley C Crater; and mare deposits of the Marsh of Decay occupy the lower half of the frame. Near the horizon, the Caucasus Mountains separate the Sea of Serenity at right from the Marsh of Mists (left), which is an arm of the Sea of Rains at far left. Autolycus (near) and Aristillus are the two large craters between the Sea of Rains and the Marsh of Mists (mapping camera frame AS15-1537).

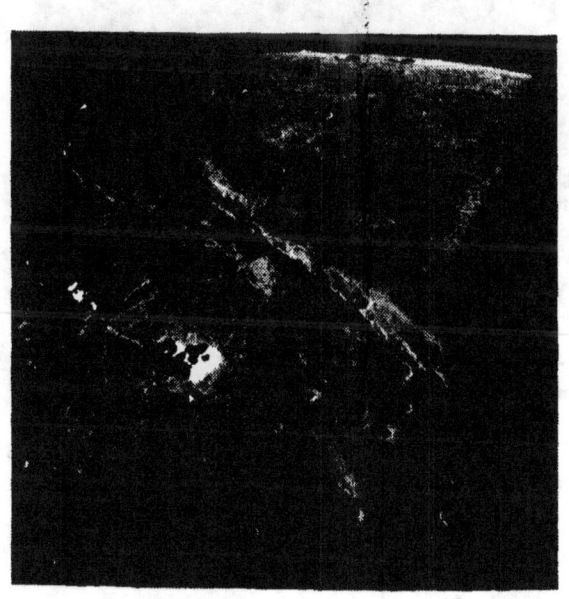

FIGURE 3-7.—The Sea of Rains illuminated by a low Sun angle. The mapping camera was used to photograph continuous strips from terminator to terminator on several revolutions. This oblique view northward across the Sea of Rains demonstrates the effectiveness of low-Sun illumination in accentuating features of low relief. Lambert Crater is just outside the lower right corner of the photograph, but ejecta from the crater extend into the field of view. Mt. Lahire casts a very long shadow across the smooth mare deposits in the left central part of the photograph; Helicon and Le Verrier are the large craters near the horizon. The Imbrian flows extend as a belt to the right and left from the prominent mare ridge. Lobate flow fronts along the north and south margins of the belt are clearly visible in the upper half of this near-terminator view. The protective cover remained within the camera field of view during revolution 35 (mapping camera frame AS15-1555).

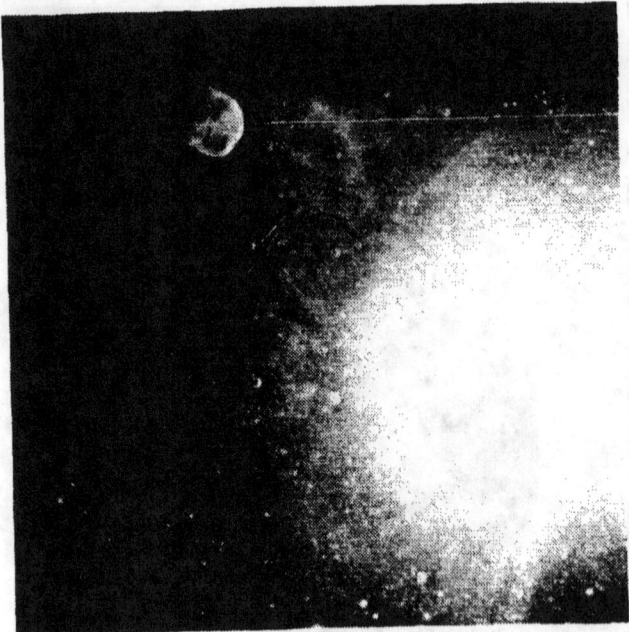

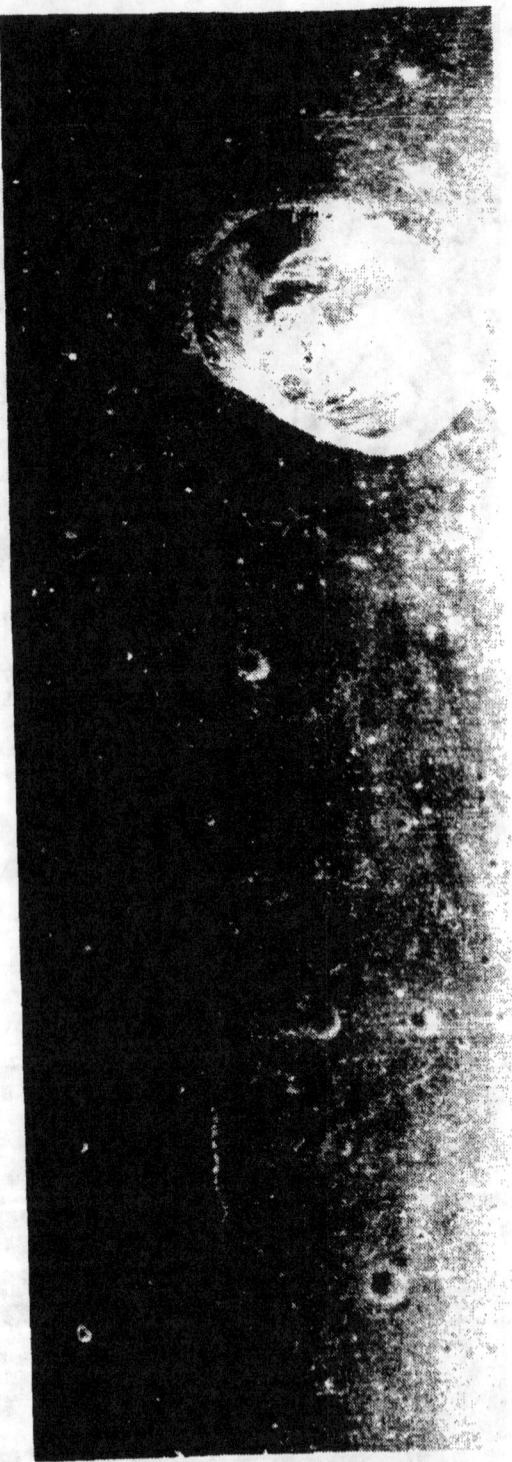

FIGURE 3-8.—Dawes Crater and mare deposits in the Seas of Serenity and Tranquillity are illuminated by a high Sun angle. Lunar-surface retroreflectivity produces the bright spot near the right edge of the photograph, which shows an area of normal mare material in the Sea of Serenity. Darker deposits surrounding Dawes Crater crop out as a ring around the southern margin of the Sea of Serenity. A sharp contact between the two units forms an arc from the upper right corner to the middle of the lower edge of the photograph. The photograph is oriented with south toward the top to facilitate comparison with figure 3-9 (mapping camera frame AS15-1658).

FIGURE 3-9.—Panoramic camera photograph of Dawes Crater. The 60-cm focal-length lens of the camera provides high-resolution photographs of the lunar surface from orbital altitudes. Approximately one-fourth of the 11.4- by 114.8-cm frame is reproduced; the area shown includes terrain south of the CSM at viewing angles between 15° and 45° off the vertical. The surface features in and near Dawes Crater can also be seen in figure 3-8. High-resolution photographs such as this will support continuing studies of lunar-surface features long after the completion of the last Apollo flight (panoramic camera frame AS15-9562).

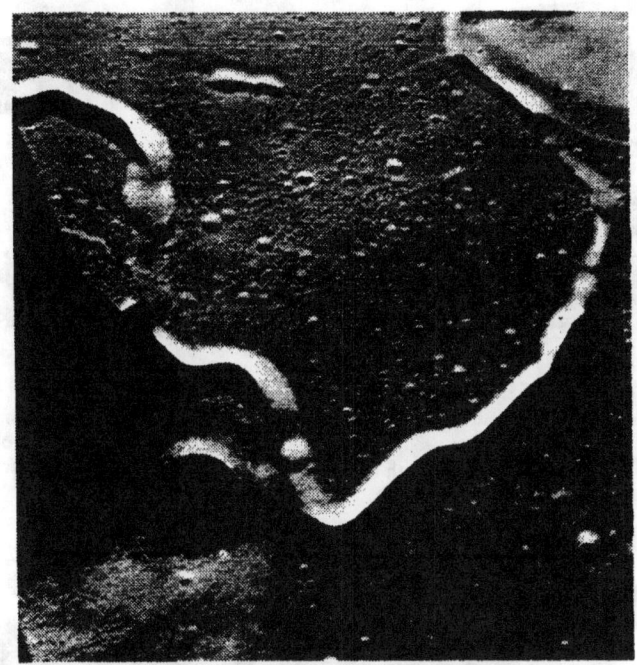

FIGURE 3-10.– The sinuous path of Hadley Rille winds along the Apennine Front. North is to the right, South Complex is at bottom center, North Complex is at the lower right, St. George Crater and Hadley Delta are in the lower left, and Hill 305 is at the upper right. At least two sets of lineations can be identified in Hadley Delta. Both sets exist in St. George Crater and on both sides of the inner slopes of Hadley Rille. Bedding is visible along some parts of the rille wall. At the upper right, the rille narrows and then widens abruptly. A smaller, less well-developed rille branches off from this enlarged area (AS15-87-11720).

FIGURE 3-11.– Hadley Delta, with Silver Spur to the left and St. George Crater to the right, is shown in this composite photograph taken during the standup EVA. Hadley C is the bright peak above St. George Crater. Photographs taken during the standup EVA show best the east-dipping lineations in Silver Spur and Hadley Delta. Close observation of the rim of St. George Crater shows the presence of a second set of lineations with an east-west strike. These two sets of lineations are somewhat obscured in photographs taken by the crew at higher Sun angles (S-71-51737).

FIGURE 3-12.—Seven of the 14 photographs taken from the left- and right-hand LM windows have been used in this composite photograph. The view covers approximately 180° from Hadley Delta in the south to the base of Mt. Hadley in the north (S-71-47078).

FIGURE 3-13.—Composite photograph forms a view to the north along Hadley Rille. Some horizontal tonal and textural differences, which can be detected in the east wall, may correlate with the horizontal bedding in the west wall of the rille. The boulder to the right center was the location of the major sampling and documentation activity at station 2, where the CDR is unloading equipment from the Rover. Fragments from the boulder and fine material from around and from under the boulder were collected (S-71-51735).

FIGURE 3-14.—The tongs are used to measure the distance between the camera and the object for these closeup photographs. The depth of field is approximately 4 cm at these camera settings. This boulder, the object of much of the activity of station 2, has a well-developed coating of glass that displays a range of vesicle sizes; some vesicles are approximately 2 cm in diameter (AS15-86-11555).

FIGURE 3-15.—The CDR places the Apollo Lunar-surface drill on the surface in preparation for adding additional core stems. The rack with core and bore stems is to his right and the solar wind spectrometer is in the foreground; the tapelike ribbons are cables that connect the central station to the individual ALSEP experiments (AS15-87-11847).

FIGURE 3-16.—The CDR drives the Rover near the LM. Material can be seen dropping from the front and rear wheels. The ALSEP is deployed between the Rover and Hill 305 on the horizon. The solar wind composition experiment appears in the background between the television camera and the antenna mast (AS15-85-11471).

FIGURE 3-17.—The LMP salutes after the flag deployment on the lunar surface. Hadley Delta forms the skyline behind the LM, which is located in a crater on a slope of approximately 10°. The Rover is parked in a north-south orientation because of thermal constraints between EVA periods. The modularized equipment stowage assembly of the LM appears just above the right front fender of the Rover (AS15-92-12446).

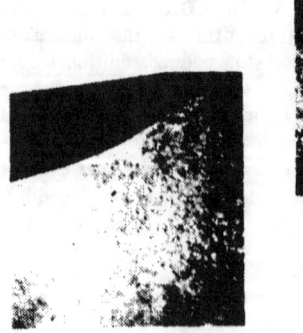

FIGURE 3-18.—During EVA-2, the CDR used the Hasselblad camera with the 500-mm telephoto lens extensively from station 6A on the side of Hadley Delta. This mosaic shows the detail in Mt. Hadley that was recorded from a distance of approximately 18 km. The largest sharp crater in this photograph is approximately 100 m in diameter. The lineations that can be resolved are approximately 10 m thick and can be traced from the summit to the base of the mountain, a distance of approximately 3000 m (S-71-48875).

FIGURE 3-19.—On the side of Hadley Delta at station 6A, the CDR took a series of photographs with the 500-mm telephoto lens. The large crater in the foreground, at a distance of approximately 2 km, is Dune; and the series of white dots to the left of the LM (approximately 5 km distant) is the deployed ALSEP. The huge crater that forms the background above the LM is Pluton at a distance of 8 km. Pluton is approximately 800 m in diameter, and the largest boulder in the crater is approximately 20 m in diameter (AS15-84-11324).

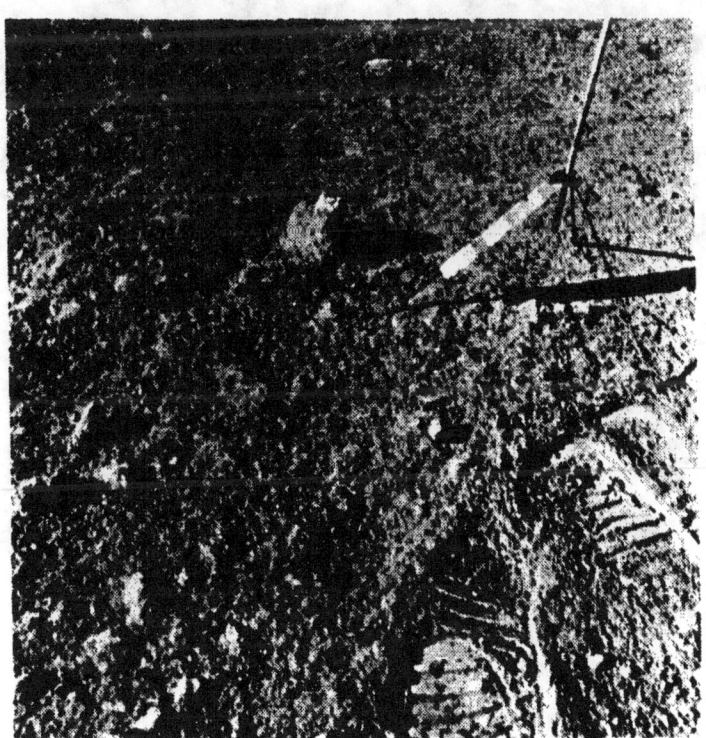

FIGURE 3-20.—The small white clast on top of the larger gray fragment located to the upper left of the gnomon is the "genesis" rock as the Apollo 15 crew found it on the lunar surface. It should be noted that this anorthositic fragment is not in situ but occurs as a clast within a breccia. This sample was collected at station 7, Spur Crater, on EVA-2 (AS15-86-11670).

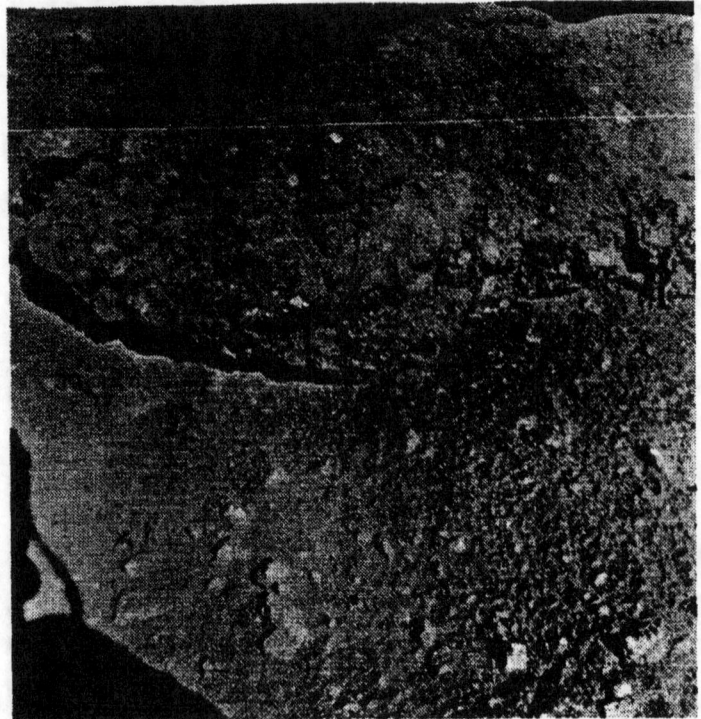

FIGURE 3-21.–During EVA-2, the Apollo 15 crew found this boulder at station 7, Spur Crater. This boulder is a breccia with a dark matrix and white clasts. A representative specimen of this type of breccia was returned to Earth as sample 15445 (AS15-86-11689).

FIGURE 3-22.–Composite photograph of Dune Crater at station 4; Dune Crater is approximately 500 m in diameter. The vesicular boulder on the edge of the crater marks the area that was sampled and documented at this location. Mt. Hadley with the spectacular lineations dominates the skyline. Two of the three terraces observed by the crew can be seen at the base of Mt. Hadley (S-71-51736).

FIGURE 3-23.—Much of the activity at station 4 was devoted to the boulders shown in this photograph. These boulders are basalts and represent material excavated during the formation of Dune Crater. The crew reported that vesicles were abundant in the boulders at this location and that some vesicles were as large as 9 cm in diameter. The abundance and size of the vesicles in these boulders suggest that this basaltic material cooled on or very near the lunar surface (AS15-87-11779).

FIGURE 3-24.—The crew made a series of closeup photographs of the large boulder at Dune Crater. The tongs are used to measure the distance from the camera to the object because of the limited depth of field. This photograph shows the size range of vesicles on this surface of the boulder; plagioclase laths are also visible (AS15-87-11773).

FIGURE 3-25.—The LMP digs a trench near the ALSEP site at the end of EVA-2. This trench, used for some of the soil-mechanics measurements, was the source of samples 15030, 15040, 15013, and the special environment sample container. At a depth of 30 to 35 cm, the LMP reported the presence of a hard layer that he could not penetrate with the scoop (AS15-92-12424).

FIGURE 3-26.—Composite photograph of the LM taken from the ALSEP location. The Apennine Front forms the background behind the LM. Wheel and foot tracks crisscross in the foreground (S-71-51738).

FIGURE 3-27.—Hadley Delta, Silver Spur, and St. George Crater form the skyline in this view toward the south of Hadley Rille from station 9A. The CDR works at the Rover to remove the Hasselblad camera for the telephotographs. It is noteworthy that the front panel on the left front fender of the Rover has been lost (AS15-82-11121).

FIGURE 3-28.—The textured pattern in the upper center of this photograph was made by the LMP with the lunar rake. The rake is used to collect a comprehensive sample—a selective collection of rocks in the 1- to 3-cm size range. Samples 15600 and 15610 were collected at station 9A (AS15-82-11155).

FIGURE 3-29.—View to the south down Hadley Rille from station 9A. Both the eastern and western sides of the rille are shown. Hadley Delta is the mountain in the background, and St. George Crater is partially visible in the upper right. The boulders in the foreground are basalts from the units that crop out along the rille (AS15-82-11147).

FIGURE 3-30.—Just below the CDR's right hand and approximately 5 cm to the right of the hammer handle is the area where samples 15595 to 15598 were chipped from this boulder. The fragments are in the bag that the CDR has in his left hand. He wears one of the EVA cuff checklists on his right wrist. The gnomon, which is used to determine the local vertical, is positioned on the sampled boulder (AS15-82-11145).

FIGURE 3-31.—The CDR walks toward Hadley Rille at station 10 to take telephotographs of the far side. He carries the Hasselblad camera with a 500-mm lens in his left hand as he walks away from the Rover (AS15-82-11168).

FIGURE 3-32.—Telephotograph of the western wall of Hadley Rille from station 10. The more obvious bedded units are overlain by a massive unit with a pronounced fracture pattern that dips to the right (north). The large white boulder at the upper right margin is approximately 8 m long. The distance along the western edge of the rille as shown in this view is approximately 150 m (AS15-89-12105).

FIGURE 3-33.—Telephotograph of Hadley Rille from station 10 showing the bedded units in situ on the western wall. This outcrop shows both massive and thin bedded units; the massive units have well-developed columnar jointing. The thin-bedded units have individual beds that are less than 1 m thick. The largest boulders on the western slope are 8 to 10 m in diameter (AS15-89-12116).

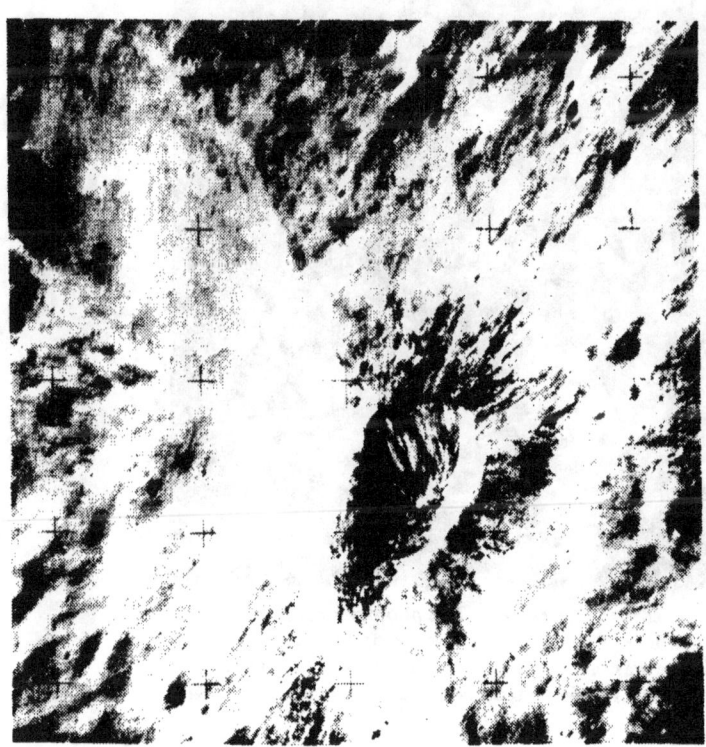

FIGURE 3-34.—Sharp raised-rim crater with rays. Rays of dark ejecta streak the bright halo surrounding this sharp crater on the crest of the northeast rim of Gibbs Crater. The rim of Gibbs, which trends from upper right to lower left through the center reseau cross and through the sharp crater, is not prominent in this telephotograph taken with the 500-mm lens. This feature was selected during premission planning for photography if the opportunity and film were available. The Apollo 15 astronauts photographed many such contingency targets during the final revolutions in lunar orbit (AS15-81-10920).

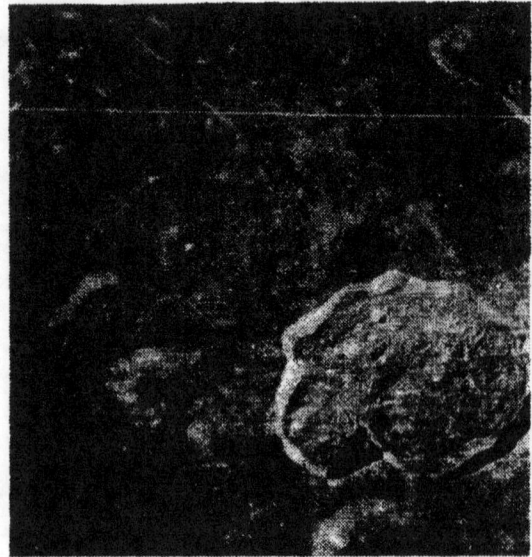

FIGURE 3-35.—Oblique view northwestward across Newcomb Crater in highland terrain northeast of the Sea of Crises near latitude 30° N, longitude 44° E. The slight embayment toward the lower left was considered (on the basis of Earth-based observations) to be another crater, which was named Newcomb A (AS15-91-12353).

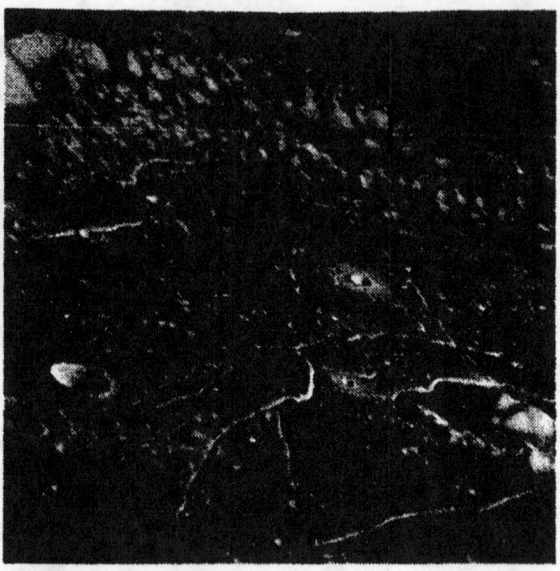

FIGURE 3-36.—Eastern floor of Humboldt Crater. The mare-type material on the floor contains radial cracks and concentric rilles. A dark-halo area is visible at the lower left corner. Low hills of material that resemble the central peak protrude through the smooth crater floor. Bright-halo craters are also evident. The "doughnut" filling of the crater at the left margin is a rare feature (AS15-93-12641).

FIGURE 3-37.—Crescent Earth low over the lunar horizon. This photograph, one of a series, was exposed through the 250-mm lens. The CSM was above a point near latitude 24° S, longitude 99° E, when this picture was taken. Steep slopes on the left horizon are the southwestern inner rim of Humboldt Crater near latitude 25° S, longitude 78° E (AS15-97-13268).

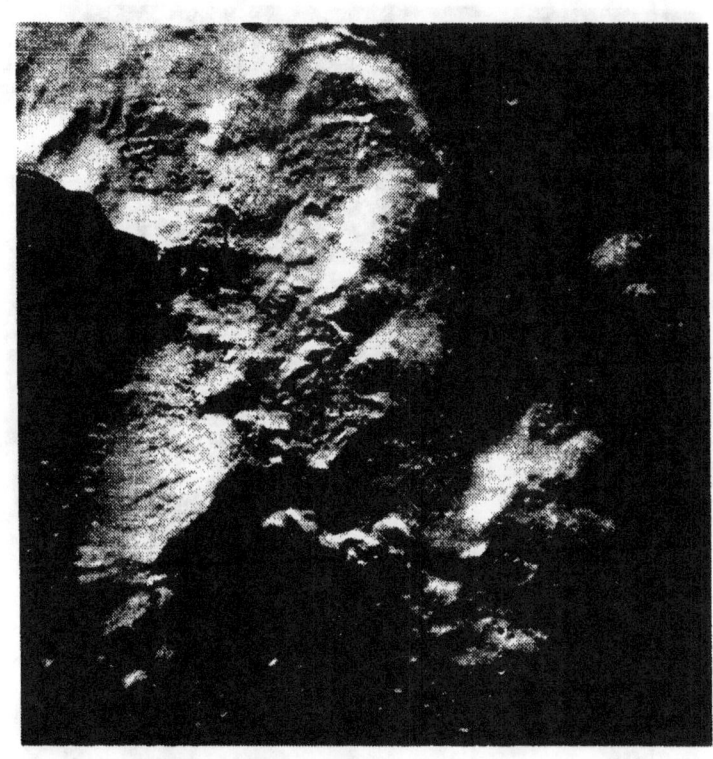

FIGURE 3-38.—East end of the central peak in Tsiolkovsky Crater. In this near-vertical view, dark crater-floor deposits contrast sharply with the high-albedo rocks of the central peak. West is toward the top of the photograph. The clearly visible outcrop along the left-facing sheer wall, near the point of the peak, is the bedding the CMP described while in orbit (AS15-96-13017).

FIGURE 3-39.—Oblique view southwestward across the far-side Crater Tsiolkovsky. The central peak of high-albedo material contrasts sharply with the dark deposit on the crater floor. The central peak is 265 km southwest of the ground track in this telephotographic view through the 250-mm lens. The far-side wall, 336 km from the camera, is much steeper and a little higher than the near rim of the crater (AS15-91-12383).

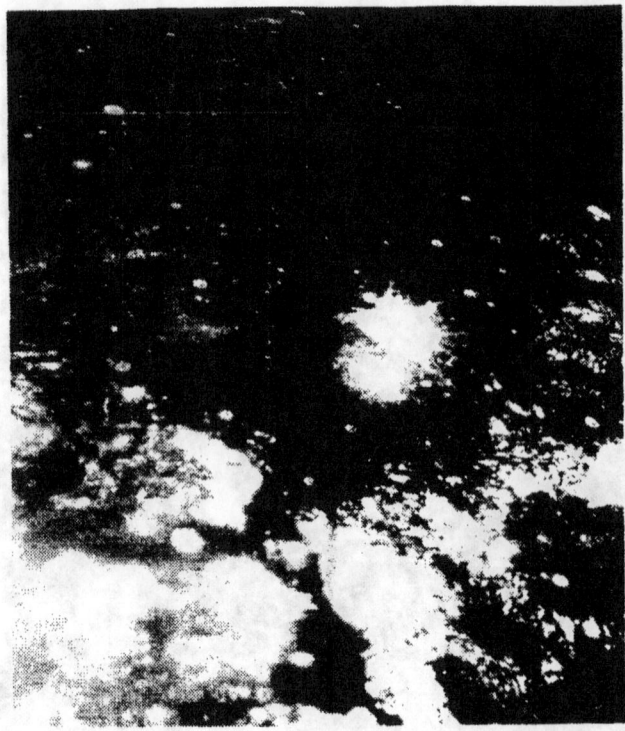

FIGURE 3-40.—Oblique view northwestward across the Littrow Rille area of the eastern Sea of Serenity. A high Sun enhances the albedo contrast. Steep slopes in the highlands, crater walls, and rays are much brighter than the normal mare materials in the central Sea of Serenity at the top of the photograph. The ring of darker mare material and the still darker mantling material of the Littrow region are near the highlands margin (AS15-94-12846).

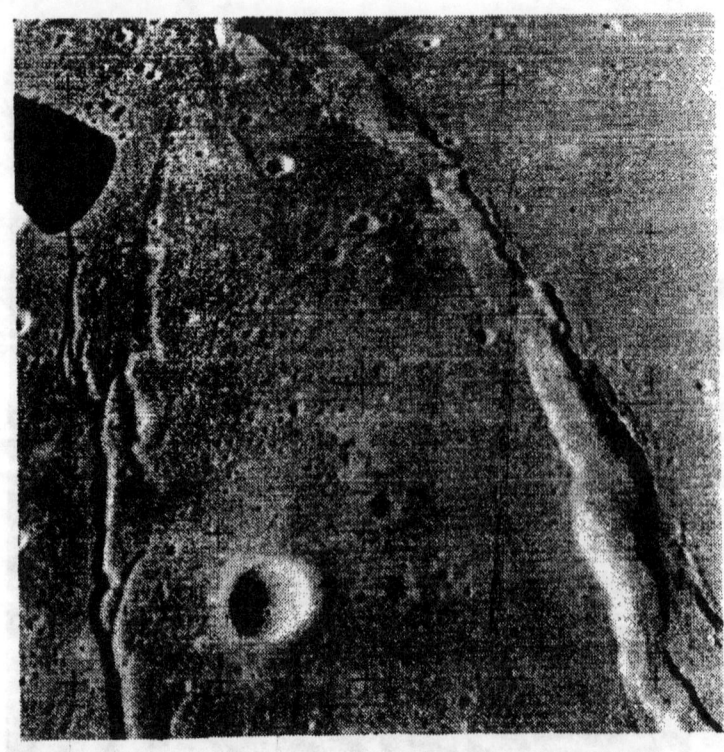

FIGURE 3-41.—South-southwestward oblique view from the LM across eastern Sea of Serenity. A branching rille (depressed center) near the left edge of the photograph contrasts sharply with the broad mare ridge (raised center) at right. The partly concealed crater to the left of the thruster is le Monnier LA. The photograph was taken with a Hasselblad camera equipped with a 60-mm lens one revolution before landing (AS15-87-11712).

FIGURE 3-42.—Oblique view northward across the Ocean of Storms. The feature that extends directly away from the camera consists of a mare ridge near the lower edge of the photograph, of a chain of elongate craters with raised rims in the central section, and of a sinuous rille near the margin of the highlands. The prominent crater near the upper right is Gruithuisen K (AS15-93-12725).

FIGURE 3-43.—View southwestward across Prinz, Aristarchus, and Herodotus Craters. The Prinz rilles are in the foreground; the Aristarchus rilles flow generally toward the camera from the Aristarchus Plateau in the background. Cobra Head and the upper end of Schröter's Valley are near the center of the upper margin. The extremely high albedo of the Aristarchus rays and halo is noteworthy (AS15-93-12602).

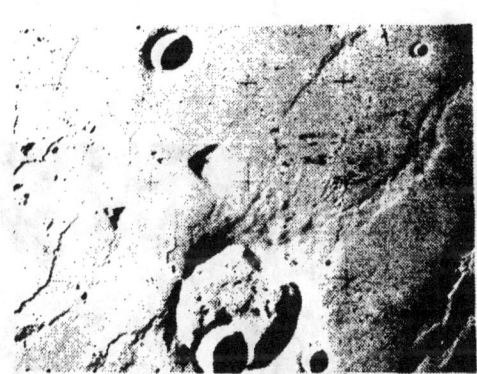

FIGURE 3-44.—View northwestward across the Ocean of Storms at the north margin of the Aristarchus Plateau. Krieger Crater, near the lower edge of the photograph, has a rim broken in two places. Krieger B, approximately one-fifth the diameter of the larger crater, is centered just inside the rim at one point. A prominent rille appears to emerge from the other break. Wollaston is the large, raised-rim crater in the upper left quadrant of the photograph. The similar orientation of mare ridges in the upper right quadrant and the ridges and rilles in the lower left quadrant are noteworthy (AS15-90-12272).

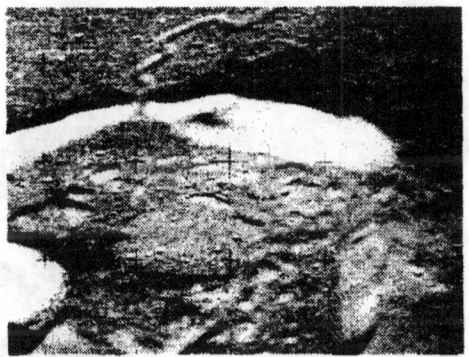

FIGURE 3-45.—Telephotographic oblique view westward across Krieger Crater. Two small dark-halo craters occupy an anomalous bench on the inner slope of the crater wall to the right of the break in the crater wall. Within the crater, no evidence exists of the sinuous channel that is prominent beyond the wall (AS15-92-12480).

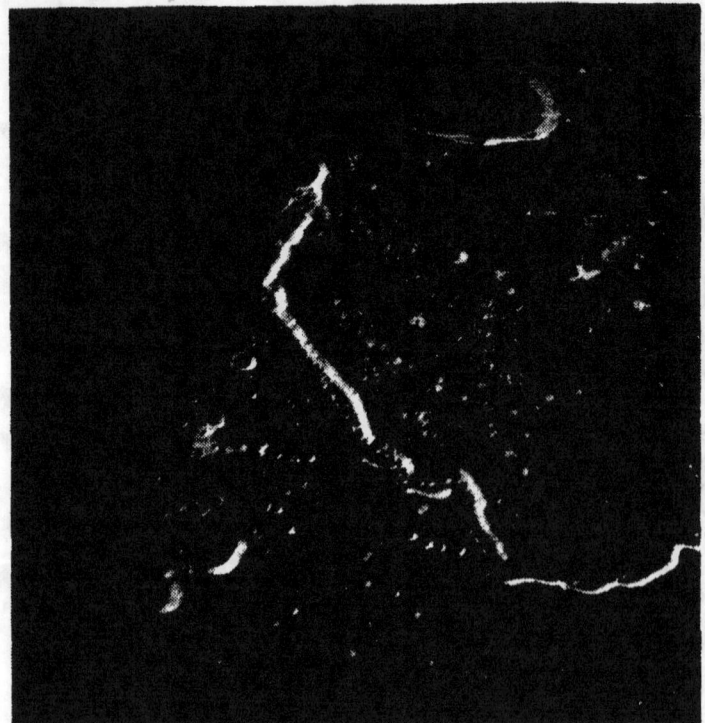

FIGURE 3-46.—Cobra Head and upper Schröter's Valley. The broad outer channel of Schröter's Valley consists of numerous more or less straight segments that join at angles, but the inner channel is truly sinuous. Cobra Head is near the high point of the Aristarchus Plateau and due north of Herodotus Crater. In this view toward the southwest, the remnants of an older, wider, and higher channel can be seen along the segment of the rille that trends toward the camera (AS15-93-12624).

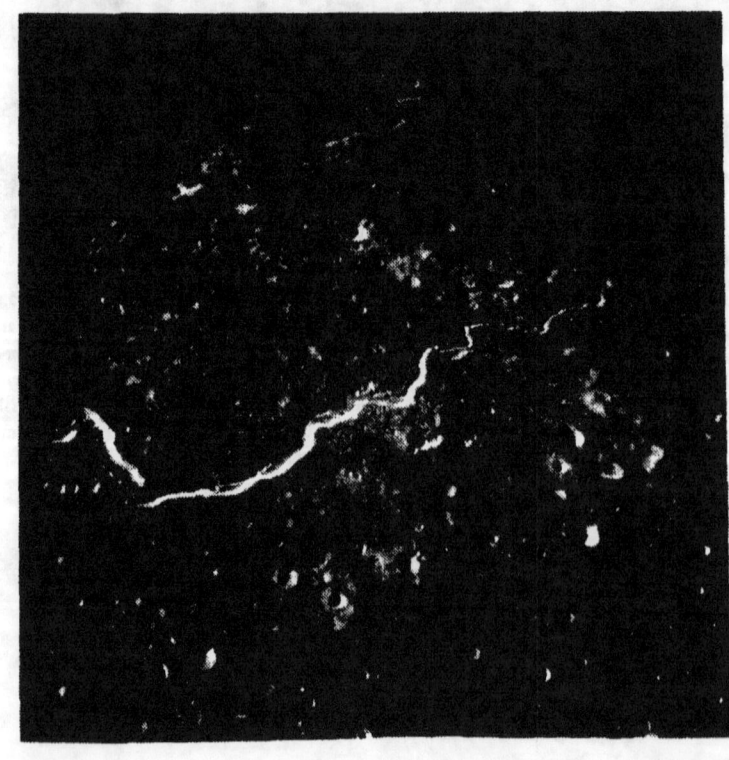

FIGURE 3-47.—This lower segment of Schröter's Valley trends southwestward down the gently sloping surface of Aristarchus Plateau. The outer channel is more segmented than sinuous, but the narrow inner channel is truly sinuous. At a point to the right of center, the narrow sinuous channel breaks through the far wall of the outer channel and winds southward toward the flatter surface of the Ocean of Storms (AS15-93-12628).

FIGURE 3-48.—A very low Sun illuminates this view of southwestern Aristarchus Plateau. The effects of changing Sun angle are best demonstrated by examining the same area illuminated by the Sun at a range of elevation angles. This eastward view includes three distinctive rilles that can be easily identified in figure 3-49 (AS15-98-13345).

FIGURE 3-49.—Telephotograph of an area of southwestern Aristarchus Plateau. The plateau was illuminated by a Sun only 15° to 20° above the horizon at the time this photograph was taken. Although this elevation is still low, a dramatic change in the surface appearance has resulted from the 10° to 15° increase in Sun angle since figure 3-48 was exposed (AS15-95-12978).

FIGURE 3-50.—Western half of Posidonius Crater on the eastern margin of the Sea of Serenity. A tightly convoluted sinuous rille crosses the raised floor of the crater, turns back, and follows the rim to a low point in the western rim (AS15-91-12366).

FIGURE 3-51.—View westward to the terminator in eastern Sea of Rains. The low Sun angle exaggerates small differences in elevation. Beer and Feuillée Craters, near the center, are at the edge of material that looks like rough mare deposits in a higher Sun angle. In this view, the affinity of this material to nearby highlands seems more likely. The chain of craters extending toward the camera from Beer Crater is Archimedes Rille I (AS15-94-12776).

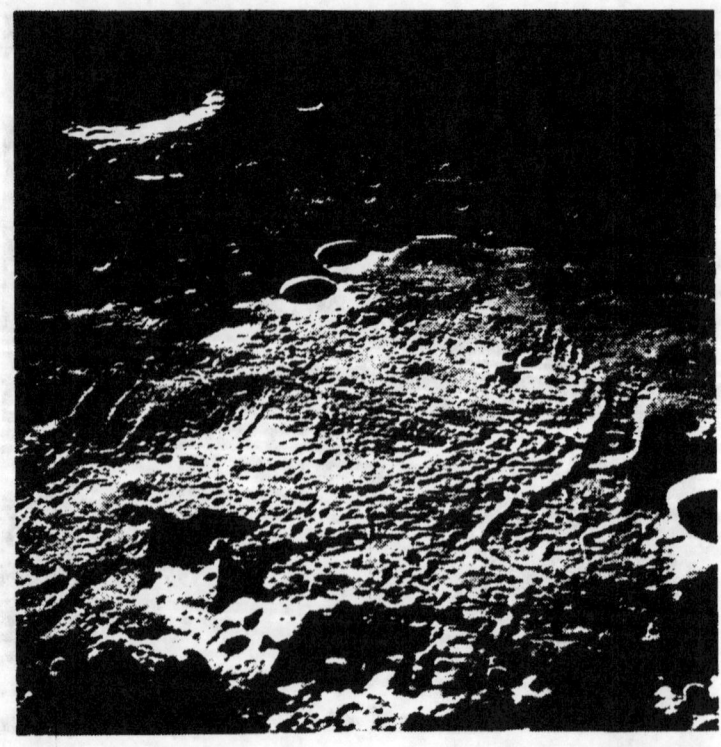

FIGURE 3-52.—View westward across southern Aristarchus Plateau. The low Sun angle accentuates mounds on the plateau and dramatically enhances the low mare ridges along the terminator in the Ocean of Storms (AS15-88-11982).

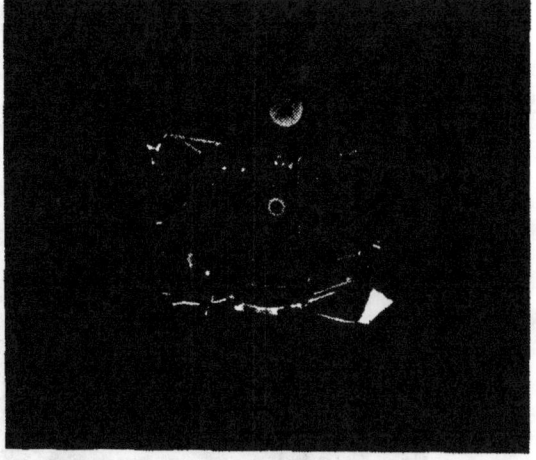

FIGURE 3-53.—The ascent stage of the LM. Between the successful rendezvous and the docking maneuver, 15 min of stationkeeping at ranges of approximately 30 m permitted close inspection of both spacecraft. Photographs like this are used to document the condition of the LM exterior. During this period, the LM crew inspected the SIM bay of the CSM from close range in an attempt to ascertain the cause of the panoramic camera V/h anomalies (AS15-96-13035).

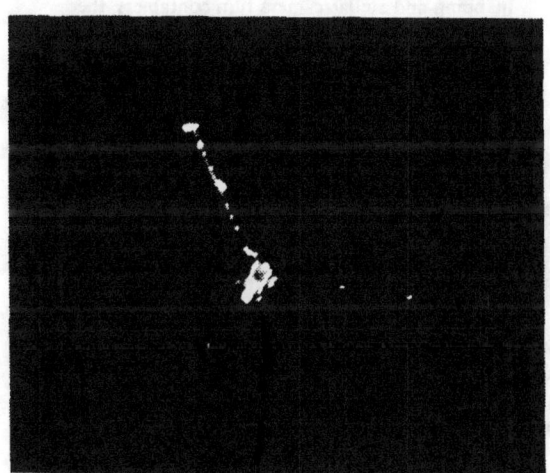

FIGURE 3-54.—The Apollo 15 subsatellite. Before transearth injection, the CSM launched a subsatellite containing three experiments: the S-band transponder, the particle shadow/boundary layer experiment, and a magnetometer. Still photographs such as this were used to document the condition of the subsatellite surfaces and deployment of the booms. The DAC photographs recorded the spin rate and wobble of the spinning subsatellite (AS15-96-13068).

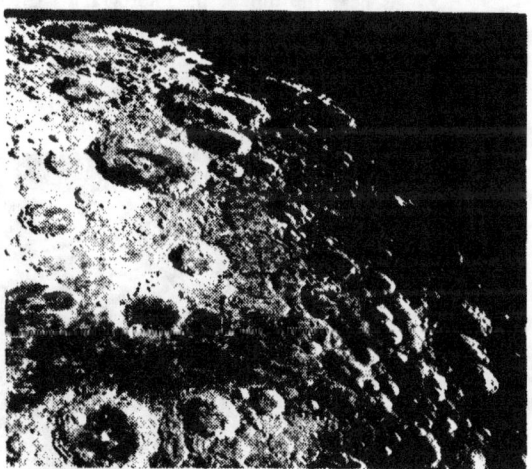

FIGURE 3-55.—View of the Moon taken after transearth injection. This photograph shows the northeast quarter of the lunar scene visible in the early portion of the transearth coast. Smyth's Sea, at the bottom, and Border Sea, just below the middle, have been observed by most of the Apollo crews; but only the Apollo 15 astronauts photographed this area with a low Sun angle. The large reseau cross at the center overlies a point near latitude 16° N, longitude 94.5° E. Fabray, the large crater near the horizon at the upper left corner, is centered near latitude 43° N, longitude 101° E (AS15-95-12998).

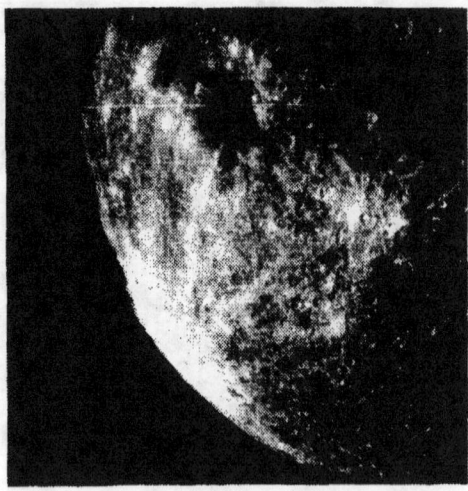

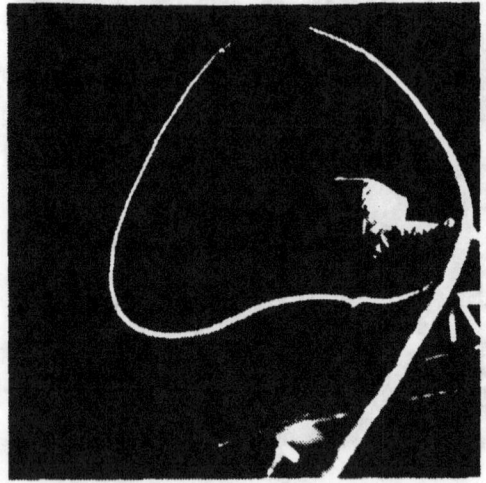

FIGURE 3-56.—This photograph, taken after transearth injection, shows the southwest quarter of the lunar scene. The bright halo and rays of Tycho Crater dominate the southwest quarter of the lunar surface. Langrenus is at the east edge of the Sea of Fertility. A narrow strip of cratered highlands separates the Sea of Fertility from the Seas of Tranquillity and Nectar. Tranquility Base, site of the Apollo 11 landing, is near the top of the photograph (AS15-94-12849).

FIGURE 3-57.—Approximately 18 hr after transearth injection, the CMP egressed the CM to retrieve film cassettes from the SIM. In this Hasselblad photograph, an open spiral of the umbilical frames the oxygen purge system mounted low on the CMP's back. He is inspecting equipment mounted in the SIM bay after handing the panoramic-camera film cassette to the crewmen in the CM. After retrieving the mapping and stellar camera film containers, the CMP ended the first transearthcoast EVA (AS15-96-13100).

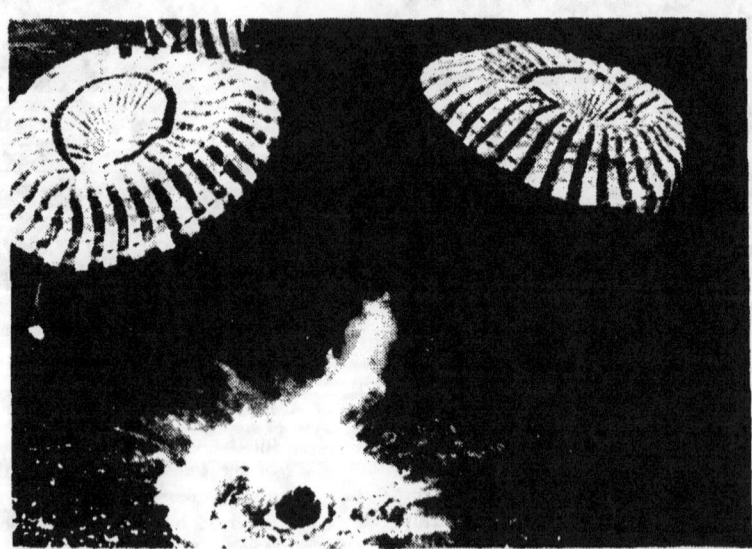

FIGURE 3-58.—Apollo 15 splashdown in the Pacific Ocean. During the final stages of descent, the parachute farthest from the camera was partially streaming. The other two parachutes are beginning to collapse as the CM rises after the initial contact (S-71-43541).

4. Crew Observations

David R. Scott,[a] Alfred M. Worden,[a] and James B. Irwin[a]

The Apollo 15 mission was the first of the series of manned lunar missions designed to extend the scientific exploration of the Moon both from lunar orbit and on the surface. On the Apollo 15 spacecraft, numerous hardware and software changes were made to the basic Apollo system; these changes necessitated new operational procedures, time lines, and techniques. Nevertheless, throughout the training cycle, we continued to think of the mission first and foremost as a scientific expedition. We were able to concentrate fully a third of our many months of training on preparing for the scientific requirements of the mission.

In this section, our observations are reported in summary form. Considerably more detailed observations were made during the course of this journey on the air-to-ground transmissions and later to the scientific teams conducting the postflight debriefing sessions. In general, these observations, impressions, and ideas are integrated into the more complete sections that follow, particularly the sections on geology, soil mechanics, and orbital science.

One aspect of the mission not covered elsewhere in this report should be particularly emphasized. We went to the Moon as trained, hopefully efficient, observers to gather data with both our scientific instruments and our minds. We spent 150 hr circling over this unique planet, exploring the Hadley Base area, and performing the scientific tasks required. Yet, in addition to making these assigned scientific observations, we left the Moon indelibly impressed with its stark, surrealistic features, its nearly overwhelming variety of landforms, and, above all, its awesome beauty. It is truly a fascinating place for exploration and study.

---

[a]NASA Manned Spacecraft Center.

TRANSLUNAR AND TRANSEARTH COASTS

With few exceptions, the scientific operations during the translunar- and transearth-coast periods were carried out as scheduled. These operations included photographic tasks, medical experiments, and, in lunar orbit and during the return to Earth, the operation of selected experiments in the scientific instrument module (SIM). The structural door covering the SIM bay was jettisoned without difficulty shortly before lunar-orbit insertion, and it could be seen drifting slowly away from the long axis of the command-service module (CSM).

LUNAR LANDING AND ASCENT

As the lunar module (LM) pitched over and gave us our first descending view of the Hadley site, the rille was the only feature we could immediately identify. The topographic relief of all local features was much less than we had anticipated from the enhanced 20-m-resolution Lunar Orbiter photography and from the preflight landing-site models. Sharp landmark features within the Hadley plain were almost nonexistent. However, first the South Cluster and then Salyut Crater and the crater adjacent to it on the north were located as we passed 5000 ft of altitude. At 2000 ft, we selected a relatively smooth area and maneuvered toward this spot. At 150 ft above the surface, we observed the first traces of blowing dust; this increased as we descended and completely obscured the lunar surface from 60 ft on down to engine shutdown.

The ascent, from the television-monitored lift-off to orbit insertion, proceeded smoothly and provided a fascinating last tour along the northern portion of Hadley Rille.

STANDUP EXTRAVEHICULAR ACTIVITY

The standup extravehicular activity (EVA) served essentially the purposes for which it was intended. We verified that the local terrain would permit good trafficability for the lunar roving vehicle (Rover); assured ourselves of suitable areas west of the LM on which to deploy the Apollo lunar surface experiments package (ALSEP); made three sets of panoramic photographs from the high vantage point; and, because of the nearly perfect Sun angle illuminating parts of the Apennine massif, we became aware of and photographed the remarkable lineations in these blocks. Although we were unable to pinpoint our location precisely from nearby landmarks, we were able to verify our touchdown point as being within the expected landing ellipse by sighting on known peaks in the surrounding mountains.

SURFACE-EXPERIMENT DEPLOYMENT

The ALSEP, solar-wind composition experiment, and laser ranging retroreflector package were deployed and properly alined without major mechanical problems. However, the emplacement of the heat-flow experiment and the subsequent collection of the deep core sample were extremely difficult, physically exhausting, and far more time consuming than had been anticipated before the flight. The operation of the hardware components of this experiment was, in general, acceptable. The major exception was the vise used for separating the drill-stem segments; its mountings had been installed backward on the geology pallet before the mission. The primary cause of the difficulties encountered with the drilling activities was the unexpected resistance of the regolith to drill penetration. Consequently, the heat-flow stem did not drill at expected rates, and it did not clear the hole cleanly by moving deep-lying material up to the surface. Because of the high torque levels on the chuck-stem interface, the drill chuck bound to the stems; in one case, it was necessary to destroy the stem itself to remove it from the chuck. Finally, the deep core could not be extracted from the uncooperative soil by normal methods; the two of us, working at the limit of our combined strength, were ultimately required to remove it. The exterior flutes contributed to this condition because the drill stem was pulled into the ground still deeper when the motor was activated. In spite of these difficulties, the heat-flow probes were ultimately emplaced, and the deep core sample was successfully collected.

LUNAR-TRAVERSE GEOLOGY

The impressive geological setting of the Hadley-Apennine landing site, bounded on three sides by the towering Apennine Mountains and on the fourth side by Hadley Rille, suggested to us that a great variety of features and samples could be expected. Also, the lack of clear, high-resolution photography of the landing site itself suggested that variation in the preflight estimates of topographic relief, surface debris, and cratering could be expected. In all cases, the actual conditions we encountered indeed exceeded these expectations.

Mare Surface

The mare surface at the Hadley site is characterized by a hummocky terrain produced by a high density of randomly scattered, rounded, subdued, low-rimmed craters of all sizes up to several hundred meters in diameter. There is a notable absence of large areas of fragmental debris or boulder fields. Uniquely fresh, 1- to 2-m debris-filled craters, with glass-covered fragments in the central portions occur occasionally, but on less than 1 percent of the mare surface which we traversed.

Apennine Mountains

The massif material comprising the Apennine Mountains is extremely surface-rounded; less than 0.1 percent has clearly exposed surface outcroppings or fresh, young craters. However, quite surprisingly, massive units of well-organized, uniform, parallel lineations appear within all blocks, and each block has a different orientation. Mt. Hadley is the most dramatic of these blocks; at least 200 bands are exposed on its southwestern slope, which dips approximately 30° to the west-northwest. Discontinuous, linear, patterned ground is visible, superimposed over these possible layers. A more definitive exposure of these units was observed at Silver Spur, where an upper unit of seven 60-m-thick layers is apparently in contact with a lower section of somewhat thinner parallel layers, which in turn show evidence of cross bedding and subhorizontal fractures. Three continuous, subhorizontal, nonuniform lineations are visible within the lower 10 percent of the Mt. Hadley vertical profile. In addition to the linear features just described, we observed (on the lower 2 percent of vertical relief of Mt. Hadley) three separate horizontal lines resembling "high water" marks. These lines are not readily apparent in the surface photography but, nonetheless, were unmistakably distinct during the traverses.

Hadley Rille

The most distinctive feature of Hadley Rille is the exposed layering within the bedrock on the upper 15 percent of the rille walls. Two major units can be identified in this region: The upper 10 percent appears as poorly organized massive blocks with an apparent fracture orientation dipping approximately 45° north, and the lower 5 percent appears as a distinct horizontal unit exposing discontinuous outcrops partially covered with talus and fines. Each exposure is characterized by approximately 10 different parallel, horizontal layers. The remainder of the slope is covered with talus, 20 to 30 percent of which is fragmental debris. There is a suggestion of another massive unit with a heavy mantle of fines at the level 40 percent from the top. The exposures at this level appear lighter in color and more rounded than the general talus debris. No significant collection of talus is apparent at any one level. The upper 10 percent of the eastern side of the rille is characterized by massive subangular blocks of fine-grained, vesicular, porphyritic basalt containing abundant plagioclase. This unit, as viewed south toward Head Valley, has the same character as the upper unit on the western wall. The bottom of the rille is gently sloping and smooth, with no evidence of flow in either direction. No accumulation of talus is evident other than occasional large boulders. Our first impression of Hadley Rille is that it results from a fracture in the mare crust.

South Cluster

South Cluster is a major feature on the Hadley plain concentrated primarily in the area easily discerned on all orbital photography of the landing site. However, because of the general lack of morphological features on the slopes of Mt. Hadley, the Swann Range, and Hadley Delta, the concentration of cratering up the slope of Hadley Delta directly south of the cluster seems to be associated with the cluster itself. This characteristic strongly suggests that a sweep of secondary projectiles from the north was indeed the origin of this major feature. A buildup of debris on the southern rims of these secondary craters is not evident, however, even though the estimated 10-percent coverage of the surface by scattered, coarse, fragmental debris in this region is unique within the Hadley area.

Sampling

The collection of representative samples was accomplished in the general vicinity of all the preplanned locations except for the North Complex/Schaber Hills region, which was unfortunately excluded by the higher priorities assigned to other scientific tasks. Nevertheless, a large variety of lunar materials was recovered from the Hadley area, including samples identified as anorthosite; basalts with vesicles of various size, orientation, and distribution and with phenocrysts of plagioclase, olivine, and pyroxene; complex breccias with an assortment of well-defined clasts; rounded glass fragments; and, finally, numbers of samples displaying subtle color differences, hints of surface-alteration products, slickensides, glass-splattered surfaces, and fractures. In addition, an assortment of lunar fines showing differences in granularity and texture was sampled, and several larger football-size rocks were returned, including the large 10-kg basalt sample.

With regard to the collection of samples on the lunar surface and, indeed, to the carrying out of lunar field geology in general, we wish particularly to emphasize our impression that our ability to identify rock types at the time of their collection seemed equal to our ability to do so during the many terrestrial field exercises of the preflight training period. We felt basically unhampered (although somewhat slowed physically) by the bulky equipment that the unique lunar environment makes necessary.

ORBITAL OBSERVATIONS

The scientific experiments, engineering tests, and wide variety of photographic tasks were completed as scheduled. Once in lunar orbit, the SIM bay was operated during the 3 days of lunar-surface activity according to a carefully detailed flight plan. Because of the complexity of the SIM bay for solo operation, a preflight decision had been made to hold to a minimum the real-time changes to the preplanned science schedule. This approach worked well, and the solo portion of the SIM operation was accomplished without major problems.

In addition to the SIM bay operation, we obtained photographs of 21 to 23 preplanned lunar targets and many unplanned targets of interest and made a number of visual observations of lunar features during our 6 days in orbit. The visual observations were made to complement photography in an attempt to obtain a better understanding of the large-scale geological processes that have formed the surface features of the Moon. Tsiolkovsky, the Aristarchus Plateau, and the Littrow region were of special interest to us. We observed Tsiolkovsky to be a large, mare-filled, impact crater uniquely placed between the large mare basins and the upland areas on the far side of the Moon. Its strikingly deep basin surrounds a prominent central peak and is rimmed with steep walls. Several faults extend through the crater walls; and, on the western side, one sees the classical flow pattern and lobate toes of a large rock avalanche that moved from the northwest rim into the subdued crater Fermi. Along the eastern side of Tsiolkovsky, we could see numerous apparent lava flows originating along fault zones and filling nearby minor craters.

The Littrow area showed distinct color banding that extended out into Mare Serenitatis. This banding resembles what we recognize terrestrially as lava flows or ash deposits (or both); and, within the darkest band, we observed many small, positive features that we identify as remarkably distinct cinder cones.

The signatures of intense volcanic constructional landforms were most noticeable around the Aristarchus Plateau region, where large numbers of flow fronts and rille features emanate from the central area. We looked particularly for signs of flow deltas at the outlets of the rilles in this region and concluded that, if indeed they had existed, these features had been inundated by subsequent flows across the mare surfaces.

Our last assigned scientific task in lunar orbit was to jettison the subsatellite. This was accomplished without difficulty, and we were able to observe and photograph the deployed satellite as it drifted outward from the CSM. Its arms were properly extended, and it was rotating with a coning angle of approximately $10°$.

5. Preliminary Geologic Investigation of The Apollo 15 Landing Site

*G.A. Swann,[a][†] N.G. Bailey,[a] R.M. Batson,[a] V.L. Freeman,[a]
M.H. Hait,[a] J.W. Head,[b] H.E. Holt,[a] K.A. Howard,[a]
J.B. Irwin,[c] K.B. Larson,[a] W.R. Muehlerger,[d] V.S. Reed,[a]
J.J. Rennilson,[e] G.G. Schaber,[a] D.R. Scott,[c] L.T. Silver,[e]
R.L. Sutton,[a] G.E. Ulrich,[a] H.G. Wilshire,[a] and E.W. Wolfe[a]*

The Apollo 15 lunar module (LM) landed at longitude 03°39'20" E, latitude 26°26'00" N on the mare surface of Palus Putredinis on the eastern edge of the Imbrium Basin. The site is between the Apennine Mountain front and Hadley Rille. The objectives of the mission, in order of decreasing priority, were description and sampling of three major geologic features—the Apennine Front, Hadley Rille, and the mare.

The greater number of periods of extravehicular activity (EVA) and the mobility provided by the lunar roving vehicle (Rover) allowed much more geologic information to be obtained from a much larger area than those explored by previous Apollo crews. A total of 5 hr was spent at traverse station stops, and the astronauts transmitted excellent descriptions of the lunar surface while in transit between stations. Approximately 78 kg of rock and soil samples were collected, and 1152 photographs were taken with the 60- and 500-mm focal-length Hasselblad cameras. Much useful information was obtained from the lunar surface television camera at eight of the 12 stations. Some information was gained from the data-acquisition (sequence) camera, and many useful photographs of the site were taken from orbit.

The geologic complexity of the site (fig. 5-1), the vast amount of returned data, and the brief time elapsed since the mission, of necessity, make this report preliminary. Many of the descriptions herein are incomplete, and most of the interpretations and conclusions are tentative. However, the facts emerging from the data reduction blend well with the premission photogeologic maps (refs. 5-1 to 5-3) from which the traverses were planned.

Much of the material in this report is taken from earlier reports that were distributed as working papers by members of the Geology Experiment Team (refs. 5-4 to 5-8).

The stations shown on the geologic map (fig. 5-2) are located at the panorama stations although much of the geologic data and many of the samples were taken from areas a significant distance from the panorama stations. All crater sizes refer to the rim-to-rim diameter unless otherwise specified. The sizes of fragments and blocks are generally given as the largest dimension of the field of view.

The topographic base (fig. 5-2) for the geologic map was compiled on an AS11 stereoplotter using Apollo 15 panoramic camera photographs AS15-9809 and AS15-9814. Frame 9809 is somewhat distorted for optimum photogrammetric use but probably did not introduce significant error into this map, because only a very small part of the frame was used in the compilation. An updated map with more rigorous horizontal and vertical control is being compiled.

The map is superimposed on an unrectified panoramic camera photograph. The contour lines and other map details do not, therefore, match equivalent details on the photograph. The map was modified to correct obvious discrepancies, but no attempt was made to match contour lines precisely to topographic details.

GEOLOGIC SETTING

The Apollo 15 crew, like the Apollo 14 crew, investigated features related to the huge multiringed Imbrium Basin. The Apollo 15 landing site is on a dark mare plain (part of the Marsh of Decay or Palus Putredinis) near the sinuous Hadley Rille and the frontal scarp of the Apennine Mountains (fig. 5-3). This scarp is the main boundary of the Imbrium Basin, which is centered approximately 650 km to the northwest (fig. 5-4, after ref. 5-9). The largest mountains of the Apennines are a chain of discon-

---
[a]U.S. Geological Survey.
[b]Bellcomm, Incorporated.
[c]NASA Manned Spacecraft Center.
[d]The University of Texas at Austin.
[e]California Institute of Technology.
[†]Principal investigator.

FIGURE 5-1.—Geologic map of the Apollo 15 landing site.

tinuous rectilinear massifs 2 to 5 km high. These mountains are interpreted as large fault blocks uplifted and segmented at the time of the Imbrium impact. Between the massifs and outward beyond them are hilly areas that merge outward with the Fra Mauro Formation, interpreted as a blanket of ejecta from the Imbrium Basin and sampled by the Apollo 14 crew. The hills appear to be jostled blocks subdued by the Fra Mauro blanket. The large massifs, however, are not subdued in this manner and so may be composed mainly of pre-Imbrian rock, perhaps thinly veneered by Imbrium ejecta. The area is near the old Serenitatis basin, which suggests that at least part of the pre-Imbrian material in the massifs is ejecta from Mare Serenitatis.

Mare material of Palus Putredinis fills lowlands at the base of the Apennines, forming a dark plain. Regional relations west of the site show that a number of events occurred between formation of the Imbrium Basin and emplacement of the mare deposits. These include deposition of the Apennine Bench Formation and the cratering event that formed Archimedes (ref. 5-1). Morphologies of craters on the mare surface at the site indicate that the mare age is late Imbrian or early Eratosthenian. It is a "red" mare, one whose spectral reflectance is enhanced in the red (ref. 5-10).

Some of the hills and mountains in the area are dark like the mare, perhaps indicating that they are coated by a thin mantle of dark material. The region contains numerous diffuse light-colored rays and satellitic clusters of secondary impact craters from the large Copernican-age craters (Autolycus and Aristillus) to the north.

Hadley Rille (fig. 5-3) follows a sinuous course through the mare and locally abuts premare massifs. The rille is one of the freshest sinuous rilles on the Moon, and rock outcrops are common along the upper part of the walls.

The LM landed within traverse distance of several important geologic features (fig. 5-1) and, during the three periods of EVA, the crew investigated most of these in detail. The landing site is on the mare surface, nearly on the crest of a very gentle ridge that trends northwest between Crescent Crater and North Complex (fig. 5-3). This part of the mare is slightly

PRELIMINARY GEOLOGIC INVESTIGATION OF THE LANDING SITE

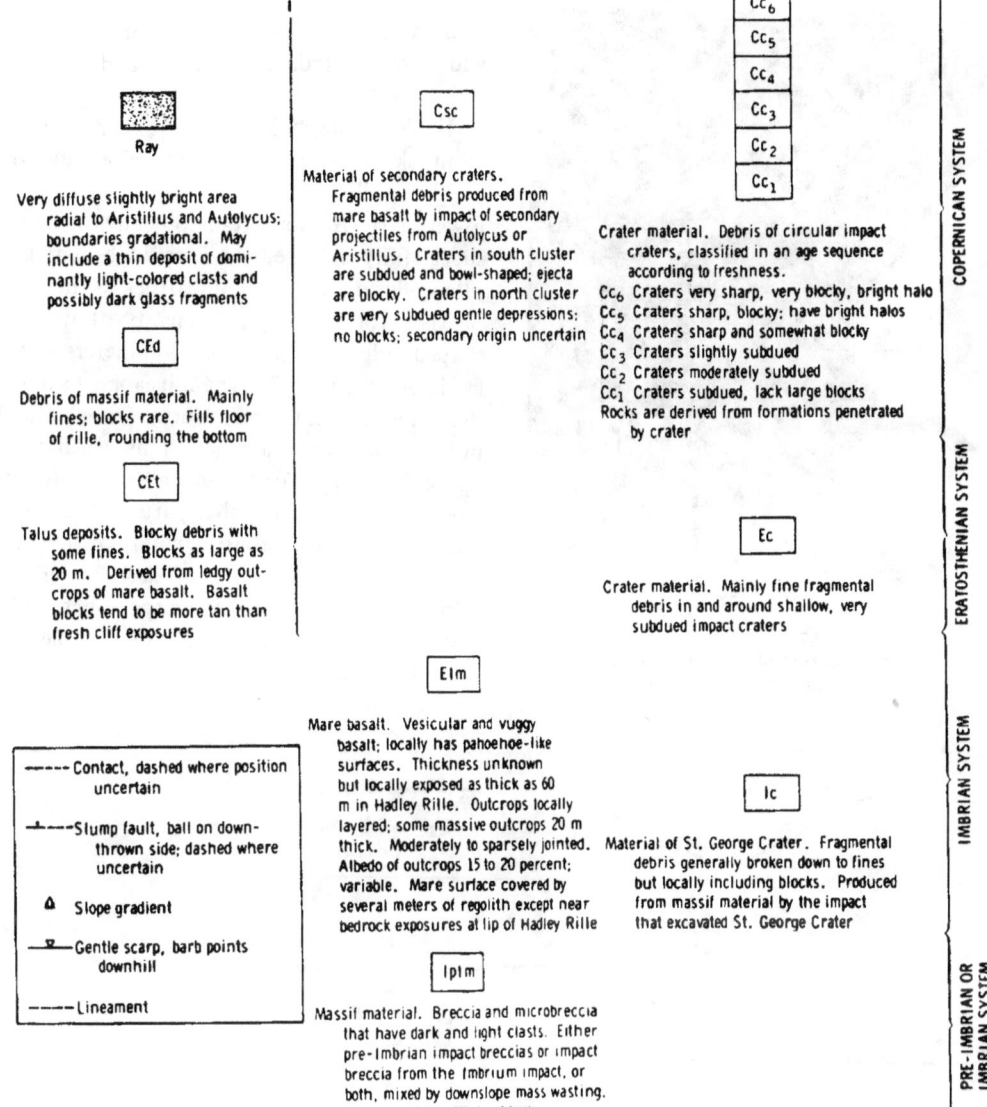

higher in albedo than elsewhere, suggesting a broad diffuse ray from Aristillus or Autolycus. Another gentle ridge in the mare is present at Elbow Crater and apparently is truncated by Hadley Rille. The mare surface contains numerous subdued craters 100 to 400 m in diameter, and many smaller ones, some of which are quite fresh. A prominent concentration of the larger craters known as the South Cluster is one of many in Palus Putredinis that form a pattern satellitic to Aristillus or Autolycus. A less distinctive cluster of more subdued craters lies immediately northwest of the landing site. The mare surface is covered with regolith approximately 5 m thick.

Two major Apennine massifs tower over the Hadley plain to heights of 4.5 and 3.5 km. These are Mt. Hadley to the northeast and Hadley Delta just south of the landing site (fig. 5-3). The face of Mt. Hadley is steep and has very high albedo. The north face of Hadley Delta, called the "Front" during the Apollo 15 mission, rises abruptly above the younger mare surface except near Elbow Crater, where the contact is gradational, apparently as a result of debris that has moved down the slopes. As elsewhere on the Moon, the steep slopes of the massifs are sparsely cratered because of rapid mixing of debris and destruction of craters by downslope movement. A

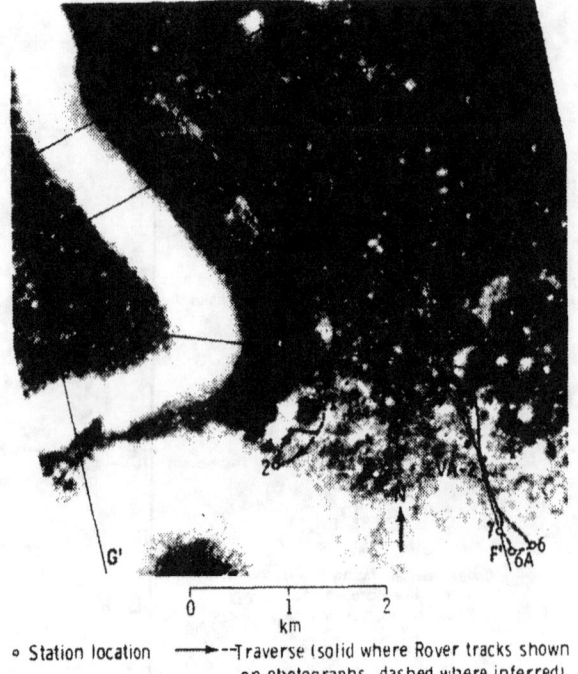

prominent exception is St. George Crater, 2.5 km wide, which predates the mare and is very subdued (fig. 5-1).

Hadley Rille is 350 m deep near the landing site. Rimrock outcrops are exposed along the upper part of the walls, and blocky talus deposits cover the lower part. An important objective of the mission, successfully achieved, was stereophotography of the rille walls and sampling of bedrock at the lip.

The North Complex consists of low irregularly shaped hills that lie a few kilometers north of the landing site (fig. 5-3). The hills appear slightly darker than the adjacent mare. North Complex and similar but slightly less dark low hills to the northwest resemble, except for the low albedo, hilly intramassif Apennine features in other parts of the region. These hills may therefore consist mainly of Imbrium ejecta, mantled by a thin layer of dark material. On the other hand, some peculiar scarps, lobes, and irregular crater chains suggest that North Complex may be a constructional volcanic form.

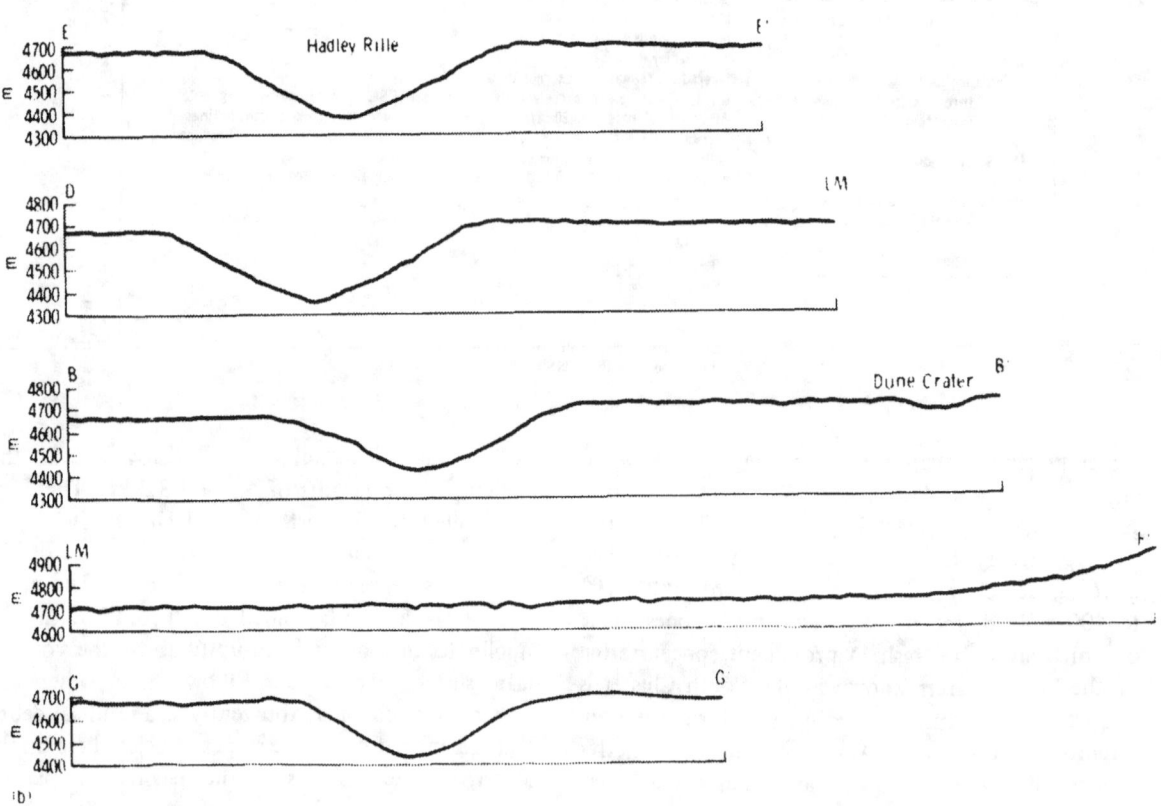

FIGURE 5-2.—Traverse map and profiles of the Apollo 15 landing site. (a) Traverse map. (b) Cross-sectional profiles.

PRELIMINARY GEOLOGIC INVESTIGATION OF THE LANDING SITE

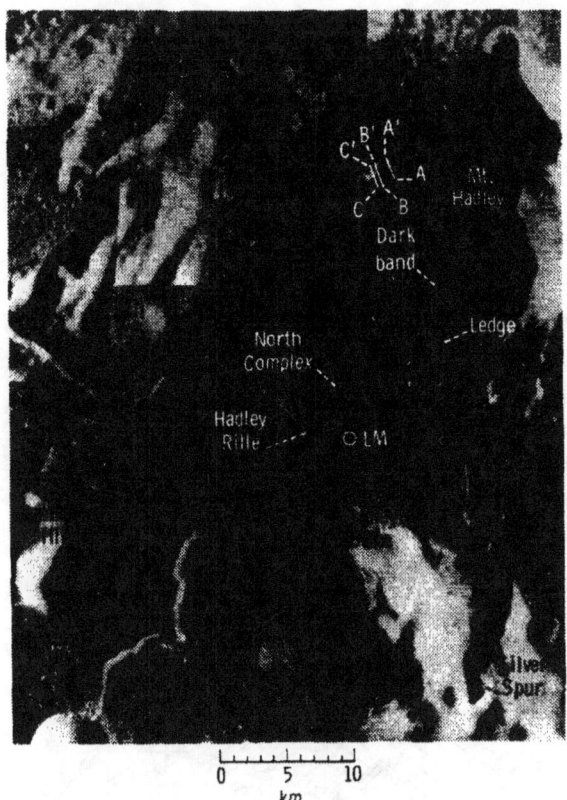

FIGURE 5-3.—Uncontrolled mosaic of Apollo 15 landing site from panoramic camera photographs.

PRELIMINARY GEOLOGIC INTERPRETATIONS

The large number of photographs returned by the Apollo 15 crew is currently being studied for a variety of purposes; the status of these studies is outlined in the following sections. Among these investigations are examination of the optical properties of surface materials determined by photometric measurements of photographs, examination of the stratigraphy of the rille walls as seen in the 500-mm photography, study of lineaments seen in photographs of the Apennine Front, and study of small-scale features such as fillets, small lineaments, and internal structures and lithologies of photographed rocks. The thorough documentation of samples and the geologic complexity of the site have further prompted a detailed examination of the local geologic environment of the samples. These details are presented herein along with preliminary interpretations of the relative ages of the sampled surfaces, gross stratigraphy of the mare and Apennine Front deposits, distribution of breccias on the mare and front areas, and comparison of the principal features of Apollo 14 and Apollo 15 breccias.

Optical Properties of Surface Materials

The combination of Apollo 15 panoramic and surface photography afforded a unique opportunity to study the optical properties of a large area of the Hadley-Apennine landing site. The additional use of a Hasselblad camera with a 500-mm lens further increased the area from which photometric measurements can be made (including the west wall of Hadley Rille and the North Complex). The photometry also benefited from the use of a black-and-white film different from that used on previous missions, reducing the halation effects and increasing the resolution and range of exposure. Preliminary results of more than 1000 measurements from the film densitometry are presented in the following ways: an estimated normal-albedo map, a range of estimated normal albedos of the fine-grain material and documented samples of each station, and the photometric properties of the rocks and fines on the rille wall and at the North Complex.

The source of the photometric data was a black-and-white second-generation master-positive film. Panoramic camera frame AS15-9814 (fig. 5-5(a)) over the Hadley-Apennine site was copied at approximately triple enlargement onto negative stock, and the relative densities of the film were measured with a Joyce-Loebl microdensitometer and recorded on digital magnetic tape. The scanning aperture was 250 μm square, which is the equivalent of a 10 m^2 integrated area of the lunar surface.

Surface photography was controlled in two ways. The first method involved the use of the gnomon and the photometric chart. The second used the film sensitometry data furnished by the Photographic Technology Laboratory at the NASA Manned Spacecraft Center. At the early deployment of the gnomon at station 1, the photometric chart data and the sensitometry data agreed within ±3 percent. Subsequent comparisons between the two disagreed by an increasing amount. The probable cause for this disagreement is the accumulation of lunar material on the chart as the mission progressed. Thus, for later periods of EVA, more reliance was placed on the film sensitometry, whereas the photometric chart aided in

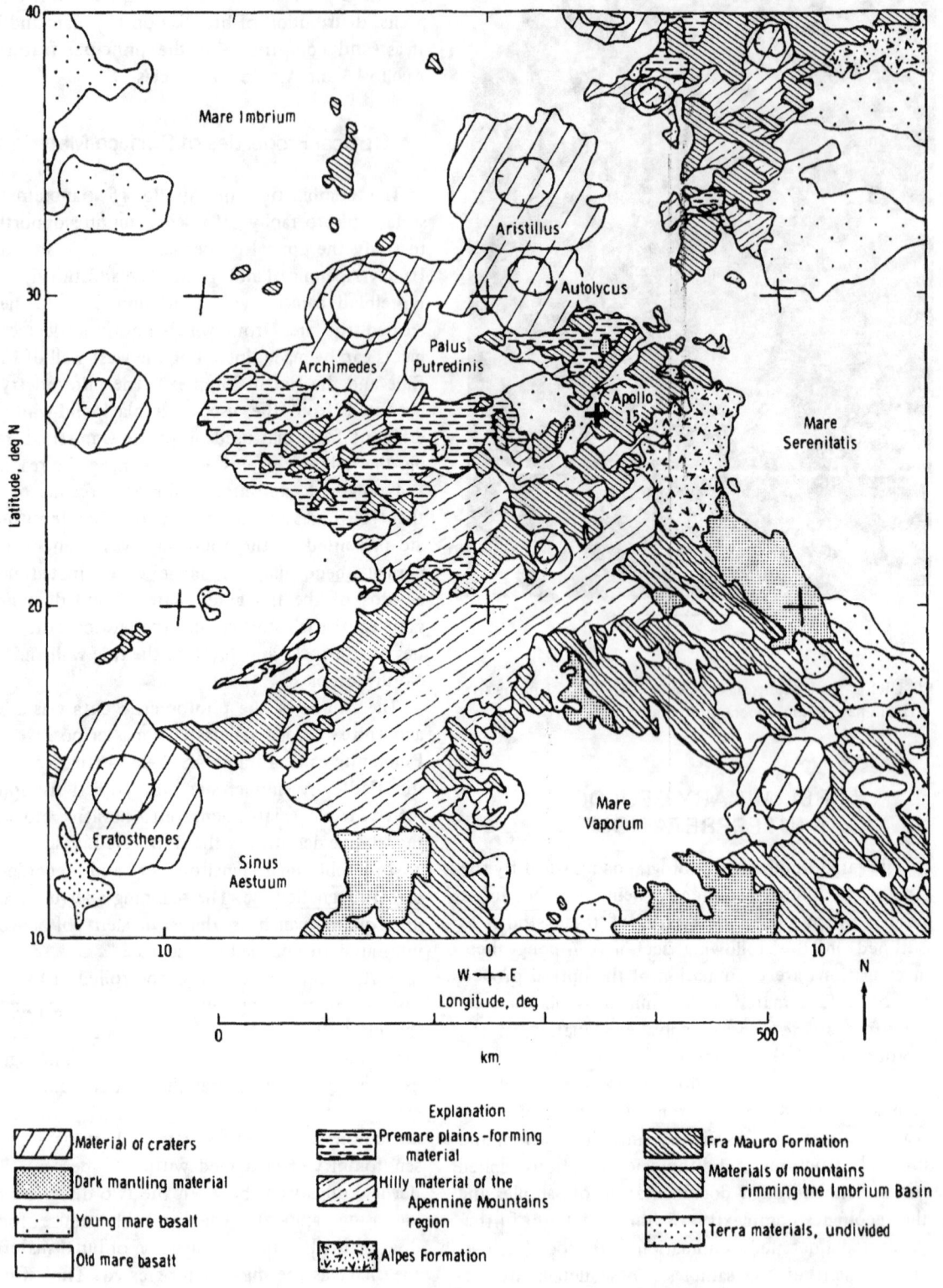

FIGURE 5-4.—Regional geologic map of the Apennine Mountains area (modified from ref. 5-9).

checking the f-stop at which the photograph was taken. The flow plan used in deriving the photometric data from the surface photography is shown in figure 5-6.

The extrapolated normal-albedo map was prepared by formatting the digital data from microdensitometry of the panoramic camera film into the VICAR image-processing system. The resulting digital numbers are a function of the film density and are thus related to the albedos of the lunar surface. The digital data were calibrated to the normal albedo at each station (except 3 and 9) by extrapolating the photometric data obtained from surface photographs of fine-grained regolith areas close to zero-phase angle (fig. 5-7). The photographs taken at stations 3 and 9 were not suitable for this type of analysis.

Each station has a variation in extrapolated normal albedos because the astronauts photographed different areas at the station. The range of albedos measured at each station is shown in figure 5-7. The averages for each station lie along an approximate line on the log-log plot, indicating that the station information was on the linear portion of the characteristic curve of the panoramic camera film. The processing of the panoramic camera frame was continued by associating a range of digital numbers with the albedo and producing a photomap of each albedo level. The resultant map (fig. 5-5 (b)) was smoothed for scanning noise. Hand corrections were made on the west side of the rille and the Apennine Front by measuring the local slope directions from the stereoscopic photopairs and profiles and by

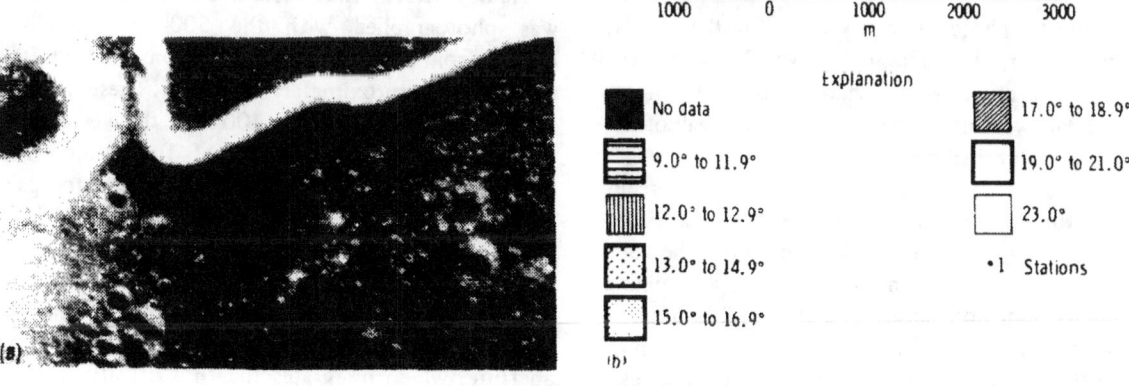

FIGURE 5-5.—Correlation of photometric data. (a) A portion of the photograph from which the extrapolated normal-albedo map was derived. The Sun was at an elevation of 43° and an azimuth of 117°, and the camera was tilted 12.5° to the right (pan AS15-9814). (b) Extrapolated normal-albedo map of the Apollo 15 landing site. The extrapolated normal albedos were corrected on the Apennine Front and west side of Hadley Rille for local slope effects. Data were not obtained from shadowed areas.

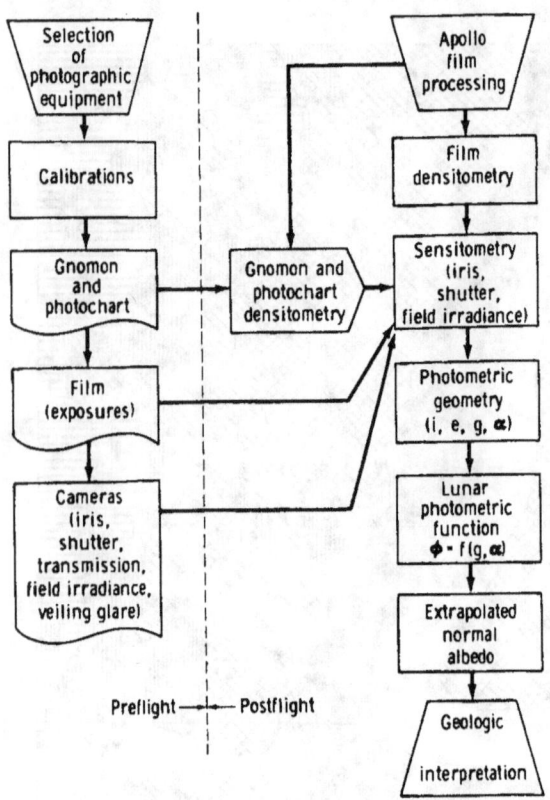

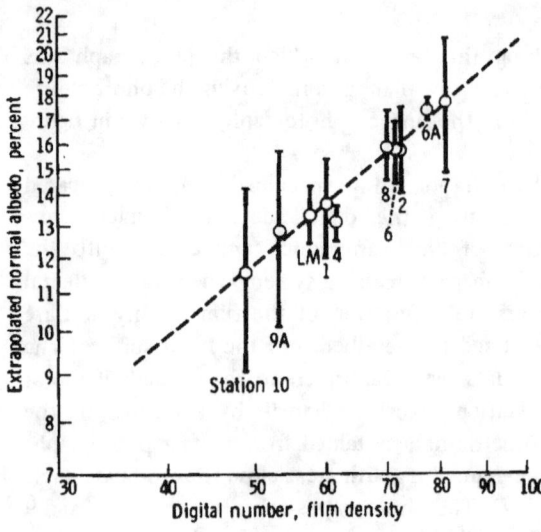

FIGURE 5-6.—Lunar geologic photometry-data flow plan showing the steps that are required to measure geological photometry from the Apollo film cameras. The lunar surface behaves in a special manner with respect to incident light i and an emergent light e reflecting an amount of light ϕ as a function of the phase angle g and radiance longitude α.

FIGURE 5-7.—Digital computer data from densitometry of panorama camera photograph (pan AS15-9814) are plotted against the extrapolated normal albedos measured from the surface photography. Bars indicate the range of albedos measured at each station.

correcting the photometric geometry. For these areas, the photometric function applied was the same as for the mare. Because of the shadows present, photometric data were not obtained on the east wall of the rille or in crater shadows.

The albedo map shows a range of 9 to 19 percent albedo for the mare area, while the Apennine Front has an albedo range of 15 to 23 percent. The zone between the Front and the mare is an area of mixing of debris from both areas. The rille wall is brighter than typical mare material because of the high proportion of the surface that is covered by blocky rock talus.

The mare surface is darker toward the east, which may indicate either areal variations in the underlying mare volcanic rocks or the presence of ray material in the western areas. The LM may have landed in an area of diffuse ray material. This lighter area can be seen in photographs AS15-84-11324 and 11325, where the lighter material has a measured albedo of 16 to 17 percent and the adjacent darker material has a measured albedo of 13 to 14 percent. Short (80-m-long) rays can be observed radiating outward from the LM on panoramic camera frame AS15-9814, suggesting that the exhaust from the descent engine exposed lighter materials.

Hadley Rille.—The western side of Hadley Rille was photographed with the 500-mm Hasselblad camera from across the rille at stations 9A and 10, a distance of approximately 1350 m. These pictures show only the upper 90 to 100 m of the western rille wall, which consists mostly of fine-grained soil, numerous talus blocks, and some massive outcrops of rock strata. Four photographs (AS15-89-12116, 12045, 12050, and 12099) were selected for photometric analysis of the massive outcrops, talus blocks, and fine-grained material. The densities of the film positives were measured through a 250-μm-diameter aperture, which integrates the film density throughout an area corresponding to a circular area on the rille wall approximately 0.75 m in diameter. The geometric orientation of the measured areas was estimated from stereoscopic study of the photographs. The luminances were then extrapolated to zero-phase luminances and expressed as normal albedos.

The photometric properties of the rock outcrops in the west side of the Hadley Rille indicate the presence of at least three rock units of different albedos—17, 19, and 21 percent. The talus blocks range in albedo from 15 to 23 percent, suggesting the presence of at least two additional rock albedo units. These rock units of different albedos indicate a buried stratigraphy of successive layers. The variation in albedo between rock units is considered to be primarily the effect of different crystal sizes and proportions between pyroxene and plagioclase and of the vesicularity of the units.

The lower discontinuous outcrop in figure 5-8 consists of eight or more layers with one layer approximately 3 m thick. The different layers have similar reflectivities, and the normal albedo is approximately 17 percent. The upper massive outcrop appears as a lighter unit and has a normal albedo of 21 percent. The massive outcrop in figure 5-9 appears to be a continuation of the upper outcrop shown in

FIGURE 5-9.—Area to the left of that shown in figure 5-8. Massive outcrop has an albedo of 21 percent and appears to be a continuation of the massive unit shown in figure 5-8. Outcrop is progressively more fractured or shattered from right to left (AS15-89-12111).

FIGURE 5-8.—Outcrops in the central part of the upper wall of Hadley Rille west of station 10. Total thickness of outcropping units is approximately 35 m. A massive light-toned unit (albedo 21 percent) with northwest-dipping fractures overlies a darker layered unit (albedo 17 percent). Talus blocks have accumulated on a bench beneath the outcrops. Subdued crater on mare surface just beyond the rille is approximately 130 m in diameter (AS15-89-12157).

figure 5-8 and has an albedo of 21 percent. The lower layered unit does not crop out, but several of the smaller talus blocks have an albedo of 17 percent, suggesting that the layered unit is present beneath the regolith.

The large blocks in figure 5-10 have a normal albedo of 19 percent. The talus blocks or partially exposed bedrock adjacent and to the left in the photograph have a normal albedo of 21 percent. A similar situation occurs in figure 5-11, in which the albedo of the lower massive outcrop is approximately 19 percent and the overlying blocks have an albedo of 21 percent.

The talus blocks in all photographs have a range in estimated normal albedo from 15 to 23 percent. Most of the larger talus blocks measured have an albedo of approximately 21 percent, which suggests that this albedo is probably representative of the most common source material for the talus blocks and probably of the most prevalent rock type in the rille wall.

The fine-grained fragmental material on the rille wall has an albedo of approximately 14.5 percent, slightly higher than the adjacent mare surface material. The brighter rille wall is most likely caused by the presence of a higher proportion of small rock fragments on the wall surface as compared with the

FIGURE 5-10.—Area to the right of that shown in figure 5-8. Large blocks (A) have an albedo of 19 percent, while the layered outcrop behind (B) has an albedo of 21 percent. A large column like block is at (C), and a platy slab is at (D) (AS15-89-12100).

mare surface. In many small areas along the rille wall, the fine-grained material appears locally brighter than 14.5 percent, probably as a result of included rock fragments as large as several centimeters across (near the resolution limit of the pictures).

North Complex.—Photometric measurements were made on four sets of 500-mm Hasselblad photographs with Sun angles of 13° to 42°. The largest block in Pluton Crater has an albedo of 18 percent. The block near the top of the north wall of Eaglecrest (fig. 5-12) has an overall albedo of 20.5 percent with a white band, too small to measure, cutting across it. Two other large blocks have albedos of 21 and 22 percent. This range of albedos was observed on the west wall of Hadley Rille and suggests that those blocks are most likely mare rock types. Five blocks scattered along the middle part of the north wall of Pluton have albedos of 25, 26, 26.5, 27, and 29 percent. This range of albedos is characteristic of white-clast breccias along the Apennine Front and around Cone Crater at the Apollo 14 site (ref. 5-11).

Rock fragments.—Thirty of the documented samples were measured for in situ photometric properties. The light anorthositic rocks and white clasts are easily differentiated from the other samples on the basis of albedos, which are as high as 25 to 32

FIGURE 5-11.—Outcrop approximately 20 m thick at left edge of the upper wall of Hadley Rille west of station 10. Albedo of outcrop is 19 percent while that of bright overlying blocks is 21 percent. Arrow points to thin layered outcrops of upper dark hackly rock (AS15-89-12115).

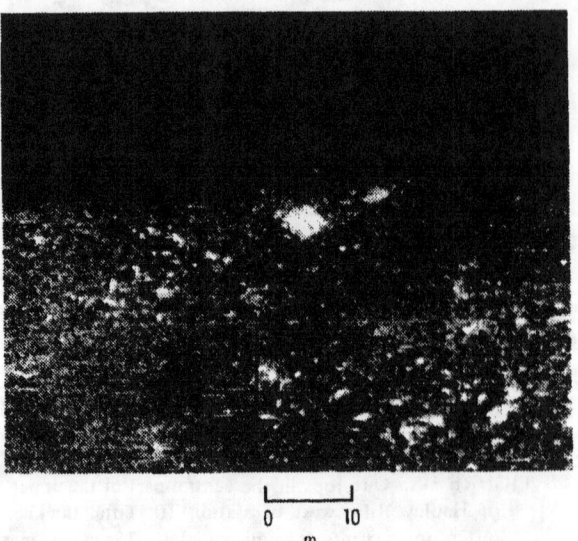

FIGURE 5-12.—Enlargement of 500-mm photograph of large block on north wall of Eaglecrest Crater (North Complex) showing possible high-albedo zone running through it (AS15-82-11209).

percent. Most of the basalts and breccias ranged from 12.5 to 23.0 percent with a mean of approximately 19 percent. Several dark clasts or dark glassy areas had albedos as low as 9.2 percent, but the integrated albedo of each of the fragments was greater than 12.5 percent. Some fragments were too small to measure effectively in the down-Sun photographs.

Apennine Front

The Apennine Front is the face of the arcuate mountain chain that borders the Imbrium Basin on the southeast. Immediately adjacent to the Apollo 15 landing site, the Front consists of the slopes of two major mountains—Mt. Hadley to the northeast and Hadley Delta to the south. These massifs rise steeply 3 to 5 km above the local mare surface.

In the premission photogeologic analyses of the Apollo 15 landing site (refs. 5-1 and 5-2), the Front massifs are interpreted as composed mainly of pre-Imbrian rocks consisting of impact breccias from the Serenitatis basin overlying a complex of impact breccias from still older basins and craters. The Imbrium impact caused faulting along lines both radial and concentric to the basin that produced the present-day arc of block-faulted mountains. Rolling hills east of the landing site between Hadley Delta and Mt. Hadley are interpreted as Imbrium-impact ejecta that may have been deposited by a base surge that arrived at the area shortly after uplift of the Apennine Mountains. These deposits of Imbrium ejecta are thought to be relatively thick in the lower intermontane areas near the basin. Locally, the deposits may form a thin mantle on the pre-Imbrian materials of the Apennine massifs.

Materials of the Front.—The Apennine Front was visited and sampled at the foot of Hadley Delta, at station 2 on the flank of St. George Crater, and at stations 6, 6A, and 7 in the vicinity of Spur Crater. In these areas, the surface material on the Front is relatively fine-grained cratered regolith that contains not only debris derived from deeper layers but also some ejecta from the nearby mare and, presumably, a small amount of exotic material from distant meteorite impacts. Geologic models derived from photointerpretation suggest that Front material sampled in the regolith may include not only pre-Imbrian massif material but also Imbrium ejecta either deposited directly in the sampled areas by the Imbrium impact or subsequently transported down the slope by mass wasting. The massif material may consist of a variety of stratigraphic units. For example, a distinctly light band is visible in the upper part of the north face of Hadley Delta in high-Sun-angle orbital panoramic camera photography.

Of 133 rock fragments collected at the Front, 106 are breccias, four are glass fragments, and 15 are crystalline rocks, including some of the distinctive mare-type basalts. Preliminary examination shows the breccias are of three types—friable, coherent, and well lithified. Friable breccias, presumably soil breccias, include a variety of clasts that are feldspathic, nonmare-type basalt, mare-type basalt, and glass spheres and fragments. Coherent breccias are characterized by a dark vitreous matrix in which feldspathic clasts, nonmare-type basalt clasts, and granulated olivines and pyroxenes are abundant. These breccias, commonly coated with frothy glass, are the dominant type at the Front and are also present on the mare. Well-lithified breccias are those in which the most abundant clasts are feldspathic and in which clasts of nonmare-type basalt are subordinate. This breccia type is characterized by the occurrence of an aphanatic crystalline matrix that intrudes the clast.

Unlike the Apollo 14 breccias, the coherent and well-lithified breccia types are notable for the absence of clasts of older breccias. Hence, the breccias may represent materials from depths below the normal levels of impact brecciation. These breccias may have been derived from large craters and basins that excavated bedrock before the Imbrium impact and deposited it as ejecta at the Imbrium site, or they may represent bedrock brecciated and uplifted by the Imbrium event itself.

Mass wasting.—In photographs taken at high Sun angle, the steep mountain slopes show prominent dark bands that extend straight downslope. These bands merge upward into the dark, low-relief, hummocky material interpreted as Imbrium ejecta (refs. 5-1 and 5-2). Excellent examples of the dark bands that presumably represent gravitational transport of debris downslope occur along the Front south of St. George Crater and east of Hadley C (fig. 5-13).

A second kind of mass-wasting process is shown in figure 5-14, a 500-mm lunar-surface photograph of the upper part of Mt. Hadley. Scarcity of blocks in the lower part of the photograph suggests that the regolith is relatively thick. In the upper half of the photograph, from the upper part of the steep lineated portion of the slope almost to the mountain top,

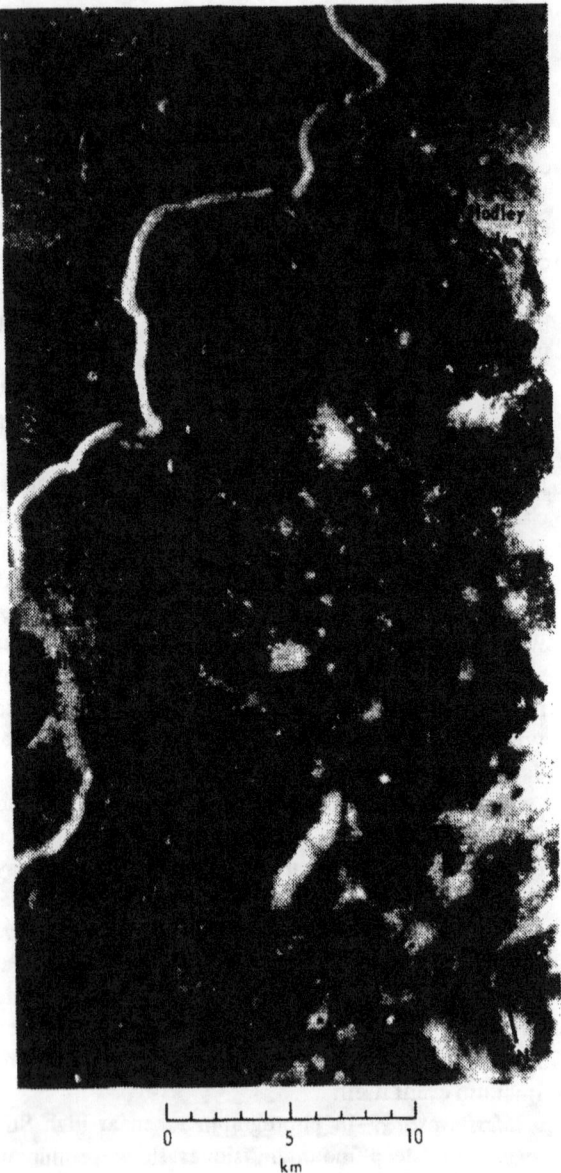

FIGURE 5-13.—Orbital photograph showing dark bands interpreted as material transported down the west slope of the Apennine Front from the mantling deposit of Imbrian ejecta (pan AS15-9814).

FIGURE 5-14.—Lunar-surface 500-mm photograph, looking northeast, of upper part of Mt. Hadley showing the blocky zone just below the hill crest. Base of blocky zone is gradational, but the zone occupies the upper one-third to one-half of the pictured slope (AS15-84-11304).

boulders are much more abundant, which indicates the presence of a zone between the mountain top and the steep main face where the regolith is preferentially thinned. Presumably, this zone occurs because ejecta from craters on the flat hilltop are distributed randomly around the source craters, whereas ejecta on the slope are distributed preferentially downslope. Hence, a zone of disequilibrium exists high on the slope where material is lost downslope more rapidly than it is replenished by impacts up on the flatter hilltop.

Lineaments — Well-developed and largely unexpected systems of lineaments suggestive of fracture and compositional layers or both were observed and photographed by the crew on both Mt. Hadley and Hadley Delta. Surface photographs taken with the 60- and 500-mm Hasselblad cameras (figs. 5-15 to 5-21) show the lineaments clearly. Orbital photographs taken with the high-resolution panoramic camera generally show the same lineament sets that the crew documented from the surface.

Previous orbital and surface photographs of the lunar surface have shown that lineaments at all scales tend to be alined in preferred directions—predominantly northwest, north, and northeast with secondary systems trending north-northwest and north-northeast. The observation that this so-called lunar grid has been recognized under very restricted lighting conditions, generally low Sun with lighting from either the east or the west, has raised the question that some of the lineaments might be artifacts produced by low-angle illumination of randomly irregular surfaces. Subsequent experiments

FIGURE 5-15.—View south toward Hadley Delta (approximately 3.5 km high) showing the northeast-trending lineament set (sloping gently left) and the north-trending lineament set (sloping more steeply to the right). The cluster of small fresh craters on lower slope of Hadley Delta at left side of photograph is noteworthy (AS15-85-11374).

FIGURE 5-16.—Lunar-surface 500-mm photograph south to rim of St. George Crater showing northeast-trending lineament set (sloping gently left) and the north-trending lineament set (sloping steeply right). The bright crater on the east rim of St. George Crater is approximately 50 m in diameter (AS15-84-11236).

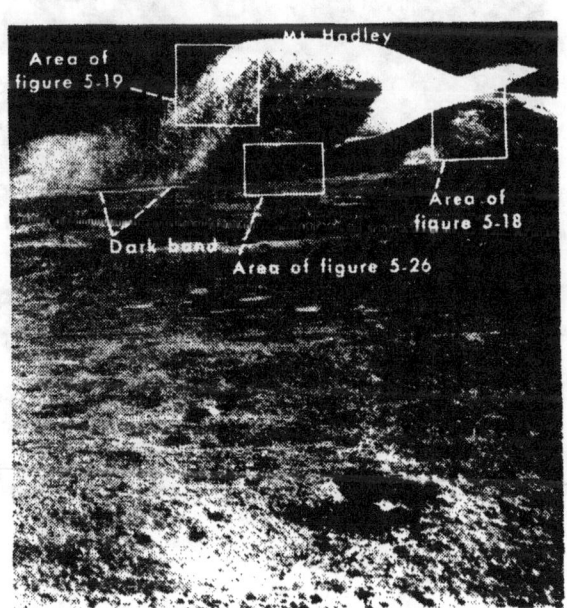

FIGURE 5-17.—Northeast view to Mount Hadley (approximately 4.5 km high). Prominent lineaments sloping steeply to the left form the northeast-trending set. Dark band near base of Mt. Hadley is noted (AS15-90-12208).

with small-scale models, still in progress,[1] show that oblique illumination on randomly irregular surfaces produces systematic sets of lineaments that resemble some of those recorded at the Hadley-Apennine site. Lineaments in the models form conjugate sets bisected by the illumination line, with which they form acute angles. This angle increases with increasing incident lighting angle.

Because of the great interpretative significance to be attached to the Apennine Front lineaments if they can be identified as the surface traces of compositional layers or of regional fracture sets, a preliminary attempt was made to evaluate the possibility that some may be lighting artifacts. Statistical analyses of lineament trends were made by measuring the orientations of approximately 1500 lineaments in six separate areas on orbital panoramic camera photographs. The results are summarized in azimuth-frequency diagrams (figs. 5-22 to 5-24). Panoramic camera photographs from three different orbits provided relatively low-, intermediate-, and high-Sun

---

[1] K. A. Howard, unpublished data.

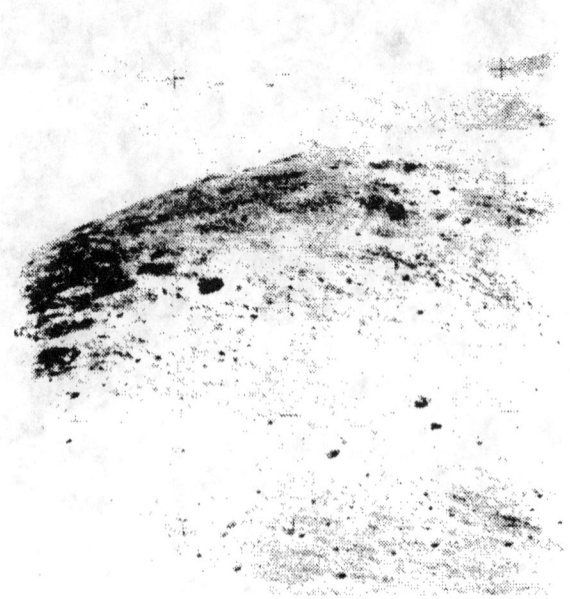

FIGURE 5-18.—Lunar-surface 500-mm photograph of area to northeast of surface of low hill south of Mt. Hadley. Area of photograph corresponds in part with area 3. Northeast-trending lineament set slopes very steeply from upper right to lower left. North-trending lineament set, sloping gently to the right, is recognizable but obscure in this view. Sharp blocky crater in left-central part of photograph is approximately 80 m in diameter (AS15-84-11317).

positions with Sun azimuths of 99°, 112.6°, and 117°, respectively, and with Sun elevations of 18°, 38.5°, and 43°.

Area 1 (fig. 5-22(b)), the west-facing slope of Hadley Delta, is illuminated only in the high-Sun orbital photographs. Lineaments on that surface are discontinuous and cluster in two groups, north and northeast, approximately 60° apart. The Sun crudely bisects the obtuse angle between the main lineament trends.

Area 2 (figs. 5-22(c) and 5-23), the northeast-facing slope of Hadley Delta, was examined in both low- and high-Sun photographs. At low Sun (fig. 5-23), the lineament distribution is characterized by major north and northeast trends approximately 40° apart, again roughly bisected by the Sun line.

In the high-Sun view of area 2 (fig. 5-22(a)), with illumination the same as for area 1, the lineament trends are similar to the area 1 trends, and the major lineament directions are again approximately bisected by the Sun line. Comparison with the low-Sun lineament plot for the same area (fig. 5-23) shows that the orientation of the north-trending set is essentially unchanged but that the northeast set has apparently shifted clockwise approximately 20°, a movement nearly equivalent in magnitude and direction to the shift in Sun azimuth from low Sun to high Sun. Despite the apparent shift with changing illumination, visual comparison of the two panoramic camera photographs shows that some of the same lineaments can be identified at either high or low Sun. (The greater number of lineaments recorded at high Sun was measured on a triple enlargement of the panoramic camera photograph, whereas the low-Sun lineaments were measured on a contact print. The scale difference in the photographs probably accounts for the differences in absolute numbers of lineaments measured at high and low Sun.)

A 60-mm photograph (fig. 5-15) of area 2 taken from the lunar surface shows the two lineament sets. The northeast set slopes gently left in the photograph and the north set, particularly evident on the rim and flanks of St. George Crater, slopes steeply to the right. A 500-mm photograph (fig. 5-16) of the St. George Crater rim taken from the lunar surface shows

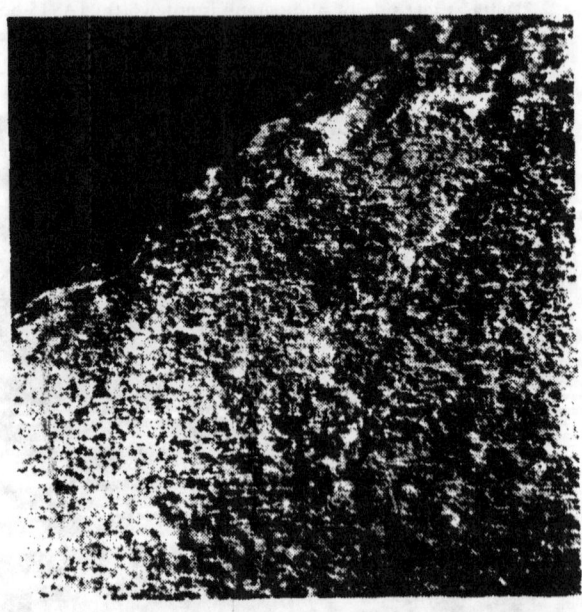

FIGURE 5-19.—Lunar-surface 500-mm photograph of area to northeast of a portion of Mt. Hadley, showing northeast-trending linears that dip steeply to left and the cross-cutting nearly horizontal regolith-covered benches. Vertical linears are present but subordinate in this view. Approximate width of photographed area is 2.4 km (AS15-84-11321).

FIGURE 5-20.—View looking southeast to Hadley Delta and Silver Spur showing the Silver Spur ledges with a gentle apparent dip to the left. Hadley Delta is approximately 3.5 km high and the distance to the crest of Silver Spur is approximately 20 km (AS15-85-11371).

more detail of the lineament patterns. Parallelism of the long edges of crater rim shadows with the north-trending (upper left to lower right) lineament set is compatible with the concept that this set may be a lighting artifact. This effect is particularly enhanced by foreshortening of the view on the more distant (south) wall of St. George Crater. The large thickness of mature regolith implied by the scarcity of blocks even near the largest and freshest crater on the northeast rim of St. George Crater (fig. 5-16) suggests that neither lineament set reflects compositional layering of the bedrock. However, if the lineaments represent the axes of fine, systematically alined ridges and troughs, they might be the surface traces of bedrock fractures propagated through the regolith according to the mechanism proposed in reference 5-12.

Area 3 is located just south of Mt. Hadley (fig. 5-24) on the southwest-sloping face of one of the low hills interpreted as Imbrium ejecta. Lineaments measured with intermediate illumination show prominent north and northeast maxima that are roughly bisected by the Sun and are separated by approximately 50°. The northeast-trending lineament set is visible in 60- and 500-mm photographs (figs. 5-17 and 5-18) taken from the lunar surface. In the 500-mm photograph, the northeast lineaments are similar to the lineaments on the rim of St. George Crater (fig. 5-16) in size, frequency, and in the occurrence of block-free regolith. Presumably the origins are similar. The north-trending set is recognizable but somewhat indistinct in photographs from the lunar surface. Brief comparison of panoramic camera photographs with different illumination angles indicates that at least some of the lineaments persist despite lighting changes.

Area 4 is located low on the south face of Mt. Hadley (fig. 5-24). Lineaments, measured with an intermediate-illumination angle cluster around north and northeast trends separated by approximately 55°, and the major lineament directions are again roughly bisected by the Sun line.

Area 5 is the entire west face of Mt. Hadley, partly shown in figures 5-17 and 5-19. Lineaments measured in an intermediate-illumination panoramic camera photograph (fig. 5-24) are grouped around three major trends that correlate with the lineaments

FIGURE 5-21.—Lunar-surface 500-mm photograph looking southeast toward Silver Spur, approximately 20 km away. View shows detail of massive ledges of the north-northeast lineament system (sloping gently left) and the possible fractures of the northwest lineament system (dark deeply shadowed depressions sloping steeply right). Slope of Hadley Delta is in the foreground (AS15-84-11250).

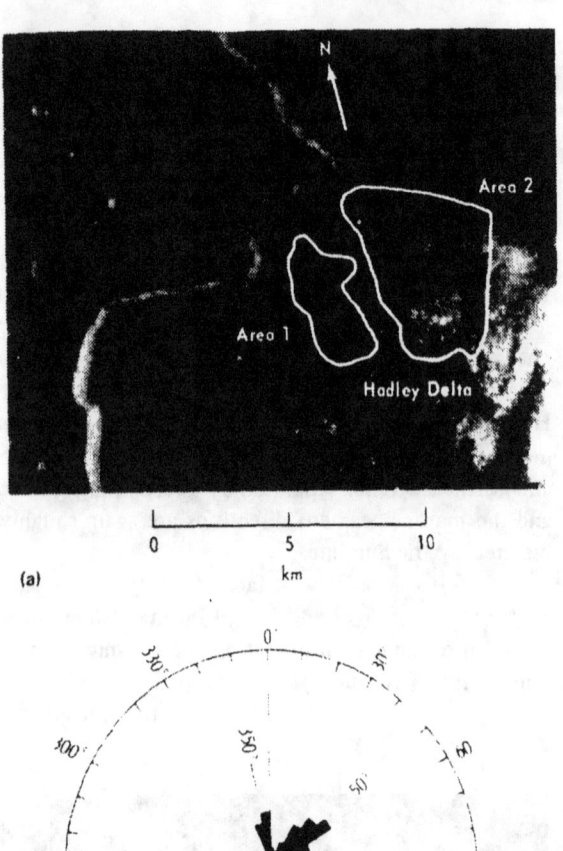

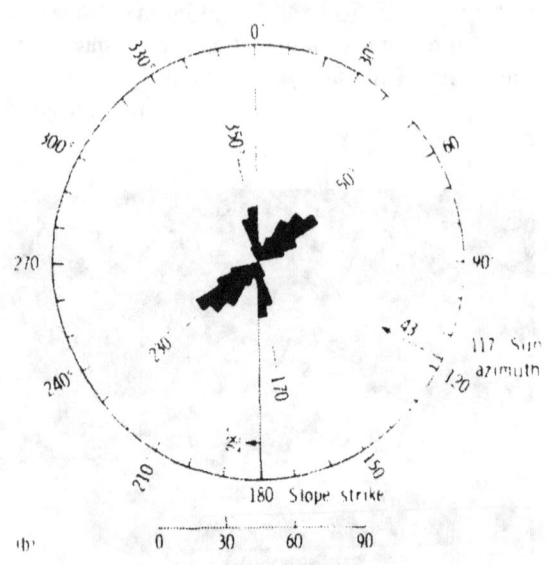

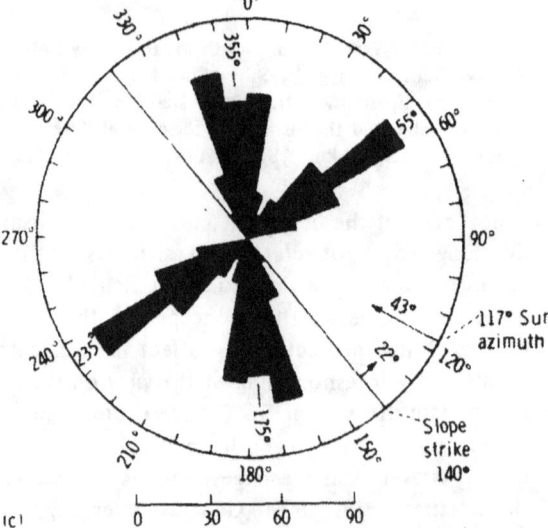

FIGURE 5-22.—Lineament azimuth-frequency plots and location maps for areas 1 and 2. Scale represents number of lineaments plotted as a function of distance from origin; lineaments are plotted in 10° increments. (a) Orbital panoramic camera view taken with high Sun (pan AS15-9809). (b) Plot of area 1 (188 lineaments). (c) Plot of area 2 (459 lineaments).

photographed from the lunar surface (figs. 5-17 and 5-19). The predominant set trends north 55° E and corresponds with the well-defined set of linears dipping steeply left in figures 5-17 and 5-19. Although no single linear can be traced across the entire outcrop face, the parallelism and crisp definition of some bright linears through distances of tens or hundreds of meters give the impression that they are the surface traces of compositional layers or of a system of well-defined, uniform fractures. The impression is heightened in a few places where the same linear apparently emerges both upslope and downslope from beneath the cover of one of the many subhorizontal, smooth, regolith-covered benches on the mountain face (fig. 5-19). In other places, faint traces of the linears extend across these benches.

A second set of linears, trending north 25° E, appears vertical in figures 5-17 and 5-19. This set is less prominent than the northeast set in both orbital and lunar-surface photographs.

The third set, trending northwest along the Sun line, is approximately parallel to the abundant subhorizontal benches prominent in figure 5-19. The benches are visible in the orbital photograph as discontinuous, narrow, somewhat sinuous bands approximately contouring the slope. The benches are distinct from the lineaments, which are straighter and more regular, and they may be slump features or benches where the slope profile has been locally flattened by cratering or downslope mass movement. The northwest-trending lineaments are a well-defined but subordinate set on the southwest mountain face, but on the northwest face they are prominent.

The face of Mt. Hadley is completely shadowed in the low-Sun orbital photographs. Comparison of high-Sun and intermediate-Sun panoramic camera photographs shows that at least some of the lineaments in the major set persist through the small illumination change.

As with areas 2 and 4, the face of Mt. Hadley is notable for the scarcity of blocks, which implies the presence of a fairly thick mature regolith. If the lineaments are not lighting artifacts, it seems more likely that they represent the traces of fractures propagated through the regolith than that they represent compositional layering in the bedrock.

Area 6 is located on the west flank of Silver Spur (figs. 5-20 and 5-24). The striking 500-mm photograph taken from the LM (fig. 5-21) shows massive ledges apparently dipping gently to the left and crossed by finer, more nearly horizontal lineaments that give the impression of crossbedding. The finer lineaments are probably caused by the foreshortened view across an undulating cratered surface, an effect similar to that on the south wall of St. George Crater (fig. 5-16). In figure 5-20, Silver Spur resembles a hogback of gently dipping stratified rock resting conformably on the apparently layered Hadley Delta massif.

The orbital photographs do not support the hogback illusion. They show two major sets of topographic lineaments that intersect in a diamond-shaped pattern with northwest- and north-northeast-trending boundaries (fig. 5-25). Continuity of lines beyond each diamond is more easily accomplished by eye than it is by drawing each linear element, because slope reversals at diamond boundaries cause alternation of shadowing and highlighting along any single linear. Furthermore, the orientation of the two sets changes slightly across the crest of Silver Spur. The linearity of each set in the vertical photographs and the relatively minor response in trend to changes in slope orientation suggest that these are lighting enhancements of surficial features that could reflect high-angle fractures or even near-vertical compositional layering along the north-northeast trend.

The north-northeast and northwest peaks of the azimuth-frequency plot (fig. 5-24) include the well-

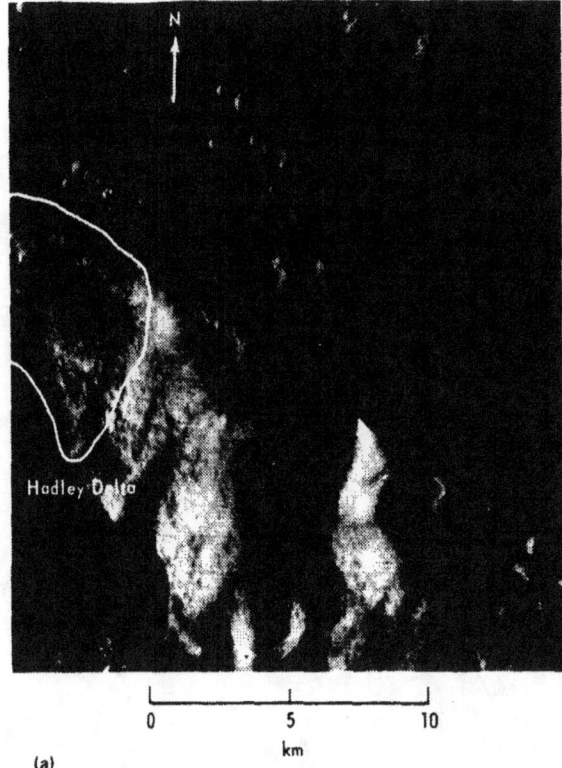

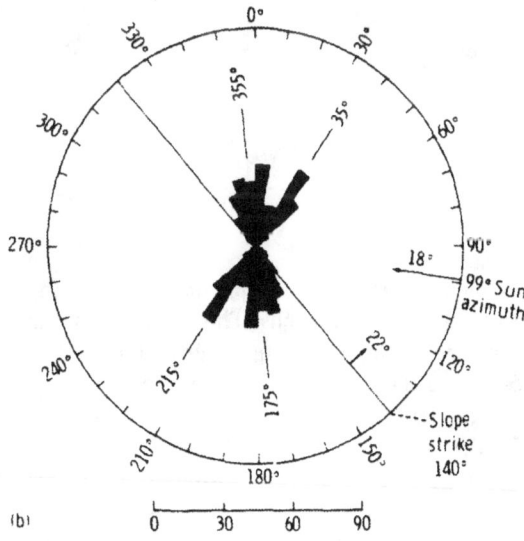

FIGURE 5-23.—Lineament azimuth-frequency plot and location map of area 2. Scale represents number of lineaments plotted as a function of distance from origin; lineaments are plotted in 10° increments. (a) Orbital panoramic camera view taken with low Sun (pan AS15-9375). (b) Plot of area 2 (252 lineaments).

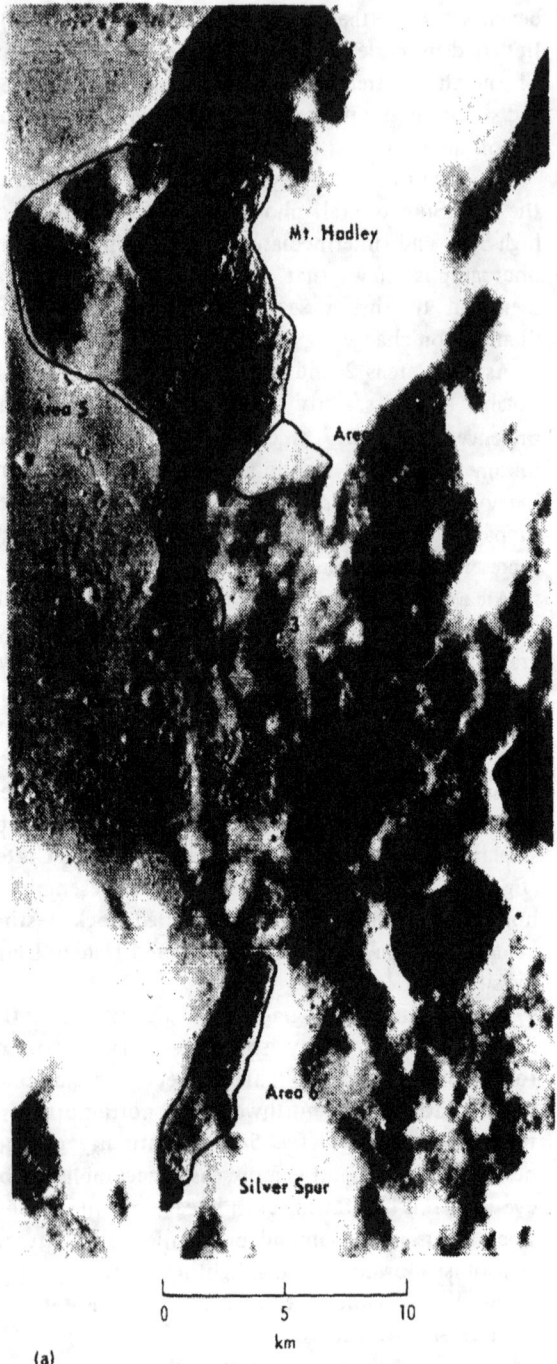
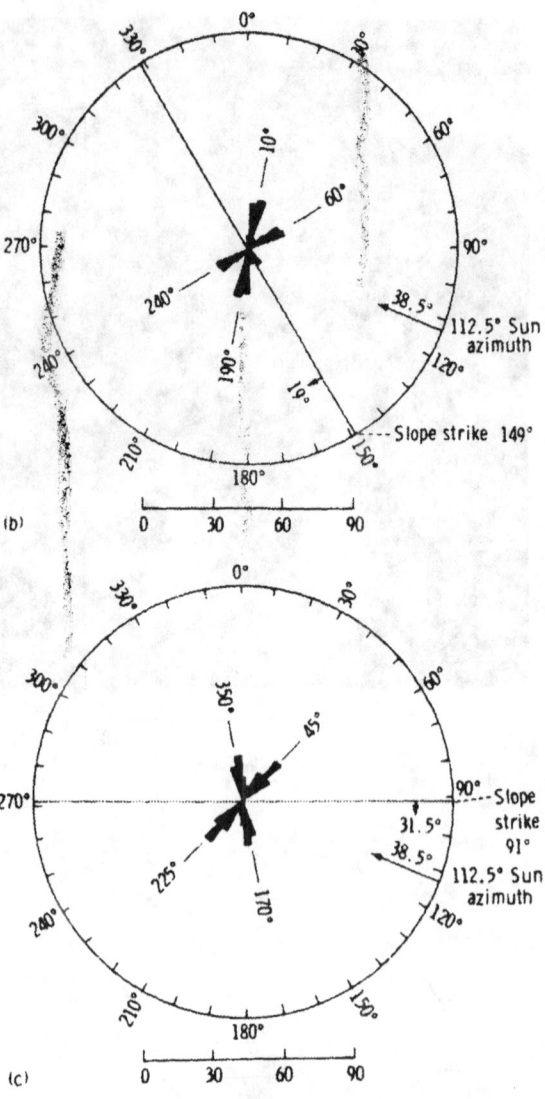

FIGURE 5-24.— Lineament azimuth-frequency plots and location maps of areas 3 to 6. Scale represents number of lineaments plotted as a function of distance from origin; lineaments are plotted in 10° increments. (a) Orbital panoramic camera view taken with intermediate Sun (pan AS15-9425). (b) Plot of area 3 (108 lineaments). (c) Plot of area 4 (107 lineaments). (d) Plot of area 5 (166 lineaments). (e) Plot of area 6 (189 lineaments).

defined topographic lineaments seen in both orbital and surface photographs. The northwest set includes the steeply right-dipping shadowed bands of figure 5-21 and the north-northeast set includes the prominent topographic ledges that appear to dip gently left. Farther north in the more heavily shadowed portion of Silver Spur, the north-northeast set of lineaments was not recognized on the panoramic camera photographs, but northwest- and north-trending lineaments were measured.

Areas 1 to 4 each occur on fairly uniform slopes of known orientation (ref. 5-13). The four slopes differ from each other greatly in orientation, but the lineament patterns in each area are similar (figs. 5-22 to 5-24). Major north and northeast trends intersect

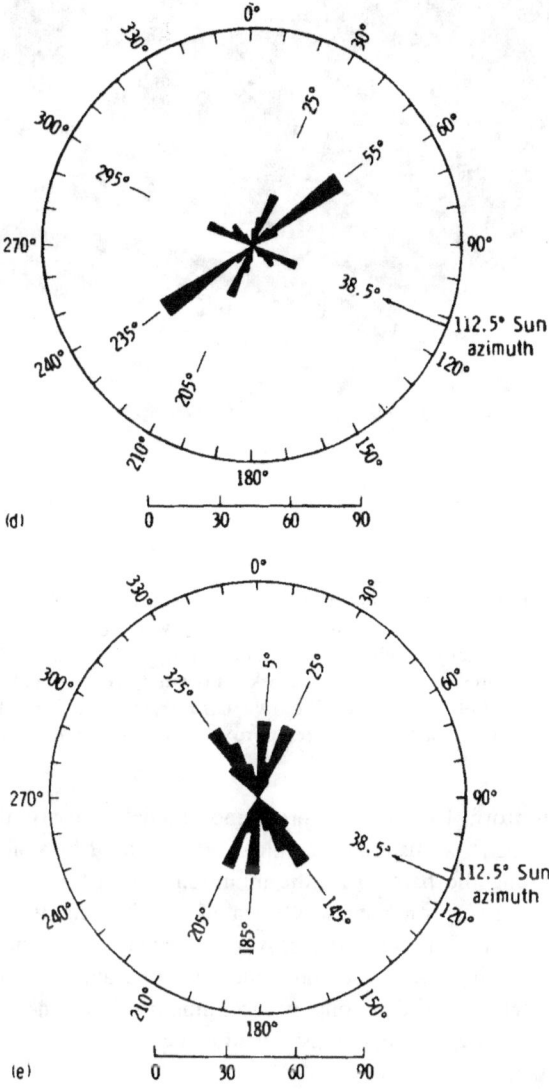

(d)

(e)

in an obtuse angle that is crudely bisected by the Sun line in each case. The lineaments have no constant relation to the slope such as gravitational effects might produce. Stereonet rotation of each slope to horizontal produces only minor changes in the lineament trends, but, in some cases, it produces major shifts in apparent Sun azimuth so that no systematic relation, related to slope, seems to exist between Sun azimuth and lineament azimuths. Apparently, the lineaments either really represent directional features of the lunar surface or, if they are lighting artifacts, they are sensitive only to the lighting direction and not to the orientation of the slope on which they occur.

In summary, the investigation to date is insufficient to assure distinction between systems of lineaments that represent closely spaced, repetitive geologic structures such as layering or regional fracture patterns and those that may be artifacts caused by oblique lighting of irregular surfaces. At Silver Spur, prominent northeast- and northwest-trending lineament sets may be directly related to steeply dipping geologic structures with distinctive topographic expression. Elsewhere, interpretation of the linears as expressions of geologic structure is equivocal.

If either the north or northeast trends prominent at stations 1 to 4 represent bedrock structure, that structure is more likely to be regional fracturing propagated through the regolith than a surface expression of compositional layering. The extensive areal distribution of the pattern and its azimuthal constancy regardless of slope orientation would suggest that, if real, it represents a regional set of nearly vertical conjugate fractures. Local variations may be superimposed on the fracture system as at Silver Spur or at Mt. Hadley, where the northeast set is prominent, but northwest and north-northeast sets also occur.

Dark band near base of Mt. Hadley.—During EVA-2, while driving to station 6, the commander (CDR) commented that the crew saw three suggestions of beddings or horizontal linear lines at the base of Mt. Hadley. He surmised that these lines might represent the high-lava mark for the basin at one time, because they were unique at the base of that mountain.

A dark band is visible along the base of Mt. Hadley in the panoramic camera photographs (figs. 5-3 and 5-24). Where photographed with the 500-mm camera in one spot (fig. 5-26), this band appears to be, at least in part, an outcropping ledge, but the photograph is of one of the less regular parts of the band. A slight hint of a similar band can be observed in a few places along the Front just south of Hadley Delta.

The top of the band is 80 to 90 m above the average level of the mare surface, and the slope of the mountain is about the same above and below the band. In a few places, the top of the band stands out as a small ledge (fig. 5-26). The surface texture of the band, and especially the slope just above the band, appears to be slightly smoother than that of the rest of the mountain face. At least three interpretations are possible for the band—an accumulation of debris at the base of the slope, a fault scarp, or a high-lava mark.

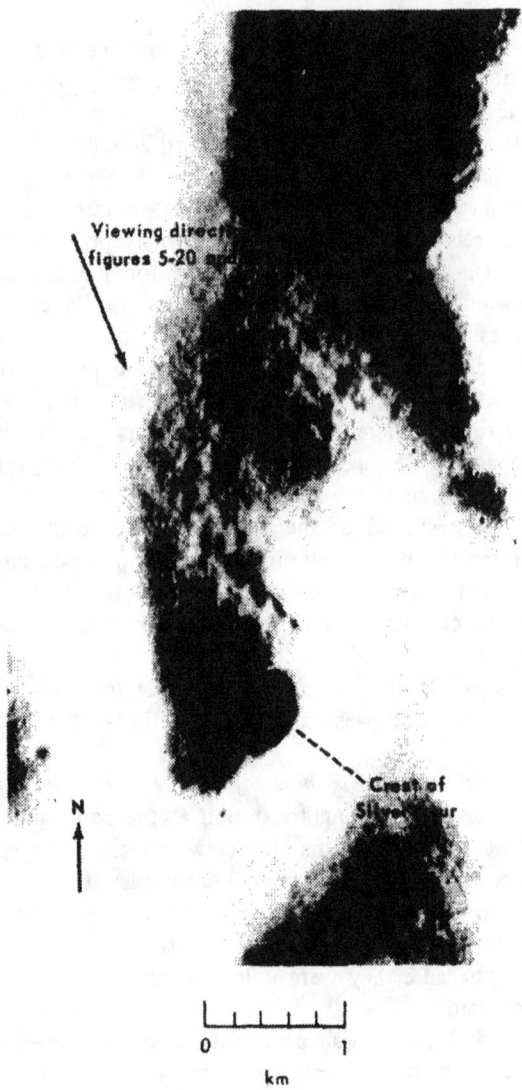

FIGURE 5-25.—Approximately vertical view of Silver Spur showing the diamond-shaped topographic pattern related to intersection of two sets of lineaments (pan AS15-9430).

FIGURE 5-26.—Lunar-surface 500-mm photograph of area to northeast of the base of Mt. Hadley where the dark band apparently coincides with an outcropping ledge that may represent a high-lava mark associated with an earlier basin-filling event. The two large fresh craters on the surface of the ledge are approximately 50 m in diameter (AS15-84-11315).

The only suggestion of a debris apron along the base of Mt. Hadley is this band, and, from analogy with other lunar slopes, it seems likely that at least some debris should have accumulated at the base of the slope. Similar bands occur near the base of other lunar mountains where they are in contact with mare. One at the Flamsteed ring was suggested as being the result of slope movement (ref. 5-14), but it has a more rounded convexity than the feature on Mt. Hadley. The band on Mt. Hadley appears remarkably uniform for a debris apron, and, if a debris accumulation, it seems that a similar feature would be visible along the base of all the mountains, with the height of the apron somewhat related to the heights and slopes of the mountains. Also, if a significant amount of debris were accumulated, the slope angle should decrease in the zone of accumulation. No marked decrease in slope exists below the top of the dark band (fig. 5-27).

The mountains in general are believed to be bounded by faults that occurred during the Imbrium impact event. Slight readjustments along the faults, with the mountain on the upthrown side, after the mare basin filling could leave a remnant of the mare basalts on the Front. The trace of the possible fault, however, is more sinuous than might be expected.

The crew's impression was that it might be a high-lava mark, left after a subsidence of the mare basalts, probably either by cooling shrinkage or, more likely, by partial drain-back into the source vent. Small ledgelike features at the margin of the mare between Mt. Hadley and Hadley Delta are associated with troughs that suggest that the lava pulled away from the mountains, as recognized from premission mapping. Features that are obviously of this type are present elsewhere on the Moon and are common in lava lakes on Earth. The outcrop ledge (fig. 5-26) supports a lava-mark interpretation, for such outcrops are rare on the mountain slope but common where mare basalt is exposed in Hadley Rille. The smoother texture may be caused by debris collecting on the small ledge. This interpretation at present appears to be the best explanation for the band; it does, however, for sufficient subsidence, require a thickness of molten lava greater than the 80 m height of the band.

Hadley Rille

Hadley Rille (fig. 5-3) is one of the freshest appearing sinuous rilles on the Moon. It is also one of the largest, being 1500 m wide, 400 m deep, and at least 100 km long. The Apollo 15 exploration produced a wealth of geologic data that provide new constraints on hypotheses of origin for this impressive canyon. Furthermore, Hadley Rille offers a new perspective into lunar geology inasmuch as it acts as a window into the subsurface and exposes strata in a cross section. The main purpose of this subsection is to provide a description of the form of the rille and the materials exposed in the rille walls.

The rille was visited or seen from four locations by the Apollo 15 crew (stations 1, 2, 9A and 10; fig. 5-2). In addition, it was photographed from the Apollo 15 command module, which transited the site many times. The new geologic information about the rille on which this report is based is of several kinds:

(1) Excellent visual observations and descriptions by the astronauts, mainly on the ground, but also from orbit

(2) Panoramic photography from the ground using Hasselblad cameras with 60-mm lenses

(3) Telephotography of the rille walls from the ground using a Hasselblad camera with a 500-mm lens

(4) Motion-picture 16-mm photography of the rille made during the LM ascent from the lunar surface

(5) Orbital photography of the rille and surrounding region at various Sun angles, using handheld Hasselblad cameras

(6) Preliminary examination of orbital photography, at various Sun angles, from the Apollo 15 panoramic camera

(7) Topography derived from the Apollo 15 panoramic camera photographs (figs. 5-1 and 5-2)

(8) Photographic documentation of collected samples

Much of the photography can be viewed stereoscopically. A wealth of additional data will be gained by detailed study of the orbital photography from the Apollo 15 panoramic and mapping cameras.

General geology of Hadley Rille.—The Apollo 15 crew visited Hadley Rille at a place where the rille changes course from a dominantly northeastward to a dominantly northwestward course (fig. 5-3). To the southwest, the rille winds through a mare-filled graben valley from its head in an elongate, cleftlike depression. To the northwest, a series of shallow septa interrupts and divides the rille into a chain of coalescing elongate bowls indicative of collapse (refs. 5-2 and 5-15). Beyond these, the rille widens again and follows the mare through a narrow gap in the mountains to the main part of Palus Putredinis.

For most of its length, the rille is incised into mare material, but locally it abuts older mountain masses, as at Hadley Delta (figs. 5-1 and 5-3). Whether the full depth of the rille is restricted to mare material or whether it extends into underlying premare material is unknown. Regional relationships indicate that the materials upon which the mare rock rests may include faulted pre-Imbrian rocks, ejecta of the Imbrium Basin, and light, plains-forming units such as the Apennine Bench Formation (ref. 5-1).

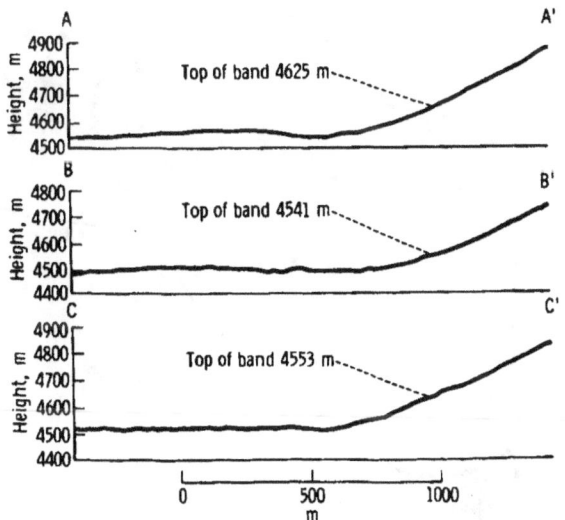

FIGURE 5-27.—Profiles through base of Mt. Hadley showing top of dark band. (Profiles compiled on an AS-11-A1 stereoplotter using two panoramic camera photographs (pans AS15-9809 and AS15-9814).)

External shape: The rille is decidedly sinuous. In any one reach, the width of the rille is constant, so that at each curve the outside has a greater radius than the inside. This observation is well illustrated by the sharp bend at Elbow Crater (fig. 5-3). This geometry cannot result from simple fracturing, because the two sides do not fit together. Instead, it must be caused by drainage or erosion in the rille, either through carrying material along its course or by back wasting of the slopes.

Subtle raised rims are locally present along the rille. Between Elbow Crater and station 9A, this gentle rise is conspicuous on the topographic map (fig. 5-2) and was described by the crew. A less prominent rim is shown on the topographic map across the rille on the southwest side.

The map also shows that the mare surface slopes gently northward away from Hadley Delta, which again confirms the observations made by the crew, who thought that the mare surface sloped gently toward the rille from the mountains east of the area of figure 5-1. The northward general slope of mare at the site is consistent with the viewpoint that Hadley Rille may have been a channel in which material was transported from south to north.

An equally important result of the topographic mapping is that the mare surface southwest of the rille is 30 to 40 m lower than that on the northeast side. This marked difference from one side of the rille to the other can also be seen in photographs taken across the rille from station 9A. These photographs, taken from a vantage point just below the northeast rim, show that the mare surface on the opposite side is visible for some distance and hence is lower. Both the higher mare surface and the higher outside rim fall on the outside of the bend in the rille. Just west of St. George Crater, an outlier of mare isolated on the east side of the rille, on the inside of a bend, is as low or lower than mare on the other side. This situation raises the question whether differences in mare elevation across the rille are systematically related to bends in the rille, as if they might be related to momentum of a fluid flowing in the rille.

Internal shape: As seen in figure 5-2, the inward slopes of the rille walls, just northwest of the Apollo 15 site, average between 25° and 30°. In the vicinity of the landing site, the regolith surface at the top of the rille generally slopes gradually down to a sharp lip approximately 5 m below the mare surface, where discontinuous outcrops form a scarp that extends as far as 35 to 60 m below the top. Beneath this is blocky talus that extends to the bottom approximately 300 m below. Where the rille cuts against a large premare massif (Apennine Front) near St. George Crater, outcrops and blocky talus are replaced by fine-grained debris similar to that on the rest of the surface of the massif. This occurrence indicates that mare rock, which forms the ledgy outcrops and the blocks derived from them, is not exposed at the rille edge along the massif.

In profile, the rille resembes a V that has a rounded bottom and slightly concave sides (fig. 5-2(b)). The walls are uneven in detail, as illustrated schematically in figure 5-28. Below the semicontinuous scarp formed by outcrops at the top, a generally flattish bench of talus gradually slopes off, commonly to one or a series of convexities and inflections in the talus. The inflections tend to be elongate or semicontinuous along the rille.

Materials in the rille.—The materials in the rille include the regolith on the mare surface, bedrock exposures in the upper 60 m, and talus deposits

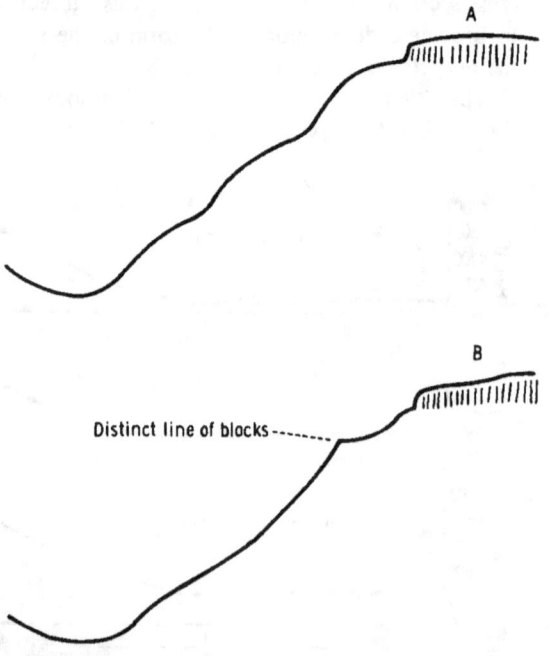

FIGURE 5-28.—Profile sketches (considerably exaggerated) to show subtle benches and inflections on the northeast wall of Hadley Rille. Point A is 2 km northwest of stations 9A and 10; point B is between station 9A and Elbow Crater.

ranging in grain size from fine-grained soil to blocks as large as 30 m across.

Lithologies: The surface geologic unit, cut along most of the length of the rille, is mare basalt. At the edge of the rille, the crew described, photographed, and sampled rocks from the top of the outcrop ledge and from the overlying regolith. Other sampling locations in the mare include Elbow and Dune Craters, both of which were excavated to depths of 50 to 100 m; stations 3 and 8 away from the rille; and a small crater (station 9) near the rim of the rille.

The local section of mare rocks consists of vuggy to vesicular porphyritic basalts. These basalts were sampled at the top of the outcrop in the rille at station 9A and also at the rims of Elbow and Dune Craters. To the extent that these crater rims consist of material excavated from the craters from depths of 50 to 100 m, this basalt may extend down that far. Mare basalt also comprises most or all of the fragments in the regolith above the lip of the rille at station 9A, and outcrops of hard basalt such as that sampled at station 9A are the obvious source of most talus blocks in the rille walls.

Material of the Apennine Front is present on the rille wall below St. George Crater and is distinguished from mare material by the general lack of blocks. This premare material was sampled near the rille at station 2 and found to be breccia. The lack of blocks denotes a large degree of disintegration and a relatively thick regolith.

Regolith: At the top of the rille above the outcrops is an excellent cross section of the lunar regolith. The vertical distance of outcrops below the general mare surface indicates that the regolith is normally approximately 5 m thick and has an irregular base. As the rille is approached from the crest of the rim, the surface slopes gently downward and the regolith thins and becomes coarser. Within approximately 25 m of the sharp lip of the rille, regolith is essentially absent, so that numerous boulders and bedrock protuberances 1 to 3 m across are exposed (fig. 5-29). This thinning is inferred to be the result of near-rim impacts that distributed material in all directions including into the rille; the narrow zone of thin regolith along the rille, however, receives material only from the east because impacts that occur within the rille to the west do not eject material up to the rim (fig. 5-30). This ejection pattern results in the loss of material toward the direction of the rille; therefore, as the rille rim

FIGURE 5-29.—Large blocks and bedrock protuberances at the lip of the rille at station 9A (AS15-82-11147).

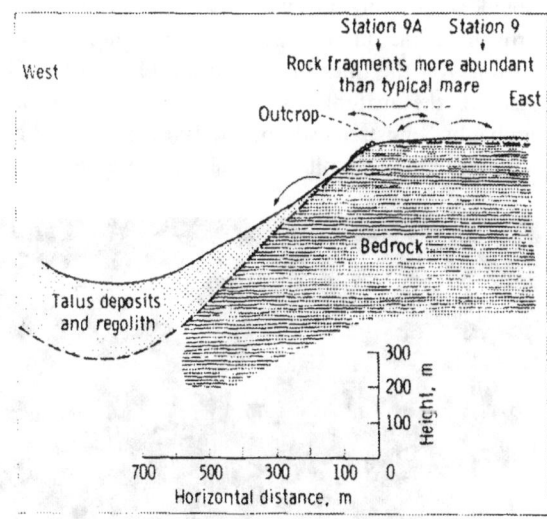

FIGURE 5-30.—Diagram to illustrate winnowing of regolith into the rille. Impacts on the rim eject material in both directions, but the rim receives ejecta only from one side. The result is a net movement of regolith material into the rille.

recedes backwards by erosion, so will the zone of thin regolith recede.

Rock fragments are more abundant in the vicinity of the rille rim than they are on the mare surface to the east. The increase becomes noticeable approximately 200 to 300 m east of the lip of the rille. Most of the fragments at a distance of 200 to 300 m are a few centimeters across. The size of the fragments increases markedly as the surface begins to slope gently down toward the rille; bedrock is reached at the lip.

The abundance of rocks in the 200- to 300-m zone along the rim of the rille is related to the nearness of outcrops to the surface in the vicinity of the rim. All craters greater than 0.5 m or so in diameter in the narrow zone along the lip penetrate the fine-grained material, and therefore the ejecta consist primarily of rock fragments. In the areas of normal regolith thickness, only those craters greater than 20 to 25 m in diameter penetrate the regolith, and even then most of the ejecta are fine-grained materials from craters approximately 100 m in diameter. Therefore, the blocky nature of the 200- to 300-m zone along the rille is due to the nearby source of rocks in the area of very thin regolith along the rille rim.

At the bottom of the rille, the fragment size has a bimodal distribution (fig. 5-31). Numerous large boulders are present, which are rocks that have broken off the outcrops above and were large enough to roll over the fines to the bottom of the rille. The rest of the material is mainly fine grained and probably consists mostly of the fines that have been winnowed into the rille by cratering processes. Filling of the bottom by these fines has rounded the V-shape. Where the rille cuts against the premare massif, the wall is made of fine-grained debris, and the bottom of the rille is shallower and flatter than elsewhere. This situation indicates a considerable fill of fine-grained debris derived from the massif.

Talus: The talus slopes that comprise the main walls of the rille are blocky compared to many lunar landscapes (figs. 5-31 and 5-32). Loose debris is approximately at the angle of repose. Recent instability of the blocks is shown by two boulder tracks visible on the slope opposite Elbow Crater (AS 15-84-11284); larger boulder tracks are evident farther south along the rille in Lunar Orbiter photographs. Talus is especially blocky where penetrated by fresh craters (fig. 5-33). The obvious source of most talus is the outcropping ledges of mare basalt near the top of the wall. Loose blocks that lie above the level of outcrops can be accounted for as blocks produced by impact gardening. Looking across the rille, the crew described loose talus blocks as slightly tanner or darker than the light-gray outcrops.

Both outcrops and blocky talus are absent below the Apennine Front and St. George Crater. A few large blocks on this slope are visible from station 1, but most of these are directly opposite Bridge Crater and probably were ejected from Bridge Crater on the opposite side of the rille. The crew described at least one boulder that had rolled down the slope. The contact in the rille wall, between the mainly fine-grained debris of the Apennine Front and the blocky talus derived from mare rock, angles under Elbow Crater (figs. 5-1 and 5-32), suggesting that at Elbow Crater the Apennine material dips shallowly under the mare fill. The section of ledgy basalt that provides blocky talus thins southward under Elbow Crater.

The blocky talus deposits are commonly poorly sorted and contain a large component of fine-grained debris as compared to talus slopes on Earth. This difference is undoubtedly caused by impact comminution of the lunar talus and to addition of fine-grained ejecta from the mare surface beyond the outcrops. Many patches of talus in the rille are so recently accumulated, however, that fine-grained debris does not fill the interstices (figs. 5-33 and 5-34). Where alined with fractured outcrops, some of these block fields give the appearance of being jumbled but not moved far from their source, similar to fields of frost-heaved blocks that cover outcrops on some terrestrial mountain peaks. In other patches,

FIGURE 5-31.— Bottom of the rille, looking north from station 2. The largest block is 15 m across (AS15-84-11287).

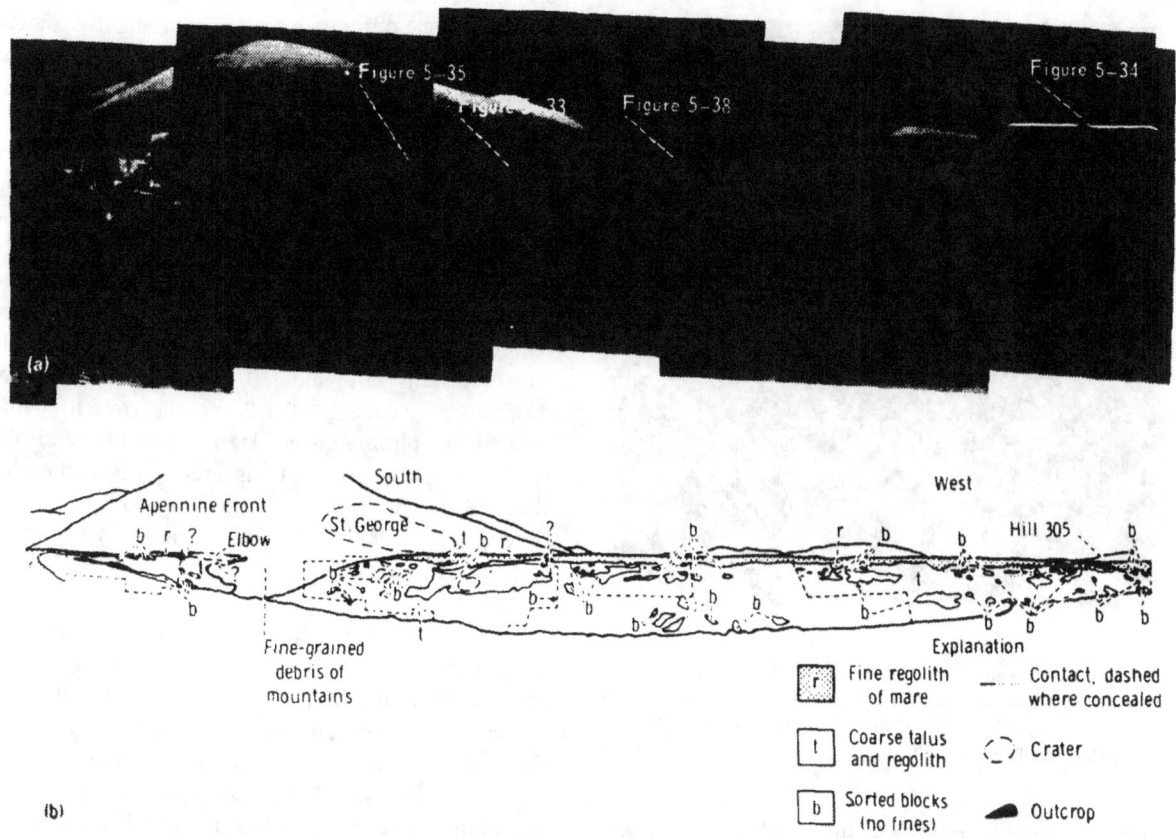

FIGURES 5-32.—Hadley Rille viewed from station 9A. (a) Photographic panorama (AS15-82-11110 to 11120). (b) Panoramic map showing outcrop distribution; the straight dashed lines show the limits of the 500-mm photography.

blocks have accumulated on gentle benches or inflections in the slope—either old craters or possibly topography that is controlled by underlying stratigraphy or structure. As seen in figure 5-32, many accumulations of blocks are elongate horizontally, either along outcrops or discontinuous benches. On the northeast wall, blocks commonly accumulate on a bench just below the outcrop scarp; farther down the wall, blocks are in places concentrated on the steep lower part of convexities in the slope. A few patches of blocks on the opposite wall are elongate down the slope like stone stripes on Earth. Near the top of the rille wall, horizontal lines of blocks underlie finer regolith in places and represent rocks that are apparently close to their bedrock source.

Several talus blocks are more than 10 m across. For example, the largest blocks in the bottom of the rille in figure 5-31 are 10 to 15 m across. The large irregular block near the top of the wall in figure 5-33 is 16 m across. The layered rock in figure 5-35, now split by fractures, is 30 m long and 8 m across the layering. A block halfway down the rille wall south-southwest across from station 9A is 15 by 18 m, and two blocks at an equivalent level to the west-southwest across from station 9A are 11 by 14 m and 10 by 11 m. The largest blocks are about the same size as, or a little thicker than, the largest unbroken outcrops (figs. 5-8 and 5-11). Unbroken blocks of basalt this large are uncommon on Earth, demonstrating that these lunar basaltic flows are both thick and remarkably unjointed compared to many terrestrial counterparts. The maximum size of talus blocks offers a potential means of estimating the minimum thickness of the source layers all along the rille.

The talus blocks show a variety of shapes and surface textures in the telephotographs. Future detailed study of these photographs offers promise of a better understanding of structures in lunar basaltic flows and also of processes of lunar weathering and

FIGURE 5-33.—Rocky 100-m crater (lower left) in the southwest wall of the rille. Lip of rille is just above the blocky areas, and beyond is the mare surface. Large block at upper right, with elliptical cavity, is 16 m across (AS15-89-12069).

erosion. For example, one subrounded boulder contains many large cavities– apparently vesicles– 20 to 50 cm in diameter. The surface of one small block has several crescentic ribs that resemble ropy pahoehoe. Many blocks are layered; one of the best examples is the large block in figure 5-35. Vesicular layering in a block was described at station 9A by the astronauts. One large block near the top of the talus appears to be bounded by columnar joints (fig. 5-10). A rimmed elliptical cavity 3 by 5 m across occurs in the large block in the upper right quadrant of figure 5-35. Many large talus blocks have split in their present location (e.g., fig. 5-35). Other evidence of lunar weathering and erosional processes on the blocks include expressions of layering and rounding of many blocks.

The thickness of the talus deposits is not known nor is the distance that the lip of the rille has receded. The present profile appears to be a consequence of wall recession by mass wasting so that the talus aprons of the two sides coalesce.

Outcrops: Rock strata in the rille near the Apollo 15 site crop out discontinuously in the upper 60 m of the walls. These outcrops were telephotographed through the 500-mm lens at three locations. The first area is a single outcrop, which may be slightly rotated out of position, seen just east of Bridge Crater across from station 2 (fig. 5-36). The outcrop, approximately 18 m long and 6 m high, consists of two prominent horizontal layers. The second area of outcrops forms a discontinuous string along the top of the northeast wall of the rille between Elbow Crater and station 9A, whence they were photographed and sampled. In side view, looking along the slope, these outcrops appear as long subhorizontal bands. The third area of outcrops is across the rille from stations 9A and 10 (fig. 5-32). Extensive outcrops along much of a 2-km-long stretch in that area were photographed stereoscopically and described from the two stations directly across the rille; these photographs provide the best observations of outcropping rocks. Most of the following description is based on the outcrops across the rille from stations 9A and 10.

An interval of massive rocks accounts for most of the outcrops. Below and above these massive rocks are discontinuous exposures that are less massive. These relationships are best seen in the area shown in figure 5-34, where dotted contacts outline the three intervals. Correlation of these exposures as continuous units between outcrops is tenuous because of the cover of talus.

The lowest unit, prominently layered (lower center of fig. 5-8), is exposed in only one small area, and the base and top are concealed by talus. The exposed thickness is approximately 8 m– 16 m if the covered interval above consists of a more erodable member of the same material. This covered interval might be fragmental but is unlikely to be an old regolith horizon considering the regularity of the exposed top of the underlying layered outcrop. At least 12 layers occur in the outcrop (fig. 5-37). Several of the more massive of these layers (1 to 3 m thick) contain less well defined internal layering or parallel banding. Thinner layers less than a meter thick occur together or separate the more massive layers from each other. These thinner layers weather out distinctly. Fractures are dominantly vertical or near vertical within this layered interval and cut distinctly across the massive units, often continuing across the thinner unit into the next massive layer. Nearly all the fractures are within 20° of vertical.

Above the overlying covered interval in figure 5-8 are prominent outcrops (15 to 20 m thick) of light-toned rock that is part of the middle interval of

PRELIMINARY GEOLOGIC INVESTIGATION OF THE LANDING SITE

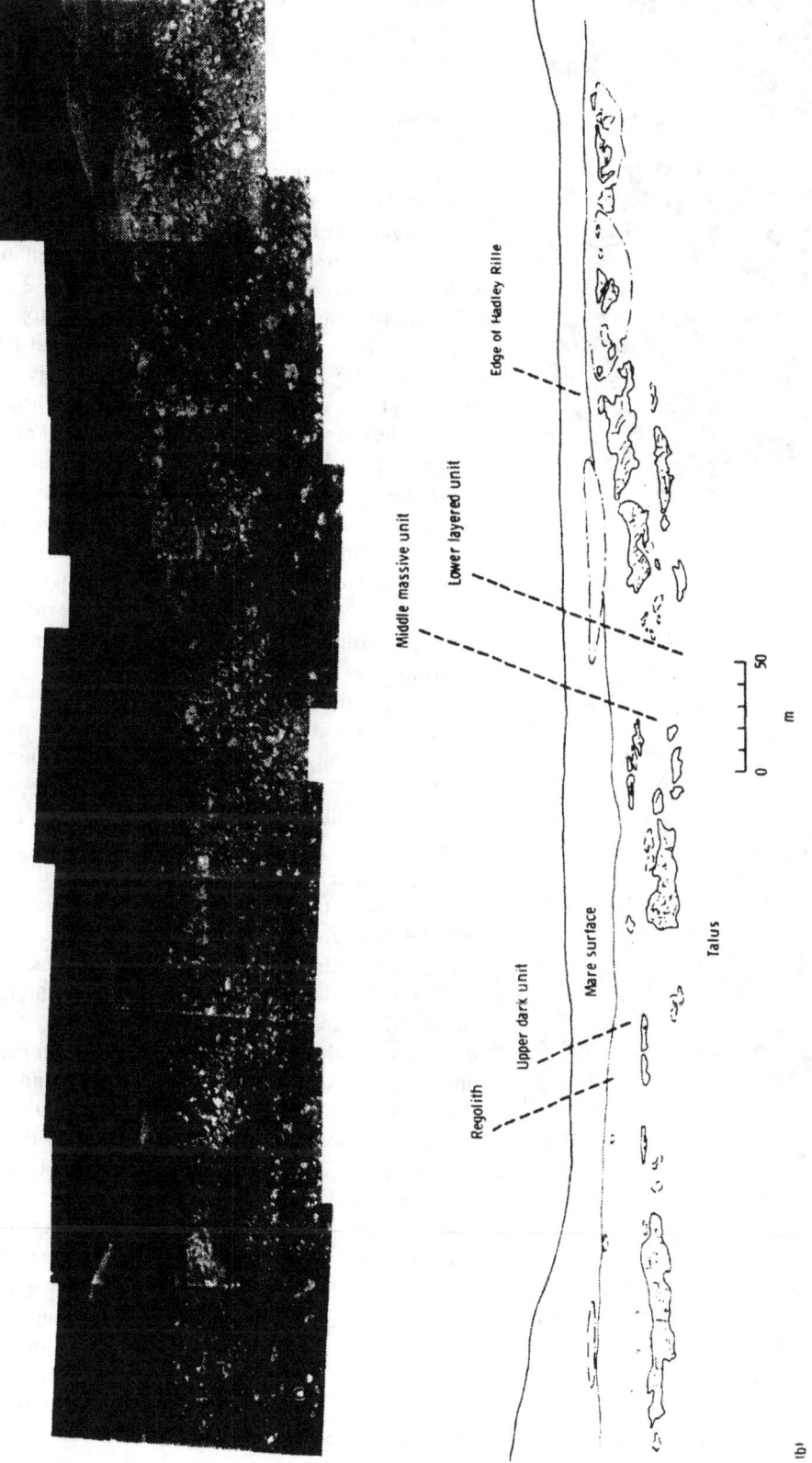

FIGURE 5-34.— Hadley Rille viewed from station 10. (a) Telephotomosaic of upper wall. Area shown is approximately 800 m wide. (b) Map of same area showing outcrops (enclosed in solid lines) and possible outcrops (enclosed in dashed lines). Lines within outcrops indicate fracturing and layering; dotted contacts indicate major units or intervals; dot-dash lines indicate craters.

FIGURE 5-35.—Wall of rille across from Elbow Crater, looking south from station 9A. Large layered block is 30 m long (AS15-89-12081).

FIGURE 5-36.—Upper part of rille wall east of Bridge Crater, photographed from station 2. Outcrop near center with a dark layer at top is 18 m long (AS15-84-11269).

the upper 4 to 6 m of the outcrops, where nearly vertical joints begin to dominate.

At the left edge of figure 5-34 is another large massive outcrop (fig. 5-11). This outcrop has more prominent vertical joints and horizontal partings and appears less bright than those just described. Whether it is a direct continuation of the massive unit shown in figure 5-8 or, instead, replaces or overlies the unit is uncertain, but both features belong to the massive interval. Clearer evidence of two superposed units within the massive interval is seen in figure 5-38. An upper unit (upper-right quadrant of the photograph) forms massive outcrops with planar faces that rarely show fine-scale pitting. Horizontal parting is developed in places, especially in areas photographed 150 m to the left of the picture. In the left-central portion of figure 5-38, below some debris, are outcrops of a lower massive unit. These lower outcrops expose irregular rounded faces that are hackly and coarsely pitted, in contrast to those above.

The interval of massive outcrops thus includes more than one unit. The massive units provide most of the outcrops as well as a high proportion of the talus blocks in the rille.

An upper nonmassive unit is tentatively identified in small intermittent exposures on the left part of figure 5-11. Only 2 to 3 m are exposed below the regolith. This upper unit of small outcrops has characteristically dark and hackly surfaces and is irregularly layered. A relatively sharp but locally irregular contact between it and lighter toned massive rocks below is tentatively identified in several places, but the difference between these two is less obvious than between the massive and the lower layered units in figure 5-8. Some parts of the massive outcrop in figure 5-11 resemble the hackly upper unit. Numerous massive light-toned blocks are present in regolith above the dark, hackly outcrops, possibly indicating that more rock such as the massive unit overlies or is interlayered with the upper hackly unit. In the right half of the area of figure 5-34, little room exists for the upper hackly unit between the massive outcrop and the mare surface; in fact, a suggestion exists that the lower two units bend upward in this area, so that the upper unit may pinch out. However, an exposure of rocks that may be correlative with the upper unit and its lower contact appears far to the south near Bridge Crater (fig. 5-36), where dark, hackly-surfaced rock overlies lighter toned rock on a prominent contact. Vertical joints there pass through both units

massive units. This massive outcrop has discontinuous thin layering or parting within it locally, averaging approximately 0.3 m thick. The most striking aspect, however, is a series of closely spaced (0.5 to 2 m) joints that dip 45° or more to the northwest. This prominent oblique jointing decreases in importance in

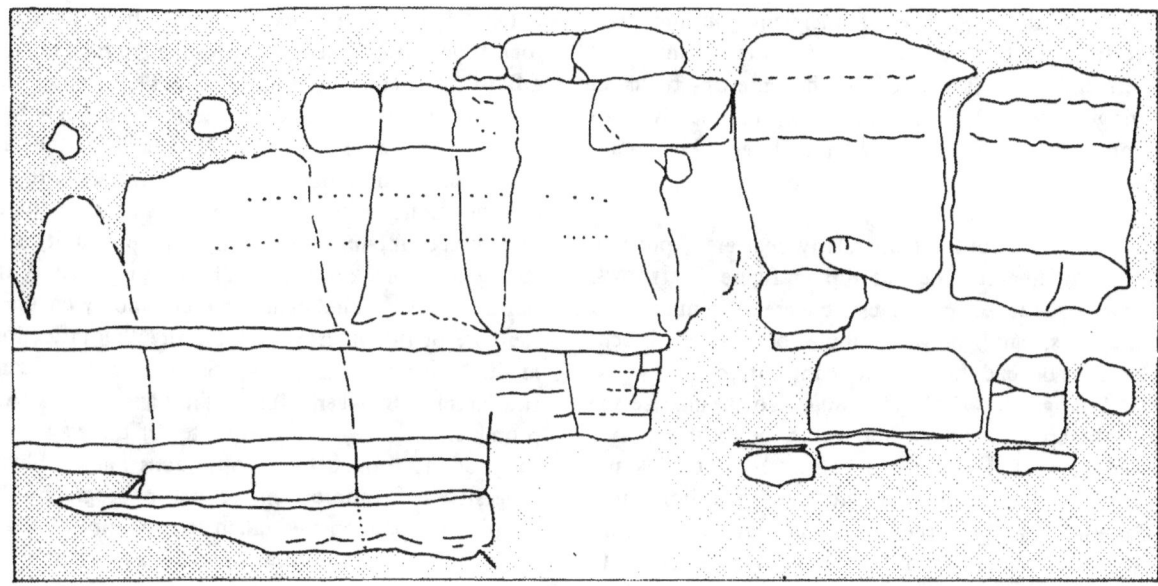

FIGURE 5-37.—Schematic drawing of layered outcrop indicated in figure 5-11. Outcrop is approximately 8 m high.

FIGURE 5-38.—Outcrops on upper rille wall across from station 9A. Width of area is approximately 150 m. Two major massive outcropping units are separated by a covered interval (AS15-89-12075).

together, suggesting no parting plane at the contact.

The attitude of all the layering is horizontal or nearly so. Slabs south of station 9A, however, slope very gently away from the rille, suggesting that the strata dip outward a few degrees. This outward dip possibly is related to the raised rim of the rille.

Small-scale internal layering is in most places expressed as horizontal notches in the rock, as in figures 5-8 and 5-11. The upper dark hackly unit, on the other hand, displays horizontal projecting ribs. Layering in the massive outcrops such as in figure 5-11 consists mainly or entirely of discontinuous parting planes or fractures. Layering in the other two intervals may, however, record successions of strata. Much of the layering is probably analogous to that seen and sampled by the astronauts at the rille edge, where rocks contain vesicular zones that weather out as layers. Most of the layering seen is inferred to be the result of erosion by repeated small impacts.

Layering of a different type is seen in an irregular slabby outcrop near the mare surface (fig. 5-11). An open crack underlies the thin slab and parallels the curving top surface, very much like a shelly pahoehoe where the fluid lava drained from beneath the cooled crust.

Most of the outcrops are jointed. Some of the northwest-dipping fractures shown in figure 5-8 are filled by a light-gray material, which may be either regolith fines or possibly veins. More pervasive than these local inclined joints are vertical joints (fig. 5-11). Stereoscopic examination of photographs of the area shown in figure 5-11 reveals that, between the more obvious fractures, are numerous vertical ribs and troughs suggestive of incipient jointing. The vertical joints may be cooling joints as in many terrestrial lava flows. Rarely the outcrops break in irregular vertical columnlike blocks; one nearly horizontal exposure is fractured into interlocking polygons approximately 3 m across that resemble columnar jointing.

Some outcrops are irregularly shattered, presumably by impact. The outcrop in figure 5-9 shows, from right to left, a progressive increase in number of fractures, until only a rubble of rotated blocks remains beyond the left end of the outcrop.

Stratigraphy of the rille wall.—To summarize the stratigraphy shown by outcrops in the southwest rille wall—several units can be recognized, most thick and massive, but some that show internal horizontal layering. Whether the units and layers represent successive strata cannot yet be said with certainty. In figure 5-34, some suggestion exists that units pinch out to the right (northwest), so that the mare surface consists of successively lower units to the right. Panoramas that include the area shown in figure 5-38 hint faintly of a similar rise of lower units to the northwest, but it is premature to speculate that this relationship is the rule.

The outcrops are limited to the upper 60 m of the wall, which may be significant in terms of regional stratigraphy. On some orbital photographs, the outcrops show up as very bright reflectors along the upper part of the sunlit slope. The bright line of outcrops can be followed to the northwest at least as far as the interruptions in the rille, confirming that the outcrop ledges are essentially continuous along the upper part of the rille.

Where photographed with sunlight just grazing the northeast wall, the outcrop ledge can be identified as forming a nearly continuous scarp just below the rim (fig. 5-1). At Elbow Crater, the ledge thins and becomes discontinuous, and the mare basalt pinches out against the Apennine massif as indicated by a change in the character of debris in the rille wall (figs. 5-1 and 5-32). The full thickness of the thinned basalt was apparently not penetrated by Elbow Crater, however, because samples from the rim and ejecta blanket are all basalt lithologies or derived from them. To the northwest, the outcrop ledge is present even adjacent to the North Complex, where map relationships suggest that a buried premare hill may project to the rille wall (ref. 5-2).

The uniform base of the outcropping ledge on either side of the rille may indicate a discontinuity in rock type at that level; beneath the hard rock may be a flat-topped stratum of more erodable material, such as thinner mare flows or perhaps the premare Apennine Bench Formation. Alternatively, the uniform base of the outcrops may only mark the top of a rather uniform talus slope and be no clue to the underlying subtalus stratigraphy. The possibility of a stratigraphic break in this area is also suggested by the flattish block-covered bench that commonly occurs at the base of the outcrops (fig. 5-28). Other inflections in the talus slope below may also be due to subtalus stratigraphy. Between Elbow Crater and station 9A, a prominent terrace, about a quarter of the way down the wall, is formed by a continuous line of blocks suggestive of outcrop in orbital photography. In telephotographic frames taken from stations 9A and 10, this line is greatly foreshortened, and whether the rocks are outcrops or merely an accumulation of blocks is uncertain.

The lack of any certain outcrops in the long talus slopes below the prominent rimrock indicates that caution must be taken in searching for subtalus stratigraphy. Other explanations are possible for benches and inflections in the talus, as shown schematically in figure 5-39. One possibility is that the benches may be rimrock slabs that have tilted and slumped into the rille as the rim receded. Some extension fractures that precede this kind of sloughing may now be present in the rim north of station 10, where some irregular troughs just back from the lip of the rille are observed (fig. 5-2). In some collapsed lava tubes, tilted blocks of this type form long hogback ridges and benches within the collapse trench (ref. 5-16); where weathering breaks the hogback into blocks, the result is an elongate train of rubble. Possibly the talus benches in Hadley Rille result from a similar process.

Another alternative is that benches and inflections can form in talus independently of underlying structure or stratigraphy. Possibly the impact-erosion process causes the rimrock outcrops to wear back faster than the rest of the talus slope. Any impact on the outcrop would immediately cause broken slabs to fall, slide off, and pile up at the base of the outcrop as a blocky bench. Downward movement along the talus slope would be less rapid because less sliding and falling would occur. The blocky bench at the top of the talus slope could conceivably act as another "outcrop," so that another scarp and another bench

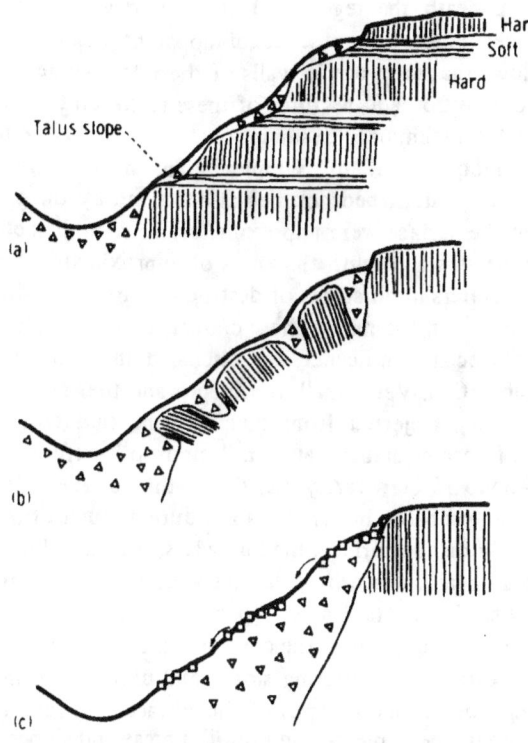

FIGURE 5-39.—Schematic illustrations of three different explanations for benches and inflections in the talus of the rille walls. (a) Alternating hard and soft layers in the subtalus stratigraphy. (b) Tilted slabs from collapse into the rille (slump blocks could be tilted backward instead). (c) Talus surface bears no relation to subtalus structure or stratigraphy; benches are formed by blocks that have fallen or slid from outcrops or block fields.

could form below it, and so on down the slope. Such an occurrence could explain the more subtle form of each succeeding convexity shown in figure 5-28.

The uniform top of the talus slope may bear no relationship to a lithologic change beneath the rimrock but may only indicate that talus deposits fill the rille to a uniform depth. Outcrops in the rille indicate that the mare basalts are at least 60 m thick; however, if the mare basalt that was sampled on the rim of Dune Crater is representative of material ejected from the estimated depth of penetration of nearly 100 m, then mare basalt is present well below the base of the rimrock outcrops in Hadley Rille.

Mare Surface

The surface of that part of the mare traversed by the Apollo 15 crew is generally a plain that slopes slightly to the northeast (fig. 5-2). To the crew, the mare appeared as a hummocky or rolling surface with subtle ridges and gentle valleys. The surface texture appeared smooth with scattered rocks on less than 5 percent of the area. Widely separated patches of roughness are the results of recent impacts that left sharp crater rims and small boulder fields. The visible ridges and valleys are probably mostly due to greatly subdued large craters, and the smoothness is caused by destruction of blocks by erosion by small impacts. Large rays crossing the mare surface were not visible to the crew either as topographic or compositional differences, but patches of high albedo were noted that may be the result of compositional differences in the mare regolith caused by an admixture of ray materials.

The observed part of the mare is bounded by Hadley Rille to the west, the North Complex to the north, and the front of Hadley Delta to the south.

The contact of the mare with the front of Hadley Delta is marked by a change of slope and a band of soft material with fewer large craters than upon the mare. The soft material of the band probably is a thicker regolith than on the mare and includes debris derived from the slope above by both cratering processes and downslope creep. This material is probably slowly extending upon the mare.

The mare surface west of Hadley Rille is lower than on the east side in the traverse area (fig. 5-2). East of the area, the mare extends to the base of the Apennine Front and, at least locally, the contact is marked by a depression on the mare side.

Preliminary study of panoramic camera photography of the landing site indicates a possible, but poorly defined, subdivision of the mare into four primary units based on crater population and general surface textural changes (fig. 5-40). Units I and III are characterized by rolling, hummocky topography, an abundance of fresh bright-halo craters, and association with the North Complex and the South Cluster, respectively. Unit II is characterized by fewer bright-halo craters and a considerably less hummocky terrain. Unit IV, located east-southeast of the South Cluster, is lower topographically than the adjacent Unit III. The contact is marked by a distinct scarp in several places. Unit IV is also more distinctly patterned with lineaments than the other three units. Preliminary topographic data indicate that Unit I may be slightly lower than Unit II and that Unit II may be slightly lower than Unit III, suggesting a possible stratified sequence of mare flows.

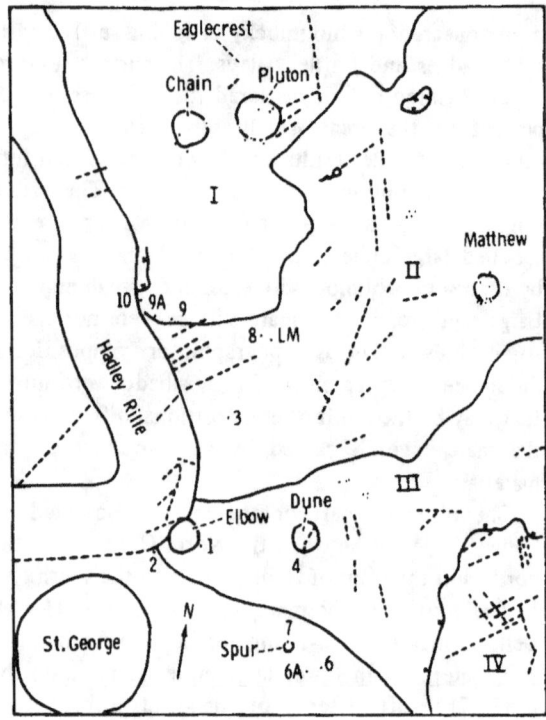

FIGURE 5-40.—Photogeologic sketch map of mare surface in the area of the Apollo 15 landing site, showing mare surface textural units and lineaments.

The LM landed near the center of a low linear northwest-trending ridge several meters high and 600 m wide, which was identified on the premission geologic maps (ref. 5-3). This low ridge extends from the North Complex area to Crescent Crater in the South Cluster but may or may not be related to either. The ridge was not observed by the crew while on the surface because of the very subtle nature of the ridge; however, the photographs taken during the standup EVA appear to indicate a slight topographic rise in the area toward the South Cluster. The low ridge trends in one of the major lunar grid directions and parallels Hadley Rille in the landing-site area. It is conceivable that the ridge is associated with the formation of the South Cluster as a deposit of fine-grained material from Autolycus or Aristillus Craters or perhaps is related to a volcanic or structural feature associated with the North Complex. Analysis of the samples collected from near the LM, Apollo lunar surface experiments package (ALSEP), and station 8 sites may help to define the nature of the ridge.

Beneath the regolith at the rille edge, the mare rocks consist of a series of apparent basaltic lava flows that crop out in walls of the rille. The detailed composition and textures of these rocks, on the basis of the preliminary examinations, are discussed in the subsection entitled "Samples." The smallest craters with apparent bedrock debris described by the crew on the surface were approximately 25 m in diameter, indicating a regolith thickness of approximately 5 m.

Craters in all stages of destruction are everywhere present on the mare. Those craters and crater clusters of special significance are discussed in a following subsection. Very small impact pits and the very small fragments ejected from them are the causes of the surface appearance called a "raindrop" texture. The Apollo 15 crew rarely mentioned this surface appearance, probably in part because, during much of their study on the surface, the Sun was sufficiently high so that small surface irregularities were not accentuated by shadows. The Hadley-Apennine landing site offers a chance to compare the development of the raindrop appearance on differing slopes and units by examining closeup photographs of the surface. The raindrop appearance is present on both flat areas and slopes, as well as on both young surfaces (such as the ejecta blanket of the station 9 crater) and older surfaces. The few photographs of the surface not showing the raindrop appearance can be accounted for by scale, Sun angle, or surface disturbance. It is concluded that the raindrop appearance forms more rapidly than it is destroyed by slope movements and that it probably is formed by the same size particles that accomplish the erosion of topographic highs such as crater rims. This size particle seems to have saturated the surface even on the very young station 9 ejecta blanket, although part of the appearance in that area may be due to small particles thrown into high trajectories at the time the crater formed. The raindrop appearance on the station 9 surface may not be the result solely of particles of the size and energy that produce zap pits on exposed rocks, because such pits appear to be rare on rocks ejected from the crater.

The panoramic camera photographs of the site taken after the LM descent show that the area disturbed during descent is an area of high albedo (fig. 5-5(a)). It seems likely that the high albedo is caused by removal of darker surface material from more compacted underlying material with a lighter color. The high-albedo area extends in the southeast quadrant to a distance of approximately 160 m from the LM. A ray extends to the north to approximately 125 m, and rays extend to the northwest and southwest to approximately 100 m. This observation suggests all the closeup stereophotographs taken at previous Apollo landing sites were within the area that was disturbed by the descent-engine exhaust.

Rocks are present on the mare but cover a very small percentage of the surface. No large boulders on the mare were visited or described except those in ejecta from large craters or along the rille. The rocks seen are generally smaller than 30 m across and appear to be proportional in abundance to the age of the local surface. No distinct rays crossing the mare can be detected by observations of rock distribution; however, the wide separation between points of photographic documentation prevents detection of broad rays. The estimated abundance of rocks at the stations on the mare in percent of surface covered is as follows. At the LM site, 1 percent; at station 8, <1 percent; at the ALSEP site, <1 percent; at station 1, 5 percent on the ejecta blanket of Elbow Crater; at station 3, 1 percent; at station 4, 5 percent on the ejecta blanket of Dune Crater; at station 9, 15 percent on the ejecta blanket of the fresh crater; at the photographic stop between the LM and station 6, 5 percent on the rim of a crater to 2 percent out from the rim on the ejecta blanket; and, at the photographic stops between station 10 and the LM, 1 percent.

North Complex

The North Complex is situated 2.5 km north-northwest at its nearest point from the LM site and covers an area of approximately 7 km² (figs. 5-3 and 5-41). The relatively broad, flat-topped crest of the complex is approximately 100 m above the LM site. The North Complex is one of many dark, low-lying, positive-relief features on the lunar maria and contains a number of mare ridgelike, lobate structures and scarps that suggest a possible volcanic constructional origin. The North Complex can also be interpreted as a premare topographic high that was surrounded and partly or totally covered by later mare volcanic rocks. The south flank is dominated by several large craters including Pluton (750 m in diameter), Chain (600 m in diameter), and Eaglecrest (400 m in diameter).

The proposed traverse to the North Complex was not accomplished largely because of a shortened third EVA (from 6 to 4.5 hr). However, thirty 500-mm

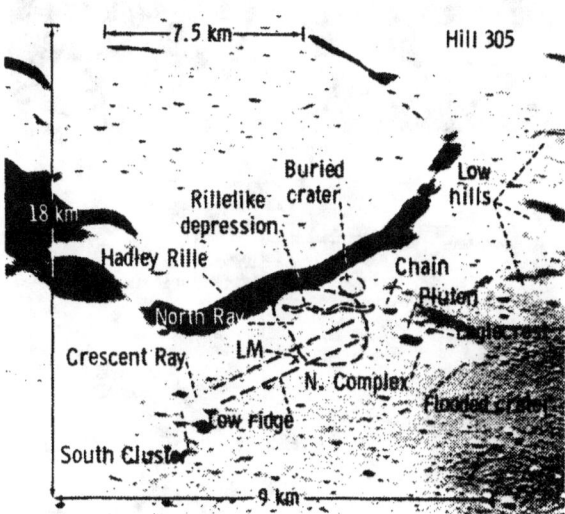

FIGURE 5-41.—Command module 500-mm photograph showing significant geologic features of the landing site (AS15-87-11718).

photographs were taken of the Pluton and Chain Craters area during the standup EVA, during station 6A activities, and from inside the LM. This series of photographs, combined with the orbital panoramic camera coverage, permits a remote geologic study of this interesting morphologic feature.

The walls of Pluton and Eaglecrest Craters (figs. 5-42 and 5-43) are covered with abundant blocks, with the largest seen on the 500-mm photographs being 9 m across. Chain Crater has fewer and smaller average-size blocks exposed on its walls (fig. 5-44). Chain Crater is apparently much older than either Pluton or Eaglecrest Craters, and the walls and floor are covered by a thick layer of fine-grained regolith. No distinct stratification is indicated by the distribution of blocks.

The question as to whether the North Complex area is a volcanic constructional feature or a feature associated with premare materials is not completely resolved. The observations, based on study of the available photographs, that suggest at least a volcanic cover are summarized in descending order of importance as follows.

(1) The low albedo of the North Complex area as compared with the slightly lighter (and more domical) hills to the north, which appear to be associated with the mountain front.

(2) The presence of lobate, flowlike scarps.

(3) Photographic evidence of the presence of very large blocks with large cavities and with albedos

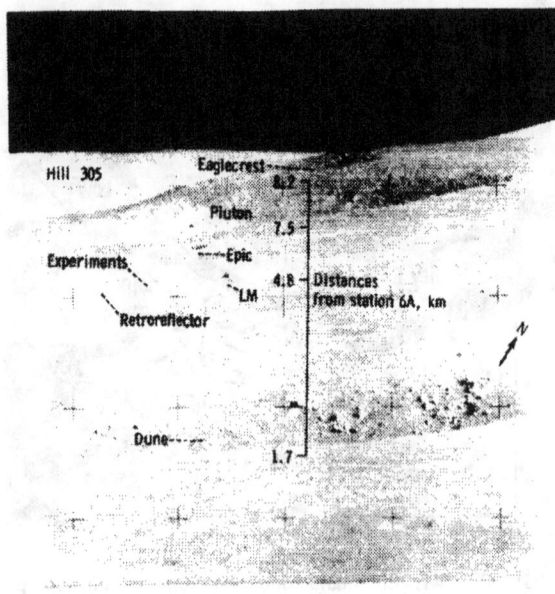

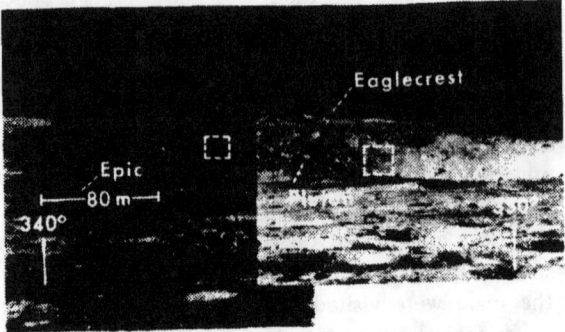

FIGURE 5-43.—Standup-EVA panorama of Pluton and Eaglecrest Craters (AS15-84-11243 and 11244).

FIGURE 5-42.—The LM and Pluton Crater are shown in this 500-mm photograph taken from station 6A during EVA-2 on the lower slopes of Hadley Delta (160 m above LM level) (AS15-84-11324).

typical of mare rocks, and distinctly layered, tabular blocks on the inner wall of Pluton Crater (figs. 5-45 and 5-46).

(4) The presence of a topographically undulating rim crest on Pluton Crater with little or no hummocky rim materials. (Impact craters characteristically have much more uniform rims even if located in undulating topography.)

(5) The possibility that the area south of Chain Crater may be a region of volcanic deposits emanating from the North Complex structure. (This area is more hummocky than other portions of the mare surface (verified by the crew during the EVA-3 traverse to the rille) and is slightly higher in elevation than the surrounding mare surface. A subtle rillelike depression extends from the southern rim of Chain Crater. A possible buried crater southwest of Chain Crater (fig. 5-41) further indicates possible volcanic activity.)

Evidence that the North Complex may be underlain by breccias includes (1) relatively high albedo values from photometric measurements made on the soil and a number of blocks on the walls on Pluton and Eaglecrest Craters (fig. 5-43), and (2) the abundance of bright, fresh craters in the North Complex and in the crater cluster area south of Chain Crater. The presence of large numbers of bright-halo craters in these areas may indicate excavation of higher albedo rocks, possibly breccias.

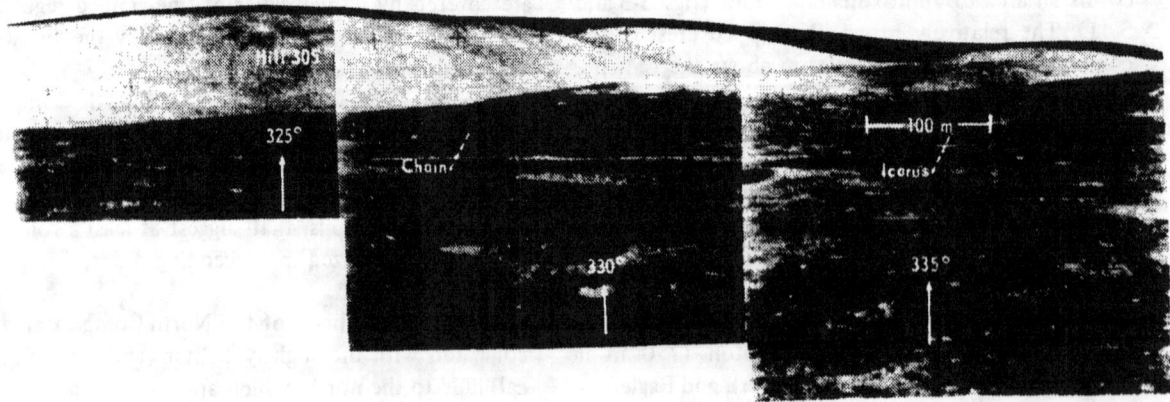

FIGURE 5-44.—Standup-EVA panorama of Chain Crater. Sun-elevation angle was 12.5° (AS15-84-11244 to 11246).

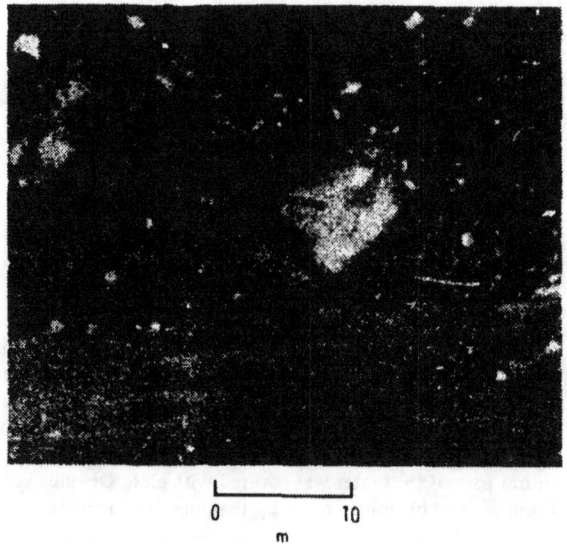

FIGURE 5-45.—Enlargement of large block with macrocavities (vesicles?) on wall of Pluton from one of the standup-EVA 500-mm photographs (AS15-82-11217).

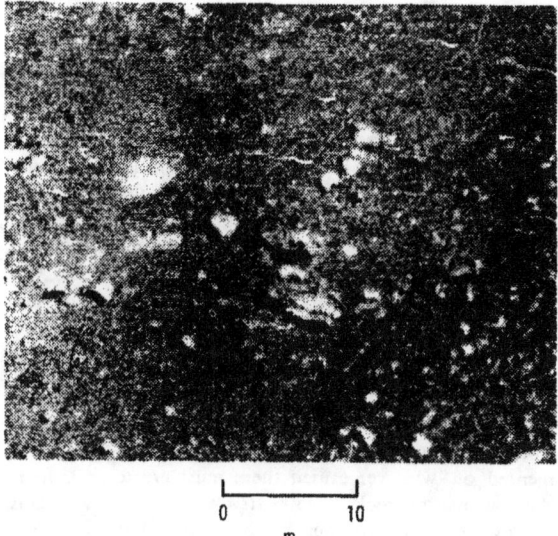

FIGURE 5-46.—Enlargement of layered block on wall of Pluton Crater (AS15-82-11215).

The eastern side of the North Complex has been overlapped by younger mare volcanics as suggested by the rather sharp contact between the two materials and the presence of a crater with a flooded floor at this contact (fig. 5-41). The 100-m height for the North Complex and the 80- to 90-m-high ledge that may be a high-lava level on the base of Mt. Hadley suggest that the entire hill may have been completely covered at least once by earlier mare volcanics, before subsidence of the mare surface to the present level.

Small-Scale Features

In addition to the major geologic features, several types of small-scale features were noted. These smaller features include craters, fillets, and lineaments other than the large lineaments on the Apennine Front.

Craters.—Craters of many sizes, shapes, and lithologic variations are visible in the photographs taken on the lunar surface and in the panoramic photographs, and many of the different craters were described and sampled by the crew. Only a few of the more interesting types are described herein.

Glassy craters: The data returned from the Apollo 15 mission provide exceptional documentation of one particular class of craters—the fresh glass-floored craters ranging from 30 cm to 15 m in diameter. The crew quickly became interested in these craters and described, photographed, and sampled them. They also developed theories of formation for craters with abundant glass and consequently were able to make observations to test the theories while they were still on the lunar surface. The air-to-ground transcript of the conversation between the CDR and the LM pilot (LMP) while they discussed glassy craters in the vicinities of the various stations is noteworthy in this respect.

LMP at LM

The crater here that I'm standing by, Joe, it's about a meter in diameter. And then, there's a smaller crater right in the center of it, and that one has fragments around in that—that have glass exposed on them, where the—the larger crater does not have any glass exposed, just the smaller crater within the large one.

CDR between LM and station 1

There's a nice little round 1-m crater with very angular fragments all over the bottom and the rims, and glass in the very center; about a meter across.

LMP at station 1

Yes, just behind you is one of those fresh craters, too, with a lot of glass in it.

LMP between stations 1 and 2

Bet there's glass in the bottom of that one.

CDR

Yes, there sure is.

CDR between stations 2 and 3

There are a lot of little craters around here—little being less than a meter—which are very rough, have a lot of debris—right up to the rim and over the top side of the rim, and no ejecta blanket to speak of, but the whole inside of the wall—take a half a meter crater and it's filled with angular, gray, fragmental debris on the order of inch size—or less, very uniformly distributed, fairly well sorted. Like—maybe they came—maybe the debris is one of our Aristillus or Autolycus friends. And there's a lot of it, so I think we'll have a chance to get it later on.

LMP between station 3 and LM

You know, Joe, these small fresh craters that we've commented on—whatever caused them must create or indurate the soil into the rocks—creates its own—own rocks, because there's just a concentration of rocks around the very fresh ones. And the small I'm talking about may be a foot to 3 feet diameter.

CDR between station 3 and LM

There's a pretty fresh one right up ahead, Jim. Looks like about 10 m across, and it's got up to 6-inch fragments around the rim—maybe 15, 20 percent of the rim has fragments in it, but nothing—no significant ejecta blanket, which I think is typical of all these around here. That one looks like it's maybe a meter and a half deep. Too bad we can't get in it, and I bet it has glass in it too.

LMP

You know, you can almost tell the ones that are going to have glass—by looking at them before you get there.

CDR

That's right, you sure can.

Mission Control (debriefing)

Do you have any feel for whether the fragments around the small fresh craters that you've called out to us are, in general, pieces of the projectile or do you think they're ejecta fragments?

CDR

Well, Joe, we're pretty sure they're projectile fragments, and that's when we really need to stop and sample.

LMP between LM and station 6

You know, on one of these trips, we ought to stop at one of these very fresh ones and really tap one. I mean these small ones, you know, just filled with rock debris and glass in the middle. Just do a systematic sampling on it. Like this one over here at 1 o'clock.

CDR at station 6

And there's one of those fresh little craters. Let's go sample that one.

LMP

Got glass in the bottom.

CDR

And it looks to me like the best thing to do—would be to—scoop the side—scoop—scoop the center where the glass is.

LMP

It all felt kind of welded together.

LMP

Like fragments all glued together. What an intricate pattern.

CDR between stations 8 and 9

I bet it chipped that hole, Jim. It went right in—it came from that—it made that crater there. And it came from 070. That angular projectile about a foot across, Joe, had made a secondary about a meter across. I bet you anything, because the—one part of the fragment was covered with glass, and the central part of the crater was covered with glass. Obviously, a secondary and obviously made by that angular fragment.

LMP at station 9

Boy, look at the fresh blocks ahead of us.

CDR

Good fresh one [crater].

LMP

It sure kicked up a lot of rocks.

CDR

Yes. Boy, it's really fresh with a lot of debris. Nice ejecta blanket. Good typical one.

CDR

And the rim is very, very soft. My boot sinks in a good—if I push on it, a good 4 inches—and the whole center part of the crater is just full of debris. Very angular, glass in the center. It's about—oh, I guess, 40 meters across and maybe 5 or 6 meters. No, not that much—3 or 4 meters deep. And a slightly raised rim. An ejecta blanket that goes out about one crater diameter, quite uniform. I don't see any rays. There's a little bench in the bottom—halfway up—about a tenth the diameter of the crater—and it seems to be all the way around, somewhat irregularly.

CDR at sampling rim

The first one I tried to pick up just fell apart. It's a clod—it's just a caked clod. This stuff is really soft.

LMP

This is sure a unique crater. Unique—that we've seen so far. Very soft on the rim. You sink in about 6 inches.

CDR

Just like big pieces of mud.

LMP

You know, this has the appearance of those small ones that we sampled, with the exception, there's no concentration of glass in the very center, except every fragment has glass on it.

CDR

That's right. Well, not every fragment; many of these clods don't have any at all. Most of them don't have any glass. Covered with dirt but it [sample] looks just like a big piece of glass.

The craters of this type have many features in common besides the limited range of sizes. They are fresh-appearing craters, noticeably blocky with glassy bottoms, and generally without a conspicuous ejecta blanket. The largest of this class of craters, at station 9, differs from the rest by having a well-developed ejecta blanket and a well-developed interior bench. The crew's comments on formation indicate that the rocks and glass in the craters result from the impacts that formed the craters rather than being excavated from the soil. The rocks and glass were considered either as part of the projectile or formed by a welding process upon impact. A final statement on the origin of the rocks and glass must await chemical and petrologic examination of the samples. However, a tentative theory, for the 1-m or smaller craters, is that the projectile arrived while still coated with molten glass that was jarred off upon impact, leaving a glassy center in the crater and partly glass-welded clods higher up the crater walls.

The crater at station 9 may have formed by the same process but is sufficiently different from smaller craters of this class to be described separately. It is exceptionally well documented photographically, but the collected samples may not be adequate to resolve problems of origin.

The station 9 crater (fig. 5-47) is situated approximately 300 m east of Hadley Rille (fig. 5-1). Approximately 75 m south of the rim of Scarp Crater, it is about the same distance southwest of the rim of an unnamed crater that is about the same size as Scarp, is younger than Scarp, and partly overlaps Scarp to the southeast. The unnamed crater and Scarp Crater are both approximately 250 m in diameter. Thus, the station 9 crater site is less than one-third crater diameter from both Scarp and the unnamed crater and can be presumed to be located upon the ejecta blankets of each. This situation would suggest that the site is topographically higher than the general surface; this observation is strengthened by the photographs taken from station 9 toward the east. The photographs taken to the west show just the top of the rille, suggesting a very slight downward slope in that direction.

The station 9 crater is approximately 15 m in

FIGURE 5-47.—Sharp, cloddy crater with bench at station 9 (AS15-82-11082).

diameter with a distinct rim, almost a meter in height, that slopes outward for a distance of approximately 2 m, where it merges with a thin ejecta blanket that extends another 15 to 20 m. The interior of the crater, 3 m deep, is very blocky with a bench slightly more than halfway down (fig. 5-47). Above the bench, the wall is steep, about at the angle of repose, except for the east side where it is very steep. The bench is nearly flat and approximately 1 m wide. Below the bench, the lower part of the crater is approximately 3 m across and a little more than 1 m deep. Within the lower part of the crater are angular blocks with a rough concentric arrangement. The largest blocks, approximately 30 cm in diameter, are arranged in a circle, about 1.5 m across, around smaller blocks.

Two types of rocks can be distinguished in the photographs of station 9 crater—an easily eroded, medium-albedo breccia (type 1) and a hard, low-albedo glass (type 2). The fine-grained material is assumed to be derived from the same material as the type 1 rocks; certain rocks of type 1 appearance, except apparently more resistant to erosion, are considered gradational from type 1 to type 2.

Easily eroded, medium-albedo type 1 rocks and fine-grained material occur on the bench within the crater, in the wall above the bench, on the rim, and in the ejecta blanket. A few of the rocks in the foreground of the photographs contain either clasts of high albedo or a bumpy surface texture suggesting clasts, in addition to faint banding. In the very steep east wall of the crater above the bench, there appear to be outcrops of the breccia in which the faint banding dips outward.

Hard, low-albedo, type 2 rocks are the only material seen below the bench and occur elsewhere in less abundance than type 1 rocks and fine-grained material. These rocks are angular and contain holes, which suggests a vesicular texture but is probably a frothy surface. Type 2 rocks are interpreted as glass or glass coating. The occurrence of only glassy rocks in the center of the crater may relate this crater to the smaller fresh glassy craters and suggests that the glass is a result of the cratering dynamics. However, the bench in this crater is unique and may mark the presence of a contact between rock types, in which case the presence of only glassy rocks in the crater center might reflect a different cratering reaction caused by a different composition of target material. The transitional rocks, which appear to be medium in albedo but resistant to erosion, could be either the result of partial conversion to glass or coating by glass (because they came from the edge of the central zone when homogeneous material was impacted) or the result of being the closest to the impact center of an upper distinct layer. In the first instance, the transitional rocks probably were derived from a depth just above the bench; in the second instance, they probably were derived from the upper layer but near the impact center.

The distribution of rocks upon the rim and ejecta blanket of the station 9 crater is not uniform. Most of the rocks appear to be mixed with the fine-grained material and probably were deposited with the fine-grained material. A significant portion of the rocks appears to rest upon the fine-grained material, as if the rocks were deposited later. Some of this distribution could result from greater resistance to erosion of specific rocks or from later churning by other impacts. However, the rim and ejecta blanket are such young surfaces that both mechanisms are considered of minor importance. The dominant factor in forming the nonuniform distribution of the rocks on the very young surfaces is interpreted as an original difference in deposition at the time of the impact that formed the crater. The bulk of the rim-surface material and of the ejecta-blanket material is considered the result of deposition from a base-surge cloud of unsorted fine material and rocks. The rocks that appear to lie on the surface are considered to be rocks that were thrown into steep trajectories by the impact and then fell upon the base-surge deposit.

The fresh, glassy craters were sampled at stations 6 and 9. Unfortunately, the crew did not have the opportunity to sample both a crater of this class and the projectile that apparently formed the crater, although they did note such an opportunity between stations 8 and 9; in addition, they observed that the projectile, as well as the center of the crater, was glassy. Sampling of both materials might have shown if the glass in the crater were derived from the projectile. At station 6, a 1-m-diameter crater with a glassy bottom was sampled both from the center and from the rim. These samples have not been studied in detail; on the basis of preliminary data, the material from the crater center is described as glassy soil and the material from the rim is described as soil. Detailed examination of these samples may determine whether the glass is derived from fusion of the local soil.

At station 9, samples were taken from the rim and from the ejecta blanket about midway from the rim to the edge of the blanket. Two rocks that appeared to lie upon the surface were taken from the rim but broke apart during transport. As photographed before sampling, these friable rocks appeared rounded. It can be concluded that they were derived from above the bench in the crater, were ejected with a steep trajectory, and were representative samples of the uppermost layer—less than 2 m thick—of the site.

One rock, one glass sphere, and soil were sampled from the ejecta blanket halfway from the rim of the edge. The soil sample, probably typical of the fine material in the base-surge deposit, has not been studied. The rock appeared to lie upon the surface and to be resistant to erosion. Laboratory examination shows the rock to be a breccia with a nearly complete vesicular glass coating. It may have been derived from the unit above the bench and may have been transported through a steep trajectory. If so, this rock represents material that is transitional from type 1 to type 2 and may have been derived from the lower part of the unit above the bench. The rock also could have been derived from the central glassy part of the crater, although it appears to be more rounded than the blocks remaining in the center. The glass sphere is an unusual sample and, although it may be representative of the glassy center of the crater, it also might have been a glass sphere in the regolith before this cratering event. The greater number of zap pits on the glass sphere indicates more total exposure time at the surface than those of the other rocks collected.

Crater clusters: Crater clusters are common features at the Apollo 15 site, but they can be discerned only if composed of large craters or if close to surface panoramic photograph stations. The clusters that occur near sampling areas are of special importance, because they may indicate a direction of origin of exotic materials, some of which may have been sampled.

The large group of secondary craters named South Cluster (station 4, fig. 5-2) is alined with and possibly related to the faint ray that crosses the mare surface from the north-northwest. The ray and the projectiles that produced the cluster are thought to have been derived from either Autolycus or Aristillus Craters. Station 4 is located on the south rim of Dune Crater, the westernmost member of the South Cluster. The soil samples or the breccia collected there could possibly include material from Autolycus or Aristillus. The size of the craters of South Cluster suggests that they, in turn, could be the sources of secondary or tertiary craters and exotic materials in nearby areas, perhaps even as far as stations 6, 6A, and 7 (see subsection entitled "Samples").

A small cluster of rimless craters occurs approximately 15 m southeast of the LM (fig. 5-48). The individual craters are approximately 1 to 2 m in diameter. The craters form a rough V with the apex to the north-northeast, which suggests a source in that direction. Although the crew traversed the cluster, none of the samples appears to have been taken from within the area affected.

A cluster of relatively fresh, 25- to 100-m-diameter craters on the flank of Hadley Delta is of special significance because three stations (6, 6A, and 7) were within the general area of the cluster (figs. 5-1 to 5-3). The crew noted the cluster went up the slope of the Front and suggested that it might be a group of secondary craters. The cluster includes Spur Crater, though Spur appears slightly more subdued and therefore may be somewhat older than the others. In photographs taken from the LM, the craters appear as a linear group oriented up the slope (fig. 5-15). Plotting those craters that can reasonably be assumed to belong to the cluster shows a more equant distribution with a subgroup forming the line. At

FIGURE 5-48.—View looking east from the LM during the standup EVA showing a cluster of small rimless craters (AS15-85-11368).

least one other cluster of craters is present on the Front. These craters are larger, more openly spaced, and alined northeast from a very bright crater on the east rim of St. George Crater. A distinct separation exists between the two clusters at the upper ends, but the clusters approach each other at the base of the slope.

The craters of the cluster that includes Spur Crater are young and show distinct rims and high-albedo halos. The linear subgroup suggests but does not prove that the projectiles came from the north or south. If so, the craters may have been made by secondary projectiles that contributed some exotic fragments to the regolith in the area of stations 6, 6A, and 7. This cluster of fresh craters is much younger and smaller than the South Cluster to the north and therefore is not related to the event that produced the South Cluster.

Craters on Hadley Rille rim: The 500-mm camera photographs taken from station 10 show two craters that are of special interest because their effects on the underlying bedrock outcrops can be studied. The first, a crater 20 to 25 m across, is directly above the right-hand side of the prominent massive outcrop in figure 5-11. No effect from the cratering is visible within the ledge, but the top of the ledge is clear of debris. Apparently, the effect of the cratering was to eject debris into the rille along the top of the resistant ledge, which served as a plane of detachment. The second crater of interest is a subdued crater approximately 130 m across on the mare surface adjacent to the rille (fig. 5-8). The eastern rim almost coincides with the lip of the rille, close to some underlying massive outcrops. The massive upper outcrop in this area has a well-developed set of northwest-dipping fractures and a lesser set of vertical fractures that are more prominent here than elsewhere along the outcrop ledge. The fractures may be the result of the cratering event.

Fillets.—Fillets were first described (ref. 5-17) as accumulations of fragments on uphill faces of rocks, as seen in photographs televised by Surveyor III. Fillets were further defined (ref. 5-18) as embankments of fine-grained material partially or entirely surrounding larger fragments. The Apollo 14 mission to Fra Mauro illustrated a greater variety of blocks and related fillets, which allowed classification on the basis of the contact relationships between the rocks and the soils on (or in) which they occur (ref. 5-11).

The Apollo 15 results provide many new examples of rock-soil contacts that can now be studied systematically, together with the samples derived from them, to determine the relative effects of rock texture or friability, rounding, surface slopes, and length of time during which the rock has been exposed to surface processes.

The three-fold classification previously presented (ref. 5-11) is expanded in this report on the basis of new evidence from the isolated boulders examined on the Apennine Front and from the abundant blocky debris near outcrops along the edge of Hadley Rille. With the exception of rocks that occur on steep slopes, forming special kinds of composite fillets, all rocks can be placed in one of the five morphologic categories (fig. 5-49). It is probable that the primary factors responsible for morphologic differences in fillets, other than slopes, are (1) the coherence (or friability) of the rock, (2) the original shape of the rock upon reaching its present location, and (3) the length of time a rock has been in the observed position. The relative friability can be determined from samples of specific boulders and from samples collected from the general surface population where rock types are locally homogeneous, as appears to be the case on the edge of Hadley Rille (stations 9A and 10), at the blocky crater east of the rille (station 9),

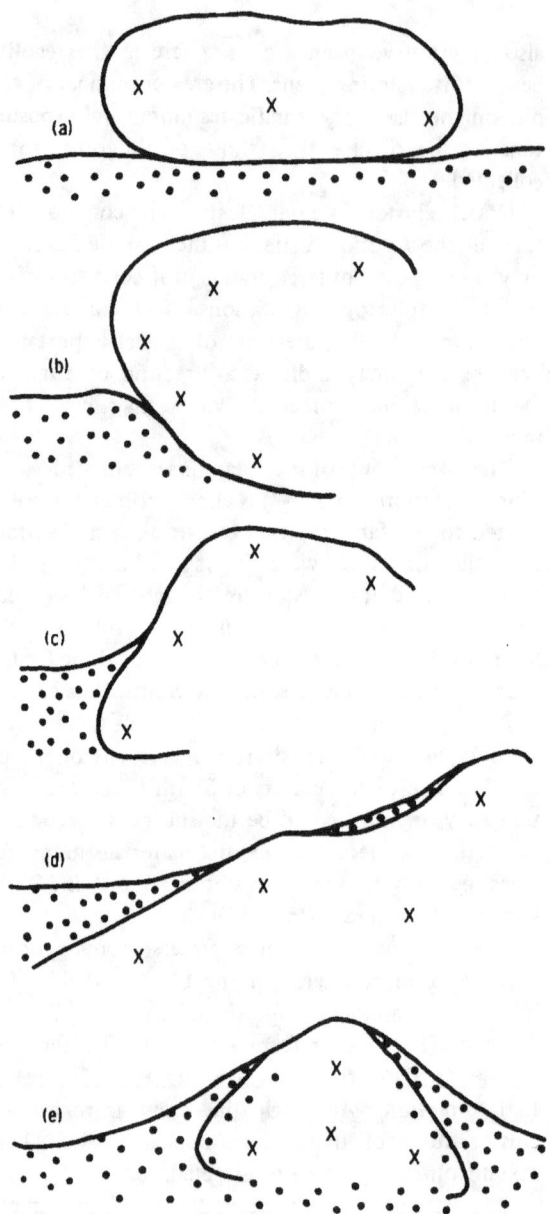

FIGURE 5-49.—Types of rock-soil contacts observed at Hadley-Apennine landing site. (a) Perched rock. (b) Overhang fillet. (c) Steep fillet. (d) Shallow fillet. (e) Mound fillet.

at Dune Crater (station 4), and at the green breccia boulder on the slopes of Hadley Delta (station 6A). The original shape of the rock can only be inferred from the coherence and present shape compared with fresh exposures of similar rocks having similar composition. The third factor, the age of the presently exposed rock surface, can be measured by cosmic radiation methods and can then be used to calibrate the relative contributions from the first two factors, coherence and shape.

Type A in figure 5-49 is a perched rock lying completely on top of the surface. This type has no observable fillet but has rather prominent overhanging edges. An excellent Apollo 15 example is the breccia boulder, approximately 1.5 m long, on the rim of Spur Crater. This rock has an irregular subrounded shape, appears to be moderately coherent, and should have an exposure age equivalent to that of the relatively fresh crater from which it came.

Many other examples of perched or unfilleted rocks occur at all stations, and these rocks can be interpreted as recent arrivals. Several perched boulders were sampled at the rille edge (basalts such as samples 15535, 15536, and 15555) and at the fresh, cloddy crater at station 9, where friable soil-breccia fragments lie high on the surface.

The overhang fillet (type B) is best illustrated by the 1-m angular boulder on the slope at station 2. This rock has a distinctive ridgelike embankment of soil and fragments parallel to the base of the east edge, but the fillet is separated from the rock face by a few centimeters. The rock may have produced this fillet by rolling or falling into the present position, banking up the soil, and rocking westward away from the fillet. Mechanical disintegration, caused by diurnal heating and cooling and by micrometeorite impacts, is discounted because a fillet does not occur on all sides of the rock. Furthermore, the lack of an uphill fillet, although the rock lies on a prominent slope, indicates a brief residence at this location.

At station 10 on the east rim of Hadley Rille, numerous tabular and subrounded basaltic blocks have overhanging postures as a result of their location on the wall of a subdued crater or, locally, on the sloping wall of the rille. These relationships are illustrated schematically in figure 5-50. Prominent uphill fillets on these very coherent rocks reflect the effect of slope combined with sufficiently long exposure. The downhill overhangs may indicate either continuing movement of fine materials downslope or interference by the boulder with the normally random redistribution of surface materials.

Steep fillets (type C) are generally narrow, concave, upward embankments of soil against a vertically to steeply inclined rock surface. At the Apollo 15 site, these fillets occur notably around large angular

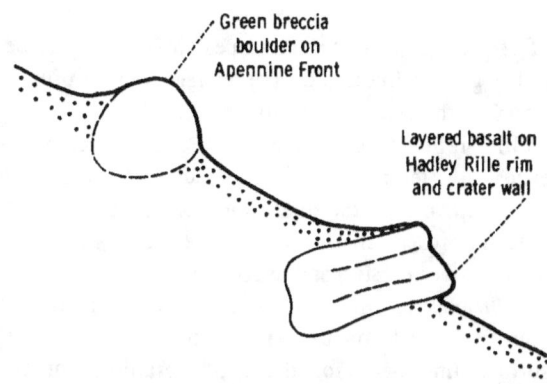

FIGURE 5-50.—Examples of uphill fillets observed by Apollo 15 crew. That the downhill contacts can be either steep or overhanging, probably reflecting the relative coherence of the boulders, is noteworthy.

basaltic blocks on the rim of Hadley Rille (stations 9A and 10) and on an isolated, rounded, vesicular, basaltic boulder at Dune Crater.

Shallow fillets (type D) occur on the presumably coherent basalts at the rille edge where the tabular, subrounded, low-profile blocks (similar to unweathered lava-flow rock forms on Earth) are overlain by broad fillets, particularly on the upslope side. In this example, rock shape and surface slopes (rather than friability) probably contribute to the morphology of the fillets, and the ages of these fillets must be greater than those of equally filleted, friable rocks elsewhere.

On the slopes of Hadley Delta, an uphill type C fillet was photographed on the green breccia boulder at station 6A; this fillet is illustrated schematically in figure 5-50. The rock is friable, and the fillet extends high on the south flank (the north side was not photographed). The possibility exists that the fillet results from mechanical disintegration of a friable rock (because samples of the high fillet (15400 to 15404) and the breccia (15405) look similar in composition) as well as from downhill movement of soil against the base of the rock.

Both of these processes—disintegration and lateral soil movement—eventually should result in a mound (type E) fillet, which should develop more rapidly for a friable rock than for a coherent rock of similar initial shape. If rocks are equally coherent, filleting and surface dust coating will occur more rapidly on a rounded or tabular flat-lying rock than on an angular, high-standing rock. If the friability and initial shapes of rocks and local surface slopes are all comparable, the degree of filleting on the average will be proportional to the exposure ages of these rocks.

Integration of the composition and physical properties of returned samples, radiation ages, and measurements of the fillet morphologies should provide improved methods of determining rates of lunar-surface erosion and deposition.

Lineaments on the mare surface.—Small-scale (centimeter widths by tens of centimeter lengths) surface lineaments of the type described for the Apollo 11 and 12 samples (ref. 5-12) are essentially absent both on the mare plains and on the lower slopes of the Hadley Delta massif in the areas visited during the periods of EVA. Large-scale (hundreds of meters length) regional lineaments mapped from the panoramic camera data of the mare surface in the vicinity of the landing site are poorly to moderately developed. They are most prevalent in the area east of the South Cluster (fig. 5-40).

The crew mentioned only once seeing what appeared to be lineaments on the mare surface during the return to the LM from station 2 (approximately 0.7 km away). The CDR thought the lineaments were oriented northwest, but the LMP was not convinced of the presence of lineaments. A few seconds later, they agreed that evident lineaments lay ahead of them, parallel to their direction of travel (essentially north). The crew confirmed after the mission that they were looking for lineaments during all travel periods but said that they did not observe any on the mare besides those mentioned.

The small-scale mare lineaments were well developed at the Apollo 11 site and in areas of the Apollo 12 site where associated with firm compacted ground on the rim crests and inner walls of 200- to 400-m-diameter subdued craters (ref. 5-12). The mare at the Apollo 15 site, however, appears more like the nonmare (Fra Mauro) Apollo 14 site where very few small-scale lineaments were evident and only a few lineaments as much as 100 m long were well developed (ref. 5-11).

A preliminary study of the larger scale lineaments on the Palus Putredinis near the Apollo 15 landing site indicates a primary northwest orientation with a secondary north-northwest trend.

Some doubt has been expressed regarding the existence of lineaments as anything more than the effects of lighting (subsection entitled "Apennine Front"). The Sun angles were about the same during the Apollo 11, 12, and 14 missions, but the small-scale lineaments trend (1) northeast and northwest and are well developed at the Apollo 11 site; (2) northeast, north, and northwest and are well developed at the Apollo 12 site (ref. 5-12); and (3) northeast, north, and northwest and are poorly developed at the Apollo 14 site (ref. 5-11). During the two Apollo 15 periods of EVA, the Sun angles were also similar to those of earlier missions, yet the lineaments are essentially absent. Therefore, the variation in trends and abundances from site to site under similar lighting conditions shows that the lineaments are not produced by lighting alone but differ from one locality to the next.

Some evidence exists that the presence or absence of the smallest scale lineaments from site to site may be affected by changes in the grain size, cohesiveness, and density of the upper tens of centimeters of the soil. These data are currently being seriously investigated along with the true relationship of solar illumination effects on the observability of the small-scale linear features.

SAMPLES

Approximately 78 kg of rock and soil samples were returned by the Apollo 15 crew. The locations of all but two large samples are known, many to within a meter or less of features seen in the panoramic camera photographs. The rather complete documentation of most of the samples makes it possible to relate the samples to the geologic environment in which they were collected. This ability is especially important in an area as complex as the Hadley-Apennine site.

The panoramic views taken during the standup EVA, in the vicinity of the LM, and at stations 1, 2, 3, 4, 6, 6A, 7, 8, 9, 9A, and 10, and at the ALSEP-deployment site are contained in appendix D at the back of this report. The panoramic views (figs. D-1 to D-16) are on extra-length foldouts for the convenience of the users.

In this subsection, detailed information is given on the rock samples and the local environments. Generalizations of particular geologic significance derived from these detailed descriptions are given first to provide a context in which any specific sample of interest may be placed. Generalizations that are pertinent to further detailed study of samples include the relative ages of the surfaces sampled and preliminary gross stratigraphic relationships of the mare basalts and breccias collected from the Apennine Front.

Estimate of Relative Ages of Sampled Surfaces

Preliminary geologic studies of the data obtained during the mission permit an evaluation of the relative ages of the surfaces visited, photographed, and sampled. The sequence was established by using crater morphology; rock-fragment size, shape, and abundance; relationship to distinct local and regional geologic and topographic (morphologic) units. Surfaces on the nearly horizontal mare tend to modify at different rates compared with those on the Apennine Front because of the complication of preferential downslope movement of debris on the Front. Similarly, mare surfaces that are influenced by proximity to the rille (e.g., station 9A) must be considered in a different context from those outside of the influence of this edge effect. The following sequence is tentative. Relative assignments should probably be correct to within one position. At all sites, surface areas that are free from smaller fresh craters are considered; recent impact features result in relationships that require special consideration.

Several fresh small crater rims and blankets were carefully sampled and documented at stations 2 and 6, which are considered the oldest sites sampled, as indicated in table 5-I. These samples provide opportunities to compare and contrast older and younger surfaces on similar materials.

In general, a correlation between the age of the surface and the mean surface exposure of the materials immediately on and underlying the surface is expectable. However, local or special circumstances can modify this relationship. For example, samples taken from the sloping surfaces of the Front represent products of transport and accumulation by impact processes modified in various degrees by downslope bias and possibly by direct downslope gravitational movement. A distinctly different mixing regimen may be the consequence of these downslope effects. At the rille margin at sites such as station 9A, a distinct rilleward slope and the absence of essentially a half space of source material provides a condition contributing to net loss (deflation) of regolith. The mean surface exposure of regolithic materials at this site follows a different function of the surface age than is found at normal mare sites.

TABLE 5-I.—*Relative Ages of Materials at Apollo 15 Sampling Sites*

[Listed in order of age, with oldest site last]

Station no.	Type location
9	Mare
9A (Rille margin)	Mare
7 (Spur)	Front
4 (Dune)	Mare
1 (Elbow)	Mare
LM	Mare
8 (ALSEP)	Mare
3	Mare
6	Front
2 (St. George)	Front

Stratigraphic Relationships of the Mare Basalts

Two main types of basalt, each with textural variants, are recognized among the mare basalts: type A, basalts rich in yellow-green to brown pyroxene, commonly in long prisms and type B, olivine-phyric basalt. Type A, with variants grading from glassy to coarsely crystalline groundmasses, is by far the dominant variety and occurs at all the three main mare sites (stations 1, 4, and 9A). Type B, with a small range in groundmass grain size, occurs mainly at station 9A.

Sampling of three widely spaced areas of mare material allows examination of lateral variations among the mare basalts. In addition, two crater sites (stations 1 and 4) were sampled, thus allowing examination of vertical variations in mare basalts to depths as great as 90 m. The prevalence of type A basalt at all three main collecting sites suggests significant lateral continuity of this type of basalt. Samples collected from the two large-crater sites on the mare are of this type. Samples from the rim of Dune Crater, possibly from a depth as great as 90 m, are porphyritic with aphanitic groundmasses in contrast to medium- or coarse-grained rocks from the flank of Dune Crater and samples from Elbow Crater. This discovery strongly suggests that more than one flow unit of pyroxene-rich basalt was sampled. Variations in granularity of the pyroxene-rich basalts as found in the radial samples taken at Elbow Crater may represent interflow variations or multiple flow units. Type B basalt was collected from a slightly higher elevation than the pyroxene-rich basalts taken from the outcrop at the edge of Hadley Rille. This type of basalt was well represented in the rake sample at station 9A but was not found at Elbow or Dune Craters. This occurrence suggests the presence of a local olivine basalt flow that is stratigraphically above the pyroxene-rich basalt in the vicinity of station 9A.

Apennine Front Samples

Three main types of breccias were collected from stations 2, 6, 6A, and 7 on the Apennine Front: (1) generally friable breccias containing fragments of glass and basalt, and debris derived from coarse-grained feldspathic rocks; some of these breccias contain sparse fragments of basalts like those collected from the mare sites; (2) generally coherent breccias with medium- to dark-gray, vitreous matrices and abundant debris derived from coarse-grained feldspathic rocks and varying proportions of nonmare-type basalt clasts; these breccias are commonly partly coated by glass; and (3) well-lithified breccias with dark-gray aphanitic matrices and clasts composed dominantly of granulated feldspathic rocks; the breccia matrices bear intrusive relationships to the clasts.

Friable-type breccia is considered to be weakly lithified regolith that locally contains a small proportion of mare basalt clasts thrown up on the Apennine Front, mostly from South Cluster. Coherent-type breccia is the most common type collected on the Apennine Front. It is also common at the LM site and at station 9 on the mare surface. Because these breccias do not contain marelike basalt clasts and are the dominant rock type collected from the Front, they are presumed to represent premare Front material. The well-lithified breccias, of which only three documented samples have been returned, are also considered to represent Apennine Front material, differing from coherent breccias mainly in proportion of the coarse-grained feldspathic clasts and in the intrusive relationships of matrix and clast shown by the well-lithified breccias. Accordingly, the Apennine Front, beneath the regolith, is considered to be composed of breccias, the clasts of which are mainly varying proportions of coarse-grained feldspathic rocks and nonmare-type basalts.

The distribution of rock types (fig. 5-51) reveals an unexpected prevalence of nonmare-type rocks on the mare surface, especially around the LM site and station 9; the best samples of mare basalts were obtained from two craters (Elbow and Dune) that penetrated the regolith on the mare and from the rille edge where the regolith is very thin or absent. However, the general geologic setting is such that contamination of the mare surface by extraneous material is greatly enhanced. Much of the sampled part of the mare surface is covered by a ray that contributed an unknown quantity of foreign material to the surface. The site is bounded on two sides by high mountains composed at least partly of breccias; dominant downslope movement of material, resulting in contamination of the mare surface, is to be expected with much less material transfer in the opposite direction. A third side of the site is bounded by North Complex, which may also contribute breccias (ref. 5-2) to the mare surface; the fourth side is bounded by Hadley Rille, which is more likely a repository than a contributor of material to the mare surface. These features, combined with the general lack of large craters on the mare surface, tend to enhance contamination of the surface by foreign debris.

An important and distinctive feature of the breccias collected from the Apennine Front is the lack of evidence of multiple brecciation that is so well displayed in the Apollo 14 samples. This observance strongly suggests that the Apollo 15 crew sampled a lower level of the lunar crust that is less reworked than that sampled by the Apollo 14 crew.

Sample Locations and Environments

Sample locations are summarized in table 5-II by Lunar Receiving Laboratory (LRL) numbers. The distribution of rock types for samples larger than 20 g is shown in figure 5-51. The rocks have been categorized only in the most general way in this figure and in table 5-III, but detailed descriptions of some individual specimens are provided in table 5-IV.

LM area.—The LM site (fig. 5-3) is on the mare surface about equidistant from North Complex and Hadley Rille. The site is within the limits of a ray (ref. 5-3) that mantles the mare surface. Distinct craters larger than a few meters across are widely separated, whereas closely spaced craters are less than 25 m across. Subdued craters nearly 1 m across overlap one another. The surface is very hummocky in detail (figs. D-1 to D-3). The surface material in the

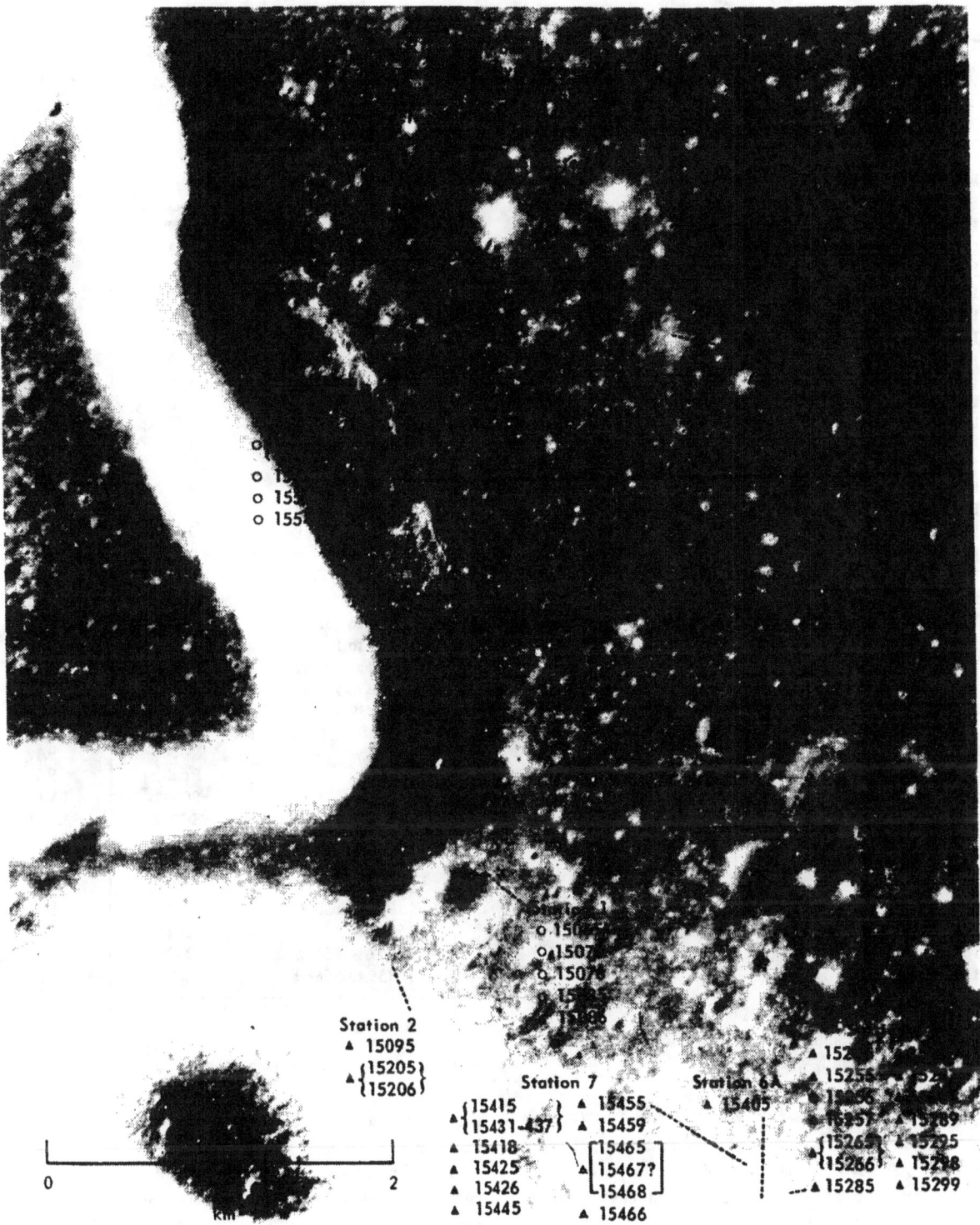

FIGURE 5-51.—Map of Hadley-Apennine landing site showing distribution of rock types larger than 20 g (rake samples excluded).

TABLE 5-II.—Index of Apollo 15 Samples by Sequential LRL Number

LRL sample no.	Container	Traverse station	LRL sample no.	Container	Traverse station
15001 to 15006	Core	8	15285 to 15289	192	6
15007 and 15008	Core	2	15290 to 15295	188	6
15009	Core	6	15297	[b]SCB-3	(?)
15010 and 15011	Core	9A	15298	Rock	6
15012	[a]SESC-1	6	15299	Rock	6
15013	[a]SESC	8	15300 to 15308	173	7
15014	[a]SESC-2	LM	15310 to 15392	172	7
15015	Rock	LM	15400 to 15405	168	6A
15016	Rock	3	15410 to 15414	194	7
15017 to 15019	Contingency	LM	15415	196	7
15020 to 15026	162	LM	15417 to 15419	194	7
15027 and 15028	162	LM	15421 to 15427	195	7
15030 to 15034	252	8	15431 to 15437	170	7
15040 to 15044	253	8	15445	171	7
15058	Rock	ALSEP	15455	198	7
15059	Rock	ALSEP	15459	Rock	7
15065	156	1	15465 to 15468	199	7
15070 to 15076	157	1	15470 to 15476	203	4
15080 to 15088	158	1	15485 and 15486	204	4
15090 to 15093 and 15095	159	2	15495	174	4
15100 to 15105	187	2	15498	Rock	4
15115 to 15119	186	2	15499	Rock	4
15125	186	2	15500 to 15508	255	9
15135	186	2	15510 to 15515	273	9
15145 to 15148	186	2	15528 and 15529	274	9A
15200 to 15204 and 15206	160	2	15530 to 15538	275	9A
15205	161	2	15545 to 15548	278	9A
15210 to 15214	180	2	15555	Rock	9A
15220 to 15224	181	2	15556	Rock	9A
15230 to 15234	182	2	15557	Rock	9A
15240 to 15245	163	6	15558	[b]SCB-2	LM(?)
15250 to 15254	164	6	15561 to 15564	[b]SCB-2	LM(?)
15255 to 15257	190	6	15565	[b]SCB-2	LM(?)
15259	192	6	15595 to 15598	281	9A
15260 to 15264	166	6	15600 to 15610	283	9A
15265 to 15267	193	6	15612 to 15630	282	9A
15268 and 15269	192	6	15632 to 15645	282	9A
15270 to 15274	167	6	15647 to 15656	282	9A
15281 to 15284	[b]SCB-3	(?)	15658 to 15689	282	9A

[a]Surface environmental sample container.

[b]Sample collection bag.

vicinity of the LM is fine grained with less than 2 percent of the surface covered by particles larger than pebble size. No lineaments of natural origin are visible, but the descent-engine exhaust caused some streaking of the surface.

Ten, possibly 11, samples were collected from an area of approximately 10 by 30 m near the LM (fig. 5-52). Of these, two are soils (surface environmental sample container (SESC) sample 15014 and contingency samples 15020 to 15024), seven or eight are breccias (15015, 15018, 15025 to 15028, and 15558(?)), and one is a hollow glass ball (15017). The breccias all lack mare-type basalt clasts. The samples were selected primarily on the basis of distinctive individual characteristics, except sample 15014, which may be considered representative of the rim material of the 6-m-diameter crater on which the LM landed.

TABLE 5-III.—*Cross Reference of Apollo 15 Lunar Rock Samples, Weights, Station Numbers, and General Rock Types*

[Comprehensive soil and fragment samples are not included: i.e., 15100 to 15148, 15300 to 15392, and 15600 to 15689.]

LRL sample no.	Weight, g	Station	Breccia	Soil breccia (a)	Basalt
15015	4515	LM	(b)		
15016	923.7	3			(c)
15025	77.3	LM		(d)	
15027	51.0	LM		(d)	
15028	59.4	LM		(d)	
15058	2672.5	ALSEP			(c)
15059	1149.2	ALSEP	(b)		
15065	1475.5	1			(c)
15075	809.3	1			(c)
15076	400.5	1			(c)
15085	471.3	1			(c)
15086	216.5	1		(d)	
15095	25.5	2	(b)		
15205	337.3 [f] [e]				
15205,1	1.6	2	(b)		
15206	92.0				
15245,1 to 90	[g]116.58	6		(d)	
15255	240.4	6	(b)		
15256	201.0	6			(b)
15257	22.5	6	(b)		
15265	314.2 [e]				
15266	271.4	6		(d)	
15285	264.2	6		(d)	
15286	34.6	6		(d)	
15287	44.9	6	(b)		
15288,0	62.6 [f]				
15288,1	7.6	6	(b)		
15289	24.	6	(b)		
15295	947.3	6	(b)		
15298	1731.4	6	(b)		
15299	1691.7	6	(b)		
15405	513.1	6A	(b)		
15415	269.4				
15431 to 15434	[f] [e]				
15435,1 to 17	[g]57.5				
15435,18	49.1				
15435,19	9.6	7	(b)		
15435,20	58.3				
15435,21 to 32	[g]32.3				
15436	3.5				
15437	1.0				
15418	1140.7	7	(b)		
15425	136.3	7	(b)		
15426	223.6	7	(b)		
15445	287.2	7	(b)		
15455,0	885.4 [f]				
15455,1 to 23	[g]53.1	7	(b)		
15459	4828.3	7	(b)		
15465	374.8 [f]				
15465,1	1.2				
15467(?)	1.1	7		(d)	

TABLE 5-III.—*Cross Reference of Apollo 15 Lunar Rock Samples, Weights, Station Numbers, and General Rock Types* - Concluded

LRL sample no.	Weight, g	Station	Breccia	Soil breccia (a)	Basalt
15468	1.3				
15466	119.2	7		(d)	
15475,0	298.2 [f]				
15475,1	85.2	4			(c)
15475,2	23.4				
15476	266.3	4			(c)
15485,0	102.2 [f][e]				
15485,1	2.7				
15486	46.8	4			(c)
15499	2024.0				
15495	908.9	4			(c)
15498	2339.8	4		(d)	
15505,0	1140.3 [f]				
15505,1	7.1				
15506(?)	22.9	9	(b)		
15508(?)	1.4				
15515,1	59.3	9		(d)	
15515,2 to 48	[g]85.4			(d)	
15529	1531.0	9A			(c)
15535	404.4 [f]				
15536	317.2	9A			(c)
15545	746.6	9A			(c)
15546	27.8	9A			(c)
15547	20.1	9A			(c)
15555	9613.7	9A			(c)
15556	1538.0	9A			(c)
15557	2518.0	9A			(c)
15558	1333.3	LM(?)		(d)	
15565,1 to 43	[g]810	9 to LM		(d)	
15595	237.6 [e]				
15596	224.8	9A			(c)
15597	145.7	9A			(c)
15598	135.7	9A			(c)

[a]Criterion for the identification of soil breccia is the presence of mare basalt clasts.
[b]Breccias without mare basalt clasts.
[c]Mare/Rille-type basalt.
[d]Breccias with mare basalt clasts.
[e]Braced numerals represent multiple samples collected from the same rock.
[f]Bracketed numerals represent the multiple samples that were formed from one sample that broke.
[g]Total weight of a sequence of fragments.
[h]Sample 15256 has the chemical composition (section 6) of a typical mare basalt, but the texture is not similar to mare basalts.
[i]Samples 15515,1 to 48 represent two separate rocks that were subsequently broken and mixed.

The lack of large samples of mare basalt, either as individual rocks or as clasts in the breccias, indicates that the coarse surface material at the LM site is foreign to the site. The breccias, which resemble those collected from the Front, may be part of the ray on which the LM landed (fig. 5-1). Alternatively, especially considering the low density of large fragments at this site, the breccias may have been ejected from impacts either on the Front or North Complex (refs. 5-1 and 5-2).

Sample 15014: Soil collected as a contaminated sample in an SESC from the northwest rim of the 6-m-diameter crater on which the LM landed. The documentary photograph (fig. 5-53) shows a faint streaking produced by alinement of centimeter-size fragments. This streaking was caused by fines being

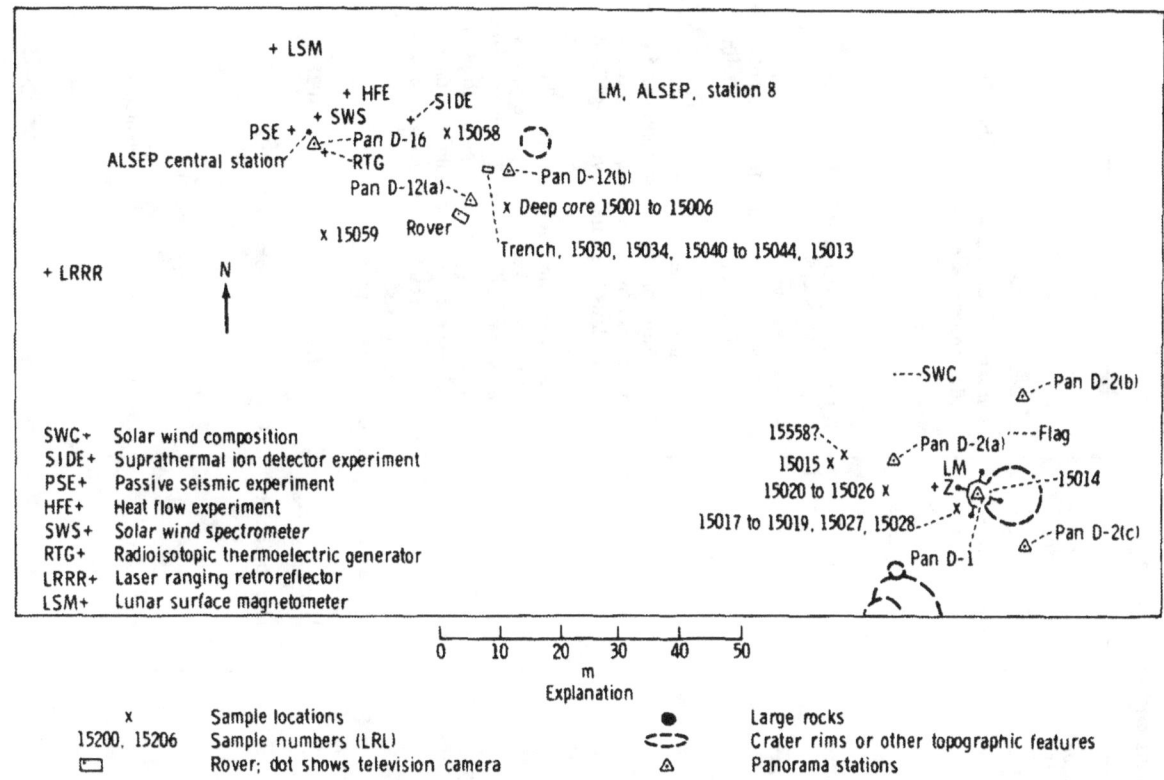

FIGURE 5-52.—Planimetric map of the LM, ALSEP, and station 8 areas.

FIGURE 5-53.—Presampling, cross-Sun photograph, looking south, of sample 15014, collected adjacent to the LM at the northeast quadrant. The faint streaking (from upper right to lower left) caused by the descent-engine exhaust is notable (AS15-88-11884).

blown away by exhaust of the LM descent engine; the sample is contaminated accordingly.

Sample 15015: A glass-covered breccia almost 30 cm long was collected approximately 20 m west of the LM +Z footpad (fig. 5-54). The area in which the sample was collected is flat and smooth. No documentary photographs showing details of the area were taken. The sample is of distinctly lower albedo than other fragments visible in the photographs, and no other fragments of comparable size occur in the vicinity (figs. D-2 and D-3).

Samples 15020 to 15026 (contingency sample): The site of the contingency sample is visible in photographs taken from the LM window (fig. D-3). The site is a small flat area between two subdued 1-m-diameter craters. Scattered coarse fragments occur in the sample area, and one sample (15025) was collected with the soil. The fine-grained material is probably representative of surface material near the LM. Sample 15025 resembles other large fragments in the vicinity, but none appears to be associated directly with any crater.

TABLE 5-IV.—*Sample Locations and Documentation*

Station no.	LRL sample no.	Lunar-surface photograph No.	Type (a)	Location status (b)	Orientation status (c)	Sample description
LM	15014	AS15-88-11884 AS15-88-11885 AS15-88-11886 AS15-88-11887 AS15-87-11838 AS15-87-11839	XSB XSB DSB XSA Pan D-2(c) Pan D-2(c)	Known	Not applicable	Soil sample, contaminated by the LM descent-engine exhaust.
LM	15015	AS15-85-11384 to AS15-85-11388 AS15-88-11931 AS15-88-11932 AS15-88-11938 to AS15-88-11945 AS15-87-11803 AS15-87-11804	Pan D-3(a) Pan D-3(b) Pan D-3(b) Pan D-3(b) Pan D-2(a) Pan D-2(a)	Known	Tentative	A blocky, angular, and largely glass-covered breccia. The dark vesicular glass is thick and rough textured on two surfaces where it encloses small fragments of soil. Where the glass coating is thin, the angular shape and blocky joint surfaces of the enclosed rock are visible. Small exposures of the enclosed block show that it is a breccia with light-colored lithic clasts in a dark-gray, fine-grained matrix.
LM	15017	AS15-86-11604 AS15-86-11605 AS15-86-11606 AS15-86-11607 AS15-86-11608 AS15-85-11388	XSE XSB XSB DSB XSA Pan D-3(a)	Known	Partial	Hollow glass sphere approximately 3 cm in diameter broken in transit. The glass is dark gray and vesicular. The surfaces are smooth except where irregular lumps of soil are enclosed by or adhere to the glass. The soil fragments contain small white clasts.
LM	15018	Same as 15017	Not identified in photographs	Known	Unknown	Glassy fragment.
LM	15019	Same as 15017	Not identified in photographs	Known	Unknown	Glassy fragment (description for 15028).
LM	15020 to 15024	AS15-85-11385 to AS15-85-11389	Pan D-3(a)	Known	Not applicable	Contingency sample (soil).
LM	15025	AS15-88-11932 AS15-88-11933	Pan D-3(b) Pan D-3(b)	Known	Unknown	Contingency sample (breccia).
LM	15026	AS15-88-11938 to AS15-88-11945	Pan D-3(b)	Known	Unknown	Contingency sample (glassy fragment).

PRELIMINARY GEOLOGIC INVESTIGATION OF THE LANDING SITE 5-51

TABLE 5-IV.—Sample Locations and Documentation - Continued

Station no.	LRL sample no.	Lunar-surface photograph No.	Type (a)	Location status (b)	Orientation status (c)	Sample description
LM	15027	AS15-86-11604 AS15-86-11605 AS15-86-11606 AS15-86-11607 AS15-86-11608 AS15-85-11388	XSB XSB XSB DSB XSA Pan D-3(a)	Known	Known	A blocky, angular, glassy breccia. Vesicular dark-gray glass containing light-colored clasts grades inward from four surfaces to breccia of light lithic clasts in a dark-gray, dense, glassy matrix. One clast 1 cm across is medium- to fine-grained basalt with brown pyroxene. Clasts between 1 and 5 mm are mostly light colored, but some pale-brown ones occur. The unresolved matrix comprises approximately 85 percent of the rock. This description also applies to 15019, which is probably a piece of 15028.
LM	15028	AS15-86-11604 AS15-86-11605 AS15-86-11606 AS15-86-11607 AS15-86-11608 AS15-85-11388	XSB XSB XSB DSB XSA Pan D-3(a)	Known	Known	This sample is a blocky, angular, glassy microbreccia containing light-colored lithic and mineral clasts in a dark-gray matrix. The dense glassy interior of the breccia grades laterally into thin vesicular glass veins within the body of the rock and to thicker frothy outer surfaces on two sides. Clasts include subordinate light-brown lithic or mineral fragments. Unresolved matrix components comprise approximately 80 percent of the matrix.
LM(?) (EVA-3)	15558	Same as 15015	Pans D-3(a) and (b)	Tentative[d]	Tentative	A blocky, subangular breccia with glass as a partial coating and fracture filling. The sample appears to have broken along subparallel fractures in a roughly orthogonal system. Lithic clasts, including basalt, up to 1 cm across are imbedded in a medium-gray matrix.
1	15065	AS15-86-11530 AS15-86-11531 AS15-85-11416 AS15-85-11417 AS15-85-11532 AS15-85-11408 AS15-85-11409	XSB XSB DSB DSB (LOC) XSA Pan D-4 Pan D-4	Known	Known	A blocky, subangular, coarse-grained basalt. A zone approximately 2.5 to 6 cm wide at one end of the specimen is conspicuously more mafic than the rest of the rock. The contact between the two domains is irregular and moderately sharp. The light side of the rock appears

TABLE 5-IV.—Sample Locations and Documentation - Continued

Station no.	LRL sample no.	Lunar-surface photograph No.	Type (a)	Location status (b)	Orientation status (c)	Sample description
1		AS15-85-11410	Pan D-4			to have a 3:1 ratio of mafic mineral to feldspar, and the dark side a ratio approximately 6:1 or 7:1. Irregular vugs to several millimeters across are concentrated in the dark domain.
1	15070 to 15074	AS15-86-11533 AS15-86-11534 AS15-85-11418 AS15-85-11419 AS15-86-11535	XSB XSB DSB DSB (LOC) XSA	Known (approx.)	Not applicable	Soil sample.
1	15075			Known	Known	A blocky, subangular, medium-grained basalt. Approximately 5 percent is vugs, to 3 mm across, that appear to be concentrated in a lens 1 by 3 cm.
1	15076	AS15-85-11408 AS15-85-11409	Pan D-4 (approx.) Pan D-4 (approx.)	Known (approx.)	Known	This sample is a blocky, angular, medium-grained basalt that closely resembles 15075. Approximately 5 percent is irregularly distributed vugs to 4 mm across. Plagioclase laths or plates appear to be alined.
1	15080 to 15084	AS15-86-11536 AS15-86-11537 AS15-85-11420 AS15-85-11421 AS15-86-11538	XSB XSB DSB DSB (LOC) XSA (fragments)	Known (approx.)	Not applicable	Fines consisting of soil scooped from beneath 15085, some soil that was described during collection as being caked on the bottom of 15085, and probably some material derived from the breccia sample 15086.
1	15085	AS15-86-11539 AS15-85-11408	XSA (soil) Pan D-4 (approx.)	Known (approx.)	Known	This description also applies to 15087, which is probably a piece of 15085. The sample is a blocky, subangular, coarse-grained basalt approximately 3 to 5 percent vugs to a few millimeters across. The texture may be subophitic, contains plagioclase to 5 mm long.
1	15086	Same as 15085	Same as 15085	Known (approx.)	Known	This description also applies to 15088, which is probably a piece of 15086. The sample is a blocky subrounded, friable, fine breccia. Lithic clasts, comprising approximately 5 percent of the rock, are both leucocratic (dominant) and melanocratic. Leucocratic types include some basaltic fragments with brown

PRELIMINARY GEOLOGIC INVESTIGATION OF THE LANDING SITE 5-53

TABLE 5-IV.—*Sample Locations and Documentation* - Continued

Station no.	LRL sample no.	Lunar-surface photograph No.	Type (a)	Location status (b)	Orientation status (c)	Sample description
1	15087	Same as 15085	Same as 15085	Known (approx.)	Known	pyroxene. Melanocratic types are very fine grained or glassy and dark gray. Clasts are seriate with 15 to 25 percent in the 0.5- to 1.0-mm size fraction.
1	15088	Same as 15085	Same as 15085	Known (approx.)	Known	Description of sample 15085.
2	15007	AS15-86-11574	XSB	Known	Not applicable	Description of sample 15086.
		AS15-86-11575	XSB			Lower core from a double drive tube.
2	15008	AS15-86-11576	XSD	Known	Not applicable	Upper core from a double drive tube.
		AS15-86-11577	XSD			
		AS15-86-11578	XSD			
		AS15-85-11443	DSB			
		AS15-85-11444	DSB			
		AS15-85-11445	DSB			
		AS15-85-11434	Pan D-5(a)			
		AS15-85-11435	Pan D-5(a)			
		AS15-85-11448	Pan D-5(c)			
2	15090 to 15093	AS15-86-11549	XSB	Known (approx.)	Not applicable	Soil sample collected with the tongs.
		AS15-86-11550	XSB			
2	15095	AS15-86-11551	XSA	Known (approx.)	Partially known (partly in shadow)	A blocky, subrounded microbreccia. The coherent breccia is approximately 85-percent glass covered, with sharp glass-breccia contacts. Clasts are less than 1 mm and are dominantly leucocratic. The breccia matrix is light gray.
		AS15-85-11435	Pan D-5(a)			
		AS15-85-14436	Pan D-5(a)			
		AS15-85-11446	Pan D-5(c)			
		AS15-85-11447	Pan D-5(c)			
2	15100 to 15104	AS15-86-11567	XSB	Known	Not applicable	Comprehensive soil sample.
		AS15-86-11568	XSB			
		AS15-85-11441	DSB			
2	15105	AS15-85-11442	DSB (LOC)	Known	Unknown	Basaltic fragment included with the comprehensive soil sample.
2	15115 to 15119	AS15-86-11572	XSA	Known	Unknown	Comprehensive rake fragments.
		AS15-86-11573	XSA			
2	15125	AS15-85-11435	Pan D-5(a)	Known	Unknown	Comprehensive rake fragments.
2	15135	AS15-85-11448	Pan D-5(c)	Known	Unknown	Comprehensive rake fragments.
2	15145 to 15148	AS15-85-14449	Pan D-5(c)	Known	Unknown	Comprehensive rake fragments.
2	15200 to 15204	Same as 15206	Same as 15206	Known	Partially known	Incidental fines collected with sample 15206.

5-54 APOLLO 15 PRELIMINARY SCIENCE REPORT

TABLE 5-IV.—Sample Locations and Documentation - Continued

Station no.	LRL sample no.	Lunar-surface photograph No.	Type (a)	Location status (b)	Orientation status (c)	Sample description
2	15205	AS15-86-11546	XSB	Known	Known	A blocky, angular breccia bounded by five joint surfaces, along two of which the rock was broken from a large boulder. Glass, in sharp contact with the breccia, coats two planar surfaces of the breccia. Clasts larger than 1 mm are dominantly leucocratic and comprise 20 to 25 percent of the rock. The matrix is coherent, light gray, and contains leucocratic clasts in seriate size arrangement from larger than 1 mm to resolution.
		AS15-86-11547	XSB			
		AS15-86-11552	XSB			
		AS15-86-11553	XSB			
		AS15-85-11439	DSB			
		AS15-85-11440	DSB			
		AS15-86-11558	XSA			
		AS15-86-11559	XSA			
		AS15-86-11560	DSA			
		AS15-85-11435	Pan D-5(a)			
		AS15-85-11436	Pan D-5(a)			
		AS15-85-11447	Pan D-5(c)			
		AS15-85-11448	Pan D-5(c)			
		AS15-85-11449	Pan D-5(c)			
2	15206	AS15-86-11546	XSB	Known	Partially known	A blocky, angular, glassy breccia with delicate protrusions; broken from a large boulder. Clasts bigger than 1 mm are leucocratic and comprise approximately 10 to 15 percent of the rock. The coherent matrix is dark gray and vuggy or vesicular with cavities to approximately 4 mm across. The cavities are larger and more abundant toward original edges of the specimen, and the outer surfaces resemble a vesicular basalt.
		AS15-86-11547	XSB			
		AS15-85-11439	DSB			
		AS15-86-11558	XSA			
		AS15-86-11559	XSA			
		AS15-86-11560	DSA			
		Pan photographs	Same as 15205			
2	15210 to 15214	AS15-86-11544	XSB	Known	Not applicable	Soil from the rim of a small crater adjacent to the north (downhill) side of the large boulder at station 2. This is the fillet sample referred to by the crew during collection.
		AS15-86-11545	XSB			
		AS15-85-11439	DSB			
		AS15-85-11440	DSB			
2	15220 to 15224	AS15-86-11548	XSD	Known	Not applicable	Soil collected from the rim of the same crater, approximately 0.5 m north of the station 2 boulder.
		AS15-86-11556	XSA			
		AS15-86-11557	XSA			
		AS15-86-11560	DSA			
		Pan photographs	Same as 15205			
2	15230 to 15234	AS15-86-11561	XSB	Known	Not applicable	Soil from beneath the boulder, collected after the rock was rolled over.
		AS15-86-11562	XSB			
		AS15-86-11563	XSB			
		AS15-86-11564	XSB			
		AS15-86-11565	XSA			
		AS15-86-11566	XSA			

PRELIMINARY GEOLOGIC INVESTIGATION OF THE LANDING SITE 5-55

TABLE 5-IV.–*Sample Locations and Documentation* - Continued

Station no.	LRL sample no.	Lunar-surface photograph No.	Type (a)	Location status (b)	Orientation status (c)	Sample description
3	15016	AS15-86-11569 Pan photographs	DSA Same as 15205	Known	Known	A blocky, subrounded, highly vesicular rock of fine-grained basalt. A slight tendency to slabby shape parallels a layering produced by variations in size and abundance of vesicles. The vesicles are smooth walled and spherical.
4	15470 to 15474	AS15-86-11579 AS15-86-11580 AS15-86-11581 AS15-86-11582 AS15-86-11583 AS15-86-11584	XSB XSB XSB XSA Pan D-7 Pan D-7	Known (approx.)	Not applicable	Soil collected with rock samples 15475 and 15476.
4	15475	AS15-87-11759 AS15-87-11760 AS15-87-11761 AS15-87-11762 AS15-87-11764 AS15-87-11763	XSB XSB DSB XSA XSA DSA (LOC)	Known (approx.)	Known	A blocky, angular, medium-coarse-grained, porphyritic basalt with pale-green prisms to 15 mm long in a slightly vuggy groundmass of plagioclase laths approximately 2 mm long and pale yellow-green pyroxene.
4	15476	Same as 15475	Same as 15475	Known (approx.)	Known	A slabby, subangular, porphyritic basalt with brown pyroxene phenocrysts to 15 mm long in a slightly vuggy groundmass plagioclase, and brown and green mafic silicates.
4	15485	AS15-87-11765 AS15-87-11766 AS15-87-11767 AS15-87-11768	XSB XSB DSB DSB	Known	Partially known	A blocky, angular basalt with brown pyroxene prisms to 7 mm long in a moderately vuggy groundmass of flow-alined plagioclase 2 to 3 mm long, and green-brown mafic silicates.
4	15486	AS15-87-11769 AS15-87-11770 AS15-87-11771 AS15-87-11772 AS15-90-12242 AS15-90-12243	OSA(?) OSA(?) OSA(?) OSA(?) Pan D-6 Pan D-6	Known	Partially known	Blocky, angular, porphyritic basalt with brown pyroxene phenocrysts to 7 mm long in a slightly vuggy groundmass of plagioclase laths and pyroxene. Light greenish-gray material coats one surface.
4	15495	Same as 15070 to 15076	Same as 15070 to 15076	Known (approx.)	Known	A blocky, subangular, vuggy basalt with dark-brown pyroxene prisms to 10 mm long in vugs.
4	15498	AS15-87-11765 AS15-87-11769 AS15-90-12242 AS15-90-12243	XSB(?) XSA(?) Pan D-6 Pan D-6	Tentative	Unknown	A blocky, angular, glass-coated breccia with approximately 10 percent leucocratic clasts bigger than 1 mm. Sizes are seriate and the rock contains 10 to 15 percent leucocratic clasts 0.5 to 1.0 mm. Dark, vesicular glass in sharp contact with the

TABLE 5-IV.—Sample Locations and Documentation - Continued

Station no.	LRL sample no.	Lunar-surface photograph No.	Type (a)	Location status (b)	Orientation status (c)	Sample description
4	15499	AS15-87-11768 AS15-87-11779 AS15-90-12242 AS15-90-12243	DSB XSA Pan D-6 Pan D-6	Known	Known	breccia covers parts of four sides of the rock. The medium-gray matrix is coherent. A blocky, angular, highly vesicular basalt. Vesicles appear to grade in size and abundance across the rock. One surface is coated by light yellowish-gray material grading into very dark brownish-gray coating at one edge. The rock is porphyritic with brown pyroxene prisms to 10 mm long in a groundmass of plagioclase laths and pyroxene.
6	15009	AS15-85-11527 AS15-85-11528 AS15-85-11529 AS15-85-11483 AS15-85-11484 AS15-85-11485 AS15-85-11511 AS15-85-11512 AS15-85-11513	DSB (LOC) DSB (LOC) DSB (LOC) Pan D-9(a) Pan D-9(a) Pan D-9(a) Pan D-9(b) Pan D-9(b) Pan D-9(b)	Known	Not applicable	Core from a single drive tube.
6	15012	AS15-85-11525 AS15-85-11526 AS15-86-11641 AS15-86-11642 AS15-86-11643 AS15-86-11644 AS15-86-11645 AS15-86-11646 AS15-85-11483 AS15-85-11484 AS15-85-11485 AS15-85-11511 AS15-85-11512 AS15-85-11513	DSB (LOC) DSB XSB XSB XSA DSA XSA XSA Pan D-9(a) Pan D-9(a) Pan D-9(a) Pan D-9(b) Pan D-9(b) Pan D-9(b)	Known	Not applicable	Bottom of trench; SESC-1 sample.
6	15240 to 15244	AS15-85-11498 AS15-85-11499 AS15-85-11500 AS15-86-11609 AS15-86-11610 AS15-86-11611 AS15-86-11612 AS15-86-11613	DSB DSB Loc XSB XSB XSB XSA XSA	Known	Not applicable	Fines collected from the center of a small glass-lined crater.

PRELIMINARY GEOLOGIC INVESTIGATION OF THE LANDING SITE 5-57

TABLE 5-IV.—Sample Locations and Documentation - Continued

Station no.	LRL sample no.	Lunar-surface photograph No.	Type (a)	Location status (b)	Orientation status (c)	Sample description
6	15245,1 to 90.	AS15-86-11614 AS15-86-11615 AS15-85-11515 AS15-85-11493 AS15-85-11494 AS15-85-11495 Same as 15240 to 15244	XSA XSA Pan D-9(b) Pan D-9(a) Pan D-9(a) Pan D-9(a) Same as 15240 to 15244	Known	Not applicable	Fragments of soil breccia, many glass covered, from the center of the same small crater as 15240 to 15244.
6	15250 to 15254	Same as 15240 to 15244	Same as 15240 to 15244	Known	Not applicable	Soil from the rim of the same small crater as 15240 to 15244.
6	15255	AS15-86-11629 AS15-86-11630 AS15-86-11631 AS15-86-11632 AS15-85-11515	XSB XSB DSB XSA Pan D-9(b)	Known	Known	A blocky, subangular, fine breccia partly coated by dark, vesicular glass. Clasts, comprising approximately 5 to 10 percent of the rock, are dominantly leucocratic. The coherent matrix is medium gray and contains leucocratic fragments from a millimeter to resolution.
6	15256	Same as 15255	Same as 15255	Known	Known	A blocky, subrounded, fine-grained, prophyritic basalt? The rock has one nearly flat surface and a fracture parallel to that surface. Breccia.
6	15257	Same as 15255	Same as 15255	Known (approx.)	Unknown	
6	15259	AS15-85-11523 AS15-85-11524 AS15-86-11633 AS15-86-11634 AS15-86-11635 AS15-86-11636 AS15-86-11637 AS15-85-11484 AS15-85-11485 AS15-85-11511 AS15-85-11512 AS15-85-11513	DSB DSB LOC LOC XSB XSB XSA Pan D-9(a) Pan D-9(a) Pan D-9(b) Pan D-9(b) Pan D-9(b)	Known (approx.)	Unknown	A small, slabby, angular breccia with 2 to 5 percent leucocratic clasts bigger than 1 mm in a light-gray, friable matrix.
6	15260 to 15264	Same as 15012	Same as 15012	Known	Not applicable	Soil from the bottom of the trench.
6	15265	AS15-85-11523 AS15-85-11524	DSB DSB	Known	Known	The sample is a slabby, subangular breccia with well-developed planar parting

TABLE 5-IV.—*Sample Locations and Documentation* - Continued

Station no.	LRL sample no.	Lunar-surface photograph No.	Type (a)	Location status (b)	Orientation status (c)	Sample description
		AS15-86-11638	XSB			parallel to the slabby direction, with 10 to 15 percent clasts bigger than 1 mm (up to 15 mm) about evenly divided among melanocratic, mesocratic, and leucocratic types. The largest clast is composed of dark-gray and chalky-white material and resembles 15455. Mesocratic clasts are light-brown aggregates of feldspar and brown pyroxene. Sizes are seriate to resolution with 5 to 10 percent in the range of 0.5 to 1.0 mm. Crude clast alinement and size sorting are suggested. The medium-gray matrix is coherent.
		AS15-86-11639	XSB			
		AS15-86-11640	XSA			
		AS15-85-11484	Pan D-9(a)			
		AS15-85-11485	Pan D-9(a)			
		AS15-85-11511	Pan D-9(b)			
		AS15-85-11512	Pan D-9(b)			
		AS15-85-11513	Pan D-9(b)			
6	15266	Same as 15265	Same as 15265	Known	Known	A blocky, angular breccia with one surface slickensided. Glass formed in places along the slickensides or was injected there. The sample contains 2 to 3 percent leucocratic to mesocratic clasts bigger than 1 mm that appear to be fragments of basalt with brown pyroxene. Sizes appear bimodal with a few percent leucocratic clasts between approximately 0.1 to 0.5 mm. The dark-gray matrix is moderately friable. Breccia.
6	15267	Same as 15259	Same as 15259	Known (approx.)	Unknown	A slabby, subrounded, coherent breccia with approximately 3 percent leucocratic clasts bigger than 1 mm. A lens (or perhaps a clast) of breccia approximately 0.5 cm thick and 2 to 3 cm long has 20 to 25 percent chalky-white clasts 1 to 5 mm across. The layer or clast closely resembles 15205.
6	15268	Same as 15259	Same as 15259	Known (approx.)	Unknown	
6	15269	Same as 15259	Same as 15259	Known (approx.)	Unknown	A slightly slabby, subangular, glassy microbreccia with 1 to 3 percent small leucocratic clasts bigger than 1 mm and 5 to 10 percent leucocratic clasts less than 1 mm across. The matrix is coherent, dark gray, and grades from a massive interior to thin, vesicular surfaces.

PRELIMINARY GEOLOGIC INVESTIGATION OF THE LANDING SITE

TABLE 5-IV.—Sample Locations and Documentation - Continued

Station no.	LRL sample no.	Lunar-surface photograph No.	Type (a)	Location status (b)	Orientation status (c)	Sample description
6	15270 to 15274	AS15-86-11656 AS15-86-11657 AS15-85-11490	XSB XSA Pan D-9(a)	Known (approx.)	Not applicable	Soil collected near the Rover.
6(?) (EVA-2)	15281 to 15284	None	None	Unknown	Not applicable	Residue fines of less than 1 cm from SCB-3, used during EVA-2. Breccia samples 15298 and 15299 were put into SCB-3 also (sample 15297).
6	15285	Same as 15259	Same as 15259	Known	Known	A blocky, angular breccia with 5 to 10 percent leucocratic clasts bigger than 1 mm. Sizes are seriate to resolution with approximately 15 percent leucocratic clasts in the size range of 0.5 to 1.0 mm. Clasts may be crudely alined. The medium-gray matrix is coherent. Vesicular, dark glass partly covers one surface.
6	15286	Same as 15259	Same as 15259	Known (approx.)	Unknown	An elongate, blocky, subrounded breccia with approximately 5 percent clasts bigger than 1 mm. Most clasts are leucocratic, dominantly chalky white, but a few are dark gray. Sizes are seriate to resolution with approximately 15 percent in the size range of 0.5 to 1.0 mm. A higher proportion of dark clasts may exist in the smaller sizes. The matrix is light gray and coherent. Dark, vesicular glass partly covers three surfaces.
6	15287	Same as 15259	Same as 15259	Known (approx.)	Unknown	A slightly slabby, subrounded to subangular breccia with 5 to 10 percent leucocratic clasts bigger than 1 mm and 1 percent melanocratic clasts. Sizes are seriate with approximately 10 percent in the size range of 0.5 to 1.0 mm. The medium-gray matrix is coherent.
6	15288	Same as 15259	Same as 15259	Known (approx.)	Unknown	A blocky, subangular, glassy microbreccia with 2 to 3 percent leucocratic clasts bigger than 1 mm. Sizes are seriate to resolution with 15 to 20 percent in the size range of 0.5 to 1.0 mm. Elongate clasts

TABLE 5-IV.- *Sample Locations and Documentation* - Continued

Station no.	LRL sample no.	Lunar-surface photograph		Location status (b)	Orientation status (c)	Sample description
		No.	Type (a)			
6	15289	Same as 15259	Same as 15259	Known (approx.)	Unknown	are well alined. The dark-gray matrix is coherent and apparently grades from a massive interior to vesiculated margins. Breccia.
6	15290 to 15294	AS15-86-11616 AS15-86-11617 AS15-85-11501 AS15-85-11502 AS15-86-11618 AS15-86-11619 AS15-86-11620 AS15-85-11495	XSB XSB DSB DSB LOC LOC LOC Pan D-9(a)	Known	Not applicable	Soil collected with breccia sample 15295.
6	15295	Same as 15290 to 15294	Same as 15290 to 15294	Known	Unknown	A blocky, subangular breccia. The rock has broken into several pieces revealing vesicular glass coatings on some interior surfaces. This sample is 1 to 2 percent leucocratic clasts bigger than 1 mm (to 15 mm). Sizes are seriate in the range of 0.5 to 1.0 mm. Clasts in this size range comprise approximately 10 to 15 percent of the rock and are dominantly leucocratic. The medium-gray matrix is moderately coherent.
6(?) (EVA-2)	15297,1 to 13	None	None	Unknown	Not applicable	Loose fragments larger than 1 cm from SCB-3 used during EVA-2. Some of these fragments may have broken from breccia samples 15298 and 15299, which were also carried loose in SCB-3.
6	15298	AS15-85-11503 AS15-85-11504 AS15-86-11621 AS15-86-11622 AS15-86-11623 AS15-85-11495 AS15-85-11515 AS15-85-11516	DSB DSB XSB XSB XSB Pan D-9(a) Pan D-9(b) Pan D-9(b)	Known	Partially known	A blocky, subangular, fine breccia with less than 1 percent leucocratic clasts bigger than 1 mm. Clast sizes are seriate, and leucocratic fragments in the range of 0.1 to 1.0 mm comprise approximately 10 to 15 percent of the rock. Fractures in the rock were injected by thin glass veins. The light- to medium-gray matrix is moderately friable.

PRELIMINARY GEOLOGIC INVESTIGATION OF THE LANDING SITE 5-61

TABLE 5-IV.–*Sample Locations and Documentation* - Continued

Station no.	LRL sample no.	Lunar-surface photograph No.	Type (a)	Location status (b)	Orientation status (c)	Sample description
6	15299	AS15-85-11505	LOC	Known (approx.)	Unknown	A blocky, angular, fine breccia with 1 to 3 percent leucocratic clasts bigger than 1 mm. Sizes are seriate to resolution, and leucocratic clasts in the size range of 0.5 to 1.0 mm comprise approximately 20 percent of the rock. The medium-gray matrix is moderately coherent.
		AS15-85-11506	DSB			
		AS15-86-11624	XSB			
		AS15-86-11625	XSB			
		AS15-86-11628	XSA			
		AS15-85-11516	Pan D-9(b)			
		AS15-85-11517	Pan D-9(b)			
6A	15400 to 15404	AS15-86-11658	XSB	Known	Not applicable	Material scraped from the upper part of the breccia boulder, and possibly some from the fillet.
		AS15-86-11659	XSB			
		AS15-86-11660	XSA			
		AS15-86-11661	XSA			
		AS15-90-12199	DS			
		AS15-90-12200	DS			
		AS15-90-12187	Pan D-10			
		AS15-90-12188	Pan D-10			
6A	15405	Same as 15400 to 15404	Same as 15404	Known	Unknown	A blocky subangular breccia cut by multiple irregular fractures. The specimen is approximately 5 percent leucocratic clasts bigger than 1 mm in a coherent dark-gray matrix and is thoroughly coated with greenish-gray soil.
7	15300 to 15304	AS15-90-12231	DSB	Known	Not applicable	Soil collected with the comprehensive sample.
		AS15-90-12232	DSB			
7	15305 to 15308	AS15-90-12233	XSA	Known	Unknown	Fragments collected with the comprehensive soil sample.
		AS15-90-12234	XSA			
		AS15-90-12217	Pan D-11 (sample area obscured by Rover)			
		AS15-90-12218				
		AS15-90-12219				
7	15310 to 15392	Same as 15300 to 15308	Same as 15308	Known	Unknown	Fragments collected in the comprehensive rake sample.
7	15410 to 15414	Same as 15417 to 15419	Same as 15419	Known	Not applicable	Soil collected with samples 15417 to 15419.
7	15415	AS15-86-11670	XSB	Known	Known	A blocky, angular to subangular rock composed largely of coarse plagioclase. Chalky-white zones or bands suggest shattering. One set of close-spaced parallel fractures exists.
		AS15-86-11671	XSB			
		AS15-90-12227	DSB			
		AS15-90-12228	DSB			
		AS15-86-11672	XSA			
		AS15-90-12201	Pan D-11			

TABLE 5-IV.—*Sample Locations and Documentation* - Continued

Station no.	LRL sample no.	Lunar-surface photograph No.	Type (a)	Location status (b)	Orientation status (c)	Sample description
7	15417	AS15-86-11662	XSB	Known (approx.)	Unknown	A small, 1.3-g, blocky, subrounded, fine breccia with 1 to 2 percent leucocratic clasts bigger than 1 mm. Sizes are seriate with 20 percent leucocratic clasts in the range of 0.5 to 1.0 mm. The medium-gray matrix is coherent.
7	15418	AS15-86-11663 AS15-90-12223 AS15-90-12224 AS15-86-11664	XSB DSB DSB (LOC) XSA 15418 (before 15419)	Known	Known	A blocky, angular breccia with 5 to 10 percent leucocratic clasts bigger than 1 mm. Sizes are seriate with 10 to 20 percent leucocratic clasts in the range of 0.5 to 1.0 mm. The light- to medium-gray matrix is coherent and has a few percent spherical vesicles to 10 mm across, most of which are concentrated at one end of the specimen.
7	15419	AS15-86-11665 AS15-90-12201	XSA 15419 Pan D-11	Known	Partially known	A blocky subangular breccia partly coated with dark vesicular glass with approximately 5 percent leucocratic clasts bigger than 1 mm (up to 6 mm); some are basaltic fragments with brown pyroxene. Sizes are seriate with approximately 15 percent clasts, many of which appear to be brownish basalts, in the size range of 0.5 to 1.0 mm. The light-gray matrix is coherent.
7	15421 to 15424 15425	AS15-86-11666 AS15-86-11667 AS15-90-12225 AS15-86-11668	XSB XSB DSB DSB (LOC) XSA	Known	Not applicable	Fines collected with or broken from fragments 15425 and 15426.
7				Known	Unknown	Blocky, subrounded fragments (probably broken in transit) of friable, light grayish-green microbreccia with a small percentage of chalky-white clasts smaller than 1 mm.
7	15426	AS15-86-11669 AS15-90-12201 AS15-90-12202	XSA Pan D-11 Pan D-11	Known	Unknown	Blocky, subrounded fragments (probably broken in transit) of friable, light grayish-green microbreccia with variable percentages of chalky-white clasts to 1 mm.
7	15427,1 to 22	Same as 15425 and 15426	Same as 15425 and 15426	Known	Unknown	Chips larger than 1-cm diameter collected with or broken from 15425 and 15426.

TABLE 5-IV.—Sample Locations and Documentation - Continued

Station no.	LRL sample no.	Lunar-surface photograph No.	Type (a)	Location status (b)	Orientation status (c)	Sample description
7	15431 to 15434	AS15-86-11670 AS15-86-11671 AS15-86-11672 AS15-90-12227	XSB XSB XSB DSB	Known	Not applicable	Fines collected with or broken from the clod beneath sample 15415.
7	15435 to 15437	AS15-90-12228 AS15-86-11673 AS15-86-11674	DSB XSD XSA	Known	Not applicable	Fragments larger than 1 cm diameter collected with or broken from the clod beneath sample 15415.
7	15445	AS15-86-11690 AS15-86-11691 AS15-86-11692 AS15-90-12201	XSB XSB XSA Pan D-11	Known	Known	A blocky, angular breccia with 20 to 25 percent very irregular, embayed leucocratic clasts to 4 cm across. Sizes are bimodal with a small percentage of leucocratic clasts smaller than 1 mm. The large clasts are mostly chalky white, but one has approximately 5 to 10 percent of a red mineral. The clasts are veined by the matrix as in sample 15455. The dark-gray matrix is coherent and has a few percent irregular vugs.
7	15455	AS15-86-11675 AS15-86-11676 AS15-86-11677 AS15-90-12229 AS15-90-12201	XSB XSB DSB XSA Pan D-11	Known	Known	A blocky, angular breccia with approximately 45 percent leucocratic clasts to 8 cm across. Sizes are bimodal with 1 percent leucocratic clasts less than 1 mm. The largest clast has very irregular embayed boundaries with the matrix and is cut by veins of matrix material that are irregular in course and width. This clast also has perhaps 25 percent 1- to 3-mm patches of light-gray material in chalky-white material. Smaller leucocratic clasts are concentrated in a band or pod. The dark-gray matrix is coherent and has a few percent vugs concentrated near one large clast.
7	15459	AS15-90-12232 AS15-90-12234 AS15-90-12235 AS15-90-12236 AS15-90-12217 AS15-90-12218 AS15-90-12219	DSB OSB XSB XSB Pan D-11 (sample area obscured by Rover)	Known (approx.)	Partially known	A blocky, angular breccia with approximately 15 percent leucocratic clasts bigger than 1 mm (to 80 mm). Sizes are seriate with approximately 15 percent leucocratic clasts in the size range of 0.5 to 1.0 mm and a small percentage of dark-gray clasts. Most clasts are chalky white, and some have a greenish-gray cast. The medium-gray matrix is coherent.

TABLE 5-IV.—*Sample Locations and Documentation* - Continued

Station no.	LRL sample no.	Lunar-surface photograph No.	Type (a)	Location status (b)	Orientation status (c)	Sample description
	15465	AS15-86-11678 AS15-86-11679 AS15-90-12230 AS15-86-11680 AS15-86-11681 AS15-90-12201	XSB XSB DSB XSB XSA Pan D-11	Known	Known	A blocky, angular breccia. The rock is composed of angular fragments, some shattered, of breccia held together by dark, vesicular glass. The breccia has approximately 5 to 7 percent leucocratic clasts bigger than 1 mm and a few melanocratic clasts. Sizes are seriate to resolution with approximately 10 percent leucocratic clasts in the size range of 0.5 to 1.0 mm. The light-gray matrix is moderately coherent.
7	15466	Same as 15465	Same as 15465	Known	Known (tentative)	Rock similar to 15465, found underneath another rock.
7	15467	Same as 15465	Same as 15465	Known (approx.)	Unknown	Glass-coated fragment similar to 15465 and 15466.
7 8 (ALSEP)	15468 15001 to 15006	Same as 15465 AS15-92-12427 AS15-92-12428 (Television shows drill site with relation to pan)	Same as 15465 Pan D-12(a) Pan D-12(a)	Known	Not applicable	Not identified in lunar-surface photographs. Deep core.
8 (ALSEP)	15013	AS15-92-12417 AS15-92-12418	XSB XSB	Known	Not applicable	Soil sample collected from bottom of trench (approximately 0.33 m below the surface); may be contaminated by soil caved in from trench sides.
8 (ALSEP)	15030 to 15034	AS15-92-12419 AS15-92-12439 AS15-92-12440	DSB XSA XSA	Known	Not applicable	Soil samples collected from bottom of trench (approximately 0.33 m below the surface); may be contaminated by soil caved in from trench sides.
8 (ALSEP)	15040 to 15044	AS15-92-12441 AS15-92-12442 AS15-92-12443 AS15-88-11872 AS15-88-11873 AS15-88-11874 AS15-88-11875 AS15-88-11876	DSA DSA DS (LOC) DSA DSA XSA to south XSA to south XSA to north	Known	Not applicable	Soil samples taken from top of trench.

PRELIMINARY GEOLOGIC INVESTIGATION OF THE LANDING SITE 5-65

TABLE 5-IV.—Sample Locations and Documentation - Continued

Station no.	LRL sample no.	Lunar-surface photograph No.	Type (a)	Location status (b)	Orientation status (c)	Sample description
		AS15-88-11877	XSA to north			
		AS15-92-12429	Pan D-12(a)			
		AS15-92-12430	Pan D-12(a)			
		AS15-92-12431	Pan D-12(a)			
		AS15-88-11947	Pan D-3(b)			
		AS15-88-11948	Pan D-3(b)			
		AS15-88-11951	Pan D-3(b)			
		AS15-88-11952	Pan D-3(b)			
8 (ALSEP)	15058	AS15-92-12410	XSB	Known	Known	A blocky, angular, medium-grained vuggy basalt. Vugs, to approximately 4 mm across, comprise 5 percent of the volume of the rock but are irregularly distributed.
		AS15-92-12411	XSB			
		AS15-92-12412	DSB			
		AS15-87-11850	Pan D-16			
		AS15-87-11851	Pan D-16			
8 (ALSEP)	15059	AS15-92-12413	DSB	Known	Known	A blocky, subangular, fine breccia that is approximately 90-percent glass covered. The exposed surface has been broken open; so the rock was originally entirely glass coated. The glass, which is in sharp contact with the breccia, is very dark gray and vesicular. The glass is coarse textured and contains tiny white clasts at one end of the rock where it is probably thicker than elsewhere. Where the glass coating is thin, the angular shape of the enclosed coherent breccia block is readily evident. The breccia contains dominantly leucocratic lithic clasts that are subrounded to well rounded. Clasts bigger than 1 mm comprise approximately 3 to 5 percent of the breccia and range up to 8 mm. Clast sizes are seriate to resolution with approximately 15 to 20 percent of the rock in the size range of 0.5 to 1.0 mm. Elongate clasts are in parallel alinement. The matrix is dark gray.
		AS15-92-12414	DSB			
		AS15-92-12415	XSB			
		AS15-87-11853	Pan D-16			
		AS15-87-11853	Pan D-16			

APOLLO 15 PRELIMINARY SCIENCE REPORT

TABLE 5-IV.—Sample Locations and Documentation - Continued

Station no.	LRL sample no.	Lunar-surface photograph No.	Type (a)	Location status (b)	Orientation status (c)	Sample description
9	15500 to 15504	AS15-82-11105 AS15-82-11106 AS15-82-11107 AS15-82-11108 AS15-82-11109 AS15-82-11090 AS15-82-11091	XSB XSB DSB LOC XSA Pan D-13 Pan D-13	Known	Not applicable	Soil collected with breccia sample 15505.
9	15505	Same as 15500 to 15504	Same as 15500 to 15504	Known	Known	A blocky, angular breccia almost entirely coated by glass. Sharply angular corners of the breccia are revealed by thin glass coating. Small exposures of the breccia have small leucocratic clasts in a dark-gray coherent matrix.
9	15506	Same as 15500 to 15504	Same as 15500 to 15504	Known	Unknown	Fragment that could have broken from 15505.
9	15507	Same as 15500 to 15504	Same as 15500 to 15504	Known	Unknown	Glassy bead collected in addition to sample 15505.
9	15508	Same as 15500 to 15504	Same as 15500 to 15504	Known	Unknown	Fragment that could have broken from 15505.
9	15510 to 15514	Same as 15500 to 15504	XSB	Known	Unknown	Fines less than 1 cm collected with or broken from two collected clods.
9	15515,1 to 48	AS15-82-11094 AS15-82-11098 AS15-82-11099 AS15-82-11100 AS15-82-11089 AS15-82-11090	XSB XSB XSB XSA Pan D-13 Pan D-13	Known	Unknown	Fragments larger than 1 cm collected with or broken from the same two collected clods as 15510 to 15514.
9(?) (EVA-3)	15561 to 15564	None	None	Unknown[d]	Not applicable	Loose fines from SCB-2 comprising incidental soil and particles broken from loose rocks in the bag.
9(?)	15565,1 to 13	None	None	Unknown	Unknown	Breccia fragments that presumably broke from a larger grab sample returned in SCB-2. (These samples originally were given LRL numbers 15565 to 15569, 15575 to 15579, and 15585 to 15587.)
9	15565,14 to 43	None	None	Unknown	Not applicable	Other small samples from SCB-2.

TABLE 5-IV.–*Sample Locations and Documentation* - Continued

Station no.	LRL sample no.	Lunar-surface photograph		Location status (b)	Orientation status (c)	Sample description
		No.	Type (a)			
9A	15010	AS15-82-11156	XSB	Known	Not applicable	Lower core from a double drive tube.
9A	15011	AS15-82-11157	XSB	Known	Not applicable	Upper core from a double drive tube.
		AS15-82-11158	DSB			
		AS15-82-11159	LOC			
		AS15-82-11160	XSD			
		AS15-82-11161	XSD			
		AS15-82-11162	XSD			
		AS15-82-11163	XSA			
		AS15-82-11123	Pan D-14			
		AS15-82-11124	Pan D-14			
9A	15528	AS15-82-11128	DSB	Known	Known	A blocky, subangular, highly vesicular, very fine grained basalt.
		AS15-82-11129	XSB			
9A	15529	AS15-82-11119	Pan D-14	Known	Known	A blocky, subangular, highly vesicular, very fine grained basalt.
		AS15-82-11120	Pan D-14			
9A	15530 to 15534	AS15-82-11139	XSB	Known	Not applicable	Soil collected near boulder chip fragments 15535 and 15536.
		AS15-82-11140	XSB			
		AS15-82-11138	DSB			
		AS15-82-11141	XSA			
		AS15-82-11126	Pan D-14			
		AS15-82-11127	Pan D-14			
9A	15535	AS15-82-11139	XSB	Known	Known	A slabby, subangular, slightly vuggy, fine- to medium-grained basalt with possible small olivine phenocrysts in a groundmass of plagioclase and green to brown mafic silicates.
		AS15-82-11140	XSB			
		AS15-82-11138	DSB			
		AS15-82-11141	XSA			
		AS15-82-11125	Pan D-14			
		AS15-82-11126	Pan D-14			
		AS15-82-11127	Pan D-14			
9A	15536	Same as 15535	Same as 15535	Known	Known	A blocky, angular, slightly vuggy, medium-grained basalt with brown and green mafic silicates and plagioclase with an equigranular texture.
9A	15537	Same as 15535	Same as 15535	Known (approx.)	Unknown	A blocky, angular, moderately vesicular, coarse-grained basalt with green and brown mafic silicates and plagioclase with an equigranular texture.
9A	15538	Same as 15535	Same as 15535	Known (approx.)	Unknown	Rock fragment similar to 15537.
9A	15545	Same as 15535	Same as 15535	Known (approx.)	Unknown	A blocky, subangular, slightly vuggy, fine- to medium-grained basalt with possible olivine phenocrysts in a groundmass of brown and green mafic silicates and small plagioclase laths.

TABLE 5-IV.—*Sample Locations and Documentation* - Continued

Station no.	LRL sample no.	Lunar-surface photograph No.	Type (a)	Location status (b)	Orientation status (c)	Sample description
9A	15546	Same as 15535	Same as 15535	Known (approx.)	Unknown	Basalt.
9A	15547	Same as 15535	Same as 15535	Known (approx.)	Unknown	Basalt.
9A	15548	Same as 15535	Same as 15535	Known (approx.)	Unknown	Basalt.
9A	15555	AS15-82-11164 AS15-82-11123 AS15-82-11124 AS15-82-11125	XSB Pan D-14 Pan D-14 Pan D-14	Known	Known	A blocky, subangular, very vuggy, coarse-grained basalt with green and brown mafic silicates and plagioclase with an equigranular texture.
9A	15556	AS15-82-11133 AS15-82-11134 AS15-82-11135	DSB DSB XSB	Known	Known	A blocky, subangular, highly vesicular, fine-grained basalt.
9A	15557	AS15-82-11117 AS15-82-11118 AS15-82-11136 AS15-82-11137	Pan D-14 Pan D-14 DSB XSB	Known	Known	Basalt.
9A	15595	AS15-82-11110 AS15-82-11111 AS15-82-11142 AS15-82-11143 AS15-82-11144 AS15-82-11145 AS15-82-11146 AS15-82-11126 AS15-82-11127	Pan D-14 Pan D-14 DSB XSB XSB XSA XSA Pan D-14 Pan D-14	Known	Known	A blocky, angular rock (broken from an outcrop) of moderately vuggy, porphyritic basalt with very irregular distribution of vugs. The specimen contains green pyroxene prisms to 6 mm in a groundmass of green and brown mafic silicates and plagioclase with a dark-gray stain or coating on one surface.
9A	15596	Same as 15595	Same as 15595	Known	Known	A blocky, angular, moderately vuggy, porphyritic basalt. Vugs are very irregularly distributed. Green pyroxene prisms to 10 mm long are set in a fine-grained groundmass of green and brown pyroxene and plagioclase(?).
9A	15597	Same as 15595	Same as 15595	Known (approx.)	Unknown	A slabby, subangular, porphyritic basalt with very few cavities. The very few green pyroxene prisms to 7 mm are set in a microcrystalline groundmass.
9A	15598	Same as 15595	Same as 15595	Known (approx.)	Unknown	A blocky, angular, slightly vuggy, fine-grained basalt with possibly a few olivine microphenocrysts.

PRELIMINARY GEOLOGIC INVESTIGATION OF THE LANDING SITE

TABLE 5-IV.—Sample Locations and Documentation - Concluded

Station no.	LRL sample no.	Lunar-surface photograph		Location status (b)	Orientation status (c)	Sample description
		No.	Type (a)			
9A	15600 to 15610	AS15-82-11151 AS15-82-11152 AS15-82-11153 AS15-82-11154 AS15-82-11155 AS15-82-11122 AS15-82-11123 AS15-82-11124	XSB XSB DSB XSA XSA Pan D-14 Pan D-14 Pan D-14	Known	Not applicable	Soil (15600 to 15604) and rock fragments (15605 to 15610) collected with the comprehensive samples.
9A	15612 to 15630	Same as 15600 to 15610	Same as 15600 to 15610	Known	Not applicable	Comprehensive rake samples.
9A	15632 to 15645	Same as 15600 to 15610	Same as 15600 to 15610	Known	Not applicable	Comprehensive rake samples.
9A	15647 to 15656	Same as 15600 to 15610	Same as 15600 to 15610	Known	Not applicable	Comprehensive rake samples.
9A	15658 to 15689	Same as 15600 to 15610	Same as 15600 to 15610	Known	Not applicable	Comprehensive rake samples.

[a] The types of documentary photographs are as follows.
DS   Down-Sun.
DSA   Down-Sun, taken after collecting sample.
DSB   Down-Sun, taken before collecting sample.
LOC   So-called "locator" photograph that shows sample location with respect to horizon or some distinctive lunar-surface feature.
OS   Oblique-to-Sun.
OSA   Oblique-to-Sun, taken after collecting sample.
OSB   Oblique-to-Sun, taken before collecting sample.
Pan   Photographs that are part of a panorama. Panoramas are provided as foldouts in appendix D.
XSA   Cross-Sun, taken after collecting sample.
XSB   Cross-Sun, taken before collecting sample.
XSD   Cross-Sun, taken during collection of sample.

[b] Samples or sample areas are considered to be known if they are recognized in photographic panoramas or can be related directly to features of known location. Locations are listed as approximately known if they can be related with confidence to general areas but are not specifically identified. Locations are considered tentative if the identity of the associated sample is in question. Locations are mostly unknown for the mixed-residue components of sample collection bags and for the group of samples numbered 15565,1 to 43 (footnote d).

[c] The lunar orientations of samples listed here are determined by two methods of study – (1) reconstruction of the documented lunar lighting characteristics using the actual sample in the LRL with oblique lighting from a nearly collimated light source and (2) for samples that are fragile or friable, comparison of LRL documentary photographs ("mugshots") with lunar-surface photographs of the same samples. A known orientation is one in which the attitude of the sample at the time of collection is certain. A partially known status is one in which the attitude of a sample is strongly suggested by comparison between LRL and lunar-surface photographs or is in some way implied by comments by the astronauts at the time of collection.

[d] Sample 15558 and the group of samples 15561 to 15565 were returned loose and undocumented in SCB-2. These samples were collected during EVA-3, probably somewhere in the traverse between station 9 and return to the LM. The crew do not remember details of their collection. Because these samples are all breccias, most or all of which include clasts of mare basalt, it is assumed that they were collected in areas of other known breccias; namely, at station 9 or near the LM. Sample 15558 is the largest single fragment and is tentatively identified as a light-colored rock near sample 15015, west of the LM, photographed from the LM window before EVA-1 and missing from photographs taken after EVA-3. Samples 15561 to 15564 are fines. Sample 15565,1 to 43 may represent a larger grab sample that disintegrated during transportation in the Rover seat pan and after transfer to SCB-2 when the crew returned to the LM. This observation supports the suggestion that sample 15565 and its fractions came from a larger sample collected at station 9.

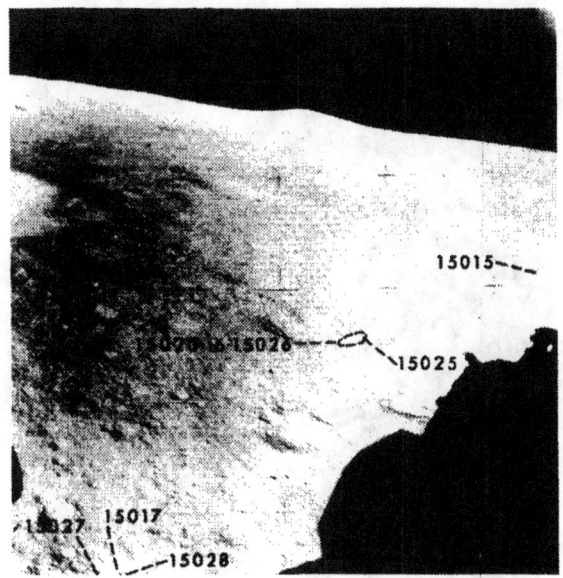

FIGURE 5-54.—Presampling, LM-window photograph, looking west, of samples 15015, 15017 to 15028, and 15558(?) collected adjacent to the LM (AS15-85-11388).

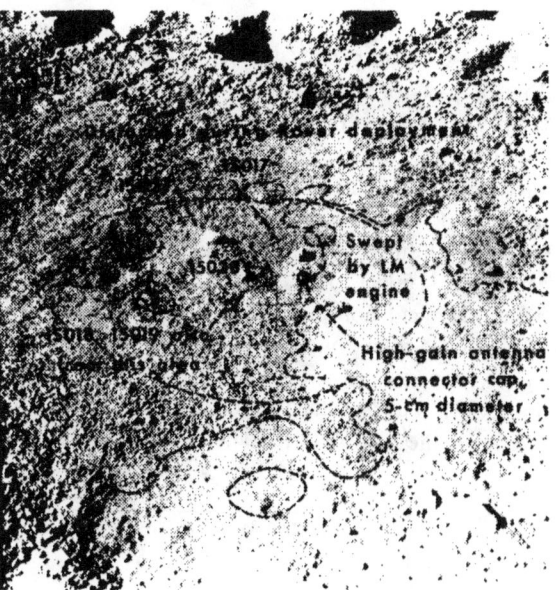

FIGURE 5-55.—Presampling, cross-Sun photograph, looking south, of samples 15017 to 15019, 15027, and 15028, collected immediately southwest of the LM (AS15-86-11604).

Samples 15017 to 15019, 15027, and 15028: Collected from a subdued 1-m-diameter crater approximately 4 m south of the LM +Z footpad (fig. 5-52), samples 15018 and 15019 have not been identified in the photographs (fig. 5-55). The surface material is fine grained with rare fragments to 4 cm across. The samples lay in a cluster distinctly apart from others of the same size, suggesting that they may be related. However, the crater appears too subdued for these fragments to be residuals of that impact. The crater containing the collected rocks adjoins two other subdued craters of similar size, and smaller, younger craters are scattered in the area. Well-developed fillets were banked against angular glassy breccia fragments (samples 15027 and 15028), but no fillet is visible against the glass sphere, 15017. The albedo of 15017 appears low and is moderate compared with 15027 and 15028. Samples 15027 and 15028 appear typical of rocks in their size range in the area. Sample 15017 is atypical in shape, but the glass itself is no different from that coating 15015.

Station 1.—Station 1 is located on the east flank of Elbow Crater, a 400-m-diameter crater that lies near the junction of the mare surface and the Apennine Front. Elbow Crater is a relatively old Copernican-age crater formed by an impact at a sharp bend in Hadley Rille (fig. 5-2). Elbow is approximately 0.5 km north of the foot of the Hadley Delta slope. Younger small craters occur on the Elbow Crater ejecta blanket. The surface is generally flat but hummocky in detail. The surface material is composed of fine-grained soil with sparse to common pebble-size and larger fragments. The larger fragments are locally concentrated around the young, fresh crater on the Elbow Crater ejecta blanket.

Nine samples were collected from three points along a line extending approximately 65 m eastward from the rim crest of Elbow Crater. Of these samples, two are soils (15070 to 15074 and 15080 to 15084), five are mare-type basalts (15065, 15075, 15076, 15085, and 15087), and two are poorly lithified soil breccias (15086 and 15088). The breccias contain much debris derived from mare-type basalts. The samples were collected radially from Elbow Crater to determine lithologic variations with depth in the mare material; excavation of the crater would ideally result in discovering ejecta, representative of approximately 75 m of material underlying the crater, distributed with the deepest material near the crater rim and the shallowest material farthest from the rim.

The basaltic samples collected from all three points of this traverse resemble one another in being coarse grained and rich in large pyroxene prisms. Each set, however, has individually distinctive characteristics, suggesting that different parts of a single flow or parts of a related group of flows have been sampled. Hence, although the younger craters have reworked the Elbow ejecta, they do not appear to have mixed the material on a scale comparable to the sample interval. The soil breccias presumably represent shock-lithified regolith material.

Sample 15065: This sample is a coarse-grained basalt fragment collected approximately 4 m east of the Elbow Crater rim crest (fig. 5-56), and approximately 1 m east of a 4-m-diameter crater (figs. 5-57 and 5-58). A few subdued craters less than 0.5 m across are near the sample site. The surrounding area is

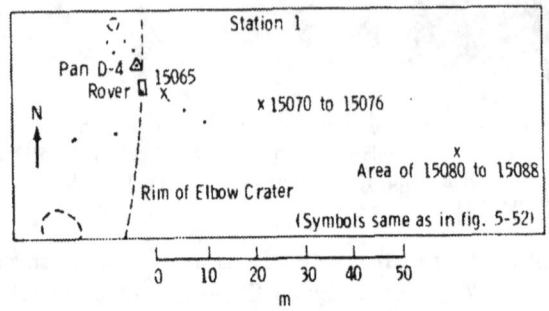

FIGURE 5-56.—Planimetric map of station 1.

FIGURE 5-57.—Presampling, oblique-to-Sun photograph, looking east-southeast, of samples 15065, 15070 to 15076, 15085, and 15086, collected at station 1 near the rim of Elbow Crater (AS15-85-11408).

FIGURE 5-58.—Presampling, cross-Sun photograph, looking south, of sample 15065, collected on east rim of Elbow Crater at station 1. This sampling point is the first of the radial sample (AS15-85-11417).

covered by fine-grained soil and pebble-size fragments, but no cobble-size fragments occur within 1 m of the sample 15065 site. A suggestion of lineaments crossing the flat area of the sample site exists. The 4-m-diameter crater has a ring of cobble-size fragments inside the rim and also contains a 1-m-wide flat block (fig. 5-58) that has a faint dark band resembling the dark compositional band crossing sample 15065. The pebbles and cobbles in the vicinity all have about the same albedo and no apparent dust cover. Sample 15065 has a small fillet less than 1 cm high; the rock was lying on a flat side and was less than one-fifth buried. The position suggests that the sample was moved after the Elbow Crater event and may have been ejected from the 4-m-diameter crater that is centered approximately 1 m from the rim crest of Elbow Crater.

Samples 15070 to 15076: These samples are soils and rocks collected from a small area approximately 25 m east of the Elbow Crater rim crest (figs. D-4 and 5-56). Scattered, more or less fresh, small craters are observed in the local area, and nearby is a group of small subdued craters just east of a cluster of rocks from which the samples were selected (fig. 5-59). The area is generally flat and has no lineaments. The rocks in the cluster all have similar surface textures and the

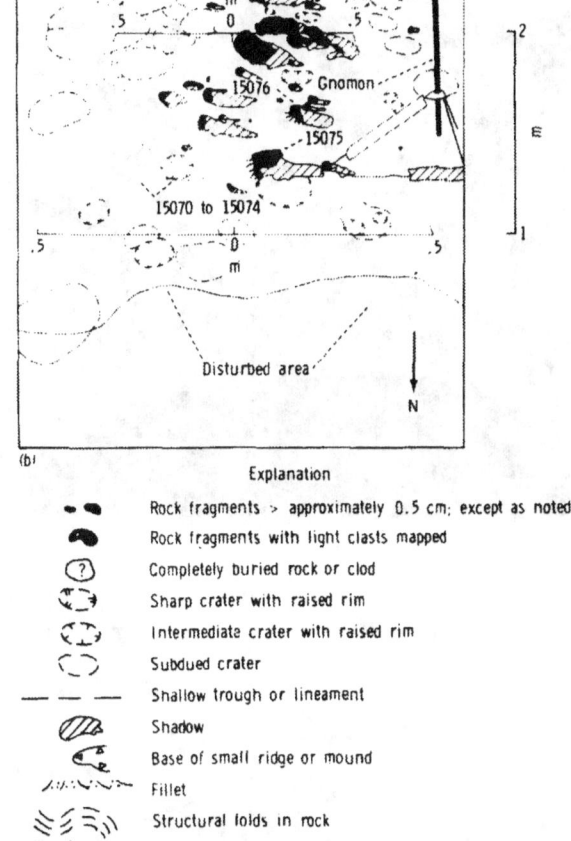

FIGURE 5-59.—Samples 15070 to 15076, collected from the east rim of Elbow Crater at station 1. This sampling point is the second in the radial sample. (a) Presampling, oblique-to-Sun photograph, looking northwest (AS15-86-11534). (b) Sample environment sketch map.

Explanation
- Rock fragments > approximately 0.5 cm; except as noted
- Rock fragments with light clasts mapped
- Completely buried rock or clod
- Sharp crater with raised rim
- Intermediate crater with raised rim
- Subdued crater
- Shallow trough or lineament
- Shadow
- Base of small ridge or mound
- Fillet
- Structural folds in rock
- Fault; arrows show apparent displacement

same albedo. Small- to moderate-sized fillets occur. Some rocks in the cluster appear to be nearly buried and others, including the samples, appear to be less than one-fourth buried. The cluster appears to be composed of related rocks. Possibly, the impact of the cluster produced the group of subdued craters 50 cm to the east. If so, the cluster was ejected from a crater younger than Elbow east of the present location of the cluster, and the samples represent Elbow ejecta farther than 25 m from the rim crest.

Samples 15080 to 15088: These samples are soils and rocks collected approximately 65 m east of the Elbow Crater rim crest (figs. D-4 and 5-53). The immediate area is generally flat and lacks lineaments (fig. 5-60). Small, fresh craters occur in the vicinity—a 10-cm-diameter crater lies approximately 20 cm west southwest of the 15086 site and a moderately fresh 15-cm-diameter crater was observed approximately 35 cm west southwest of the 15085 site. The surface material is fine grained with many pebble-sized fragments. Cobble-size fragments such as samples 15085 and 15086 are distinctly spaced (about one every 2 m^2). The albedo of all the fragments appears the same, and no dust is visible on the larger ones. Fillets less than 1 cm high are banked against some fragments, including 15085, but 15086 had no fillet. The collected rocks were about one-fourth buried. The positions of associated fresh craters suggest that samples 15085 and 15086 were moved to the sampled positions from the west-southwest and that the original positions in the Elbow blanket were less than 60 m from the rim crest.

Station 2.—Station 2 is located on the northeastern flank of St. George Crater near the base of the Apennine Front (figs. D-5, 5-2, and 5-61). St. George is a subdued crater of Imbrian age somewhat larger than 2 km in diameter, located at the foot of Hadley Delta. Smaller craters, 60 m and less in diameter, pock the crater walls and flanks of St. George. The station 2 photographs suggest that the flank of St.

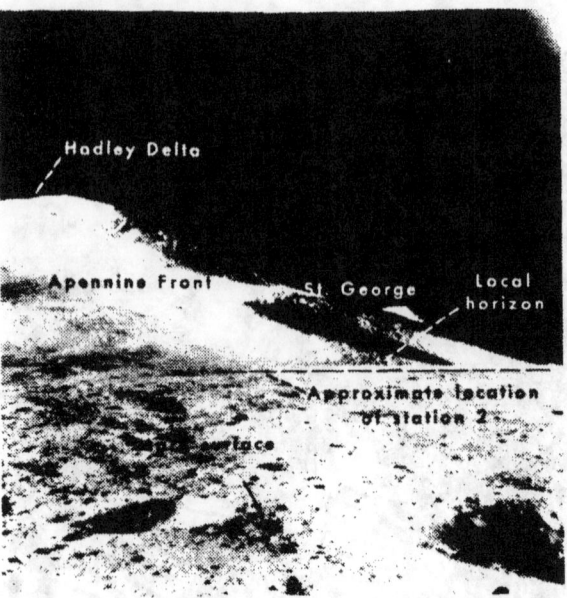

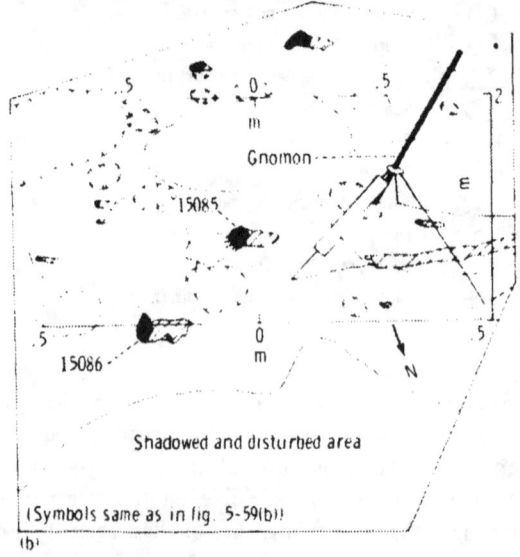

FIGURE 5-60.— Samples 15085 to 15088, collected from the east of the rim of Elbow Crater. This is the third sampling point (farthest) of the radial sample. Samples 15087 and 15088 were apparently collected incidentally with 15085 and 15086. (a) Presampling, cross-Sun photograph, looking south (AS15-86-11536). (b) Sample environment sketch map.

George Crater is underlain by a thick mature regolith. The surface material in the vicinity is composed of fine-grained material with widely scattered fragments a few centimeters across. Except for two blocks associated with apparent secondary impact craters

FIGURE 5-61.— Standup-EVA photograph, looking approximately south, of location of station 2. Exact position of station 2, near the base of the Apennine Front, is concealed by the local horizon on the most distant visible part of the mare surface (AS15-85-11375).

FIGURE 5-62.— Lunar-surface photograph of cratered, undulating surface and fine-grained regolith with widely scattered pebbles and cobbles at station 2. The large block beside the astronaut, ejected by a distant impact, and soil near the block and beneath it, were sampled (AS15-85-11435).

(fig. 5-62), no fragments larger than 10 cm were seen. The notably fine grain size, scarcity of blocks, and extremely undulatory character of the surface suggest that the area is one of intensely cratered and gardened regolith. The regolith is apparently cohesive; numerous small (<5 cm) clods are visible in areas where the surface has been kicked up (fig. 5-63) or broken by the tongs (fig. 5-64). Clods are also visible near small, fresh craters.

Twenty-two samples were collected from an area of approximately 10 by 15 m at station 2 (fig. 5-65). Of these samples, two are cores, five are soils, eight are breccias, and seven are mare-type basalts. The breccias lack mare-type basalt clasts. Two breccia samples were collected from a large boulder, and soil samples were collected to provide representatives of the general character of the soil, of the fillet adjacent to the boulder, and of the soil from beneath the boulder. The soil sample collected as "fillet" is probably soil ejected from the small shallow crater adjacent to the boulder. The remaining rock samples are the rake samples, designed to give a statistical sampling of rocks in the size range between soil and the average documented sample. Basaltic fragments in the rake sample include some that are virtually identical with basalt ejected from the nearby Elbow Crater as well as two fragments that closely resemble

FIGURE 5-64.—Presampling, cross-Sun photograph, looking south, of samples 15090 to 15093 and 15095, collected from near large rock at station 2 (AS15-86-11549).

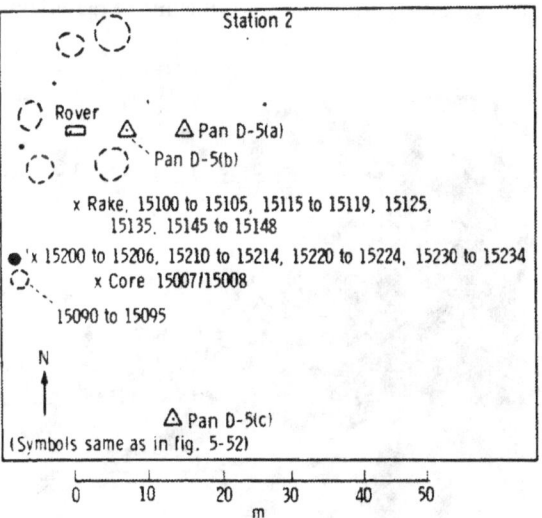

FIGURE 5-65.—Planimetric map of Station 2.

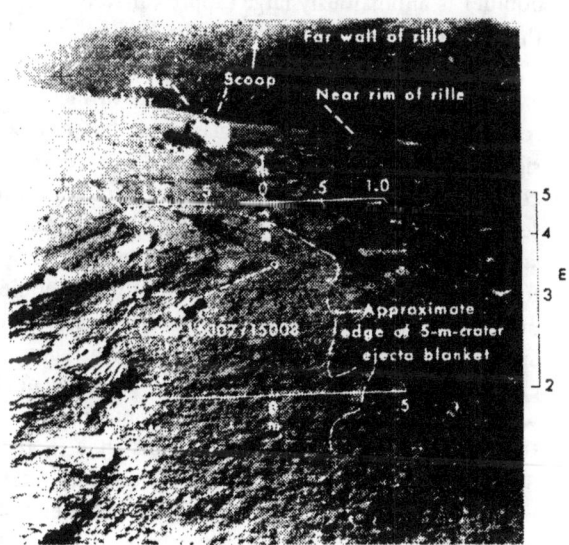

FIGURE 5-63.—Presampling, cross-Sun photograph, looking northwest, of samples 15007 (upper) and 15008 (lower), double core, collected at station 2. The large boulder has been rolled so that the former base now faces the camera (AS15-85-11443).

the basalts collected at station 9A. Breccias in this collection have an unusual assemblage of fine-grained gabbroic clasts that are not like either the mare basalts or other gabbroic samples collected from the Apennine Front.

Soil samples collected at station 2 should provide representatives of thoroughly gardened, mature regolith developed on the ejecta blanket of St. George

Crater. Admixtures of mare basalt are likely, considering the proximity of Elbow Crater and the occurrence of mare-type basalts in the rake samples.

Samples 15007 and 15008: These samples comprise a double core-tube sample taken approximately 5 m southeast of the large boulder shown in figure 5-63. The core-tube was driven close to the rim and within the continuous ejecta blanket of a 10-m-diameter crater; it likely penetrated the buried contact between the local ejecta blanket and the previously exposed regolith surface.

Samples 15090 to 15093 and 15095: Samples 15090 to 15093 are soil samples and 15095 is a glass-coated breccia sample collected close to the large boulder at station 2. In the immediate vicinity are scattered coherent fragments a few centimeters across (fig. 5-64) on generally fine-grained soil. In size and appearance, sample 15095 is typical of fragments scattered on the surface throughout the station 2 area.

Samples 15100 to 15105: Samples 15100 to 15104 are soils collected from the rake-sample site (fig. 5-66) 5 m east of the large boulder at station 2. The samples were taken from the pile of material left by the raking operation at the edge of the raked area. Sample 15105 is a small fragment of fine-grained mare-type basalt from the soil.

Samples 15115 to 15119, 15125, 15135, and 15145 to 15148: These rake samples were collected 5 m east of the large boulder at station 2 (fig. 5-66). As shown in the figure, no surface fragments larger than 10 cm in the rake area exist, and fragments coarser than 1 cm are rare. Crew comments made during collection of the rake sample verify the scarcity of coherent fragments large enough (<1 cm) to be retained by the rake. The rake-sample fragments are typical in size and distribution of the coarse fraction in the upper 15 cm of the regolith throughout the station 2 area. Samples 15115 to 15119 and 15125 are mare-type basalts similar to those collected at Elbow Crater and station 9A. Sample 15135 is a glassy breccia lacking clasts of mare-type basalt. Samples 15145 to 15148 are probably soil breccias, but they contain an unusual assemblage of fine-grained gabbroic clasts.

Samples 15205 and 15206: These samples are pieces of breccia broken from the large boulder at station 2 (figs. 5-67 and 5-68). Sample 15206 came from a corner on the south (uphill) side of the rock, and sample 15205 came from a corner at the intersection of the top and east faces of the rock. The boulder is anomalously large (approximately 1 m) for the area. Coherent fragments on the surface at station 2 are scattered in occurrence and generally do not exceed 10 cm in diameter. Much of the surface of the rock is covered by bubbly glass that has not been significantly abraded (fig. 5-69). Rock edges and corners, controlled in part by surface-fracture intersections, are sharp (fig. 5-68). The boulder lies on the south rim of a small, fresh crater that appears to open to the northwest (fig. 5-67). Apparently, the boulder impacted at a low angle from the north or northwest and rolled uphill onto the rim of its own secondary crater. A similar crater, apparently elongate northwest-southeast, with a large boulder on or near the southeastern rim is visible in the distance in figure 5-62. The two features may be secondary craters produced by two blocks thrown from a crater to the northwest. A small fillet occurs where the block lies on the high part of the crater rim (figs. 5-68 and 5-70), but the underside of the block forms an overhang at the south end. With the boulder removed, the fillet resembles the slight ridge that might be expected to ring the boulder where it compressed the

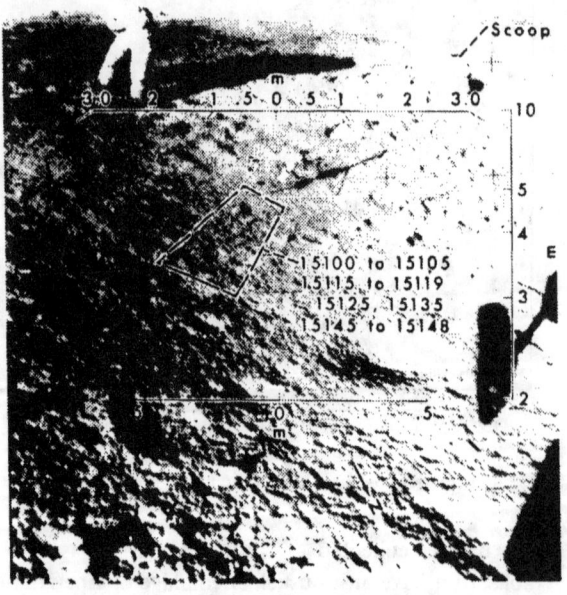

FIGURE 5-66.—Presampling, oblique-to-Sun photograph, looking southwest, of samples 15100 to 15105 (soil) and samples 15115 to 15148 (rake), collected as a comprehensive sample at station 2. The large boulder has been rolled so that the former bottom side now faces the camera (AS15-85-11442).

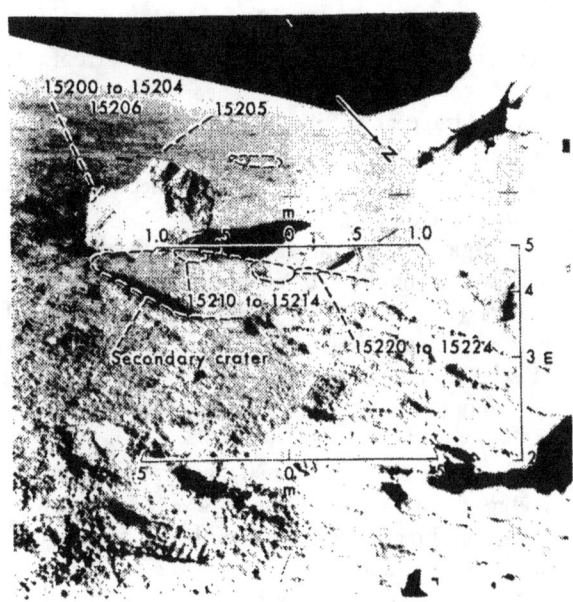

FIGURE 5-67.—Presampling, oblique-to-Sun photograph, looking southwest, of near-field view of the large rock at station 2 and its secondary crater showing collection sites for samples 15200 to 15206, 15210 to 15214, and 15220 to 15224. Samples 15230 to 15234 were collected from under boulder, which was rolled to the west (AS15-85-11440).

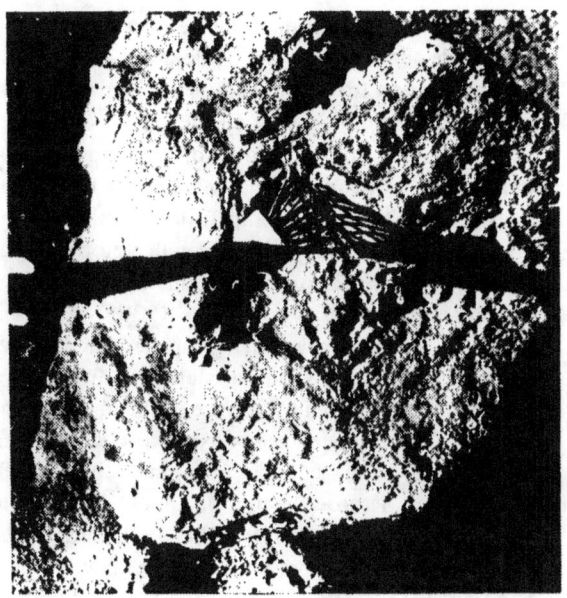

FIGURE 5-69.—Cross-Sun photograph, looking south, of large block with well-preserved coating of bubbly glass at station 2 (AS15-86-11554).

FIGURE 5-68.—Postsampling, oblique-to-Sun photograph, looking northwest, of collection sites for samples 15200 to 15206, 15210 to 15214, and 15220 to 15224 (AS15-86-11560). The inset is from a postsampling, cross-Sun photograph looking north (AS15-86-11558).

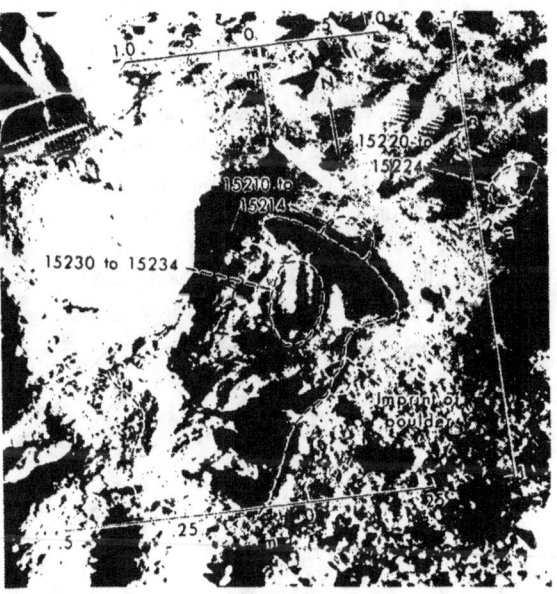

FIGURE 5-70.—Postsampling, cross-Sun photograph, looking north, of underside of large rock at station 2, showing the boulder imprint in the crater rim and the collection site for samples 15230 to 15234 in the soil beneath the boulder. Collection sites for samples 15210 to 15214 and 15220 to 15224 are also indicated (AS15-86-11565).

soft secondary crater rim. Absence of a fillet on the uphill side suggests that the boulder has been in the present position for a relatively brief time. It appears likely that the block was ejected from a post-St. George Crater position to the northwest, as perhaps was another block in a similar setting—the block that is visible in figure 5-62. Comparison of soil collected from under the boulder with soil collected from near the boulder may yield useful information. If the glass that coats the fractures of sample 15205 was mobilized by the same impact that propelled the boulder to the presampling position at station 2, the date of the impact as indicated by the age of the glass may be equivalent to the difference in exposure ages between soil collected from under the rock and soil collected from nearby. Because the breccia samples taken from the boulder have no mare-type basalt fragments, the boulder presumably represents premare material.

Samples 15210 to 15214, 15220 to 15224, and 15230 to 15234: These samples are soils collected from the rim of the secondary impact crater adjacent to the large boulder at station 2 (figs. 5-66, 5-68, and 5-70). Samples 15210 to 15214 were collected near the base of the boulder; samples 15220 to 15224 were collected from the crater rim approximately 1 m northwest of the boulder; and samples 15230 to 15234 were collected from the south crater rim beneath the boulder.

Station 3.—Station 3 is on the mare surface near the boundary between two subdued craters, each approximately 125 m west-southwest of Rhysling Crater (figs. D-7, 5-2, and 5-71). The surface is undulating on large and small scales because of the intersection of numerous craters. Within the area visible in figure D-7, many 5- to 25-m-diameter subdued craters contribute to surface irregularities. Scattered over the surface are fragments from a few to 25 cm across (fig. 5-72, near where the Rover tracks disappear to the south). The station 3 area appears in general to be a mature surface. One mare-type basaltic fragment was collected at station 3, an unscheduled stop.

Sample 15016: This sample was collected from an area with abundant 10-cm- to 1-m-diameter subdued craters (fig. 5-72). The sample is a very vesicular basalt, rounded by surface erosion. Lithologically, it closely resembles samples 15535 and 15536, as well as a number of fragments in the rake samples from station 9A.

In detail, the nearby mare surface is pitted with craters on the order of one to several centimeters in diameter. Fragments from less than 1 mm to 15 cm across are scattered sparsely over the surface with no noticeable concentrations. Moderately well developed lineaments in the area of sample 15016 trend mainly northwest with a few that trend northeast. Fillets are not well developed against 15016 or its vesicular neighbor (fig. 5-73), and most of the larger fragments farther from 15016 (fig. 5-72) also lack fillets. The apparent depth of burial of the fragments is typically less than one-fourth the height. Sample 15016 was

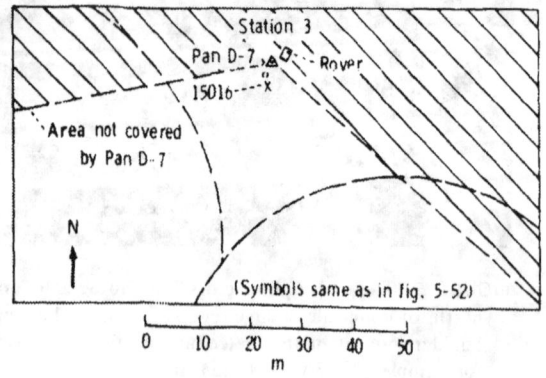

FIGURE 5-71.—Planimetric map of station 3.

FIGURE 5-72.—Postsampling, cross-Sun photograph, looking south, of area from which sample 15016 was collected. The Rover tracks are approximately 2 m apart (AS15-86-11584).

Station 4.—Station 4 is on the south rim of Dune Crater, a 460-m-diameter crater in South Cluster and just west of a slightly younger 100-m-diameter crater that has been inset into the Dune rim (figs. 5-2 and 5-74). Photographs of the northern wall of Dune Crater indicate that the upper one-third has a rather uniform exposure of boulders, probably representative of the upper mare flows (fig. D-6). The smoother, lower slopes of the crater wall and floor are covered with fine-grained infilling materials (fig. 5-75). The relatively low abundance of fragments away from the rim crest of Dune Crater indicates a relatively mature

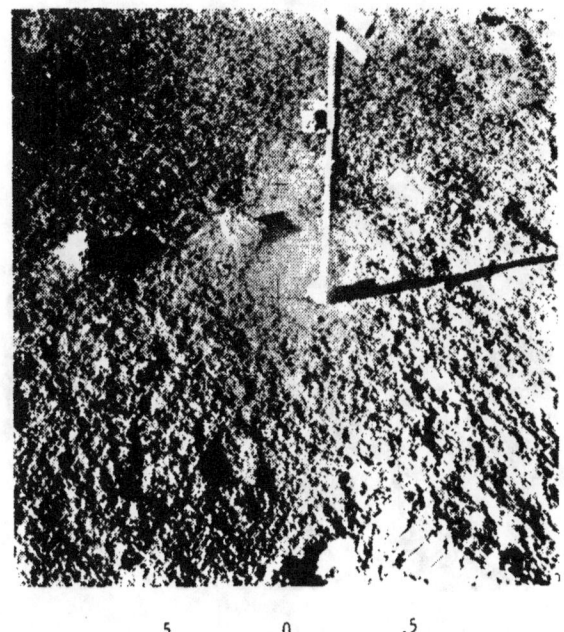

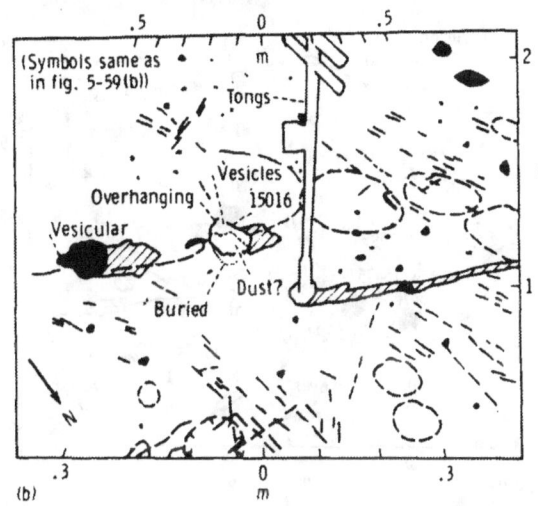

FIGURE 5-73.—Sample 15016, collected from station 3. (a) Presampling, cross-Sun photograph, looking south (AS15-86-11579). (b) Sample environment sketch map.

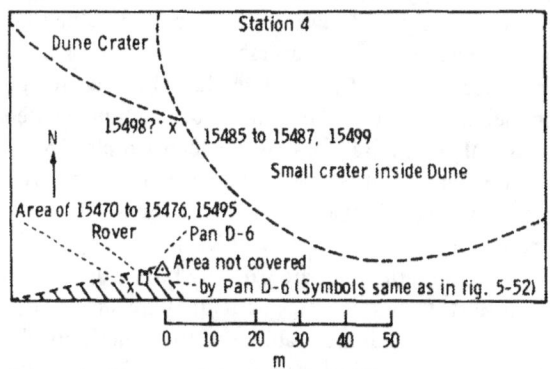

FIGURE 5-74.—Planimetric map of station 4.

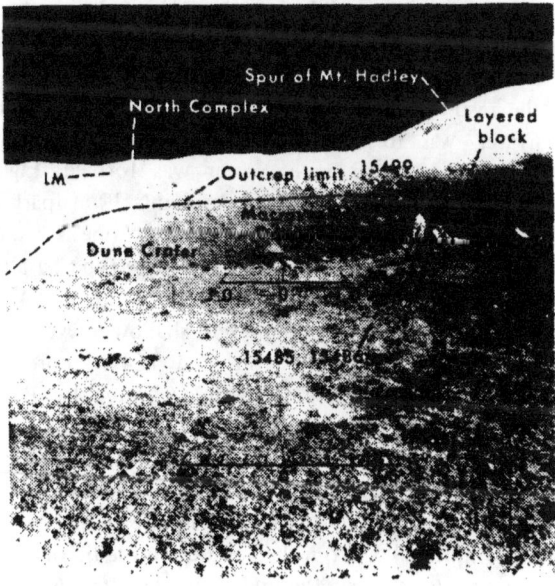

FIGURE 5-75.—Presampling photograph, looking north, of the boulders at the south rim of Dune Crater, showing locations of samples 15485, 15486, 15498, and 15499 (AS15-90-12242).

collected from the rim of a 50-cm-diameter crater. Only the northeast bottom corner of the sample was embedded in soil, and the south end was suspended above the inner wall of the crater. The rock was free of dust when collected. Comparison of figures 5-72 and 5-73 suggests that sample 15016 is representative in shape, burial, and filleting of the large fragments in the area. The rounded corners and edges suggest that 15016 has been on or near the surface for a long time.

surface and a fairly early Copernican age for the South Cluster secondary cratering event. Two areas, speparated by approximately 30 m, were sampled (fig. 5-74). The surface immediately surrounding the site of samples 15470 to 15478 and 15495 has a moderate cover of fragments (fig. 5-76). Small nearby craters are sparse. Lineaments are not visible in the documentary photographs, and the slope in the sample vicinity is negligible. Rock fragments greater than 6 cm across in the sample area are not filleted. The 17-cm rock partly hidden by the gnomon staff (fig. 5-76) appears to be the only large fragment that is significantly buried. Nearly all the rocks appear to be moderately dust covered. The apparent albedo of the fragments is low to average, and no photographic evidence exists of any light-clast breccia in the immediate vicinity. The local surface immediately south of the basalt block from which samples 15485, 15486, and 15499 were taken (fig. 5-75) is a well-developed fillet banked against the block, with a moderate abundance of rock fragments (fig. 5-77). The rocks on the fillet are partially buried and coated with dust. Only a few fragments that are smaller than 2 cm or so are resolvable on the fillet, possibly because of the downslope movement of fines covering them. Possibly for the same reason, no small craters exist on the fillet.

The proximity of the large boulder to the junction of the two craters suggests that the boulder position and the associated prominent boulder group may be the result of a double excavation—once when Dune Crater was formed and again when the slightly younger 100-m-diameter crater was formed. Eight samples were collected from two points 30 m apart at station 4 (fig. 5-74). Of these samples, one is a soil, one is a soil breccia, and six are mare-type basalts. The sample locations are such that samples 15475, 15476, and 15495 may represent shallower levels and that samples 15485, 15486, and 15499 represent deeper levels in mare material excavated by Dune Crater.

The dimater of Dune Crater is such that blocks near the rim may have come from depths as great as 90 m below the mare surface. If so, the chilled nature of samples taken from the boulder on the lip of Dune Crater suggests that the underlying basalts consist of more than one flow unit. The breccia, apparently a soil breccia containing mare-type basalt clasts, and the soil samples may show evidence of flow units in addition to those represented by the large samples.

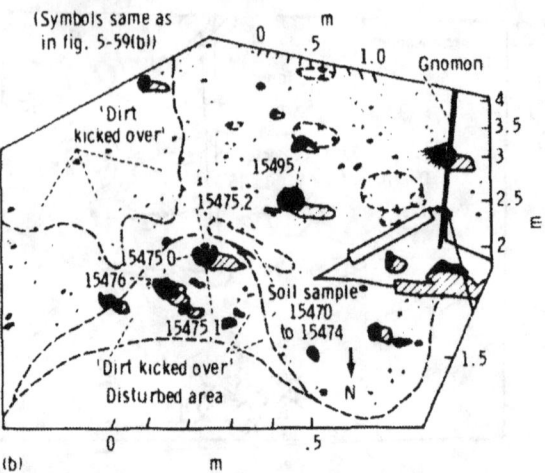

FIGURE 5-76.—Samples 15470 to 15476, 15478, 15479, and vicinity. (a) Presampling photograph, looking south (AS15-87-11759). (b) Sample environment sketch map.

Samples 15470 to 15476, and 15495: Soil and mare-type basalts were collected approximately 30 m south-southeast of the rim crest of Dune Crater. Soil samples 15470 to 15474 were taken from between rock samples 15475 and 15495 (fig. 5-76), but these samples may contain a small percentage of disturbed (kicked) material resulting from astronaut activities in the area before sampling. The soil was collected from very near the surface, probably only the upper 3 to 4 cm. Basalt samples 15475 and 15495 are similar in size, shape, texture, and general lack of filleting or extensive burial. Sample 15476 is slightly tabular and

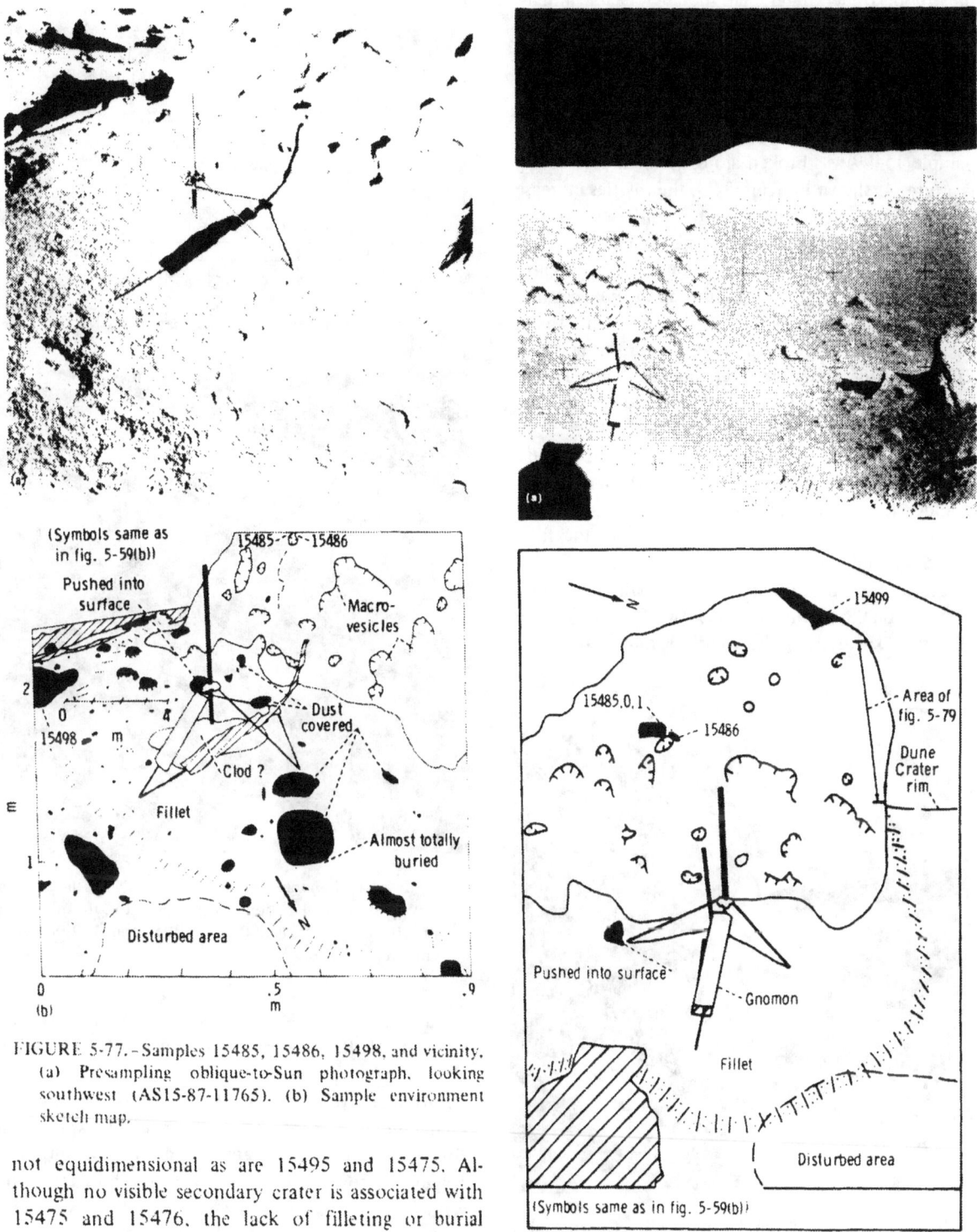

FIGURE 5-77.—Samples 15485, 15486, 15498, and vicinity. (a) Presampling oblique-to-Sun photograph, looking southwest (AS15-87-11765). (b) Sample environment sketch map.

not equidimensional as are 15495 and 15475. Although no visible secondary crater is associated with 15475 and 15476, the lack of filleting or burial suggests that they may have been recycled onto the surface at least once since the Dune Crater event.

Samples 15485, 15486, and 15499: These samples are basaltic fragments broken from a large boulder on

FIGURE 5-78.—Samples 15485, 15486, 15499, and vicinity. (a) Presampling, down-Sun photograph, looking west (AS15-87-11768). (b) Sample environment sketch map.

the lip of Dune Crater. The boulder has strikingly large, vesicular cavities exposed on the east face (fig. 5-78). Samples 15485 and 15486 were taken from the overhanging edges of one of the cavities. The latter broke on a preexisting fracture, part of a set characterized by a distinctive olive-gray alteration. Sample 15499 was broken along a similar preexisting fracture. As shown in figure 5-79, the cavities increase in size from left to right on the north end of the boulder (the width of the tongs where they rest against the rock is 9.8 cm). Comparison of figures 5-77(a), 5-78(a), and 5-79 suggests an overall vesicular zonation of the large boulder. It appears that a finely vesicular zone extends across the area from which samples 15485 and 15486 were collected, and a more coarsely vesicular zone extends up through the area from which sample 15499 was collected.

Sample 15498: Sample 15498 is a soil breccia collected near the base of the fillet on the large basalt boulder from which 15485 was collected. The fragment was one-third to one-half buried in the fillet (fig. 5-77) and was moderately dust covered.

Station 6. Station 6, the easternmost point sampled on the Apennine Front (figs. 5-2 and 5-80) is nearly 3 km east of St. George Crater and 5 km

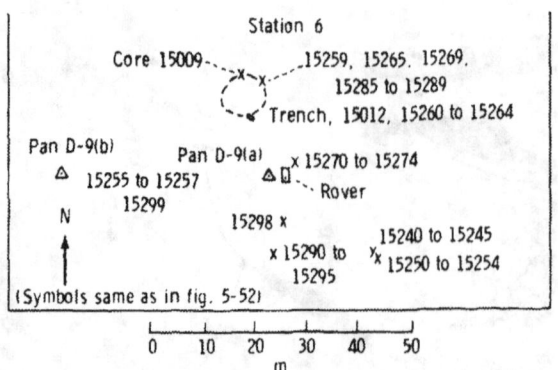

FIGURE 5-80.—Planimetric map of station 6.

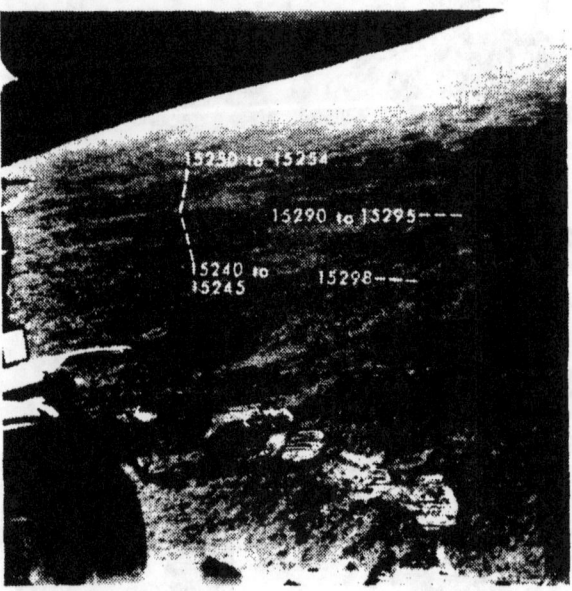

FIGURE 5-81.—Panorama view (fig. D-9(a)), southeast from Rover, showing locations of samples 15240 to 15245, 15250 to 15254, 15290 to 15295, and 15298 before sampling (AS15-85-11495).

FIGURE 5-79.—Photograph, looking south, of north end of large basalt boulder, showing increase in vesicle size from left to right. The largest holes do not seem to be in zones. Sample 15499 is off camera at the top of the picture (AS15-87-11773).

southeast of the LM. The site is on the north-facing slope of Hadley Delta, approximately 90 to 100 m above the mare surface. The surface slopes from approximately 5° to 15°. A subtle difference in soil characteristics at the sample site may be expressed in the fine granular appearance and deeper burial of fragments toward the east (fig. 5-81) as compared with coarser textures (higher concentrations of centimeter-size fragments) and less burial west and north of the Rover (fig. 5-82). Fragments greater than 10 cm across are uniformly but sparsely distributed in the intercrater areas, with notably higher density

FIGURE 5-82.—Panorama view (fig. D-9(b)), looking east, showing locations of samples 15240 to 15245, 15250 to 15254, 15290 to 15295, 15298, and 15299 after collection. Samples 15255 and 15256 are shown before collection; samples 15270 to 15274 were behind the Rover (AS15-85-11515).

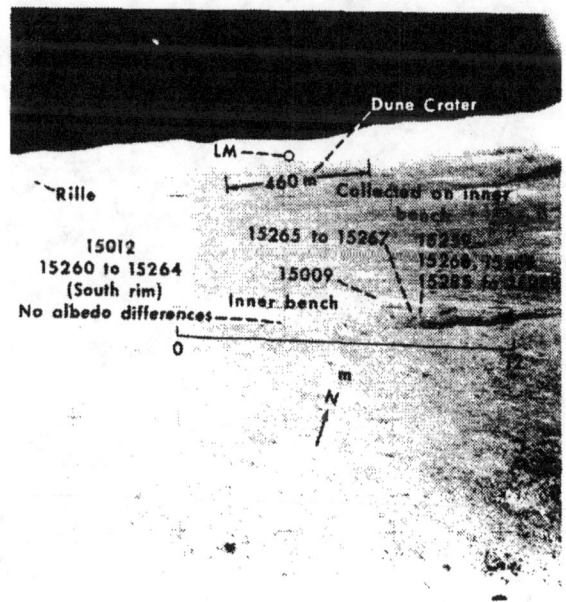

FIGURE 5-83.—Presampling, panorama photograph (fig. D-9(a)), looking northwest, of 12-m-diameter crater north of Rover at station 6. Samples 15009, 15012, 15259, 15260 to 15264, 15265 to 15269, and 15285 to 15289 are located (AS15-85-11485).

around a 12-m-diameter crater (fig. 5-83). The only fresh craters present are less than 1 m in diameter, and from these only clods of soil or soil breccias were excavated. Some samples were collected from the 12-m-diameter crater north of the Rover (fig. 5-83); they are considered to be recycled from underlying regolith. Other samples are not associated with this crater. Hence, two types of environments from which samples were taken are defined at station 6—the 12-m-diameter crater environment and the Hadley Delta slope environment.

The rim of the 12-m-diameter crater is subdued but distinctly raised approximately 0.5 m above the surrounding slope. Several 1-m-diameter sharp craters are on the rim of the crater (fig. 5-84). Parts of the rim have well-developed benches, particularly the north and northeast sides, and a bench is observed on the inner wall on the northeast side of the crater. Small parallel east-trending lineaments are visible on the north wall of the inner bench near the sample site. The slope in the immediate vicinity is 15° to 20° toward the crater floor, but the regional slope is approximately 10° N. Filleting and burial of the rock fragments are less than might be expected for undisturbed ejecta from the surface by the small, fresh, 1-m-diameter craters. Dust cover on many of the rock fragments seems higher than normal, probably the result of ejecta from craters in the immediate area. The north rim of the 12-m-diameter crater has a higher-than-average albedo, and light-colored soil was kicked up during sampling activities.

Lineaments occur locally in the Hadley Delta slope environment. The soil is finely granular with centimeter-size fragments comprising 1 to 15 percent of the surface layer. Larger fragments are sparse and are not directly associated with specific craters larger than a meter.

Forty-five samples were collected from an area approximately 40 by 55 m at station 6 (fig. 5-80). Of these, one is a core sample, 16 are soil samples, 27 are breccias, and one is a basalt. At least four of the breccias contain mare-type basalt clasts. The basaltic fragment is a Front-type basalt. The drive tube and 17 other samples (fig. 5-83) were collected at various points in and around a 12-m-diameter crater. Other samples were selected on the basis of distinctive individual features.

The detailed sampling in and around the 12-m-diameter crater will allow study of vertical and lateral variations in the local regolith. The core was taken

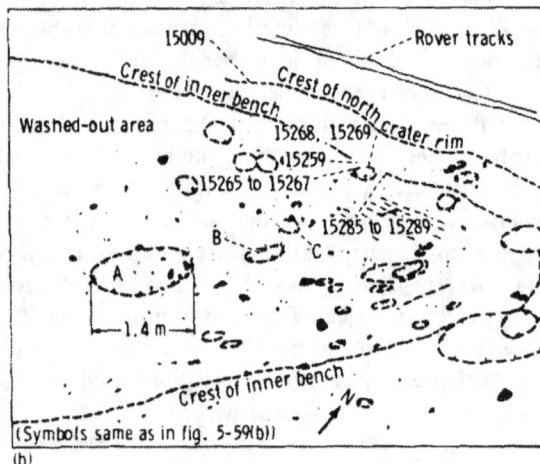

FIGURE 5-84.—East end of 12-m-diameter crater at station 6, showing area of samples 15259, 15265 to 15269, and 15285 to 15289. (a) Presampling, cross-Sun photograph, looking north (AS15-86-11634). (b) Sample environment sketch map, showing distribution of fragments in the sample area.

from the north rim of the crater, and a trench sample was taken from the south rim. Additional samples were taken from both inside and outside this crater. Most of the breccias collected at this station appear to be lithified regolith. They contain abundant feldspathic clasts as well as a few mare-type basalt clasts. The clast population should bear on the lithologic character of the Front beneath the regolith. Other samples are of glass-cemented soil from a 1-m-diameter crater; these appear typical of a great many such craters in the area. The one Front-type basaltic sample is the largest rock of this type collected.

Sample 15009: Sample 15009 is a core sample collected from the north rim of the 12-m-diameter crater (fig. 5-83), 10 to 15 m downslope from the Rover (fig. 5-80). The surface into which the single core was driven (fig. 5-85) is fine grained but littered with larger angular fragments approximately 0.5 to 2 cm across. The crew noted that the north rim of this crater was more "granular" than the south rim. A few very subdued 10- to 20-cm-diameter craters pock the surface.

Samples 15259, 15265 to 15269, and 15285 to 15289: These samples were collected from an inner bench on the northeast wall of the 12-m-diameter crater, approximately 15 m downslope from the Rover (figs. 5-80 and 5-83). The distribution of rock fragments in the vicinity of the sampling area is shown in figure 5-86. The largest single rock fragment collected in the area is labeled in figure 5-86(b) as the source of specimens 15265 to 15267. The fragment was too large for a sample bag and was broken with a hammer. The two largest pieces (15265 and 15266) from the rock are illustrated in figure 5-87 in the positions in which they landed after breaking. Sample 15267 has not been identified in the photographs. The original large breccia fragment was one-third

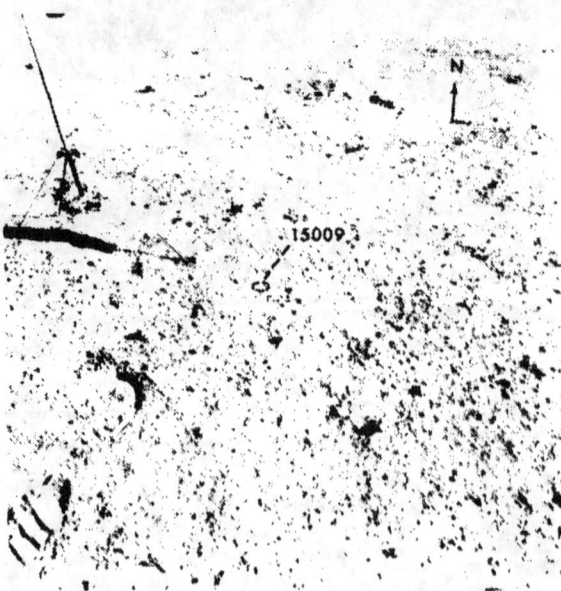

FIGURE 5-85.—Presampling, cross-Sun photograph of site of core-tube sample 15009 on north rim of 12-m-diameter crater at station 6 (AS15-86-11648).

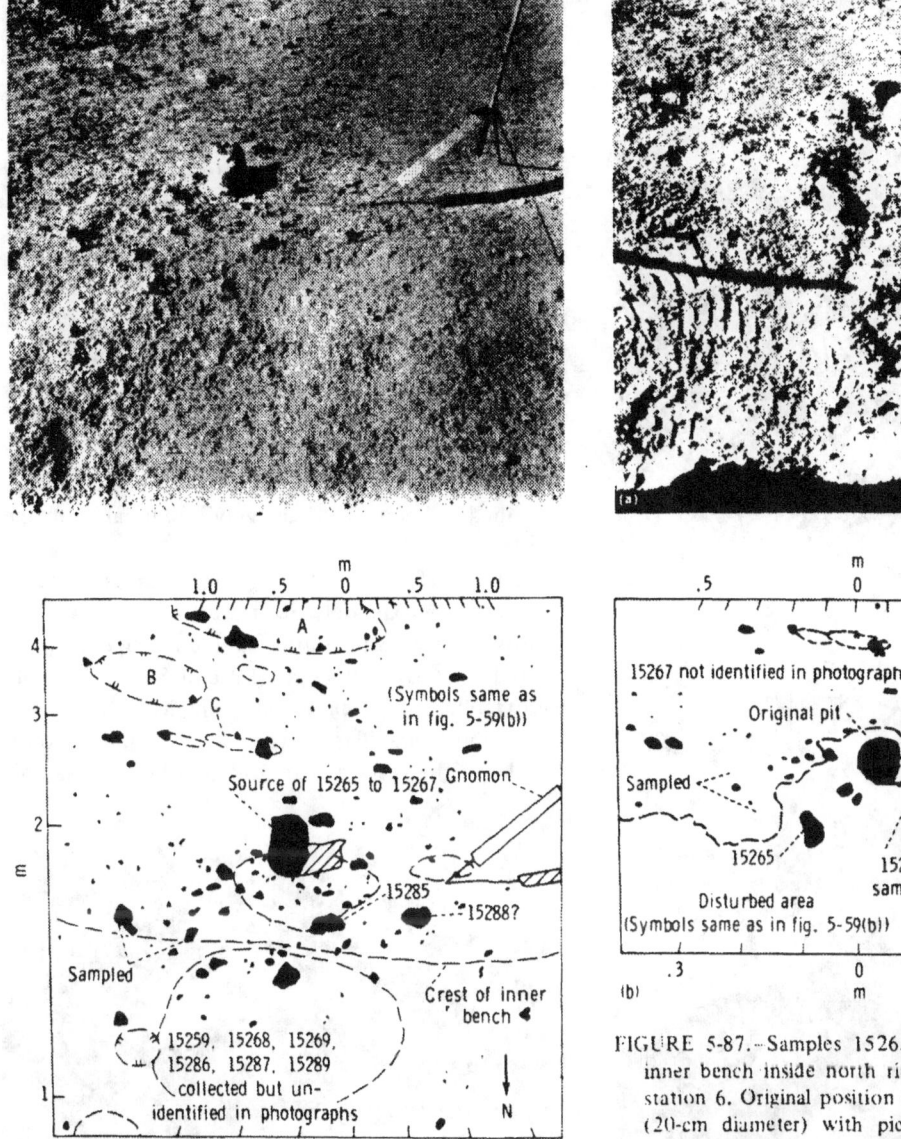

FIGURE 5-86.—Samples 15259, 15265 to 15269, and 15285 to 15289 collected on inner bench, inside the north rim of 12-m-diameter crater at station 6. (a) Presampling, cross-Sun photograph, looking south (AS15-86-11635). (b) Sample environment sketch map.

FIGURE 5-87.—Samples 15265 to 15267 and vicinity on inner bench inside north rim of 12-m-diameter crater at station 6. Original position of rock was in shallow crater (20-cm diameter) with pieces surrounding it. (a) Presampling, cross-Sun photograph, looking south (AS15-86-11638). (b) Sample environment sketch map.

buried and had a very small fillet. Most of the burial may have been due to impact of the fragment. The 15265-to-15267 rock was found lying on the north rim of a 40-cm-diameter crater, most likely the result of a low-velocity impact of the rock. The source of the rock may have been the 1.4-m-diameter crater labeled A in figures 5-84 and 5-86. The smaller craters labeled B and C are probably too small to have been the source of the rock. The abundant small fragments immediately surrounding the larger breccia appear to be pieces of it, possibly broken off upon impact. Samples 15259, 15268, and 15286 to 15289 are samples of these small pieces surrounding the 15265-to-15267 rock.

Samples 15012, and 15260 to 15264: These

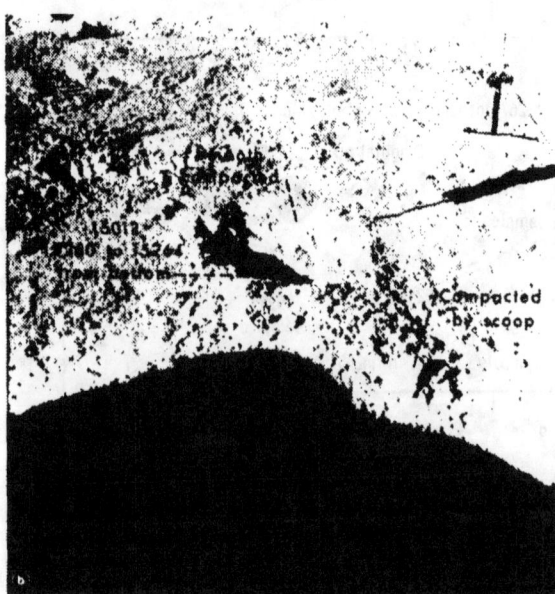

FIGURE 5-88.—Site of trench dug on south rim of the 12-m-diameter crater. (a) Presampling, cross-Sun photograph, looking north (AS15-86-11641). (b) Postsampling photograph, looking southwest, of trench from which samples 15012 (SESC-1) and 15260 to 15264 were collected (AS15-86-11643).

distinctly fewer than on the north rim of the same crater (fig. 5-85).

Samples 15240 to 15245, and 15250 to 15254: Samples 15240 to 15245 were collected from the floor, and samples 15250 to 15254 were taken from the east rim of a 1-m-diameter crater approximately 20 m southeast and upslope from the Rover (fig. 5-81). The crater is marked by a concentration of fragments, primarily clods up to 10 cm across, on an otherwise smooth, finely granular surface (fig. 5-89). The crater is superposed on the south wall of a subdued 3-m-diameter crater that has no visible ejecta blanket. No lineaments were seen in the photographs, and no variations in albedo are observed. Samples 15240 to 15245 represent two scoops of "welded" splash glass, broken soil breccias, and soil from the floor of the crater; samples 15250 to 15254 were described as "very fine light gray" material from the east rim.

Samples 15255 to 15257: These samples are rocks collected approximately 30 m west of the Rover and approximately 25 m southwest and upslope from the 12-m-diameter crater (figs. D-9 and 5-80). The only basalt collected at station 6 is sample 15256, which has a more granular surface texture and a lower profile on the surface than 15255, a breccia. Sample

samples are trench samples from the south rim of the 12-m-diameter crater, 10 to 15 m downslope from the Rover (figs. 5-80 and 5-83). The surface where the trench was dug (fig. 5-88) is littered with fragments ranging from 0.5 to 2 cm, but they are

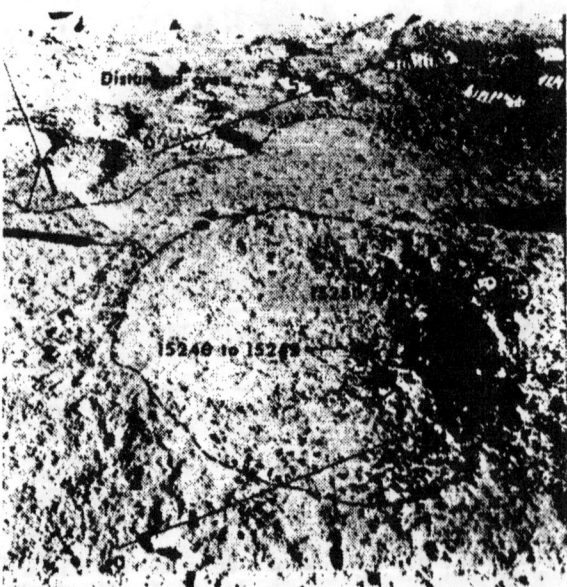

FIGURE 5-89.—Presampling, cross-Sun photograph of "fresh little crater" at station 6 from which samples 15240 to 15245 and 15250 to 15254 were collected (AS15-86-11610).

15257, a breccia, has not yet been identified in the photographs. The general surface texture around samples 15255 to 15257 is coarsely granular with 10 to 15 percent of the undisturbed area covered by centimeter-size clods. Several craters 10 to 30 cm across with slightly raised rims but no pronounced ejecta blankets occur in the vicinity. No distinctive lineaments are apparent, but a very subtle pattern trending north-south occurs near the gnomon (fig. 5-90). Two 3- to 4-cm fragments near the gnomon appear partly buried and slightly filleted on the uphill side. These fragments are candidates for identification as sample 15257. No albedo differences in the rocks or soil are visible down-Sun.

Samples 15270 to 15274: These soil samples were collected from the compressed wheel track behind the Rover (figs. D-9, 5-80, and 5-91). The adjacent undisturbed soil surface appears typical of the coarsely granular texture to the west and north at station 6. Five to 7 percent of the immediate vicinity is covered by centimeter-size clods. The area is lacking in fresh craters or coherent rock fragments. Before reaching the Rover, the crew commented how the chevrons of the wheels compacted the soil, but they inferred that their bootprints were much deeper than the wheel tracks.

Samples 15290 to 15295: These rock and soil samples were collected upslope 10 to 15 m south of the Rover (figs. D-9 and 5-80). Sample 15295 is the single large angular rock seen in figure 5-81. The rock is distinctive in the area because of the large size (15 cm long), angularity, and the presence of a fillet developed on the uphill side (fig. 5-92). Samples 15290 to 15294 are soil samples representative of the local regolith.

Sample 15298: This sample is a fractured soil breccia collected approximately 10 m south and upslope from the Rover. Five to 7 percent of the general surface is covered by centimeter-size clods (fig. 5-93). Other fragments larger than 10 cm are rare. Two small fresh craters several meters north of the sample have excavated centimeter-size clods and may be the source of similar ones near the sample. An unexplained linear pattern formed by these clods follows a curved path east of the sample, trending northeast and bending left toward the northwest as shown in figure 5-93. Sample 15298 has a slightly higher apparent albedo than the surrounding soil and was one-fourth to one-third buried before collection. Penetrating fractures can be seen on both the east and west sides of the rock parallel to the length (figs. 5-81 and 5-93). Sample 15298 has no obvious relation to

FIGURE 5-90.—Presampling, cross-Sun photograph of samples 15255 and 15256 at station 6. Sample 15257 was also collected in this area but has not been identified in the photographs (AS15-86-11630).

FIGURE 5-91.—Postsampling, cross-Sun photograph of location of soil samples 15270 to 15274 from the Rover track (AS15-86-11657).

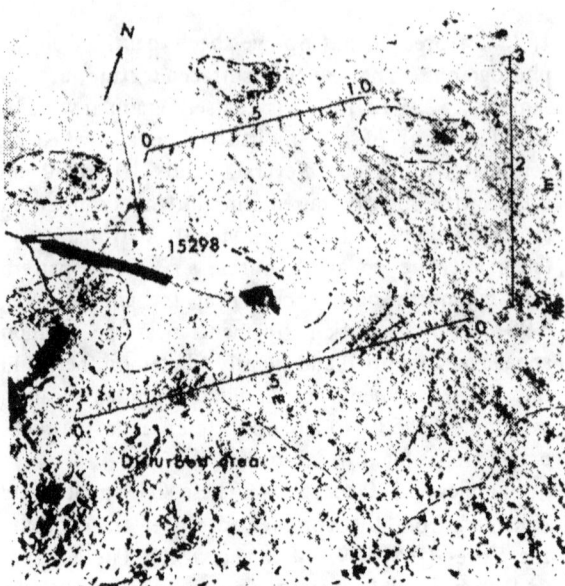

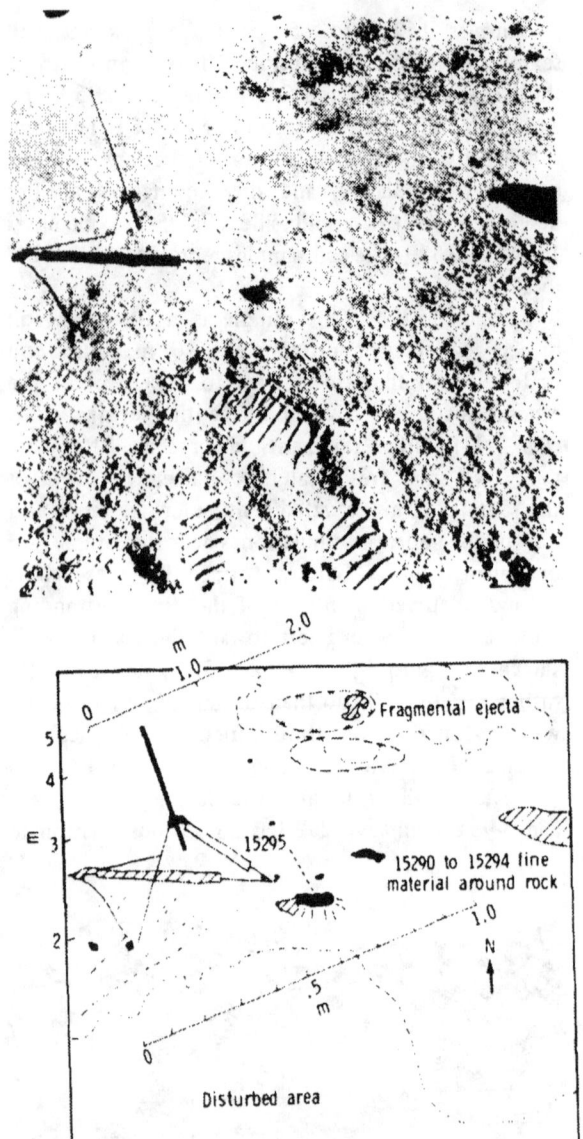

FIGURE 5-92.—Samples 15290 to 15295 collected at station 6. The location of the soil (15290 to 15294) is inferred to be from around the rock (15295). (a) Presampling, cross-Sun photograph, looking north (AS15-86-11617). (b) Sample environment sketch map.

nearby craters or other features and the source is unknown.

Sample 15299: This sample was collected approximately 25 m west-southwest of the Rover (figs. D-9 and 5-80). In figure 5-94, the rock appears to lie on the surface with no evidence of burial or filleting. The

FIGURE 5-93.—Presampling, cross-Sun photograph of sample 15298 approximately 10 m upslope from the Rover. The largest rock, a breccia, was collected at station 6 (AS15-86-11622).

rock appeared to the astronauts to have struck the surface approximately 30 cm east of the collection site. Stereophotographs of the inferred impact site (fig. 5-94) do not show a depression from the impact; however, a light area that is anomalously bright in the down-Sun view occurs at the impact spot cited by the crew. The general surface is coarsely granular, and 10 to 15 percent is covered by centimeter-size clods distributed uniformly around the area. The population of craters greater than 10 cm across is small in the immediate vicinity, and none is obviously related to the sample. Five lineaments on the south side of the sample trend north, parallel to the area slope. Sample 15299 has some irregular patches of high albedo in the down-Sun photograph, indicative of light-clast components in the breccia.

Station 6A.—Station 6A is the highest location explored on the Apennine Front (figs. D-10, 5-2, and 5-95). It was an intermediate stop made at a distinctive isolated boulder en route to Spur Crater (station 7). The station is approximately 130 m above the mare and 250 m south-southeast of Spur Crater (fig. D-10). The boulder was sighted as a distinctive landmark and sampling target during the traverse to station 6, and, upon returning to it, the crew described it as a "big breccia" and as the "first green rock" they had seen.

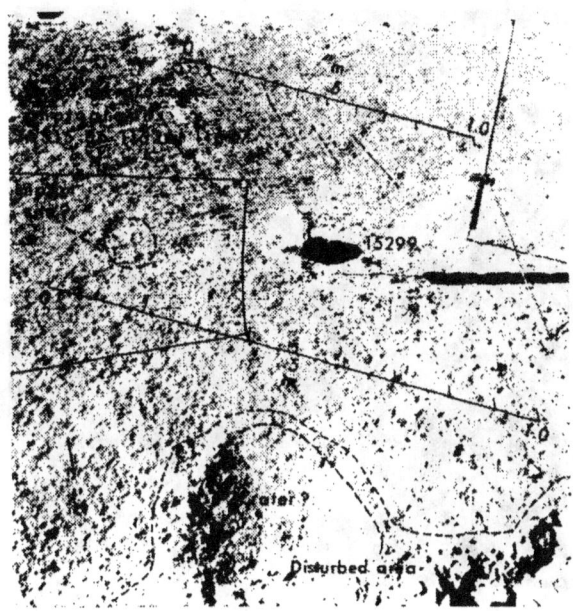

FIGURE 5-94.—Presampling, cross-Sun photograph of sample 15299 at station 6 approximately 25 m west-southwest of the Rover. The second largest rock, a breccia, was collected in this area (AS15-86-11624).

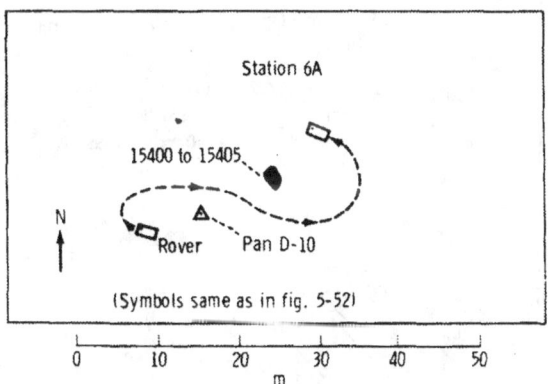

FIGURE 5-95.—Planimetric map of station 6A.

Two samples were collected from a small area on and near the boulder (fig. 5-95). The soil samples (15400 to 15404) were collected from a green, stripelike area on the boulder, and sample 15405 is a fragment of breccia representing the boulder. The breccia lacks mare-type basalt clasts.

Although the boulder lies 250 m from Spur Crater, it closely resembles samples 15445 and 15455, which were collected at Spur Crater, and the greenish-gray soils resemble friable, green rocks and green soils collected at Spur Crater. Hence, it is possible that the boulder was ejected from Spur Crater, or, at least, that it was derived from the same type of material that was excavated by the Spur Crater event.

Samples 15400 to 15405: Samples 15405 (breccia) and 15400 to 15404 (soil) were collected from the singular, rounded, 3-m-long boulder and from a green soil-filled linear depression developed high on the south side of the boulder (fig. 5-96). A prominent fillet forms a limited apron on the south side of the boulder and may be related to the soil sample collected at the top of the boulder. Relatively few fragments occur in the vicinity of the rock. Most are angular and lie on the fillet, but sparse, randomly distributed fragments occur in all directions. A small blocky crater, from which presumably indurated soil was excavated, lies downslope to the northeast. No local crater is associated with the boulder, and local smaller craters are generally obscured by the hummocky slopes of the area. Visible lineaments are close-spaced grooves in the near field, trending approximately northwest; they are several millimeters in width, several meters in length, and are roughly parallel to much larger linear-appearing hummocks. The slope is to the north, probably the steepest one (estimated at approximately 15°) traversed during the mission. The fillet occurs upslope from the boulder, but the high position on the boulder and the presence of larger fragments suggest, possibly, that it may be derived at least partly from the rock against which it lies. Among other possibilities for consideration is that the fillet and the soil stripe on the boulder are plastered on and against the boulder as soil ejecta from a nearby crater. The down-Sun view (fig. 5-97) illustrates the relative albedo differences between white irregular clasts in the light matrix and the darker, clast-poor surface. Similar clasts are visible in fragments photographed on the slope northwest of the boulder.

Except for the larger size (1 by 3 m exposed) and greater depth of burial, the boulder is similar to the few white clast-bearing fragments on the surrounding surface. The large boulder at Spur Crater (station 7) has a similar surface texture but contains larger clasts and has no apparent fillet.

Station 7. Station 7 is located at the rim of the 100-m-diameter Spur Crater on the lower slope of Hadley Delta, approximately 60 m above the level of the LM (figs. D-11, 5-2, and 5-98). The regional slope is approximately 10° N toward the mare surface. The slopes of Spur Crater are pocked by many small

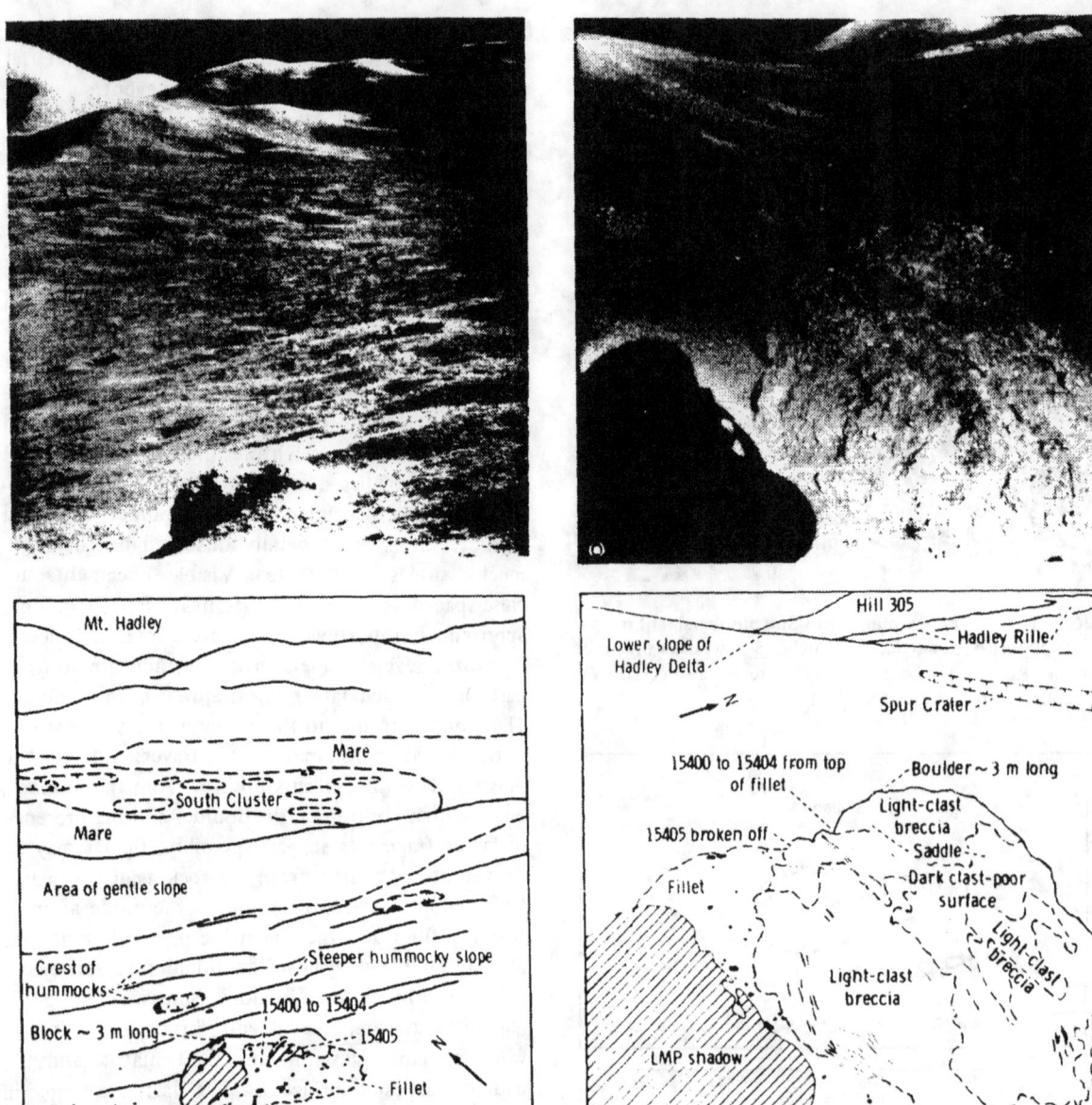

FIGURE 5-96.— View from station 6A, highest vantage point reached on any traverse. (a) Photograph looking northeast across 3-m-wide filleted boulder (AS15-90-12188). (b) Sketch map showing locations of samples 15400 to 15405 collected from filleted boulder.

FIGURE 5-97.— View at station 6A showing albedo variations in clasts and matrix. (a) Presampling, down-Sun, closeup photograph of 3-m-wide breccia boulder; scale is not shown because of the steep slope (AS15-90-12199). (b) Sketch map of surface textures on boulder showing locations of samples 15400 to 15405 from east end.

craters that have reworked Spur ejecta. Rock fragments from centimeter size to 1.5 m are abundant on the upper slopes of Spur Crater.

Ninety-three samples were collected in a rim-circumferential area approximately 20 by 60 m, along the north rim of Spur Crater. Of these, four are soils, 70 are breccias, 15 are crystalline rocks, and four are small glass fragments. Seventy-eight of these are small fragments collected in the rake sample, including some breccias and all the crystalline rocks. Two of

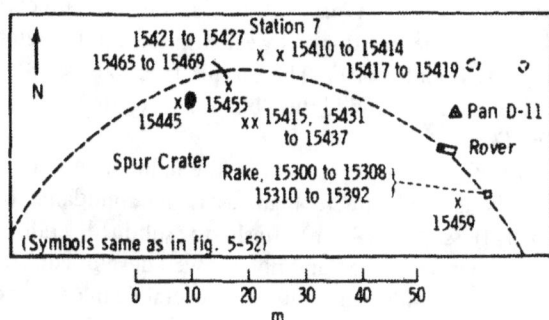

FIGURE 5-98.—Planimetric map of station 7.

the large breccia samples have mare-type basalt clasts, but most are characterized by chalky-white clasts and lack mare-type basalt clasts.

Samples 15410 to 15414 and 15417 to 15419: Soil samples 15410 to 15414 and breccia samples 15417 to 15419 were collected from the summit of the subdued rim crest of Spur Crater and approximately 18 m northeast of the 1.5-m-wide breccia block on the north-northwest rim of Spur Crater (figs. D-11 and 5-98). The local surface in the sample vicinity is moderately well strewn with rock fragments (figs. 5-99 and 5-100) up to several tens of centimeters across. The sample area is characterized, however, by the abundance of less than 1-cm-size fragments. Small craters less than 1 m are generally common and range from very fresh to moderately subdued. Lineaments are not present. Because the sample locality is just on the summit of the Spur Crater rim crest, no local surface slope exists.

It is probable that the diverse rock materials sampled at station 7 represent the major components of the Spur Crater ejecta from a maximum depth of approximately 20 m below the surface of the Hadley Delta front. A few small crystalline rocks collected in the rake sample are mare-type basalts and a very small proportion of the breccia samples contains mare-type basalt clasts, probably derived from the South Cluster impact area. Most or all of the sampled materials, including the anorthosite, 15415, have undergone at least two and possibly more episodes of intense shock, including pre-Imbrian (pre-Imbrium Basin ejecta deposits), Imbrian (Imbrium event), and Copernican (Spur Crater event) shocks.

Most of the rock fragments more than 10 cm across have moderate to well-developed fillets (possibly because of the nearly level surface in the near sample environment). The immediate sample locality

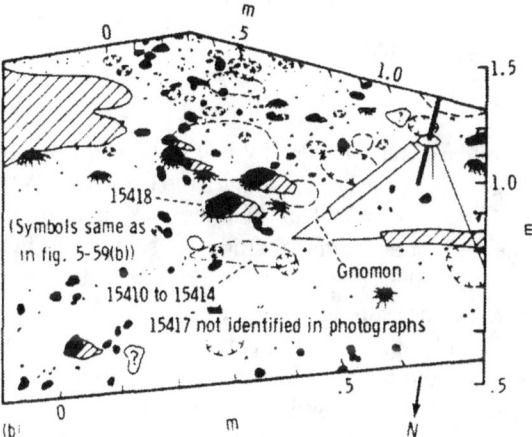

FIGURE 5-99.—Sample 15418 and vicinity of samples 15410 to 15414, 15417, and 15419. (a) Presampling, cross-Sun photograph looking south (AS15-86-11662). (b) Sample environment sketch map.

is of much lower albedo than that of the sample 15415 environment, although a few scattered high-albedo rock fragments or breccia "clasts" are seen in the area (AS15-90-12223). Most of the large fragments appear to have a significant dust cover, and many are buried to approximately one-fourth to one-half their height.

As shown in figure 5-99, sample 15418 is associated with several other rocks larger than 10 cm and with three moderately fresh to moderately subdued craters. The largest crater (0.3 to 0.4 m diameter) due south of the sample is subdued and may possibly be a

FIGURE 5-100.—Cross-Sun photograph, looking south, of areas from which samples 15410 to 15414, 15417, and 15418 were collected; sample 15419 before collecting (AS15-86-11664).

secondary crater formed by one or more of the three larger rocks (including sample 15418). The filleting and degree of burial of all three rocks indicate that they have been in this position for some time, but it is likely that they all have been recycled to the surface at least once, perhaps by the small crater previously mentioned. Sample 15419 is a small, glass-coated breccia fragment that was found in the hole left by removal of sample 15418. The soil sample taken just north of the site of 15418 probably was the source of the small (1.3 g) breccia sample 15417.

As stated, all sampled breccias appear to have been recycled to the surface (at least once) by small cratering events; however, they likely represent Spur Crater ejecta derived from as deep as 20 m. High-albedo soil was kicked up adjacent to this sampling site, indicating a possible abundance of light-clast breccia in the immediate vicinity. The crew also mentioned that the soil here "caked" on the surface (upper layer). (This phenomenon was responsible for their decision to take the soil sample.) Soil was also apparently caked on top of the 15418 breccia sample when collected.

Samples 15415, and 15431 to 15437: Sample 15415 (anorthosite) was collected at station 7 approximately 10 m east of the 1.5-m-wide breccia block on the north-northwest rim of Spur Crater (figs. 5-98 and 5-101). The sample location was 3 m east of sample 15455, the "black and white breccia" (fig. D-11).

The fragment population in the immediate sample vicinity is moderate to high, as is the abundance of small (less than 1 m) fresh to subdued craters. Lineations are sparse and none is well developed. The sample rests on a gentle slope of several degrees to the south toward the bottom of Spur Crater. Filleting on rock fragments in the immediate vicinity of the sample is moderate, and none of the rocks larger than 6 cm across is entirely free of fillets (fig. 5-101).

The area surrounding sample 15415 is generally of higher albedo than most of the Spur Crater northern rim; however, isolated patches of light material in the soil occur in other parts of the Spur Crater sampling area. Dust coating on fragments in the vicinity of sample 15415 appears to be greater than that at the nearby sampling site of sample 15445. The depth of burial of most fragments greater than 6 cm across is between one-fourth and one-half of the rock heights.

The size and shape of sample 15415 are not unique on the northern rim of Spur Crater, but the material (sample 15435, shock-lithified soil breccia or extremely weathered breccia) from which it was plucked is one of the largest "fragments" (approximately 17 cm across) in the immediate vicinity. The texture and albedo of sample 15415 were described by the crew as unique to the area, probably by virtue of the abundance of reflective plagioclase surfaces. However, numerous rock fragments with similar-appearing albedos are scattered around the north rim of Spur Crater in the sample area as seen in the down-Sun photographs (fig. 5-101). (This observation was also verified by the crew.) These fragments, including sample 15455, may also be of similar origin and composition as sample 15415, but apparently they have sustained more shock and do not have the crystalline plagioclase that 15415 contains. This difference may be the only major one between 15415 and the other light-clast fragments observed in the immediate vicinity of Spur Crater.

The east side of samples 15431 to 15437, from which sample 15415 was plucked as viewed in figure 5-101, does not appear to contain any observable high-albedo clasts other than sample 15415. Samples 15431 to 15437 are filleted (likely the result primarily of disintegration of the poorly indurated

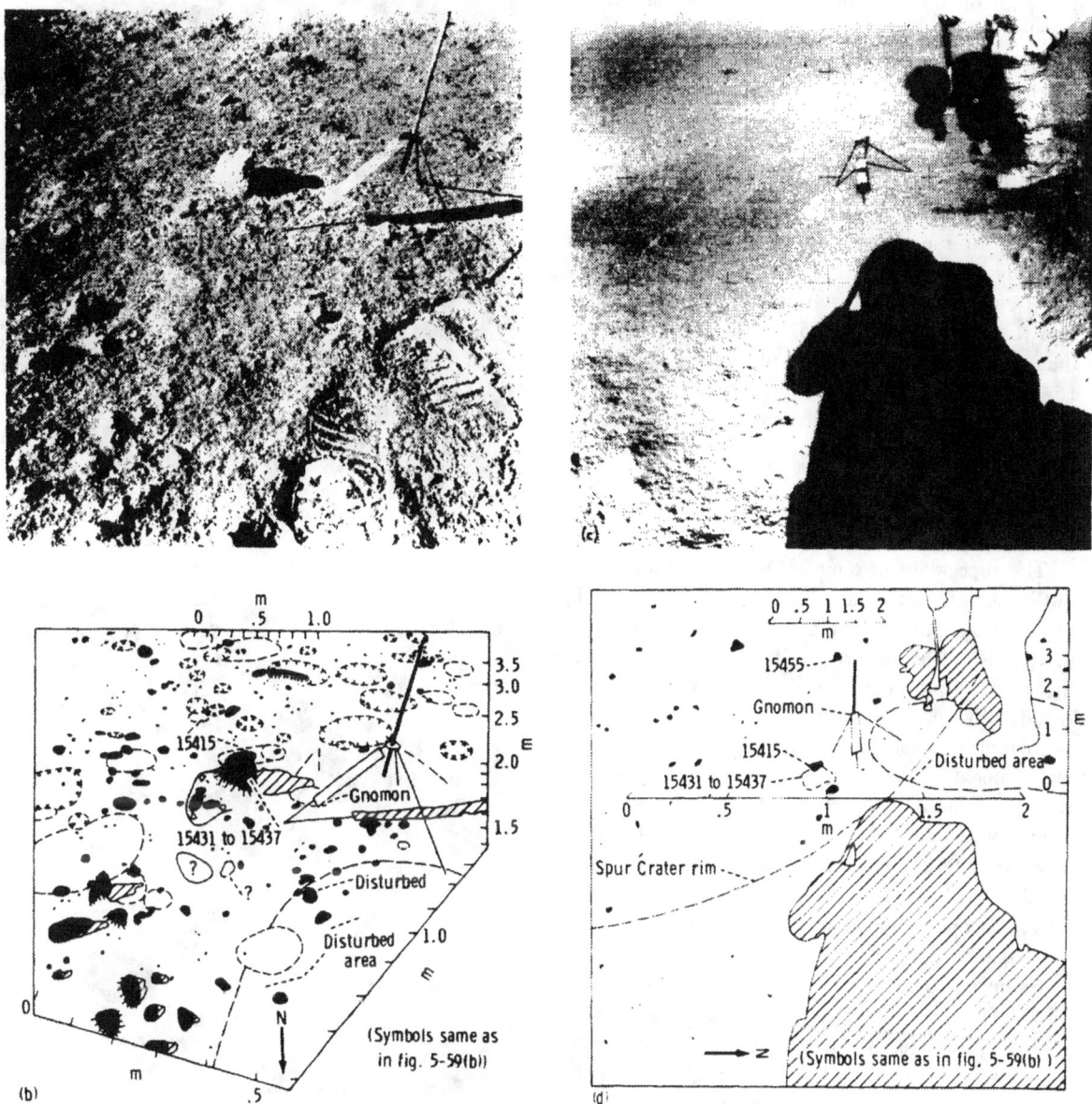

FIGURE 5-101.—Samples 15415 and 15435. (a) Presampling, cross-Sun photograph, looking south (AS15-86-11670). (b) Sample environment sketch map. (c) Presampling, down-Sun photograph, looking west (AS15-90-12227). (d) Sample environment sketch map of vicinity, showing distribution of only the light clasts.

breccia) and may be buried as much as one-fourth of their height. All evidence indicates that sample 15415 and the associated samples 15431 to 15437 have been in this position for a long time. No closely associated secondary crater exists from which the breccia may have been thrown.

In summary, sample 15415 is a very high albedo, plagioclase-rich clast, plucked from the lunar top of a much lower albedo, poorly consolidated breccia (of presently unknown genesis) on the northern rim crest of Spur Crater. The sample was described by the crew as exotic to the local area by virture of the shiny, high-albedo plagioclase content and the "come and sample me" position on top of sample 15435. Sample 15415 was described by the CDR as being easily lifted off with the tongs and therefore poorly attached to sample 15431 to 15437 (possibly weathered loose). He confirmed personally later that it was his opinion that the anorthosite was a part of samples 15431 to 15437.

It is difficult to state positively the depth from which samples 15415 and 15431 to 15437 were derived, but the abundance and distribution of similar light-clast breccia in the vicinity indicate the original source in this area to be Spur Crater ejecta from as deep as 20 m. The samples may, however, have been recycled to the surface by at least one post-Spur cratering event. Sample 15415 has undergone at least two fragmentation events (one before incorporation into the breccia (samples 15431 to 15437) and one that produced the fragment of breccia containing 15415), and at least one relithification event (which produced the breccia). This observation suggests a rather complex history, probably associated with cratering, which decreases the probability that 15415 is indigenous to the bedrock in the immediate vicinity of Spur Crater.

The appearance of sample 15415 as a "weathered out" clast from a friable breccia matrix makes the genetic association of the anorthositic clast with the matrix also rather uncertain. The clast may have been incorporated into a soil breccia by lithification of the regolith by the Spur Crater event or an event that postdated the Spur Crater event (but in the local environment), or it may have been earlier incorporated in a poorly indurated tectonic breccia ejected from Spur Crater and weathered to the present state since that time. A detailed comparison of samples 15431 to 15437 matrix with other more coherent breccia samples should clarify this uncertainty regarding origin.

Samples 15421 to 15427: Samples 15421 to 15424 (fines), 15425 and 15426 (rocks), and 15427 (chips)—"green rocks"—were collected on the crest of the north rim of Spur Crater approximately 15 m northeast of the 1.5-m-across breccia block on the crater north-northwest rim (fig. 5-98). The collection area is approximately 8 m north of the site of the anorthositic sample 15415. The rock samples are part of a small cluster of fragments that are as much as 25 cm across (fig. 5-102). Craters in the near-sample environment are sparse with only two subdued 35-cm-diameter craters in the mapped area. A small (7 to 8 cm) secondary crater, with an associated rock fragment in the bottom, is present in the near vicinity (fig. 5-102), but none is clearly related to the rock cluster sampled.

Lineaments are not present in the immediate area.

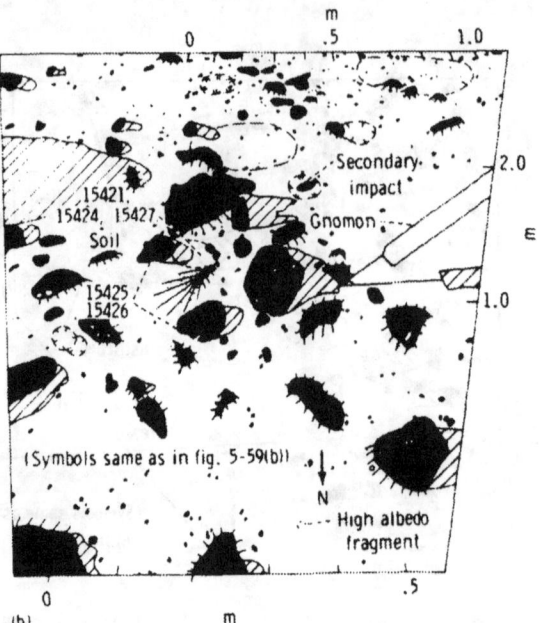

FIGURE 5-102.- Samples 15421 to 15427. (a) Presampling, cross-Sun photograph, looking south (AS15-86-11666). (b) Sample environment sketch map.

The slope in the sample environment is gently to the south (1° to 2°) toward the bottom of Spur Crater, but the regional surface slopes approximately 10° to the north down the Hadley Delta front. Fillets are not well developed on the rocks in the sample fragment cluster, but nearly all the fragments (except the two samples) are partially buried.

The general albedo of the rocks in the sample cluster appears to be typical of other fragments in the area. Abundant light clasts are visible (fig. 5-102). Dust coating on the samples is low to average, but moderate to high on the rest of the fragments forming the sample cluster.

Samples 15425 and 15426 were collected by the crew because these samples appeared to have a green tint similar to other materials they had seen previously on the Hadley Delta traverse at station 6A. The crew mentioned that the samples appeared to be part of a cluster of rocks that all seemed to be "roughly the same." The CDR suggested that possibly all the fragments in the cluster were part of a "big fragment, but it broke when it hit." This observation is favored by the general abundance of light-clast breccias in the cluster. As mentioned, the dust cover on the samples also appears to be less than that of the other fragments. The albedo of the samples appears to be average in the documentary photographs, but many rocks in the cluster have mappable light clasts. Samples 15425 and 15426 are similar in size and shape to the rest of the fragments in the rock cluster, but the apparent lack of extensive burial and dust cover makes them appear rather unique in the near-sample environment, and they may have been emplaced later than the remainder of the partially buried cluster fragments. The samples are, however, probably part of the Spur Crater ejecta from a depth of perhaps 20 m beneath the Hadley Delta front at this point. The samples are unique, however, by virtue of the soft, green coating and the general friable nature.

Breccia block (not directly sampled): The largest rock fragment on the Spur Crater rim is a 1.5- by 0.6-m, light-clast breccia block lying on the crater north-northwest rim crest (figs. D-11, 5-98, and 5-103). This block has many interesting features, and, although not directly sampled, a smaller rock fragment next to it was collected (sample 15445) and described as being likely of the same material as the large breccia block and probably derived from it.

Light clasts are abundant within the breccia block, and similar rocks are common on the nearby surface. The breccia block also has well-developed joints or fractures, and apparent folds, indicative of a complex history of intense shock and deformation. This particular block probably represents a larger version of most of the smaller breccia samples collected at station 7, and, because of the size, shows shock history information and general textural details over a much larger area than will be found on the returned samples.

The block is broken in several places along what appears to be a set of fractures with faces parallel to the long dimension of the block. A second, equally well developed, major fracture set is perpendicular to the other. Three or four additional minor sets are also present.

Well-developed folds appear to be present on the boulder. The folds are visibly enhanced on the block by what appears to be a differential weathering or erosion of the individual fold "layers" and the interface between them (fig. 5-103). The fold axis appears to be parallel to the fracture faces mapped on the block and is oriented parallel to the long dimension of the block.

Some areas on the base of the breccia block appear to have a vesicular texture and may be a glassy outer covering on the rock resulting from the latest shock event.

The block lies on the crater rim with no fillet buildup and with minimal burial. The southern end of the block is actually off the surface where the crater wall slopes toward the bottom of the crater (fig. 5-103(a)). The situation of the block on the top of the inner-crater-wall slope to the south permits most surface fines to migrate toward the center of the crater and inhibits buildup around the block itself. It is unlikely that the block has been moved from the present position since being excavated by the Spur Crater event. The condition of the Spur Crater boulder illustrates the extreme deformation that perhaps all the sampled light-clast breccias have undergone.

Sample 15445: Sample 15445 was located approximately 0.6 m west-southwest of the 1.5-m-across breccia block on the north-northwest rim crest of Spur Crater (figs. D-11 and 5-98). The local surface is notably free of large (greater than 6 cm) partially buried rocks. Most of the small rocks on the surface have little or no filleting (fig. 5-104). Many of the small rocks lie on small topographic highs, indicating the possibility that they may be recently derived from the breccia boulder by spalling. Most rock fragments surrounding sample 15445 are less than 1 cm across; very few are more than 2 to 3 cm across. The local surface slopes moderately to the south-southeast toward the floor of Spur Crater.

Lineaments are not present in the vicinity of

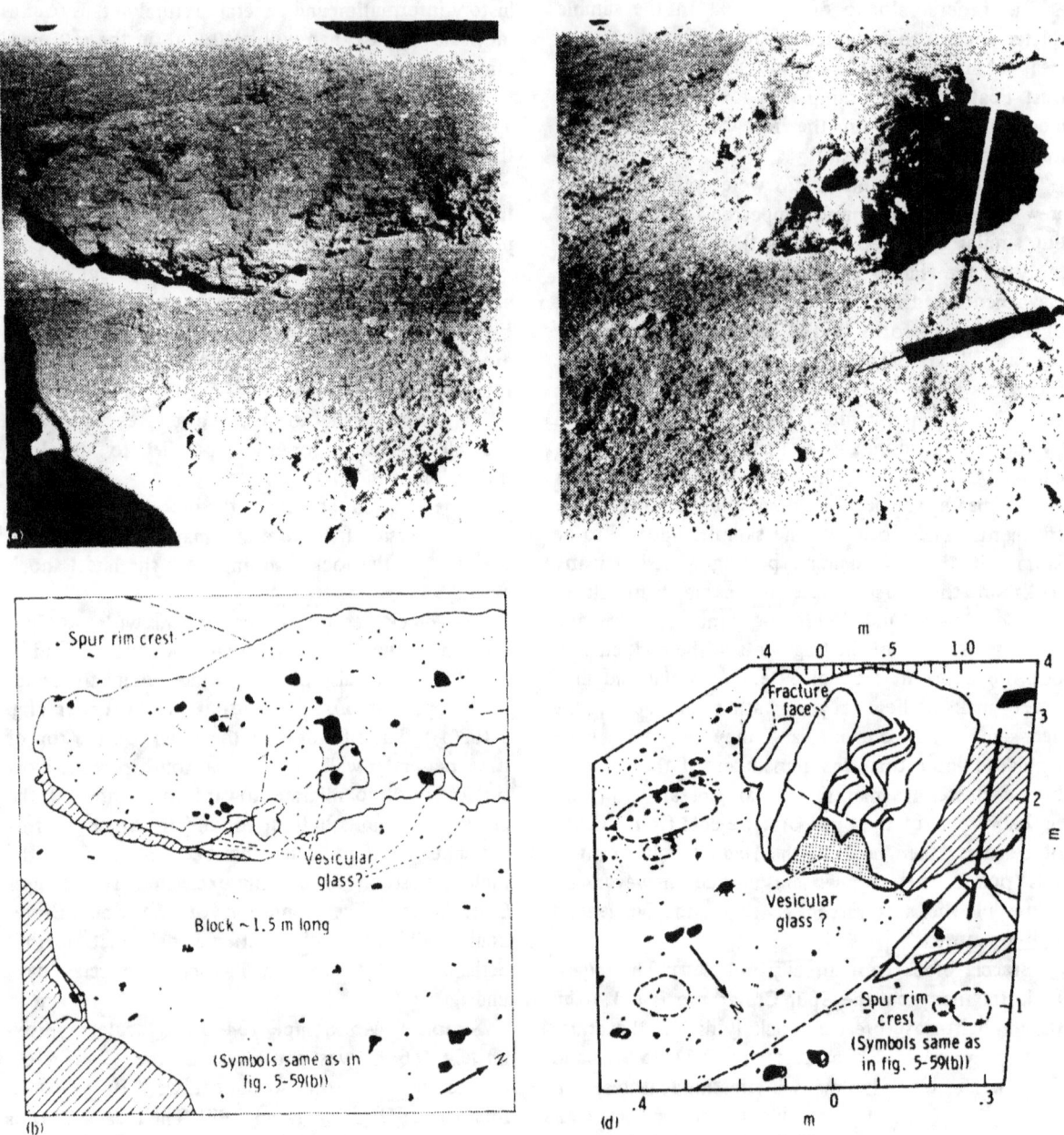

FIGURE 5-103.—Block on north-northwest rim of Spur Crater. (a) Down-Sun photograph of breccia block 1.5 m across (AS15-86-11689). (b) Sketch map of surface of breccia block, showing light clasts in and around block, fractures and glass coatings on block. (c) Cross-Sun photograph looking south, showing fold structures near center of north face of block (AS15-86-11682). (d) Sketch map of breccia block, showing traces of fold planes on surface near center of block.

sample 15445. A few scattered subdued craters (less than 0.3 m) were near sample 15445, but none is close enough to the sample to be associated with it as a secondary crater. The sample does not appear to be clearly representative of the rocks in the immediate environment of the 1.5-m-across breccia block, especially in shape and texture, but 15445 does resemble them in regard to the general lack of fillet buildup. Most larger (greater than 3 to 4 cm) rocks in the sample area are rough textured and irregular in shape, unlike sample 15445, which is equidimensional and relatively smooth but with angular corners.

The sample and the breccia block are probably representative of subregolithic material characteristic of the Hadley Delta front at the 60-m elevation (above the plains) where sampled. The position of the large breccia block (and possibly sample 15445) on the rim crest of Spur Crater indicates that this material may have been derived from the maximum sample depth (approximately 20 m) from within the crater.

Sample 15455: Sample 15455 was collected approximately 10 m east of the 1.5-m-across breccia block and 3 m west of the anorthositic sample (15415) on the north rim of Spur Crater on the Hadley Delta front (figs. D-11 and 5-98).

The surface in the immediate vicinity is not characterized by the cluster or grouping of rock debris that was typical of several other sample sites at station 7. Fewer rocks of all sizes are observed, and sample 15455 (approximately 10 cm across) is the largest fragment visible in presampling photography (fig. 5-105). No sharp, fresh, primary or secondary small craters are seen in the sample area. The surface in the sample vicinity slopes to the south several degrees toward the floor of Spur Crater.

Fillets on the rocks at the sample site are small to average in size, and some larger rocks, including sample 15455, lie on small topographic highs above the immediate surface level. Albedo of the local rocks and associated soil can be best observed on the down-Sun photograph (fig. 5-105(a)). The undisturbed surface next to the sample site appears to have an albedo typical of the area, but an area immediately to the north disturbed by the astronauts' boots contains higher albedo material, perhaps indicating the presence of abundant light-clast breccia material in the general vicinity underlying the upper few millimeters of surface soil (fig. 5-105). The presence of numerous rock fragments in the area of samples 15415 and 15455 with high-albedo clasts or large phenocrysts is illustrated in figure 5-101. This local area appears to be heavily strewn with such light-clast breccias, all of which may be related to a single ejecta stream from Spur Crater and perhaps from the same depth in the Hadley Delta front.

Dust covering on the local rocks surrounding sample 15455 is probably less than average, especially on the perched rocks. Burial is less than average in the immediate sample vicinity.

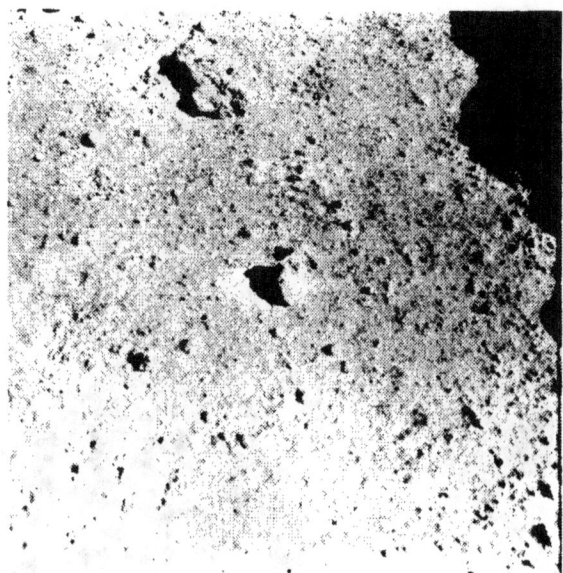

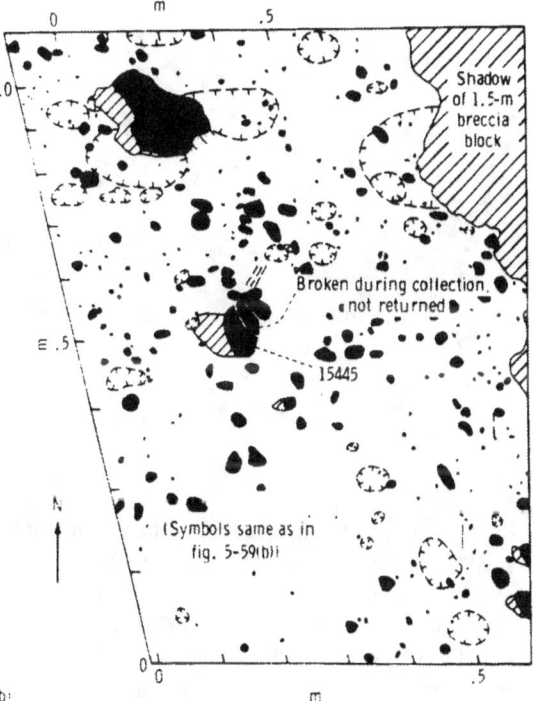

FIGURE 5-104.—Sample 15445. (a) Presampling, cross-Sun photograph, looking north (AS15-86-11691). (b) Sample environment sketch map.

Sample 15455 is closely associated with a subdued, shallow 1-m-diameter crater from which the sample reexcavated to the surface. The local slopes associated with sample 15455 are gently to the south

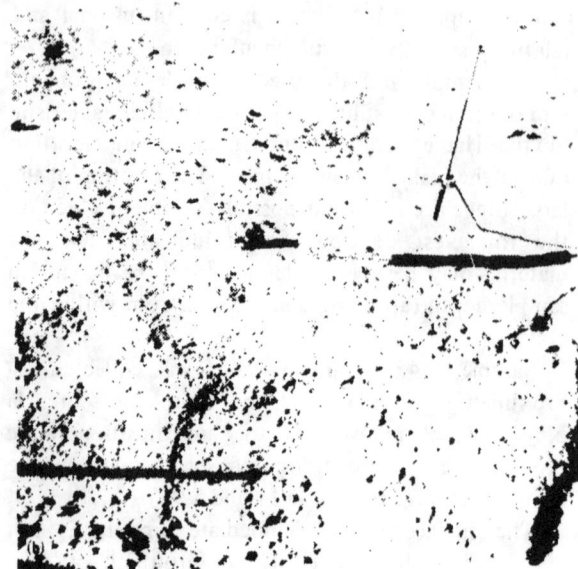
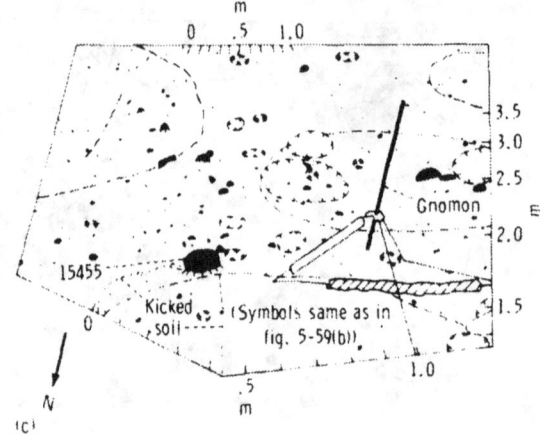

FIGURE 5-105.—Sample 15455 and vicinity. (a) Presampling, down-Sun photograph, looking west (AS15-90-12229). (b) Presampling, cross-Sun photograph, looking south (AS15-86-11675). (c) Sample environment sketch map.

on the south side and gently to the north (into the small 1-m-diameter crater) on the north side of the sample. The fact that the sample is lying exactly on the rim crest of the small recent crater where fines should be preferentially moved into the crater by small impacts may account for the lack of burial.

The sample is perhaps slightly larger than other rocks in the immediate vicinity, but it is certainly not unusual in respect to the area encompassing samples 15415, 15455, and the 1.5-m-across breccia block. The texture and structure of sample 15455 are not particularly unique, but the down-Sun photograph (fig. 5-105(a)) does reveal the presence of larger-than-average high-albedo clasts on the northeast side.

The sample is obviously one fragment of many, in the general area of 100 m², that are characterized by the presence of high-albedo clasts. The clasts are associated with high-albedo soils exposed by cratering on the surface or from below the upper few millimeters of the present surface. All these fragments may have been excavated from approximately 20 m below the original Spur Crater surface and possibly represent either layered deposits or debris from pre-Imbrian or Imbrium (Fra Mauro) ejecta deposits comprising the Hadley Delta front.

Sample 15459: Sample 15459 was collected on the northeast inner wall of Spur Crater (above the steepest part of the wall slope) approximately 6 m southwest of the rim crest and 50 m east-southeast of the 1.5-m-across block on the north-northwest rim of the crater (figs. D-11 and 5-98). The rock fragments in the immediate vicinity of the sample form a group and range in size from a few centimeters to approximately 1.0 by 0.5 m (fig. 5-106). The rake-sample area is 3 m northeast of the 15459 site. Two main types of rocks are distinguishable by shape and texture in the immediate sampling area. The first, characteristic of rock sample 15459, as well as the large 1.0- by 0.5-m block adjacent to it, has at least three or four directions of jointing or parting giving the rocks a distinct rectangular, "slabby" appearance. These jointed rocks are all subrounded, at least on the exposed surfaces. The only diverse rock visible in the photographs is a 20- to 30-cm-long, very rough, possibly vesicular fragment lying just to the east of the large block (fig. 5-106). This rough rock may be an exotic thrown up to this position by a local impact in the mare plains to the north; however, it is, more likely, a vesiculated glass and breccia fragment aggregate similar to samples 15465 and 15466.

The crater population in the near-sample environment is unusually low with a definite lack of small fresh craters. This situation is probably a result of the location of the sample area over the inner wall of Spur Crater, with downslope movement quickly destroying craters in this size range.

A few well-developed lineaments are present in the near field of figure 5-106(a). The presence of even these few lineaments is considered unusually high at the Apollo 15 site when compared with all other sample-documentation areas studied and may be related to fracturing under the regolith at the rim crest of Spur Crater. Fillets are poorly developed on almost all rocks in the sample area. Most of the fragments are buried from one-third to one-half the height. No postsampling photograph is available to verify the exact burial depth.

The albedo of the rocks and soil surrounding the group of rocks appears average to slightly below average for the Spur Crater rim sample sites. No light-clast material is visible, and, where the soil has been disturbed during sampling activities, no higher-than-average albedo soil is observed.

Sample 15459 was taken from the center of a 30- by 30- by 15-cm rock that had two well-developed internal fractures parallel to the larger external surfaces. The CDR said that it looked like "layering in it." The sample was broken loose along these fractures with the scoop.

Both the sampled rock and the large fractured

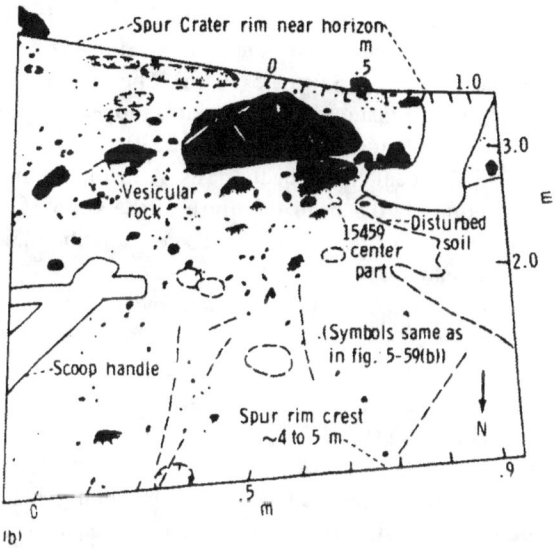

FIGURE 5-106.– Sample 15459. (a) Presampling, cross-Sun photograph, looking south (AS15-90-12235). (b) Sample environment sketch map.

block immediately adjacent to the southeast are buried from one-third to one-half their height and apparently have been in these relative positions for quite some time, probably since the formation of Spur Crater. The external form and internal fracture patterns are roughly parallel, and they may be parts of an original single block. Dust cover on the sampled block appears minimal.

The jointed or fractured sample block and the

larger adjacent block are probably the same material and have undergone similar histories. They probably represent highly fractured Apennine Mountain material that has been excavated by the Spur Crater event from a depth approaching 20 m and have not been moved significantly or buried since that time.

Samples 15465 to 15466: Samples 15465 and 15466 were taken about 3 m east-northeast of the 1.5-m-across breccia block on the north-northwest rim crest of Spur Crater (figs. D-11 and 5-98). The sample location is just inside the Spur Crater rim crest on a surface sloping several degrees to the south-southeast toward the crater floor.

The surface in the immediate vicinity of the sample is rather blocky with fragment sizes as large as 12 cm across (fig. 5-107). The rock types are unusually diverse for such a limited area and include (1) equidimensional subangular fragments, (2) rough macrovesicular-appearing fragments, and (3) foliated ropy-to-rough fragments. The largest fragments in the near-sample environment are grouped into a 1-m² area as illustrated in figure 5-107(b). The rocks in the small group characterizing the area appear in the down-Sun photograph (AS15-90-12230) to contain rather abundant high-albedo clasts. A disturbed area approximately 0.8 m to the north of the sample site (over the Spur Crater rim summit) contains abundant high-albedo soil (fig. 5-105).

Small craters, less than 1 m in diameter, are of low to average abundance. Lineaments are sparse, but two rather distinct linear troughs are present in a 12-cm-diameter crater approximately 20 cm east of the sample site (fig. 5-107).

All the larger rock fragments have moderately well-developed fillets and appear to be about equally buried to one-fourth to one-third their height. Dust cover on the rocks is difficult to evaluate from the available photography.

Samples 15300 to 15308 and 15310 to 15392: Samples 15300 to 15308 are scooped soil (including rock fragments 15305 to 15308). Samples 15310 to 15314 are soil collected with comprehensive rake fragments 15315 to 15392. The samples were collected on the northeast rim of Spur Crater (figs. 5-2 and 5-98).

The raked area was slightly disturbed in the eastern part by footprints before raking (fig. 5-108), but this perturbation should not degrade the quality of the samples. The soil samples were taken from approximately the center of the raked area.

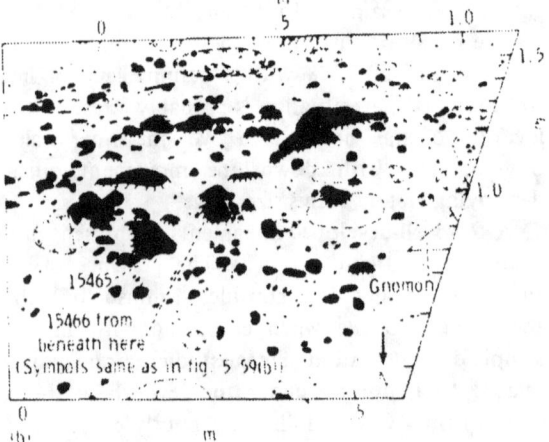

FIGURE 5-107.— Samples 15465 and 15466. (a) Presampling, cross-Sun photograph, looking south (AS15-86-11678). (b) Sample environment sketch map.

As previously described, small fragments are moderately abundant on and near the rim crest of Spur Crater. Most of the fragments collected are breccias and soil breccias, which is consistent with the interpretation that most of the material in the vicinity is breccia that was originally derived from the event that formed the Imbrium Basin, some of which has been recycled by later, local cratering events such as the Spur Crater event.

Station 8-ALSEP site. — The station 8-ALSEP site is located approximately 125 m northwest of the LM on ray-mantled mare surface like that at the LM site (fig. 5-2). The surface is generally flat and smooth,

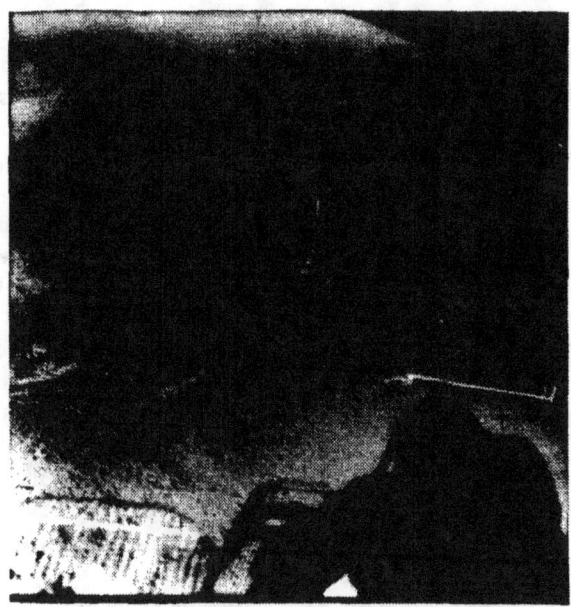

FIGURE 5-108.—Presampling, down-Sun photograph, looking west, of vicinity of scooped samples 15300 to 15308 and of raked samples 15310 to 15392 collected at station 7 (AS15-90-12232).

but it is hummocky in detail. Large craters in the area are subdued; small, fresh craters also appear smooth owing to a lack of clods on their rims. In places, the surface has a "raindrop" appearance resulting from scattered subdued craters less than 1 cm across. The surface material is fine grained with cobble-size fragments covering less than 1 percent of the surface. No clusters or linear arrangements of the larger rocks are visible on the photographs (figs. D-12 and D-16).

Five samples were collected from an area approximately 25 by 50 m near the ALSEP (fig. 5-52). Of these samples, three are soils taken from a trench (fig. 5-109), one is a mare-type basalt (fig. 5-110), and one is a breccia that does not contain mare-type basalt clasts (fig. 5-111). The trench was dug and sampled to determine the shallow stratigraphy of the regolith. The other two samples were collected on the basis of size and distinctive individual features.

The breccia fragment (sample 15059) closely resembles sample 15015 and, like the LM site breccias, may be part of the ray (fig. 5-1) or it may have been ejected from impacts on the Front or North Complex. The basaltic fragment is a mare-type basalt, but the fact that it is the only large basaltic fragment collected from the LM-ALSEP area suggests that it is not indigenous to the immediate area.

FIGURE 5-109.—Deep trench from which samples 15013, 15030 to 15034, and 15040 to 15044 were taken. (a) Presampling, cross-Sun photograph, looking north, of approximate area of trench (AS15-92-12417). (b) Cross-Sun photograph, looking south, of trench (AS15-92-12439).

Samples 15013, 15030 to 15034, and 15040 to 15044: The samples are soils collected from the trench (fig. 5-109). Samples 15013 and 15030 to 15034 were collected from the bottom of the trench (approximately 30 cm below the surface), but they may be contaminated by soil caved from the sides of

FIGURE 5-110.—Presampling, cross-Sun photograph, looking south, of sample 15058, collected near the ALSEP (AS15-92-12410).

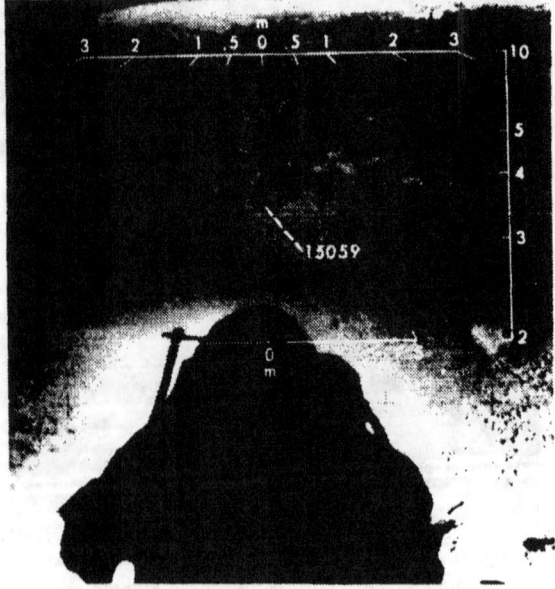

FIGURE 5-111.—Presampling, down-Sun photograph, looking west, of sample 15059, collected near the ALSEP (AS15-92-12413).

the trench. Samples 15040 to 15044 were taken from the top of the trench.

Sample 15058: This sample is a basaltic fragment collected approximately 30 m east-northeast of the ALSEP central station (fig. 5-52). The fragment (fig. 5-110), angular and rough surfaced, had no visible dust coating when collected. The rock projected above the surface in a manner suggesting that it was less than one-fourth buried and no fillets were banked against it. No other rocks of comparable size occur in the area.

Sample 15059: This sample is a glass-covered breccia collected approximately 15 m south of the ALSEP central station (fig. 5-52). The rock is angular and rough surfaced with a low albedo. The rock, which appears to be dust covered, had no definite fillet and was lying on the surface (fig. 5-111). No other fragments of comparable size occurred in the vicinity.

Station 9.—Station 9 is located on the mare surface approximately 1400 m west of the LM and 300 m east of Hadley Rille (figs. 5-2 and 5-112). The site lies just outside the mapped limits of a diffuse ray. The surrounding area is broadly flat and smooth but hummocky in detail. The sample station is located at a 15-m-diameter crater of very fresh appearance and is surrounded by abundant friable clods and glass. The surface material near the rim of the crater was observed to be noticeably soft. Because the crater does not penetrate what appears to be true bedrock, the regolith is greater than 3 m thick at station 9, as opposed to the very thin regolith at nearby station 9A.

Four samples were collected at two points approximately 15 m apart. These samples were collected as two rocks (sample 15507 and one that broke in transit into samples 15505, 15506, and 15508) and two clods (also broken in transit: samples 15510 to 15515). Samples 15505 to 15508 were collected from the ejecta blanket approximately 15 m from the crater rim, and the others were collected from the rim.

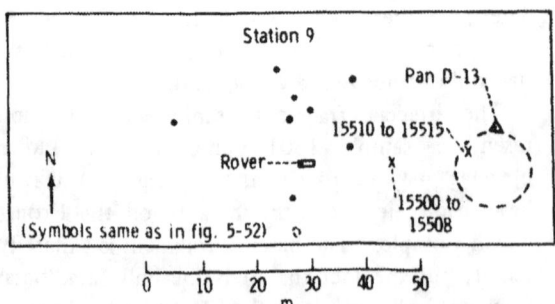

FIGURE 5-112.—Planimetric map of station 9.

The ejecta blanket of this crater probably represents the freshest surface of any significant extent that has been sampled by the Apollo missions. This exposure, however, is probably not the first exposure to the lunar surface of many of the particles in the samples. Glassy samples 15505 to 15508, possibly derived from the coherent dark fragments below the bench, lack clasts of mare-type basalt, whereas the clod samples contain clasts of mare-type basalt. This observation indicates very considerable additions of nonmare material to the mare surface.

Samples 15505 to 15508: Samples 15505 and 15507 (as well as samples 15506 and 15508, which probably broke off 15505) were collected about halfway from the rim of the 15-m-diameter crater on the ejecta blanket that is approximately 25 m across (figs. D-13 and 5-112). The surface is saturated with raindrop depressions and fragments less than 1 cm in diameter; cobble-size fragments and craters are sparse (fig. 5-113). The samples are probably part of the ejecta blanket and not related to smaller features in the field of view. Sample 15505 was less than one-half buried, other nearby fragments are mostly buried to slightly buried, and only a few have fillets. The nearby fragments are rounded to subangular (fig. 5-113(a)). Albedo differences on local fragments are not apparent; however, considering the crater and the ejecta blanket as a whole, two types of fragments are present—fragments of easily eroded material and darker (glassy?) hard fragments. The local slope is gentle.

Sample 15505 is among the largest 2 percent of fragments on the ejecta blanket at this distance from the rim. It is also more angular than most samples and is less buried than most. It was apparently free from dust and without a fillet. The sample probably is part of the ejecta blanket but may have been one of the last fragments to come to rest. The small crater just up-Sun from the sample (fig. 5-113) may have been made by the sample.

Sample 15507 is a hollow, dark, glass ball and may be representative of the hard dark material that occurs abundantly in the center of the crater but in lesser amount in the ejecta blanket.

Samples 15510 to 15515: Samples 15510 to 15515 were collected from the northwest rim crest of the 15-m-diameter cloddy crater at station 9 (fig. 5-112). The two clods shown in figures D-13 and 5-114 disintegrated to form the parts that are numbered 15510 to 15515. The crater itself is

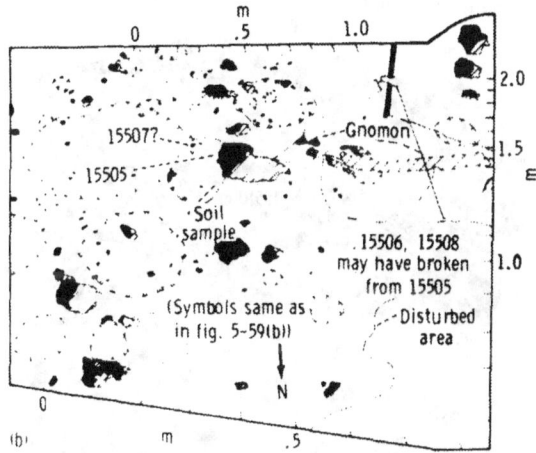

FIGURE 5-113.—Sample 15505, 15506, 15508, and vicinity of unidentified sample 15507. (a) Presampling cross-Sun photograph (AS15-82-11105). (b) Sample environment sketch map.

approximately 2 m deep with a distinct bench approximately 1 m or slightly more above the crater floor (fig. 5-47). Below the bench, the crater walls and floor are covered by a jumble of dark blocks with no visible fine-grained debris matrix. Fragments are abundant on the rim crest and on the walls of the crater. On the rim, where these samples were collected, most of the fragments are easily eroded breccias similar to samples 15510 to 15515, mixed with approximately 10 percent of hard glassy fragments resembling samples 15505 to 15508. Closely spaced but poorly-developed north-trending linea-

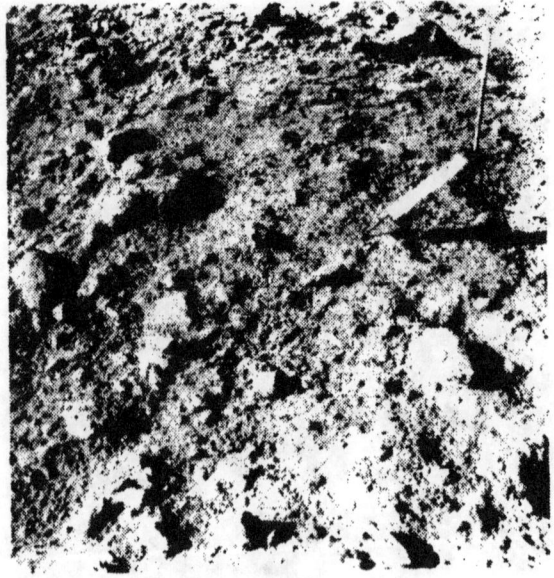

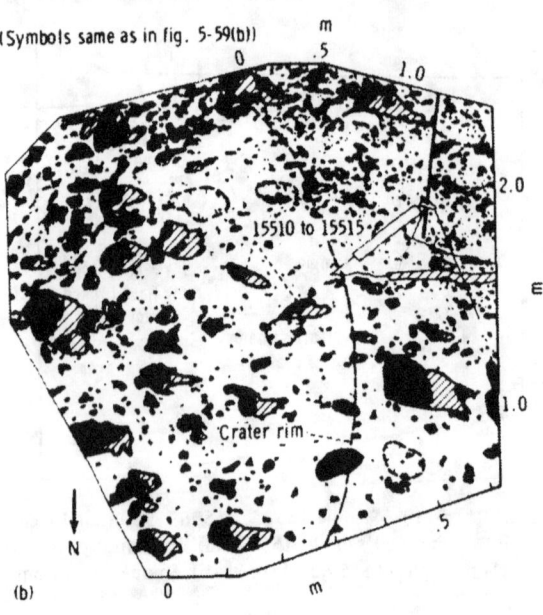

FIGURE 5-114.—Clods that were collected and broke in transit to form samples 15510 to 15515. (a) Presampling, cross-Sun photograph, looking south (AS15-82-11098). (b) Sample environment sketch map.

ments are visible. The local slope is a few degrees southeast into the crater. Fillets are common next to the easily eroded breccias, but these fillets are small and merge with the fine-grained matrix of the ejecta and were not mapped (fig. 5-114). Most of the glassy fragments and some breccia fragments appear to lie on the ejecta blanket. Only those fragments that are mostly buried appear dusty on top. Some of the dust may be derived by erosion of the fragment itself; however, more may have been deposited along with the partially buried fragments. The clod samples are definitely more rounded and lighter in color than the dark blocky material and lack the abundant vesicles. They lie upon the ejecta blanket and probably fell back upon the rim after most of the blanket was deposited. Lacking fillets and appearing dust free, the samples likely originated from a depth just less than the depth of the bench in the crater.

Station 9A.—Station 9A is located at the edge of Hadley Rille approximately 1800 m west of the LM (figs. D-14, 5-2, and 5-115). The surface generally slopes toward the rille rim from the station site. The only visible fresh crater in the vicinity of station 9A that is larger than 2 m in diameter is the 3-m-diameter crater where samples 15535 and 15536 were collected. The undulating surface is probably largely a result of old subdued craters. Small surface lineaments are essentially absent. Many of the rock fragments have well-developed fillets banked against their sides, and in general the more rounded fragments have the best-developed fillets (fig. D-14). Slightly protruding rock fragments with well-developed fillets are commonly dust covered. The general absence of fresh craters, the rounding of rock fragments, and the well-developed fillets suggest that the sampled surface is mature.

Rock fragments are more abundant within approximately 300 m of the rille rim than they are on the mare surface to the east. Most of the fragments at a distance of 200 to 300 m are a few centimeters across. The size of the fragments increases markedly toward the rille, and near the rim are numerous boulders and bedrock protuberances from 1 to 3 m across (fig. D-14).

As described in the subsection entitled "Hadley Rille," the regolith thins in the vicinity of the rille and, within approximately 25 m of the rim, the regolith is essentially absent. The abundance of rocks in the 200- to 300-m-wide zone along the rille is related to the nearness of outcrop to the surface in the vicinity of the rim.

One hundred and three samples were collected from an area approximately 30 by 75 m at station 9A. Of these samples, two are core samples, six are breccias, 92 are mare-type basalts, and three are soils. The breccias contain mare-type basalt fragments. Samples were collected from outcrops at the edge of

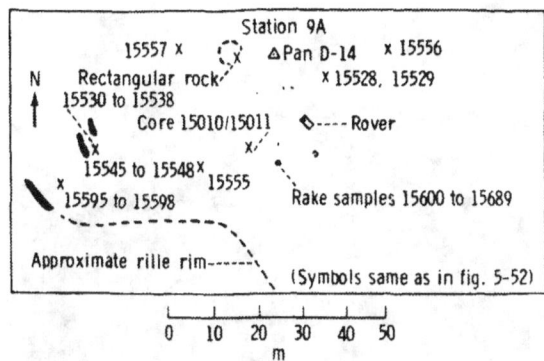

FIGURE 5-115.—Planimetric map of station 9A.

the rille and from loose debris a little above outcrop level. Other loose samples were collected from the surface still farther above outcrop level (fig. 5-115). Eighty-six small samples were collected in the rake sample, also taken from above outcrop level.

Because of the near-surface source for bedrock fragments adjacent to the rille rim, it is concluded that most of the samples collected are probably representative of the local bedrock in the vicinity of station 9A. Samples 15535, 15536, 15595, and 15596 are almost certainly representative of bedrock at the spots where collected. Samples 15535 and 15536 were collected from the ejecta of the fresh crater that penetrates the thin regolith cover. Samples 15595 and 15596 were collected from a large rock in the group of rocks exposed along the rille rim and were identified by the crew as samples of bedrock from the rim. Samples 15529 and 15556, because they lay on the surface, do not appear to have been in place for a long time, and samples 15555 and 15557 appear to have been in place for somewhat longer periods.

Samples 15010/15011: Samples 15010/15011 are from the double core tube taken approximately 20 m north of the rim of Hadley Rille (figs. 5-1 and 5-115). The surface in the vicinity of the core is generally level and no fresh crater is apparent in the immediate vicinity of the sample site. Fragments as much as approximately 20 cm across are common in the area, and boulders greater than 1 m across are sparsely scattered. Small subdued craters ranging from raindrop-size depressions to 20 cm across are fairly common. Small fillets are banked against many of the rock fragments in the sample vicinity (fig. 5-116).

The surface in the immediate vicinity of the core appears to have been undisturbed by footprints or

FIGURE 5-116.—Presampling, oblique-to-Sun photograph, looking southwest, of vicinity of samples 15010 and 15011 (double core) (AS15-82-11159).

Rover tracks. Therefore, the uppermost part of the core, except for the disturbance caused by driving the core and subsequent handling, should be representative of the undisturbed lunar surface.

Samples 15528 and 15529: Samples 15528 and 15529 were collected approximately 60 m northeast of the rim of Hadley Rille (figs. 5-1 and 5-115). The general surface description for sample 15529 applies also to sample 15528, which has not been identified on the photographs.

The surface is fairly level in the vicinity of the sample site (fig. 5-117). Sample 15529 was collected from an area where rock fragments are fairly abundant in the size range from 20 cm down to the limit of resolution of the photograph. Fragments are more abundant in this area than at the LM site, but less abundant than nearer the rille rim. No fresh craters are in the immediate vicinity of the sample; a few subdued craters a few centimeters to 0.5 m in diameter are present. Raindrop-size depressions are abundant. No lineaments are visible in the photographs (fig. 5-117(a)).

Fillets are well developed on all the rock fragments in the near vicinity of sample 15529. A fillet approximately 3 cm high was banked against sample 15529. Dust coatings are common on many of the

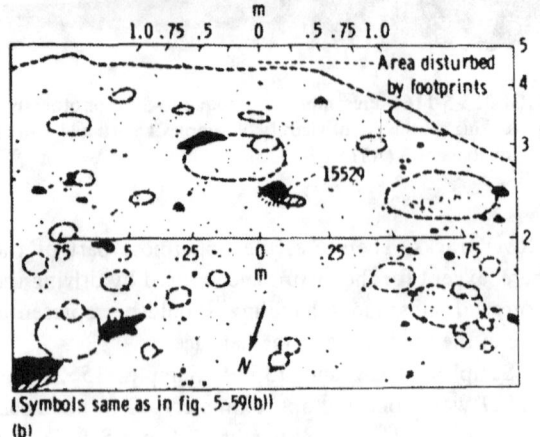

FIGURE 5-117.—Sample 15529 and vicinity. (a) Presampling, oblique-to-Sun photograph, looking southeast (AS15-82-11129). (b) Sketch map; rock fragments and clods approximately 2 cm across are shown.

FIGURE 5-118.—Presampling, cross-Sun photograph, looking south, of samples 15530 to 15538, 15545 to 15548, and vicinity. Samples 15537, 15538, and 15545 to 15548 are not recognized in photographs (AS15-82-11140).

fragments in the area; no dust coating is visible, however, on sample 15529 (fig. 5-117(a)).

All the rock fragments in the vicinity of the sample are well rounded. Sample 15529 is more spherical than most of the other nearby fragments, but it is rounded by approximately the same amount.

Samples 15530 to 15538 and 15540 to 15548: Samples 15535 and 15536 were collected approximately 20 m east of the rim of Hadley Rille, from the north rim of a moderately fresh, blocky, 3-m-diameter crater (figs. 5-115, 5-118, and 5-119). The two samples were chipped from a 0.75-m-diameter boulder. In addition, rock and soil samples 15530 to 15534, 15537, 15538, and 15540 to 15548 were collected in the vicinity; the general surface description also applies to these samples though they have not been identified on the photographs.

Boulders as large as 1 m across are abundant on the rim of the 3-m-diameter crater and are moderately abundant in the general vicinity. Smaller fragments, down to the limit of resolution of the photographs, are also abundant. No other fresh craters are visible on the photographs in the immediate vicinity of the sample site. A linear trough approximately 5 cm deep, 10 cm wide, and 1 m long that trends approximately west is adjacent to the sampled rock (fig. 5-118). No other linear features are apparent.

Fillets are generally poorly developed or absent in the sample site. The rocks appear to be free from dust coatings except where covered by dust kicked by the astronauts.

The rock from which samples 15535 and 15536 were chipped appears to be representative of the other rocks in the area. Most of the rocks are subangular with a few planar faces that may have been joint surfaces in the bedrock from which the rocks were derived. Most of the rocks were probably derived from within the 3-m-diameter crater. The angularity of the rocks, the lack of fillet develop-

FIGURE 5-119.—Photograph, looking southwest, of rocks from which samples 15535, 15536, 15595, and 15596 were chipped. Rock A is the same as that identified in figure 5-29 (AS15-82-11126).

ment, and the blocky nature of the crater rim suggest that the crater was formed rather recently. This apparently young ejecta surface has probably not had many fragments from foreign surfaces deposited upon it. This observation, plus the locally thin regolith, suggests that the rocks including samples 15535 and 15536 are probably representative of local bedrock.

Sample 15555: Sample 15555 was collected approximately 12 m north of the rim of Hadley Rille (figs. D-14, 5-1, and 5-115). Rock fragments are more abundant in the area of the sample site than in the LM site, but less abundant than nearer the rille rim. The fragments range in size from 35 cm down to the limit of resolution of the photographs. Small fragments or clumps of fines less than a centimeter in diameter are abundant (fig. 5-120). No fresh craters are in the immediate vicinity of the sample; a few subdued craters a few centimeters in diameter are present. Raindrop-size depressions are abundant. A few poorly developed lineaments several centimeters long are present.

Fillets appear to be fairly well developed on all the rock fragments in the near vicinity of sample 15555. However, the steep inclination of the camera axis and the lack of stereocoverage make it impossible to define the extent and height of the fillets specifically.

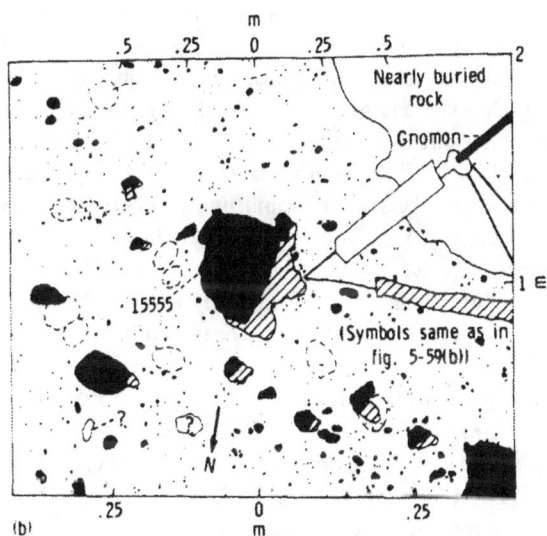

FIGURE 5-120.—Sample 15555 and vicinity. (a) Presampling, cross-Sun photograph, looking south (AS15-82-11164). (b) Sketch map; rock fragments and clods approximately 1 cm across are shown.

Dust coatings appear to be present on all the fragments in the photograph except sample 15555 (fig. 5-120(a)), probably because the sample is larger and extends farther above the surface than the other fragments in the vicinity.

A large, very subdued, dust-coated, and nearly buried rock appears present in the lower left corner of figure 5-120(a). A few very small rock fragments lie on the surface of the buried rock, but these fragments are much less abundant than on the normal surface of fine-grained material. Zap pits are common on the surface of this partially buried rock. All the rock fragments in the vicinity of the sample are well rounded. The surface texture of most of the fragments is somewhat rough, probably as a result of zap pits.

Sample 15555, in relationship to the surface, appears in the photographs to be representative of the other fragments in the immediate vicinity; however, it is much larger than any other feature except the partly buried rock. Fragments as large and larger than sample 15555 are common in the general area, however (fig. D-14).

Sample 15556: Sample 15556 was collected approximately 60 m northeast of the rim of Hadley Rille. Rock fragments as large as 20 cm across are fairly common in the sample site, and small fragments or clumps of fines less than a centimeter across are abundant. One small fresh crater approximately 20 cm in diameter occurs approximately 1 m east of the sample. This crater is surrounded by ejected clumps of fines, but the crater is probably not related to the sample. No other fresh craters are present in the vicinity of the sample. Raindrop-size depressions are abundant. No lineaments are apparent in the photographs (fig. 5-121).

Fillets as high as 3 cm are fairly well developed on all the rock fragments in the near vicinity of sample 15556. Dust coatings appear to be present on all the fragments in the photo except for sample 15556.

Sample 15557: Sample 15557 was collected approximately 40 m north of the rim of Hadley Rille (figs. 5-1 and 5-115). Rock fragments greater than 2 cm across are generally sparse within a few meters of the sample. Several subdued craters less than 0.5 m in diameter are present in the vicinity of the sample. One crater approximately 0.25 m in diameter with cloddy ejecta is approximately 0.5 m to the southwest of the sample. The freshness of this crater suggests that it is not related to the sample. No small lineaments are present (fig. 5-122).

Sample 15557 is well rounded and has a well-developed fillet as does the 3-cm-wide fragment just south of the sample. The 5-cm-wide fragment in the lower right-hand corner of figure 5-122 is angular and does not have a well-developed fillet. This observation supports the contention that well-developed fillets are generally associated with the well-rounded rocks.

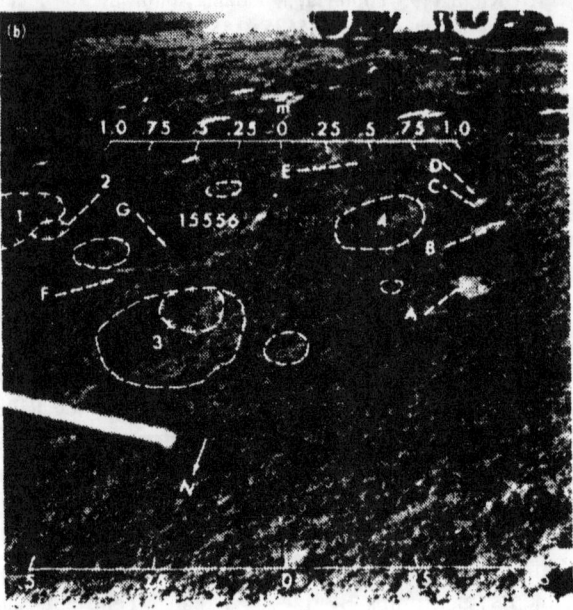

FIGURE 5-121.—Sample 15556 and vicinity. (a) Presampling, down-Sun photograph, looking west (AS15-82-11133). (b) Presampling, oblique-to-Sun photograph, looking southeast (AS15-82-11135).

Sample 15557 is the largest rock fragment in the immediate vicinity; because of the size, it is difficult to compare this sample with other nearby rock fragments. It is similar in appearance to other well-rounded and filleted rocks around station 9A and probably had the best-developed fillet of any of the documented samples collected at this station,

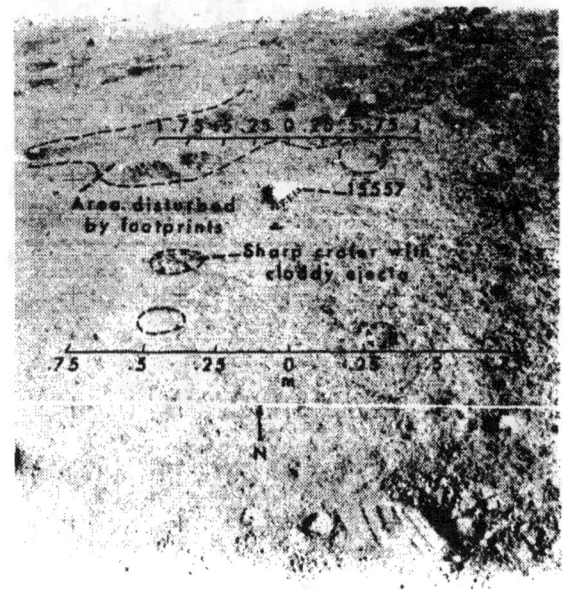

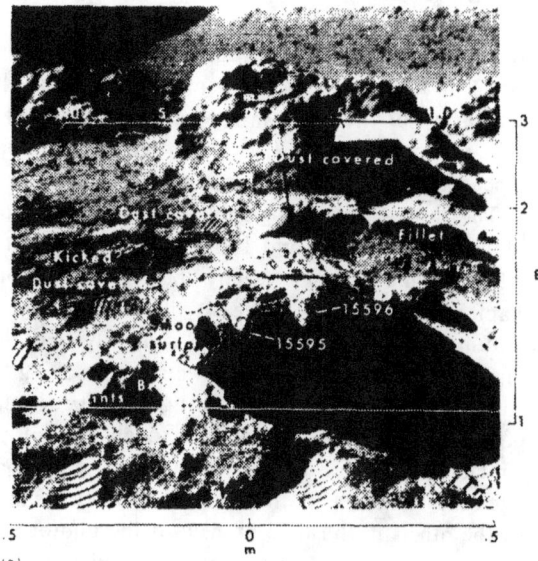

FIGURE 5-122.—Presampling, cross-Sun photograph, looking north, of sample 15557 and vicinity (AS15-82-11137).

which suggests that it may have been in place longer than the other samples.

Samples 15595 to 15598: Samples 15595 and 15596 were collected approximately 8 m east of the rim of Hadley Rille. These two samples were chipped from the surface of a large rock (fig. 5-123) identified by the crew as bedrock exposed along the rim of Hadley Rille. The rock from which the samples were chipped is one of hundreds of such rocks exposed along both sides of the rille. In addition, samples 15597 and 15598 were collected at this site, and the general surface description applies to these samples, although they have not been identified on the photographs.

The ground slopes gently from the sample site toward the rim of Hadley Rille. Rock fragments on the surface become increasingly abundant as the rille rim is approached; large boulders or exposures of bedrock as much as several meters across are abundant along the rille rim (fig. 5-29). The rock from which samples 15595 and 15596 were chipped is one of these large exposures (figs. 5-29 and 5-119). Smaller fragments are also abundant from small boulder size down to the limit of resolution of the photographs.

Most of the rock fragments have well-developed fillets banked against the sides, except beneath overhanging parts of the rocks. Fillets as high as 10

FIGURE 5-123.—Views of rock from which samples 15595 and 15596 were chipped and vicinity. (a) Presampling, oblique-to-Sun photograph, looking southeast (AS15-82-11143); samples 15597 and 15598 cannot be identified in this view. (b) Presampling, down-Sun photograph, looking west (AS15-82-11142).

cm are common on the large exposed rocks. Many of the fragments are partially coated with dust, especially those the tops of which are near the surface and those that have gentle slopes low on the sides. Some of the smaller fragments, however, appear to lie on small topographic highs of fine-grained materials.

The rock from which samples 15595 and 15596 were chipped appears to be similar to other large rocks exposed in the area. Approximately 50 cm high and 2 m across, the rock is somewhat more tabular than many of the rocks in the same size range. It is well rounded with a hackly, pitted surface and appears to have through-going planar structures.

The down-Sun photograph of the boulder from which samples 15595 and 15596 were collected (fig. 5-123(b)) and, to some extent, the cross-Sun photograph (fig. 5-123(a)) suggest that the northeast protuberance of the boulder has a slightly smoother surface. Sample 15596 was collected from along the contact between this smooth portion and the rougher portion of the rock, from the rough side of the contact. Generally, the surface texture, gray tone, size, and rounding of the sampled boulder is similar to the others in the field of view of the photograph.

The sampled boulder appears to be one of a group of large boulders or bedrock protuberances exposed all along the upper rim of the rille. These rocks are probably in place or very nearly in place, and therefore the two samples almost certainly are from the highest bedrock unit exposed along the east edge of the rille.

Samples 15600 to 15610, 15612 to 15630, 15632 to 15645, 15647 to 15656, and 15658 to 15689: These rake samples were collected approximately 20 m northeast of the rim of Hadley Rille (figs. 5-1 and 5-115). The surface in the immediate vicinity of the samples is smooth, level, and free of rock fragments greater than 5 cm across (fig. 5-124). The general vicinity is littered with fragments commonly as large as 20 cm across and with a few sparsely scattered boulders greater than 1 m across (fig. D-19).

Several moderately subdued to subdued craters as large as approximately 2 m in diameter are present in the general vicinity, but the sampled area is sufficiently far from craters greater than 1 m in diameter to be in their continuous ejecta (fig. 5-124). The surface was undisturbed by footprints or wheeltracks before the raking.

A few subdued craters ranging from raindrop-size depressions to 20 cm across are present in the raked area. Many of the rock fragments have small fillets banked against the sides.

Station 10. – Station 10, located approximately 200 m north-northwest of station 9A, was a stop for photography (figs. D-15 and 5-125). The offset distance from station 9A provided a base for the

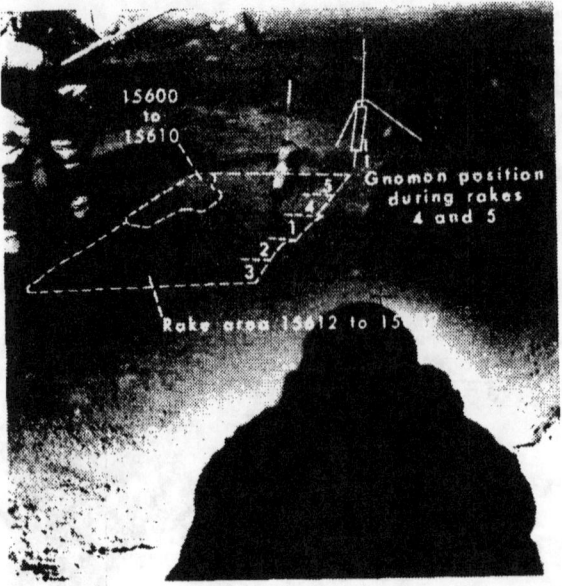

FIGURE 5-124. – Presampling, down-Sun photograph, looking west, of samples 15600 to 15610 and 15612 to 15689 collected with rake at station 9A (AS15-82-11153).

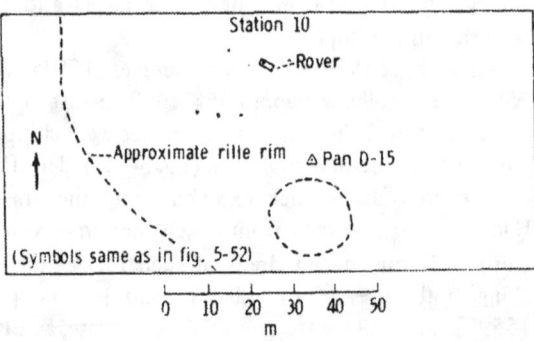

FIGURE 5-125. – Planimetric map of station 10.

CDR's stereoscopic 500-mm photography of the west wall of Hadley Rille and of the St. George Crater rim. In addition, stereoscopy was obtained from the LMP's full wide-angle panoramas taken at both stations. Geologic descriptions by the crew at station 10 provide annotation of the photography, including that of several large vesicular basalt boulders that are near the north rim of a 60-m-diameter crater where the Rover was parked.

SUMMARY

The Apollo 15 crew landed near the sinuous Hadley Rille on mare material at the base of the

Apennine Mountains. The objectives of the mission, set by premission photogeologic study of the site, were description and sampling of the Apennine Front, Hadley Rille, and mare materials of the Imbrium Basin.

The Apennine Mountains, composed of rectilinear massifs rising 3.5 to 4.5 km above Hadley Rille, are interpreted as fault blocks uplifted and segmented at the time of the Imbrium impact. The mountain front, therefore, probably exposes a section of pre-Imbrian lunar crust, the sampling and description of which were highest priority during the mission. This objective had added significance because of the successful sampling in breccias of higher levels of pre-Imbrian crust by the Apollo 14 crew.

The mountains around the site are characterized by gentle to moderate slopes and rounded outlines. Large blocks are scarce on the mountain front, suggesting the presence of a thick regolith. Linear ribs present on Silver Spur appear to represent nearly vertical structure. Two prominent sets of parallel lineaments on many of the mountain slopes may be artifacts produced by low oblique lighting or may represent sets of steeply dipping fractures propagated through the regolith.

Rocks collected from the Front consist mainly of breccias. Clasts in these coherent breccias are predominantly debris derived from coarse-grained feldspathic rocks, one of the largest samples of which is nearly pure plagioclase. Excellent samples of this material will allow detailed study of this most important element of the pre-Imbrian crust.

An important and distinctive feature of the breccias collected from the Apennine Front is the absence of clasts of older breccias that are so abundant in the Apollo 14 samples. This absence of clasts indicates a less reworked part of the lunar crust than was sampled by the Apollo 14 crew; samples of these lower levels of the crust provide material with less complicated histories than those obtained by the preceding crew.

A striking and common feature of the breccias collected from the Front is partial to nearly complete coating of sharply angular blocks by glass. Such glass-coated breccias not only are common along the Hadley Delta front but also at the LM site and at station 9 on the mare surface. The presence on the mare surface of breccias lacking mare-type basalt clasts is ascribed to erosion of the highlands surrounding the mare embayment, possible contributions of foreign debris from a ray across the mare surface, and general lack of fresh large craters of the mare that would recycle subregolithic material to the surface in the traversed areas.

Hadley Rille is one of the freshest of the sinuous rilles on the Moon; the origin of these features has long been controversial. At the edge of the rille, the first extensive observations of lunar outcrops and their stratigraphy were made. Outcrops were described, sampled directly, and documented by closeup and long-focal-length photography. The photographs reveal outcrops to depths of 60 m in the rille walls and thus yield the local minimum thickness of mare material. The outcrops are distinctly layered and are composed of more than one unit as revealed by varying surface textures and albedos. Some of the rocks cropping out in the rille walls are so massive and poorly jointed that the talus contains enormous blocks. The rille has slightly raised rims, and the east side is substantially higher than the west side.

The mare surface generally is smooth to rolling because of numerous subdued craters. The regolith is approximately 5 m thick. Rocks account for approximately 1 percent of the surface except at craters, where fragments cover as much as 15 percent of the local surface. Small-scale lineaments are rare compared with the mare areas visited by the Apollo 11 and 12 crews. The mare at the site is tentatively subdivided into four areas on the basis of crater density and general surface texture. The mare has lower albedo to the east, possibly because of a ray from Aristillus or Autolycus Craters.

Mare material collected at three widely spaced points—Elbow Crater, Dune Crater, and the rille edge—consists mainly of basalt with abundant coarse pyroxene, indicating considerable lateral continuity of this rock type. Olivine basalt forms a more restricted unit that, where sampled at the edge of the rille, probably overlies the pyroxene-rich basalts.

The North Complex, an irregular dark hilly area north of the landing site, was not explored but was photographed extensively (particularly the wall of Pluton Crater). Blocks on this wall include some (1) that are as bright as breccias on Hadley Delta and others (2) that have albedos typical of mare basalts, have large cavities, or are distinctly layered. This observation suggests that both major rock types may be present in the North Complex.

Several types of fillets occur where rocks are embedded in soil. Fillet development is seen to be an evolving process that is dependent on surface slope, rock shape, rock friability, and time.

The surfaces at different sampling sites vary in age, from old at stations 2 and 6 and at the LM-ALSEP area to young at the rille edge and at crater rims. A 15-m-diameter crater with glassy ejecta at station 9 is probably the youngest surface of any significant extent yet sampled. Good documentation of numerous small glass-floored craters in the site suggests that the glass is related to impact processes.

The increased capability of the extended-stay (J-series) missions over the H-missions that preceded Apollo 15 was shown to be a significant improvement in the geologic exploration of the Moon. Particularly significant to geologic exploration was the performance of the Rover, the greater number and longer duration of the periods of EVA, and the demonstrated capability to land in mountainous regions of the Moon. The Apollo 15 mission clearly demonstrated the scientific benefits of manned space exploration.

REFERENCES

5-1. Carr, M.H.; and El-Baz, Farouk: Geologic Map of the Apennine-Hadley Region of the Moon–Apollo 15 Pre-Mission Map. U.S. Geol. Survey Misc. Geol. Inv. Map I-723 (sheet 1 of 2), 1971.

5-2. Howard, K.A.: Geologic Map of Part of the Apennine-Hadley Region of the Moon–Apollo 15 Pre-Mission Map. U.S. Geol. Survey Misc. Geol. Inv. Map I-723 (sheet 2 of 2), 1971.

5-3. Schaber, G.G.; and Head, J.W.: Surface Operational Map of the Apennine-Hadley Landing Site–Apollo 15. U.S. Geol. Survey Open File Rept., July 1971.

5-4. Apollo Lunar Geology Investigation Team: Preliminary Report on the Geology and Field Petrology at the Apollo 15 Landing Site. U.S. Geol. Survey Interagency Rept. 32, Aug. 5, 1971.

5-5. Batson, R.M.; Larson, K.B.; Reed, V.S.; et al.: Preliminary Catalog of Pictures Taken on the Lunar Surface During the Apollo 15 Mission. U.S. Geol. Survey Interagency Rept. 35, Aug. 27, 1971.

5-6. Sutton, R.L.; Hait, M.H.; Wolfe, E.W.; et al.: Preliminary Documentation of the Apollo 15 Samples. U.S. Geol. Survey Interagency Rept. 34, Aug. 26, 1971.

5-7. Swann, G.A.; Hait, M.H.; Schaber, G.G.; Freeman, V.L.; et al.: Preliminary Description of Apollo 15 Sample Environments. U.S. Geol. Survey Interagency Rept. 36, Sept. 1971.

5-8. Howard, K.A.; Head, J.W.; and Swann, G.A.: Geology of Hadley Rille: Preliminary Report. U.S. Geol. Survey Interagency Rept. 41, 1971.

5-9. Wilhelms, D.E.; and McCauley, J.F.: Geologic Map of the Near Side of the Moon. U.S. Geol. Survey Misc. Geol. Inv. Map I-703, 1971.

5-10. Soderblom, L.A.; and Lebofsky, L.A.: A Technique of Rapid Determination of Relative Ages of Lunar Areas from Orbital Photography. J. Geophys. Res., vol. 76, 1971.

5-11. Swann, G.A.; Bailey, N.G.; Batson, R.M.; Eggleton, R.E.; et al.: Preliminary Geologic Investigations of the Apollo 14 Landing Site. Sec. 3 of Apollo 14 Preliminary Science Report, NASA SP-272, 1971.

5-12. Schaber, G.G.; and Swann, G.A.: Surface Lineaments at the Apollo 11 and 12 Landing Sites. Proceedings of Second Lunar Science Conference, vol. 1, A.A. Levinson, ed., MIT Press (Cambridge, Mass.), 1971, pp. 27-38.

5-13. Anon.: NASA Lunar Topographic Map, Rima Hadley (sheets A and B). Topocom, Jan. 1971.

5-14. Milton, Daniel J.: Slopes on the Moon. Science, vol. 156, no. 3778, May 26, 1967, p. 1135.

5-15. Greeley, Ronald: Lunar Hadley Rille: Considerations of Its Origin. Science, vol. 172, no. 3984, May 14, 1971, pp. 722-725.

5-16. Howard, K.A.: Lava Channels in Northeastern California. Am. Geophys. Union Trans., vol. 50, no. 4, 1969, p. 341.

5-17. Gault, D.; Collins, R.; Gold, T.; Green, J.; et al.: Lunar Theory and Processes. Scientific Results. Part II of Surveyor III Mission Report, sec. VIII, JPL Tech. Rept. 32-117, 1967, pp. 195-213.

5-18. Morris, E.C.; and Shoemaker, E.M.: Fragmental Debris. Geology. Television Observations from Surveyor. Science Results. Part II of Surveyor Project Final Report, sec. III, part. G3, JPL Tech. Rept. 32-1265, 1968, pp. 69-86.

6. Preliminary Examination of Lunar Samples

The Lunar Sample Preliminary Examination Team[a]

The primary scientific objective of lunar exploration is the characterization of the body as a planet. In particular, the present chemical and physical structure of the Moon and the endogenic and exogenic processes that have shaped the planet must be determined in order that the initial state of the Moon can be inferred. The prospect of achieving this objective within the scope of the lunar-exploration program depends on both the complexity of the Moon and the skill and imagination with which the resources of the program have been and are being used.

A very important element in planning the exploration of the lunar surface is the choice of sites to be studied during manned landings. To achieve the most comprehensive coverage of the lunar surface, each landing site should facilitate the exploration of terrain with a variety of structural, temporal, and chemical characteristics. The Apollo 15 landing site is located among mountains that rise more than 4000 m above the elevation of the site and a valley that cuts more than 300 m into the surface on which the spacecraft landed. This vertical relief provided exploration opportunities that may expand the knowledge of the Moon in both time (the wide range of stratigraphic units sampled) and space (the additional information about the subsurface structure).

The morphology, mineralogy, petrology, and chemistry of the samples returned from the Apollo 15 landing site are discussed in this section. A selenological description of the area from which the samples were taken is given in section 5 of this report.

The diversity of the samples and the variety of sample environments found at the Apollo 15 site have resulted in several hypotheses that relate individual samples to local situations after even a preliminary examination of the samples. Several somewhat speculative hypotheses that relate individual samples to a geological framework are discussed in this section in the hope that the hypotheses will provide guidelines for more detailed studies of the samples to arrive at an integrated understanding of the selenology of the Apollo 15 site.

A total of 77 kg of samples, which consists of more than 350 individual samples from 10 sampling areas, was collected during three periods of extravehicular activity (EVA) on July 31, August 1, and August 2, 1971. These samples consist of rock specimens that range in weight from 1 or 2 g to more than 9.5 kg, core tubes, and a variety of soil specimens. The samples come from two distinct selenologic regions, the mare plain and the foot of a 3000-m-high mountain known as Hadley Delta. A brief description of the individual specimens that weigh more than 10 g is given in table 6-I. The rocks from the mare plain consist of two basic types, (1) the extrusive and hypabyssal basaltic rocks and (2) the glass-coated and glass-cemented breccias. The breccias are apparently concentrated in the region around the lunar module (LM) (fig. 5-2 in sec. 5 of this report). The rock samples from the foot of the mountain front comprise a very diverse set of rock types that range from breccias to possible metaigneous rocks. Thirteen soil samples, with an aggregate weight of 4.5 kg, were collected from the mare plain; and 18 soil samples, with an aggregate weight of 8.5 kg, were collected from the mountain front.

Small rock specimens with average dimensions that range from 0.5 to 6 cm were obtained at three sampling sites (St. George Crater, Spur Crater, and the edge of Hadley Rille). These samples, which were obtained by raking or sieving the soil at these locations, are statistically representative of the smaller rock fragments found at the sites. These samples are listed and described by individual rock type in table 6-II.

---

[a]The team composition is listed in "Acknowledgments" at the end of this section.

TABLE 6-I.–Inventory of Rocks That Weigh More Than 10 g[a]

Sample no.	Mass, g	Location	Description	Sample no.	Mass, g	Location	Description
15065	1575	Station 1	Gabbro or very coarse basalt	15418	1141	Station 7	from large boulder
15075	809	Station 1	Gabbro or very coarse basalt	15419	18	Station 7	Shock-melted gabbroic anorthosite
15076	400	Station 1	Gabbro or very coarse basalt	15425	126	Station 7	Glass-coated breccia
15085	471	Station 1	Gabbro or very coarse basalt	15426	223	Station 7	Friable green clod
15086	216	Station 1	Soil breccia	15415	269	Station 7	Friable green clod
15095	25	Station 1	Glass-coated microbreccia	15435	400	Station 7	Anorthosite
15206	92	Station 2	Glassy breccia chipped from large boulder				Clods of breccia from "pedestal," which was the matrix of the anorthosite
15205	337	Station 2	Glassy breccia chipped from large boulder	15455	938	Station 7	Large white norite clasts in black breccia
15016	924	Station 3	Scoriaceous basalt	15465	375	Station 7	Glass-coated breccia
15475	406	Station 4	Gabbro or very coarse basalt	15466	123	Station 7	Glass-coated breccia
15476	266	Station 4	Gabbro or very coarse basalt	15445	287	Station 7	Breccia
15495	909	Station 4	Gabbro or very coarse basalt	15459	4828	Station 7	Breccia with large clasts up to 8 cm
15499	2025	Station 4	Porphyritic vitrophyre chipped from boulder	15058	2672	Station 8	Basalt
15485	105	Station 4	Porphyritic vitrophyre chipped from boulder; has white alteration crust on fractures	15059	1149	Station 8	Glass-coated breccia
				15515	143	Station 9	Clods of soil from rim of fresh crater
				15505	1171	Station 9	Glass-coated microbreccia
15486	47	Station 4	Porphyritic vitrophyre chipped from boulder	15529	1531	Station 9A	Vesicular basalt
				15556	1538	Station 9A	Vesicular basalt
15498	2340	Station 4	Microbreccia with partial coating of glass	15557	2518	Station 9A	Vesicular basalt
				15535	404	Station 9A	Olivine basalt
15245	115	Station 6	Many glassy agglutinates from center of 1-m-diameter fresh crater	15536	321	Station 9A	Olivine basalt
				15545	747	Station 9A	Olivine basalt
				15546	28	Station 9A	Basalt
15295	947	Station 6	Breccia with partial coating of glass	15547	20	Station 9A	Olivine basalt
				15595	237	Station 9A	Porphyritic basalt chipped from "outcrop" boulder
15298	1731	Station 6	Microbreccia				
15299	1692	Station 6	Breccia with glassy surface	15596	225	Station 9A	Porphyritic basalt chipped from "outcrop" boulder
15255	240	Station 6	Glass-coated breccia				
15256	201	Station 6	Shock-melted basalt	15597	146	Station 9A	Olivine basalt
15257	22	Station 6	Microbreccia with partial coating of glass	15598	136	Station 9A	Olivine basalt
				15555	9614	Station 9A	Basalt
15268	11	Station 6	Microbreccia with white-clast layer	15558	1333	Unknown	Breccia with glass in fractures
15285	264	Station 6	Breccia with partial coating of glass	15565 to 15587	735	Unknown	Thirteen fragments of broken-up breccia
15286	40	Station 6	Glassy breccia				
15287	45	Station 6	Soil breccia				

TABLE 6-I.—Inventory of Rocks That Weigh More Than 10 g[a] – Concluded

Sample no.	Mass, g	Location	Description	Sample no.	Mass, g	Location	Description
15015	4735	LM area	Glass-coated breccia	15288	63	Station 6	Glassy breccia
15025	77	LM area	Fine breccia	15289	24	Station 6	Breccia with partial coating of glass
15017	17	LM area	Hollow glass sphere	15265	314	Station 6	Fine breccia
15027	51	LM area	Glass-coated breccia	15266	271	Station 6	Fine breccia with glass coating
15028	59	LM area	Glass-coated breccia	15405	513	Station 6A	Slightly recrystallized breccia chipped

[a]Rake samples have not been included.

The descriptions reported in this section are based on both a careful macroscopic study of all the rock samples and more detailed petrographic and chemical studies of a smaller set of the samples. Most individual soil samples have been examined in thin section and by microscopic examination of several hundred grains of each soil. At this stage, core-tube studies have been limited to nondestructive X-radiographs that have resulted in gross textural characterizations of the cores.

Samples that were selected for chemical and isotopic studies and thin-section analyses have been restricted to a representative sample set that has been chosen by the Lunar Sample Preliminary Examination Team and the Lunar Sample Analysis Planning Team for preliminary characterization of the entire Apollo 15 collection.

CHEMISTRY AND PETROGRAPHY OF ROCK SAMPLES

Mare Basalts

The vertical section exposed in the wall of Hadley Rille (fig. 6-1) indicates that the subsurface structure in the mare region probably consists of a series of basalt flows with thicknesses ranging from 1 or 2 m to tens of meters. When examined with the naked eye, the majority of the rocks from the mare regions seems to be very fresh igneous rocks with macroscopic textures that range from massive, dense basalts or gabbros to scoriaceous basalts. Many vuggy rocks contain large, clearly zoned pyroxene phenocrysts that extend into the vugs.

The chemical compositions of ten individual basalt samples are listed in table 6-III. The basalt compositions listed in the table include the compositions of representative samples from all of the mare sampling points, along with a single metabasalt (sample 15256), which was collected at station 6 at the foot of the mountain front. The range of compositions observed in these ten analyses is remarkably small. The only significant variation is found in the magnesium oxide (MgO), iron oxide (FeO), and titanium dioxide (TiO_2) contents. The overall chemical characteristics observed for these basaltic rocks are similar to the chemical characteristics of the basaltic rocks found at Statio Tranquillitatis and at the Apollo 12 and Luna 16 sites. In particular, the high iron (Fe) content and the correspondingly high FeO/MgO ratio should be noted. Also, as in previously examined mare basalts, the low sodium monoxide (Na_2O) content distinguishes the lunar samples from all terrestrial basalts. The average TiO_2 content is somewhat lower than the content observed in samples from previous mare sites. A more detailed comparison of the Apollo 15 mare basalts with terrestrial and other lunar samples is given in figures 6-2 and 6-3. The FeO and MgO concentrations shown in figure 6-2 reveal the systematic and remarkable difference in Fe content of the mare basalts and the terrestrial and lunar-highland basalts. Both the content of Fe and magnesium (Mg) and the FeO/MgO ratio in a large group of mare basalts are similar to those found in eucritic meteorites. The FeO/MgO ratio for four different mare regions falls in a rather narrow range between 2.2 and

TABLE 6-II.—*Samples Obtained With the Rake*[a]

Group[b]	Sample no.	Description
Samples from station 2 (St. George Crater)[c]		
1	15115 to 15117	Slightly vuggy, coarse-grained, porphyritic-zoned, green-to-brown, clinopyroxene basalt
2	15118, 15119	Slightly vuggy, medium-grained, porphyritic-zoned, clinopyroxene basalt (finer-grained variant of group 1)
3	15105	Vesicular-to-vuggy, fine-grained, microporphyritic-olivine, brown-pigeonite basalt
4	15125	Fine-grained, microporphyritic-olivine, gray-pyroxene basalt with no vesicles or vugs
5	15135	Dark-gray, vesicular, glassy microbreccia with white clasts
6	15145 to 15148	Light-gray soil breccia with clasts of group 1 type rocks and minerals
Samples from station 7 (Spur Crater)[d]		
1	15315 to 15320	Soil breccias distinguished by clasts of basalt that contain cinnamon-brown pyroxene; other clasts of the anorthosite-gabbro group, the gray-pyroxene-basalt group, the green-glass-microbreccia group, and the group 7 basalts also may be present
2	15306, 15321 to 15355	Soil breccias distinguished by a lack of basalt containing cinnamon-brown pyroxene; otherwise, clasts similar to the group 1 clasts
3	15308, 15356 to 15360	Tough microbreccias with light-colored lithic and mineral clasts in a dark-gray, aphanitic, subvitreous matrix (similar to the black part of sample 15455)
4	15361	Pale-green, tough, microcrystalline rock; probably a finely granulated gabbro or norite
5	15362 to 15364	Anorthosite; one, finely granulated; other two, coarser grained
6	15365 to 15377	Friable green-glass microbreccia; one with clasts of green-pyroxene-bearing gabbro in a matrix of green glass and pyroxene
7	15378 to 15384	Fine-grained basalts with very light brown to light-yellow-brown pyroxenes and a plagioclase-to-mafic-mineral ratio of 1:1; no vugs or vesicles
8	15385 to 15388	Slightly vuggy, coarse grained, zoned, green-to-red-brown pyroxene basalt with plagioclase-to-mafic-mineral ratio of approximately 1:2
9	15307, 15389 to 15393	Glass
Samples from station 9A (Hadley Rille)[c]		
1	15606, 15612 to 15630	Highly vesicular (40 to 50 percent), red-brown-pyroxene, porphyritic-olivine, equigranular-to-subophitic basalt
2	15610, 15632 to 15645	Vuggy, medium grained, porphyritic-zoned green-to-red-brown clinopyroxene (and olivine?) basalt
3	15608 to 15609, 15647 to 15656	Fine-grained subophitic basalt with deep-brown pyroxene and a few to no olivine phenocrysts, no vesicles, and only a very few vugs; sample 15647 slightly recrystallized

TABLE 6-II.—*Samples Obtained With the Rake*[a] — Concluded

Group[b]	Sample no.	Description
4	15605, 15658 to 15664, 15670	Vuggy, medium-grained, porphyritic-zoned green-to-red-brown pyroxene, equigranular-to-subophitic basalt
5	15607, 15665, 15668, 15669	Moderately vesicular to vuggy, fine-grained-porphyritic-olivine, equigranular basalt
6	15666, 15667	Vuggy, medium-grained, highly porphyritic-zoned, green-to-brown pyroxene; probably subophitic basalt
7	15671 to 15673	Highly vesicular to partly vuggy, medium-grained, porphyritic-zoned, green-to-brown-pyroxene, subophitic basalt
8	15674 to 15681	Slightly vuggy, porphyritic-olivine, fine-grained, equigranular basalt with reddish-brown pyroxene in matrix
9	15682	Porphyritic-zoned, green-to-brown, pyroxene basalt with plumose-plagioclase reddish-brown-pyroxene intergrowth as matrix
10	15683	Slightly vuggy, fine-grained, porphyritic-olivine and plagioclase, intergranular basalt
11	15684 to 15686	Breccia fragments cemented by black glass (agglutinates)
12	15687, 15688	Granular igneous rocks (basalts) cemented by vesicular black glass
13	15689	Breccia with unique, bright, orange-brown, sugary clasts (pyroxene?); also clasts of anorthosite

[a]The samples range in size from 0.5 to 6 cm.
[b]The groupings are based on morphological and mineralogical characterization of the sample surfaces by microscopic examination.
[c]The 12 samples range in mass from 1.0 to 27.6 g.
[d]The 81 samples range in mass from 0.3 to 64.3 g.
[e]The 81 samples range in mass from 0.2 to 336.7 g.

FIGURE 6-1.—Layering in the west wall of Hadley Rille, as viewed from station 9. The top layer appears to be relatively thick (approximately 25 m), but the next series of layers is relatively thin. The total width of the wall section shown in the photograph is 150 m (AS15-89-12104).

3.8, if the high-Mg basalts (which are probably enriched in olivine) from the Apollo 12 site are excluded.

The calcium oxide (CaO) and aluminum oxide (Al_2O_3) concentrations (fig. 6-3) indicate that the mare and highland basalts are clearly separated by aluminum (Al)/calcium (Ca) ratio of chondritic meteorites. Moreover, when all mare basalts are considered as a group, a significant correlation appears to exist between Ca and Al. Major terrestrial basalt types also appear to fall on either side of the meteoritic Ca/Al ratio. Oceanic-ridge basalts and island-arc basalts are generally enriched in Al relative to Ca; that is, they have Al/Ca ratios in excess of the meteoritic ratio. Conversely, nepheline normative basalts tend to be depleted in Al relative to Ca. The comparisons shown in figure 6-3 suggest that both lunar and terrestrial basalts differ from most meteorites, which have a Ca/Al ratio that is essentially constant (ref. 6-2). The variable Ca/Al ratios observed in samples from the lunar surface are evidence against the hypothesis that the origin of achondrites is the lunar surface. The trace-element abundances (including potassium (K), uranium (U), and thorium (Th)) for the Apollo 15

TABLE 6-III.—*X-Ray Fluorescence Analysis of Igneous and Metaigneous Rocks*[a]

(a) Chemical-compound abundance

Compound	Sample no. (station no.)									
	15016 (3)	15058 (8)	15076 (1)	[b]15256 (6)	15499 (4)	15555 (9A)	15556 (9A)	[c]15415 (7)	15418 (7)	[d]14310
	Abundance, percent (by weight)									
SiO_2	43.97	47.81	48.80	44.93	47.62	44.24	45.11	44.08	44.97	47.19
TiO_2	2.31	1.77	1.46	2.54	1.81	2.26	2.76	.02	.27	1.24
Al_2O_3	8.43	8.87	9.30	8.89	9.27	8.48	9.43	35.49	26.73	20.14
FeO	22.58	19.97	18.62	22.21	20.26	22.47	22.25	.23	5.37	8.38
MnO	.33	.28	.27	.29	.28	.29	.29	.00	.08	.11
MgO	11.14	9.01	9.46	9.08	8.94	11.19	7.73	.09	5.38	7.87
CaO	9.40	10.32	10.82	10.27	10.40	9.45	10.83	19.68	16.10	12.29
Na_2O	.21	.28	.26	.28	.29	.24	.26	.34	.31	.63
K_2O	.03	.03	.03	.03	.06	.03	.03	<0.01	.03	.49
P_2O_5	.07	.08	.03	.06	.08	.06	.08	.01	.03	.34
S	.07	.07	.03	.08	.07	.05	.08	.00	.03	.02
Cr_2O_3	–	–	–	–	–	.70	–	–	.11	.18
Total	98.54	98.49	99.08	98.66	99.08	99.46	98.94	99.95	99.41	98.87
	Abundance, ppm									
Sr	83	101	99	100	105	92	107	184	152	189
Zr	95	98	50	90	112	78	91	–	67	847

(b) Mineral abundance

Mineral	Sample no. (station no.)									
	15016 (3)	15058 (8)	15076 (1)	[b]15256 (6)	15499 (4)	15555 (9A)	15556 (9A)	[c]15415 (7)	15418 (7)	[d]14310
	Normative abundance									
Quartz	0.00	1.15	1.61	0.00	0.38	0.00	0.00	0.10	0.00	0.16
Orthoclase	.18	.18	.18	.18	.36	.18	.18	.06	.18	2.90
Albite	1.78	2.37	2.20	2.37	2.45	2.03	2.20	2.88	2.62	5.25
Anorthite	21.97	22.86	24.12	22.91	23.82	21.97	24.48	95.29	71.46	50.73
Diopside	20.25	23.35	24.56	23.27	22.90	20.51	24.26	1.22	6.73	6.58
Ferro-hypersthene	32.14	44.97	43.55	35.76	45.48	32.12	37.53	.00	12.38	29.93
Olivine	17.61	.00	.00	9.14	.00	17.47	4.70	.00	5.33	.00
Ilmenite	4.39	3.36	2.77	4.82	3.44	4.29	5.24	.04	.51	2.36

[a]Major-element content (except for sodium (Na)) was determined on a 280-mg aliquot of sample prepared according to the procedure outlined in reference 6-1. The Na content was determined by atomic absorption spectrometry on a 20-mg sample.

[b]Metabasalt.

[c]Anorthosite.

[d]Apollo 14 control sample.

basaltic rocks are, in general, similar to the abundances found in the Apollo 12 igneous rocks.

Petrographic thin sections of 13 different Apollo 15 basalt samples have been examined as a part of this preliminary study. The results indicate that all basalts from the mare regions have primary igneous textures and can be grouped to suggest that widespread layers underlie the mare region. A single

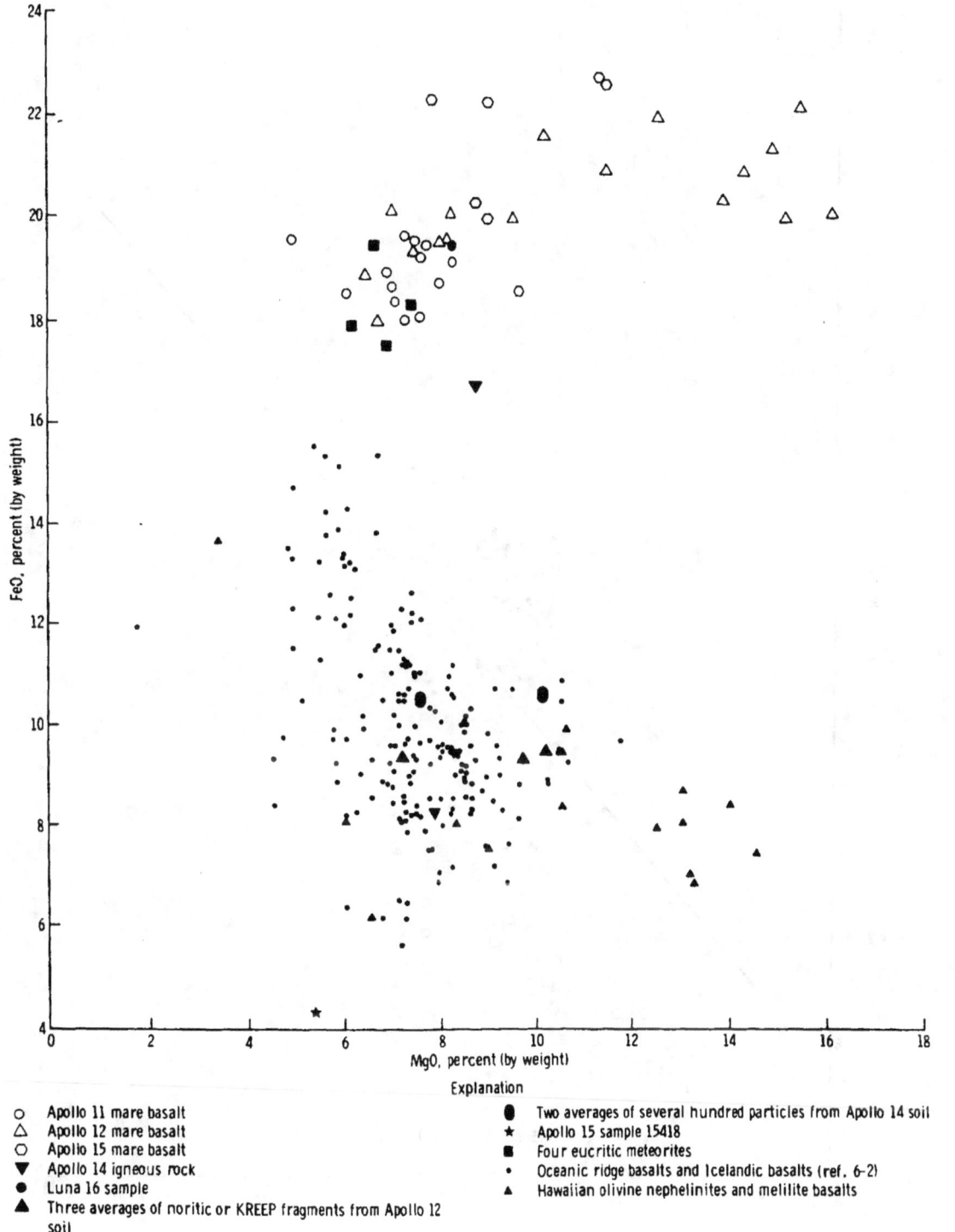

FIGURE 6-2.—The MgO and FeO content of lunar rocks.

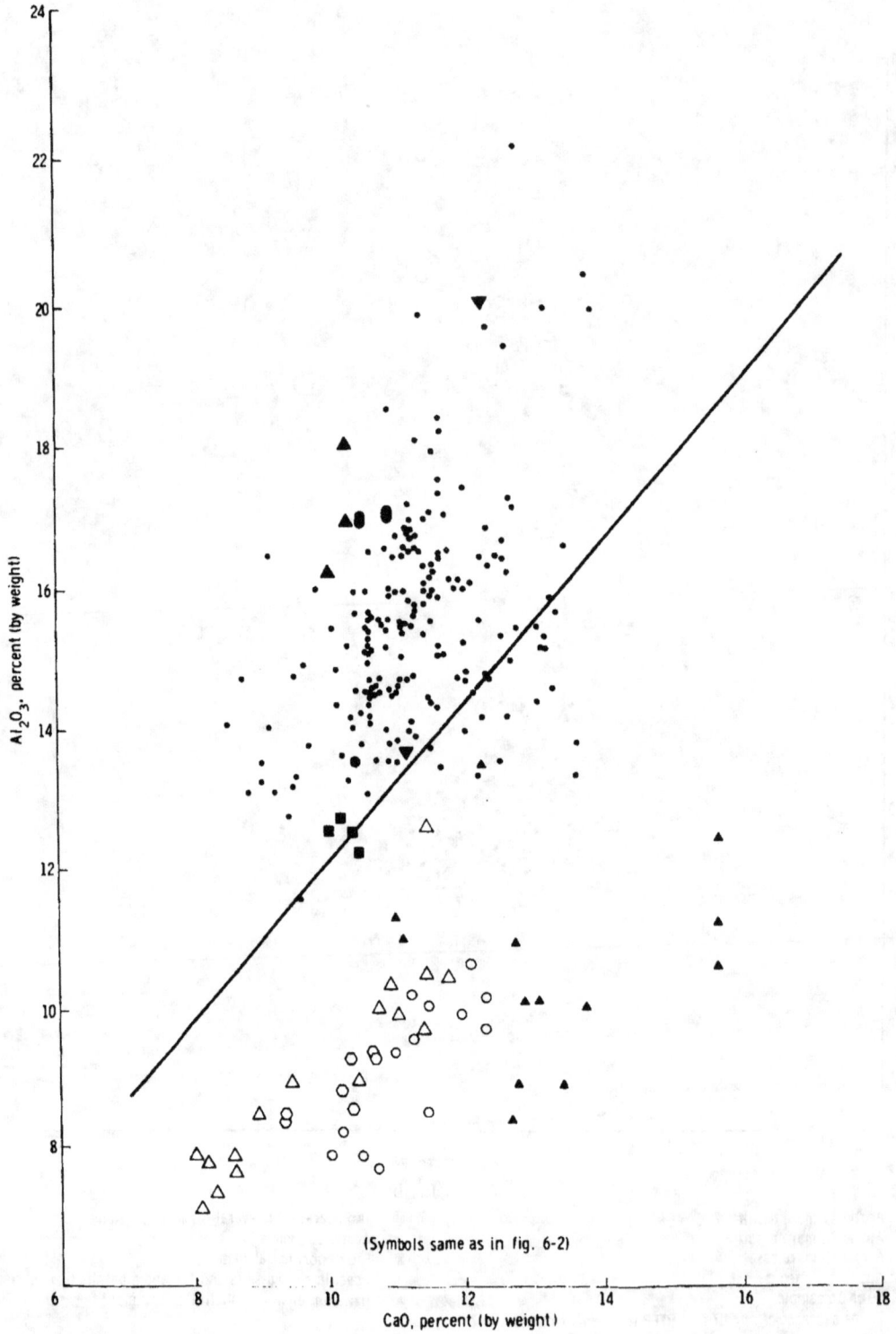

FIGURE 6-3.—The CaO and Al_2O_3 content of lunar and terrestrial basaltic rocks. The line is the locus of CaO and Al_2O_3 contents of all chondritic, eucritic, and howarditic meteorites.

basaltic rock (sample 15256) from the mountain front appears to be extensively shock-metamorphosed.

Four distinctly different textural types of mare basalts have been discerned from thin-section studies.

(1) Porphyritic clinopyroxene basalt (samples 15058, 15076, 15085, and 15475) is typified by 3- to 9-mm-long prisms with cores of yellow-green pigeonite and rims of brown augite as phenocrysts in a subophitic matrix that contains moderately to strongly zoned plagioclase with a-axial inclusions of pyroxene. Tridymite occurs in these thin sections (fig. 6-4).

(2) Porphyritic clinopyroxene basalt vitrophyre (samples 15485, 15499, and 15597) is typified by 1- to 7-mm-long skeletal prisms that have hollow cores

FIGURE 6-5.—Thin section of sample 15485,3 (plane light) showing skeletal clinopyroxene crystals in glassy maroon matrix. The vesicle contains fragments of crystals and glass.

surrounded by pigeonite and thin rims of augite as phenocrysts in a glassy to devitrified matrix that consists of 3-μm plagioclase and pyroxene in plumose intergrowths. The phenocrysts have preferred orientations; one specimen has a prism of skeletal olivine in addition to pyroxene (fig. 6-5).

(3) Porphyritic olivine basalt (samples 15535, 15545, and 15555) is typified by approximately 10-percent euhedral-olivine phenocrysts in an inequigranular matrix of poikilitic plagioclase and a wide range of sizes of weakly zoned clinopyroxene (pigeonite cores to augite rims). Cristobalite occurs in all three thin sections (fig. 6-6).

(4) Highly vesicular (scoriaceous) basalt (samples 15016 and 15556) is typified by over 50-percent weakly zoned clinopyroxene (pigeonite cores to augite rims) as part of a subophitic to intergranular

FIGURE 6-4.—Thin section of sample 15076,12 (crossed polars) showing zoned clinopyroxene and plagioclase that contain cores of pyroxene. The inner zone is pigeonite, and the outer zone is augite; the opaque mineral is ilmenite.

FIGURE 6-6.—Thin section of sample 15535,11 (crossed polars) showing olivine phenocrysts in groundmass of pyroxene and poikilitic plagioclase.

observation was the occurrence of a very fine grained, white, 0.1-mm-thick crust along fractures of this rock type in the vicinity of Dune Crater. This is the first lunar sample with evidence of fluid deposition of this type.

The type 2 mare basalt was found in fragments collected at the edge of Hadley Rille and in the form of a large boulder at Dune Crater. Two identical chips, taken 1 m apart, were obtained from this boulder. The two sampling points where this basalt type was found occur at the extremities of the region that was sampled, which thus suggests that this basalt type may also be a widespread unit in the underlying bedrock.

Most samples of the type 3 and 4 mare basalts come from the station at the edge of Hadley Rille. One type 3 sample was found halfway between

texture. Plagioclase is weakly to strongly zoned (fig. 6-7).

The type 1 mare basalt is quite ubiquitous. It was found in thin sections of samples from Elbow Crater, Dune Crater, and the LM landing site. Careful macroscopic examination of samples from the Hadley Rille station 9A suggests that the same rock type also occurs there. The most detailed collection of this rock type was obtained at three sampling points along a 100-m-long radial traverse across the rim of Elbow Crater. The type 1 mare basalts have a wide range of grain sizes, which range from gabbroic at the rim of Elbow Crater (sample 15065) to much finer grained porphyritic rocks observed at Hadley Rille station. The tentative hypothesis is that this variation indicates that the samples were ejected from different depths in a thick, widespread unit that is near the top of the underlying bedrock in the area. An interesting

FIGURE 6-7.—Thin section of sample 15556,15 (plane light) showing large vesicles and possible plagioclase phenocrysts in a matrix of granular pyroxene, olivine, and opaque minerals.

Elbow Crater and the LM. In addition, some fragments of the type 3 basalt may occur in the rake sample taken at Spur Crater. At this stage of the examination, the distribution and occurrence of type 3 and 4 basalts do not indicate any clear-cut hypotheses that relate them to the underlying bedrock.

Nonmare Basalts

Seven rake fragments from Spur Crater and many clasts in breccias from the mountain-front area are basalts that are not represented in the suite from the mare areas. Although no thin sections of these basaltic fragments are available, they are distinctly different from the mare basalts in several respects: (1) the plagioclase-to-mafic-mineral ratio is approximately 1, whereas, in mare rocks, the ratio is closer to 0.5; (2) the pyroxene is light brown or tan, whereas, in mare rocks, the color is deep cinnamon brown; (3) the pyroxenes show no zoning, whereas, in mare rocks, a green core with a red-brown rim generally can be found; (4) a yellow-brown to yellow-green phase (possibly another pyroxene) occurs in roughly equal proportions to the light-brown pyroxene; (5) the opaque-mineral population tends to be platy, whereas, in the mare rocks, the opaque-mineral population is more lathlike to equidimensional; (6) the grain size of all phases is less than 1 mm, whereas, in the mare suite, some grains are as large as 1 cm; and (7) no vugs or vesicles occur, whereas the mare basalts are slightly to highly vesicular or vuggy (or both).

Thin sections of breccia from the mountain front contain basalt clasts with textures and mineral proportions significantly different from the mare basalts. One of the basalt clast types is similar to sample 14310 of the material returned from the Apollo 14 site (ref. 6-3). It is tempting to speculate that these nonmare basalts that were found in the breccia and rake samples are from a sequence of premare basalt strata that may be represented by the several-kilometer-thick sequence of layers observed in the mountains to the south and east of the landing site (sec. 5 of this report).

Clastic and Metamorphosed Rocks

A wide variety of rocks with textural characteristics that clearly indicate a multiple-event origin has been returned from the Apollo 15 site. Although the greatest variety of these rocks is concentrated along the foot of Hadley Delta, breccias of various types occur throughout the area. The individual samples from these regions are listed and briefly described in table 6-I. The samples collected at the mountain front contain a number of unique specimens that cannot be more generally characterized. Several individual specimens that deserve particular mention, even in this preliminary report, will be discussed in detail.

Much of the material collected at the mountain front is ultimately derived from basic igneous rocks (i.e., extrusive basalts of plagioclase-enriched cumulates). Unlike the mare samples, however, all of the samples from the mountain front have undergone substantial shock metamorphism or brecciation, followed by lithification, in the time interval between the crystallization of the igneous source rock and the deposition of these rocks at the present site. At the present time, it is not clear whether most of the shock-metamorphism and brecciation history observed in these rocks is post-Imbrian (i.e., followed the formation of the mountain front), or whether some of the breccias existed in essentially the present form before the Imbrian impact. In other words, it is not known whether any of the igneous rocks represented in the breccias are derived from igneous structures that underlie the debris layer that now covers the mountain front where the samples were collected. Much more careful study of these samples is required before this question can be answered.

One fragment of extremely pure anorthite (sample 15415) was recognized as an anorthosite during the lunar-surface traverse. A macroscopic photograph of this specimen is shown in figure 6-8. The bulk chemical composition and the normative composition are listed in table 6-III. Thin-section studies of this sample reveal that the rock consists of over 99-percent plagioclase grains (An_{95} to An_{98}) from 1 to 3 cm across with approximately 1-percent clinopyroxene (0.1 mm across), which consists of interstitial augite and pigeonite and augite inclusions in large plagioclase crystals. The wide range of crystal dimensions found in this rock makes it difficult to characterize the texture in a single photomicrograph. Furthermore, thin sections of the rock reveal a variety of cataclastic textures, including at least two distinctly different types of postcrystallization deformation. Small patches of highly fractured and shattered plagioclase surrounded by glass are observed in one or two sections, which suggests that this anortho-

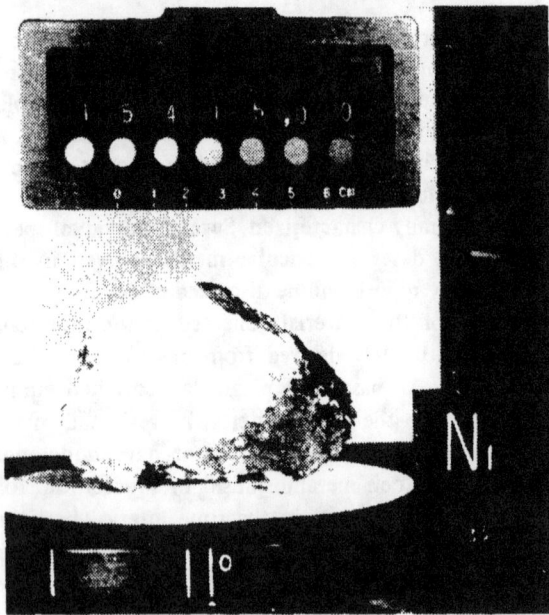

FIGURE 6-8.—Anorthosite (sample 15415) found as a large clast in breccia at the rim of Spur Crater. The darker part of the sample is a dust coating on a portion of the surface embedded in the breccia matrix (S-71-42951).

site fragment was involved in a substantial impact subsequent to crystallization. In addition, polygonal plagioclase grains that range from 0.1 to 1.3 mm in average dimension often occur along fractures within a single crystal of plagioclase (fig. 6-9). Similar polygonal grains also occur along crystal boundaries. The polygonal plagioclase grains suggest that an older deformation and annealing took place during either the original cooling of the anorthosite or a subsequent high-temperature period. Photographs from the lunar surface show that the rock was perched on a pedestal of breccia, which suggests that the anorthosite was a clast in a breccia. Examination of a thin section of the pedestal material shows a complex set of flow-banded glass, glassy breccia, anorthosite fragments, and basaltic material, which is further evidence that the anorthosite was a clast.

A second unique rock (sample 15455) consists of one-half white material and one-half black material, with inclusions of white in black along the contact (fig. 6-10). Thin-section study indicates the white portion to be a severely shattered norite, with approximately 75 percent plagioclase and 25 percent orthopyroxene. Reconstruction of the relict grains indicates coarse plagioclase grains with interstitial orthopyroxene. The black material, which occurs as

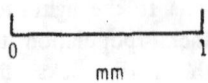

FIGURE 6-9.—Thin section of anorthosite (sample 15415) showing parallel trains of polygonal plagioclase grains that have recrystallized along old fractures in a large plagioclase grain, as evidenced by displaced twin lamellae.

black veins in the norite, is a breccia that contains clasts of norite, anorthosite, and nonmare basalts.

Two rocks (samples 15256 and 15418) appear to be shock-melted igneous rocks. Rock sample 15256 contains a mixture of clasts (or inclusions) in fine-grained laths of clinopyroxene and plagioclase alined along a curving flow-type structure. The chemical composition (table 6-III) is nearly identical to that of the high-Fe-content mare basalts. That this rock was ejected from the mare plain onto the mountain front by a collision of sufficient magnitude to metamorphose a dense crystalline rock seems likely. Rock sample 15418 is unique both in texture and in chemical composition. The chemical composition is that of a very anorthite-rich gabbroic anorthosite (table 6-III). The FeO/MgO ratio of this rock (fig. 6-2) clearly indicates that the rock was not derived from liquids with the high FeO/MgO ratio of the mare basalts. In fact, the FeO/MgO ratio of both samples 15418 and 15415 (the anorthosite) relates these plagioclase-rich rocks to the high-Al and low-Fe basalt found at the Apollo 14 site and in the Apollo 12 soil fragments.

The thin sections of rock sample 15418 reveal a banded structure of devitrified glass and microcrystalline basalts (fig. 6-11), as well as patches of devitrified maskelynite, annealed pyroxene, and microcrystalline basalt. This structure suggests that the fragment was completely shock-melted and subsequently crystallized or annealed to produce fine-grained crystalline patches or devitrified-glass patches.

The K content of this rock, determined by gamma-ray counting, is less than 100 parts per million (ppm). This value suggests an original cumulative origin for the rock, with very little admixture of soil or other rocks during impact metamorphism.

The largest sample returned from the mountain front was a coarse breccia (sample 15459) with clasts as large as 8 cm across (fig. 6-12). This sample, which was collected at the rim of Spur Crater, is a tough, dense breccia with a matrix that has a substantial proportion of glass. The most abundant clasts consist of light-colored basalt and anorthosite. A small

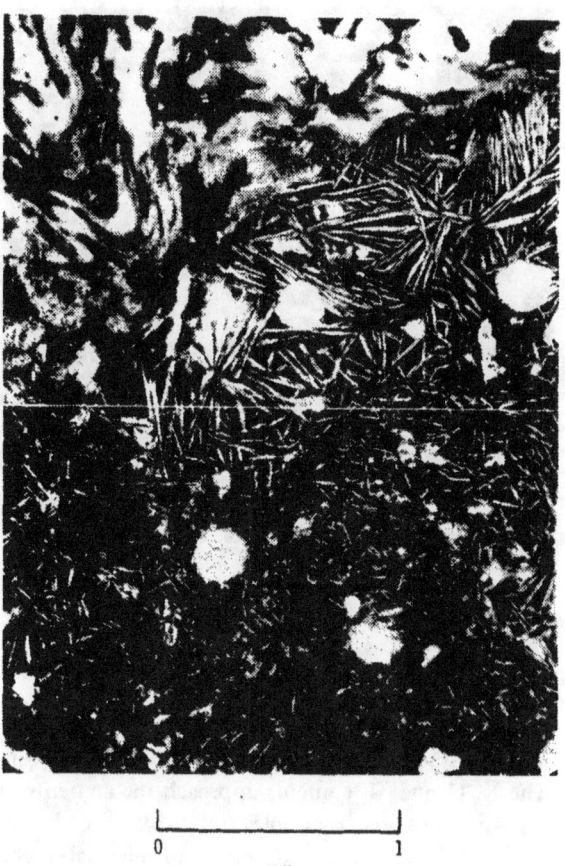

FIGURE 6-11.—Thin section of sample 15418,8 (plane light) showing several vesicles; outer zone of swirled, devitrified glass; and two inner zones of intergrown plagioclase needles (light gray) and pyroxene (dark gray).

proportion of the clasts is composed of green-to-cinnamon-brown pyroxene, which, thus far, has been found only in mare basalts. However, overall clast characteristics indicate that this breccia sample is probably derived from a nonmare terrain in which a small amount of mare material is present.

The largest rock sample closely observed at the mountain front was found at station 2 on the flank of St. George Crater. Two oriented samples (15205 and 15206) from this 1- to 2-m block were returned. In addition, the block was overturned to obtain a soil sample from underneath the boulder. Both the surface photographs and the two returned samples indicate that this rock is a typical breccia sample. The rock consists of a matrix with large amounts of devitrified glass in which clasts of nonmare basalt and gabbro are set. The K, U, and Th contents of sample

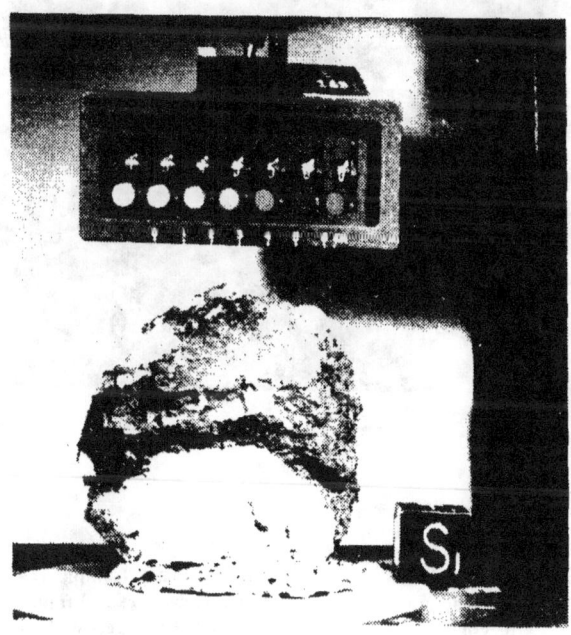

FIGURE 6-10.—Coarse breccia (sample 15455) from the rim of Spur Crater. White areas are highly shattered norite clasts in black, flow-banded breccia (S-71-43891).

FIGURE 6-12.—Coarse breccia (sample 15459) from the rim of Spur Crater. The largest visible clast is 8 cm across; the glassy matrix makes up more than 50 percent of the breccia (S-71-45977).

structure and composition of many of these rocks are not well known at this time because they are obscured by extensive glass coatings. Most of the breccias contain undevitrified glass, either in the matrix or as clasts. Of the 15 thin sections of breccias that have been examined, 13 have a significant amount of brown glass as matrix; some of the thin sections are composed of as much as 50-percent brown glass. Of the remaining two thin sections, one contains clear glass clasts, and partly devitrified glass in the matrix, and the other (from sample 15405) is the only thin section with a clearly recrystallized (or annealed) matrix. However, even in the case of the

15206 are higher than the contents of any other samples studied during the preliminary examination. The K, U, and Th contents approach the contents of the Apollo 14 breccia samples (ref. 6-3).

During the surface traverse, an unusual green material was noticed at the rim of Spur Crater. Several friable fragments of this material were returned (sample 15426). Microscopic studies of these samples indicate that the fragments are breccias that consist of more than 50-percent green glass, which occurs as spheres and fragments of spheres (fig. 6-13). Many of the green-glass spheres have partially devitrified to crystals composed of orthopyroxene. In a few cases, these crystals occur in a radiating pattern similar to the pattern seen in many meteoritic chondrules. Chemical analyses of both this material (sample 15923) and associated soil samples (sample 15301) indicate an unusually high MgO concentration. Similar green-glass spheres are abundant in the Spur Crater soil samples and are present in smaller amounts in most other soil samples.

The breccias collected from the mare plain are clearly different from the breccias collected from the mountain front, in that the mare-plain breccias have a high abundance of mare basalts and opaque minerals, and the mountain-front breccias have a high abundance of nonmare basalts and gabbroic rocks and a low abundance of opaque minerals. The internal

FIGURE 6-13.—Thin section of green rock (sample 15426) showing green-glass spheres of various sizes and fragments of green glass (medium-gray areas). The large bright areas are devitrified green-glass spheres with small patches of nondevitrified green glass remaining (center of photograph).

thin section with a recrystallized matrix, undevitrified clasts of clear glass were found.

Although essentially all the breccias are unrecrystallized, many indicate a complex history of events. In several cases, glass veinlets crosscut other breccia components and transect the preferred orientation, which indicates remobilization of matrix material. In other cases, faulting of glass spheres and other clasts indicates a mechanical disruption of previously existing breccia. Apparently, multiple stages of melting and fracturing can occur without any evidence of recrystallization.

SOIL AND CORE-TUBE SAMPLES

Soil Samples

The 31 different soil samples returned from the Apollo 15 mission represent at least three distinctly different lunar regions, which are mountain slopes, typical mare regolith, and the ray or ridge that trends approximately 30° west of north through the landing site. Four large soil samples, which weigh approximately 1 kg each, were returned for detailed characterization of these regions. All soil samples, except for those returned in the surface environmental sample container, were sieved into four size fractions (less than 1 mm, 1 to 2 mm, 2 to 4 mm, and 4 to 10 mm). The three largest size fractions, which were designated as coarse fines, made up 2 to 10 percent of the soil collected. A 100-mg portion from 18 of the fractions with a particle size less than 1 mm has been examined microscopically to characterize the different soil types. These samples were sieved into standard size fractions that were then individually examined by using both binocular and petrographic microscopes. The size distribution of the finest fractions was determined by direct measurement of individual grains in a computer-controlled optical measuring system. The grain size distributions for a number of soil samples are summarized in figure 6-14.

The particle types in the Apollo 15 soils are similar to the soil from the previous missions in most respects. The major difference is the presence of green-glass spheres that have an index of refraction of 1.65. The soil is composed of five major particle types.

(1) *Agglutinates (or glazed aggregates) plus brown-glass droplets and fragments.*—These particles

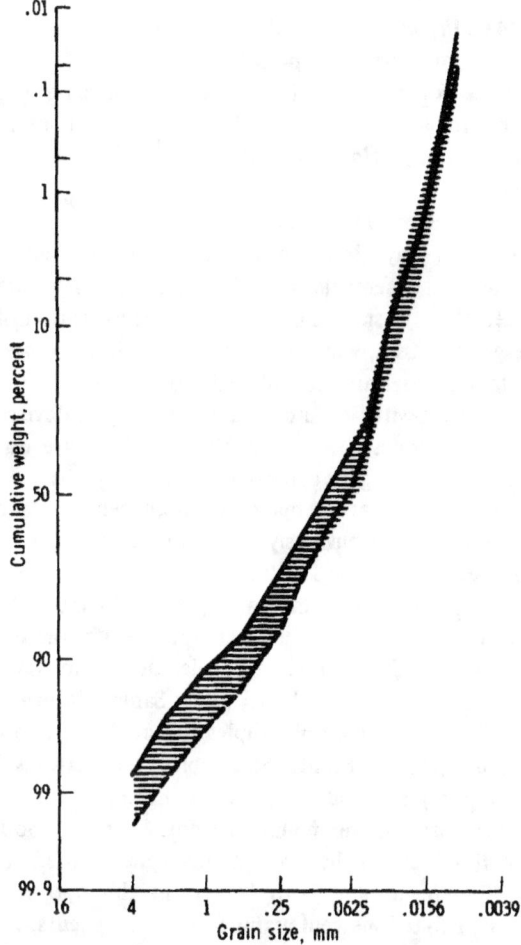

FIGURE 6-14.—Grain size distributions of representative soil samples from stations 2, 6, 7, 8, 9, 9A, and the LM site, plotted on a probability scale. Most of the curves are very similar, and the range is plotted as a shaded band. The solid line represents the coarsest grained soil from station 9A; the dashed line represents soil from station 8. Median grain sizes range from 42 to 98 μm; all soil samples are poorly sorted. Most of the soils are slightly finer grained than soils from other landing sites.

consist of detrital fragments of pyroxene, plagioclase, and glass, bonded by dark-brown to black inhomogeneous glass. The agglutinates are often vesicular and quite irregular in shape.

(2) *Basalt fragments.*—Several different textures are present, granular, ophitic, and hyalocrystalline. Mineral assemblages are usually clinopyroxene, plagioclase, olivine, and opaque minerals.

(3) *Mineral fragments.*—These include mainly augite, pigeonite, plagioclase, orthopyroxene, and olivine.

(4) *Microbreccias.*—Both glassy and recrystallized breccia matrices are present. The vitric (glassy) breccias contain primarily clasts of plagioclase and clinopyroxene. The recrystallized breccias contain mineral clasts, clasts of older breccias, and basalt clasts.

(5) *Glasses.*—The glasses include mainly brown, green, and gray glasses in the form of both droplets and angular fragments. Colorless and red glasses are present in very small amounts. With the exception of the green glasses, which are clear, most of the glasses contain detrital and crystal inclusions. Many of the glasses are devitrified and exhibit a variety of devitrification textures. The green glasses, which are different from any glass component previously observed in lunar soils, are remarkably homogeneous and nonvesicular and obviously are identical to the green glass found in sample 15426.

For purposes of comparison, the soil-sample descriptions used in this section are for one size fraction (0.125 to 0.250 mm). More detailed descriptions will be available in the Apollo 15 Lunar Sample Information Catalog. Of the soil samples collected at the base of the Apennine Front, the samples from stations 2 (four samples) and 7 (three samples) are most characteristic of the mountain front. Soils from both sampling locations have a high green-glass content (7 to 40 percent) and 7- to 14-percent clinopyroxene grains, 1.5- to 7-percent granular basalt fragments, 13- to 16-percent vitric and recrystallized microbreccia fragments, and 40 to 63 percent agglutinates (plus brown glass).

Soils collected from station 6 (five samples), the LM area (one sample), and the Apollo lunar surface experiments package (ALSEP) site (two samples) are similar in composition. The samples from these stations contain 4- to 5-percent basalt fragments that have a variety of textures (granular, ophitic, subophitic, and hyalocrystalline). Also found were 2- to 7-percent recrystallized microbreccias (also, 1- to 3-percent vitric microbreccias in the station 6 soil), 8- to 13-percent clinopyroxene, and 58 to 71 percent agglutinates (plus brown glass). The main difference within this group of samples is in the amount of green glass (0 to 3 percent for the soils from the LM and ALSEP sites and 7 to 11 percent for the soil from station 6).

Two soil samples collected at the edge of Hadley Rille (station 9A) contain abundant clinopyroxene grains (30 to 35 percent); 7- to 16-percent basalt fragments, which have ophitic and granular textures; 5-percent vitric and recrystallized microbreccia fragments; and 30 to 38 percent agglutinates (plus brown glass). The soil collected at station 9 is a hybrid and consists of what ideally would be a mixture of the soils from the LM area and the Hadley Rille edge.

The LM and ALSEP sites are located on a narrow ridge that trends south-southeast to the South Cluster of craters (identified from the Apollo 15 orbital photography). By continuing the trend of the ridge, the ridge would intercept station 6. With the exception of the differences in the green-glass content, the soils from station 6 are very similar to the soils from the LM area. The ridge has been interpreted by Carr and El Baz (ref. 6-4) as a ray from Aristillus or Autolycus. The similarity of soil samples collected along this trend may be another bit of evidence for the presence of this ray.

The soils from stations 2 and 7 appear to be representative of the Apennine Front. The station 7 soil samples, which were collected close to station 6, are from the edge of the relatively fresh, 100-m-diameter Spur Crater, which may have penetrated any surficial ray material to excavate debris from the Apennine Front. The soils appear to have been derived from several types of microbreccia, granular basalts, and clastic rocks composed of green glass, all of which could be rock types derived from the stratified rocks visible on the upper slopes of Hadley Delta. The variable abundance of the unique green-glass component and the concentration of this component in both soil and breccia samples from Spur Crater indicate that green glass may provide a particularly important tracer in the study of the regolith at this site. The rille soils are what would be expected from ground-up basalt flows. The microbreccia component may have been derived from the debris that extends out from the Apennine Mountains or from the well-stratified rocks that are interbedded with basalt flows in the Hadley Rille wall.

Soil Chemistry

The chemical compositions of seven individual soil samples, as determined by standard X-ray fluorescence procedures, are listed in table 6-IV. These compositions are systematically different from the composition of the rocks that presumably underlie the soil samples, particularly in the mare regions. In these regions, the soil has a much higher Al_2O_3

TABLE 6-IV.—X-Ray Fluorescence Analysis of Soil and Breccia Samples[a]

Compound	Soil										Breccia		
	Sample no. (station no.)												
	[b]10084	[c]12070	[d]14163	[e]15021	[f]15101 (2)	15271 (6)	[f]15301 (7)	15471 (4)	15501 (9)	[f]15601 (9A)	15265 (6)	15558 (9A)	[g]15923 (6)
Abundance, percent													
SiO_2	41.86	45.91	47.17	46.56	45.95	46.70	45.91	46.10	46.21	45.05	46.94	46.31	45.18
TiO_2	7.56	2.81	1.79	1.75	1.27	1.47	1.17	1.58	1.81	1.98	1.40	1.89	1.14
Al_2O_3	13.55	12.50	17.22	13.73	17.38	16.51	14.53	12.91	12.20	10.20	16.71	12.40	15.06
FeO	15.94	16.40	10.35	15.21	11.65	12.15	14.05	16.24	16.72	19.79	11.18	16.54	13.72
MnO	.21	.22	.14	.20	.16	.16	.19	.21	.22	.26	.15	.22	.18
MgO	7.82	10.00	9.37	10.37	10.36	10.55	12.12	11.11	10.80	10.89	9.95	10.51	12.14
CaO	12.08	10.43	10.95	10.54	11.52	11.29	10.70	10.42	10.25	9.87	11.19	10.18	11.11
Na_2O	.40	.41	.66	.41	.39	.43	.35	.32	.37	.29	.51	.42	.36
K_2O	.13	.25	.58	.20	.17	.21	.16	.12	.16	.10	.25	.19	.11
P_2O_5	.11	.27	.46	.18	.13	.21	.15	.12	.17	.11	.25	.21	.09
S	.15	.08	.08	.06	.06	.08	.04	.07	.07	.06	.08	.09	.06
Cr_2O_3	.32	.43	.22	--	--	.38	--	.47	.49	.56	.33	.51	.40
Total	100.13	99.71	98.99	99.21	99.04	100.14	99.37	99.67	99.47	99.18	98.94	99.47	99.55
Abundance, ppm													
Sr	169	136	186	135	142	144	113	124	122	109	150	123	111
Zr	312	529	978	410	313	382	260	229	317	199	169	356	152
Nb	19	33	65	24	19	23	17	15	20	13	29	22	10
Rb	3.3	6.9	15	6.1	4.9	5.6	3.8	3.0	4.7	3.1	7.8	5.3	2.7
Th	2.2	6.7	13	3.8	3.3	4.4	4.2	3.0	3.1	<2	4.8	3.6	<2
Ni	238	276	322	288	260	--	268	--	--	--	--	--	--
Y	105	110	213	91	69	84	60	54	72	47	100	78	39

[a]Trace-element abundances were determined for a 1-g powder by using the procedure outlined in reference 6-5.
[b]Apollo 11 sample.
[c]Apollo 12 sample.
[d]Apollo 14 sample.
[e]LM contingency sample.
[f]Comprehensive sample.
[g]Green clod.

content and a much lower FeO content than the associated igneous rocks; furthermore, the niobium (Nb), K, U, and Th contents of all soils are systematically greater than the contents determined for igneous-rock sample 15555. In addition, individual soil samples from the mare region differ in composition from one another, particularly in the characteristics that distinguish the soil from the mare basalts.

The soil collected from the edge of the Hadley Rille is, in all respects, more like the mare basalts than the soil collected in the vicinity of the LM. The three soil samples from the mountain front are consistently higher in Al_2O_3 and lower in FeO than the mare soils; however, the concentrations of other elements (e.g., K, yttrium (Y), and Nb) do not enable mountain-front soils to be distinguished from mare soils. The higher Al_2O_3 and lower FeO content of the mountain-front soil again suggests that Al-rich units must occur in abundance in the mountain front from which this soil was at least partially derived. The inverse correlation of FeO and Al_2O_3 for the soil and breccia samples from this site is shown in figure 6-15. This correlation is nearly linear between 8 and 20 percent for Al_2O_3 and between 10 and 22 percent for FeO. That neither sample 15415 nor sample

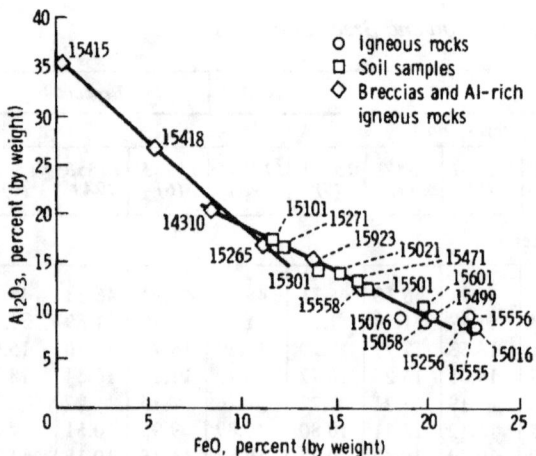

FIGURE 6-15.—The Al_2O_3 and FeO content of Apollo 15 soil and rock samples.

15418 falls on the linear correlation observed for soil and breccia samples is significant; however, Apollo 14 sample 14310 falls on the linear relationship. Thus, both the clast in the mountain-front breccia and the average composition found in the mountain-front soil suggest that the dominant material found in the front may be a plagioclase-rich basalt. In other words, the two end members that are responsible for the linear correlation in figure 6-15 may be an Fe-rich mare basalt and an Al-rich, Fe-poor premare basalt.

Deep-Drill-Core and Core-Tube Samples

A substantial portion of the soil returned from the Apollo 15 landing site was in the form of regolith cores. A rotary percussive drill that was used for the first time during the Apollo 15 mission enabled a deep core to be extracted from a depth of 2.4 m. The length of this core may represent a significant fraction of the total regolith thickness at the Apollo 15 site. Moreover, the unique location of the sample (in a region where both detritus from the degradation of the mountain front and the Autolycus or Aristillus ray material make substantial contributions to the regolith) will greatly enhance the scientific value of the sample. That this single sample will extend the knowledge of recent lunar stratigraphy (as well as solar history) more than any other single sample returned from the Moon is quite likely. In addition, the lowest parts of the core extend well below the penetration depth of cosmic rays; thus, the first lunar samples unaffected by the recent cosmic-ray exposure history of the lunar surface are a part of the core.

The deep-drill tube consists of six sections, each approximately 40 cm long. All the drill sections, except for the lower section, were returned completely full. The total mass of this core material is 1.33 kg. Both the ease of drilling experienced on the lunar surface and the density variations observed for different sections of the core tube suggest that the upper portion of the regolith may consist of layers with different densities. The lowest section of the deep-drill core has an unusually high density, 2.15 g/cm^3, which is the highest density for any soil sample so far returned from the lunar surface.

Only nondestructive examinations by X-radiographs and gamma-ray counting of the deep-drill core have been undertaken to date. Stereo X-radiographs of the entire length of the core reveal significant variations in the abundance of small pebbles; and, more importantly, they show that significant density variations exist along the length of the core. Fifty-eight individual layers that range from 0.5 to 21 cm in thickness have been delineated by using these somewhat subjective criteria. Both the layer contacts and the orientation and occurrence of individual rock fragments are shown in figure 6-16. Nondestructive gamma-ray measurements made at three different depths in the deep-drill core show that the ^{26}Al and ^{22}Na activity decreases markedly in the lowest sections, as would be expected.

Three drive tubes, with maximum penetrations up to 70 cm and containing a total sample weight of 3.3 kg, also were returned from the Apollo 15 site. At present, none of the cores has been opened for examination along the length of the tube; however, X-radiographs have been obtained over the entire length of all tubes. Textural changes within the cores have been made with stereopairs of these X-radiographs. These descriptions are somewhat subjective, with grain sizes and sorting estimated by comparisons with soils of known grain size. Rock fragments larger than 2 mm can be measured, and the shapes can be described. Layering and the distribution of rock fragments in several individual cores are shown in figure 6-16.

A double core was collected at station 2 (St. George Crater) along the outer edge of a 10-m-diameter crater. The core that was collected is 63.9 cm long and weighs 1.279 kg. The bulk density of the soil in the upper half of the double core is 1.36 ± 0.05 g/cm^3, and the density in the lower half is 1.66

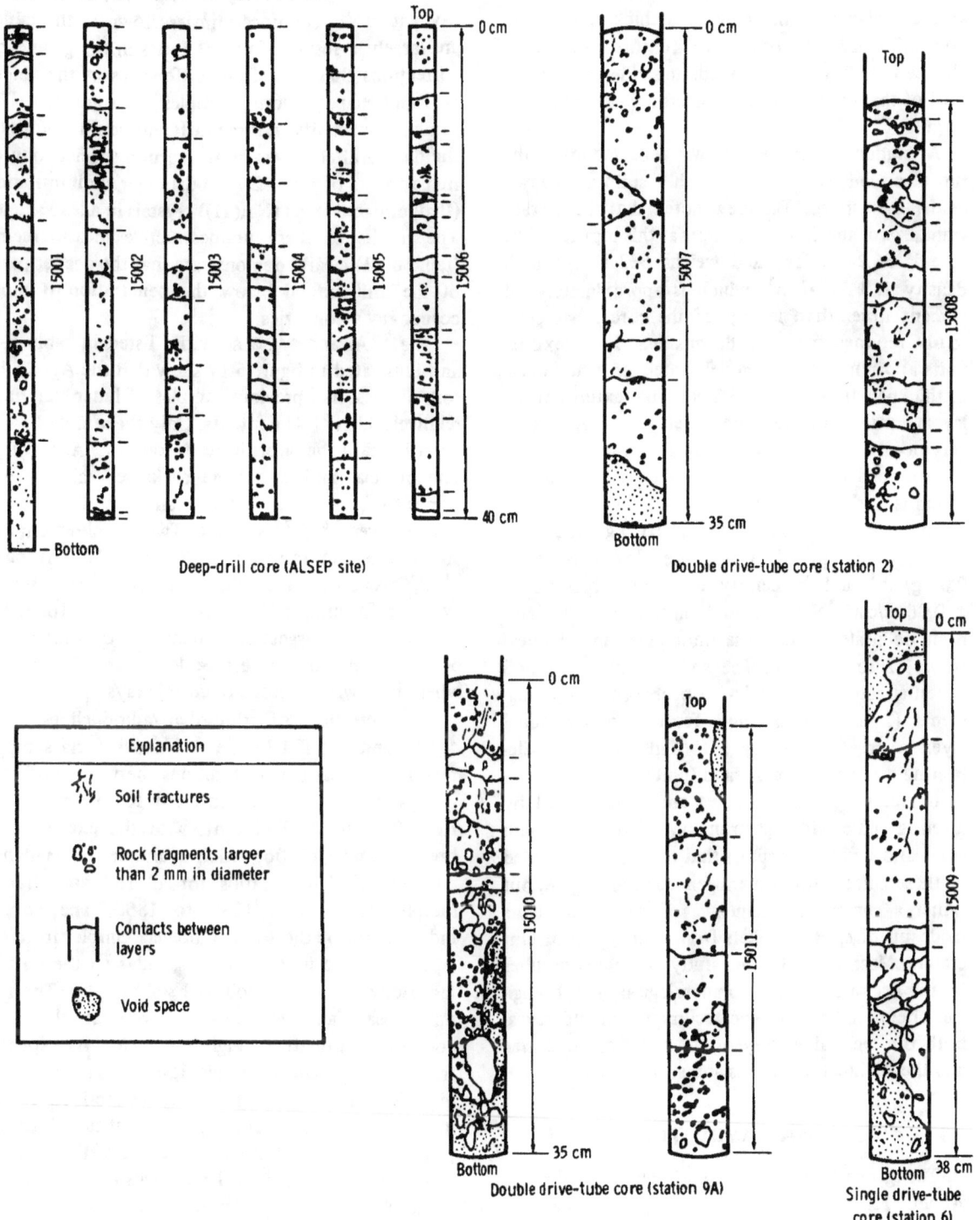

FIGURE 6-16.—Sketches of X-radiographs of the six sections of the deep-drill core collected with a rotary-percussion drill (samples 15001 to 15006) and of the drive-tube cores (samples 15007 to 15011).

± 0.02 g/cm³. Based on textural changes 14 layers were found within the core; bed thicknesses range from 1.2 to 22.0 cm (fig. 6-16). Textures range from silt size (unit 2) to pebbly, medium-sand size (unit 4). Most of the contacts between layers appear to be very irregular.

At station 6, a single core was collected inside the rim of a subdued crater where white soil was exposed below the surface. The crew noted that the average grain size of the soil was coarser at this location. The core is 36.2 cm long and weighs 622 g. The bulk density is 1.35 g/cm³, which is approximately 14 percent lower than in any of the core tubes previously returned from Apollo missions. Four textural units apparently have been delineated, and they range in thickness from 1.6 to 15.2 cm. The textures of the layers range from moderately well sorted silt size to very poorly sorted, granule-bearing, fine-sand size.

A second double core was taken at station 9A, which is at the edge of Hadley Rille. The double core is 65.5 cm long and weighs 1.401 kg. The bulk density in the upper half of the double core is 1.69 ± 0.03 g/cm³, and the density in the lower half is 1.85 ± 0.06 g/cm³. Nine textural units apparently have been delineated, and these units range in thickness from 1.2 to 13.0 cm. The soil is very coarse and contains abundant granule- to pebble-size rock fragments (2 to 48 mm long). Textures of individual layers range from sandy silt (unit 2) to granule-bearing medium to coarse sand (unit 9).

In summary, all the cores are characterized by layering and a wide spectrum of soil textures, grain sizes, and sorting. It appears that the regolith sampled in these cores is not a homogeneous rubble pile, but rather a complicated sequence of clastic sediments made up of ejecta blankets from nearby and distant craters. Much more detailed study of these core tubes by the use of a variety of mineralogical, morphological, chemical, and isotopic techniques should reveal both the regional history of the soil from this area and the specific local events and structure.

GAMMA-RAY ACTIVITY

The gamma-ray spectra of 19 samples from the Apollo 15 site have been measured by using the low-level counting technique described in reference 6-6. The principal gamma-ray activity is a result of the natural radioactivity of K, U, and Th and the cosmic-ray-induced radioactivity of ^{26}Al and ^{22}Na.

The results of this survey are summarized in tables 6-V and 6-VI. The uncertainties shown in the tables are largely a result of uncertainties in the geometric corrections that arise from differences in the thickness and shape of actual samples and standards. The gamma-ray activity in three different depth regions of the deep-drill core was measured in a special counter that consists of two 12.7-cm-diameter sodium iodide (thallium activated) (NaI(Tl)) crystals in a lead shield. The results of these nondestructive measurements (table 6-VI) clearly demonstrate that the deeper parts of the drill extend below the penetration of most cosmic-ray secondaries.

The ^{26}Al and ^{22}Na activities listed in table 6-V and illustrated in figure 6-17 show that the Apollo 15 soil samples, like previous samples of lunar soil, have relatively high ^{26}Al contents. Only the sample 15431 from beneath the anorthosite is low in ^{26}Al activity. The breccias and crystalline rocks generally have a somewhat-lower ^{26}Al activity and, on the average, have a lower ^{26}Al/^{22}Na ratio. Two samples from the Apollo 15 site (samples 15206 and 15086) have ^{26}Al/^{22}Na ratios of less than 1. In the absence of any evidence for unique chemical compositions for these specimens, the suggestion is made that they may have been brought to the surface less than 10^6 yr ago, where they were exposed to cosmic rays.

Concentrations of primordial radionuclides (^{40}K, ^{238}U, and ^{232}Th) in the soil and breccia samples reveal no clear-cut distinctions between different groups of samples. The inferred K contents range from 0.12 to 0.19 percent, with the exception of breccia sample 15206, which is unique for its high K content. The K/U ratios for all soil and breccia samples range from 1230 to 1850. The natural radioactivity of the soil and breccia samples from the Apollo 15 site is, thus, quite similar to the natural radioactivity of the Apollo 12 soil samples. The two mare basalts analyzed thus far have relatively low K contents and relatively high K/U ratios. As expected from the mineralogy, sample 15415 has an unusually high K/U ratio. This fact can be ascribed to the inability of quadravalent U to substitute into the plagioclase lattice. With this exception, the range of K/U ratios in the Apollo 15 samples is similar to the range observed in samples from previous sites. With these additional observations, little doubt appears to exist that the K/U ratio of the Moon is different from that of the Earth, which is 1.0×10^4 to 1.2×10^4 (ref. 6-7).

TABLE 6-V.—*Gamma-Ray Analysis of Lunar Samples*

Sample no.	Mass, g	K, percent (by weight)	Th, ppm	U, ppm	K/U ratio	^{26}Al, dpm/kg	^{22}Na, dpm/kg
Soils (<1mm)							
15431	145.4	0.19 ± 0.02	4.8 ± 0.7	1.1 ± 0.2	1730	68 ± 14	36 ± 5
15021	132	0.16 ± 0.02	5.1 ± 0.7	1.3 ± 0.2	1230	175 ± 25	50 ± 7
15271,16	527.9	0.16 ± 0.03	4.2 ± 0.8	1.2 ± 0.2	1330	130 ± 20	34 ± 5
15211,2	104.2	0.15 ± 0.03	3.8 ± 0.8	0.96 ± 0.20	1530	130 ± 20	57 ± 9
15301	557.2	0.12 ± 0.02	3.2 ± 0.5	0.88 ± 0.15	1360	104 ± 20	40 ± 10
Breccias and metaigneous rocks							
15206	92.0	0.45 ± 0.06	11 ± 2	3.0 ± 0.6	1500	38 ± 15	45 ± 10
15265	314.2	0.19 ± 0.03	5.1 ± 1.0	1.3 ± 0.2	1460	79 ± 15	38 ± 9
15558	1333.3	0.17 ± 0.02	3.4 ± 0.4	1.0 ± 0.1	1700	84 ± 15	36 ± 10
15466	118.0	0.15 ± 0.03	3.5 ± 0.7	0.93 ± 0.20	1610	84 ± 15	40 ± 9
15086	172.1	0.14 ± 0.03	3.2 ± 0.5	0.76 ± 0.11	1840	39 ± 15	40 ± 15
15455	881.1	0.090 ± 0.020	1.9 ± 0.4	0.50 ± 0.10	1800	65 ± 20	39 ± 15
15426,1	125.7	0.082 ± 0.01	1.9 ± 0.4	0.43 ± 0.10	1900	59 ± 12	38 ± 8
15418	1140.7	0.0086 ± 0.0010	0.13 ± 0.04	0.04 ± 0.01	2150	120 ± 40	25 ± 10
Crystalline rocks							
15085	471.3	0.041 ± 0.005	0.51 ± 0.10	0.13 ± 0.03	3150	71 ± 15	33 ± 10
15256	201.0	0.034 ± 0.004	0.46 ± 0.10	0.15 ± 0.02	2260	95 ± 15	36 ± 8
15415	269.4	0.012 ± 0.002	0.007 ± 0.030	0.0024 ± 0.007	--	115 ± 15	36 ± 5

TABLE 6-VI.—*Gamma-Ray Analysis of the Deep-Drill Core*

Deep-drill section	Depth, cm	K, percent (by weight)	Th, ppm	U, ppm	K/U ratio	^{26}Al, dpm/kg	^{22}Na, dpm/kg	β, epm/kg
6-3	11 to 21	0.19 ± 0.03	4.7 ± 1.0	1.3 ± 0.3	1460	57 ± 20	33 ± 18	77 ± 15
4-2	106 to 117	0.17 ± 0.03	4.3 ± 1.0	1.1 ± 0.3	1550	16 ± 18	34 ± 15	45 ± 10
1-2	234 to 245	0.19 ± 0.03	3.7 ± 1.0	1.1 ± 0.3	1730	<11	<10	<9

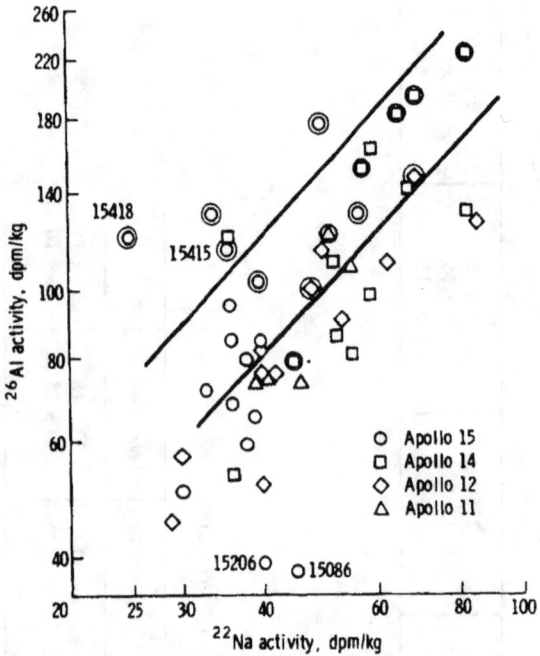

FIGURE 6-17.—Cosmic-ray-induced activities of ^{26}Al and ^{22}Na in the lunar samples. Soil samples are shown by enclosed symbols. These data show that soil samples generally have higher ^{26}Al/^{22}Na disintegration ratios. Within the bounded area, the ^{26}Al/^{22}Na disintegration ratio is between 2 and 3. Sample 15415 is the anorthosite, and sample 15418 is the shock-melted gabbroic anorthosite.

NOBLE-GAS AND CARBON CONTENT

Abundances and isotopic compositions of the five noble gases, helium (He), neon (Ne), argon (Ar), krypton (Kr), and xenon (Xe) were determined by mass spectrometry on five soil and four fragmental-rock samples and are listed in table 6-VII and shown on figure 6-18. Soil samples 15301, 15021, 15101, and 15601 were sampled over a wide area from locations near the LM, St. George Crater, Spur Crater, and Hadley Rille, respectively. Sample 15923, which is soil that contains abundant green-glass spheres, was collected with rock samples 15425 and 15426 near Spur Crater. An almost pure glass fraction and a soil fraction were analyzed separately for this sample. Breccia sample 15558 was collected near the LM, while the other fragmental rocks were collected near the mountain front.

Concentrations of noble gases in most of the Apollo 15 samples that have been examined are quite similar to noble-gas concentrations previously determined in bulk soils returned from the Apollo 12 and 14 sites and are somewhat lower than the concentrations in the Apollo 11 soils and breccia. These noble gases are predominantly of a solar-wind origin, with smaller amounts of a nuclear-reaction-produced component. Elemental abundance ratios fall within the range of values determined previously for lunar soils and have average values (with the exception of sample 15923) of ^{4}He/^{20}Ne = 45, ^{20}Ne/^{36}Ar = 6.3, ^{36}Ar/^{132}Xe = 18 000, and ^{84}Kr/^{132}Xe = 7.5. Individual values for the various ratios vary by a factor of 2 as a result of the differences in efficiencies for solar-wind entrapment and retention. In general, the isotopic composition of the solar-wind component is essentially identical to values determined previously for lunar material returned from the Apollo 11 and 14 sites. For the soil samples, the trapped ^{20}Ne/^{22}Ne ratio is 12.66, and the trapped ^{36}Ar/^{38}Ar ratio is 5.38.

Variations exist among a number of samples in the ^{4}He/^{3}He and ^{40}Ar/^{36}Ar ratios, which apparently cannot be explained by differences in radiogenic ^{4}He and ^{40}Ar contents. Thus, the distinct ^{4}He/^{3}He value of approximately 2430 that is characteristic of all soil samples, except for sample 15601, and the value of 2063 that is characteristic of breccia samples 15265 and 15558 may represent an actual isotopic difference in the trapped solar He in these samples. Likewise, the trapped ^{40}Ar/^{36}Ar ratios vary among samples by a factor as large as 3, which persists even after corrections are made for radiogenic ^{40}Ar based on measured K contents and an age of 3.35×10^9 yr. This variation in the ^{40}Ar/^{36}Ar ratio exists even for those soil samples that have similar ^{36}Ar contents; this phenomenon was not noticed for soils returned from the Apollo 11 and 12 sites. Indications are that the variation in this ratio is a result of differing contents of lunar-atmosphere ^{40}Ar. Three out of four breccias examined have identical measured ^{40}Ar/^{36}Ar values. Whether the variable ^{40}Ar component in the soils is a result of retention efficiencies, geographical distribution of the samples, or some other factor is not apparent. The low ^{40}Ar/^{36}Ar ratios measured on all these samples preclude accurate determinations of radiogenic ^{40}Ar from in situ decay of K.

The isotopic compositions of Kr and selenium (Se) in these various samples reflect a solar-wind origin

TABLE 6-VII.—Noble-Gas Contents of Soil and Breccia Samples[a]

Sample no.	Description of sample	Mass, mg	^3He, cm^3/g (b)	^4He, cm^3/g (b)	^{22}Ne, cm^3/g (b)	^{36}Ar, cm^3/g (b)	^{84}Kr, cm^3/g (b)	^{132}Xe, cm^3/g (b)	$^4He/^3He$	$^{20}Ne/^{22}Ne$	$^{22}Ne/^{21}Ne$	$^{36}Ar/^{38}Ar$	$^{40}Ar/^{36}Ar$
15301,1	Fines	6.14	18.8 × 10^{-6}	45 800 × 10^{-6}	110 × 10^{-6}	233 × 10^{-6}	79.9 × 10^{-9}	16.2 × 10^{-9}	2435	12.60 ± 0.02	28.03 ± 0.16	5.37 ± 0.01	1.78 ± 0.01
15021,4	Fines	23.71	30.9	74 600	114	229	87.2	11.7	2419	12.68 ± 0.04	27.91 ± 0.06	5.41 ± 0.02	0.747 ± 0.005
15101,2	Fines	23.06	24.3	59 450	103	204	70.2	8.68	2444	12.71 ± 0.06	28.56 ± 0.29	5.42 ± 0.01	1.172 ± 0.005
15601,3	Fines	7.62	16.8	35 600	58.3	95.1	44.3	5.51	2123	12.61 ± 0.02	25.42 ± 0.14	5.32 ± 0.01	0.932 ± 0.005
15923,2	Glass	25.14	1.17	1 756	7.54	3.94	1.63	.332	1458	11.65 ± 0.02	10.93 ± 0.10	4.59 ± 0.005	4.40 ± 0.04
15923,2	Fines	8.92	2.48	5 970	14.5	5.92	4.39	1.54	2413	12.72 ± 0.03	25.38 ± 0.12	5.08 ± 0.20	2.71 ± 0.10
15265,3	Breccia	7.67	13.2	27 200	45.9	105	48.8	4.98	2062	12.57 ± 0.01	24.50 ± 0.08	5.33 ± 0.02	1.87 ± 0.01
15298,3	Breccia	8.83	22.1	50 600	89.8	203	125	14.4	2289	12.59 ± 0.02	27.71 ± 0.13	5.36 ± 0.01	1.08 ± 0.01
15498,2	Breccia	18.93	5.70	15 670	44.4	88.8	22.4	3.27	2748	12.24 ± 0.05	21.00 ± 0.09	5.30 ± 0.02	1.88 ± 0.01
15558,3	Breccia	7.47	23.8	49 150	74.2	135	78.9	13.3	2064	12.50 ± 0.02	24.25 ± 0.12	5.31 ± 0.02	1.89 ± 0.01

[a] All abundances are ± 5 to 10 percent based on multiple standard gas analyses. Uncertainties in isotopic ratios represent one standard deviation of multiple measurements. Abundance blank corrections typically were approximately 1 percent and, in no case, greater than 5 percent. Isotopic-ratio blank correctives were significant only for the $^{40}Ar/^{36}Ar$ ratio and have been applied.
[b] At standard temperature and pressure.

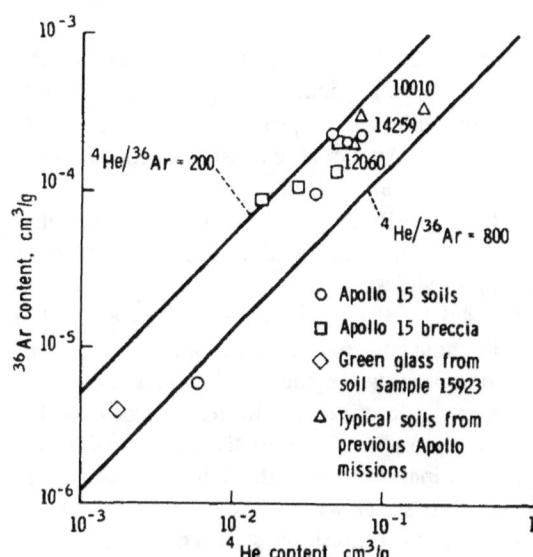

FIGURE 6-18.—Comparison of ^4He and ^{36}Ar contents measured in Apollo 15 soils and breccias and in typical soils from previous Apollo missions. The majority of lunar soils and breccias analyzed exhibit $^4He/^{36}Ar$ ratios between 200 and 800 and show variations in concentrations of these isotopes by a factor of 100. Implantation during solar-wind bombardment is the chief source of these elements.

with small variable amounts of nuclear-reaction-produced gases. The measured ratios for Kr and Xe in all of these samples will be published at a later date. Xenon-136 (relative to mass 132) appears to be slightly more abundant; but, otherwise, the trapped Kr and Xe composition is essentially identical to that measured in previous lunar soils. No evidence of excess ^{129}Xe of a radiogenic origin was seen. Concentrations of cosmic-ray spallation-produced ^{21}Ne, ^{80}Kr, and ^{126}Xe have been calculated for all of these samples. Differences in exposure ages appear to exist, even when the variations in target chemistry are considered, and fall in the range of 50×10^6 to 500×10^6 yr.

Total Carbon Content

The total carbon (C) content of 16 samples has been determined by direct combustion of soil or pulverized rock samples and by using the technique described in reference 6-8. The results of these analyses are shown in figure 6-19, along with previous results obtained for Apollo 11, 12, and 14 samples. The C content of Apollo 15 soil and breccia samples is systematically higher than the C content of the Apollo 15 igneous rocks and is similar to the C content found in soil samples from other sites. The C content of the Apollo 15 extrusive igneous rocks is significantly lower than that reported for igneous rocks from previous sites. This difference probably results from improved handling and contamination-control procedures.

The shock-melted anorthosite (sample 15418) from the mountain front has an even lower C content than the extrusive igneous rocks. This difference perhaps can be ascribed to the removal of C during shock metamorphism. Alternatively, the difference may be an indication that the gabbroic anorthosite from which this rock was derived had an even lower C content than the extrusive basalts. The low C content of this rock further substantiates the earlier assertion that the rock underwent shock metamorphism with negligible admixture of soil or regolith to the parent rock.

The substantiation of the systematic difference in the C content of the soil and igneous rocks appears to confirm the hypothesis (ref. 6-8) that most of the C in the lunar soil may be ascribed to the solar wind. From the data given in this section, the inference may be made that the C content of the average Moon is much lower than was previously inferred from analyses of lunar soil or lunar igneous rocks.

DISCUSSION AND CONCLUSIONS

The primary objective of the Apollo 15 mission was to sample and return to Earth premare materials exposed in the steep mountain front that borders the Hadley Plain. A second objective, somewhat lower in priority, was to sample and examine the edge of Hadley Rille. Even the present cursory examination of the returned samples and the rille-edge photography indicates rather clearly that the mare structure under the Hadley Plain is similar to the structure of many terrestrial lava fields that have been built up by a series of successive lava flows. The textures and bulk chemical compositions of the mare lavas examined in this study confirm previous conclusions that the lunar mare is composed of a series of extrusive volcanic rocks that are rich in Fe and poor in Na.

The suite of samples returned from the base of Hadley Delta is remarkably variable. Several unique rock specimens that were obtained from this sampling site provide significant insights into lunar processes, such as shock metamorphism, and into the pre-Imbrian igneous history. At present, these samples do not provide a clear picture of the kinds of rocks and rock units that make up the mountains that surround the Hadley Plain. The samples clearly do not exclude the intriguing possibility that unmodified extrusive and intrusive igneous rocks may be exposed in the mountain front. Two specimens provide further and quite clear evidence for extensive separation of plagioclase from relatively large igneous intrusions in pre-Imbrian time.

Several specific conclusions can be drawn from the preliminary examination of the lunar samples returned from the Apollo 15 site.

(1) The K/U ratio of both the mare basalts and the soil further establishes the earlier conclusion (ref. 6-7) that the K/U ratio clearly distinguishes the Moon from the Earth.

(2) The soil chemistry and lithology of fragments in the rake sample and in breccia clasts indicate that plagioclase or Al-rich basalts are a common component of the premare lunar surface.

(3) Examination of the 2.5-m-deep-drill core indicates that the lunar regolith probably has a substantial stratigraphic history.

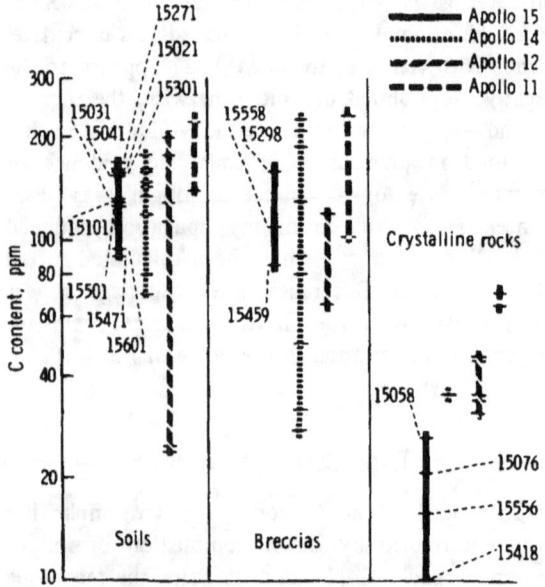

FIGURE 6-19.— Comparison of the total C abundances for the Apollo 15 sample types with those of previous Apollo missions. Previous total C abundance data are taken from references 6-2 and 6-8.

(4) An unusual green-glass component, which is rich in the mountain-front soil samples, suggests that a Mg- and Fe-rich rock type that has not been sampled at previous Apollo sites exists in the vicinity of the Apollo 15 site.

(5) Most of the C found on the lunar surface was implaced there by the solar wind.

(6) Variations in the ^{40}Ar abundance in several soil samples further support the hypothesis that ^{40}Ar may escape into the lunar atmosphere and be recaptured from the atmosphere.

Additional studies of the samples described in this section will almost certainly lead to additional interpretations of the history and selenology of the Apollo 15 site. In summary, several of the outstanding questions that deserve serious attention during the detailed study of samples returned from the Apollo 15 site will be restated.

(1) What is the nature of the stratified material that presumably underlies the regolith that blankets the mountain front? What time span is represented by the layers that appear to make up the mountains?

(2) How many extrusive units that underlie the mare are represented in the returned samples? Can these units be resolved in terms of absolute ages?

(3) Can the Autolycus or Aristillus ray be uniquely associated with some of the returned samples? If it can, what is the age of deposition of the ray?

(4) What are the times of deposition of the individual layers found in the different core tubes?

REFERENCES

6-1. Norrish, K.; and Hutton, J.T.: An Accurate X-ray Spectrographic Method for the Analysis of a Wide Range of Geological Samples. Geochim. Cosmochim. Acta, vol. 33, Apr. 1969, pp. 431-453.

6-2. Ahrens, L.H.; and Von Michaelis, H.: The Composition of Stony Meteorites III, Some Inter-Element Relationships. Earth Planet. Sci. Lett., vol. 5, 1969, pp. 395-400.

6-3. Lunar Sample Preliminary Examination Team: Preliminary Examination of Lunar Samples from Apollo 14. Science, vol. 173, no. 3998, Aug. 20, 1971, pp. 681-693.

6-4. Carr, M.H.; Howard, K.A.; and El-Baz, Farouk: Geologic Maps of the Apennine-Hadley Region of the Moon, Apollo 15 Pre-Mission Maps, U.S. Geol. Survey Map I-723, 1971.

6-5. Norrish, K.; and Chappell, B.W.: X-ray Fluorescence Spectrography. Physical Methods in Determinative Mineralogy, ch. 4, J. Zussman, ed., Academic Press (London and New York), 1967, pp. 161-214.

6-6. O'Kelley, G.D.; Eldridge, J.S.; Schonfeld, E.; and Bell, P.R.: Cosmogenic Radionuclide Concentrations and Exposure Ages of Lunar Samples from Apollo 12. Proceedings of the Second Lunar Science Conference, vol. 2, A.A. Levinson, ed., MIT Press (Cambridge, Mass.), pp. 1747-1756.

6-7. Lunar Sample Preliminary Examination Team: Preliminary Examination of Lunar Samples from Apollo 11. Science, vol. 165, no. 3899, Sept. 19, 1969, pp. 1211-1227.

6-8. Moore, C.B.; Gibson, E. K.; Larimer, J.W.; Lewis, C.F.; et al.: Total Carbon and Nitrogen Abundances in Apollo 11 Lunar Samples and Selected Achondrites and Basalts. Proceedings of the Apollo 11 Lunar Science Conference, vol. 2, A.A. Levinson, ed., Pergamon Press (New York), 1970, pp. 1375-1382.

ACKNOWLEDGMENTS

The Preliminary Examination Team serves three purposes: (1) presentation of geologic, petrographic, and chemical data about the returned samples to the Lunar Sample Analysis Planning Team as an aid in planning sample allocations to principal investigators; (2) compilation of specimen descriptions for a sample catalog that is prepared by the curator's office and distributed to principal investigators; and (3) preparation of a preliminary report about the sample return to be made available to the general public through publication in a scientific journal.

The members of the Lunar Sample Preliminary Examination Team are P.W. Gast, NASA Manned Spacecraft Center (MSC); W.C. Phinney, MSC; M.B. Duke, MSC; L.T. Silver, California Institute of Technology; N.J. Hubbard, MSC; G.H. Heiken, MSC; P. Butler, MSC; D.S. McKay, MSC; J.L. Warner, MSC; D.A. Morrison, MSC; F. Horz, MSC; J. Head, Bellcomm, Inc.; G.E. Lofgren, MSC; W.I. Ridley, MSC; A.M. Reid, MSC; H. Wilshire, U.S. Geological Survey (USGS); J.F. Lindsay, Lunar Science Institute (LSI); W.D. Carrier, MSC; P. Jakes, LSI; M.N. Bass, MSC; P.R. Brett, MSC; E.D. Jackson, USGS; J.M. Rhodes, Lockheed Electronics Corp. (LEC); B.M. Bansal, LEC; J.E. Wainwright, LEC; K.A. Parker, LEC; K.V. Rodgers, LEC; J.E. Keith, MSC; R.S.Clark, MSC; E. Schonfeld, MSC; L. Bennett, MSC; M. Robbins, Brown and Root-Northrop (BRN); W. Portenier, BRN; D.D. Bogard, MSC; W.R. Hart, BRN; W.C. Hirsch, BRN; R.B. Wilkin, BRN; E.K. Gibson, MSC; C.B. Moore, Arizona State Universtiy (ASU); and C.F. Lewis, ASU.

The members of the Lunar Sample Analysis Planning Team are A.J. Calio, MSC; M.B. Duke, MSC; J.W. Harris, MSC; A.L. Burlingame, University of California at Berkeley; D.S. Burnett, California Institute of Technology; B. Doe, USGS: D.E. Gault, NASA Ames Research Center; L.A. Haskin, University of Wisconsin; W.G. Melson, U.S. National Museum, Smithsonian Institution; J. Papike, State University of New York; R.O. Pepin, University of Minnesota; P.B. Price, University of California at Berkeley; H. Schnoes, University of Wisconsin; B. Tilling, NASA Headquarters; N.M. Toksoz, Massachusetts Institute of Technology; and J.A. Wood, Astrophysical Observatory, Smithsonian Institution.

7. Soil-Mechanics Experiment

J.K. Mitchell,[a][†] L.G. Bromwell,[b] W.D. Carrier, III,[c] N.C. Costes,[d] W.N. Houston,[a] and R.F. Scott[e]

INTRODUCTION

The purpose of the soil-mechanics experiment is to obtain data on the physical characteristics and mechanical properties of the lunar soil at the surface and subsurface and the variations of these properties in lateral directions. The characteristics of the unconsolidated surface materials provide a record of the past influences of time, stress, and environment. Of particular importance are such properties as particle size and shape; particle-size distribution, density, strength, and compressibility; and the variations of these properties from point to point. An additional objective is to develop information that will aid in the interpretation of data obtained from other surface activities or experiments and in the development of lunar-surface models to aid in the solution of engineering properties associated with future lunar exploration.

The Apollo 15 soil-mechanics experiment has offered greater opportunity for study of the mechanical properties of the lunar soil than previous missions, not only because of the extended lunar-surface stay time and enhanced mobility provided by the lunar roving vehicle (Rover), but also because four new data sources were available for the first time. These sources were (1) the self-recording penetrometer (SRP), (2) new, larger diameter, thin-walled core tubes, (3) the Rover, and (4) the Apollo lunar-surface drill (ALSD). These data sources have provided the best bases for quantitative analyses thus far available in the Apollo Program.

For the first time, quantitative measurement of forces of interaction between a soil-testing device and the lunar surface has been possible. The diversity of the Hadley-Apennine area, the traverse capability provided by the Rover, and the extended extra-vehicular-activity (EVA) periods compared with the earlier missions have provided opportunity for study of the mechanical properties of the soil associated with several geologic units.

Although many of the analyses and results presented in this report are preliminary in nature and more detailed analyses and simulations are planned, the following main results have been obtained.

(1) Although the surface conditions appear quite similar throughout the Hadley-Apennine site, considerable variability exists in soil properties, both regionally and locally, as well as with depth.

(2) In situ densities range from approximately 1.36 to 2.15 g/cm^3, a range that indicates very great ranges in strength and compressibility behavior.

(3) No evidence of deep-seated slope failures has been noted, although surficial downslope movement of soil has occurred, and the soil on steep slopes along the Apennine Front is in a near-failure condition.

(4) Quantitative data provided by the SRP and the soil-mechanics trench have indicated a density of almost 2 g/cm^3, a friction angle of approximately 50°, and a cohesion of 1 kN/m^2 for the soil at station 8 (fig. 5-2, section 5). These values are higher than those deduced for sites studied in earlier missions.

(5) New core tubes developed for this mission performed very well, and subsequent studies should enable a reliable estimation of in situ densities from the returned samples.

These and a number of other conclusions have emerged from the data and analyses presented in this report.

---

[a] University of California at Berkeley.
[b] Massachusetts Institute of Technology.
[c] NASA Manned Spacecraft Center.
[d] NASA Marshall Space Flight Center.
[e] California Institute of Technology.
[†] Principal Investigator.

SUMMARY OF PREVIOUS RESULTS

Observations at five Surveyor landing sites and at Mare Tranquillitatis (Apollo 11) and Oceanus Procellarum (Apollo 12) indicated relatively similar soil conditions, although Apollo 12 core-tube samples showed a greater variation in grain-size distribution with depth than had been found in the Apollo 11 core-tube samples. On the basis of data from these missions, it was established (refs. 7-1 and 7-2) that the lunar soil is generally composed of particles in the silty-fine-sand range and that the material possesses a small cohesion and a friction angle estimated to be 35° to 40°. Best estimates of the in-place density of the soil range from approximately 1.5 to 2.0 g/cm^3. Simulation studies (ref. 7-3) have shown that both the cohesion and angle of internal friction are likely to be very sensitive functions of density.

Fra Mauro, the Apollo 14 landing site, represented a topographically and geologically different region of the Moon than had been visited previously. At that site, a greater variation in soil characteristics, both laterally and within the upper few tens of centimeters, was observed (ref. 7-4). Much coarser material (medium- to coarse-sand size) was encountered at depths of only a few centimeters at some points, and the soil, in some areas, was much less cohesive than the soil observed from previous missions. The results of measurements using the Apollo simple penetrometer suggested that the soil in the vicinity of the Apollo 14 Apollo lunar surface experiments package (ALSEP) may be somewhat stronger than soil at the landing sites of Surveyor III and VII as reported in reference 7-5. However, computations of soil cohesion at the site of the Apollo 14 soil-mechanics trench yield lower bound estimates (0.03 to 0.10 kN/m^2) considerably less than anticipated (0.35 to 0.70 kN/cm^2) from the results of earlier missions. Available data suggested also that the soil at the Fra Mauro site generally increases in strength with depth and is less dense and less strong at the rims of small craters than in level intercrater regions.

METHODS

Quantitative analyses of the mechanical properties of the lunar soil in situ are made using two main approaches, singly and in combination. The approaches are (1) simulations, wherein terrestrial measurements are made using appropriately designed lunar-soil simulants to provide a basis for prediction of probable behavior before the mission and replication of actual behavior after the mission and (2) theoretical analyses, which can be used to relate observed behavior to soil properties and imposed boundary conditions. Because of the difference between lunar and terrestrial gravity, theoretical adjustment of the results of simulations usually is required.

Houston and Namiq (ref. 7-6) and Costes et al. (ref. 7-7) have described simulation studies for the prediction of the penetration resistance of lunar soils and the evaluation of lunar-soil mechanical properties from in-place penetration data. Mitchell et al. (ref. 7-3) relate footprint depth to soil density. Houston and Mitchell (ref. 7-8) and Carrier et al. (ref. 7-9) describe how simulations can be used to determine the influences of core-tube sampling on the original properties of the lunar soil.

Theories of soil mechanics are reasonably well established, although the inherent variability of most soils and difficulties in determination of stresses in the ground require judgment in the application of these theories. Scott (ref. 7-10) and other soil-mechanics texts present these theories in detail. The theory of elasticity is used for computation of stresses and displacements, and the theory of plasticity is used to relate failure stresses and loads to soil-strength parameters. For these failure analyses, the Mohr-Coulomb strength theory is used. According to this theory, which has been shown to be sufficiently accurate for most terrestrial soils, the shear strength s can be represented by

$$s = c + \sigma \tan \phi \qquad (7\text{-}1)$$

where c is unit cohesion, σ is normal stress on the failure plane, and ϕ is the angle of internal friction. It has been assumed, on the basis of extremely limited laboratory data, that the same approach can be applied to lunar-soil behavior.

DESCRIPTION OF DATA SOURCES

As has been the case for the three previous Apollo missions, observational data provided by crew commentary and debriefings and by photography have been useful for deduction of soil properties. The excellent quality of the television, coupled with the fact that video coverage was available for most of the stations visited by the crew, has made detailed study of some of the activities of interest to the soil-mechanics experiment possible. Interactions between the astronauts and the lunar surface, as indicated by their footprints, and interactions of the small scoop, tongs, core tubes, and flagpole with the lunar surface, have provided valuable soil-behavior information. Quantitative data have been obtained from the following sources.

Soil-Mechanics Trench

During EVA-2, the lunar module pilot (LMP) excavated a trench at station 8 (fig. 5-2, section 5) with a near-vertical face to a depth of approximately 28 cm. This trench provides data on soil conditions with depth and a basis for computation of soil cohesion, as described subsequently in this section.

Self-Recording Penetrometer

The SRP, available for the first time on Apollo 15, was used to obtain data on penetration compared to force in the upper part of the lunar soil. The SRP (fig. 7-1) weighs 2.3 kg, can penetrate to a maximum depth of 76 cm, and can measure penetration force to a maximum of 111 N. The record of each penetration is scribed on a recording drum contained in the upper housing assembly.

The lunar-surface reference plane, which folds for storage, rests on the lunar surface during a measurement and serves as datum for measurement of penetration depth. Three penetrating cones, each of $30°$ apex angle and base areas of 1.29, 3.22, and 6.45 cm^2, are available for attachment to the penetration shaft, as well as a 2.54- by 12.7-cm bearing plate. The 3.22-cm^2 (base area) cone and the bearing plate were used for a series of six measurements at station 8. The SRP is shown in use during a premission simulation at the NASA Kennedy Space Center in figure 7-2.

Core Tubes

Core tubes of a different design than those previously available were used during the Apollo 15 mission. These thin-walled tubes made of aluminum are 37.5 cm long, 4.13 cm inside diameter, and 4.38 cm outside diameter. Individual tubes can be used singly or in combination. The components of a double-core-tube assembly are shown in figure 7-3; a double-core-tube sampling at station 9A during EVA-3 is depicted in figure 7-4.

The new core-tube designs were developed to satisfy three objectives: (1) to reduce the amount of sample disturbance, (2) to increase the size of the sample, and (3) to facilitate ease of sampling by the crew. These considerations are discussed in references 7-8 and 7-9. Preliminary evaluations based on crew comments and on Lunar Sample Preliminary Examination Team (LSPET) examination of the Apollo 15 cores indicate that these objectives were achieved.

Rover

The Rover is a four-wheeled surface vehicle with a double-Ackerman steering system. Each wheel is powered by an electric motor. The wheel "tires" are

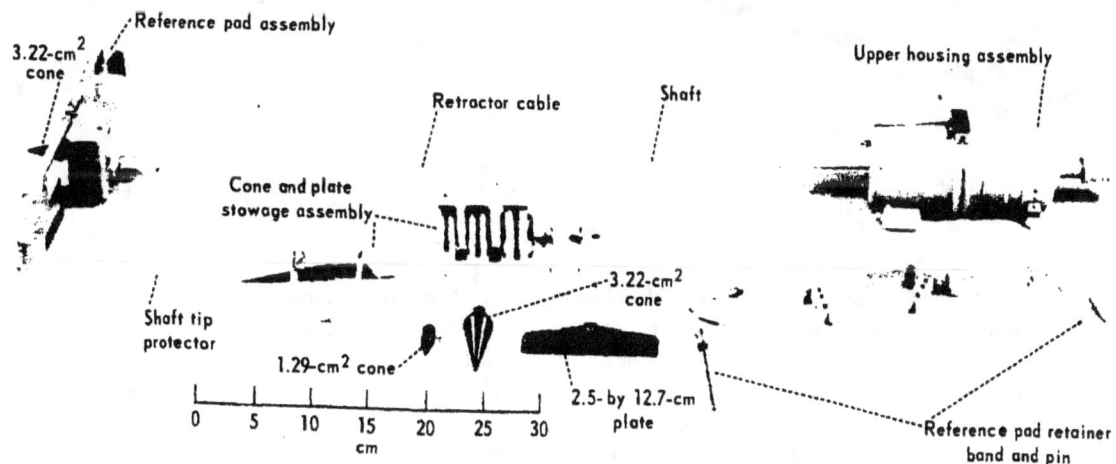

FIGURE 7-1.—Self-recording penetrometer.

FIGURE 7-2.—Self-recording penetrometer in use during premission simulation.

approximately 290 N. At this load, the average unit pressure exerted by the wheel on the soil is approximately 0.7 N/cm^2 and the tire deflection is 5.1 cm. At wheel loads of 178 N and 377 N, corresponding to wheel-load transfer at slope angles of 20°, the wheel deflections are 3.6 cm and 5.6 cm, respectively. The Rover is shown in the vicinity of the ALSEP site during EVA-1 in figure 7-5.

SOIL CHARACTERISTICS AT THE HADLEY-APENNINE SITE

Soil cover is present at all points in the Hadley-Apennine Region except for the bedrock exposures visible on the Hadley Rille wall. The soil layer appears to become thinner going down over the rim of the rille. Away from the rille, a soil depth of 3 to 4.5 m was estimated by the commander (CDR) on the basis of a crater observed during EVA-2. The surface appears similar in color (i.e., shades of gray and gray-brown) to that seen at the other Apollo sites, although wider variations were observed. Surface textures are also similar, ranging from smooth areas free of rock fragments through patterned ground to

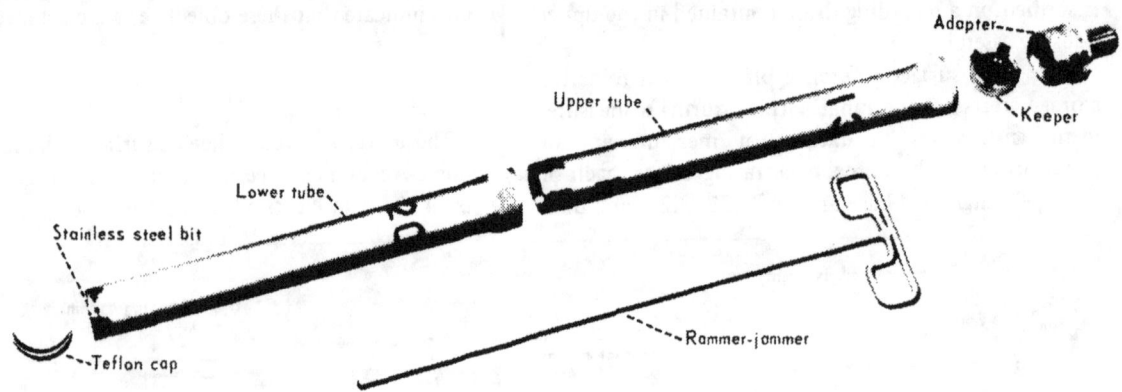

FIGURE 7-3.—Apollo 15 double core tube as used on EVA-1 and EVA-3. The single tube taken on EVA-2 was an upper tube.

made of thin, steel, piano-wire mesh, and 50 percent of the contact area with the lunar surface is covered with a chevron tread. The unloaded wheel has a diameter of 81.5 cm, a section width of 23.2 cm, and a section height of 18.6 cm. The average wheel load on level ground in lunar gravity, including the weight of the vehicle, the payload, and two crewmen, is

areas heavily populated by larger rocks and fragments. Of considerable interest and importance is the fact that the soil strength and compressibility (and, therefore, almost certainly, the density) vary significantly, not only on a large scale from station to station but also locally within short distances, as will be shown later.

FIGURE 7-4.—Double core tube at station 9A pushed to a depth of 22 cm. The tube was driven to a final depth of 68 cm by application of approximately 50 hammer blows (AS15-82-11161).

FIGURE 7-5.—Rover near ALSEP site during EVA-1 (AS15-85-11471).

Textural and Compositional Characteristics

Grain-size-distribution curves have been obtained by the LSPET for samples from several locations. Some are shown in figure 7-6, and bands indicating size ranges for samples from the previous Apollo sites (refs. 7-11 and 7-12) are also indicated. It is of interest that the samples examined thus far do not exhibit as much variability in grain-size distribution as that observed for different samples from the Apollo 12 and 14 sites. Available distributions indicate the Apollo 15 soils to be well-graded, silty, fine sands and fine, sandy silts. The sample from bag 194 (station 7 near Spur Crater) is one of the coarsest samples returned. No data are available on size distributions of particles finer than 0.044 mm. Photomicrographs of four size ranges from a sample taken at the bottom of the soil-mechanics trench are shown in figure 7-7. It may be seen that most particles are subrounded to angular, with occasional spherical particles. Gross particle shapes are typical of those in terrestrial soils of similar gradations. However, the surface textures of many of the particles (e.g., the agglutinates and the microbreccias) are more irregular than in common terrestrial soils. The influences of these unusual characteristics on mechanical properties are yet to be determined.

Study of the soil fraction finer than 1 mm by the LSPET has shown that soils from different areas have different compositions (table 7-I). It is reasonable to expect that some of the physical-property differences observed in different areas reflect these compositional differences.

Soil Profiles

Data on the variability of lunar-soil properties with depth below the surface are available from four sources: the core tubes, the deep core sample obtained using the ALSD, the soil-mechanics trench, and the SRP. The LMP reported no signs of layering while excavating the trench to a depth of 30 cm at station 8, and no layering is visible in the photographs of the trench. However, the LMP did report encountering some small white and black fragments. The trench bottom was reported to be of much firmer material than the overlying soil. Samples from the trench bottom were chipped out in platy fragments approximately 0.5 cm in length.

However, the results of X-ray examination of the core tubes and deep drill samples have led the LSPET to conclude that many different units exist with depth. The presence of a large number of units indicates a very complex soil structure, which implies a high local variability in properties.

Core Samples

Drive tubes.—More than three times as much lunar soil and rock was returned in the Apollo 15 drive core tubes than from the three previous missions com-

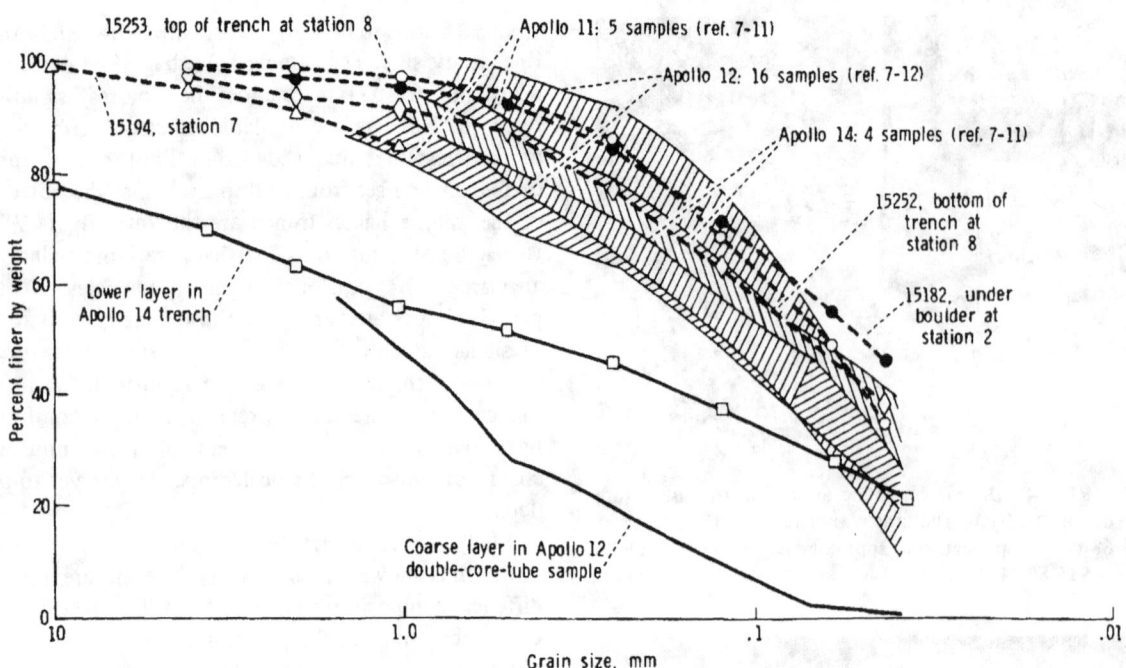

FIGURE 7-6.—Grain-size-distribution curves for several Apollo 15 samples compared with curves for samples from other Apollo sites.

bined (3302 g compared to 932 g). The core samples also appear to be less disturbed than the earlier samples. These improvements are a direct result of a new core tube designed on the basis of soil-mechanics considerations. The new tubes (fig. 7-3) reflect four important changes compared with the tubes designed for use in the previous missions: (1) inside diameter increased from 1.97 to 4.13 cm (the geometry of the Apollo 11, 12 to 14, and 15 core tubes are compared in fig. 7-8), (2) decreased wall thickness, (3) elimination of the Teflon follower and the introduction of the keeper, and (4) redesign of the bit.

The previous core tubes used a follower that was pushed up inside the core tube by the soil column during sampling. The follower was intended to resist movement of the soil inside the tube until it could be returned to Earth. Unfortunately, the follower also exerted a force of approximately 13 N to the soil during sampling, which adversely affected the recovery ratio.[1] Simulations performed by Carrier et al. (ref. 7-9) indicated that the follower reduced the recovery ratio from 80 percent to 55 percent for an Apollo 12 to 14 single-core-tube sample and from 70 percent to 63 percent for a double-core-tube sample. The new keeper, shown in the exploded view of the Apollo 15 core tube in figure 7-3, is stored in the adapter until after the sample has been obtained. The astronaut then inserts the "rammer-jammer" through a hole in the top of the adapter and pushes the keeper down until it comes into contact with the soil. The keeper has four leaf springs that dig into the wall of the core tube and resist movement in the opposite direction, thereby containing and preserving the core sample.

Drive core samples.—One core-tube sample was recovered on each of the Apollo 15 EVA periods. Data for these samples are given in table 7-II. A double-core-tube sample was taken at station 2 (fig. 5-2, section 5) on the rim of a 10-m crater between Elbow and St. George Craters at the Apennine Front. The crew pushed the first tube to the full depth, and 35 hammer blows were required to sink the upper tube. A single core was taken at station 6 inside the rim of a 10-m crater, approximately 500 m east of Spur Crater, also at the Apennine Front. The tube was pushed to full depth and no hammering was necessary. A double-core-tube sample was recovered at station 9A at the edge of Hadley Rille, approxi-

---

[1] Ratio of length of sample obtained to depth tube driven × 100 percent.

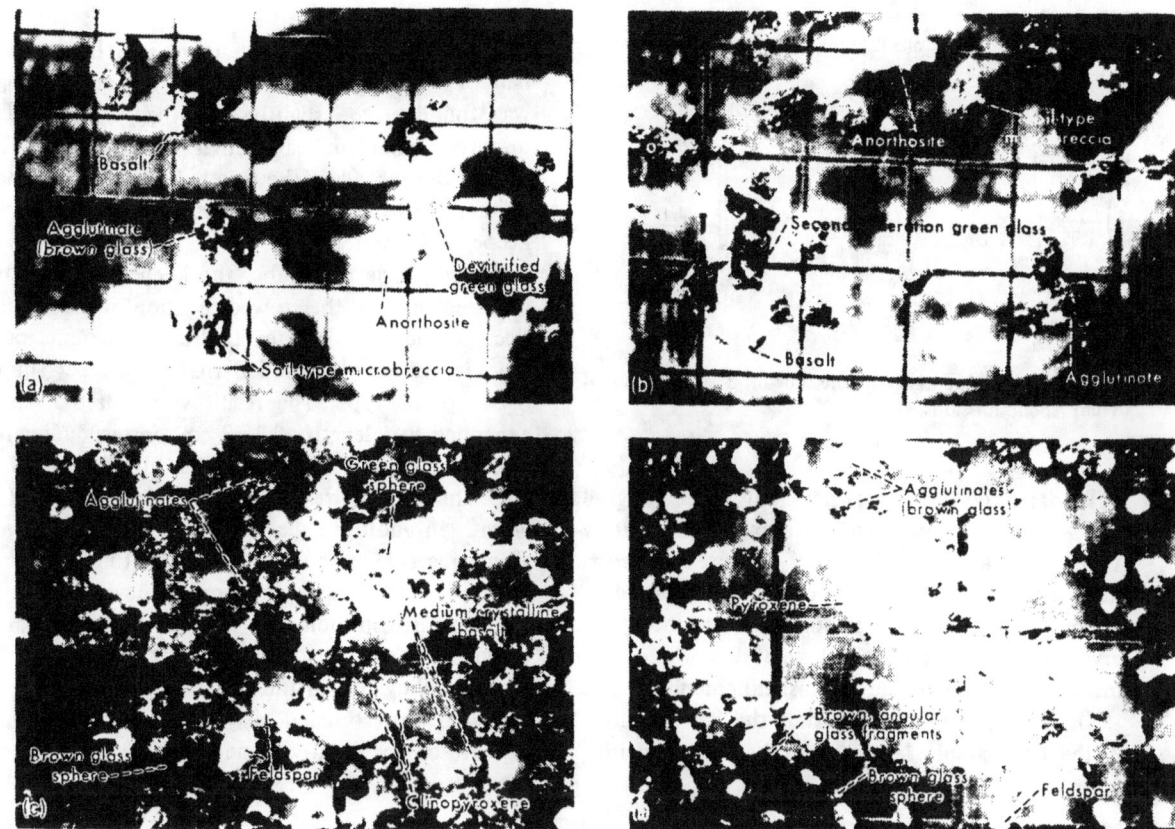

FIGURE 7-7.—Photomicrographs of four particle-size ranges from sample taken at the bottom of the soil-mechanics trench. Grid lines in photographs are 1 by 1 mm. (a) 0.5 to 1 mm (S-71-45452). (b) 0.25 to 0.5 mm (S-71-45446). (c) 0.125 to 0.25 mm (S-71-45450). (d) 0.0625 to 125 mm (S-71-45444).

TABLE 7-I.—*Compositional Characteristics of Different Soil Samples*[a]

Type of material	Composition, percent, at—				
	Apennine Front area			Lunar module area	Hadley Rille area
	Station 2	Station 6	Station 7		Station 9
Agglutinates and brown glass	~25	~46	~18	High	16 to 35
Clear green glass	12	4 to 6	High	None	<2
Mafic silicates	~18	10 to 20		15 to 20	10 to 30
Feldspar	30 to 40	18 to 20	16	6 to 10	20 to 35
Anorthosite		0 to 10	5 to 8	4 to 10	
Microbreccia		5 to 30		Trace	
Crystalline basalt			5 to 8	5 to 6	5 to 25

[a]Determined by the Lunar Sample Preliminary Examination Team.

mately 200 m west of Scarp Crater. The crew was able to push the tube to a depth of only two-thirds of the length of the bottom tube, and approximately 50 hammer blows were required to drive the tube to full depth. This additional driving effort was undoubtedly attributable to a higher soil density and strength at this location (as discussed later) as well as to the presence of rock fragments in the soil matrix.

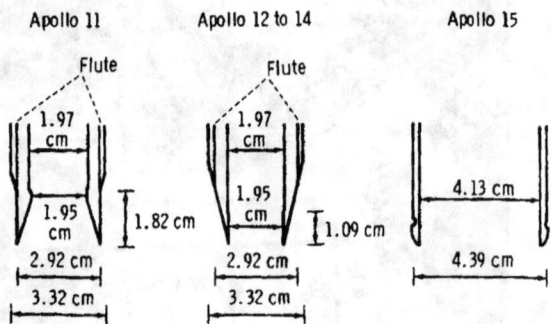

FIGURE 7-8.—Comparison of core-tube-bit designs for different Apollo missions.

To date, the core tubes have only been weighed and X-rayed in the Lunar Receiving Laboratory (LRL). A detailed description of the core samples on the basis of these X-radiographs is presented in section 6. Considerable stratigraphy has been observed as noted earlier, and careful study of the drive-tube samples should be most enlightening.

The X-radiographs also permit the determination of the core-sample lengths and the bulk densities, which are also presented in table 7-II. In the lower half of the sample from station 2, the sample length was found to be slightly less than nominal. This discovery would indicate either that the sample fell out of the top when the two halves were unscrewed or that the sample was compressed slightly when the keeper was inserted.

In the single core tube, the keeper was found to have remained in the stowed location in the adapter. Because the crew inserted the rammer-jammer properly, it has been concluded that the keeper slipped back up the tube. The result was that the sample expanded to a length of 36.2 cm, corresponding to a bulk density of 1.28 g/cm^3. If a nominal length of 34.9 cm is used, the calculated bulk density is 1.33 g/cm^3. In addition, the X-radiographs reveal a void along one side at the bottom of this tube. The crew described this sample location as having a coarser grain-size distribution than at other points at station 6, and this situation may account for part of the sample falling out of the tube before it was capped. The void was estimated to occupy 6 cm^3 (less than 2 percent of the total volume), and the bulk density

TABLE 7-II.—*Preliminary Data on Apollo 15 Core Samples*

Serial no.	Sample no.	Station	Weight, g	Length, cm	Bulk density, g/cm^3	Tube depth (pushed), cm	Total depth (pushed and driven), cm	Hammer blows, no.	Core recovery, percent
Drive tube (4.13 cm inside diameter)									
EVA-1									
[a]2003	15008	2	510.1	28 ± 1	1.36 ± 0.05	34.6	70.1	35	88 to 93
[a]2010	15007		768.7	[b]33.9 to 34.9	1.64 to 1.69				
EVA-2									
2007	15009	6	622.0	[c]36.2 to 34.9	1.35	34.6	34.6	0	101 to 105
EVA-3									
[a]2009	15011	9A	660.7	29.2 ± 0.5	1.69 ± 0.03	22.4	67.6	~50	91 to 96
[a]2014	15010		740.4	[b]32.9 to 34.9	1.79 to 1.91				
Drill stem (2.04 cm inside diameter)									
022 (top)	15006	8	210.6	32.9 to 39.9	1.62 to 1.96	[d]~236	--	--	100 to 102
023	15005		239.1	39.9	1.84				
011	15004		227.9	39.9	1.75				
020	15003		223.0	39.9	1.79				
010	15002		210.1	39.9	1.62				
027 (bottom)	15001		232.8	[e]33.2 ± 0.5 by 42.5	2.15 ± 0.03				

[a]Double.
[b]Sample either fell out of top of lower half of tube or was compressed when keeper was inserted.
[c]Nominal length is 34.9 cm; keeper slipped out of position.
[d]Drilled full depth.
[e]Sample fell out of the bottom of the drill stem.

was corrected to 1.35 g/cm³ accordingly. This density and that of the top half of the double-core-tube sample from station 2 are approximately 15 percent lower than the density of any of the samples previously returned.

As determined from the X-radiograph of the returned sample tube, approximately 54 cm³ of soil fell out of the bottom of the tube taken at station 9A before the tube was capped. In addition, the sample length was found to be less than nominal. This discovery would indicate either that the sample fell out of the top when the two halves were unscrewed or that the sample was compressed when the keeper was inserted. The high relative density at this location contradicts the latter interpretation and supports the former. Until further studies can be made, a range of possible densities is indicated as shown in table 7-II.

Drill-stem samples.—Characteristics of the ALSD and the deep drill-sampling procedure are described in section 11. The sample lengths shown in table 7-II were determined from X-radiographs that are discussed in detail in section 6. The sample length for the top section (serial number 022) was difficult to determine accurately, and a range of values is indicated. Some of the core (approximately 9.3 cm) fell out of the bottom of the drill stem (serial number 027). The bulk density of the remaining portion is approximately 2.15 g/cm³, which is 8 percent higher than the density of any previously returned core sample.

Soil Variability

One of the most striking characteristics of the soils in the Hadley-Apennine region is the great variability in properties from point to point, both regionally and locally. Vertical variability is indicated by the different units and densities observed in the core samples.

A series of footprints from different stations is shown in figure 7-9. In general, the deeper the footprint, the less dense, less strong, and more compressible the soil. Simulations (ref. 7-3) have shown that only small differences in the depth of footprints correspond to relatively large differences in soil properties. On the average, the soil on the Front was less strong and less dense than that by the lunar module (LM) and at the ALSEP site, and the surface was free of significant numbers of large fragments. In general, near Hadley Rille, the soil was relatively strong and less compressible than in other areas. Large fragments were abundant on the surface. The holes remaining after core-tube sampling at stations 6 and 9A are shown in figure 7-10. Bulging of the ground surface around the hole at station 9A indicates a stronger, less compressible soil than at station 6. As noted earlier, the single core tube at station 6 was pushed easily to the full depth, whereas the bottom tube of the double core at station 9A could be pushed only to two-thirds of the depth. These findings were somewhat surprising, because premission expectations had been that the Apennine Front would be firm with abundant coarse fragments and that the maria areas would be soft.

Local variations in strength and compressibility are common as well; an example of these variations in the vicinity of the LM is shown in figure 7-11. Footprints several centimeters deep may be seen in the foreground, whereas very little sinkage is seen in the middle ground area of the photograph.

Dust and Adhesion

Numerous instances of dust adherence to equipment, astronauts' suits, and lunar rocks were reported during the Apollo 15 EVA periods. The quantity of dust adhering to objects and the number of instances where brushing and cleaning were necessary were much more frequent than on previous missions, with the possible exception of the Apollo 12 mission.

The Rover kicked up quantities of dust during acceleration and when passing through the rims of soft craters. Little of the dust impacted on the Rover itself or on the astronauts, and it did not cause any problems with visibility or operation of the vehicle, although frequent cleaning of the lunar communications relay unit (LCRU) was required to prevent overheating of the television camera circuits. No dust accumulation was noted in the wire wheels, but a thin layer of dust eventually covered most of the vehicle.

Minor operational problems were caused by thin layers of dust on the camera lenses and dials, gnomon color chart, navigation maps, and LCRU mirror. As on previous missions, the adhering dust was brushed off easily. However, the dust was so prevalent that, during part of the mission, the astronauts reported that, to set the lens, dust had to be wiped from the camera settings every time they took a picture.

SLOPE STABILITY

A preliminary study of the 70- and 500-mm

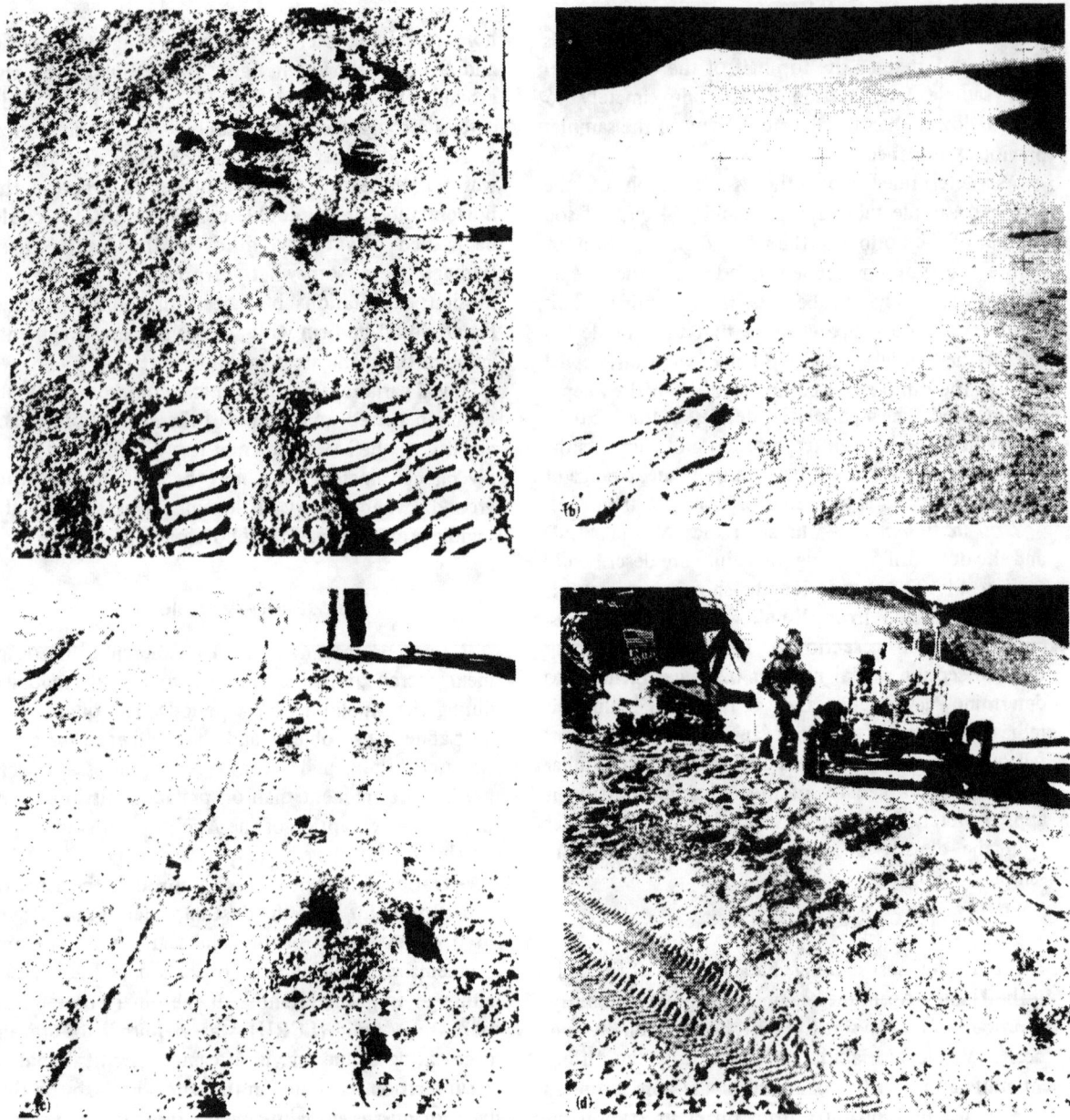

FIGURE 7-9.—Footprints from several locations illustrating soils of different strength. (a) Moderately firm soil at station 1 (AS15-86-11534). (b) Soft soil at station 2 (AS15-85-11424). (c) Very soft to soft soil at station 6 (AS15-86-11654). (d) Medium-strong soil at the LM (AS15-86-11599).

photography available thus far has been made for evidence of slope instability and past slope failures. No indications exist of previous deep-seated slope failures of the type that have been suggested by Lunar Orbiter photos of some areas of the Moon.

The near-surface zones of some slopes may be near incipient failure, however. The foreground of figure 7-12 shows failure under footprints as one of the astronauts traversed the slope in the vicinity of station 6A. Detailed analysis of conditions in this area must await more precise determination of the slope angle, which is estimated to be $10°$ to $20°$.

FIGURE 7-9.—Concluded. (e) Moderately firm to firm soil at station 9A (AS15-82-11121). (f) Firm soil at station 10 (AS15-82-11168).

FIGURE 7-10.—Core-tube holes at two sampling sites. (a) Core-tube hole at station 6 (AS15-86-11651). (b) Core-tube hole at station 9A. The raised ground surface around the station 9A hole indicates stronger, less compressible soil than at station 6 (AS15-82-11163).

Downslope movement of surficial material on the rille walls is evident. The movement of fine-grained material has left bedrock exposed on the upper slopes in some areas. Fillets are seen on the uphill side of many rocks, indicating soil movement around the rock. Other rocks without fillets can be seen, which suggests that (1) the rock itself may have rolled or slid downhill relative to the soil or (2) the soil in the vicinity of the rock has not undergone movement. Because no boulder tracks are visible, any rock movements must have occurred sufficiently long ago for subsequent soil movement to fill in any tracks formed initially. But if tracks have been filled in, then the associated rocks would be expected to be filleted as a result of the soil movement. Thus, the second hypothesis appears to be more tenable.

FIGURE 7-11.— Local variability in soil strength and density as indicated by shallow and deep footprints in the vicinity of the LM (AS15-92-12445).

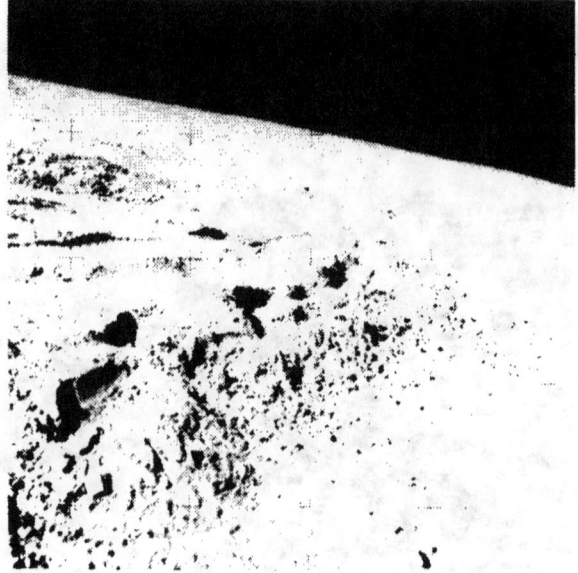

FIGURE 7-12.— Incipient slope failure as indicated by slipping out of soil beneath astronauts' feet (AS15-90-12197).

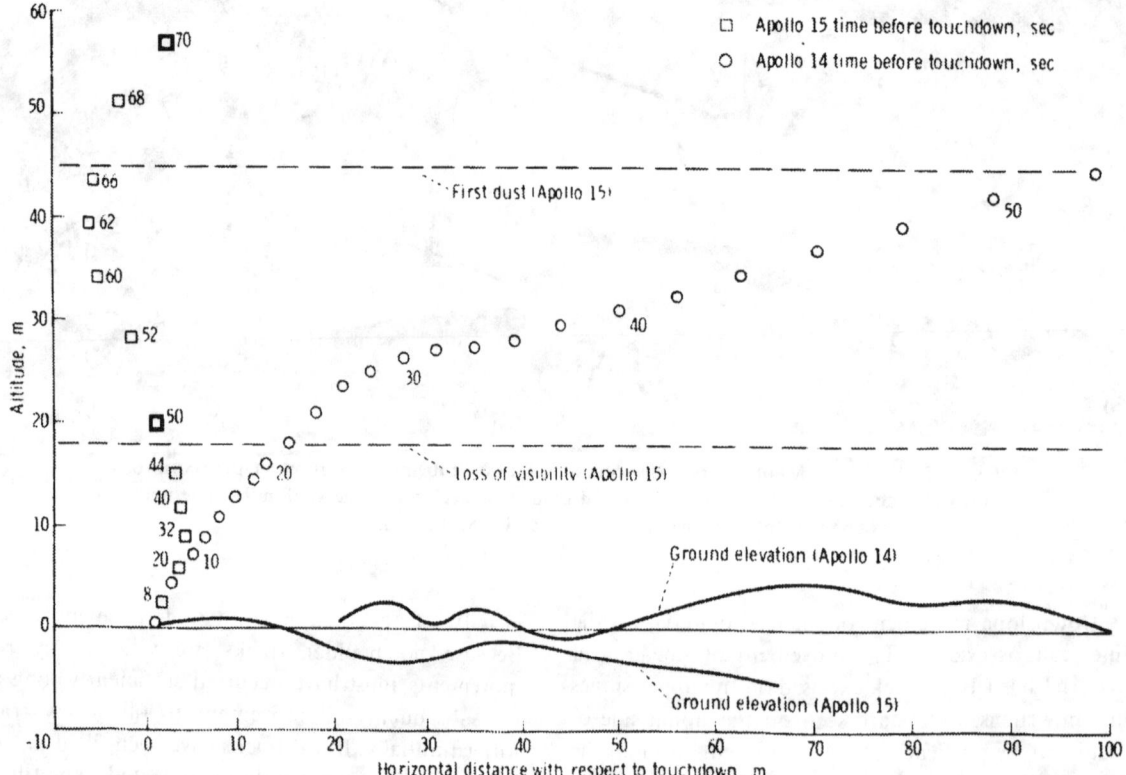

FIGURE 7-13. Apollo 14 and Apollo 15 descent trajectories.

SOIL BEHAVIOR DURING LM DESCENT AND LANDING

The Apollo 15 descent was much steeper and considerably slower than those of previous Apollo landings. The Apollo 14 and 15 descent trajectories are compared in figure 7-13. The final 30 m of descent occurred essentially vertically in a period of approximately 60 sec. In earlier landings (refs. 7-1, 7-2, and 7-4), only the last 3 to 6 m of descent were more or less vertical and occupied about half the time required for the Apollo 15 LM to descend through the same distance. The crew commented that they observed the first lunar-surface dust movement resulting from their landing at a height of approximately 46 m and noted that the last 18 m of descent were accomplished under conditions of no surface visibility as a result of the quantity of lunar soil being eroded by the descent engine. These were, therefore, the poorest visibility conditions during any Apollo landing. Previously, blowing dust had caused major difficulties only in the Apollo 12 descent and then only in the final 6 m. The dust problem may be related to the nature of the descent path and vertical velocity as well as to the local soil and the Sun-angle conditions.

Once again, from the photographs of the landing gear taken on the lunar surface, no stroking of the shock absorbers is evident, indicating only small, dynamic impact forces during landing. Only nominal penetration of the footpads into the lunar surface to a depth of several centimeters has occurred. However, in the landed position (fig. 7-14), the LM is tilted up to the west approximately 8° and up to the north through the same angle because of the lunar-surface topography. The +Z and +Y footpads appear to have landed on a slight rise, whereas the −Z footpad rests in a shallow crater 5 or 6 m in diameter. The −Y footpad is also in a slight depression. The LM is oriented with the +Z axis (the leg with the ladder) pointing due west. In the landing, principally as a consequence of the topographic relief, the descent-engine bell contacted the surface, crushing the bell slightly. The Apollo 15 mission is the first on which this has occurred and may have resulted, in part, from the fact that the Apollo 15 LM engine bell is larger than those used in earlier missions. No photographic indications are visible showing any lateral translation of the footpads during the final stages of descent. Because the underside of the LM so closely ap-

FIGURE 7-14.—The LM in the landed position is tilted up approximately 8° to the northwest because of surface topography (AS15-86-11600).

proached the lunar surface, the surface area below the spacecraft is largely in shadow, and signs of the erosion that took place in descent are not evident. In addition, on this mission, the photographs of the area around the landed LM were not taken soon enough after landing to show the surface undisturbed by the astronauts' surface operations. On photograph AS15-85-11364, taken from the top of the LM before astronaut egress, some signs of possible erosion tracks across the surface can be seen.

SOIL-ROVER INTERACTION

The use of Rover-performance data and the interaction of the Rover wheels with the lunar surface as indicators of variability in the consistency and mechanical properties of the surficial material in the Hadley-Apennine region can be made in several ways, including the following.

(1) Differences in the mean depth, shape, and surface texture of tracks developed by the chevron-covered Rover wire-mesh wheels

(2) Extent and shape of a "rooster tail," developed by fine-grained material ejected as a result of wheel-soil interaction, and characteristic speeds at which such a rooster tail is developed or astronaut visibility is degraded (or both)

(3) Net accumulation of fine-grained material inside the open wire-mesh wheel

(4) Variations in mobility performance or power consumption under constant throttle for a given slope and surface roughness

(5) Variations in the ability of the vehicle to climb slopes of the same inclination

(6) Vehicle immobilization resulting from wheel spin-out or skidding at different areas

No quantitative information exists regarding the interaction of the Rover with the lunar surface while the vehicle was in motion on level or sloping ground. Also, inasmuch as the mission profile was well within the expected capabilities of the Rover and the vehicle was never operated under performance-limiting conditions or under degraded operating modes (except for the front-steering failure during EVA-1), no direct quantitative information exists regarding the limiting mobility-performance capabilities at the Hadley-Apennine region.

The only semiquantitative and qualitative information from the interaction of the vehicle with the lunar surface can be extracted from (1) crew descriptions; (2) photographic coverage of the EVA periods, including a short 16-mm movie taken with the data-acquistion camera while the vehicle was in motion along segments of the EVA-2 traverse; and (3) Rover A-h integrator, odometer, and speedometer read-outs.

Because of the low pressure exerted by the wheels on the lunar soil, caused in part by the light wheel load (approximately 290 N on level terrain) and in part by the wheel flexibility, the average depth of the wheel tracks was only approximately 1-1/4 cm and varied from near zero to 5 cm. High wheel sinkage was usually developed when the vehicle was traversing small fresh craters. On one occasion, because of its light weight, the Rover had the tendency to slide sideways down a rather steep slope as soon as the astronauts stepped off the vehicle. Detailed knowledge of the exact circumstances that led to the tendency of the vehicle to slide downslope may be used to estimate the shear-strength characteristics of the surficial material at that location. Therefore, this particular behavior of the vehicle will be examined further in subsequent analyses.

The 50-percent chevron-covered, wire-mesh Rover wheels developed excellent traction with the lunar surficial material. In most cases, a sharp imprint of the chevron tread was clearly discernible, indicating that the surficial soil possessed some cohesion and that the amount of wheel slip was minimal. The latter observation is also corroborated by data from the Rover odometer and navigation systems, both of which were calibrated with a constant wheel-slip bias of 2.3 percent. An average wheel sinkage of approximately 1-1/4 cm at a wheel slip of 2.3 percent agrees with the data obtained from Rover wheel-soil interaction tests on lunar-soil simulants performed at the facilities of the U.S. Army Engineers Waterways Experiment Station (WES), Vicksburg, Mississippi, before the mission (ref. 7-13).

In one instance at the ALSEP site, the wheels attained a 100-percent slip while the vehicle was being started. While spinning out, the wheels dug into the lunar soil to a depth of approximately 13 cm (i.e., to the lower part of the wheel rim). The apparent looseness of the soil at this location can be attributed to a local variation in the material consistency, because information relating to the mechanical properties of lunar soil at the ALSEP site (obtained from other sources and discussed in other sections of this report) suggest that the material in this area is, in general, firm.

Driving on previously developed Rover tracks did not materially change the performance of the vehicle, although the LMP commented that, in some instances, the vehicle speed tended to increase. On the basis of crew debriefings and photographic coverage, it appears that the Rover was operated on slopes ranging from 0° to 12°. Because of its light weight and the excellent traction developed by the Rover wire-mesh wheels on the lunar soil, the general performance of the vehicle on these slopes was reported to be satisfactory. On the basis of wheel-soil interaction tests performed on lunar-soil simulants before the mission, the maximum slope angle that could be negotiated by the Rover had been estimated to be approximately 20°. Therefore, it appears that the slopes that were actually negotiated at the Hadley-Apennine region represented, at most, 60 percent of the estimated maximum slope-climbing capability of the vehicle.

Manuevering the vehicle on slopes did not present any serious problems. It was reported that the vehicle could be controlled more easily upslope than downslope; and, when the vehicle was traversing along slope contours, the wheels on the downslope side tended to displace the soil laterally and to sink a greater amount than the wheels on the upslope side.

This soil behavior again should be interpreted as being local and related to the surficial material rather than to any deep-seated material instability.

Based on crew observations, it appears that no perceptible amount of soil was collected inside the wheel when the vehicle was in motion. This observation is in agreement with the behavior of the lunar-soil simulant used in the WES wheel-soil interaction tests within the range of wheel slip realized during the Rover operation on the lunar surface.

At high vehicle accelerations, a rooster tail was developed by fine-grained material ejected from the wheels. During the performance of the wheel-soil interaction task (Grand Prix), the maximum height of the trajectory of the ejected material was estimated to be 4.5 m. It appears that, because of the presence of the fenders, the material was being ejected forward from the uncovered sides of the wheels. The CDR reported that the ejected dust was below the level of his vision.

In anticipation of local or regional variations in the mechanical properties of the lunar soil traversed by the Rover, extensive wheel-soil interaction studies were performed at the Waterways Experiment Station using a lunar-soil simulant of crushed basalt similar to the one used by Mitchell et al. (ref. 7-3) and Costes et al. (refs. 7-7 and 7-14) for lunar-soil-mechanics simulation studies. For the WES tests, the lunar-soil simulant, designated as LSS (WES mix), had been placed in five consistencies, with the following ranges in properties: specific gravity of solids, 2.69; void ratio, 0.90 to 0.69; and bulk density, 1.52 to 1.71 g/cm^3.

If the specific gravity of the solid particles of the soil at the Hadley-Apennine area is the same (3.1) as that for the single samples tested from the Apollo 11 and Apollo 12 landing sites, the bulk density of the lunar soil at the same void ratios as those for the LSS (WES mix) would range from 1.63 to 1.83 g/cm^3. The angle of internal friction of the soil, obtained from triaxial compression tests on air-dry specimens at normal stresses of approximately 0.7 N/cm^2, ranged between 38.5° and 41.0° (ref. 7-13); cohesion of the soil ranged between 0 and 0.29 N/cm^2; and the penetration-resistance gradient ranged between 0.2 and 5.9 N/cm^3. It appears that the range of cohesion and penetration resistance gradient in the soil simulants encompassed the known and calculated range of lunar-soil conditions in the Hadley-Apennine region. Therefore, the apparent agreement between the observed behavior of the Rover on the lunar surface with its expected behavior (based on the WES wheel-soil interaction studies) is an indirect indication of the mechanical properties of the surficial material at the Hadley-Apennine region. More detailed evaluations of Rover wheel-soil interactions at the Apollo 15 site are planned.

QUANTITATIVE ANALYSES OF SOIL-MECHANICS-TRENCH AND PENETROMETER EXPERIMENTS

Lunar-surface activities unique to the soil-mechanics experiment were conducted at station 8 (fig. 5-2, section 5). From analyses of the soil-mechanics trench and data obtained using the SRP, estimates of the in-place density, cohesion, and angle of internal friction are possible.

Penetrometer Measurements

The LMP used the SRP for six penetrations—four with the 3.22-cm^2 (base area) cone and two with the 2.54- by 12.70-cm bearing plate. The force-penetration records were scribed on the data drum, which has been returned for analysis.

The penetration curves for tests using the 3.22-cm^2 cone adjacent to the soil-mechanics trench and in a fresh Rover track are shown in figures 7-15(a) and 7-15(b), respectively. It is difficult to determine precisely the depth of penetration from the curves for the other four penetrations because the surface-reference pad of the penetrometer apparently rode up on the shaft during the tests. The surface-reference pad tended to ride up on the shaft when the SRP was vibrated because, although the weight of the reference pad was essentially balanced by the force on the retractor spring, the friction between the reference-pad bushing and the shaft was less than had been anticipated. In each case, however, the stress-penetration curves provide an upper bound on the depth of penetration for an applied force of 111 N, which gives a lower bound on the slope G of the stress-penetration curve.

The average slope G of the stress-penetration curve has been correlated with soil porosity, and this correlation can be used to estimate porosity at station 8 from the stress-penetration curves in figure 7-15. The average slope G was determined (dashed lines in fig. 7-15). Lower bound values of G were determined

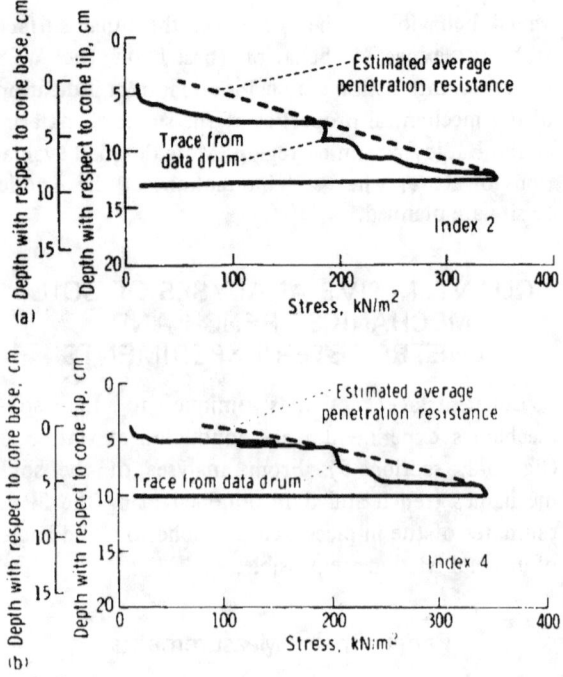

FIGURE 7-15.—Stress compared to penetration records. (a) Adjacent to soil-mechanics trench. (b) In fresh Rover track.

approximately 22 to approximately 16, indicating that the penetrating cone passed into a slightly softer layer below approximately 2 cm. This observation is consistent with a slight compaction of the upper few centimeters under the Rover wheel.

From the data in table 7-III, it appears that a reasonable average value of G for station 8 in uncompacted soil is approximately 4060 to 4360 $kN/m^2/m$. To compare these values with those obtained on Earth, an account of the effect of gravity must be made. The ratio of G under terrestrial gravity to G under lunar gravity for a soil deposit at a given porosity is defined as the gravity-reduction factor. Theoretical and experimental analyses (ref. 7-6) have shown that the gravity-reduction factor ranges from almost 6 for loose soils to approximately 4 for very dense soils. Using the factor for relatively dense soil (behavior of soil from station 8 was characteristic of dense soil), a value of G that is equivalent to that for the soil at station 8 under terrestrial gravity may be computed to be approximately 1.6×10^4 $kN/m^2/m$ for soil at the same porosity. For soil with a gradation of that of station 8 material, the corresponding porosity ranges from approximately 35 to approximately 38 percent (refs. 7-6, 7-7, and 7-13). To

TABLE 7-III.—*Summary of Cone-Penetration-Test Results*[a]

Index no.	Location	Penetration at 34 N/cm^2, cm	Gradient, G, $kN/m^2/m$	Penetration depth Base diameter
2	Adjacent to trench	8.25	4060	4.06
3	Bottom of trench	<10.25	>3250	--
4	In Rover track	5.25		2.58
	Upper 2 cm		7590	
	Lower 4 cm		4360	
5	Adjacent to Rover track	<11.25	>2980	--

[a]Cone with base area of 3.22 cm^2.

similarly from the other two cone tests not shown in figure 7-15. All values are listed in table 7-III.

The data from the SRP test in the Rover track (fig. 7-15(b)) show a slight decrease in slope at a depth of approximately 2 cm (with respect to the base of the cone).[2] The slope decreases from a G-value of

[2]From independent analyses using soil-cohesion values determined from the soil-mechanics trench and penetrometer data, it was determined that the intercepts at zero penetration with respect to the base of the cone must be no larger than the values shown in figure 7-15.

convert these parameters to density, a value of specific gravity G_s is required. Because a value of G_s for Apollo 15 soil has not yet been obtained, the value of 3.1 obtained for single samples of Apollo 11 and 12 soils may be used as an estimate. Porosity, void ratio (ratio of void volume to solid volume), and density for soil with a specific gravity of 3.1 are related in figure 7-16. The estimated range in soil porosity at station 8, as derived from SRP data, is summarized in table 7-IV.

Correlations between porosity and angle of internal friction ϕ have been developed for lunar-soil

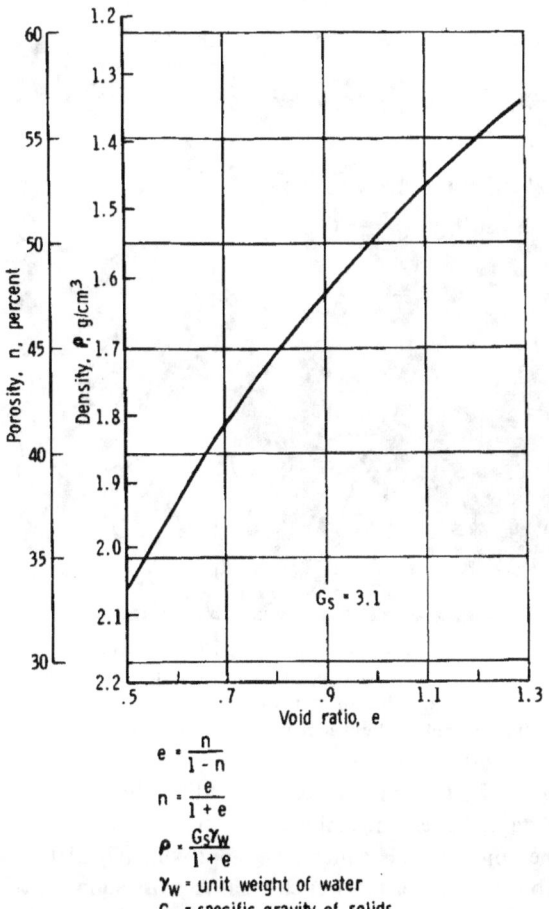

FIGURE 7-16.—Relationships between void ratio, porosity, and density for a soil with a specific gravity of 3.1.

simulants (refs. 7-6, 7-13, and 7-14). From these correlations, ϕ is estimated to be $49.5° \pm 2°$.

The estimated densities in table 7-IV are considered appropriate for the upper 10 to 20 cm at station 8. These values are significantly higher than the density of 1.84 g/cm³ measured from the uppermost section of the returned drill cores (table 7-II) obtained in the same area. One or more of the following factors may be responsible for this apparent inconsistency.

(1) The station 8 soil may have a specific gravity significantly less than the assumed value of 3.1. If so, the computed density values in table 7-IV would be lower, although the porosities and void ratios would be unchanged. This question cannot be resolved until specific gravity is measured for Apollo 15 soil.

(2) The drill core may have been loosened during sampling. As a part of the analyses for this report, a series of medium dense to dense deposits of lunar-soil simulant was prepared. Tests on the prepared simulant included driving Apollo 15 prototype core tubes with a hammer. For an initial porosity of approximately 38 percent, no significant change in density was observed during sampling. However, for an initial porosity as low as 35 percent, core-tube studies by Houston and Mitchell (ref. 7-8) indicate that the soil may loosen appreciably during sampling.

(3) Both the estimate of 1.97 g/cm³ from the penetration tests and 1.84 g/cm³ from the drill core may be correct and reflect local variability.

Soil-Mechanics Trench

Near the end of EVA-2, the soil-mechanics trench was excavated by the LMP at a point approximately 55 m east-southeast of the ALSEP central station. The lunar surface at the trench site (fig. 7-17) was approximately level except for two small, shallow craters just east of the gnomon. Excavation of the trench was accomplished by using the small scoop attached to the extension handle. Analysis of the television film and commentary by the LMP indicated that excavation proceeded smoothly and without difficulty to a depth reported at the time to be approximately 36 to 41 cm, where a much harder layer was encountered. Subsequent analysis of the television film and the Hasselblad electric data camera photographs has shown that the actual depth was probably somewhat less.

TABLE 7-IV.—*Estimated Ranges in Porosity and Friction Angle ϕ for Station 8 Soil as Determined from SRP Data*

Value	Porosity, n, percent	Void ratio, e	Density, ρ, g/cm³ (a)	Friction angle, ϕ, deg
Best estimate	36.5	0.575	1.97	49.5
Range	35 to 38	0.54 to 0.61	1.92 to 2.01	47.5 to 51.5

[a] G_s = 3.1.

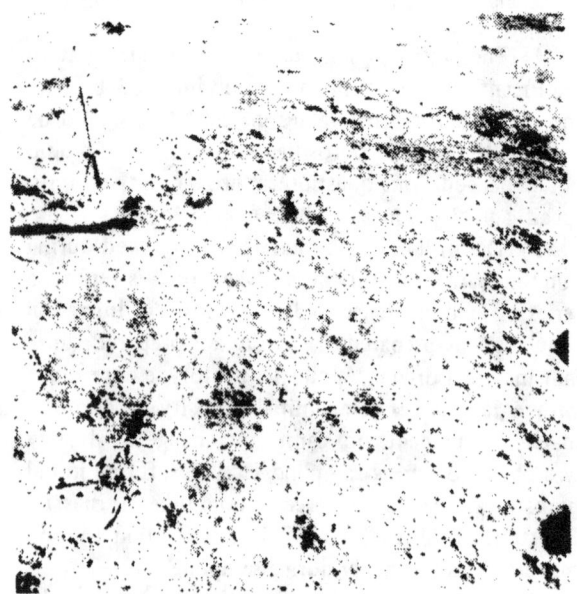

FIGURE 7-17.— Undisturbed lunar surface before excavation of the soil-mechanics trench at station 8. Two small, shallow craters may be seen just to the east of the gnomon (AS15-92-12417).

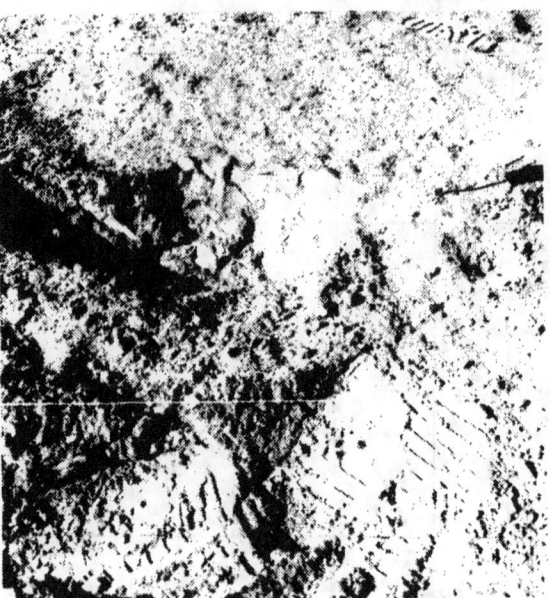

FIGURE 7-18.— Cross-Sun photograph from the north of the completed soil-mechanics trench excavated by the lunar module pilot. Scoop marks on near-vertical face reflect fine-grained, cohesive character of the soil (AS15-92-12440).

No evidence exists of layering in the trench wall. The soil was fine grained and cohesive, and a vertical face could be maintained without difficulty. A cross-Sun photograph from the north of the completed trench is shown in figure 7-18. The excavated soil was distributed to the north (foreground of photograph). The smooth scoop marks in the trench wall are evidence of the fineness and cohesiveness of the soil. The footprints in the foreground show the characteristics of recompacted, disturbed material.

The material at the bottom of the trench was reported to be much harder than that above. The LMP indicated that a smooth, flat bottom could be made easily and that further excavation necessitated chipping out the material, which came out in platy fragments approximately 0.5 cm long. However, a sample returned from the trench bottom was dark gray and very cohesive and gave no evidence of hardpan upon examination in the LRL. The cohesion was not destroyed by remolding even after prolonged exposure to an atmosphere. A sample from the top of the trench was similar in behavior to the sample from the bottom, although its grain size was slightly finer (fig. 7-6).

After sampling and photographic documentation of the completed trench, failure of the vertical side wall was induced by loading at the top with the 2.5- by 12.7-cm bearing plate attached to the SRP. The plate was oriented parallel to the trench wall and with the longitudinal center line approximately 10 cm from the top of the trench wall. A cross-Sun view of the failed trench is shown in figure 7-19. The imprint of the lunar reference plane is clearly visible in the photographs. The imprint is 35.6 cm long and 7.9 cm wide.

Detailed photogrammetric analysis of the trench photography is not yet complete. However, sufficiently accurate determination of the trench dimensions has been made to permit some estimates of soil-strength parameters. Failure of the trench wall required the application of a force to the penetrometer bearing plate in excess of the 111-N spring measuring capacity of the SRP. The LMP estimated that he applied an additional 44 N before failure occurred. Collapse was sudden and complete.

It has been shown that the values of soil-strength parameters required for equilibrium of a near-vertical, homogeneous slope are insensitive to the assumed shape of the failure surface (e.g., plane surface of sliding, circular arc, or log spiral). If a planar failure surface is assumed and the shear surfaces at the ends of the failure zone are neglected, the forces and geometry needed for analysis are as shown in figure 7-20.

FIGURE 7-19.—Cross-Sun view of soil-mechanics trench at failed vertical wall (AS15-88-11874).

For this case, the analysis is insensitive to the soil unit weight; a density value of 1.8 g/cm³ is assumed, which gives a unit weight on the Moon of 0.00294 N/cm³. Equilibrium of the forces shown in figure 7-20 can be expressed in terms of force components parallel to the failure plane; that is

$$F_D = \left(W_s \cdot 12.25\right) \cos\left(45 - \frac{\phi}{2}\right) \quad (7\text{-}2)$$

$$F_R = 11.1 \csc\left(45 - \frac{\phi}{2}\right) c + \left(W_s \cdot 12.25\right) \sin\left(45 - \frac{\phi}{2}\right) \tan\phi \quad (7\text{-}3)$$

where

F_D = driving force
F_R = resisting force
c = unit cohesion
ϕ = angle of internal friction
W_s = weight of the failure wedge per unit length

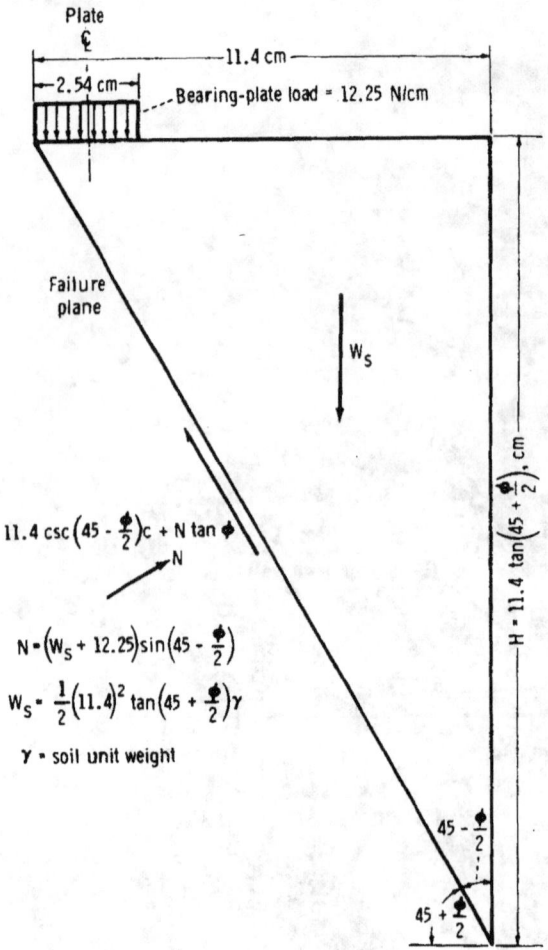

FIGURE 7-20.—Plane-failure analysis of soil-mechanics trench.

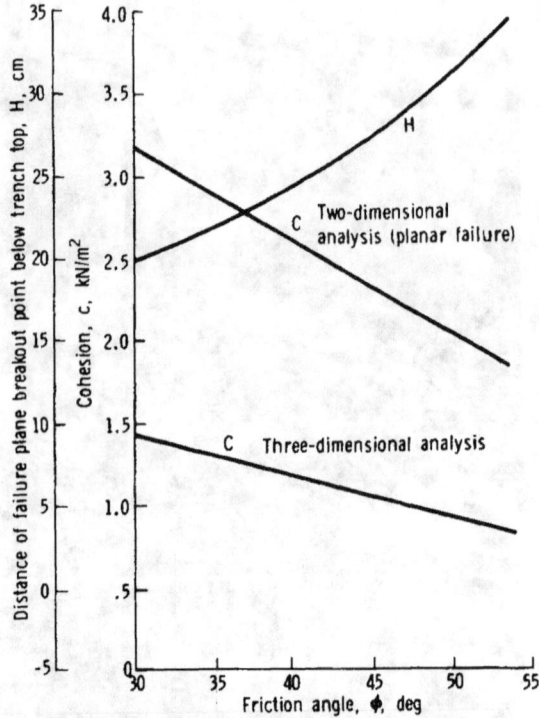

FIGURE 7-21.—Critical values of soil cohesion for trench-wall stability.

After equating F_D and F_R and rearranging, the required cohesion may be expressed in terms of the friction angle as

$$c = \frac{W_s + 12.25}{22.8}\left(\frac{1 - \sin\phi}{\cos\phi}\right) \quad (7\text{-}4)$$

By the theory of plasticity, this same expression may be derived as an upper bound solution. It may be shown that the selected failure surface is kinematically admissible.

Values of cohesion have been determined as a function of friction angle for the assumed conditions with the results shown in figure 7-21. Also shown in figure 7-21 are the corresponding values of H, which represent the distance below the top of the trench face at which the failure surface should break out.

Reliable determination of this distance is not possible from the available trench photography; however, the original trench depth appears to be sufficient to accommodate failure in accordance with any of the indicated values of ϕ.

In the analysis that resulted in the cohesion values shown in figure 7-21, end effects were neglected (i.e., plane strain, two-dimensional behavior was assumed). As shown in figure 7-19, the failure wedge involved significant shear areas at the ends. Because of this situation, the values of cohesion shown in figure 7-21 are too high. A preliminary estimate of this shape effect may be made by reducing the computed cohesion values in proportion to the ratio of the area of the assumed planar failure surface to the area of a failure surface that includes the ends of the failure wedge. In the present case, the ratio is computed to be 45 percent for all values of ϕ, resulting in the reduced values of cohesion indicated by the lower curve for cohesion as a function of friction angle in figure 7-21. Although this correction improves the accuracy of the computed cohesion, some uncertainty remains concerning the magnitude of the force required to cause failure. If it is assumed that the LMP's estimate of 44 N more than the 111-N capacity of the SRP spring is accurate to ±22 N, the cohesion values in figure 7-21 will be correct within approximately ±15 percent.

Two important features of the trench experiment are that the computed cohesion is not a sensitive function of the friction angle and that the calculation is virtually independent of the value used for soil density. Thus, even when values of friction angle and density are uncertain, the trench experiment provides a reliable basis for determination of soil cohesion.

Strength Parameters Deduced from Penetration Resistance

One of the most surprising findings at station 8, during measurements with the SRP, was the very high resistance offered by the soil to penetration by the cone with a base area of 3.22-cm². Because of the tendency of the lunar-surface reference plane to ride up on the penetrometer shaft, precise values of penetration are not known for each of the penetration tests, and the exact shape of the curve (force as a function of depth) was not obtained. However, estimates of penetration are possible (table 7-III).

The resistance to penetration q_p can be calculated by

$$q_p = cN_c\xi_c + \gamma B N_{\gamma q}\xi_{\gamma q} \qquad (7\text{-}5)$$

where

c = unit cohesion
γ = unit weight = ρg
B = width or diameter of loaded area
$\xi_c, \xi_{\gamma q}$ = shape factors
$N_c, N_{\gamma q}$ = bearing capacity factors = $f(\phi, \delta/\phi, \alpha, D/B)$
ϕ = angle of internal friction
δ = friction angle between penetrometer cone and soil
α = half the cone apex angle
ρ = density
D/B = ratio of penetration depth to cone diameter

An appropriate penetration-failure mechanism has been assumed for dense soil to enable calculation of the bearing-capacity factors. Durgunoglu[3] has substantiated this failure mechanism by means of model tests and has derived the equations needed for determination of $N_{\gamma q}$, $\xi_{\gamma q}$, and $N_c\xi_c$. The value of δ/ϕ has been taken as 0.55 based on the results of friction measurements between a ground-basalt lunar-soil simulant and hard anodized aluminum similar to that used for the SRP cones. Results from the bearing-capacity equation are insensitive to the value assumed for γ; 0.00294 N/cm³ has been assumed here, corresponding to a density ρ of 1.8 g/cm³. Values of $N_{\gamma q}$, $\xi_{\gamma q}$, and $N_c\xi_c$ have been calculated for different values of D/B and ϕ, and q_p = 34.5 N/cm² (fig. 7-22).

The D/B ratios for the tests at station 8 fall in the range of approximately 2.5 to 4.1 (table 7-III). Thus, the curve in figure 7-22 for D/B = 3 may represent the actual conditions reasonably well.

The relationship of cohesion compared to friction angle for the trench has been superimposed on figure 7-22. The intersections between this curve and the curves of c as a function of ϕ for the penetration tests give conditions that can satisfy both trench failure and the penetration test simultaneously. For D/B = 3, the required cohesion is 0.94 kN/m², and the angle of internal friction is 51.7°. This value compares favorably with that obtained by comparison of the observed penetration behavior with that of the terrestrial simulants (table 7-IV). Although the average friction angle (50.6°), computed by the two

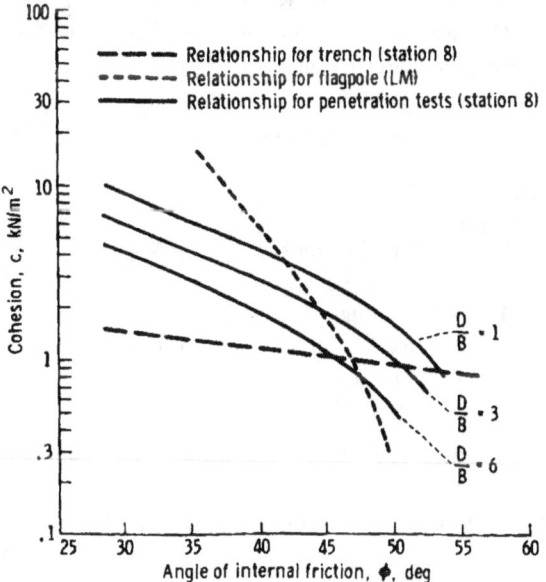

FIGURE 7-22.—Cohesion as a function of friction angle for different penetrations and an 111-N force applied to the SRP.

---

[3]Durgunoglu, H.T.: Ph. D. dissertation (in preparation). Dept. of Civil Engineering, Univ. of Calif. at Berkeley, 1971.

TABLE 7-V.—*Comparison of Estimated Cohesion Values for the Apollo Landing Sites*

Mission	Location	Cohesion, kN/m^2
11	Mare Tranquillitatis	0.35 to 0.70
12	Oceanus Procellarum	0.35 to 0.70
14	Fra Mauro	0.03 to 0.10
15	Hadley-Apennine	0.9 to 1.1

methods, is higher than has been estimated at other Apollo sites, it is consistent with the high soil density at station 8. Similarly, a cohesion of almost 1.0 kN/m^2 is higher than previously measured; but this value, too, can be accounted for by the high density and the relatively fine-grained soil consistency. Table 7-V compares estimates of soil cohesion for the four Apollo landing sites.

A third relationship between c and ϕ may be deduced from the penetration of the flagpole into the soil near the LM. The flagpole, made of chrome-anodized aluminum, is a hollow tube with an outside diameter of 2.226 cm and a wall thickness of 0.089 cm. From study of the television tapes, it was deduced that the 119.05-cm-long lower section of the pole was pushed to a depth of approximately 51 cm before requiring hammering. The LMP was observed to apply his full weight to the pole because both feet were off the ground simultaneously. His suited weight in the lunar gravity field is approximately 27 kg.

The force of penetration F is resisted by end bearing and skin friction according to

$$F = q_p A_p + f_s A_s \qquad (7\text{-}6)$$

where

q_p = unit end-bearing capacity = $cN_c \xi_c + \gamma B N_{\gamma q} \xi_{\gamma q}$
A_p = end-bearing area
A_s = surface area in contact with the soil
f_s = unit skin friction

If the unit skin friction is assumed to increase linearly from zero at the ground surface to a maximum at the bottom of the pole, depth D, then f_s is given by

$$f_s = \frac{\gamma D K \tan \delta}{2} \qquad (7\text{-}7)$$

where K is the coefficient of lateral Earth pressure $\cong 0.5$, and $\tan \delta$ is the friction coefficient between soil and pole $\cong 0.5$.

With the aid of these relationships and the assumption that the flagpole behaved in a manner similar to that of the core tubes and did not plug during penetration, values of c have been computed and plotted on figure 7-22 as a function of ϕ. This relationship defines smaller values for c and ϕ than are required to satisfy the behavior at station 8. This difference could be attributed to a lower soil density at the flagpole location. From examination of the LM and photographs (e.g., fig. 7-11) it is assumed that this may be the case. The flagpole appears to have been placed in the rim of a small crater, and the soil at small crater rims is generally softer than in intercrater regions.

DISCUSSION

Lunar-Soil Density

The bulk density of the lunar soil has been the subject of speculation since early in the lunar-exploration program. Table 7-VI summarizes some of the estimates that have been made since that time.

A density of 0.3 g/cm^3 (corresponding to a porosity of 90 percent) was assumed by Jaffe (refs. 7-15 and 7-16) in an effort to calculate lower bound bearing capacities for the design of unmanned and manned lunar-landing craft. Halajian (ref. 7-17) also used a very low density, 0.4 g/cm^3, but believed that the strength of the lunar surface was similar to that of pumice. The grain-size distribution and the lunar-soil/footpad interaction observed on Surveyor I (June 1966) suggested a value of 1.5 g/cm^3 (ref. 7-18). In December 1966, the Russian probe, Luna 13, provided the first in-place measurement of soil density on the Moon by means of a gamma-ray device. Unfortunately, the calibration curve for this device was double valued, and it was necessary to choose between a value of 0.8 and 2.1 g/cm^3. Cherkasov et al. (ref. 7-19) chose the lesser value. Based on the results from the soil-mechanics surface-sampler experiments on Surveyors III and VII, Scott and Roberson (refs. 7-5 and 7-20) confirmed the Surveyor I value of 1.5 g/cm^3 and argued (ref. 7-21) that the Russian investigators had chosen the wrong portion of their calibration curve.

Ironically, the drive-tube data from Apollo 11 also were ambiguous, because of the shape of the bit. The

TABLE 7-VI.—*Estimates of Lunar-Soil Density*

Bulk density, ρ, g/cm^3	Investigator	Landing site	Reference
0.3	Jaffe	--	7-15 and 7-16
0.4	Halajian	--	7-17
1.5	Christensen et al.	Surveyor I	7-18
0.8	Cherkasov et al.	Luna 13	7-19
1.5	Scott and Roberson, and Scott	Surveyors III and VII	7-20, 7-5, and 7-21
1.54 to 1.75	Costes and Mitchell	Apollo 11	7-22
0.75 to >1.75	Scott et al.	Apollo 11	7-23
[a]1.81 to 1.92	Costes et al.	Apollo 11	7-7
1.6 to 2.0	Scott et al.	Apollo 12	7-23
[a]1.80 to 1.84	Costes et al.	Apollo 12	7-7
1.55 to 1.90	Houston and Mitchell	Apollo 12	7-8
1.7 to 1.9	Carrier et al.	Apollo 12	7-9
1.2	Vinogradov	Luna 16	7-24
1.35 to 2.15	Mitchell et al.	Apollo 15	(b)

[a]Upper bound estimates.
[b]This report.

bulk densities of the soil in the two core tubes were 1.59 and 1.71 g/cm^3 (ref. 7-1) or 1.54 and 1.75 g/cm^3 as later reported by Costes and Mitchell (ref. 7-22) by taking into account possible differences in core-tube diameter. These densities could have indicated an in situ density from 0.75 g/cm^3 to more than 1.75 g/cm^3 (ref. 7-23).

The shape of the Apollo 12 drive-tube bits reduced the uncertainty, and the density at this site was estimated to be 1.6 to 2.0 g/cm^3 (ref. 7-23). Core-tube simulations performed later by Houston and Mitchell (ref. 7-8) and Carrier et al. (ref. 7-9) yielded additional estimates of 1.55 to 1.90 g/cm^3 and 1.7 to 1.9 g/cm^3, respectively. Based on penetration-resistance data from the Apollo 11 and 12 landing sites, Costes et al. (ref. 7-7) gave upper bound estimates of the density at the two sites of 1.8 to 1.94 g/cm^3 and 1.81 to 1.84 g/cm^3, respectively. Vinogradov (ref. 7-24) estimated a value of 1.2 g/cm^3 from a rotary-drill sample returned by Luna 16.

Density of the lunar soil at the Apollo 15 site.—The early estimates of lunar-soil density were intended as lower bounds for the entire lunar surface. When returned core-tube samples became available, it was possible to estimate a range of densities for a given landing site. The new core tubes on Apollo 15 have permitted estimates of the in situ density for different locations within the site.

The density at each of the drive-tube locations is estimated by correcting the bulk density in the tubes for disturbance caused by sampling. These corrections must await detailed core-tube-simulation studies, which will be performed later. In the meantime, the high percent core recoveries (table 7-II) suggest that the corrections will be small, and a preliminary estimate can be made of density as opposed to depth at the three core-tube locations (fig. 7-23). The top 25 to 35 cm of soil along the Apennine Front (stations 2 and 6) have very similar, low average values of density, ~1.35 g/cm^3. The soil density evidently increases rapidly with depth. The soil density measured at the Apennine Front is approximately 10 percent less than the density at any previous Surveyor or Apollo site and approaches that of the Luna 16 site (1.2 g/cm^3). The average soil density at Hadley Rille (station 9A) is significantly higher in the top 30 cm (~1.69 g/cm^3) and increases less rapidly with depth. If the density is assumed to increase linearly with depth, the station 2 data would yield a density of 1.2 g/cm^3 at the surface, increasing to 1.8 g/cm^3 at a depth of 63 cm. The station 9A data would yield a value of 1.6 g/cm^3 at the surface (33 percent higher than at station 2) and 1.9 g/cm^3 at a depth of 64 cm. Densitometric analyses of the X-radiographs are planned in an effort to develop detailed relationships of density as a function of depth for the Apollo 15 core tubes.

The in situ density at the soil-mechanics trench,

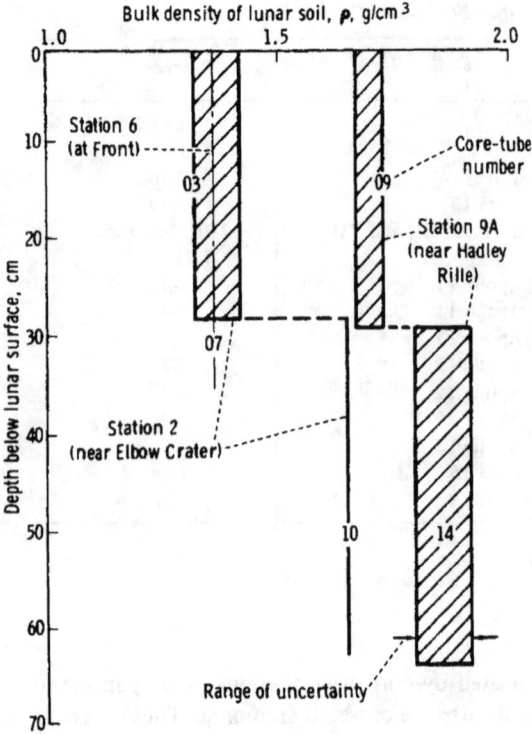

FIGURE 7-23.—Preliminary density compared to depth estimates at the three Apollo 15 core-tube sites.

station 8, has been estimated to be in the range of 1.92 to 2.01 g/cm³, based on the penetration test results. The data in table 7-II indicate a density range of 1.62 to 2.15 g/cm³ for the samples in the deep drill stem obtained from the same area. Average density of these samples is approximately 1.8 g/cm². Possible explanations for the differences according to the two methods have been discussed in the subsection entitled "Penetrometer Measurements." Because significant variations in density exist both regionally and locally on the lunar surface, further study is required to relate these differences in a consistent manner to surface-material composition, history, and lunar processes.

Relative and absolute density.—Now that more accurate values of the absolute density of lunar-soil deposits are becoming available, it is important that the relative density be determined, because mechanical properties are strongly dependent on relative density as well as on absolute density and porosity.

Relative density D_R is defined by

$$D_R = \frac{e_{max} - e}{e_{max} - e_{min}} \times 100 \text{ percent} \quad (7\text{-}8)$$

where

e_{max} = maximum void ratio (corresponding to minimum density) at which the soil deposit can exist

e_{min} = minimum void ratio (corresponding to maximum density) at which the soil deposit can exist

It can be shown easily that relative density can also be calculated in terms of bulk density according to

$$D_R = \frac{\rho_{max}}{\rho} \frac{\rho - \rho_{min}}{\rho_{max} - \rho_{min}} \times 100 \text{ percent} \quad (7\text{-}9)$$

If $\rho = \rho_{min}$ (or $e = e_{max}$), $D_R = 0$ percent and the deposit is exceptionally loose; if $\rho = \rho_{max}$ (or $e = e_{min}$), $D_R = 100$ percent and the soil is very compact.

Compressibility, disturbance during sampling, penetration resistance, and shear strength are far more dependent on the relative density than on the absolute density of a given soil deposit. Because ρ_{max} and ρ_{min} can vary from soil to soil (depending primarily on the specific gravity, grain-size distribution, grain-shape distribution, and particle surface texture), different deposits can have different absolute densities but similar relative densities. The behavior of such deposits would then be similar. It is also possible to have similar absolute densities and quite different relative densities, and this will result in significantly different behavior. The lunar soils at the Apollo 15 site have been deposited at different absolute and relative densities. Determination of the relative densities would contribute significantly to the evaluation of the data from many experiments. Until determinations of specific gravity and minimum and maximum density are made on returned lunar samples, definitive conclusions are difficult to reach.

Penetrability Considerations

Penetration of the lunar surface for purposes of measuring in situ properties, obtaining core-tube samples, or emplacing probes (as in the heat-flow experiment (HFE)) may be limited by the presence of obstructing large particles or by excessive penetration resistance. A probability analysis was made to determine the likelihood that the three core tubes, four cone penetrations, two heat-flow drill holes, and the deep core could be made to their respective depths without encountering an obstructing particle or rock fragment. The method of analysis has been described previously by Mitchell et al. (ref. 7-4). The Apollo 12 size-frequency distribution curve was used for this preliminary analysis because complete Apollo 14 and Apollo 15 size-frequency distribution curves are not yet available.

The analysis indicated a probability of 0.9 that the four cone-penetration tests would reach their respective depths without striking a particle equal to or larger than the cone diameter of 2.03 cm. The probability that the three core tubes could be driven to their respective depths without striking a particle equal to or larger than the tube diameter (4.39 cm) was 0.75. From another analysis, also based on Apollo 12 particle-size distributions, it was predicted that the chance of the core-tube material containing one particle of approximately 1.3-cm diameter is approximately 50 percent. The probability of finding a rock between 2 and 4 cm in the core tubes is only 0.2. In actuality, the Apollo 15 cores contain several rock fragments in the 1.3-cm size range, and the double-core-tube sample taken at station 9A near Hadley Rille contained a rock fragment 2.2 by 2.6 by 4.8 cm near the bottom of the core.

The probability that the HFE drill core would reach full depth (2.36 m) without striking a fragment equal to or greater than the outside diameter of the drill (2.62 cm) was 0.6. The probability that both HFE holes would reach depths of 178 and 175 cm, respectively, without striking a particle equal to or greater than the drill bit diameter (2.856 cm) was 0.5. If the HFE holes had reached full design depth of 3 m, the calculated probability is 0.7 that a particle greater than or equal to the drill diameter would be encountered and 0.5 that a particle more than twice the drill diameter would be encountered.

In areas of high density, penetration to depths of more than a few centimeters using a penetrometer or core tube may not be possible without mechanical assistance (e.g., drill or jack) to aid the astronaut. To investigate this possibility, a soil simulant was prepared to provide behavior comparable to that observed at station 8, and penetration and core-tube-sampling studies were made. For the soil simulant, the stress-penetration curves obtained were very similar in shape, slope G, and appearance to those obtained at station 8.

It was necessary to make the porosity of the simulant greater than that estimated for the lunar soil at station 8 to account for the effect of gravity. Because the resistance to penetration with the SRP was essentially the same, quantitatively, for the lunar soil as for the simulant, it is reasonable to conclude that the resistance to core-tube penetration would also be similar. Two separate core tubes similar to those used on Apollo 15 were first pushed and then hammered to the depth of a single core tube. For an applied vertical static force of approximately 245 N, the average depth of penetration was approximately 10 cm. A total of 60 blows with a hammer similar to that used on Apollo 15 were then required to drive the tube the rest of the way.

These data indicate that considerable difficulty would have been encountered in obtaining a single-core-tube sample at station 8 if it had been attempted. Driving a double core tube probably would have been impossible. Thus, it appears from these studies that if the total depth of penetration with the SRP using the 3.22-cm^2 cone is approximately 7.5 cm or less, core-tube sampling may not be practical.

Core-Tube and Borehole Stability

Figure 7-10 shows open holes that remained after removal of the core tubes from the ground. The crewmen reported that the deep drill hole also remained open after the drill stem was removed. Some soil cohesion is required to prevent collapse of the holes, and a simplified analysis of this condition is possible using the theory of elasticity. The maximum principal stress difference $(\sigma_\theta - \sigma_r)_{max}$ is at the surface of the hole, where radial stress σ_r is zero, and tangential stress σ_θ equals $2p$, where p is the lateral pressure in the ground away from the zone of influence of the hole. If the soil adjacent to the hole is disturbed or yields, a plastic zone will form around the hole; however, it may be shown (ref. 7-25) that the maximum shear stress in the zone will be less than that for the purely elastic case. Determination of values of c and ϕ to withstand the maximum applied shear stress $\tau_{max} = \dfrac{(\sigma_\theta - \sigma_r)_{max}}{2}$ will enable determination of the depth to which stresses will be elastic. The appropriate equation is

$$c = p\left(\frac{1-\sin\phi}{\cos\phi}\right) = \frac{\nu}{1-\nu}\rho g z\left(\frac{1-\sin\phi}{\cos\phi}\right)$$
(7-10)

where ν is Poisson's ratio, g is acceleration caused by gravity, and z is depth of elastic zone.

For a Poisson's ratio of 1/3 (which corresponds to an Earth pressure coefficient of 0.5) and a density of 1.8 g/cm³, the relationship between c,ϕ, and depth of elastic zone is as shown in figure 7-24. Below this depth, a plastic zone will exist extending to a distance r_e from the centerline of the hole. For any finite values of c and z, the value of r_e is finite, and a failure of the walls should not occur. However, as the hole becomes deep and the plastic zone becomes large, extensive lateral straining of the soil may occur, eventually causing a closure of the hole by inward squeezing of the soil. This phenomenon would not be expected to occur for the relatively shallow depths being drilled on the lunar surface and for the values of cohesion and friction angle that have been determined.

CONCLUSIONS

More extensive opportunities for detailed study of the mechanical properties of lunar soil have been provided by the Apollo 15 mission than by previous missions; and, for the first time in the Apollo Program, quantitative measurement of forces of interaction between a soil-testing device and the lunar surface has been possible. Preliminary conclusions can be drawn from the analyses completed to date.

(1) The lunar surface of the Hadley-Apennine site is similar in color, texture, and general behavior to that at the previous Apollo sites.

(2) Variability between grain-size distributions of different samples from the Apollo 15 site does not appear to be as great as at the Apollo 12 and 14 sites.

(3) Considerable variability exists in soil properties, as reflected by density, strength, and compressibility, both with depth and laterally. Lateral variations are both regional (as characterized by conditions ranging from soft, compressible soil along the Apennine Front to firmer, relatively incompressible soil near the rim of Hadley Rille) and local as can be observed from variable footprint depths visible in many photographs.

(4) Through the use of new core-tubes, designed on the basis of soil-mechanics considerations and used for the first time on the Apollo 15 mission, 3302 g of relatively undisturbed lunar soil were returned. The performance of these tubes was excellent.

(5) In situ soil densities that were deduced from the core-tube and drill-stem samples vary considerably (from 1.36 to 2.15 g/cm³). These results reinforce the evidence for soil variability available from other sources (e.g., photography and crew commentary).

(6) No evidence exists of past deep-seated slope failures, although the surface material may be in a near-failure condition along the Apennine Front, and there is evidence of the downslope movement of surficial material on the Hadley Rille walls.

(7) Blowing dust caused greater visibility degradation during LM landing than in previous missions. This situation may be related to the descent path, vertical velocity, Sun angle, and local soil conditions.

(8) Limited amounts of quantitative data are available on Rover-soil interaction. The apparent agreement between the observed Rover behavior on the lunar surface and the expected behavior, based on premission simulation studies, provides an indirect measure of the mechanical properties of the surficial soil in the Hadley-Apennine region.

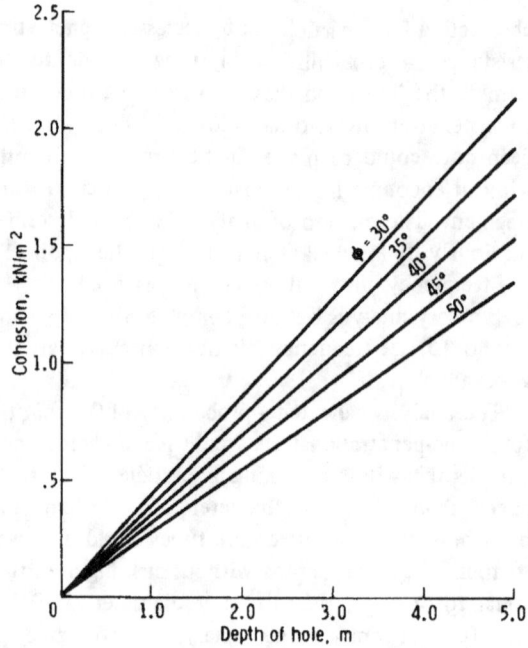

FIGURE 7-24.–Depth to bottom of elastic zone in an open bore hole.

(9) The SRP, used for the first time on this mission, has provided quantitative information on the penetration resistance of the lunar surface. Penetration data obtained at station 8 have indicated a soil of high density (1.97 g/cm^3), high strength, and low compressibility. Both theoretical analyses and the behavior of terrestrial simulants indicate an angle of internal friction at this site of approximately 50°. This high value is consistent with the high density.

(10) Analysis of the soil-mechanics trench-wall failure and the SRP data lead to an estimate for soil cohesion at station 8 of approximately 1.0 kN/m^2. This represents a cohesion greater than that apparent at the Apollo 11, 12, and 14 sites. This cohesion would be expected on the basis of the fine grain size and high density.

(11) A consideration of the variability of soil density on the lunar surface in conjunction with the strong dependence of other properties (strength and compressibility) on density, porosity, and relative density reinforces the need for determinations of the specific gravity and maximum and minimum densities for lunar-soil samples, if proper interpretation of lunar-soil behavior is to be made.

(12) The results of terrestrial simulations have indicated that it is unlikely that a core tube could have been pushed or hammered to its full length into the lunar surface at station 8. The data provide a basis for estimating the feasibility of core-tube sampling from the depth of penetration obtainable using the SRP.

(13) The stability of open core-tube holes and boreholes on the lunar surface has been analyzed, and collapse would not be expected for the shallow depths being drilled.

(14) The methods used to obtain soil-mechanics data have worked well, with the exception of the tendency of the SRP reference plane to ride up (which can be corrected easily). The quantitative values for soil properties deduced from the test results are considered reliable. The close correspondence between properties deduced using simulants and from theoretical analyses is particularly significant.

(15) Additional analyses are needed to relate the properties of lunar soil deduced herein and the variability of such properties to compositional and geological conditions on the lunar surface and to the processes that have shaped their history.

REFERENCES

7-1. Costes, N.C.; Carrier, W.D.; Mitchell, J.K.; and Scott, R.F.: Apollo 11 Soil Mechanics Investigation. Sec. 4 of Apollo 11 Preliminary Science Report, NASA SP-214, 1969.

7-2. Scott, R.F.; Carrier, W.D.; Costes, N.C.; and Mitchell, J.K.: Mechanical Properties of the Lunar Regolith. Sec. 10 of Apollo 12 Preliminary Science Report, Part C, NASA SP-235, 1970.

7-3. Mitchell, J.K.; Houston, W.N.; Vinson, T.S.; Durgunoglu, T.; et al.: Lunar Surface Engineering Properties and Stabilization of Lunar Soils. Univ. of Calif. at Berkeley, Space Sciences Laboratory, NASA Contract NAS 8-21432, 1971.

7-4. Mitchell, J.K.; Bromwell, L.G.; Carrier, W.D.; Costes, N.C.; and Scott, R.F.: Soil Mechanics Experiment. Sec. 4 of Apollo 14 Preliminary Science Report, NASA SP-272, 1971.

7-5. Scott, R.F.; and Roberson, F.I.: Soil Mechanics Surface Sampler. Surveyor Project Final Report, Part II: Science Results, Technical Report 32-1265, JPL, Pasadena, Calif., June 15, 1968, pp. 195-206.

7-6. Houston, W.N.; and Namiq, L.I.: Penetration Resistance of Lunar Soils. J. Terramechanics, vol. 8, no. 1, 1971, pp. 59-69.

7-7. Costes, N.C.; Cohron, G.T.; and Moss, D.C.: Cone Penetration Resistance Test – An Approach to Evaluating the In-Place Strength and Packing Characteristics of Lunar Soils. Proceedings of the Second Lunar Science Conference, vol. 3, A.A. Levinson, ed., Pergamon Press (Cambridge, Mass.), 1971, pp. 1973-1987.

7-8. Houston, W.N.; and Mitchell, J.K.: Lunar Core Tube Sampling. Proceedings of the Second Lunar Science Conference, vol. 3, A.A. Levinson, ed., MIT Press (Cambridge, Mass.), 1971, pp. 1953-1958.

7-9. Carrier, W.D., III; Johnson, S.W.; Werner, R.A.; and Schmidt, R.: Disturbance in Samples Recovered with the Apollo Core Tubes. Proceedings of the Second Lunar Science Conference, vol. 3, A.A. Levinson, ed., MIT Press (Cambridge, Mass.), 1971, pp. 1959-1972.

7-10. Scott, R.F.: Principles of Soil Mechanics. Addison-Wesley (Reading, Mass.), 1963.

7-11. Lunar Sample Preliminary Examination Team: Preliminary Examination of Lunar Samples from Apollo 14. Science, vol. 173, no. 3998, Aug. 20, 1971, pp. 681-693.

7-12. Lindsay, J.F.: Sedimentology of Apollo 11 and 12 Lunar Soils. J. Sed. Petr., Sept. 1971.

7-13. Green, A.J.; and Melzer, K.J.: Performance of Boeing-GM Wheels in a Lunar Soil Simulant (Basalt). Tech. Rept. M-70-15, USAE WES, Vicksburg, Miss., 1970.

7-14. Costes, N.C.; et al.: Lunar Soil Simulation Studies in Support of the Apollo 11 Mission. Geotechnical Research Laboratory, Space Sciences Laboratory, Marshall Space Flight Center, Huntsville, Ala., 1969.

7-15. Jaffe, Leonard D.: Depth and Strength of the Lunar Dust. Trans. Amer. Geophys. Union, vol. 45, Dec. 30, 1964, p. 628.

7-16. Jaffe, Leonard D.: Strength of the Lunar Dust. J. Geophys. Res., vol. 70, no. 24, Dec. 15, 1965, pp. 6139-6146.

7-17. Halajian, J.D.: The Case for a Cohesive Lunar Surface Model. Report ADR 04-40-64.2, Grumman Aircraft Engineering Corp., Research Dept., Bethpage, New York, June 1964.

7-18. Christensen, E.M.; Batterson, S.A.; Benson, H.E.; Chandler, C.E.; et al.: Lunar Surface Mechanical Properties. Surveyor I Mission Report, Part II: Scientific Data and Results, Technical Report 32-1023, JPL, Pasadena, Calif., Sept. 10, 1966, pp. 69-85.

7-19. Cherkasov, I.I.; Vakhnin, V.M.; Kemurjian, A.L.; Mikhailov, L.N.; et al.: Physical and Mechanical Properties of the Lunar Surface Layer by Means of Luna 13 Automatic Station. Moon and Planets II, A. Dollfus, ed., North-Holland Pub. Co. (Amsterdam), 1968, pp. 70-76.

7-20. Scott, R.F.; and Roberson, F.I.: Soil Mechanics Surface Sampler: Lunar Tests, Results, and Analyses. Surveyor III Mission Report, Part II: Scientific Data and Results, Tech. Rept. 32-1177, JPL, Pasadena, Calif., June 1, 1967, pp. 69-110.

7-21. Scott, R.F.: The Density of the Lunar Surface Soil. J. Geophys. Res., vol. 73, no. 16, Aug. 15, 1968, pp. 5469-5471.

7-22. Costes, N.C.; and Mitchell, J.K.: Apollo 11 Soil Mechanics Investigation. Proceedings of the Apollo 11 Lunar Science Conference, vol. 3, A.A. Levinson, ed., Pergamon Press (New York), 1970, pp. 2025-2044.

7-23. Scott, R.F.; Carrier, W.D., III; Costes, N.C.; and Mitchell, J.K.: Apollo 12 Soil Mechanics Investigation. Geotechnique, vol. 21, no. 1, Mar. 1971, pp. 1-14.

7-24. Vinogradov, A.P.: Preliminary Data on Lunar Ground Brought to Earth by Automatic Probe "Luna-16." Proceedings of the Second Lunar Science Conference, vol. 1, A.A. Levinson, ed., MIT Press (Cambridge, Mass.), 1971, pp. 1-16.

7-25. Obert, L.; and Duvall, W.I.: Rock Mechanics and the Design of Structure in Rock. John Wiley & Sons, Inc. (New York), 1967.

ACKNOWLEDGMENTS

The assistance of the individuals who have contributed to the preparation for, execution of, and data analysis from the soil-mechanics experiment is gratefully acknowledged. In particular, the authors wish to express appreciation to Astronauts David R. Scott and James B. Irwin, who were able to complete the station 8 soil-mechanics activities in an excellent manner despite a greatly compressed time line.

The basic design concept of the SRP was developed at the Geotechnical Research Laboratory of the NASA Marshall Space Flight Center Space Sciences Laboratory with the support of Teledyne-Brown Engineering Company, Huntsville, Alabama. Dr. Rolland G. Sturm, Roland H. Norton, George E. Campbell, and G.T. Cohron were instrumental in the development of this concept. The final design, construction, and qualification of the flight article were carried through by W.N. Dunaway and W. Lyon of the NASA Manned Spacecraft Center and W. Young of the General Electric Company.

Dr. H. John Hovland, L.I. Namiq, H. Turan Durgunoglu, Donald D. Treadwell, and Y. Moriwaki, all of the soil-mechanics research staff at the University of California, were responsible for many of the simulation studies, data analyses, and computations discussed in this report.

Dr. Stewart W. Johnson, National Research Council Senior Postdoctoral Fellow from the Air Force Institute of Technology, and Richard A. Werner, Lisimaco Carrasco, and Ralf Schmidt of Lockheed Electronics Company participated in work at the Manned Spacecraft Center.

Fred W. Johnson participated in the probability analyses and photographic studies done at the Massachusetts Institute of Technology.

8. Passive Seismic Experiment

*Gary V. Latham,[a][†] Maurice Ewing,[a] Frank Press,[b]
George Sutton,[c] James Dorman,[a] Yosio Nakamura,[d] Nafi Toksoz,[b]
David Lammlein,[a] and Fred Duennebier[c]*

With the successful installation of a geophysical station at Hadley Rille and the continued operation of the Apollo 12 and Apollo 14 stations approximately 1100 km southwest, the Apollo Program has for the first time achieved a network of seismic stations on the lunar surface, a network that is absolutely essential for the location of natural events on the Moon. The establishment of this network is one of the most important milestones in the geophysical exploration of the Moon.

Four major discoveries have resulted from the analysis of seismic data from this network for the 45-day period represented by this report.

(1) The Moon has a crust and a mantle, at least in the region of the Apollo 12 and Apollo 14 stations. The thickness of the crust is between 55 and 70 km and may consist of two layers. The contrast in elastic properties of the rocks that comprise the major structural units is at least as great as that existing between the crust and mantle of the Earth.

(2) Although present data do not permit a completely unambiguous interpretation, the best solution obtainable places the most active moonquake focus at a depth of 800 km, slightly deeper than any known earthquake. These moonquakes occur in monthly cycles and are triggered by lunar tides.

(3) In addition to the repeating moonquakes, moonquake "swarms" have been discovered. During periods of swarm activity, events may occur as frequently as one event every 2 hr during intervals lasting several days. The source of these swarms is unknown at present.

(4) Most of the seismic energy from a surface source is efficiently confined or trapped for a long time in the near-source region by efficient scattering near the lunar surface. The seismic energy slowly leaks to distant parts of the Moon, probably by more efficient seismic radiation within the lunar interior. Meteoroid impact signals are probably received from all parts of the Moon by efficient interior propagation.

The purpose of the passive seismic experiment (PSE) is to detect vibrations of the lunar surface and to use these data to determine the internal structure, physical state, and tectonic activity of the Moon. Sources of seismic energy may be internal (moonquakes) or external (meteoroid impacts and manmade impacts). A secondary objective of the experiment is the determination of the number and the masses of meteoroids that strike the lunar surface. The instrument is also capable of measuring tilts of the lunar surface and changes in gravity that occur at the PSE location.

Since deployment and activation of the Apollo 15 PSE on July 31, 1971, the instrument has operated as planned, except as noted in the following subsection entitled "Instrument Description and Performance." The sensor was installed west of the lunar module (LM) 110 m from the nearest LM footpad.

Signals were recorded from astronaut activities, particularly the movements of the lunar roving vehicle (Rover), at all points along the traverses (maximum range, approximately 5 km). The variation with range of Rover-generated seismic signals provides a measure of the amplitude decay law for seismic signals generated at close range.

The velocity of sound in the lunar regolith at the Apollo 15 landing site, as determined by timing the

---
[a]Lamont-Doherty Geological Observatory.
[b]Massachusetts Institute of Technology.
[c]University of Hawaii.
[d]General Dynamics.
[†]Principal investigator.

seismic signal generated by the LM ascent propulsion engine, is approximately 92 m/sec. This value is remarkably close to the regolith velocities of 104 and 108 m/sec measured at the Apollo 12 and Apollo 14 sites, respectively. The uniformity of the regolith velocities measured at widely separated sites indicates that the process of comminution by meteoroid impacts has produced a layer of remarkably uniform mechanical properties over the entire surface of the Moon.

Seismic signals from 75 events believed to be of natural origin were recorded by the long-period (LP) seismometers at one or more of the seismic stations during the 42-day period after the LM ascent. Of these events, at least two were moonquakes that originated in the region of greatest seismic activity, previously identified from recordings at the Apollo 12 and Apollo 14 stations; four were moonquakes from other locations; and 16 others were possible moonquakes. Of the remaining 53 events, 35 were probable meteoroid impacts, and 18 were too small or indistinct to be classified. The moonquakes occurred most frequently near the times of minimum (perigee) and maximum (apogee) distance between the Earth and the Moon during each monthly revolution of the Moon about the Earth, suggesting that the moonquakes are triggered by tidal stresses.

The A_1 epicenter (the point on the lunar surface directly above the moonquake source) is estimated to be nearly equidistant from the Apollo 12 and Apollo 14 sites and at a range of approximately 600 km south of the Apollo 12 station. Except for the absence of the surface-reflected phases usually seen on the records from deep earthquakes, the seismic data suggest that A_1 events, and perhaps all moonquakes, are deep. The best solution obtained to date places the A_1 moonquake focus at a depth of approximately 800 km. This depth is somewhat greater than that of any known earthquake. If this location is verified by future data, the result will have fundamental implications relative to the present state of the lunar interior. The nature of the stresses that might generate moonquakes at this depth is unknown at present. However, a secular accumulation of strain is inferred from the uniform polarity of the signals. In any case, if moonquakes originate at this depth, the deep lunar interior must possess sufficient shear strength to accumulate stress to the point of rupture. This condition places an upper bound on the temperatures that can exist at these depths.

A new method of data processing that enhances very small signals has revealed that episodes of greatly increased seismic activity occur. The events are very small and occur in some cases at average rates of one every 2 hr during intervals of several days. Individual events of these swarms appear to be moonquakes. Six such swarms occurred during April and May. These moonquake swarms do not appear to correlate with the monthly tidal cycle, as do the repeating moonquakes described previously. The locations and focal mechanisms of the sources of moonquake swarms are as yet unknown.

Seismic signals were recorded from two manmade impacts (the SIVB stage of the Saturn launch vehicle and the LM ascent stage) during the Apollo 15 mission. The LM-impact signal was the first event of precisely known location and time recorded by three instruments on the lunar surface. Data from these impacts combined with data from the impacts accomplished during the Apollo 12, 13, and 14 missions, and the lunar-surface magnetometer results are the main sources of information about the internal structure of the Moon. The general characteristics of the recorded seismic signals suggest that the outer shell of the Moon, to depths no greater than 20 km, is highly heterogeneous. The heterogeneity of the outer zone results in intensive scattering of seismic waves and greatly complicates the recorded signals. The structure of the scattering zone is not precisely known, but the presence of craters must contribute to the general complexity of the zone. The entire Moon is probably mantled by such a layer. The layered structure photographed at the Apollo 15 site suggests a sequence of thin lava flows. Each flow may have been highly fractured by thermal stresses while cooling and by meteoroid impacts before the overlying layer was deposited. A highly heterogeneous structure many kilometers thick may have been built up in this manner.

The LM ascent stage struck the surface 93 km west of the Apollo 15 station. The characteristic rumble from this impact spread slowly outward and was detected at the Apollo 15 station in approximately 22 sec, and at the Apollo 12 and Apollo 14 stations, 1100 km to the southwest, in approximately 7 min. The fact that this small source of energy was detected at such great range strongly supports the earlier hypothesis that the lunar interior, beneath the scattering zone, transmits seismic energy with extremely high efficiency. This observation strengthens the belief that meteoroid impacts are being detected from the entire lunar surface.

The Apollo 15 SIVB impact extended the depth of penetration of seismic rays to approximately 80 km. From these additional data, it now appears that a change in composition must occur at a depth of between 55 and 70 km. If so, this would be strong evidence of the presence of a lunar crust, analogous to the crust of the Earth and of about the same thickness. In the crustal zone, competent rock with a velocity for compressional waves of approximately 6 km/sec is reached at a depth of 15 to 20 km. The velocity begins to increase from 6 km/sec at a depth of between 20 and 25 km and reaches 9 km/sec at a depth of between 55 and 70 km. The presence of a secondary compressional-wave (P-wave) arrival suggests the possible existence of an intermediate layer that has a velocity of 7.5 km/sec. Thus, the lunar crust, in the region of the Apollo 12 and Apollo 14 stations, may consist of two layers: a surface layer with a velocity of 6 km/sec and a thickness of 20 to 25 km, overlying a layer with a thickness of between 30 and 50 km and a velocity of 7.5 km/sec, with a sharp increase in velocity to 9 km/sec at the base of the lower layer. Alternately, the transition from 6 to 9 km/sec may be gradual. Velocities between 6.7 and 6.9 km/sec are expected for the feldspar-rich rocks found at the surface. Thus, the measured velocities are within the range expected for these rocks.

The 9-km/sec material below a depth of 55 to 70 km may be the parent material from which the crustal rock has differentiated. If so, the Moon has undergone large-scale magmatic differentiation similar to that of the Earth.

INSTRUMENT DESCRIPTION AND PERFORMANCE

A seismometer consists simply of a mass that is free to move in one direction and that is suspended by means of a spring (or a combination of springs and hinges) from a framework. The suspended mass is provided with damping to suppress vibrations at the natural frequency of the system. The framework rests on the surface, the motions of which are to be studied, and moves with the surface. The suspended mass tends to remain fixed in space because of its own inertia, while the frame moves in relation to the mass. The resulting relative motion between the mass and the framework can be recorded and used to calculate original ground motion if the instrumental constants are known.

The Apollo 15 PSE consists of two main subsystems: the sensor unit and the electronics module. The sensor unit, shown schematically in figure 8-1, contains three matched LP seismometers (with resonant periods of 15 sec) alined orthogonally to measure one vertical (Z) and two horizontal (X and Y) components of surface motion. The sensor unit also includes a single-axis short-period (SP) seismometer (with a resonant period of 1 sec) that is sensitive to vertical motion at higher frequencies. The instrument is constructed principally of beryllium and weighs 11.5 kg, including the electronics module and thermal insulation. Without insulation, the sensor unit is 23 cm in diameter and 29 cm high. The total power drain varies between 4.3 and 7.4 W.

Instrument temperature control is provided by a 6-W heater, a proportional controller, and an aluminized Mylar insulation. The insulating shroud is spread over the local surface to reduce temperature variations of the surface material.

The LP seismometer detects vibrations of the lunar surface in the frequency range from 0.004 to 2 Hz. The SP seismometer covers the band from 0.05 to 20 Hz. The LP seismometers can detect ground motions as small as 0.3 nm at maximum sensitivity in the flat-response mode; the SP seismometer can detect ground motions of 0.3 nm at 1 Hz.

The LP horizontal-component (LPX and LPY) seismometers are very sensitive to tilt and must be leveled to high accuracy. In the Apollo system, the seismometers are leveled by means of a two-axis, motor-driven gimbal. A third motor adjusts the LP vertical-component (LPZ) seismometer in the vertical direction. Motor operation is controlled by command. Calibration of the complete system is accomplished by applying an accurate increment or step of current to the coil of each of the four seismometers by transmission of a command from Earth. The current step is equivalent to a known step of ground acceleration.

A caging system is provided to secure all critical elements of the instrument against damage during the transport and deployment phases of the Apollo mission. In the present design, a pneumatic system is used in which pressurized bellows expand to clamp fragile parts in place. Uncaging is performed on command by piercing the connecting line by means of a small explosive device.

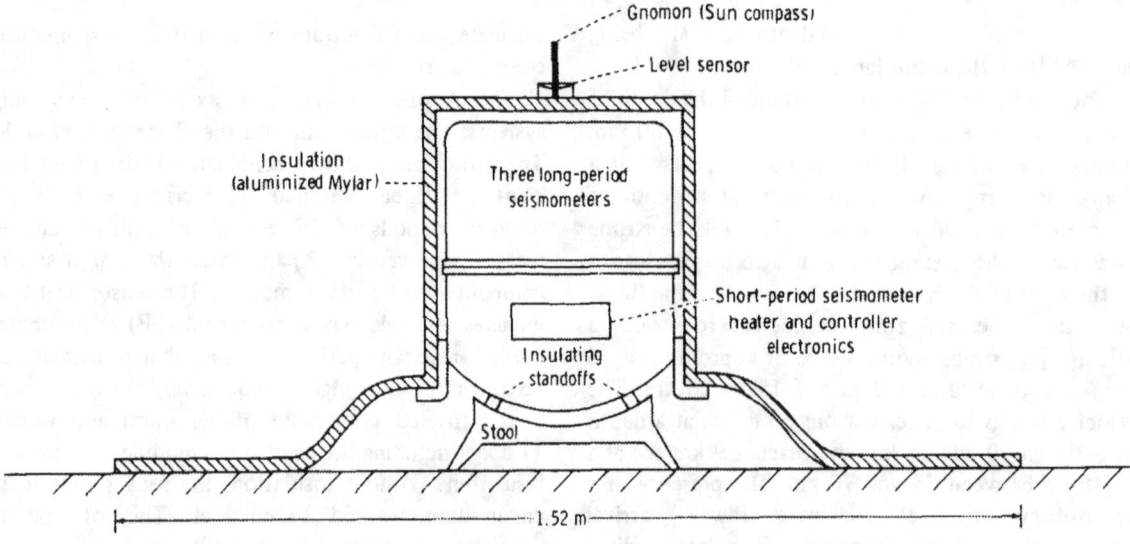

FIGURE 8-1.—Schematic diagram of PSE.

The seismometer system is controlled from Earth by a set of 15 commands that govern functions such as speed and direction of leveling motors, and instrument gain and calibration. In figure 8-2, the seismometer is shown fully deployed on the lunar surface.

Two modes of operation of the LP seismometers are possible: the flat-response mode and the peaked-response mode. In the flat-response mode, the seismometers have natural periods of 15 sec. In the peaked-response mode, the seismometers act as underdamped pendulums with natural periods of 2.2 sec. Maximum sensitivity is increased by a factor of 6 in the peaked-response mode, but sensitivity to low-frequency signals is reduced. The response curves for both modes are shown in figure 8-3.

The PSE was deployed 3 m west of the central station. No difficulty was experienced in deploying the experiment. Since initial activation of the PSE, all elements have operated as planned, with the exception of the sensor thermal-control system. The sensor temperature fell below the design setpoint of 126° F to a minimum of 112° F during the first lunar night and rose to a maximum of 133° F during the second lunar day. Preliminary examination of the instrument photographs suggests that these temperature variations may be a result of unevenness in the thermal shroud. The shroud skirt appears to be raised off the surface at several places to the extent that significant reduction in the effective insulation might occur.

As recorded at the Apollo 12 and Apollo 14 sites, episodes of seismic disturbances are observed on the LP seismometers throughout the lunar day. These disturbances are most intense near times of terminator passage and are believed to be caused by thermal contraction and expansion of the Mylar thermal shroud that blankets the sensor.

FIGURE 8-2.—Seismometer after deployment on the lunar surface (AS15-86-11590).

PASSIVE SEISMIC EXPERIMENT

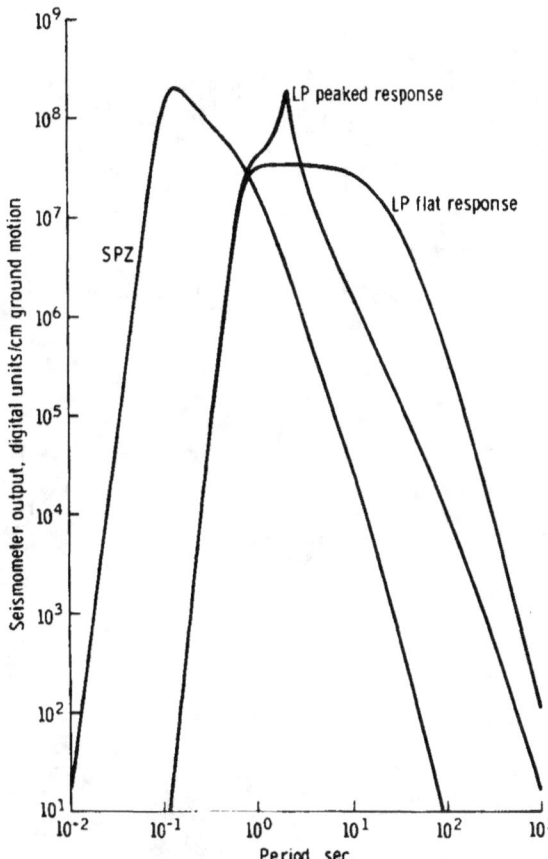

FIGURE 8-3.—Response curves for the LP and SP vertical-component seismometers. The ordinate scale is in digital units (DU) per centimeter ground motion amplitude. A DU is the signal variation that corresponds to a change in the least significant bit of the 10-bit data word.

RESULTS

The Apollo 15 PSE is a continuation of observations made during the Apollo 11, 12, 13, and 14 missions (refs. 8-1 to 8-8).

Preascent and Ascent Period

Before the LM ascent, many signals corresponding to various astronaut activities within the LM and on the surface were recorded, primarily on the SP vertical-component (SPZ) seismometer. Signals generated by the astronauts' footfalls and the motions of the Rover were detected at all points along the traverse (maximum range, approximately 5 km). A signal of particular interest during this period was generated by the thrust of the LM ascent engine.

The signal began 0.69 sec after the burn began and lasted approximately 4.5 min. As shown in figure 8-4, a second arrival can be recognized that occurs 0.51 sec after the first arrival. By comparison with Apollo 14 active-seismic-experiment data and previous LM ascent signals, the first arrival is interpreted as a wave refracted along an interface with higher velocity material at depth. The second arrival has an apparent velocity of 92 m/sec and is interpreted as a wave traveling through the top (regolith) layer. This value is remarkably close to the regolith velocities of 104 and 108 m/sec measured at the Apollo 12 and Apollo 14 sites, respectively. The uniformity of results from widely separated sites indicates that the process of comminution by meteoroid impacts has produced a layer of remarkably uniform mechanical properties over the entire surface of the Moon. If a regolith depth of 5 m is assumed at the Apollo 15 site, the velocity of seismic waves in the underlying material is 185 m/sec.

Signals generated by movements of the Rover were detected by the SPZ seismometer to ranges of approximately 4 to 5 km. The maximum range of detection was limited during extravehicular activity by the large seismic-background level originating within the LM. The maximum range of detection would be approximately 10 km under normal quiescent conditions.

Starting and stopping of the Rover produced only gradual buildup and decay of the seismic-signal amplitude rather than abrupt changes in signal amplitude. This gradual change in amplitude can be explained as resulting from intensive scattering of seismic waves in the upper layer of lunar material. Measurement of the amplitude of the signal as a function of range for various frequency components will provide an estimate of the statistical distribution of scatterers in this surface zone. This analysis has not yet been completed.

Signals From Impacts of the SIVB and LM Ascent Stages

Signals from two manmade impacts (the SIVB stage and the LM ascent stage) were recorded as part of the Apollo 15 mission. The SIVB impact preceded emplacement of the Apollo lunar surface experiments package (ALSEP) used on the Apollo 15 mission and was recorded at both the Apollo 12 and the Apollo

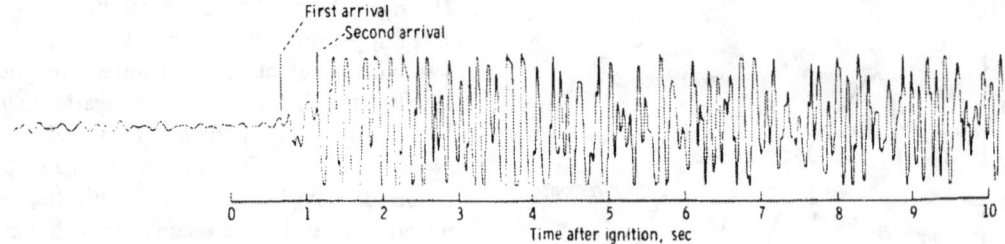

FIGURE 8-4.—Signal recorded by the SPZ seismometer from the lift-off of the Apollo 15 LM ascent stage.

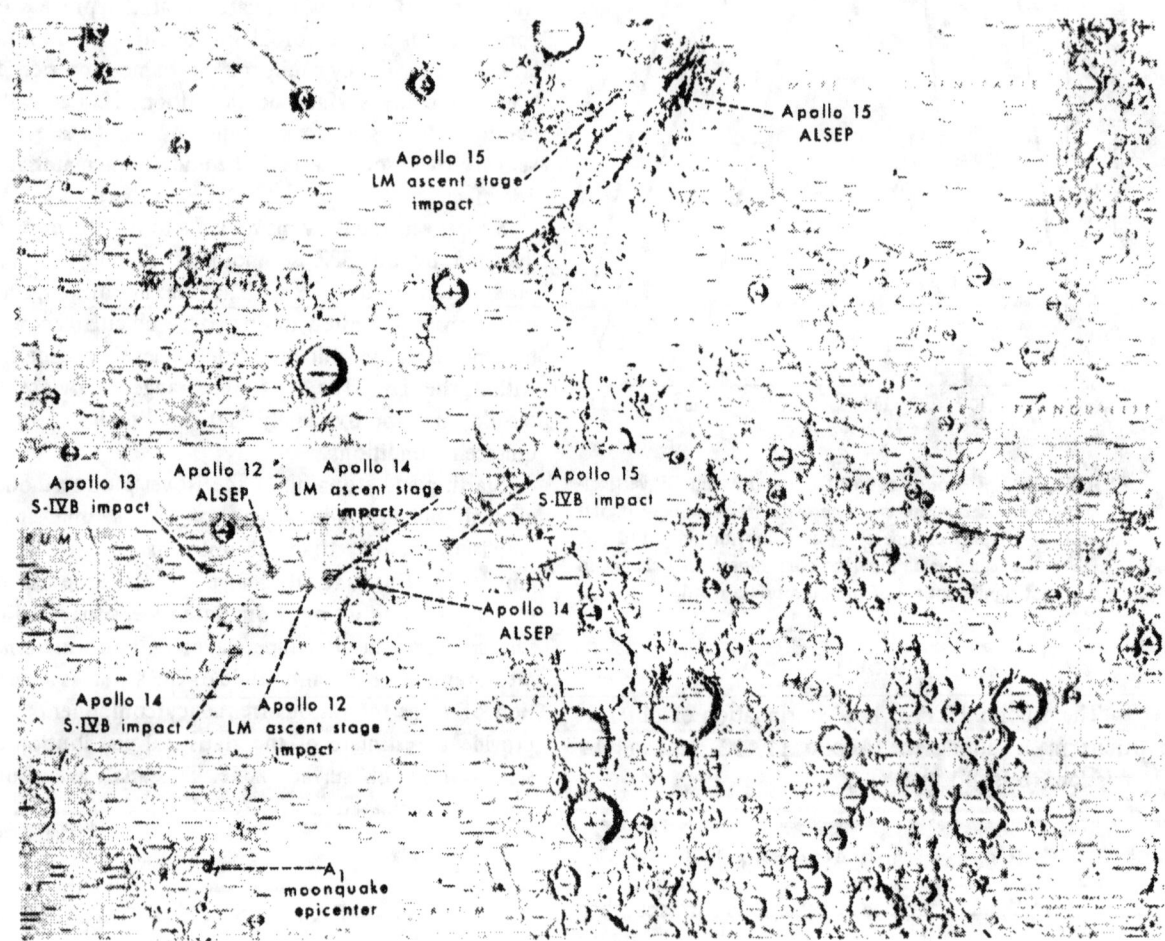

FIGURE 8-5.—Locations of ALSEP stations, LM, and SIVB impacts, and the epicenter of the most active source of moonquakes (A_1 zone).

14 stations, while the LM ascent-stage impact was also recorded at the Apollo 15 station. These impacts have aided greatly in understanding the lunar structure to a depth of approximately 80 km.

The locations of the three operating stations and all artificial impacts to date are listed in table 8-I and are also shown on the lunar map in figure 8-5. Relevant distances between these locations are also

TABLE 8-I.—*Coordinates of Seismic Stations, Impact Points, and Relevant Distances*

Location	Coordinates	Distance, km, from —		
		Apollo 12 site	Apollo 14 site	Apollo 15 site
Apollo 12 site	3.04° S, 23.42° W	--	--	--
Apollo 14 site	3.65° S, 17.48° W	181	--	--
Apollo 15 site	26.08° N, 3.66° E	1188	1095	--
Apollo 12 LM impact point	3.94° S, 21.20° W	73	--	--
Apollo 13 SIVB impact point	2.75° S, 27.86° W	135	--	--
Apollo 14 SIVB impact point	8.00° S, 26.06° W	170	--	--
Apollo 14 LM impact point	3.42° S, 19.67° W	114	67	--
Apollo 15 SIVB impact point	1.51° S, 11.81° W	355	184	--
Apollo 15 LM impact point	26.36° N, 0.25° E	1130	1049	93

listed in table 8-I. Pertinent parameters for the two impacts of the Apollo 15 mission are given in table 8-II.

The seismic signals from the last two artificial impacts, in compressed time scales, are shown in figure 8-6. These new impact signals are similar in character to previous impact signals. The signals are extremely prolonged, with very gradual increase and decrease in signal intensity and with little correlation between any two components of ground motion except at the first motion of the SIVB signal. Various distinctive pulses can be seen in the early parts of the records, but the only seismic phase that is identifiable with certainty is the initial P-wave arrival. These characteristics are believed to result from intensive scattering of the seismic waves in a highly heterogeneous outer shell, combined with low dissipation in the regions of the impact points and the seismic stations.

The initial portions of the new impact signals are shown in figure 8-7 on expanded time scales. The arrivals of the first P-waves and a tentatively identified shear wave (S-wave) are indicated in figure 8-7. The S-wave arrival has been identified on low-pass-filtered records and is not conspicuous in the broadband recording shown in figure 8-7. Traveltimes of these phases, together with those of previous manmade impact signals, are listed in table 8-III and are shown in figure 8-8. Detailed interpretation of the traveltime curves and related data are given in the section entitled "Discussion."

Short-Period Events

Several thousand signals with a great variety of shapes and sizes were recorded on the SPZ seismometer during the first 45 days of operation of the Apollo 15 PSE. The general level of recorded activity gradually subsided through the first lunar night after the initial activation of the PSE and increased abruptly at sunrise. Most of the events are attributed to venting or circulation of fluids and thermoelastic "popping" within the LM descent stage. The PSE is located approximately 110 m from the nearest footpad of the LM. The level of such activity was even higher during the operation of the Apollo 11 PSE but lower during the initial operating period of the Apollo 14 PSE. These relationships are explained by the smaller separation between the LM and the PSE in the Apollo 11 deployment (16.8 m) and by the greater separation between the LM and the PSE in the Apollo 14 deployment (178 m). These results cannot be compared with data obtained from the Apollo 12 SPZ seismometer because of the failure of that seismometer.

TABLE 8-II.—*Parameters of Apollo 15 Manmade Impacts*

Impact parameters	SIVB	LM
Day, G.m.t.	July 29	August 3
Range time,[a] G.m.t., hr:min:sec	20:58:42.9	03:03:37.0
Real time, G.m.t., hr:min:sec	20:58:41.6	03:03:35.8
Velocity, km/sec	2.58	1.70
Mass, kg	13 852	2385
Kinetic energy, ergs	4.61×10^{17}	3.43×10^{16}
Angle from horizontal, deg	62	3.2
Heading, deg	97	284

[a]Range time is the time that the signal of the associated event was observed on Earth.

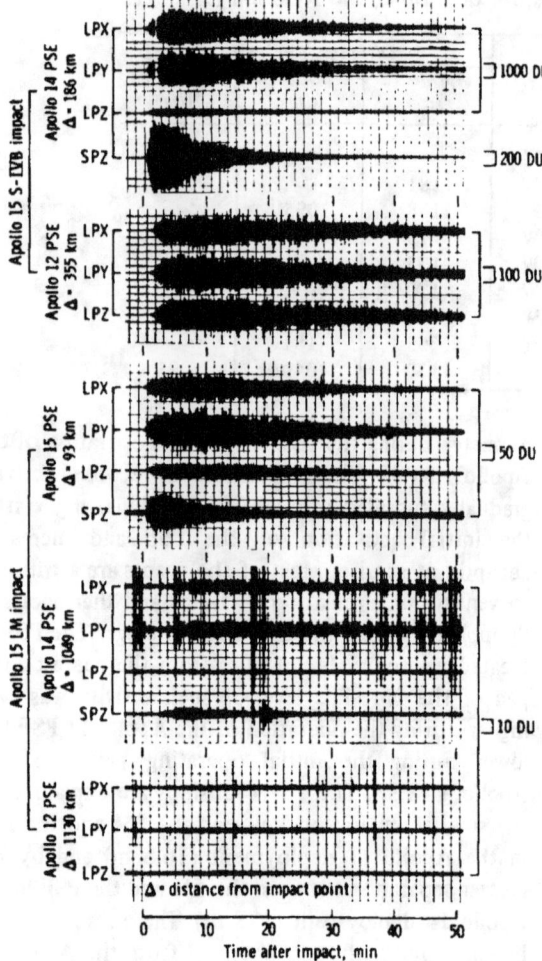

FIGURE 8-6.—Compressed time-scale records of the seismic signals received from the SIVB and LM impacts of the Apollo 15 mission at the Apollo 12, 14, and 15 ALSEP stations. The LPX, LPY, LPZ, and SPZ represent two long-period horizontal, long-period vertical, and short-period vertical components, respectively. For the amplitude scale DU, see figure 8-3.

The variation of SPZ seismometer activity is clearly related to the solar cycle at both the Apollo 14 and the Apollo 15 sites. The observed SPZ seismometer activity at the Apollo 14 site during the seventh and eighth lunations after deployment is shown in figure 8-9. Only SPZ seismometer events that show a gradual buildup and decay are included in figure 8-9. Events that have impulsive beginnings or very small rise times are probably generated by the LM or other equipment left on the Moon and are not included. Approximately 40 hours after sunrise at the Apollo 14 site, the number of observed SPZ seismometer events abruptly increases from a nighttime rate of four to five events per day to a daytime rate of between 30 and 50 events per day. This level of activity remains fairly constant until approximately 80 hours after sunset. At this time, the rate decreases abruptly to approximately 10 events per day, followed by a slower decrease to a nearly constant rate of four to five events per day approximately 5 days before sunrise. A similar pattern of activity is observed at the Apollo 15 site, but the level of activity is much higher than was detected during the first lunation at the Apollo 14 site, even though the LM and PSE separations at the two stations do not differ greatly. A rate of approximately 30 events per day was observed during the first lunar night after deployment of the Apollo 15 PSE as compared to a rate of four to five events per day for a nighttime period at the Apollo 14 site. This difference may indicate the presence of natural sources of seismic activity at the Apollo 15 site that are not present at the Apollo 14 site.

Possible sources of the observed SPZ seismometer activity at both the Apollo 14 and the Apollo 15 sites are thermal effects on the LM, the PSE thermal shroud, the cable connecting the sensor to the central station, other ALSEP instruments, the lunar soil, and nearby rocks. The phase lags noted between sunrise and the initiation of activity, and between sunset and the decrease in activity possibly reflect the thermal time constant of the source. Also, the continuing activity observed during the seventh lunar night at the Apollo 14 site suggests natural sources. Close meteoroid impacts and micromoonquakes would be possible sources for some of the signals. It is expected that more definite conclusions will be reached concerning the sources of signals recorded on the SPZ seismometers at the Apollo 14 and Apollo 15 sites, as the contribution from the LM descent stage at each station decreases during succeeding lunations. Then, the nighttime rate of SPZ seismometer activity common to both stations will set an upper limit on the possible meteoroid contribution, while a large discrepancy between the observed rates at the two stations would indicate a contribution from other sources near the more active station.

Natural Long-Period Events

Data on distant natural events come primarily from the LP seismometers. After an initial period of

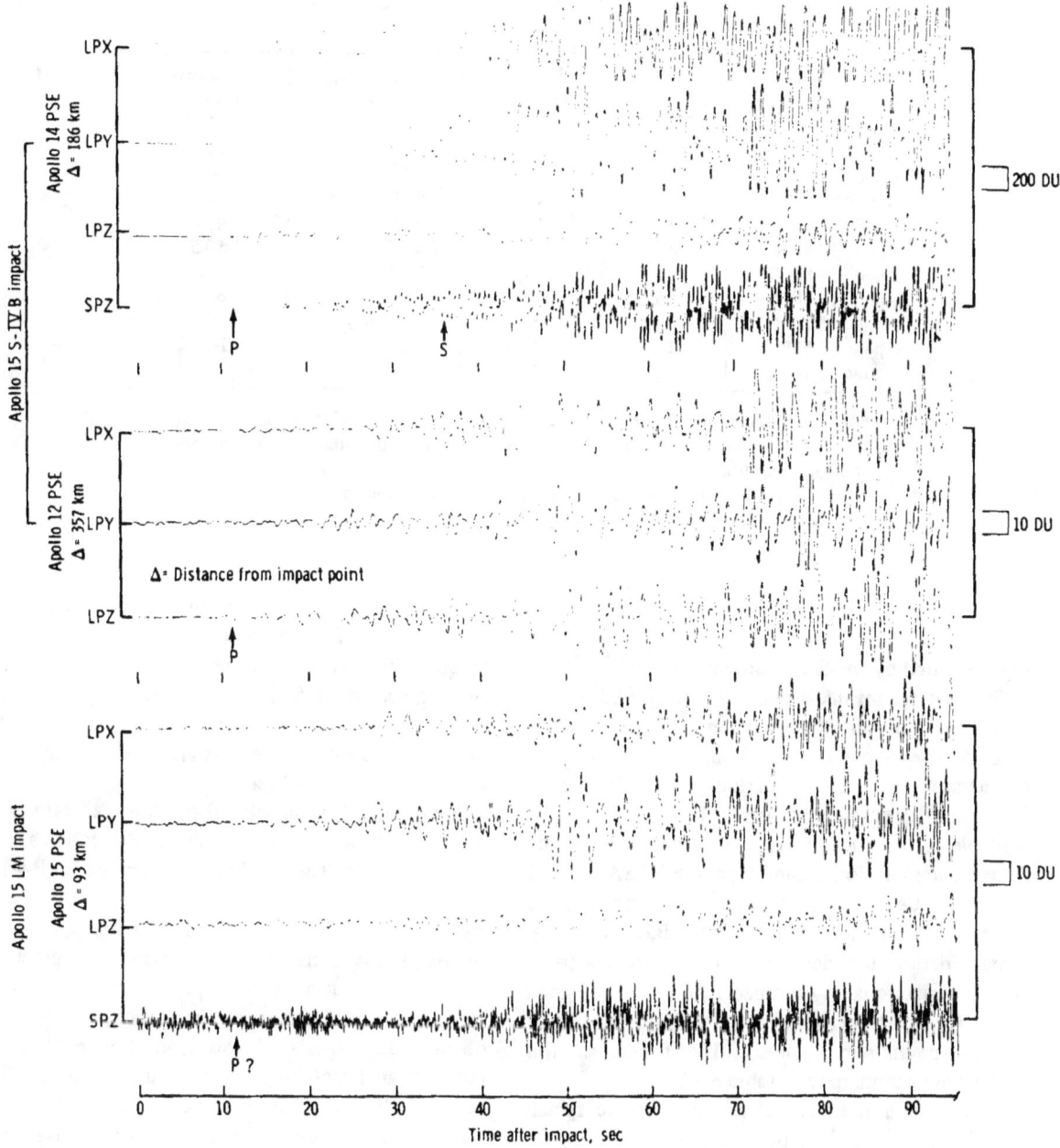

FIGURE 8-7.—Initial portions of the seismic signals from the Apollo 15 SIVB impact recorded at the Apollo 12 and Apollo 14 stations and from the Apollo 15 LM impact recorded at the Apollo 15 station. Zero on the time scale corresponds to 24.9 sec after impact for the top four traces, 44.1 sec after impact for the middle three traces, and 10.2 sec after impact for the bottom four traces. The earliest detectable signal is identified is the P-wave arrival, indicated by an arrow with the letter P. A later arrival, tentatively identified as the S-wave arrival from low-pass-filtered records and particle motion diagrams, is indicated by an arrow with the letter S.

TABLE 8-III.—*Observed Traveltimes of P-Wave and S-Wave Arrivals From LM and SIVB Impacts*

Impacting vehicle	Station	Distance, km	P-wave traveltime, sec	S-wave traveltime, sec[a]
Apollo 14 LM	14	67	17.8	31
Apollo 12 LM	12	73	ND[b]	NI[c]
Apollo 15 LM	15	93	[d]22?	NI
Apollo 14 LM	12	114	25?	45.0
Apollo 13 SIVB	12	135	28.6	50.5
Apollo 14 SIVB	12	170	35.7	55.7
Apollo 15 SIVB	14	186	36.6[e] (37.6)	60.5
Apollo 15 SIVB	12	357	55.5 (61)	NI
Apollo 15 LM	14	1049	ND	ND
Apollo 15 LM	12	1130	ND	ND

[a]All S-wave traveltimes are tentative because of uncertainties in identifying S-wave arrivals.
[b]ND = not detectable, signal amplitude below threshold of instrument detection capability.
[c]NI = not identifiable.
[d]Question marks indicate uncertain picks caused by noise background.
[e]Figures in parentheses indicate strong second arrivals.

days or weeks during which LM-generated high-frequency noise gradually subsides, these data can be supplemented by the SPZ seismometer data.

Seventy-five seismic signals were identified on the recordings from the Apollo 14 and Apollo 15 stations that were available from real-time data acquisition throughout the 42-day period after LM ascent. Continuous data from the Apollo 12 station were not available for this period because of limitations in the data decommutation capability at the NASA Manned Spacecraft Center (MSC). These data were retained on magnetic tape for future analysis. Based on signal characteristics, the detected events were classified either as moonquakes or impacts. Some of the signals were too small to be classified by the procedures used in this preliminary analysis. The results of this analysis are summarized in table 8-IV.

The criteria used for classifying seismic signals have been discussed in other reports (refs. 8-6 to 8-8), in which moonquakes and meteoroid impacts were identified as the sources of the recorded signals. Briefly, the signal characteristics that are useful in distinguishing a moonquake from a meteoroid impact are a short rise time and the presence of an H-phase. The H-phase of a moonquake is preceded only by low signal amplitudes and may be the first detectable motion of a weak event. The H-phase is a strong group of waves with a relatively sharp beginning, generally stronger on the horizontal components than on the vertical component, suggesting a shear-wave mechanism. The H-phase often contains lower frequencies than other portions of the record, peaking near the 0.5-Hz natural frequency of the LP seismometers. The maximum amplitude of the moonquake signal envelope is either at or only a few minutes after the H-phase.

Well-recorded signals also show an abruptly beginning, low-amplitude train that precedes and continues into the H-phase, suggesting a P-wave precursor. Numerous signals with these characteristics are found to fall into groups of matching signals (A_1 to A_{10}); members of each group have precisely identical waveforms. This property indicates a repeating moonquake focus to be the source of each group of matching signals. However, in this preliminary report, signal matching has been used in only a few cases to classify events because expanded time-scale playouts are required for the detailed phase comparisons.

In contrast with events containing an H-phase are numerous events (designated C-events) that have very emergent beginnings, smoothly varying envelope amplitudes, and no abrupt changes in signal frequency and amplitude. Because these are also the characteristics of artificial impact signals, C-events are believed to be meteoroid impacts. Unclassified events in table 8-IV are mainly those recorded too weakly to show diagnostic criteria on the drum seismograms.

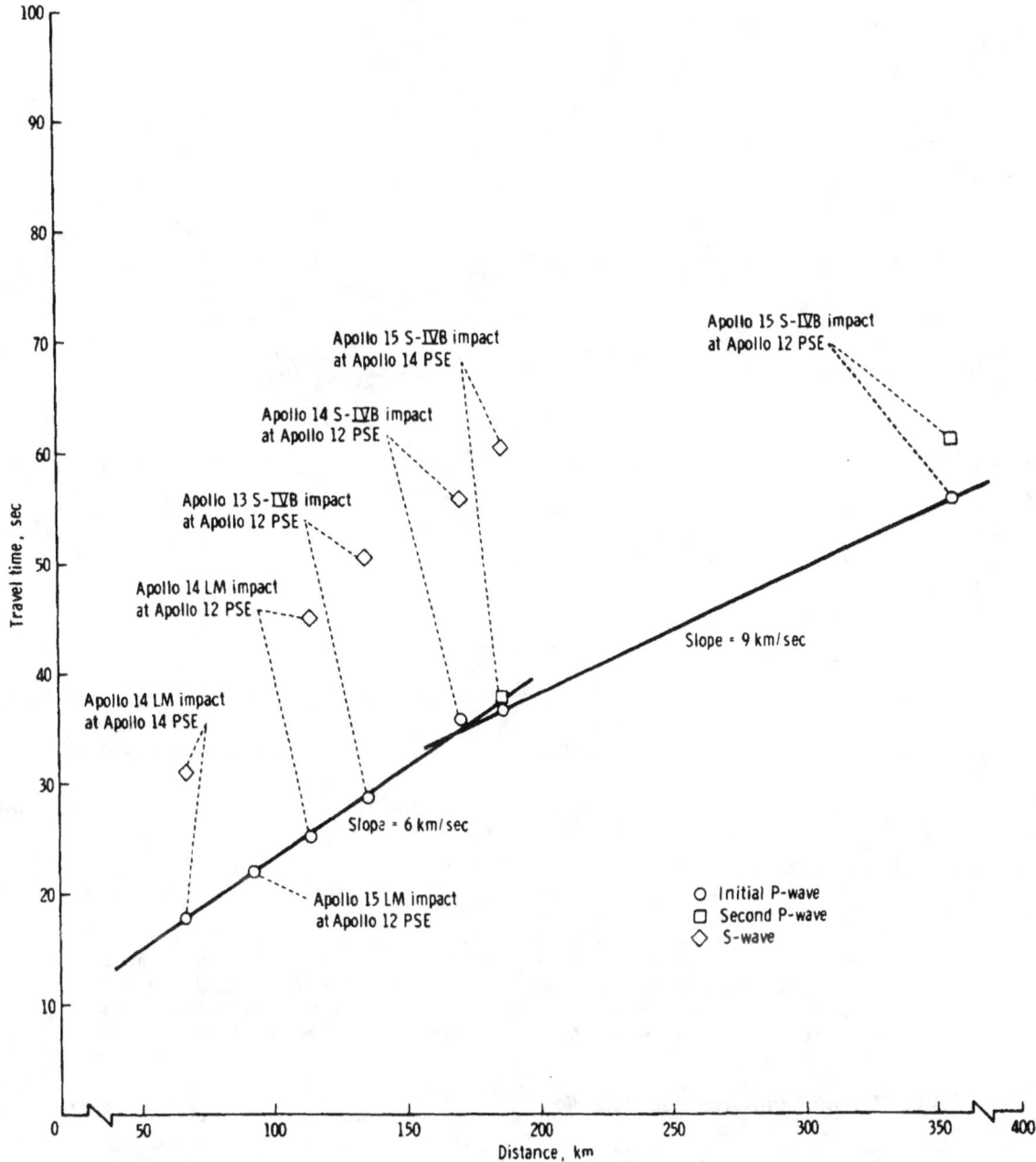

FIGURE 8-8.—Traveltimes of seismic waves recorded from the LM and SIVB impacts of the Apollo 12, 13, 14, and 15 missions. Circles indicate the first detectable arrivals, identified as P-wave arrivals. Squares are used to identify strong second arrivals. The S-wave arrival times are identified with much less certainty than the P-wave arrivals.

In table 8-IV, the number of moonquakes, impacts, and unclassified events during the 42-day period after LM ascent is given according to the stations at which the events were detected. Data in the three columns in table 8-IV are mutually exclusive; that is, an event counted at both the Apollo 14 and the Apollo 15 stations is not counted among those detected at only those stations. Fifty of the signals detected by the LP seismometers were also detected by the SP seismometer. These signals are

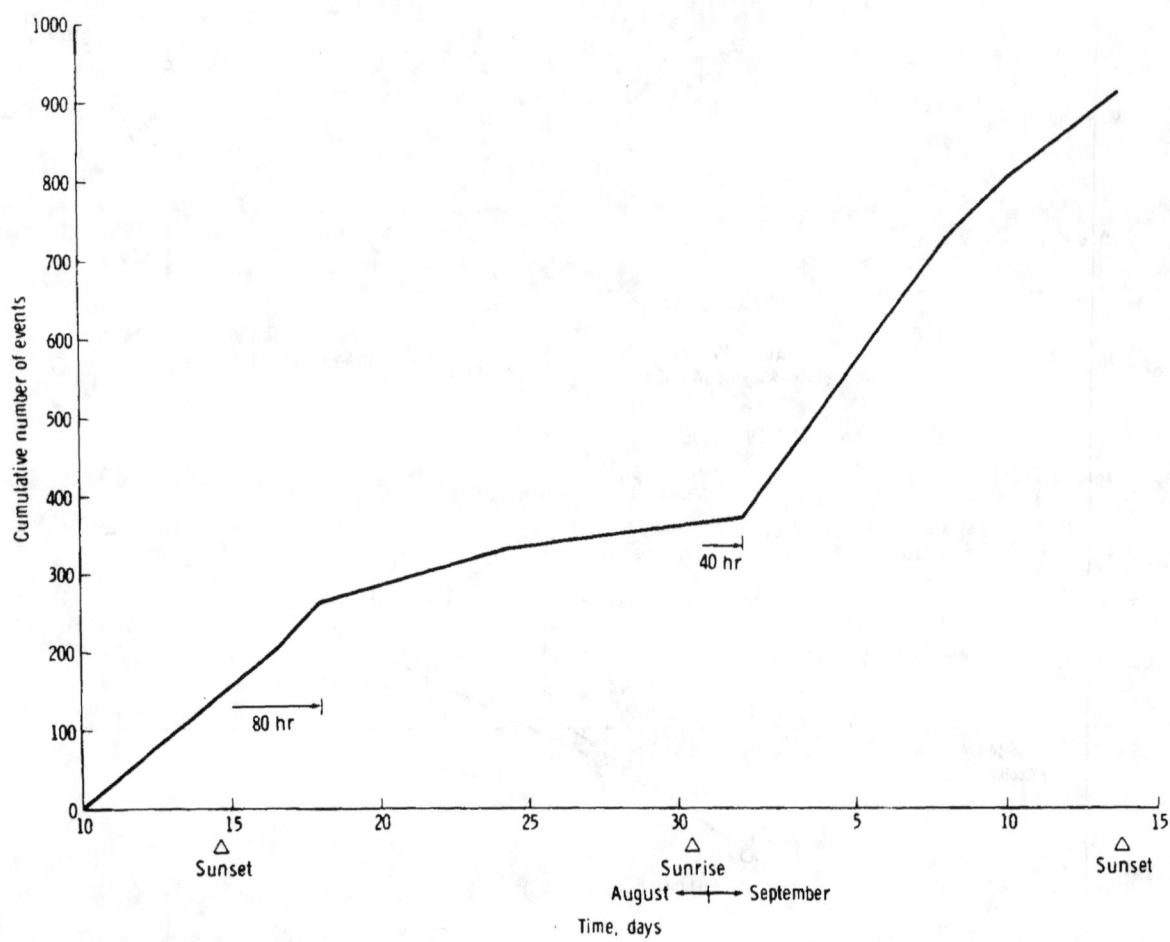

FIGURE 8-9.—Cumulative number of SPZ events observed at the Apollo 14 site as a function of time for the period August 10 to September 15. This time interval contains parts of the seventh and eighth lunations observed since deployment of the Apollo 14 PSE. Only SPZ signals that show a gradual buildup and decay are included. Signals that have impulsive beginnings or very short rise times are probably generated by the LM or other ALSEP instruments and are not included.

TABLE 8-IV.—*Seismic Events Recorded at Apollo 14 and 15 Sites*

Events	Sites			
	14	15	14 and 15	Total
LP event type				
Moonquake	11	0	11	22
Impact	26	4	5	35
Unclassified	16	1	1	18
Total LP	53	5	17	75
SP event type				
Moonquake	15	0	0	15
Impact	19	3	3	25
Unclassified	10	0	0	10
Total SP	44	3	3	50

counted in the lower part of table 8-IV. Signals detected by the SPZ seismometer only were not included in the listing. Classification of the SP events is based primarily on the character of the corresponding LP signals.

The cumulative distributions of the maximum amplitudes of the signal envelopes for the various categories of events are plotted in figure 8-10. The slope of such curves is a measure of the relative abundance of large and small events. The separate curves, which show the contributions of all events, all moonquakes, and all impacts for each station, are discussed in the following paragraphs.

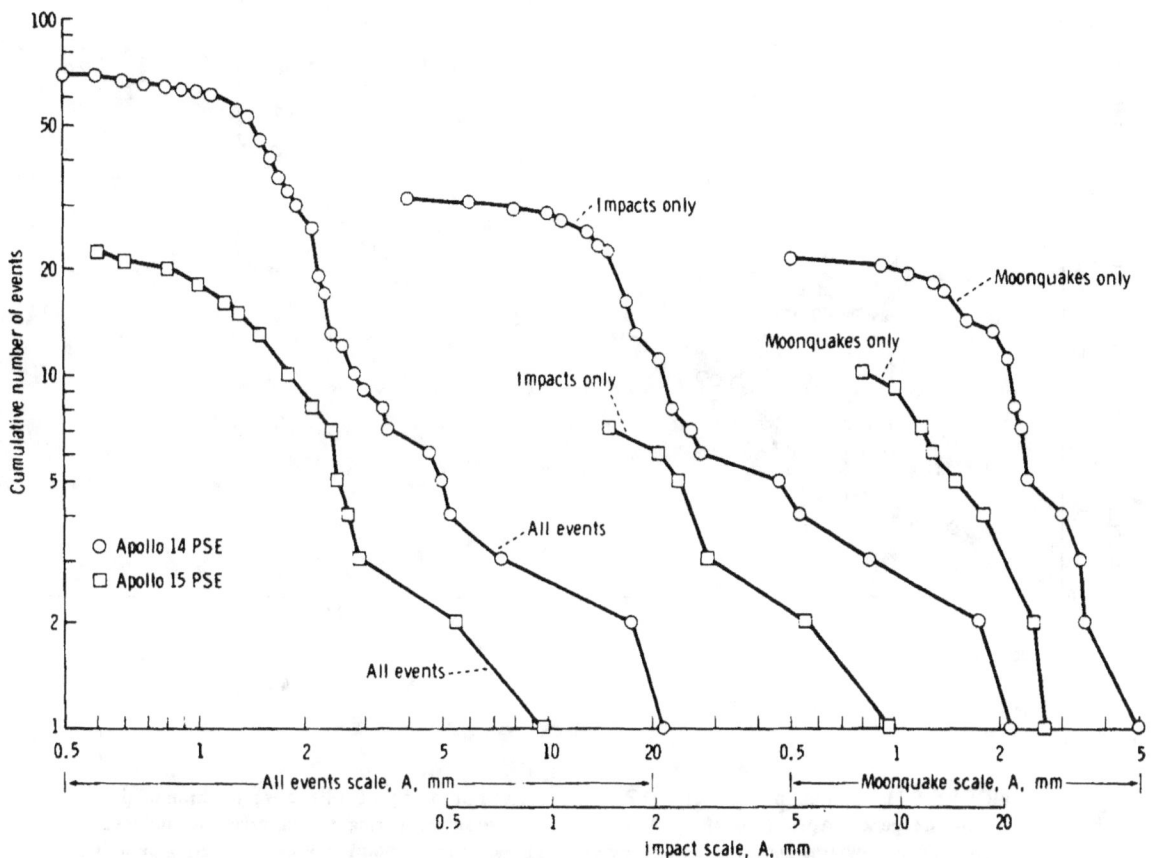

FIGURE 8-10.—Cumulative curves of LP amplitude data for natural events. Data used are amplitude measurements of the events represented in table 8-IV covering a 42-day period from LM lift-off August 2, 17:11 G.m.t., to September 13, 17:58 G.m.T. The amplitude variable $A = (X^2 + Y^2 + Z^2)^{1/2}$ where X, Y, and Z are the maximum peak-to-peak amplitudes of the smoothed envelopes of seismic signals recorded by the LP seismometers. Each plotted point (n,A) represents the number of events that have amplitude A or greater in the group represented by the particular curve. Unclassified events are included with the "all events" data but not with the impact or moonquake data.

A new data-processing technique that enhances very small signals has revealed that episodes of frequent small moonquakes occur. These episodes begin and end abruptly with no conspicuously large event in the series. By analogy with similar Earth phenomena, these signals are referred to as moonquake swarms. Examples of swarm activity are shown in the curve of figure 8-11. The slope of this curve is proportional to the rate of occurrence of moonquakes detected at the Apollo 14 station as a function of time from April 7 to May 10. Four intervals of increased activity occurred during this period. Two similar swarms occurred from May 16 to 20 and from May 25 to 29. Three swarms were detected at the Apollo 12 and Apollo 14 stations during the month after activation of the Apollo 15 station. Signals from moonquake swarms are not detectable by visual inspection of the real-time recordings from the Apollo 15 station for this period. However, additional data processing will be required to reach a definite conclusion on this point.

DISCUSSION

Scattering of Seismic Waves and Near-Surface Structure of the Moon

The characteristic long duration of lunar seismic

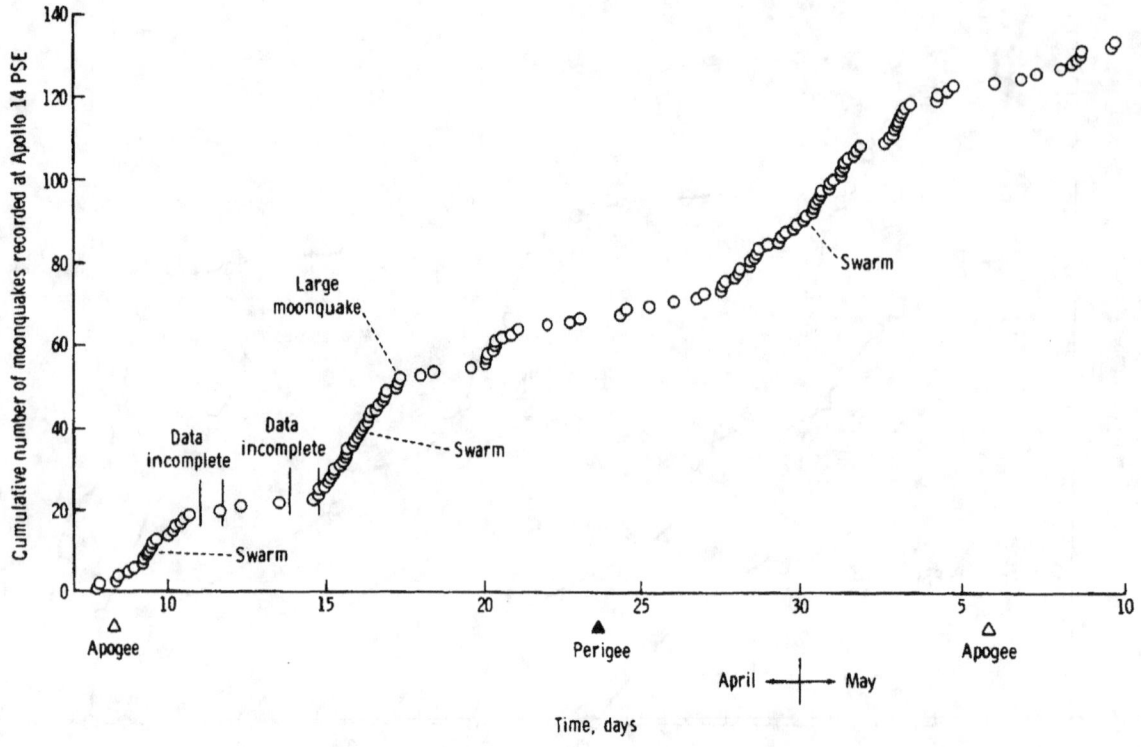

FIGURE 8-11.—Cumulative number of LP events observed at the Apollo 14 site as a function of time for the period April 7 to May 10. All LP events observed during this interval are included. Moonquake swarms appear as abrupt increases in the cumulative number of events. During periods of swarm activity, a rate of eight to 12 events per day is observed as compared to the normal rate of one to two events per day. Many of the swarm events are also detected at the Apollo 12 station.

wave trains has been interpreted as resulting from intensive scattering of seismic waves in a heterogeneous surface layer that probably blankets the entire Moon (refs. 8-1 to 8-7). Transmission of seismic energy below the scattering zone is believed to be highly efficient. This hypothesis is further confirmed by the data obtained during the Apollo 15 mission.

In figure 8-12, the rise times of artificial impact signals measured on narrowband-filtered seismograms are plotted against distance. The rise times increase with distance (and decreasing frequency) at near ranges, reach a maximum value at 100 to 150 km, and remain nearly at this level at greater distances. Seismic rays emerging at 100 to 150 km penetrate approximately 15 to 20 km into the Moon. Thus, the scattering zone must be of this thickness or less. However, a zone a few kilometers thick may account for the observed scattering. Below this level, the lunar material must behave more nearly as an ideal transmitter of seismic waves to produce no further significant increase in rise time at far distances.

In such a lunar structure, seismic waves generated by an impact are intensively scattered near the source and are observed as a scattered wave train in the near ranges. A part of this energy leaks into the lunar interior as a prolonged wave train, propagates through the lunar interior without significant scattering, undergoes further scattering when it returns to the surface, and is observed as a prolonged wave train at a distant seismic station. Therefore, the degree of scattering does not depend on the distance at far ranges. The relatively short rise times of moonquake signals are explained by the fact that seismic energy from a deep source must propagate through the scattering zone only one time en route to a seismic station.

The rise-time variation in the far ranges is irregular. This irregularity may reflect geographic differences in

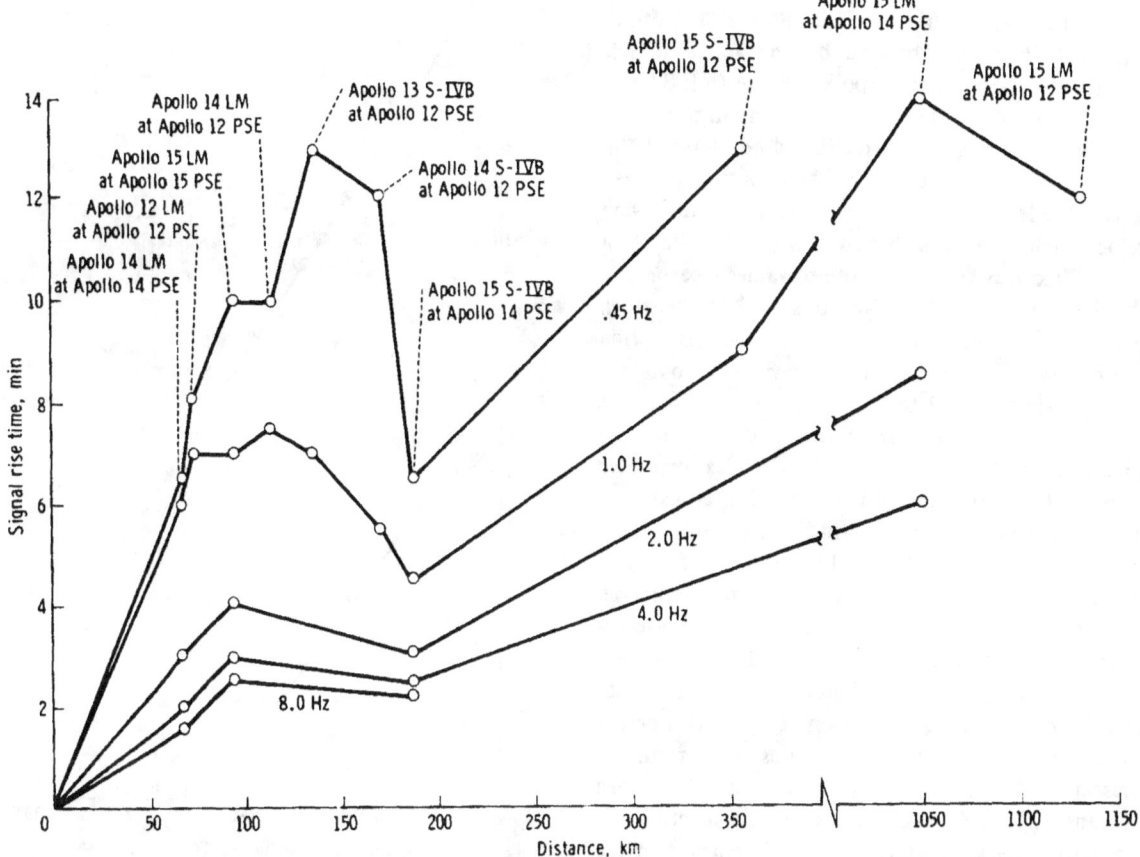

FIGURE 8-12.—Rise times of LM- and SIVB-impact signals measured on narrowband-filtered LPZ and SPZ seismograms. Four-pole Butterworth filters were used to produce the filtered seismograms. The rise time is measured from the time of impact to the approximate peak of the signal strength. The apparent trough in the rise-time curve at approximately 180 km coincides with the focusing of seismic energy at this range evidenced by enhancement of the P-wave amplitudes and the amplitudes of signal envelope maximums.

the complexity of the scattering zone. Important details of the scattering zone have not yet been determined; however, it is evident that abundant craters of various sizes and associated structural disturbances near the surface of the Moon must be elements of this unusual structure. Further study of the rise-time data of a large number of lunar events will eventually lead to a more quantitative description of the lunar scattering zone.

Lunar Structure Below the Scattering Zone

Information about the lunar structure can be obtained from the traveltimes, amplitudes, and angles of emergence of P-waves. The traveltimes provide the most directly interpretable data, whereas the other information supplements these data for determining the finer features of the velocity profile.

As shown in figure 8-8, the observed traveltime curve between distances of 67 and 186 km can be represented by a nearly constant apparent velocity of 6 km/sec. Deviations from a straight line in this range are less than 1 sec, despite the fact that these data represent traveltimes over paths with widely varying azimuths. This observation indicates that the outer 25 km of the Moon is fairly uniform in various directions in the region of the Apollo 12 and Apollo 14 stations.

The nearly constant apparent velocity of 6 km/sec in this distance range means that the P-wave velocity is nearly constant at 6 km/sec in a certain depth range in the lunar interior. How the P-wave velocity increases from approximately 300 m/sec at a depth of less than 76 m, as observed by the active seismic experiment (ASE) at the Apollo 14 site (ref. 8-9), to 6 km/sec is uncertain because of the gap in traveltime data between ranges of a few hundred meters (ASE range) and 67 km (Apollo 14 LM impact). However, it can be deduced from these data that the P-wave velocity must equal or exceed 6 km/sec at a depth of 15 to 20 km; 6 km/sec is close to values measured in the laboratory on lunar igneous rock under high pressure (refs. 8-10 and 8-11). Thus, the assumption can be made that the seismic wave velocity to depths of approximately 20 to 25 km is about the same as the velocity found at corresponding confining pressures in the laboratory. Traveltimes predicted from such a model of the lunar interior, adjusted to match the near-surface ASE findings, are found to be in close agreement with the observed traveltimes to a distance of 170 km. Thus, the simplest model consistent with the data would consist of loosely consolidated material in the near-surface zone, self-compacted to yield a strong increase in velocity with depth, with a gradual transition to more competent rock below. The velocity increases as a result of pressure and reaches 6 km/sec at a depth of between 15 and 20 km. The effects of compaction by meteoroid impacts, which may be important in the upper few kilometers, have been ignored.

An important new finding of the Apollo 15 mission is that an apparent velocity of 9 km/sec is observed between distances of 186 and 357 km as determined from recordings of the Apollo 15 SIVB impact at the Apollo 14 and Apollo 12 stations, respectively. Explanations of this high apparent velocity in terms of lateral inhomogeneities and local structural effects in material with a velocity of 6 km/sec or less appear to be unsatisfactory. Thus, it is concluded that a layer with a P-wave velocity of approximately 9 km/sec exists below a depth of 25 km in the lunar interior. Whether the transition from 6 to 9 km/sec is gradual or sharp cannot be determined from the available traveltimes alone.

More detailed information on lunar structure can be obtained when other associated data, such as signal-amplitude variations and second arrivals, are also taken into account. The observed variation of the maximum P-wave amplitudes for the LM and SIVB impact signals is shown in figure 8-13. From this plot, it can be seen that the LM data must be adjusted

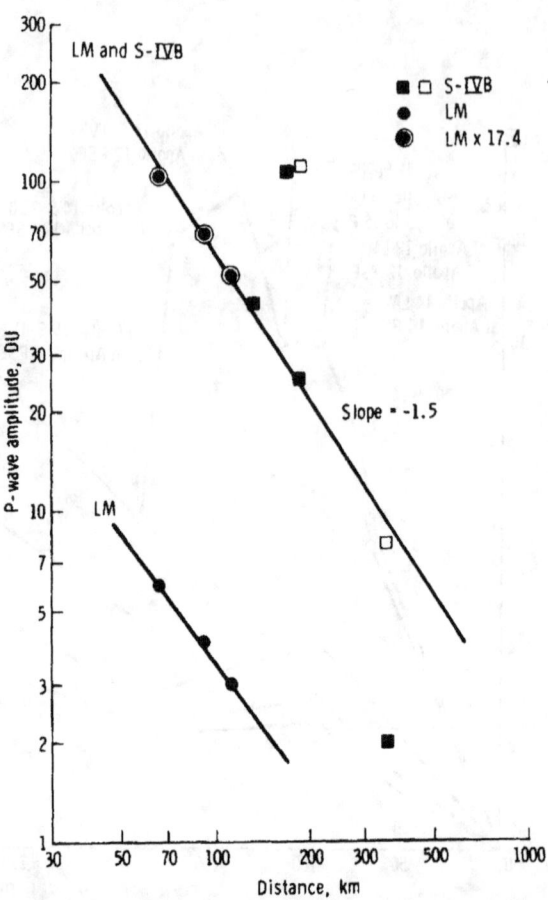

FIGURE 8-13.—Maximum peak-to-peak amplitudes, in digital units, of P-wave arrivals from the LM and SIVB impacts of the Apollo 12, 13, 14, and 15 missions, recorded by the LP seismometers. Each LM had nearly the same kinetic energy of approximately 3.3×10^{16} ergs at impact; each SIVB had nearly the same kinetic energy of approximately 4.6×10^{17} ergs at impact. The LM signal amplitudes are adjusted upward by a factor of 17.4 to give a smooth fit to the SIVB data. This empirically determined factor is required to compensate for the lower kinetic energy and shallower angle of the LM impacts relative to the SIVB impacts. The digital unit is defined in figure 8-3.

upward by a factor of 17.4 to yield values that fit smoothly on the curve for the SIVB data. This factor is required to compensate for the lower kinetic energy and much shallower angle of the LM impact. Except for the two data points near 180 km, the P-wave amplitude decreases monotonically in approximate proportion to the -1.5 power of the distance. The P-wave amplitudes at 170 and 186 km are greater than the base-level amplitude by a factor of 4.5, corresponding to an energy concentration of a factor of 20. This observation indicates that the lunar structure is such that focusing of seismic energy occurs in this range interval. A change in the slope of the travel-time curve, from 6 km/sec to a higher velocity, also occurs in this range interval. Both amplitude and travel-time variations require that the P-wave velocity begin to increase at a depth of between 20 and 25 km, to higher velocities below. Thus, a change in composition or phase at a depth of between 20 and 25 km is inferred.

Strong secondary arrivals from the Apollo 15 SIVB impact, as plotted in figure 8-8, can be interpreted as resulting from the presence of an intermediate layer with a P-wave velocity approaching 7.5 km/sec and a thickness of between 30 and 50 km. Arrivals after the initial P-wave must be treated with caution, however, because of the possibility that they may correspond to multipaths through an irregular medium. This ambiguity can be resolved only by obtaining additional impact data in this region, as is presently planned for the Apollo 16 mission.

Additional data to aid in the determination of detailed lunar structure are horizontal-to-vertical amplitude ratios of P-wave arrivals (fig. 8-14). The horizontal-to-vertical amplitude ratio is a direct measure of the P-wave incident angle and thus is an indication of the relative velocity at the deepest point where each seismic ray penetrates. This ratio decreases with increasing distance, indicating a gradual increase of velocity from the penetration depth of the ray emerging at 67 km (approximately 13 km) to the penetration depth of the ray emerging at 186 km (approximately 25 km). A quantitative interpretation of the data requires a knowledge of the velocity structure within the upper few kilometers of the lunar interior. Conversely, because the velocities at depth are known from the traveltime curve, the velocity structure in the upper few kilometers of the lunar interior can be deduced from the horizontal-to-vertical amplitude ratio. This latter step has not yet been completed.

Two interpretations of the P-wave arrival data representing the range of likely models for the upper 80 km of the lunar interior are shown in figure 8-15. In model 1, the transition in P-wave velocity from 6 to 9 km/sec is gradual; model 2 introduces two abrupt changes. In both models, the thickness of the surface crustal layer is between 20 and 25 km. The P-wave velocity reaches approximately 6 km at the

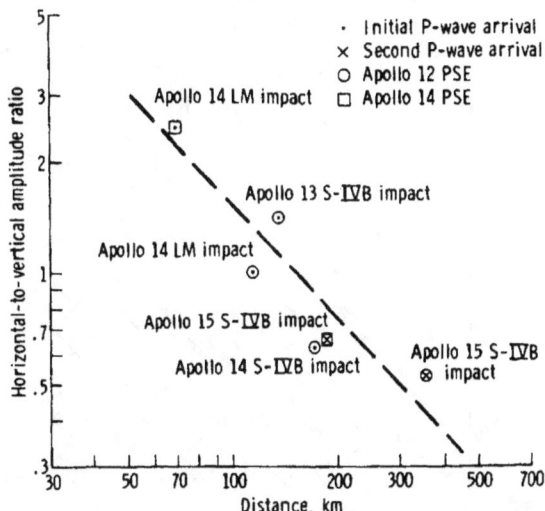

FIGURE 8-14.—Horizontal-to-vertical amplitude ratios of major P-wave arrivals from LM and SIVB impacts of the Apollo 12, 13, 14, and 15 missions, as recorded by the LP seismometers. The ratio is greater than 1 when the ground-particle motion is more nearly horizontal than vertical. Aside from the use of this diagram in interpreting velocity structure in the lunar interior, the range dependence seen in this diagram may also be used for estimating distances to large meteoroid impacts in this distance range.

base of this layer. The top zone of the layer (the scattering zone) is a zone in which the velocity increases rapidly with depth as a result of self-compaction and in which seismic waves are intensively scattered. The velocity below layer 1 may increase gradually, reaching 9 km/sec at a depth of between 60 and 70 km (model 1); or it may reach a constant velocity of 7.5 km/sec, with a sharp transition to 9 km/sec at a depth of between 55 and 70 km (model 2). Thus, the zone below the top layer may be a transition zone or it may be a second crustal layer with a thickness of between 30 and 50 km and a P-wave velocity of 7.5 km/sec, overlying the higher velocity mantle. The presence of deeper discontinuities is not precluded by present data.

The identification of shear waves in the impact signals is sufficiently uncertain that discussion of structural interpretation based upon their traveltimes must be deferred until further analysis can be completed. Similarly, it is expected that further study of body-wave arrivals from moonquakes and meteoroid impacts will aid in further elucidation of lunar structure.

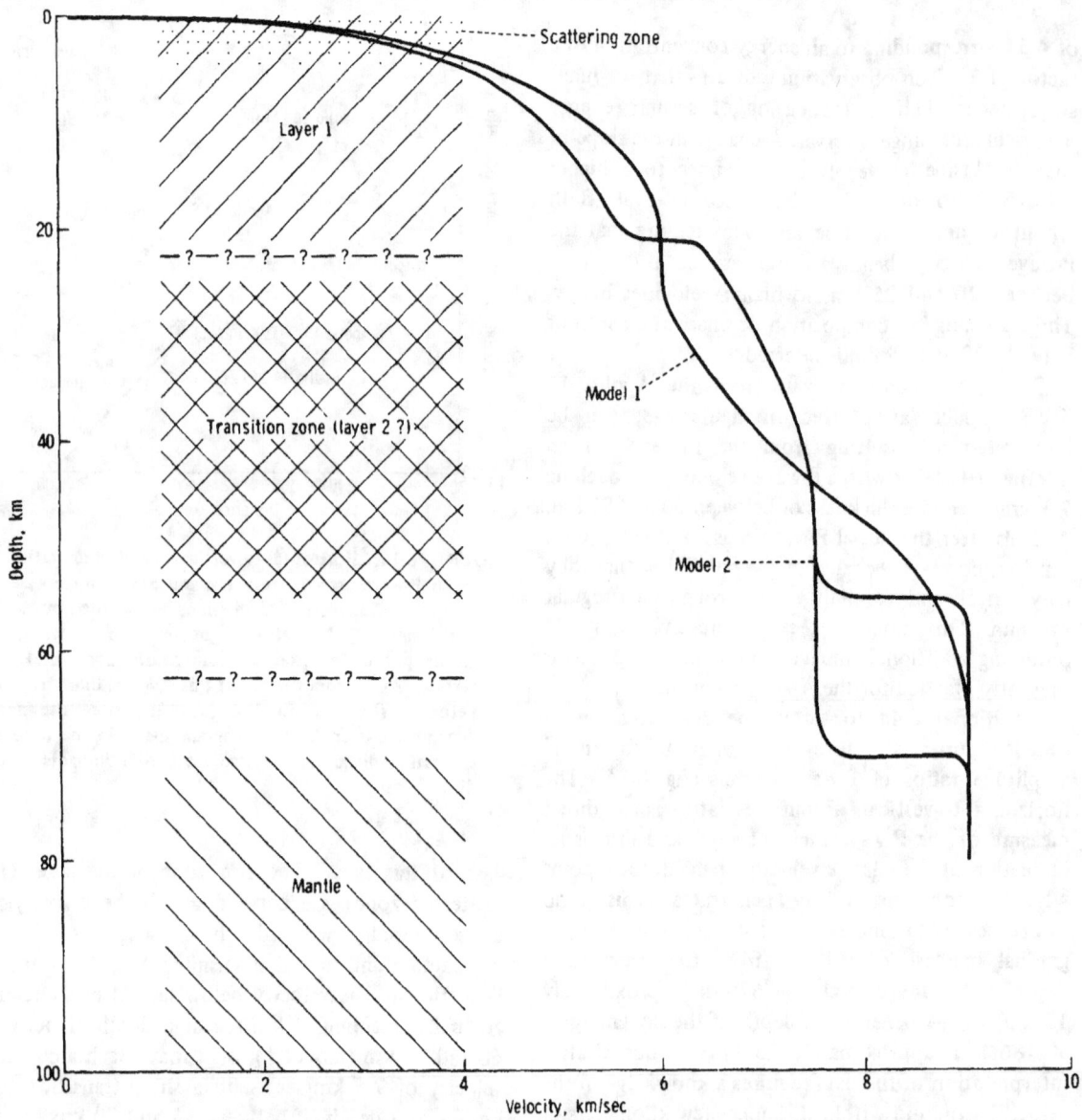

FIGURE 8-15.–A tentative P-wave velocity profile in the upper 80 km of the lunar interior deduced from P-wave arrivals from the LM and SIVB impacts of the Apollo 12, 13, 14, and 15 missions.

No P-wave traveltime data are available at present beyond a distance of 357 km corresponding to a depth of penetration of approximately 80 km. However, other features of the signal wave train, such as the peak-to-peak amplitude of the signal envelope and the rise time of the signal, can be used to interpret the structure of the lunar interior below this depth.

The maximum peak-to-peak amplitude of the signal envelope shows an almost monotonic decrease with increasing range in figure 8-16 for both the SIVB and the LM-impact data. The only exception is the amplitude generated by the Apollo 15 SIVB at the Apollo 14 station (range, 186 km), which is somewhat larger than the amplitude of the Apollo 14 SIVB impact recorded at the Apollo 12 station

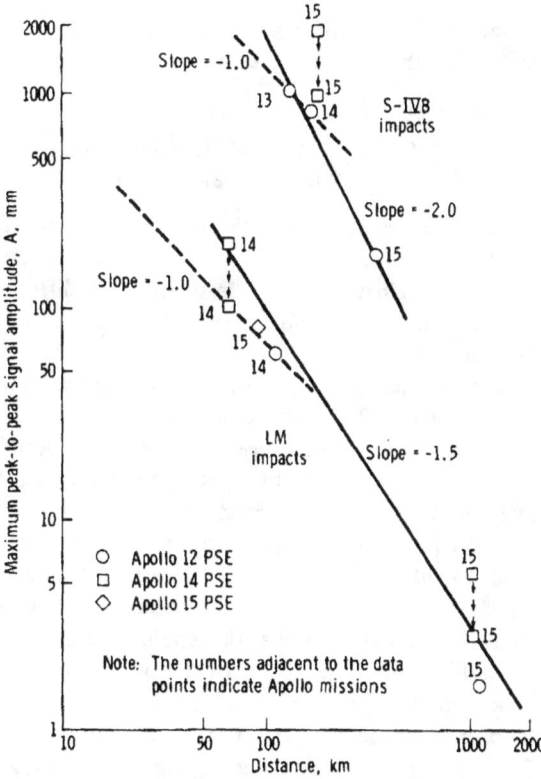

FIGURE 8-16.—Peak signal amplitudes A as a function of ranges r of artificial-impact recordings. The amplitude given is $A = (X^2 + Y^2 + Z^2)^{1/2}$, where X, Y, and Z are the maximum peak-to-peak amplitudes (mm) of unfiltered signal envelopes recorded at full gain on the three LP component seismograms. Data of the Apollo 13, 14, and 15 SIVB impacts and of the Apollo 14 and 15 LM impacts are shown. The SIVB-impact data form an upper group of points, and LM-impact data fall in a lower group. The Apollo 14 station amplitudes are plotted at their observed values and also as reduced by a factor of 2.0 to correspond to the observed average difference in sensitivity between the Apollo 12 and Apollo 14 stations. The slopes of lines drawn through various sets of data indicate a range variation of amplitude between $r^{-1.0}$ and $r^{-2.0}$.

(range, 170 km), even after allowing for the higher sensitivity of the Apollo 14 station. This difference is probably a result of the focusing effect at this range. Also, as for P-wave amplitudes, the maximum signal-envelope amplitudes for SIVB impacts cannot be compared directly with those of LM impacts for two reasons: the shallow impact angle of LM vehicles (3.3° to 3.7° from the horizontal) introduces uncertainty as to the seismic coupling efficiencies of LM impacts; also, the coupling efficiency of an SIVB impact may greatly exceed that of an impact of the LM ascent stage because the greater kinetic energy causes penetration into harder material. At ranges of as much as 170 km, an amplitude variation as $r^{-1.0}$ is satisfactory for LM and SIVB data, separately. The amplitude of the Apollo 15 LM impact at the Apollo 15 station also fits this variation. Stronger falloff, at least for distances beyond 186 km, is indicated by the Apollo 15 longer range data.

In figure 8-17, all SIVB-generated amplitudes that are shown in figure 8-16 have been divided by a factor of 20. This conversion factor provides a smooth overlap between the LM and SIVB amplitude data. The data can be fit by two straight-line segments intersecting at approximately 200 km. The simple, near-range variation as r^{-1} was noted in reference 8-7. It is apparent that a distance variation of approximately $r^{-1.8}$ is required at greater ranges. The range dependence of $r^{-1.8}$ and the LM/SIVB adjustment factor of 20 determined for the maximums in the signal envelopes both agree reasonably well with the values of $r^{-1.5}$ and 17.4 found independently from the P-wave amplitudes.

The Absence of Surface Waves and Surface-Reflected Phases

The major differences between terrestrial and lunar seismic signals now appear to be explained by the presence of a heterogeneous surface layer that blankets the Moon to a probable depth of several kilometers, with a maximum thickness of 20 km. Seismic waves with the observed wavelengths are intensively scattered within this zone. Seismic wave velocities and absorption of seismic energy are both quite low in this zone. The acoustic properties of this zone, the scattering zone, can probably be explained as the result of meteoroid impacts and the nearly complete absence of fluids. The lower boundary of the scattering zone is the depth below which the rock, conditioned by pressure, transmits seismic waves without significant scattering.

The presence of the scattering zone probably accounts for two features of lunar seismic signals that earlier puzzled the experiment team: (1) the absence of normal surface waves (Love waves and Rayleigh waves) and (2) the poor definition or total absence of identifiable surface-reflected phases, particularly from deep moonquakes.

Surface waves, with wavelengths of approximately the thickness of the scattering zone and smaller, will

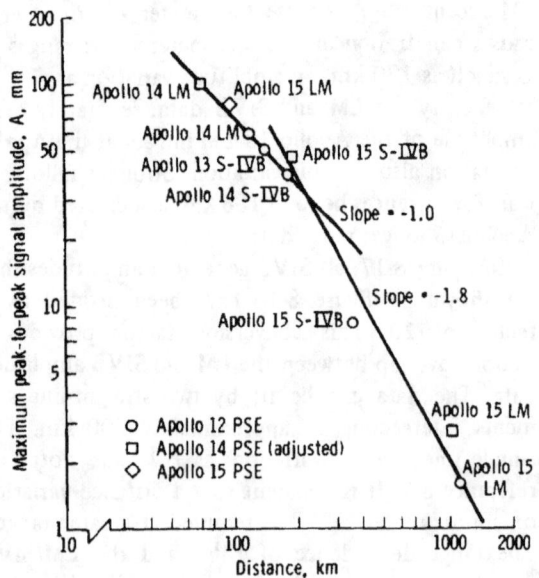

FIGURE 8-17.—Adjusted variation of artificial-impact amplitude with range. The data points of figure 8-16 are replotted with the same abscissas. Ordinates of SIVB amplitude data are plotted after division by a factor of 20. All Apollo 14 station data are adjusted for relative station sensitivity by dividing amplitude values by 2.0. With these adjustments, an amplitude variation as $r^{-1.0}$ for ranges 67 to approximately 200 km and as $r^{-1.8}$ at greater ranges is a satisfactory representation of the available data.

scatter quickly into body waves within the scattering zone and thus will propagate only to very short ranges. Surface waves long enough to propagate coherently (signal periods greater than approximately 5 sec) would be generated only by events larger than any that have occurred during 20 months of observation. Such events must be rare.

A thin surface zone of intensive scattering, coupled with a rapid increase of seismic velocity with depth, will make the lunar surface a very poor reflector for seismic waves incident on the surface from the lunar interior. In such a structure, a small proportion of the incident seismic energy follows the simple ray path of total reflection. Most of the energy is trapped in the surface region and leaks out slowly as from a primary surface source. As a result, no sharp increase of seismic energy is observed at a time when a surface-reflected seismic phase is expected from an event at a large distance. Phases corresponding to surface-reflected P- and S-waves have been tentatively identified in several of the LM and SIVB signals, but they are poorly defined.

Location and Focal Mechanism of Moonquakes

Moonquakes detected at the Apollo 12 and Apollo 14 stations before the Apollo 15 mission are believed to have originated at no less than 10 different focuses, although some of these may be quite close to one another. However, a single focus (A_1 zone) accounts for nearly 80 percent of the total detected seismic energy. Two moonquakes from the A_1 zone were recorded by the three stations of the Apollo seismic network during the first two perigee periods after activation of the Apollo 15 station, with one moonquake at each perigee. Both moonquakes were small to intermediate in size relative to the range of A_1 signal amplitudes detected thus far. The P-waves from these events arrived at the Apollo 12 station 1.8 sec earlier than at the Apollo 14 station but could not be detected at the Apollo 15 station. Shear waves (H-phase) arrived first at the Apollo 12 station, and 3.4 sec and 113.9 sec later at the Apollo 14 and Apollo 15 stations, respectively. Using these arrival times and the lunar models shown in figure 8-15, the epicenter (point on the surface directly above the moonquake focus) is located at latitude 21° S and longitude 28° W, approximately 600 km south-southwest of the Apollo 12 station, as shown in figure 8-5. The depth of the focus is approximately 800 km, somewhat deeper than any known earthquake. The remaining moonquake focuses have not yet been located.

The source of strain energy released as moonquakes is not known; but, if significant depth of focus for all moonquakes is verified by future data, these data will have profound implications concerning the lunar interior. In general, this result would require that, unlike the Earth, the shear strength of the lunar material at a depth of 800 km must be large enough to sustain appreciable stress and that maximum stress differences originate at this depth. These conditions place strong constraints on the temperature distribution in the deep lunar interior.

The nearly exact repetition of moonquake signals from a given focal zone over periods of many months requires that the focal zones be small, 10 km in diameter or less, and fixed in location over periods approaching 2 yr. If moonquake focuses were separated by as much as 1 wavelength, larger differences would be observed among moonquake signals.

As noted previously (refs. 8-5 to 8-8), the moonquakes occur in monthly cycles near times of apogee and perigee. This phenomenon suggests that the moonquakes are triggered by lunar tides. This hypothesis is strengthened by the observation that the total seismic-energy release and the interval between the times of occurrence of the first moonquakes each month and times of perigee both show 7-month periodicities which also appear in the long-term gravity variations. With a few possible exceptions, the polarities of signals belonging to a set of matching events are identical. This observation implies that the source mechanism is a progressive dislocation and not one that periodically reverses in direction. Conceivably, detectable movements in one direction may be compensated by many small, undetectable movements in the opposite direction. A progressive source mechanism suggests a secular accumulation of strain periodically triggered by lunar tides. Whether this strain is local, regional, or moonwide is an intriguing problem for further study. Several possible sources are slight expansion of the Moon by internal radiogenic heating or slight contraction on cooling, a gradual settling of the lunar body from an ellipsoidal form to a more nearly spherical form as the Moon gradually recedes from the Earth, localized strains caused by uncompensated masses, or localized thermal stresses.

For samples of earthquake data, the cumulative amplitude curves often have a nearly linear slope known as the b-value. The b-values measured for tectonic earthquakes are normally close to 1. The b-values of moonquake data in figure 8-10 are approximately 2. Higher b-values, as measured for moonquakes, are typical of one class of earthquakes —those associated with volcanic activity, which are presumably generated by subsurface movements of magma.

Laboratory experiments have demonstrated that high b-values are associated with microfracturing in rock samples subjected to small mechanical stresses (ref. 8-12) and with cracking from thermal stresses induced by heating and cooling of samples (ref. 8-13). High b-values are also measured in laboratory tests when two surfaces are rubbed together under high pressure (ref. 8-14). Thus, although no definite conclusion regarding the focal mechanism of moonquakes can be based on b-value data alone, comparison of laboratory experimental data and seismic measurements for earthquakes suggests that moonquakes may be generated by thermal stresses, possibly of volcanic origin; tectonic stresses at low stress levels; or dislocations along preexisting fractures. These are tentative suggestions, but they serve as hypotheses against which future data will be tested.

Moonquake Swarm Activity

To date, one of the most significant discoveries of lunar seismology is the observation of moonquake swarms. Each swarm is a distinctive sequence of moonquakes closely grouped in space and time, generally containing no conspicuous event. The lunar seismic activity detected between April 7 and May 10 by the Apollo 12 and Apollo 14 LP seismometers is shown in figure 8-11. Two major and two or three minor moonquake swarms occurred during this interval. Other swarms were observed between May 16 to 20 and between May 25 to 29. A moonquake swarm is characterized by an abrupt beginning and ending of activity. Events are recorded at a nearly constant rate of eight to 12 per day during the swarm, as compared to one or two per day between periods of swarm activity. Swarms, unlike other moonquake activity observed to date, do not appear to correlate with lunar tides.

Swarm events appear to be moonquakes because they have some of the characteristics of category A (matching) moonquakes, such as prominent H-phases, low-frequency spectra, and relatively small rise times. Also, the pattern of activity suggests that individual events of a particular swarm are not isolated or unrelated but that they represent an extended process occurring within a limited source region. However, swarm moonquakes differ from previously identified moonquakes. Unlike category A moonquakes, the swarm moonquakes do not appear to have matching waveforms. The waveforms of six of the largest swarm moonquakes do not match in detail among themselves nor with any other event, although the three largest moonquakes of the May 16 to 20 swarm show some similarities. The smaller swarm events have not yet been examined in detail. By contrast, no large differences in the category A_1 moonquake signals recorded at the Apollo 12 station have been observed during a period of 20 months. Thus, although the dimension of the category A_1 focal zone is thought to be approximately 1 wavelength (10 km) or less, the nonmatching swarm events must be separated by at least 1 wavelength or more. A swarm of 30 moonquakes must be distributed throughout a minimum volume of approximately 10^4 km^3, or the moonquakes may occur within a planar or linear zone of larger dimensions.

The b-values, or slopes, of the cumulative amplitude distributions for the swarms discussed are in the range 2.1 to 2.4, which is near the average value of 2 measured for the larger, periodic moonquakes. (See the section entitled "Statistics of Long-Period Events.")

An unusual swarm began on April 14 with recorded events occurring at a nearly constant rate of 12 per day as the average amplitude of the signals increased with time for a period of approximately 2 days. On April 17, after a 9-hr break in activity, three large moonquakes occurred, the last being the largest ever recorded. No significant aftershock activity was observed. This sequence of activity is represented in figure 8-11. The largest moonquake, though well recorded by the Apollo 12 and Apollo 14 stations, preceded deployment of the Apollo 15 station, and the exact location therefore remains uncertain. However, by assuming that this large moonquake occurred at a depth of 800 km, as estimated for category A_1 moonquakes, an epicenter of latitude 20° N and longitude 72° E is obtained. This location is approximately 2700 and 2900 km from the Apollo 14 and Apollo 12 sites, respectively. The Richter magnitude of this moonquake is 2 to 3 depending on the method of calculation used. The largest category A_1 moonquake observed to date had a magnitude of approximately 1 to 2. The average swarm moonquake in this sequence was comparable in magnitude to a small to intermediate category A_1 moonquake. The buildup in the amplitude of events during the swarm suggests that they are indeed related. However, large moonquakes are not associated with any of the other swarms observed to date.

Similar swarms are common in volcanic regions of the Earth where they often occur before, during, and after eruptions. Swarms are also observed in areas of geologically recent, but not current, volcanism. Sykes (ref. 8-15) reports that earthquake swarms frequently occur along the crustal zones of midoceanic ridges, which are centers of sea-floor spreading and abundant submarine volcanism. Swarms also precede most of the major volcanic eruptions in the Lesser Antilles (ref. 8-16). They may also be observed in nonvolcanic areas where they are believed to represent minor adjustment of crustal blocks to local stress conditions. Sometimes, a swarm precedes a large tectonic earthquake. Whether related to volcanism or not, all earthquake swarms are thought to be of shallow origin. With such relationships in mind, the data on moonquake swarms and matching moonquakes will be examined with great interest as clues to present lunar tectonism.

Moonquakes and Lunar Tectonism

It now appears certain that seismic-energy release related to lunar tides does occur within the Moon. However, the magnitudes and numbers of these events are small in comparison to the total seismic activity that would be recorded by an equivalent seismic station on Earth. Estimated seismic-energy release from the largest moonquakes ranges between 10^9 and 10^{12} ergs. The total energy released by moonquakes, if the Apollo 12 region is typical of the entire Moon, is approximately 10^{11} to 10^{15} ergs/yr. This value compares with approximately 5×10^{24} ergs/yr for total seismic-energy release within the Earth. Considerable uncertainty exists in this estimate, primarily because of the difficulty of estimating the ranges of natural events. However, the average rate of seismic-energy release within the Moon is clearly much less than that of the Earth. Thus, internal convection currents leading to significant lunar tectonism are probably absent. Further, the absence of conspicuous offset surface features and of compressional features such as folded mountains is evidence against significant lunar tectonic activity, past or present. Presently, the outer shell of the Moon appears to be relatively cold, rigid, and tectonically stable compared to the Earth, except for the minor disruptive influence correlated with lunar tides. However, the presence of moonquake swarms suggests continuing minor adjustment to crustal stresses. The occurrence of deep-focus moonquakes leaves open the possibility of slow convection currents at great depth beneath a rigid outer shell.

Statistics of Long-Period Events

Data on the incidence and coincidence of LP events at the Apollo 14 and Apollo 15 stations illustrate some basic properties of the lunar seismic environment. The cumulative amplitude curves (fig. 8-10) characterize the size distribution of events. The LP events are clearly a mixture of different source types. Therefore, care was taken in classifying events as moonquakes or impacts on the basis of the seismogram characteristics. As indicated in table 8-IV, almost 80 percent of all detected LP events were classified by the criteria discussed previously, although some of the identifications remain in doubt pending further analysis.

Separate cumulative curves of moonquakes and impacts (fig. 8-10) have different slopes, which indicate that the selection process has separated the events into categories representing different source mechanisms. Representing the curves of figure 8-10 by equations of the form

$$\log n = a - b \log A \qquad (8\text{-}1)$$

the b-value of moonquake distributions is approximately 2 and, for impact distributions, approximately 1. All curves flatten in the low-amplitude range where events become undetectable because of small size. Further, the curves for impact data have a possible break in slope near the midpoints. This observation may indicate an otherwise undetected effect of mixed source types. The suggestion of a break in slope is especially strong for the Apollo 14 station impact data, which are more numerous. Possibly, the impact data may be contaminated with data from swarms of small moonquakes that are not identified as such. Alternatively, impact data may be a composite of two different meteoroid populations, such as stony meteoroids and cometary meteoroids (ref. 8-17).

From table 8-IV (LP signals only), 26 impacts were detected at the Apollo 14 station only, four impacts were detected at the Apollo 15 station only, and five impacts were detected at both stations. From these data, the maximum ranges at which the impacts were detected at each station can be estimated. Assuming that the events identified as impacts are randomly distributed meteoroid impacts, the ratio of the areas from which impacts are detected by the two stations (A_{14}/A_{15}) must approximately equal the ratio of the number of recorded events, or

$$\frac{A_{14}}{A_{15}} = \frac{r_{14}^2}{r_{15}^2} = \frac{31}{9} \qquad (8\text{-}2)$$

and

$$\frac{A_{OV}}{\pi r_{14}^2} = \frac{5}{31} \qquad (8\text{-}3)$$

where A_{OV} is the overlap between the two areas, and r_{14} and r_{15} are the radii of the areas of perceptibility at the Apollo 14 and Apollo 15 stations, respectively. Using the additional fact that the stations are 1095 km apart, equations (8-2) and (8-3) can be solved to give r_{14} = 1362 km and r_{15} = 732 km, respectively. Using the same method for impacts recorded at the Apollo 12 station, r_{12} = 951 km is obtained. Thus, a meteoroid that is barely detectable at the Apollo 14 station at a range of 1362 km will be detected only to ranges of 951 and 732 km at the Apollo 12 and Apollo 15 stations, respectively. The instrument sensitivities at the respective stations are matched to within 5 percent. Thus, the differences in the ranges of detectability must be a consequence of differences in local structure at the three sites. These estimates are subject to considerable uncertainty at present, because of the relatively small number of events available for comparison. However, the method illustrated in this report will eventually lead to accurate estimates as data accumulate.

In contrast with the meteoroid-impact signals, most moonquakes are recorded at all three seismic stations. This fact supports the hypothesis that moonquake focuses are deep within the Moon and, hence, more nearly equidistant from the stations than are randomly distributed meteoroid impacts.

Using amplitude data from seismic signals recorded during the Apollo 14 mission (ref. 8-7), the meteoroid flux was estimated assuming that a distribution of meteoroid masses (individual source-receiver distances being unknown) produces the observed distribution of signal amplitudes. The key to such an analysis is a knowledge of the amplitude falloff law, or variation of amplitude A with range r. The amplitude-range calibration from 63 to 172 km, then available from artificial impacts prior to Apollo 15, was used in deriving the previous estimate. In that range, the amplitude appears to vary as r^{-1}, which would be appropriate at all ranges for a Moon with constant seismic velocity throughout the interior. From data on the more distant artificial Apollo 15 impacts, the amplitude appears to fall off more rapidly in the range between 188 and 1100 km. This variation will become better known after the artificial impacts of the Apollo 16 and Apollo 17 missions.

Considerations of this sort may increase the flux estimate by an order of magnitude. The lower b-value now obtained for the observed cumulative amplitude curve may also modify the estimate of flux, though by a lesser amount. This assumption follows from relationships expressing the calibration between meteorite mass and seismic amplitude as well as the integration of flux contributions over the surface of the Moon, which both depend on the observed b-value. Therefore, it seems best to postpone a revision of the earlier flux estimate until these factors are better known.

CONCLUSIONS

Natural lunar seismic events detected by the Apollo seismic network are moonquakes and meteoroid impacts. The moonquakes fall into two categories: periodic moonquakes and moonquake swarms. All the moonquakes are small (maximum Richter magnitudes between 1 and 2). With few exceptions, the periodic moonquakes occur at monthly intervals near times of perigee and apogee and show correlations with the longer term (7-month) lunar-gravity variations. The moonquakes originate at not less than 10 different locations. However, a single focal zone accounts for 80 percent of the total seismic energy detected. The epicenter of the active zone has been tentatively located at a point 600 km south-southwest from the Apollo 12 and Apollo 14 stations. The focus is approximately 800 km deep. Each focal zone must be small (less than 10 km in linear dimension). Changes in record character that would imply migration of the focal zone or changes in focal mechanism have not been observed in the records over a period of 20 months. Cumulative strain at each location is inferred. Thus, the moonquakes appear to be releasing internal strain of unknown origin, the release being triggered by tidal stresses. The occurrence of moonquakes at great depths implies that the lunar interior at these depths is rigid enough to support appreciable stress and that maximum stress differences occur at these depths. If the strain released as seismic energy is of thermal origin, strong constraints are placed on acceptable thermal models for the deep lunar interior.

Episodes of frequent small moonquakes, called moonquake swarms, have been observed. The occurrence of such swarms appears to bear no relation to the lunar tidal cycle, although present data are not sufficient to preclude this possibility. The source of moonquake swarms has not been determined.

The average rate of seismic-energy release within the Moon is much less than that of the Earth. Thus, internal convection currents leading to significant lunar tectonism appear to be absent. Presently, the outer shell of the Moon appears to be relatively cold, rigid, and tectonically stable compared to the Earth. However, the occurrence of moonquakes at great depth suggests the possibility of very deep convective motion. Moonquake swarms may be generated within the outer shell of the Moon as a result of continuing minor adjustments to crustal stresses.

Seismic evidence of a lunar crust has been found. In the region of the Apollo 12 and Apollo 14 stations, the thickness of the crust is between 55 and 70 km. The velocity of compressional waves in the crustal rock varies between 6.0 and 7.5 km/sec. This range brackets the velocities expected for the feldspar-rich rocks found at the surface. The transition to the subcrustal material may be gradual, beginning at a depth of 20 to 25 km; or sharp, with a major discontinuity at between 55 and 70 km. In either case, the compressional-wave velocity must reach 9 km/sec at a depth between 55 and 70 km. The past occurrence of large-scale magmatic differentiation, at least in the outer shell of the Moon, is inferred from these results.

REFERENCES

8-1. Latham, Gary; Ewing, Maurice; Press, Frank; Sutton, George; et al.: Passive Seismic Experiment. Sec. 6 of Apollo 11 Preliminary Science Report. NASA SP-214, 1969.

8-2. Latham, Gary; Ewing, Maurice; Press, Frank; Sutton, George; et al.: Apollo 11 Passive Seismic Experiment. Science, vol. 167, no. 3918, Jan. 30, 1970, pp. 455-467.

8-3. Latham, Gary; Ewing, Maurice; Press, Frank; Sutton, George; et al.: Apollo 11 Passive Seismic Experiment. Proceedings of the Apollo 11 Lunar Science Conference, Vol. 3, A. A. Levinson, ed., Pergamon Press (New York), 1970, pp. 2309-2320.

8-4. Latham, Gary; Ewing, Maurice; Press, Frank; Sutton, George; et al.: Passive Seismic Experiment. Sec. 3 of Apollo 12 Preliminary Science Report. NASA SP-235, 1970.

8-5. Latham, Gary; Ewing, Maurice; Press, Frank; Sutton, George; et al.: Seismic Data From Man-Made Impacts on the Moon. Science, vol. 170, no. 3958, Nov. 6, 1970, pp. 620-626.

8-6. Ewing, Maurice; Latham, Gary; Press, Frank; Sutton, George; et al.: Seismology of the Moon and Implications of Internal Structure, Origin, and Evolution. Highlights of Astronomy, D. Reidel Pub. Co. (Dordrecht, Holland), 1971.

8-7. Latham, Gary; Ewing, Maurice; Press, Frank; Sutton, Geroge; et al.: Passive Seismic Experiment. Sec. 6 of Apollo 14 Preliminary Science Report. NASA SP-272, 1971.

8-8. Latham, Gary; Ewing, Maurice; Press, Frank; Sutton, George; et al.: Moonquakes. Science, vol. 174, 1971.

8-9. Kovach, Robert; Watkins, Joel; and Landers, Tom: Active Seismic Experiment. Sec. 7 of Apollo 14 Preliminary Science Report. NASA SP-272, 1971.

8-10. Schreiber, E.; Anderson, O.; Soga, N.; Warren, N.; et al.: Sound Velocity and Compressibility for Lunar Rocks 17 and 46 and for Glass Spheres from the Lunar Soil. Science, vol. 167, no. 3918, Jan. 30, 1970, pp. 732-734.

8-11. Kanamori, H.; Nur, A.; Chung, D.; Wones, D.; et al.: Elastic Wave Velocities of Lunar Samples at High Pressures and Their Geophysical Implications. Science, vol. 167, no. 3918, Jan. 30, 1970, pp. 726-727.

8-12. Scholz, C. H.: The Frequency Magnitude Relation of Microfracturing in Rock and Its Relation to Earthquakes. Seismol. Soc. Amer. Bull., vol. 58, no. 1, 1968, pp. 399-415.

8-13. Warren, Nicholas W.; and Latham, Gary: An Experimental Study of Thermally Induced Microfracturing and Its Relation to Volcanic Seismicity. J. Geophys. Res., vol. 75, no. 23, Aug. 10, 1970, pp. 4455-4464.

8-14. Nakamura, Y.; Veach, C.; and McCauley, B.: Research Rept. ERR-FW-1176, General Dynamics Corp., 1971.

8-15. Sykes, Lynn R.: Earthquakes and Sea-Floor Spreading. J. Geophys. Res., vol. 75, no. 32, Nov. 10, 1970, pp. 6598-6611.

8-16. Robson, G. R.; Barr, K. G.; and Smith, G. W.: Earthquake Series in St. Kitts-Nevis, 1961-62. Nature, vol. 195, no. 4845, Sept. 8, 1962, pp. 972-974.

8-17. Hawkins, Gerald S.: The Meteor Population. NASA CR-51365, 1963.

9. Lunar-Surface Magnetometer Experiment

*P. Dyal,[a] C. W. Parkin,[a,b]
and C. P. Sonett[a,†]*

INTRODUCTION

The Apollo 15 lunar-surface magnetometer (LSM) is one of a network of magnetometers that have been deployed on the Moon to study intrinsic remanent magnetic fields and global magnetic response of the Moon to large-scale solar and terrestrial magnetic fields. From these field measurements, properties of the lunar interior such as magnetic permeability, electrical conductivity, and temperature can be calculated. In addition, correlation with solar-wind-spectrometer data allows study of the solar-wind plasma interaction with the Moon and, in turn, investigation of the resulting absorption of gases and accretion of an ionosphere. These physical parameters and processes determined from magnetometer measurements must be accounted for by comprehensive theories of origin and evolution of the Moon and solar system.

The Apollo 12, 14, and 15 magnetometer network has yielded unique information about the history and present physical state of the Moon. The measured remanent magnetic fields vary considerably from site to site: 38 ± 3 gammas at Apollo 12, 103 ± 5 and 43 ± 6 gammas at two Apollo 14 sites separated by 1.1 km, and 6 ± 4 gammas at Apollo 15. The strengths and variety of these field magnitudes imply that the field sources are local rather than global in extent. Analyses of samples from the Apollo 12 source region indicate that a magnetizing field $>10^3$ gammas existed when the source material cooled below its Curie temperature. The large field and sample remanence measurements indicate that material located below the randomly oriented regolith was uniformly magnetized over large areas.

The global magnetic response of the Moon to solar and terrestrial fields varies considerably with the lunar orbital position (fig. 9-1). During times when the Moon is immersed in the steady geomagnetic-tail field, the bulk relative lunar magnetic permeability is calculated to be $\mu/\mu_0 = 1.03 \pm 0.13$. When the Moon is located in the free-streaming solar wind, measurements from an LSM on the nighttime side of the Moon can be analyzed to determine the lunar electrical-conductivity profile. Apollo 12 LSM data have been analyzed for solar-wind magnetic-field step-transient events to calculate conductivity $\sigma_1 \sim 10^{-4}$ mhos/m for a shell of radial thickness $\Delta R = R_1 - R_2$ where $0.95R \leq R_1 < R_M$ and $R_2 \sim 0.6 R_M$, and conductivity $\sigma_2 \sim 10^{-2}$ mhos/m for the core bounded by R_2. (R_M is the radius of the Moon.) The temperature of the lunar interior can be calculated for assumed material compositions; for an olivine Moon, temperatures are calculated to be approximately 810 K for the shell and approximately 1240 K for the core. Comparison of preliminary Apollo 15 data with the Apollo 12 data indicates that the lunar response to solar-wind transients is similar at the two sites and that Apollo 15 data, when fully processed and analyzed, will allow calculation of both horizontal and radial conductivity profiles.

Measurements made by an LSM while on the lunar daytime side can also be used to estimate electrical conductivity and temperature (ref. 9-1); furthermore, correlation with solar-plasma measurements yields information concerning the plasma interaction with lunar remanent and induced fields. The 38-gamma remanent field at Apollo 12 is found to be compressed by the solar wind during times of high solar-plasma density. Since the Apollo 15 remanent field is

---

[a]NASA Ames Research Center.
[b]National Research Council postdoctoral associate.
[†]Principal investigator.

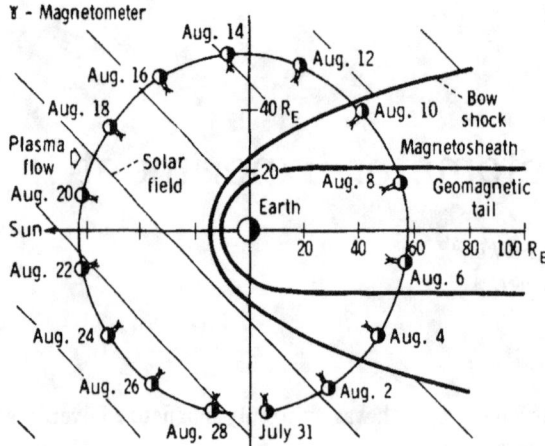

FIGURE 9-1.—Lunar orbit projection onto the solar ecliptic plane, showing the Apollo 15 magnetometer during the first postdeployment lunation. During a complete revolution around the Earth, the magnetometer passes through the bow shock of the Earth, the magnetosheath, the geomagnetic tail, and the interplanetary region dominated by solar-plasma fields.

much lower than 38 gammas, detailed analysis of Apollo 15 data should allow investigation of the plasma interaction with induced lunar fields alone.

THEORY

The total magnetic field B_A measured on the surface by an Apollo LSM is a vector sum of the following possible fields.

$$B_A = B_E + B_S + B_\mu + B_P + B_T + B_D + B_F \quad (9\text{-}1)$$

where B_E is the total external (solar or terrestrial) driving magnetic field measured by the Explorer 35 and Apollo 15 subsatellite lunar-orbiting magnetometers while outside the antisolar lunar cavity; B_S is the steady remanent field at the surface site; B_μ is the magnetization field induced in permeable lunar material; B_P is the poloidal field caused by eddy currents induced in the lunar interior by changing external fields; B_T is the toroidal field corresponding to unipolar electrical currents driven through the Moon by the $V \times B_E$ electric field; B_D is the field associated with the diamagnetic lunar cavity; and B_F is the total field associated with the hydromagnetic solar-wind flow past the Moon. It is noted that the fields of equation (9-1) can also be classed as external

(B_E), permanent (B_S), induced (B_μ, B_P, B_T), and solar wind interaction field (B_D, B_F).

During one complete lunar orbit around the Earth, the electromagnetic environment of the Moon varies considerably with the lunar position (fig. 9-1). The relative importance of the fields in equation (9-1) also varies with orbital position, and, therefore, different magnetic fields can be investigated during different times of each lunation.

When the Moon is passing through the geomagnetic tail, the ambient plasma-particle density is significantly reduced and the plasma-associated terms B_T, B_D, and B_F can be neglected. In quiet regions of the tail, B_E is constant, and the eddy-current induction field $B_P \to 0$. Equation (9-1) then reduces to

$$B_A = B_E + B_\mu + B_S \quad (9\text{-}2)$$

The magnetization field B_μ (fig. 9-2) is proportional to the external driving field B_E.

$$B_A = (1 + K)B_E + B_S \quad (9\text{-}3)$$

(the proportionality constant K in turn depends upon the permeability and the dimensions of the permeable region of the Moon); therefore, when the tail field B_E is zero (e.g., during neutral sheet crossings), $B_A = B_S$ and the Apollo surface field measures the steady remanent field alone. After B_S has been determined, the equation can be solved for the proportionality

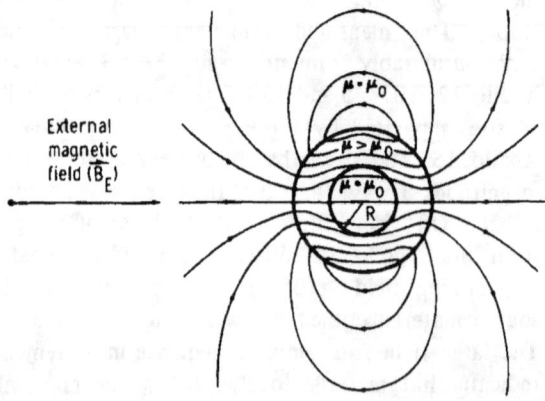

FIGURE 9-2.—Induced magnetization field B_μ. A global permeable shell placed in the uniform geomagnetic tail tends to concentrate the field lines inside the shell.

constant K, and the magnetic permeability μ can be calculated.

During times when the LSM is located on the dark (antisolar) side of the Moon, another combination of terms in equation (9-1) can be neglected to a first approximation. While in the darkside cavity, the LSM is isolated from the plasma flow; hydromagnetic effects such as surface field compression by the solar wind will not occur and $\mathbf{B}_F$ can be disregarded. Subsequent analysis will show that the bulk lunar magnetic permeability is close to that of free space, and $\mathbf{B}_\mu$ is henceforth neglected. From time-series step-transient analysis, $\mathbf{B}_T$ and $\mathbf{B}_D$ have been found to be negligible (ref. 9-2) compared to the eddy-current induction field $\mathbf{B}_P$ (fig. 9-3), and equation (9-1) reduces to

$$\mathbf{B}_A = \mathbf{B}_P + \mathbf{B}_E + \mathbf{B}_S \qquad (9\text{-}4)$$

After $\mathbf{B}_S$ has been calculated using equation (9-3), only the poloidal field $\mathbf{B}_P$ is unknown. Equation (9-4) can then be solved for certain assumed lunar models, and curve fits of data to the solution determine the model-dependent conductivity profile $\sigma(R)$. Furthermore, electrical conductivity is related to temperature, and the lunar-interior temperature can be calculated for assumed lunar-material compositions.

A different set of field terms in equation (9-1) is dominant for LSM data obtained during lunar daytime: $\mathbf{B}_D \rightarrow 0$ outside the cavity, and the global fields $\mathbf{B}_\mu$ and $\mathbf{B}_T$ can again be neglected in comparison to $\mathbf{B}_P$ because of empirical findings expressed previously. For the lunar daytime data, the LSM is exposed directly to the solar wind (ref. 9-3) and the associated wave modes and plasma-field interaction effects; therefore, the interaction term $\mathbf{B}_F$ will not be assumed negligible in general, and equation (9-1) becomes

$$\mathbf{B}_A = \mathbf{B}_P + \mathbf{B}_E + \mathbf{B}_S + \mathbf{B}_F \qquad (9\text{-}5)$$

The interaction field $\mathbf{B}_F$ has been found to be important during times of high solar-wind-particle density (ref. 9-4); at the Apollo 12 site, the local 38-gamma remanent field is compressed by high-pressure plasma flow as shown conceptually in figure 9-4. Plasma properties must therefore be taken into account for analysis of lunar daytime magnetometer data.

EXPERIMENTAL TECHNIQUE

The experimental technique used to measure the magnetic field required the astronauts to deploy, on the lunar surface, a magnetometer that would continuously measure and transmit information by radio to Earth for a period of 1 yr. The LSM was deployed at longitude 3°29' E and latitude 26°26' N at 18:29 G.m.t. on July 31, 1971. A photograph of the LSM fully deployed and alined at Hadley Rille is shown in figure 9-5, and a list of the Apollo 15 instrument characteristics is given in table 9-I. A detailed description of the LSM can be found in reference 9-5.

DESCRIPTION OF THE LSM

Fluxgate Sensor

The three orthogonal vector components of the magnetic field are measured by three fluxgate sensors (refs. 9-6 and 9-7). Each sensor, shown schematically in figure 9-6, consists of a flattened toroidal Permalloy core that is driven to saturation by a sinusoidal current of frequency 6000 Hz. The sense winding detects the superposition of the drive-winding magnetic field and the total lunar-surface

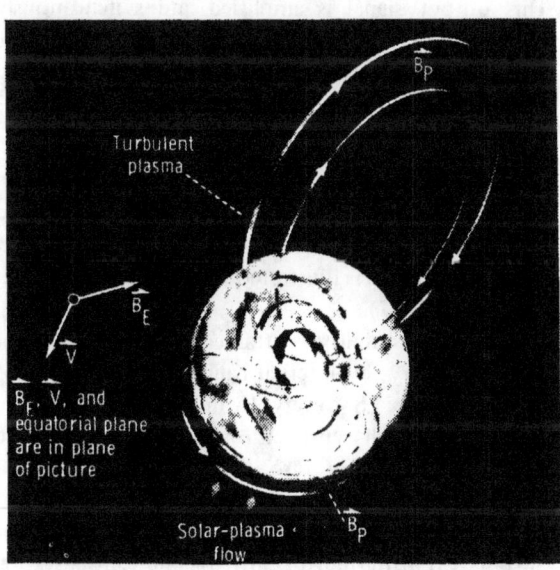

FIGURE 9-3.—Induced eddy-current magnetic field $\mathbf{B}_P$. A poloidal field $\mathbf{B}_P$ is induced by time-dependent fluctuations in the solar-wind magnetic field. The lunar equatorial plane, the solar magnetic field $\mathbf{B}_E$, and the velocity $\mathbf{V}$ of the Moon with respect to the solar wind are in the plane of the page.

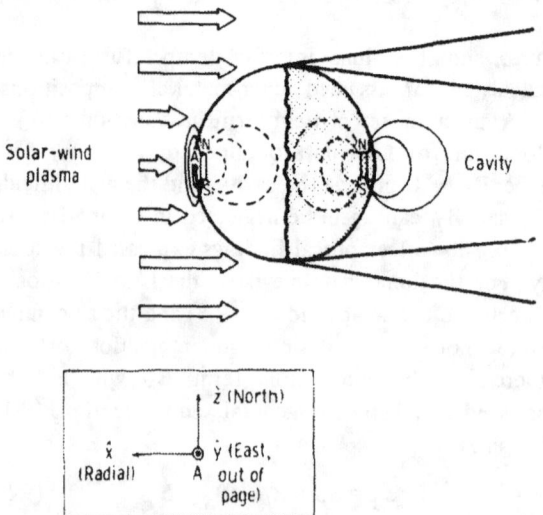

FIGURE 9-4.—Compression of a local remanent magnetic field by a high-density solar-wind plasma. The remanent field is unperturbed during lunar night (antisolar side), while on the sunlit side the horizontal components are compressed. The insert shows the coordinate system used in this paper.

FIGURE 9-5.—The Apollo 15 lunar-surface magnetometer deployed on the Moon near Hadley Rille. Sensors are at the top ends of the booms and approximately 75 cm above the lunar surface (AS15-87-11845).

TABLE 9-I.—*Apollo 15 Magnetometer Characteristics*

Parameter	Value
Ranges, gamma	0 to ±200
	0 to ±100
	0 to ±50
Resolution, gamma	±0.1
Frequency response, Hz	dc to 3
Angular response	Proportional to cosine of angle between magnetic-field vector and sensor axis
Sensor geometry	3 orthogonal sensors at ends of 100-cm booms; orientation determination to within 1° in lunar coordinates
Commands	10 ground and 1 spacecraft
Analog zero determination	180° flip of sensor
Internal calibration, percent	0, ±25, ±50, and ±75 of full scale
Field bias offset capability, percent	0, ±25, ±50, and ±75 of full scale
Modes of operation	Orthogonal field measurements, gradient measurement, internal calibration
Power, W	3.5, average in daytime
	9.4, average in nighttime
Weight, kg	8.9
Size, cm	25 × 28 × 63
Operating temperature, °C	−50 to +85

This output signal is amplified and synchronously demodulated to drive a voltage to the analog-to-digital converter and then through the central-station radio to Earth.

Electronics

The electronic components for the LSM are located in a thermally insulated box. The operation of the electronics is illustrated in the functional block diagram (fig. 9-7).

Long-term stability is attained by extensive use of digital circuitry, by internal calibration of the analog portion of the LSM every 18 hr, and by mechanical rotation of each sensor through 180° in order to determine the sensor zero offset. The analog output of the sensor electronics is internally processed by a low-pass digital filter and a telemetry encoder; the output is transmitted to Earth via the central-station S-band transmitter. A typical internal flip-calibration sequence is shown in figure 9-8.

The LSM has two data samplers: the analog-to-digital converter (26.5 samples/sec) and the central-

field; as a result, a second harmonic of the driving frequency is generated in the sense winding with a magnitude that is proportional to the strength of the surface field. The phase of the second harmonic signal with respect to the drive signal indicates the direction of the surface field with respect to the sensor axis.

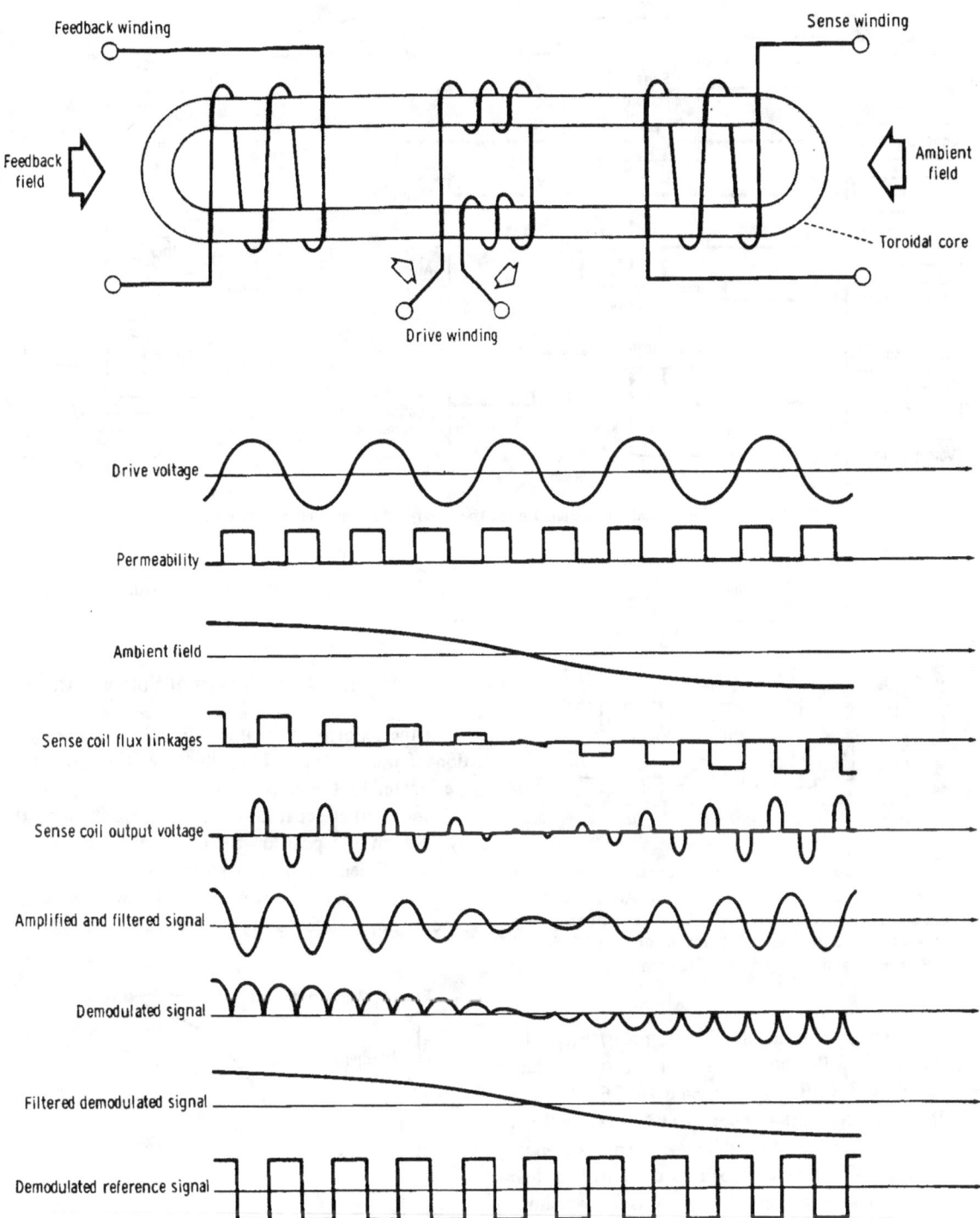

FIGURE 9-6.—Schematic outline and operation of the Ames fluxgate sensor; signal amplitudes are plotted as functions of time.

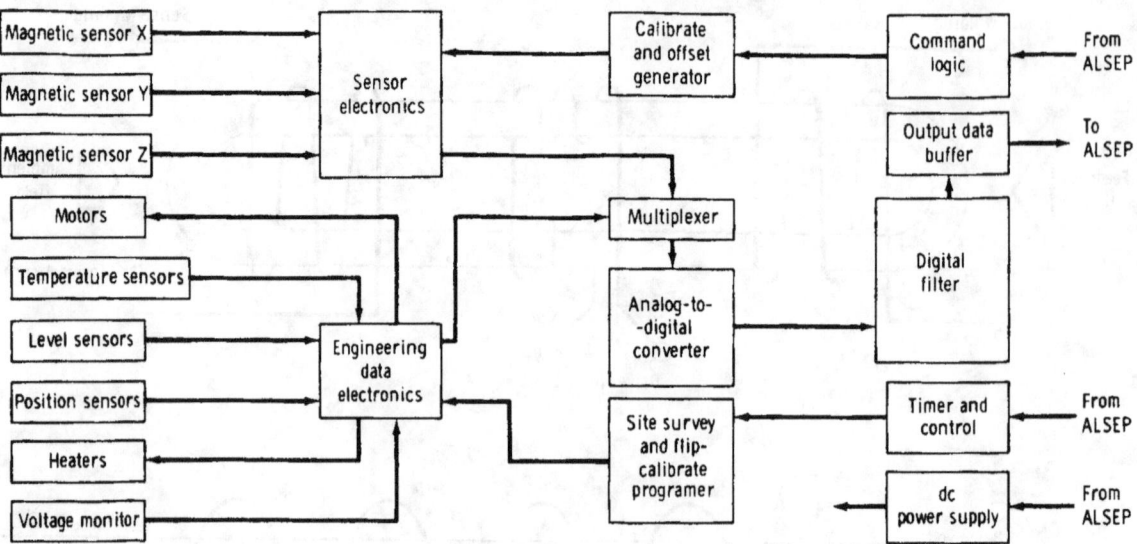

FIGURE 9-7.—Functional block diagram for the lunar-surface magnetometer electronics.

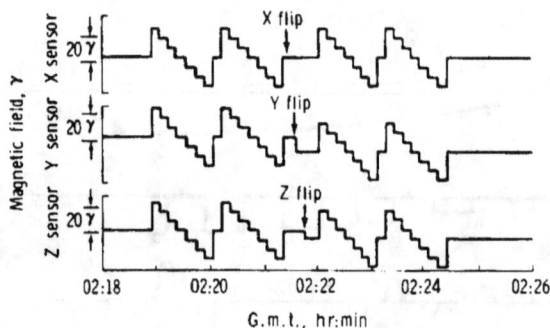

FIGURE 9-8.—Magnetometer data obtained during an internal flip-calibration sequence on August 5. This sequence is repeated every 18 hr by timer command and involves internally generated biases in steps of 75, 50, 25, 0, −25, −50, and −75 percent of full scale.

digital filter can be bypassed by ground command in order to pass on higher frequency information.

Mechanical and Thermal Subsystems

In the exterior mechanical and thermal configuration of the Apollo 15 LSM, the three fluxgate sensors are located at the ends of three 100-cm-long orthogonal booms that separate the sensors from each other by 150 cm and position them 75 cm above the lunar surface. Orientation measurements with respect to lunar coordinates are made with two devices. A shadowgraph and bubble level are used by the

station telemetry encoder (3.3 samples/sec). The prealias filter following the sensor electronics had attenuations of 3 dB at 1.7 Hz, 64 dB at 26.5 Hz, and 58 dB at the Nyquist frequency (13.2 Hz), with an attenuation rate of 22 dB/octave. The four-pole Bessel digital filter limits the alias error to less than 0.05 percent and has less than 1 percent overshoot for a step-function response. This filter has an attenuation of 3 dB at 0.3 Hz and 48 dB at the telemetry-sampling Nyquist frequency (1.6 Hz), and it has a phase response that is linear with frequency. The response of the entire LSM measurement system to a step-function input is shown in figure 9-9. The

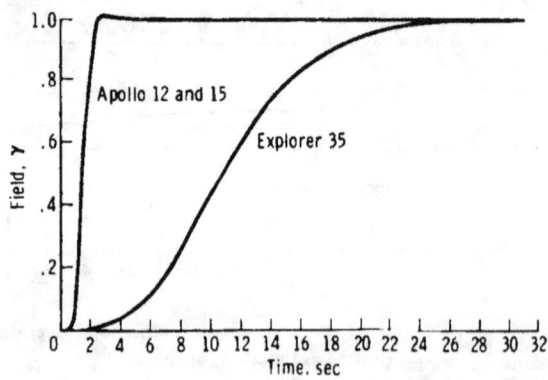

FIGURE 9-9.—Laboratory measurements comparing responses of the Apollo 12 and 15 and Explorer 35 magnetometers to a 1.0-gamma magnetic step input.

astronaut to aline the LSM and to measure azimuthal orientation with respect to the Moon-to-Sun line to an accuracy of 0.5°. Gravity-level sensors measure instrument-tilt angles to an accuracy of 0.2° every 4.8 sec.

In addition to the instrument normal mode of operation in which three vector field components are measured, the LSM has a gradiometer mode in which commands are sent to operate three motors. These motors rotate the sensors such that all simultaneously aline parallel first to one of the three axes, then to each of the other two boom axes in turn. The rotating alinement permits the vector gradient to be calculated in the plane of the sensors and also permits an independent measurement of the magnetic-field vector at each sensor position.

The thermal subsystem is designed to allow the LSM to operate over the complete lunar day-night cycle. Thermal control is accomplished by a combination of insulation, control surfaces, and heaters that collectively operate to keep the electronics between 278 K and 338 K. A plot of the temperature of the X-sensor and the electronics for the first postdeployment lunation are shown in figure 9-10. It is apparent that the thermal control subsystem holds the temperature well within the operating range of 223 to 358 K.

Data Flow and Mission Operation

The LSM experiment is controlled from the NASA Manned Spacecraft Center (MSC) by commands transmitted to the Apollo lunar surface experiments package (ALSEP) from remote tracking stations. The data are recorded on magnetic tape at the remote sites and are also sent directly to MSC for real-time analysis in order to establish the proper range, offset, frequency response, thermal control, and operating

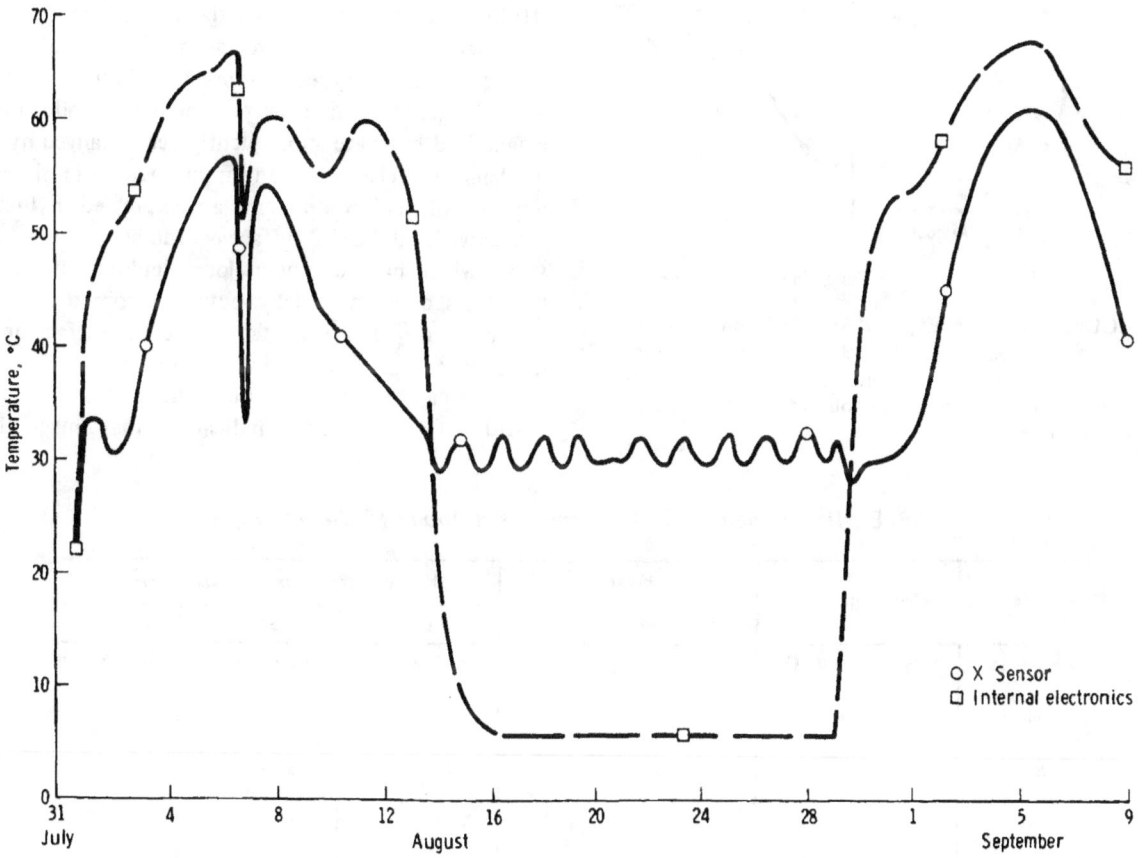

FIGURE 9-10.–Temperature inside one of the fluxgate sensors and inside the electronics box during the first postdeployment lunation of Apollo 15. The decrease in temperature during the first lunar day is due to an eclipse occurring on August 6.

mode. The one-time gradiometer-mode sequence of commands was successfully executed on August 6.

Explorer 35 Magnetometer

The ambient steady-state and time-dependent magnetic fields in the lunar environment are measured by the Explorer 35 satellite magnetometer. The satellite has an orbital period of 11.5 hr, an apolune of 9390 km, and a perilune of 2570 km (fig. 9-11). The Explorer 35 magnetometer measures three magnetic-field vector components every 6.14 sec and has an alias filter with 18-dB attenuation at the Nyquist frequency (0.08 Hz) of the spacecraft data-sampling system. The instrument has a phase shift linear with frequency, and its step-function response is slower than that of the Apollo 12 instrument (fig. 9-9). Reference 9-8 contains further information about the Explorer 35 magnetometer. Figure 9-11 also shows the orbit of the Apollo 15 particles and fields subsatellite that also carries a magnetometer (sec. 22 of this report).

RESULTS AND DISCUSSION

Remanent Magnetic Fields at the Apollo Sites

Local steady fields B_S at four sites have been calculated by least-squares fits of measurements to equation (9-3). Table 9-II and figure 9-12 list the steady fields for the Apollo 12 and 15 sites and for two separate sites of Apollo 14. These fields are all attributed to remanent magnetization in the nearby subsurface material. The remanent fields could be caused by various types of sources, including nearby platelike regions that were originally uniformly magnetized but have subsequently been changed by a mechanism such as volcanism or meteoroid shock impact. All field components are expressed in their respective local ALSEP surface-coordinate systems ($\hat{x}$, $\hat{y}$, $\hat{z}$), which have origins at local deployment sites; each $\hat{x}$ is directed radially outward from the local surface, and $\hat{y}$ and $\hat{z}$ are tangent to the surface and directed eastward and northward, respectively.

The measured remanent fields differ by at least an order of magnitude, indicating that magnetic

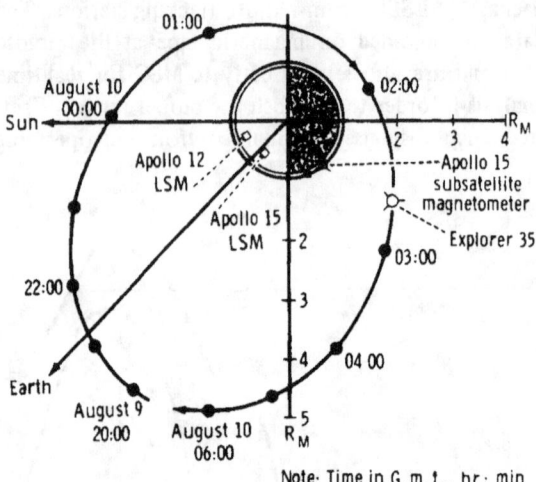

FIGURE 9-11.—The Explorer 35 orbit around the Moon, projected onto the solar ecliptic plane. The period of revolution is 11.5 hr. The Apollo 12 and 15 surface-instrument positions and Apollo 15 subsatellite trajectory are shown.

TABLE 9-II.—*Magnetic-Field Measurements at Apollo 15, 14, and 12 Sites*

Site	Coordinates, deg	Field magnitude, gammas	Magnetic-field components, gammas		
			Up	East	North
Apollo 15	26.1° N, 3.7° E	6 ± 4	+4 ± 4	+1 ± 3	+4 ± 3
Apollo 14	3.7° S, 17.5° W				
Site A[a]		103 ± 5	−93 ± 4	+38 ± 5	−24 ± 8
Site C'[a]		43 ± 6	−15 ± 4	−36 ± 5	−19 ± 5
Apollo 12	3.2° S, 23.4° W	38 ± 3	−24.4 ± 2.0	+13.0 ± 1.8	−25.6 ± 0.8

[a]These sites are shown in Figure 3-1, section 3, of the Apollo 14 Preliminary Science Report, NASA SP-272.

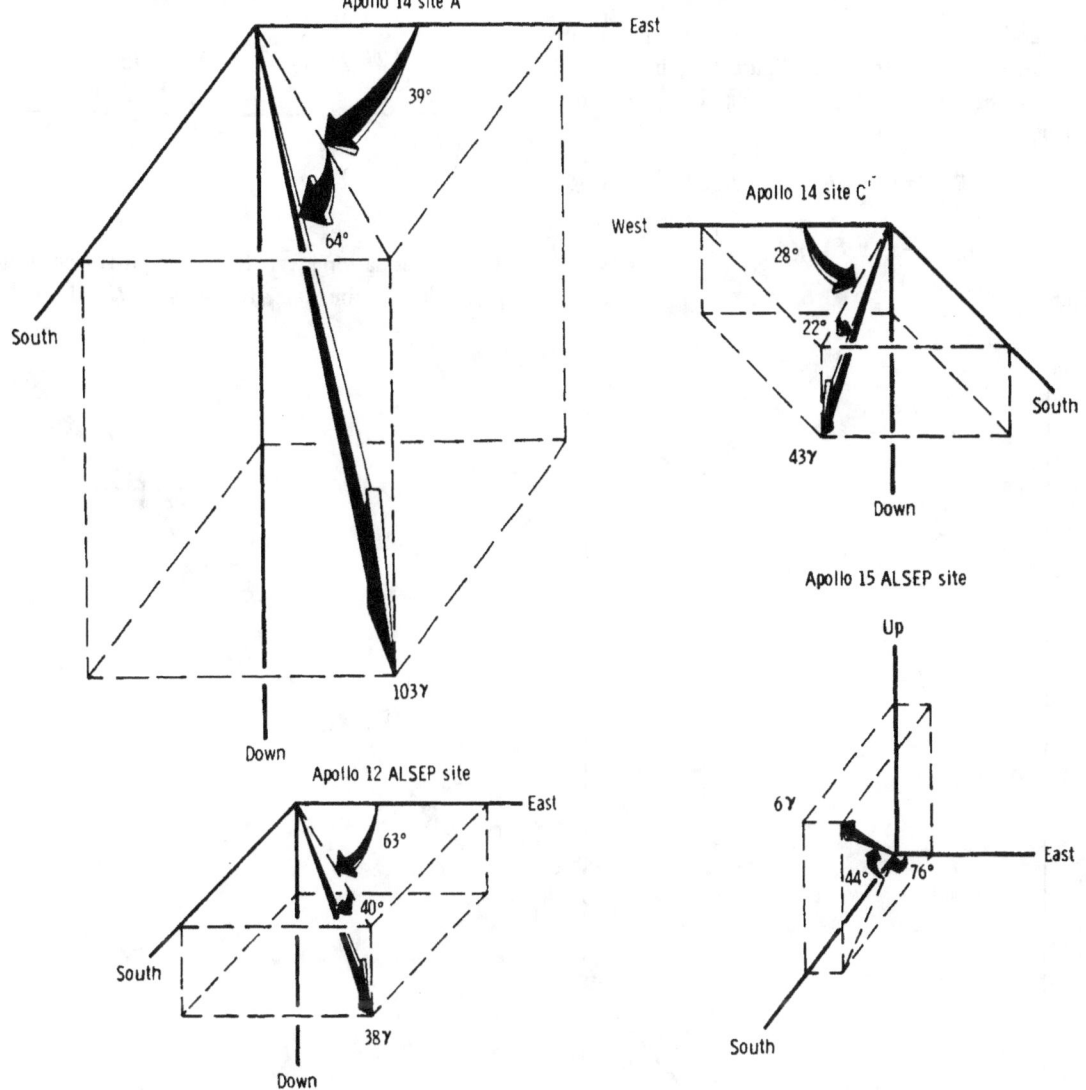

FIGURE 9-12.—Magnitudes and orientations of the vector magnetic fields measured at the Apollo 12, 14, and 15 landing sites.

concentrations (magcons) exist at widely separated regions of the Moon (ref. 9-9). The three high field readings (at the Apollo 12 and 14 sites) were taken at separations of no more than 180 km, whereas the Apollo 15 reading, taken 1200 km distant, showed comparatively little field. The large differences between the remanent magnetic fields of the Apollo 15 region and the fields of the region common to the Apollo 12 and 14 sites suggest the possibility that the two regions were formed at different times under different ambient-field conditions or that the magnetic properties of the materials in the two regions differ substantially. The Apollo 15 measurement, moreover, was obtained on the edge of the Mare Imbrium mascon basin; the fact that little or no magnetic field exists at that site leads to the preliminary conclusion that mascons are not highly magnetic.

Relative Magnetic Permeability of the Lunar Interior

During times when the Moon is immersed in steady regions of the geomagnetic tail, equations

(9-2) and (9-3) apply. Solutions for the spherically symmetric case illustrated in figure 9-2, in terms of local ALSEP surface coordinates $(\hat{x}, \hat{y}, \hat{z})$ described earlier, are as follows (ref. 9-2).

$$B_{Ax} = (1 + 2F)B_{Ex} + B_{Sx} \quad (9\text{-}6)$$

$$B_{Ay,z} = (1 - F)B_{Ey,z} + B_{Sy,z} \quad (9\text{-}7)$$

where

$$F = \frac{(2k_M + 1)(k_M - 1)\left[1 - \left(\dfrac{R}{R_M}\right)\right]^3}{(2k_M + 1)(k_M + 2) - \left(2\dfrac{R}{R_M}\right)^3 (k_M - 1)^2} \quad (9\text{-}8)$$

In this equation, k_M is the relative permeability μ/μ_o; R_M is the lunar radius; and R is the radius of

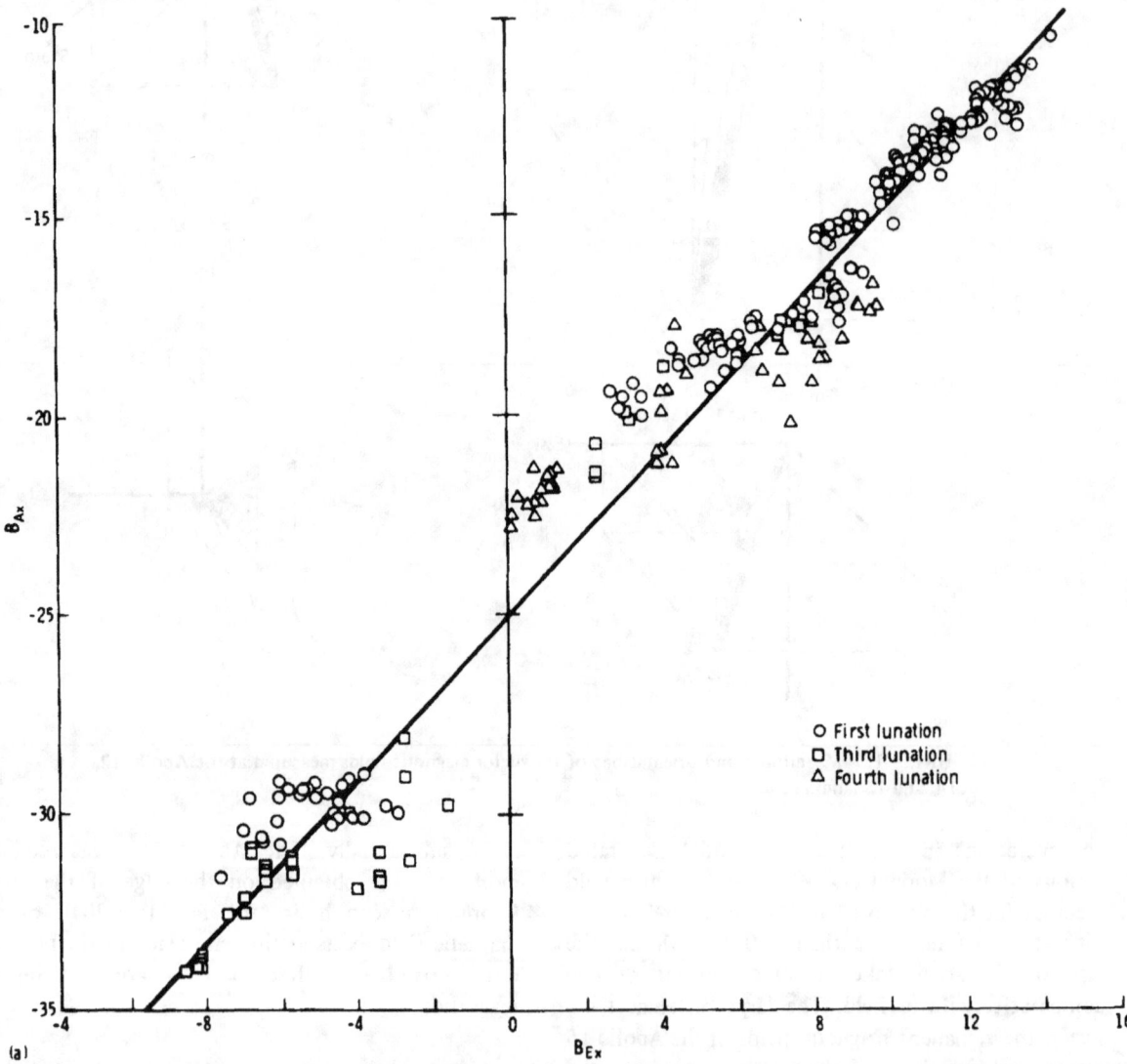

FIGURE 9-13.—Graphical representations of equations (9-6) and (9-8). (a) Radial component of Apollo 12 total surface magnetic field B_{Ax} as a function of the radial component of external driving field B_{Ex}. Data points consist of measurements in quiet regions of the geomagnetic tail taken during three different postdeployment lunations. The B_{Ax} intercept of the least-squares best-fit solid line gives the radial component of the Apollo 12 remanent field; the best-fit slope corresponds to a value of 1.03 ± 0.13 for the bulk relative permeability μ/μ_o of the Moon.

the boundary which encloses lunar material with temperature above the Curie point.

Figure 9-13(a) shows a plot of radial components of Apollo 12 LSM fields (B_{Ax}) versus the geomagnetic-tail field (B_{Ex}) measured by Explorer 35. A least-squares fit and slope calculations determine the factor F, which is used to determine the relative magnetic permeability for an assured inner radius R, as shown in figure 9-13(b). For the bulk permeability of the Moon (the case $R = 0$), $\mu/\mu_o = 1.03 \pm 0.13$, a value very close to that of free space.

Electrical Conductivity and Temperature of the Lunar Interior

Analysis of lunar-dark-side step-transient events has provided a means of calculating electrical-conductivity and temperature profiles of the interior of the Moon (refs. 9-2, 9-10, and 9-11). The transient magnetic response of a three-layer model of the Moon to a moving solar field discontinuity is shown qualitatively in figure 9-14. Figures 9-15 and 9-16 show samples of real-time data from the Apollo 12 and 15 lunar surface magnetometers in instrument coordinates (i.e., the X, Y, and Z data curves of these two figures along the respective instrument X, Y, and Z sensor-arm directions should not be confused with the ALSEP coordinate system described earlier). Figure 9-15 includes daytime data for two separate transient events, while figure 9-16 shows samples of nighttime data. There are drastic differences in magnitudes between the daytime and nighttime data; these differences stem from two primary causes: (1) daytime data were taken while the magnetometers were located in the turbulent magnetosheath region

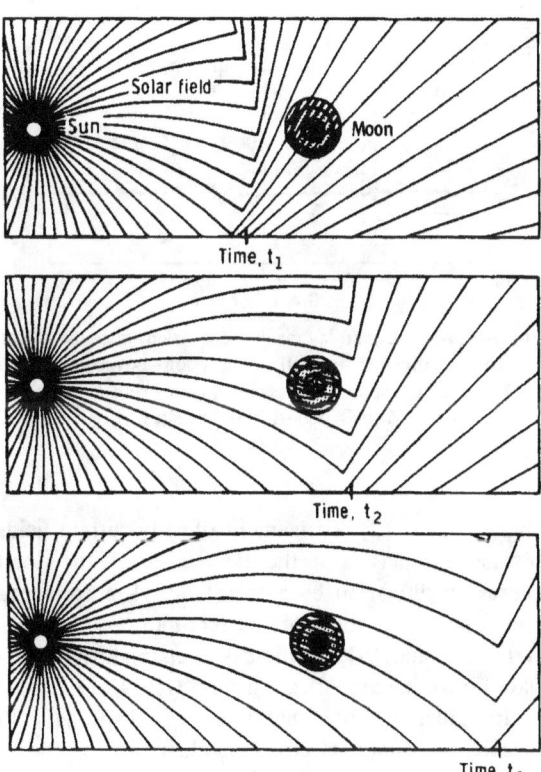

FIGURE 9-14.—Transient magnetic response of a three-layer conductivity model of the Moon for a case in which a directional discontinuity in the solar-wind magnetic field travels outward past the Moon. The three layers are characterized by successively increasing electrical conductivity with depth into the Moon. Eddy currents persist longest in the deep-core region of highest conductivity.

FIGURE 9-13(b).—The function F is related to relative magnetic permeability $k_M = \mu/\mu_o$ for various values of R/R_M. The terms R and R_M are internal and external radii, respectively, of a global permeable shell.

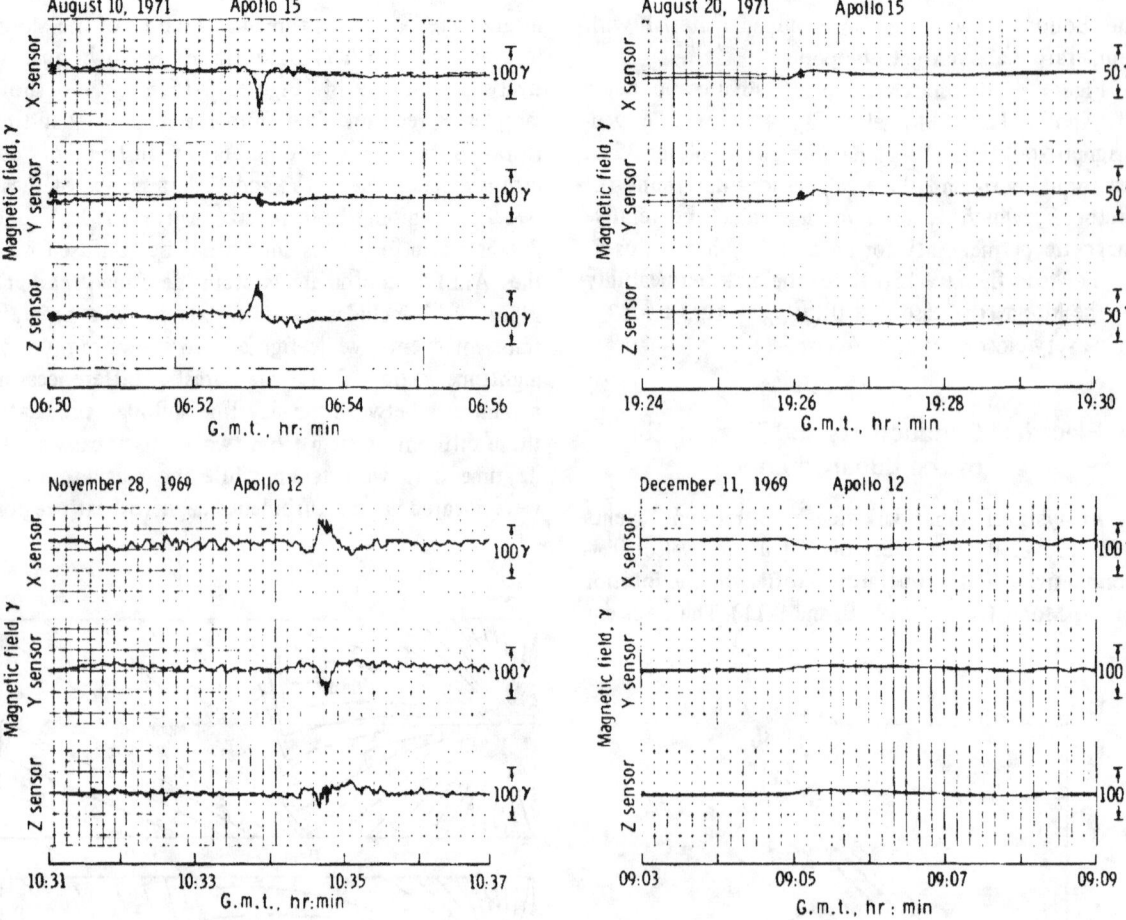

FIGURE 9-15.—Lunar-day magnetic transient response of the Moon, expressed in instrument boom-axis coordinates. The Moon responds the same qualitatively at both the Apollo 12 and the Apollo 15 landing sites.

FIGURE 9-16.—Lunar night magnetic transient response of the Moon, expressed in instrument boom-axis coordinates. Qualitatively, the Moon responds the same at both the Apollo 12 and the Apollo 15 landing sites.

(fig. 9-1); and (2) confinement of eddy-current fields by the solar plasma on the daytime side of the Moon causes amplification by a factor of ~4 in horizontal components of the eddy-current-induced field B_P (refs. 9-1 and 9-12), whereas deep nighttime data have horizontal amplification ~1.5 (ref. 9-11).

In order to use nighttime transient data to calculate a lunar electrical-conductivity profile, a simple two-layer model of the Moon is considered as a first approximation.

The two-layer model has the following assumed properties: the spherically symmetric model has a homogeneous inner core of scalar electrical conductivity σ_1 and radius R_1 surrounded by a nonconducting outer shell of outer radius R_M (the lunar radius). The sphere is in a vacuum, and permeability is everywhere that of free space ($\mu = \mu_0$). Conduction currents dominate displacement currents within the sphere, and dimensions of external-field transients are large compared to the diameter of the sphere. A transient event is assumed to affect all parts of the lunar sphere instantaneously, because the solar wind transports a step discontinuity across the entire Moon in less than 9 sec. This period has been found to be much less than the decay time of lunar-induced eddy currents.

The vector components of the magnetic field at the lunar surface are listed subsequently for the case of an external magnetic-field step transient of magnitude $\Delta B_E = B_{Ef} - B_{Eo}$ applied to the lunar sphere at time $t = 0$. Explorer 35 initial and final external fields are B_{Eo} and B_{Ef}, respectively. The solutions for the components of the vector field measured on the lunar surface ($B_A = B_P + B_E + B_S$) can be expressed as follows.

$$B_{Ax} = -3\left(\frac{R_1}{R_M}\right)^3 (\Delta B_{Ex}) F(t) + B_{Ex} + B_{Sx} \quad (9\text{-}9)$$

$$B_{Ay,z} = \frac{3}{2}\left(\frac{R_1}{R_M}\right)^3 (\Delta B_{Ey,z}) F(t) + B_{Ey,z} + B_{Sy,z} \quad (9\text{-}10)$$

where $\Delta B_{Ei} = B_{Eif} - B_{Eio}$; $i = x, y, z$; and R_1 and R_M are radii of the conducting core and the Moon, respectively. The initial and final external applied field components are B_{Eio} and B_{Eif}, respectively, and are both measured by Explorer 35. Total surface fields measured by the Apollo 12 LSM are B_{Ai}, and the time dependence of the magnetic field is expressed as

$$F(t) = \frac{2}{\pi^2} \sum_{s=1}^{\infty} \frac{1}{s^2} \exp\left(\frac{-s^2 \pi^2 t}{\mu_0 \sigma_1 R_1^2}\right) \quad (9\text{-}11)$$

Solutions for equations (9-9) and (9-10) for radial and tangential transients are shown graphically in figure 9-17. A plot of simultaneous Apollo 12 and Explorer 35 data is shown in figure 9-18, and qualitative agreement between theory and data is easily seen. More thorough analysis has shown that the data better fit a three-layer model (refs. 9-2 and 9-11), yielding the conductivities shown in figure 9-19. Assuming a material composition for the lunar interior such as pure olivine, the calculated conductivity profile can be related to a temperature profile (refs. 9-13 to 9-15); temperatures for three assumed material compositions are listed in figure 9-19.

Lunar Field Compression by the Solar Wind

As discussed earlier, equation (9-1) reduces to equation (9-5) for daytime LSM data in the magnetosheath or free-streaming solar wind. Regrouping the field terms, equation (9-5) is rewritten as follows.

$$\Delta B \equiv B_A - (B_E + B_S) = B_P + B_F \quad (9\text{-}12)$$

For frequencies $\leq 3 \times 10^{-4}$ Hz, B_P should approach zero (refs. 9-1 and 9-10), allowing an experimental opportunity to consider the flow field B_F alone.

If the 38-gamma remanent field at the Apollo 12

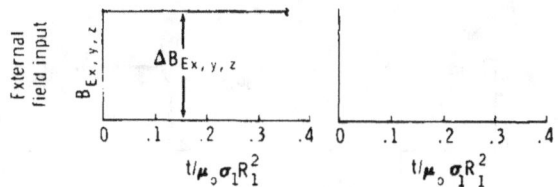

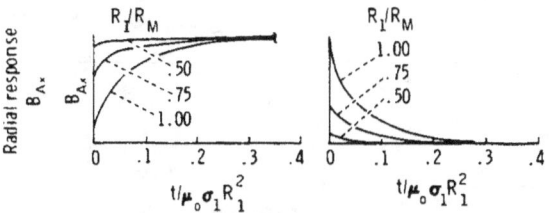

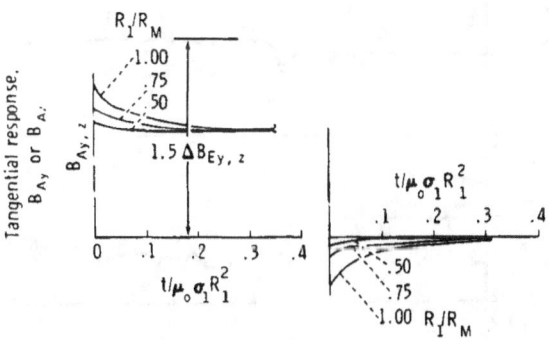

FIGURE 9-17.—Theoretical solutions for the lunar night vacuum poloidal magnetic-field response of a homogeneous conducting lunar core of radius R_1 to a step-function transient in the driving solar-wind magnetic field. For a step-function change ΔB_E in the external driving field (measured by Explorer 35), the total magnetic field at the surface of the Moon B_A (measured by the Apollo 12 and 15 magnetometers) will be damped in the radial (B_{Ax}) component and will overshoot in the tangential (B_{Ay} and B_{Az}) components. The initial overshoot magnitude is limited to a maximum value $B_{Ey}/2$ or $B_{Ez}/2$ for the case $R_1 \rightarrow R_M$. A family of curves is shown for different values of the parameter R_1/R_M.

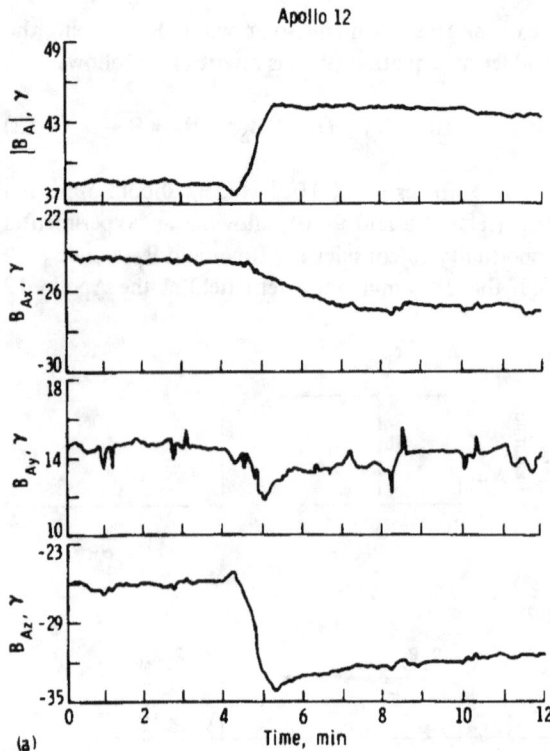

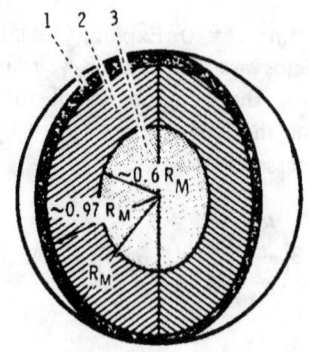

Region	Electrical conductivity, σ, mhos/m	Temperature, °K		
		Olivine	Peridotite	Apollo 11 surface sample
1	$<10^{-9}$	<440	<430	<300
2	$\sim 10^{-4}$	~810	~890	~530
3	$\sim 10^{-2}$	~1240	~1270	~740

FIGURE 9-19.—Conductivity (σ) and temperature contours for a three-layer Moon. Temperature calculations are based on σ as a function of temperature for pure olivine and peridotite (ref. 9-14) and an Apollo 11 surface sample (ref. 9-15).

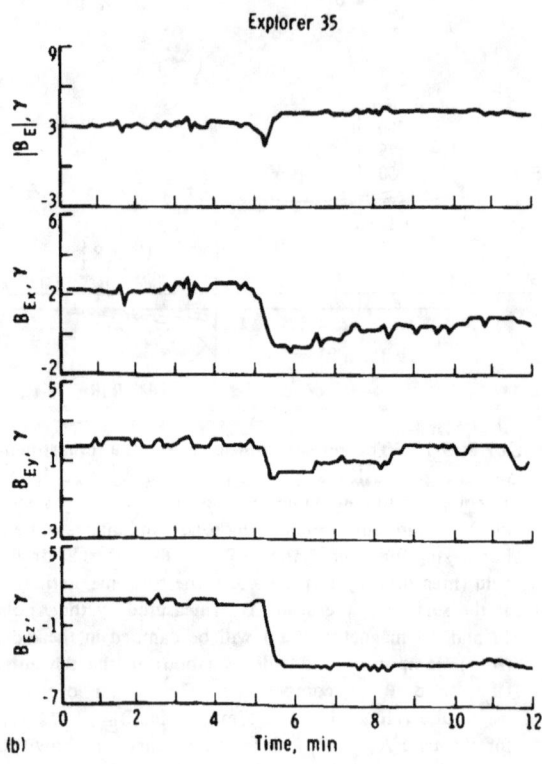

site of a sufficiently large-scale size, a high-density solar wind should compress the remanent field (ref. 9-9). To investigate the possibility that the net-interaction field $\Delta \mathbf{B} \rightarrow \mathbf{B}_F$ is a field-compression phenomenon, plots are shown in figure 9-20 for low frequency (1-hr averages) of the energy-density increase of the horizontal magnetic field ($\Delta B_\theta^2/8\pi$) and simultaneous energy density of the solar wind directed normal to the Apollo 12 site ($\rho v^2 \cos^2 \alpha$ where α is the angle between the average solar-wind velocity vector $\mathbf{v}$ and the normal lunar surface at the Apollo 12 site). A definite correlation exists between

FIGURE 9-18.—Response to step-function transients in all three vector components. The components are expressed in an ALSEP coordinate system, which originates on the lunar surface at the Apollo 12 site. The x-axis is directed radially outward from the lunar surface; y and z are tangential to the surface, directed eastward and northward, respectively. Damping on the radial x-axis and overshoot on the tangential y and z axes are apparent. Apollo 12 and Explorer 35 data scales differ because of the existence of a 38 ± gammas steady field at the Apollo 12 site. (a) Apollo 12 data; experiment begun December 8, 1969, 13:58:02 G.m.t. (b) Explorer 35 data; experiment begun December 8, 1969, 13:57:59 G.m.t.

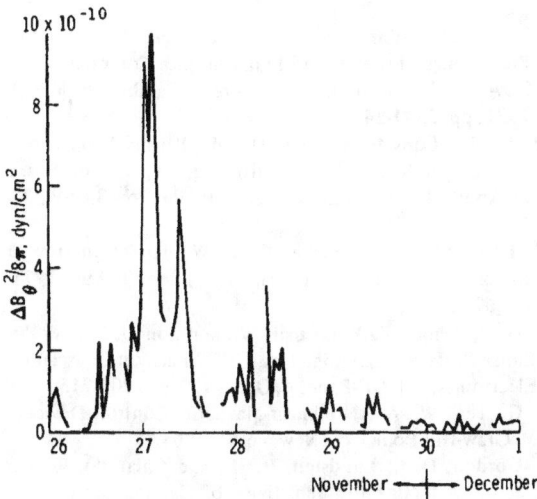 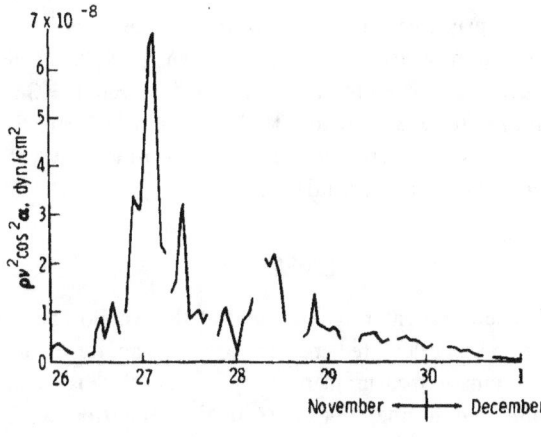

FIGURE 9-20.—Simultaneous plots of the square of horizontal magnetic field difference $\Delta B_\theta^2/8\pi = 1/8\pi\,(\Delta B_y^2 + \Delta B_z^2)$ where $\Delta B_i = B_{Ai} - (B_{Ei} + B_{Si}), i = y, z$ and vertical solar-wind pressure at the Apollo 12 site, showing the correlation between magnetic pressure and plasma pressure. The angle α is the angle between the average solar-wind velocity v and the normal to the lunar surface at the Apollo 12 site. (Plasma data are obtained from ref. 9-3.)

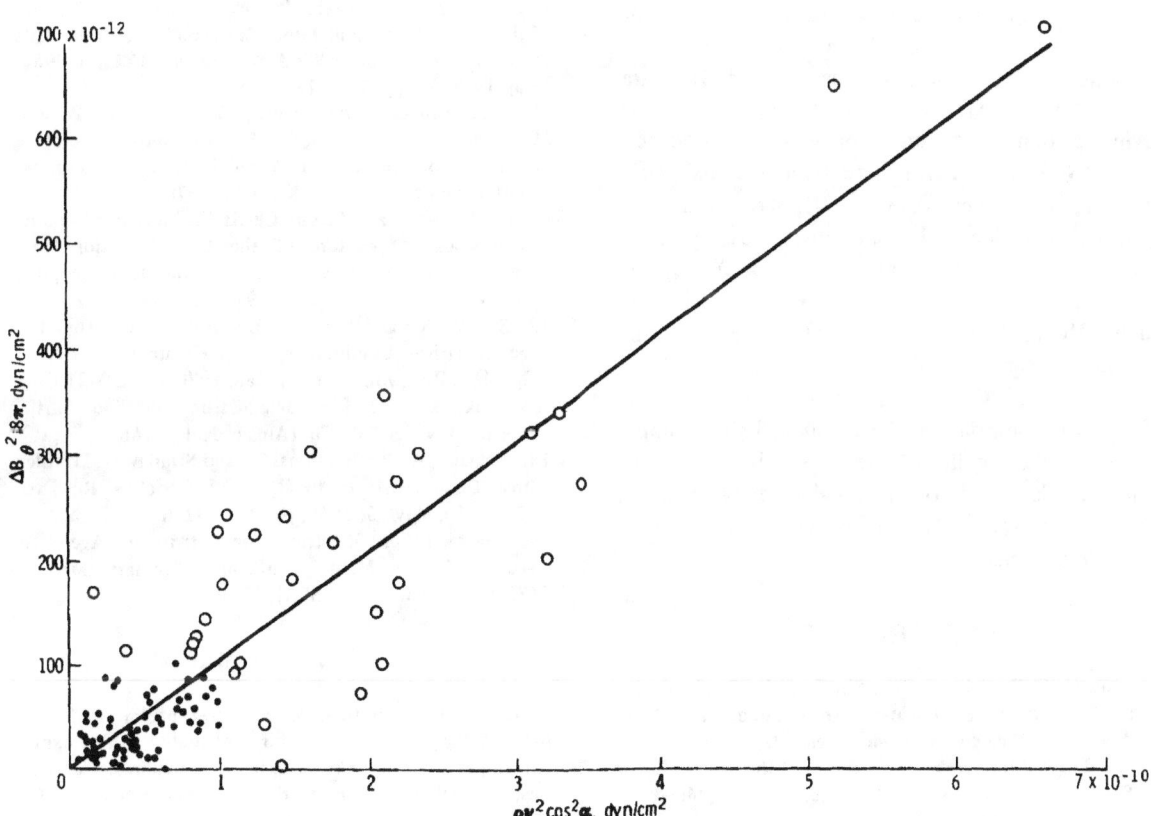

FIGURE 9-21.—Horizontal magnetic-field energy density $\Delta B_\theta^2/8\pi = \Delta B_y^2 + \Delta B_z^2$ plotted as a function of solar-wind pressure at the Apollo 12 site.

these quantities; the relationship between magnetic- and plasma-energy densities is shown to be linear in figure 9-21. Therefore, it is concluded that the 38-gamma remanent field at the Apollo 12 site is compressed on the lunar solar side during times of high solar-wind plasma density.

SUMMARY

The remanent magnetic field at the Apollo 15 site has been calculated to be 6 ± 4 gammas from preliminary measurements. The field value is small compared to the 38-gamma field at the Apollo 12 site, and the 103 ± 5 and 43 ± 6 gamma fields at two Apollo 14 sites. The Apollo 15 site lies near the edge of the Mare Imbrium mascon basin; the fact that little or no remanent field exists at that site leads to the preliminary conclusion that mascons are not highly magnetic.

The bulk relative magnetic permeability of the Moon has been calculated from measurements obtained in the geomagnetic-tail region to be $\mu/\mu_0 = 1.03 \pm 0.13$. A radial electrical-conductivity profile of the lunar interior has been determined from magnetic-field step-transient measurements. The data fit a spherically symmetric three-layer lunar model having a thin outer crust of very low electrical conductivity. The intermediate layer of radial thickness $R_1\ R_2$, where $0.95 R_M \leq R_1 < R_M$ and $R_2 \sim 0.6 R_M$, has electrical conductivity $\sigma_1 \sim 10^{-4}$ mhos/m; the inner core has radius $R_2 \sim 0.6 R_M$ and conductivity $\sigma_2 \sim 10^{-2}$ mhos/m. For the case of an olivine Moon, the temperatures of the three layers are as follows: crust <440 K, intermediate layer ~810 K, and core ~1240 K. Qualitatively, the inductive eddy-current response at the Apollo 15 site is similar to that at the Apollo 12 site. It has been observed that the solar wind compresses the steady remanent field at the Apollo 12 site during periods of high solar-plasma density.

REFERENCES

9-1. Sonett, C. P.; Schubert, G.; Smith, B. F.; Schwartz, K.; and Colburn, D. S.: Lunar Electrical Conductivity from Apollo 12 Magnetometer Measurements: Compositional and Thermal Inferences. Proceedings of the Second Lunar Science Conference, vol. 3, A. A. Levinson, ed., MIT Press (Cambridge, Mass.), 1971, pp. 2415-2431.

9-2. Dyal, Palmer; and Parkin, Curtis W.: The Apollo 12 Magnetometer Experiment: Internal Lunar Properties from Transient and Steady Magnetic Field Measurements. Proceedings of the Second Lunar Science Conference, vol. 3, A. A. Levinson, ed., MIT Press (Cambridge, Mass.), 1971, pp. 2391-2413.

9-3. Snyder, Conway W.; Clay, Douglas R.; and Neugebauer, Marcia: The Solar-Wind Spectrometer Experiment. Sec. 5 of Apollo 12 Preliminary Science Report, NASA SP-235, 1970.

9-4. Dyal, Palmer; and Parkin, Curtis W.: The Magnetism of the Moon. Scientific American, vol. 1, no. 2, Aug. 1971, pp. 62-73.

9-5. Dyal, Palmer; Parkin, Curtis W.; and Sonett, Charles P.: Lunar Surface Magnetometer. IEEE Trans. on Geoscience Electronics, vol. GE-8, no. 4, Oct. 1970, pp. 203-215.

9-6. Geyger, W. A.: Nonlinear-Magnetic Control Devices. McGraw-Hill Book Co. (New York), 1964.

9-7. Gordon, D. I.; Lundsten, R. H.; and Chiarodo, R. A.: Factors Affecting the Sensitivity of Gamma-Level Ring-Core Magnetometers. IEEE Trans. on Magnetics, vol. MAG-1, no. 4, Dec. 1965, pp. 330-337.

9-8. Sonett, C. P.; Colburn, D. S.; Currie, R. G.; and Mihalov, J. D.: The Geomagnetic Tail: Topology, Reconnection and Interaction with the Moon. Physics of the Magnetosphere, R. L. Carovillano, J. F. McClay, and H. R. Radoski, eds., D. Reidel Publishing Co. (Dordrecht, Holland), 1967.

9-9. Barnes, Aaron; Cassen, Patrick; Mihalov, J. D.; and Eviatar, A.: Permanent Lunar Surface Magnetism and its Deflection of the Solar Wind. Science, vol. 172, no. 3984, May 14, 1971, pp. 716-718.

9-10. Dyal, Palmer; Parkin, Curtis W.; Sonett, C. P.; and Colburn, D. S.: Electrical Conductivity and Temperature of the Lunar Interior from Magnetic Transient Response Measurements. NASA TM X-62012, 1970.

9-11. Dyal, Palmer; and Parkin, Curtis W.: Electrical Conductivity and Temperature of the Lunar Interior from Magnetic Transient-Response Measurements. J. Geophys. Res., vol. 76, no. 25, Sept. 1, 1971, pp. 5947-5969.

9-12. Sill, W. R.; and Blank, J. L.: Methods for Estimating the Electrical Conductivity of the Lunar Interior. J. Geophys. Res., vol. 75, no. 1, Jan. 1970, pp. 201-210.

9-13. Rikitake, T.: Electromagnetism and the Earth's Interior. Elsevier Pub. Co. (Amsterdam), 1966.

9-14. England, A. W.; Simmons, G.; and Strangway, D.: Electrical Conductivity of the Moon. J. Geophys. Res., vol. 73, no. 10, May 15, 1968, pp. 3219-3226.

9-15. Nagata, Takesi: Electrical Conductivity and Age of the Moon. Paper M.14, Thirteenth Planetary Meeting, COSPAR (Leningrad, USSR), 1970.

ACKNOWLEDGMENTS

The authors wish to express their appreciation for the efforts of the many people who contributed to this experiment. In particular, thanks are extended to John Keeler, Fred Bates, Kenneth Lewis, and Marion Legg whose diligent effort contributed to the success of this experiment. Special thanks are also extended to Dr. David S. Colburn for the use of Explorer 35 data.

10. Solar-Wind Spectrometer Experiment

*Douglas R. Clay,[a] Bruce E. Goldstein,[a]
Marcia Neugebauer,[a] and Conway W. Snyder[a][†]*

With the deployment of the Apollo 15 lunar surface experiments package, two identical solar-wind spectrometers (SWS), separated by approximately 1100 km, are now on the lunar surface. The spectrometers provide the first opportunity to measure the properties of the solar plasma simultaneously at two locations a fixed distance apart. It is hoped that these simultaneous observations will yield new information about the plasma and its interaction with the Moon and the geomagnetic field. At the time of preparation of this report, magnetic tapes of only 20 hr of simultaneous data had been received. These data are discussed in this preliminary report.

The SWS experiment was designed with the objective of measuring protons and electrons at the lunar surface. Solar-wind and magnetic-field measurements by the Lunar Orbiter Explorer 35 spacecraft (refs. 10-1 to 10-5) have established that no plasma shock is present ahead of the Moon and that the solar wind is not deflected significantly by the Moon. Since then, the SWS experiment of Apollo 12 (ref. 10-6), the charged-particle lunar environment experiment of Apollo 14 (ref. 10-7), and the suprathermal ion detector experiments of Apollo 12 and 14 (refs. 10-8 and 10-9) have established that solar-wind protons reach the lunar surface without major deflection. In addition, magnetometer experiments at the Apollo 12 and 14 sites (refs. 10-10 and 10-11) detected local magnetic fields with strengths of 30 to 100 gammas.

The scientific objective of the Apollo 15 SWS experiment is an investigation of the following phenomena.

(1) *Lunar photoelectric layer.*—measure electron-charge fluxes at the lunar surface.

(2) *Interaction of the solar wind with the local magnetic field.*—determine the change in proton direction and bulk velocity by comparison with interplanetary data.

(3) *Lunar limb shocks.*—determine if lunar limb shocks are detectable at the experiment site during dawn or dusk.

(4) *Solar-wind monitor.*—use the experiment to measure solar-wind conditions.

(5) *Motion and thickness of bow shock and magnetopause.*—make direct, simultaneous comparisons with data taken at the Apollo 12 site.

(6) *Plasma fluctuations.*—study time-dependent phenomena such as waves and plasma discontinuities modified by the local magnetic field and lunar photoelectron layer.

INSTRUMENT DESCRIPTION

The basic sensor in the SWS is a Faraday cup, which measures the charged-particle flux entering the cup by collecting the ions and by using a sensitive current amplifier to determine the resultant current flow. Energy spectra of positively and negatively charged particles are obtained by applying fixed sequences of square-wave ac retarding potentials to a modulator grid and by measuring the resulting changes in current. Similar detectors have been flown on a variety of space probes (ref. 10-3).

To be sensitive to solar-wind plasma from any direction (above the horizon of the Moon) and to ascertain the angular distribution of the solar-wind plasma, the SWS has an array of seven cups. Because the cups are identical, an isotropic flux of particles would produce equal currents in each cup. If the flux is not isotropic but appears in more than one cup, analysis of the relative amounts of current in the collectors can provide information on the direction of plasma flow and its anisotropy. The central cup faces

---

[a] Jet Propulsion Laboratory, California Institute of Technology.
[†] Principal investigator.

the vertical, and the remaining six symmetrically surround it, each facing 60° from the vertical. The combined acceptance cones of all cups cover most of the upward hemisphere. Each cup has a circular opening, five circular grids, and a circular collector. The function of the grid structures is to apply an ac modulating field to incoming particles and to screen the modulating field from the inputs to the sensitive preamplifiers (fig. 10-1). The entrance apertures of the cups were protected from damage or dust by covers which remained in place until after the departure of the ascent portion of the lunar module (LM). The response of this cup to ions and electrons has been measured by laboratory plasma calibrations at various plasma conditions. The angular dependence for ions, averaged over all seven cups, is shown in figure 10-2 and agrees quite well with the calculated response. The result for electrons is similar at small angles, but has a large-angle tail caused principally by secondary electrons ejected from the modulator grid. This mode of response can occur only when the modulator voltage is negative (i.e., the electron-measuring portion of the instrument operation).

The electronics package for the instrument is in a temperature-controlled container, which hangs below the sensor assembly. The electronics package includes power supplies, a digital programer that controls the voltages in the sensors as required, the current-measuring circuitry, and data-conditioning circuits.

The instrument operates in an invariable sequence in which a complete set of plasma measurements is made every 28.1 sec. The sequence consists of 14 energy steps spaced by a factor of $\sqrt{2}$ for positive ions and seven energy steps spaced by a factor of 2 for electrons. In high range, the energy varies from 50 to 10 400 eV for ions and from 10 to 1480 eV for electrons; and, in low range, each level is reduced by a constant factor of 1.68. The range is changed by ground command. A large number of internal calibrations is provided, and every critical voltage is read out at intervals of 7.5 min or less.

The sequence of measurements starts at the lowest-energy-proton step. Eight measurements—each of seven cups and the combined output of all cups—are taken during the 1.22 sec at each step. Following the 17.1 sec required for all 14 proton steps in ascending energy, 2.5 sec of calibration measurements are made; and, finally, 8.5 sec of electron data are taken sequentially from lowest to highest energy levels.

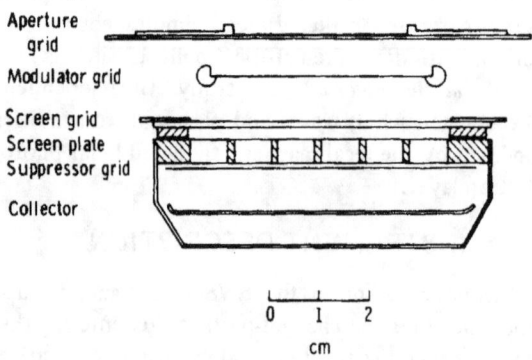

FIGURE 10-1.—Faraday cup sensor.

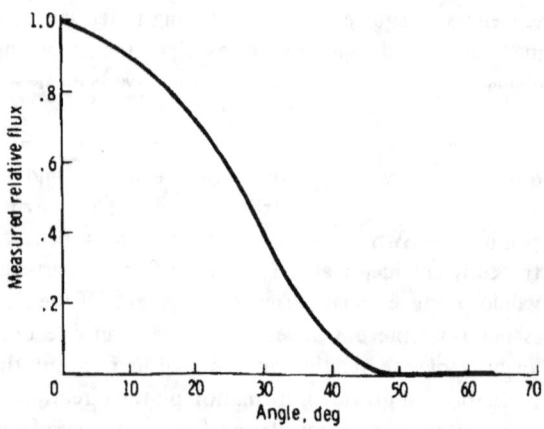

FIGURE 10-2.—Angular response of Faraday cup.

INSTRUMENT DEPLOYMENT

The SWS was deployed without difficulty near the end of the first extravehicular activity period on July 31, 1971. The spectrometer is shown on the lunar surface prior to dust-cover removal in figures 10-3 and 10-4. A portion of the deployment procedure provided for proper orientation of the instrument. The orientation was facilitated by a shadowing device (to aline in an easterly direction) and a pendulum suspension from the supporting legs (for self-leveling in one dimension). A study of the shadowing as seen in the photographs indicates preliminary values of 1° clockwise rotation (as seen from above), 2° slope to the east, and 2-1/2° slope to the north. All values are well within the specified tolerance of 5°.

Cups numbered 1 to 6 are ordered in a clockwise (as seen from above) manner, with cup 1 pointing to

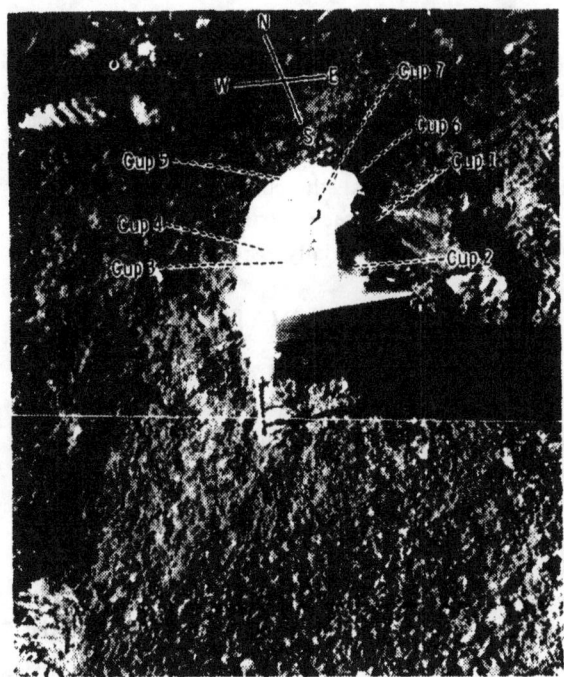

FIGURE 10-3.—Solar-wind spectrometer experiment deployed on the Moon as viewed from the north prior to dust-cover removal (AS15-86-11594).

FIGURE 10-4.—Solar-wind spectrometer experiment as viewed from the south prior to dust-cover removal. Shadowing patterns, especially on the surface of cup 3, give information about orientation of the deployed instrument (AS15-86-11593).

the east. The cup centered in the vertical direction is designated number 7. As a consequence of the SWS orientation and the 26.07° N latitude of the Apollo 15 site, the particles coming directly from the Sun enter cup 1 during lunar dawn. The direction of the average solar plasma is below cup 1 at sunrise and traverses a path between cups 1 and 2; between cups 1, 2, and 7 at midmorning; between cups 2 and 7; and is nearly midway between cups 2, 3, and 7 at midday. In a symmetric manner, the plasma passes between cups 3 and 7; between cups 3 and 4; and, finally, below cup 4 at sunset.

Shortly after deployment, at 19:37:10 G.m.t. on July 31, the SWS was turned on to provide background data with sensor covers in place. Approximately 1 hr after LM ascent, at 18:07:32 G.m.t. on August 2, the covers were removed by command from Earth, and detection of solar plasma began.

INSTRUMENT PERFORMANCE

Sampling of the housekeeping and calibration data for the first 2 months has indicated that the SWS has operated properly. The thermal control of the instrument is completely adequate, with electronics-package temperatures ranging from 254 to 328 K and sensor temperatures from 163 to 344 K. These values agree well with those of the SWS on Apollo 12 and provide confidence that this instrument also will have a long life of stable operation.

METHOD OF ANALYSIS FOR POSITIVE-ION SPECTRA

The usual method of obtaining solar-wind parameters (bulk velocity including direction, proton density, and most-probable thermal speed) is described in the following paragraphs.

The digital numbers for the 14 energy windows and the seven cups are converted to currents at the collectors. From the sums of current over all energies for each cup, the cups that have collected plasma are identified; and, if more than one cup, angles of plasma incidence are estimated. The current ratios from two cups can give only one angle; therefore, current in three cups is required to define the two angles of bulk velocity. For a plasma of velocity between Mach 4 and 20 (as interplanetary solar wind usually is), the geometry and the sensor array typically provide significant currents in two cups, with measurable currents in one or three cups occurring less often. For times when one or more angles are unknown, the assumption is made that the undetermined angle is that of the average solar wind; the average angles used are dependent upon time (i.e., the position of the Sun in the lunar sky corrected by 5° in the ecliptic plane for aberration caused by the motion of the Moon about the Sun).

The angles and the spectrum of currents then are combined to get a differential flux distribution for the cup with the largest total current, including corrections for transparency as a function of angle. A curve-fitting program then fits (in a least-squares sense) two gaussian curves to the data in the six to eight energy windows about the peak. The six degrees of freedom for the fitting are reduced to five by assuming that the bulk velocity of hydrogen and helium ions is the same. This analysis provides estimates of the hydrogen density, hydrogen velocity, hydrogen most-probable thermal speed, helium density, and helium most-probable thermal speed. Because the angular response of the cup is somewhat dependent upon the thermal speeds (higher thermal speeds yielding greater transparency for an incoming plasma at large angles), the curve fitting is reiterated when the thermal-speed results are significantly different from those initially assumed.

ION OBSERVATIONS

The SWS at the Apollo 15 site has detected solar plasma with parameters characteristic of both interplanetary solar wind and magnetosheath plasma. (See fig. 10-5 for nominal plasma regions near the Earth and the Moon.) A time plot of bulk speed and proton density for a portion of August 2 is shown in figure 10-6. The upper two curves for Apollo 15 SWS measurements show plasma detected at the time of the dust-cover removal (18:07 G.m.t.). Shown on the same time scale are SWS measurements from the Apollo 12 site. There is good agreement in relative changes for all parameters at each site. The plasma properties at the Apollo 12 site seem to fluctuate

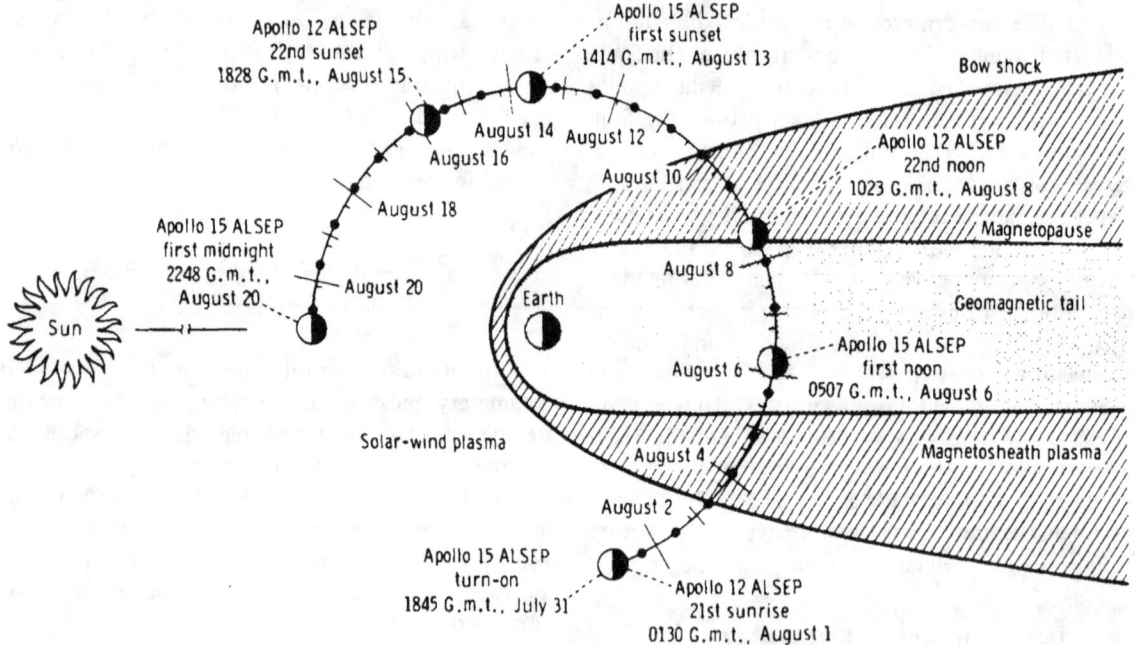

FIGURE 10-5.—Solar plasma regions in the vicinity of the Earth and the Moon. Moon positions are shown relative to the Earth-Sun line.

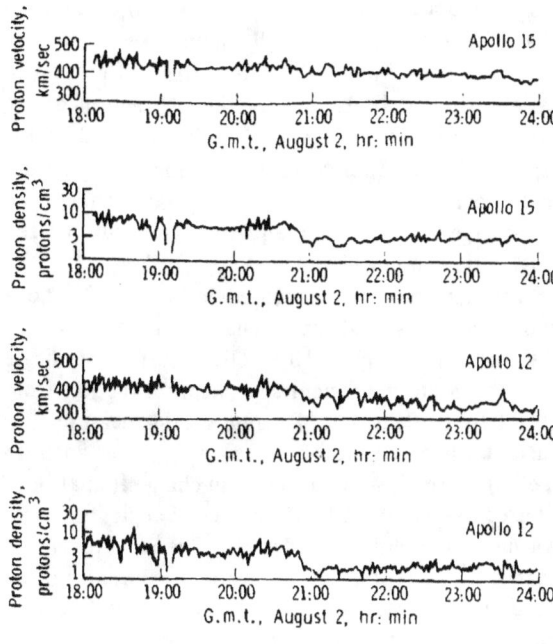

FIGURE 10-6.—Proton velocity and density at the Apollo 12 and 15 sites for 6 hr of August 2.

somewhat more. The spectra at both sites for the period between 18:00 and 19:00 G.m.t. show typical magnetosheath plasma, with large and frequent changes in velocity, density, and thermal speed. The mean thermal speed of protons was approximately 82 km/sec during this period. This relatively high value is characteristic of magnetosheath plasma. A gap can be seen in data plotted from 19:05 to 19:10 G.m.t., when the density became unusually low. This condition was observed at both Apollo sites and subsequently will be discussed in more detail.

Several crossings of the bow shock of the Earth are apparent in the interval from 19:00 to 21:00 G.m.t. The spectra between 21:00 and 24:00 G.m.t. are typical of interplanetary solar wind, with relatively small fluctuations of parameters and smaller thermal speeds (approximately 63 km/sec for this interval). The identification of these regions is corroborated by the signatures of the magnetic field as seen by the lunar surface magnetometer.

The orientation of the Apollo 12 SWS is similar to that of the Apollo 15 SWS, but the Apollo 12 SWS is closer to the equator (2.97° S latitude). Near 18:00 G.m.t. on August 2, the solar-wind direction (corrected for aberration) was 11-1/2° below the axis of the easterly facing cup of the Apollo 12 SWS. At this position, plasma enters only one cup; thus, all angles used in the analysis are those of the assumed average plasma. At the same time, the assumed average solar-wind direction at the Apollo 15 SWS is midway between cups 2 and 1 (29-1/2° from the normal of each) and 41° off the vertical (from the normal of cup 7). On many occasions, enough plasma is detected in cups 2, 1, and 7 to determine the two angles of the plasma velocity. These results show a deviation from the assumed angle of approximately 15° west but still near the ecliptic plane.

Both the proton velocity and the density appear to be lower at the Apollo 12 SWS. A detailed investigation of a few cases has shown that the density discrepancy can be reduced to 30 percent by assuming that the solar-wind direction at the Apollo 12 site is the same as that at the Apollo 15 site.

The obvious need arises for all angles of solar plasma to be known simultaneously at both spectrometers. As more data become available for analysis, it is anticipated that such times will occur. This condition is most likely to occur during the afternoon when the Moon has passed the bow shock and when the Apollo 12 SWS has data in both the vertical and westerly cups. In the past, there have been many occasions when a third cup is also detecting plasma so that all angles are calculable. At these fortunate times, it is further anticipated that the deflection of the plasma by the local magnetic or electric fields will be discernible.

The general features of plasma speed and density as the Moon approached the geomagnetic tail on August 4 are shown in figure 10-7. There are several gaps in the graph where the density dropped well below the threshold of detection of the SWS (less than 0.15 proton/cm^3). This response is typical in the geomagnetic-tail region. Several such passages are

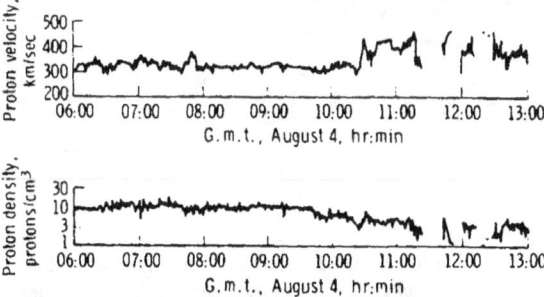

FIGURE 10-7.—Proton velocity and density for 7 hr of August 4 when Apollo 15 site first entered the geomagnetic-tail region.

shown near 11:30 and 12:15 G.m.t. The Moon passed into the geomagnetic tail at 23:02 G.m.t. on August 4, 1971.

When discontinuities in plasma parameters occur (such as at bow-shock and magnetopause crossings), it is hoped that one velocity component for the discontinuity can be determined by comparisons of simultaneous measurements at the Apollo 12 and 15 sites.

ELECTRON AND PROTON ENERGY SPECTRA

Between 19:05 and 19:10 G.m.t. on August 2, a sharp drop in solar-wind density was observed, as shown in figure 10-6. An expanded version of the data (fig. 10-8) shows proton and electron energy spectra for the period 19:00 to 19:12 G.m.t. The third and fourth rows of this figure display ion energy spectra measured by the Apollo 15 and 12 instruments, respectively, derived from the measurements in which all seven Faraday cups are connected electronically. (The gaps in the data are caused by periodic calibration measurements.) Raw-data numbers are plotted; these are approximately proportional to charge flux. Both instruments show a simultaneous drop in proton charge flux at 19:04 G.m.t. and a recovery at 19:11 G.m.t. Also, the total charge flux summed over all channels seems to vary more at the Apollo 12 site. This difference in variation might be caused by the interaction of the plasma with the strong local magnetic field at the Apollo 12 site. Alternatively, directional fluctuations in the ecliptic plane would yield greater changes in the charge-flux measurements at the Apollo 12 site than at the Apollo 15 site.

The data numbers measured in cups 1, 2, and 7 and summed over all energy channels are displayed at the bottom of figure 10-8. These currents indicate that the protons substantially change direction from 19:03 to 19:05 G.m.t. Because a cup measures less current when the protons enter at a greater angle to the cup normal, some decrease in charge flux at both sites can be attributed to a change of angle; but most is caused by a change in density. The relative lack of change of current in cup 1 lends no support to the possibility of ecliptic directional changes causing proton flux fluctuations at the Apollo 12 site.

The electron energy spectra measured at the Apollo 15 site are shown in figure 10-8; the first row represents cup 4, and the second row represents cup 5. Both cups point well over 90° away from the solar direction and are, therefore, uncontaminated by secondary electrons generated inside the cup by protons or by photons. The data numbers plotted

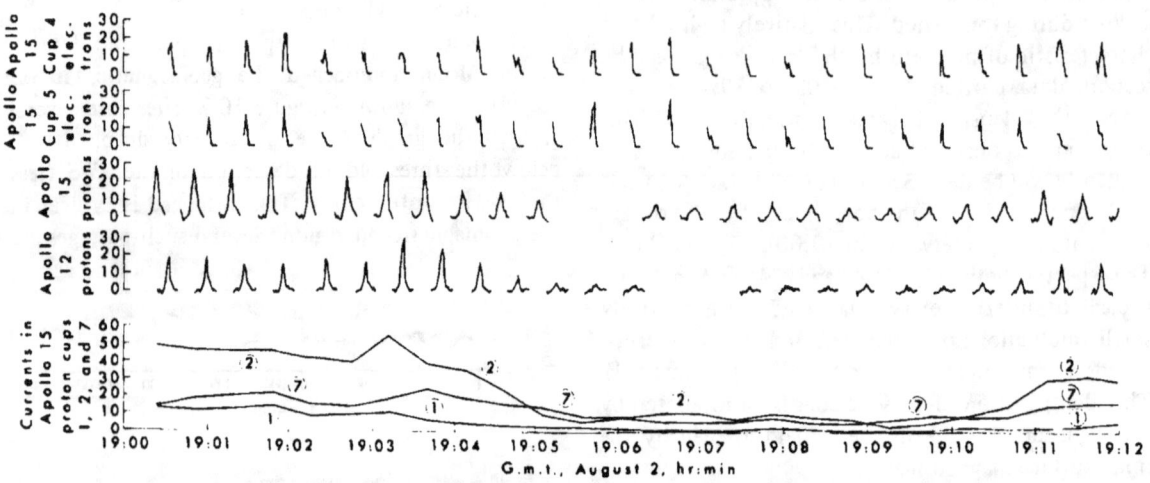

FIGURE 10-8.—Apollo 15 and Apollo 12 SWS data from August 2. The first two rows are electron charge-flux spectra measured by cups 4 and 5 of the Apollo 15 instrument. The next two rows are total ion charge-flux spectra obtained at the Apollo 15 and Apollo 12 sites. The three lines at the bottom are charge flux summed over energy channels measured in cups 1, 2, and 7 of the Apollo 15 instrument.

have not been corrected for the response of the cup as a function of incident electron energy. Therefore, the shapes of the electron spectra plotted are somewhat distorted. The error is almost the same for all electron spectra, so electron spectra can be compared directly to each other.

It should be noted that, in many cases, the spectra observed in cups 4 and 5 are similar in shape and magnitude. In the high-density proton region, cup 4 has a tendency to observe more electrons than cup 5; but, in the low-density region, this relation is reversed. This phenomenon might be caused by changes in orientation of the magnetic field. Also, electron fluxes and proton fluxes do not seem to be strictly proportional from spectrum to spectrum. This fact suggests that a large portion of the electrons observed may be lunar photoelectrons.

Some electron spectra plotted have two peaks. Also, in some cases, the magnitude and shape (energy distribution) measured in cups 4 and 5 are considerably different. Because the time between measurements for cups 4 and 5 in a given energy channel is 0.151 sec, fluctuations with frequencies near 5 Hz are implied. The spectra of protons from the summed cups and from cups 1, 2, and 7 (not shown) show no rapid oscillations. With the aid of magnetic-field data, it may be possible to attribute electron fluctuations to changes in field direction.

SUMMARY AND RESULTS

This report is based on examination of 4 days of Apollo 15 data and 1 day of Apollo 12 data within the same time interval. The Apollo 15 instrument is performing as expected. Solar-wind plasma, magnetosheath plasma, and magnetopause crossings have been observed. Proton measurements at the Apollo 15 and Apollo 12 sites show similar behavior. When angular estimates at both instruments become available, detailed comparisons of velocity and density estimates will be made.

Rapid fluctuations (5 Hz) have been observed in the electron spectra. Interpretation of this phenomenon requires knowledge of magnetic-field direction changes. Complete data and detailed comparison with other solar-wind experiments will be required before comprehensive conclusions can be drawn.

REFERENCES

10-1. Colburn, D. S.; Currie, R. G.; Mihalov, J. D.; and Sonett, C. P.: Diamagnetic Solar-Wind Cavity Discovered Behind Moon. Science, vol. 158, no. 3804, Nov. 24, 1967, pp. 1040-1042.

10-2. Sonett, C. P.; Colburn, D. S.; and Currie, R. G.: The Intrinsic Magnetic Field of the Moon. J. Geophys. Res., vol. 72, no. 21, Nov. 1, 1967, pp. 5503-5507.

10-3. Ness, N. F.; Behannon, K. W.; Taylor, H. E.; and Whang, Y. C.: Perturbations of the Interplanetary Magnetic Field by the Lunar Wake. J. Geophys. Res., vol. 73, no. 11, June 1, 1968, pp. 3241-3440.

10-4. Ness, N. F.; Behannon, K. W.; Searce, C. S.; and Cantarno, S. C.: Early Results from the Magnetic Field Experiment on Lunar Explorer 35. J. Geophys. Res., vol. 72, no. 23, Dec. 1, 1967, pp. 5769-5778.

10-5. Lyon, E. F.; Bridge, H. S.; and Binsack, J. H.: Explorer 35 Plasma Measurements in the Vicinity of the Moon. J. Geophys. Res., vol. 72, no. 23, Dec. 1, 1967, pp. 6113-6117.

10-6. Snyder, Conway W.; Clay, Douglas R.; and Neugebauer, Marcia: The Solar-Wind Spectrometer Experiment. Sec. 5 of Apollo 12 Preliminary Science Report, NASA SP-235, 1970.

10-7. O'Brien, Brian J.; and Reasoner, David L.: Charged-Particle Lunar Environment Experiment. Sec. 10 of Apollo 14 Preliminary Science Report, NASA SP-272, 1971.

10-8. Freeman, J. W., Jr.; Balsiger, H.; and Hills, H. K.: Suprathermal Ion Detector Experiment (Lunar Ionosphere Detector). Sec. 6 of Apollo 12 Preliminary Science Report, NASA SP-235, 1970.

10-9. Hills, H. K.; and Freeman, J. W., Jr.: Suprathermal Ion Detector Experiment (Lunar Ionosphere Detector). Sec. 8 of Apollo 14 Preliminary Science Report, NASA SP-272, 1971.

10-10. Dyal, Palmer; Parkin, Curtis W.; and Sonett, Charles P.: Apollo 12 Magnetometer: Measurement of a Steady Magnetic Field on the Surface of the Moon. Science, vol. 169, no. 3947, Aug. 21, 1970, pp. 762-764.

10-11. Dyal, P.; Parkin, C. W.; Sonett, C. P.; DuBois, R. L.; and Simmons, G.: Lunar Portable Magnetometer Experiment. Sec. 13 of Apollo 14 Preliminary Science Report, NASA SP-272, 1971.

11. Heat-Flow Experiment

Marcus G. Langseth, Jr.,[a][†] Sydney P. Clark, Jr.,[b] John L. Chute, Jr.,[a] Stephen J. Keihm,[a] and Alfred E. Wechsler[c]

The purpose of the heat-flow experiment is to determine the rate of heat flow from the lunar interior by temperature and thermal-property measurements in the lunar subsurface. Heat loss is directly related to the internal temperature and the rate of internal heat production; therefore, measurements of these quantities enable limits to be set on long-lived radioisotopic abundances (the chief source of interior heating), the internal temperature, and the thermal evolution of the Moon.

Preliminary analysis of the data from one heat-flow probe indicates that the heat flow from depth below the Hadley Rille site is 3.3×10^{-6} W/cm^2 (± 15 percent). This value is approximately one-half the average heat flow of the Earth. Further analysis of data over several lunations is required to demonstrate that this value is representative of the heat flow at the Hadley Rille site. Subsurface temperature at a depth of 1 m below the surface is approximately 252.4 K at one probe site and 250.7 K at the other site. These temperatures are approximately 35 K above the mean surface temperature and indicate that the thermal conductivity in the surficial layer of the Moon is highly temperature dependent. Between 1 and 1.5 m, the rate of temperature increase as a function of depth is 1.75 K/m (± 2 percent) at the probe 1 site. In situ measurements indicate that the thermal conductivity of the regolith increases with depth. Thermal-conductivity values between 1.4×10^{-4} and 2.5×10^{-4} W/cm-K were determined; these values are a factor of 7 to 10 greater than the values of the surface conductivity. Lunar-surface brightness temperatures during the first lunar night have been deduced from temperatures of thermocouples above and on the lunar surface. The cooldown history after sunset suggests that a substantial increase in conductivity occurs at a depth on the order of several centimeters. Temperature measurements were also recorded during the total eclipse on August 6, 1971.

EXPERIMENT CONCEPT AND DESIGN

Heat Budget in the Surface Layer of the Moon

The temperature and the heat flux at the surface of the Moon are determined mainly by the solar energy impinging on the surface during one-half of the 29.5-day lunation cycle. During the lunar day, the surface temperature rises to approximately 380 K, which results in heat flow into the subsurface. After lunar sunset, the surface temperature drops to nearly 100 K, and heat flows out of the subsurface and is lost by radiation into space. These very large temperature excursions, in part, are a result of the extremely low thermal conductivity and volumetric heat capacity of the fine rock powders that mantle most of the lunar surface (ref. 11-1) and, in part, are a result of the very tenuous atmosphere of the Moon. The low thermal conductivity of the bulk of the regolith (ref. 11-2) strongly inhibits the flow of energy into and out of the subsurface. At a depth of approximately 50 cm, the large surface variation of 280 K is attenuated to a nearly undetectable amplitude.

At low lunar latitudes, the surface temperature, averaged over one lunation, is approximately 220 K. This mean surface temperature is determined by the balance of solar energy flowing into and energy radiated out of the surface during a complete lunation. The mean temperature in the subsurface (at decimeter depths) may be higher than the mean

---

[a] Lamont-Doherty Geological Observatory.
[b] Yale University.
[c] Arthur D. Little, Inc.
[†] Principal investigator.

surface temperature by a few tens of degrees. As will be shown in more detail later in this section, the mean subsurface temperature at the Hadley Rille site is considerably higher than the mean surface temperature. This increase in mean temperature is a result of the important role of heat transfer by radiation in the fine powders on the surface. The efficiency of radiative heat transfer between particles varies in proportion to the cube of the absolute temperature (ref. 11-3); consequently, during the lunar day, heat flows more readily into the Moon than it flows out during the night. The subsurface heat balance, over one lunation, requires that a substantial steady-state heat loss be maintained in the upper several centimeters of lunar material to eliminate the excess heat that penetrates during the day (ref. 11-4). This outward heat flow causes a steep downward increase in mean temperature that extends to a depth of a few decimeters, the depth to which diurnal waves penetrate.

At depths greater than those to which diurnal waves penetrate, the thermal regime is dominated by heat flow from the lunar interior. This flow results from high interior temperatures and, in the subsurface, is directly proportional to the increase of temperature with depth (the vertical temperature gradient) and to the thermal conductivity. These quantities are related by the equation

$$F_z = -k \frac{dT}{dz} \quad (11\text{-}1)$$

where F_z is the vertical component of the heat flow, k is the thermal conductivity, T is the temperature, and z is the depth. The average heat flow of the Earth has been determined to be 6.2×10^{-6} W/cm² by numerous measurements (ref. 11-5). Estimates of the lunar heat flow, based on microwave emission from the Moon, have ranged from 1.0×10^{-6} W/cm² (ref. 11-6) to 3.3×10^{-6} W/cm² (ref. 11-7), or one-sixth to one-half the Earth heat flow. Thermal-history calculations, based on chrondritic- and terrestrial-isotope abundances for the Moon (ref. 11-8), result in heat-flow estimates of 1×10^{-6} to 2.5×10^{-6} W/cm² for this period in the lunar history. Because of the extremely low conductivity of the regolith, even these very low heat flows would result in gradients ranging from a few tenths of a degree per meter to a few degrees per meter.

Instrument Design

The essential measurements for determining heat flow are made by two slender temperature-sensing probes that are placed in predrilled holes in the subsurface, spaced approximately 10 m apart. Two probes enable two independent measurements of the heat flow to be made in order to gain some knowledge of the lateral variation of heat flow at the Hadley Rille site. Each probe consists of two nearly identical 50-cm-long sections (fig. 11-1). Each section of each heat-flow probe has two accurate (±0.001 K) differential thermometers that measure temperature differences between points separated by approximately 47 and 28 cm. With these thermometers, a measurement (with an accuracy of ±0.05 K) of absolute temperature at four points on each probe section also can be made.

Additional temperature measurements are provided by four thermocouple junctions in the cables that connect each probe to the electronics unit. The thermocouple junctions are located at distances of approximately 0, 0.65, 1.15, and 1.65 m from the topmost gradient sensors (fig. 11-1). The reference junction for these thermocouples is thermally joined to a platinum-resistance reference thermometer, which is mounted on the radiator plate of the electronics unit. The temperature measurements obtained from the heat-flow experiment are summarized in table 11-I.

The differential thermometers consist of four platinum resistance elements wired in a bridge configuration (fig. 11-2). The bridge is excited by successive 2.6-msec, 8-V pulses, first of one polarity and then of the other. The output voltage, the excitation voltage, and the bridge current are measured and used to determine the absolute temperature and the temperature difference. The ratio of bridge output voltage to excitation voltage and the bridge resistance are calibrated for 42 different pairs of temperature and temperature-difference values. The accuracy of these calibrations is better than the accuracies specified in table 11-I. The thermocouples are calibrated at four temperature points, and the reference bridge is calibrated at five temperature points.

Conductivity measurements are made by means of heaters that surround each of the eight gradient-bridge sensors. The experiment is designed to measure conductivity in two ranges: a lower range of $1 \times$

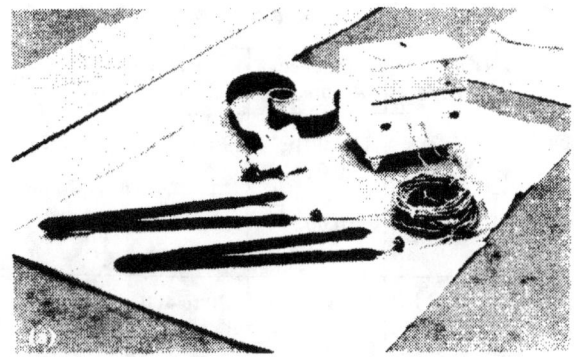
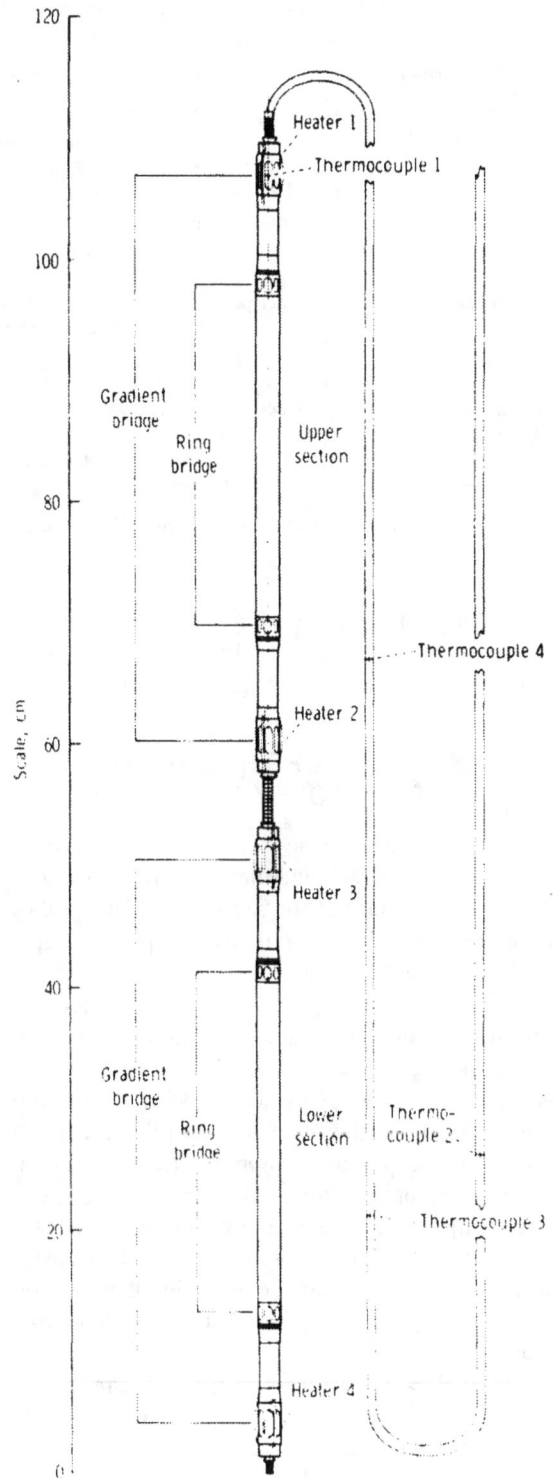

FIGURE 11-1.—Heat-flow experiment and dual-purpose heat-flow probe. (a) Heat-flow-experiment equipment. (b) Schematic of heat-flow probe.

10^{-5} to 5×10^{-4} W/cm·K and a higher range of 2×10^{-4} to 4×10^{-3} W/cm·K. To enable measurements in the lower range to be made, a heater is energized at 0.002 W, and the temperature rise of the underlying gradient sensor is recorded as a function of time for a period of 36 hr. The temperature rise and the rate of temperature rise can be interpreted in terms of the conductivity of the surrounding lunar material. Measurements in the higher range of conductivities are made by energizing the same heater at 0.5 W and monitoring the temperature rise at the ring sensor 10 cm away for a period of approximately 8 hr.

Operation of the Experiment

During normal operation of the experiment (mode 1 operation), temperatures of all gradient bridges, thermocouples, and the reference bridge (as well as temperature differences of all gradient bridges) are sampled every 7.2 min. When a heater is turned on at 0.002 W to enable measurements to be made in the lower conductivity range, the experiment is said to be operating in mode 2. The mode 3 operation is designed for the measurement of conductivity in the higher range. In this mode, temperature and temperature difference at a selected ring bridge are read every 54 sec. These modes of operation and heater turnon are controlled by commands transmitted from Earth.

The detection circuitry for measuring bridge voltages and thermocouple outputs is contained in a housing separate from the Apollo lunar surface experiments package (ALSEP) central station (fig. 11-1(a)) and is designed to compensate for amplifier offset and gain change. This compensation is achieved

TABLE 11-I.—*Summary of the Heat-Flow-Experiment Temperature Measurements*

Thermometer	Number	Location	Temperature difference, K		Absolute temperature, K	
			Range	Accuracy	Range	Accuracy
Gradient bridge[a]	1 per section	Sensors separated by 47 cm	±2 ±20	±0.001 ±0.01	190 to 270	±0.05
Ring bridge	1 per section	Sensors separated by 28 cm	±2	±0.002	190 to 270	±0.05
Thermocouple	4 per probe	Spaced 65 cm apart in the first 2.5 m of the cable above the probe	--	--	70 to 400	±0.07
Thermocouple reference bridge	1 per experiment	Mounted on the radiation plate of the electronics box	--	--	253 to 363	±0.01

[a]Gradient-bridge measurements of temperature difference are made at 2 sensitivities, with a ratio of 10 to 1.

by making all bridge measurements with bipolar excitation and by measuring the ratio of the output voltage to the excitation voltage (fig. 11-2).

EMPLACEMENT OF THE EXPERIMENT AT THE HADLEY RILLE SITE

Drilling of the holes to emplace the heat-flow probes was more difficult than had been expected. The resistant nature of the subsurface at the Hadley Rille site prevented penetration to the planned depth of 3 m. Instead, at the probe 1 site, the borestem penetrated 1.62 m (fig. 11-3); and, at the probe 2 site, the borestem penetrated approximately 1.60 m. The configuration of the probe in each hole is shown in figure 11-3. An obstruction, which was probably a break in the stem at a depth of approximately 1 m, prevented probe 2 from passing to the bottom of the borestem. Because of the very large temperature differences over the upper section, which extends above the surface, no valid temperature measurements were obtained by the ring and gradient bridges on the probe 2 upper section during most of the lunation cycle.

The shallow emplacement of the probes resulted in five of the cable thermocouples lying on, or just above, the lunar surface. These cable thermocouples come into radiative balance with the lunar surface and space, and the measured temperatures can be interpreted in terms of lunar-surface brightness temperatures. A sixth thermocouple is in the borestem projecting above the lunar surface at the probe 1 site.

When this section was written, surface-temperature and subsurface-temperature data had been recorded for nearly one and a half lunation cycles. During the first lunar noon (August 6), a full eclipse of the Sun by the Earth occurred. The thermocouples recorded surface-temperature data at 54-sec intervals during this eclipse. Six in situ conductivity measurements for the low range of values also have been conducted. Only three of these measurements are reported herein.

SUBSURFACE TEMPERATURES

The surface-temperature measurements during the lunar night and during the August 6 eclipse indicate that the surface layer surrounding the probes has an extremely low thermal conductivity. The subsurface measurements reveal that the conductivity must increase substantially with depth, and values of approximately 1.5×10^{-4} W/cm-K are found at a depth of 1 m. With these values of conductivity, it is unlikely that any measurable time variation of temperature as a result of the diurnal cycle existed at depths below 50 cm before the borestem and the heat probe were emplaced. However, after emplacement, the relatively high thermal conductance of the borestem and the radiative transfer along the inside of the stem allowed surface-temperature variations to penetrate to greater depths. After several lunations, a new periodic steady-state condition will be established

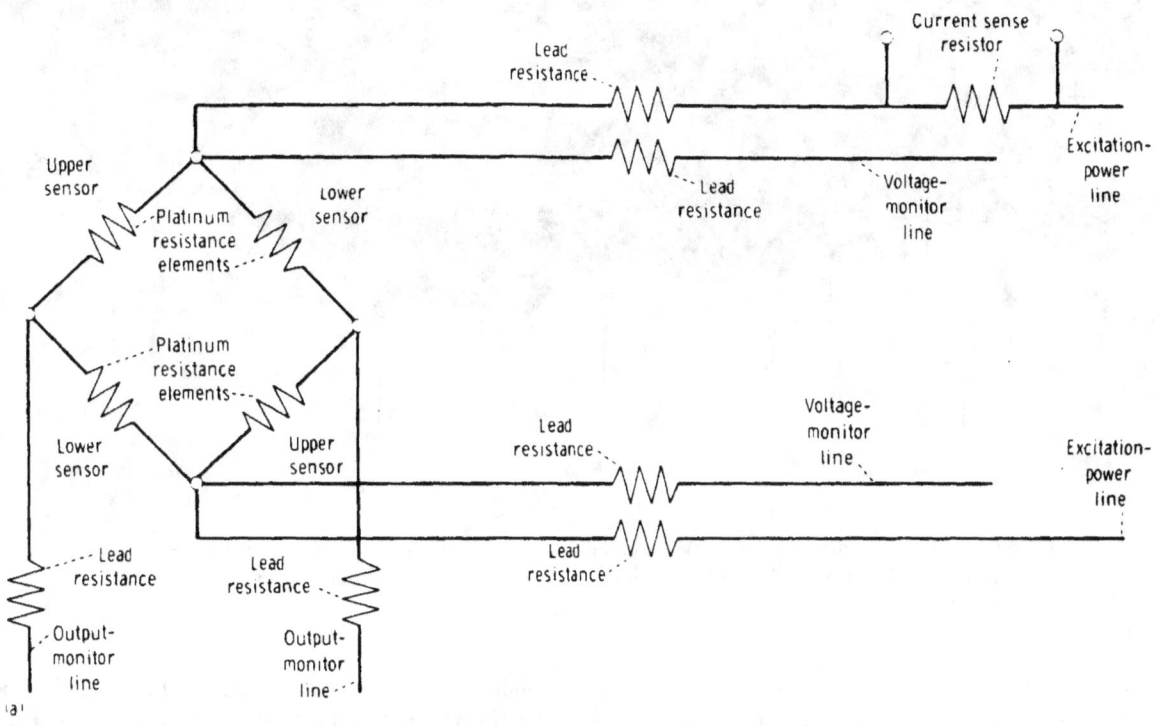

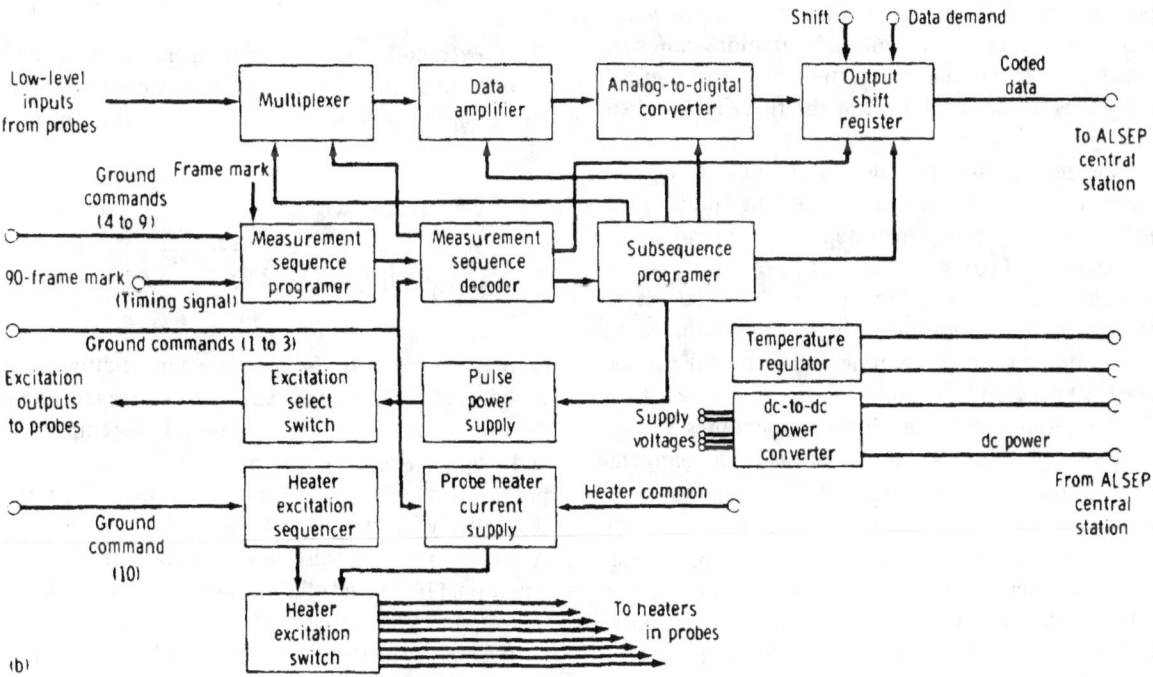

FIGURE 11-2.—Heat-flow experiment. (a) Circuit diagram of the differential thermometer. (b) Block diagram of the experiment electronics.

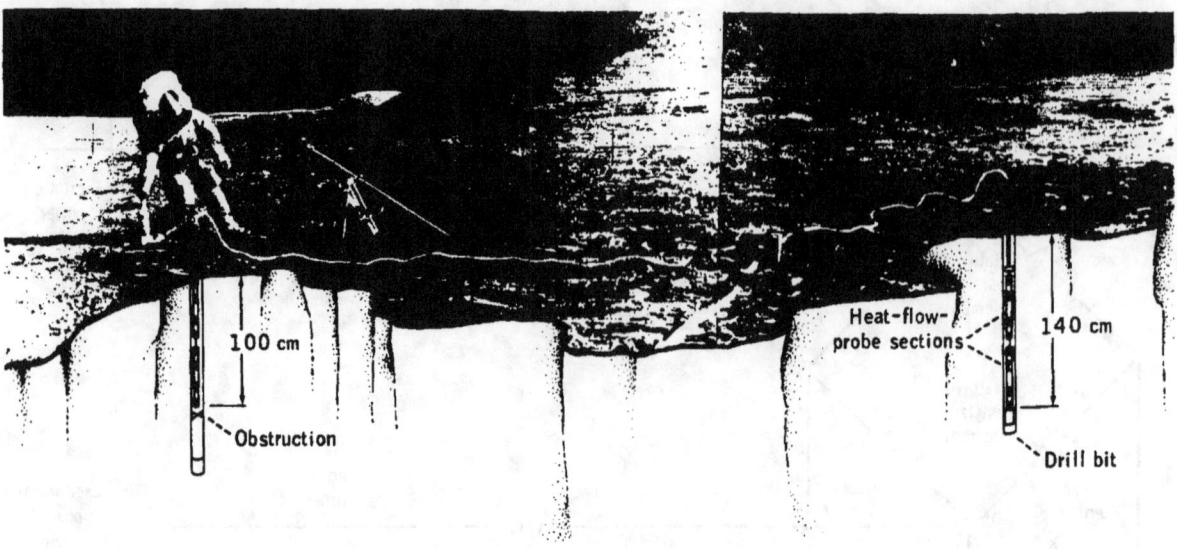

FIGURE 11-3.—Panorama of the heat-flow-experiment emplacement site (composite of photographs AS15-87-11847 and AS15-87-11848) and a cutaway drawing of the heat-probe and thermocouple locations.

around the borestem; initially, however, the borestem and probes will equilibrate toward temperatures that existed in the subsurface before emplacement. By applying the theory of the cooling of a cylinder, which is discussed in the appendix of this section, it is possible to deduce the undisturbed temperature profile at depths where diurnal variations are negligible and to obtain some estimate of the thermal properties of the regolith from the first few hundred hours of equilibration.

The temperature histories of all subsurface thermometers and the evolution of the profiles of temperature as a function of depth for probes 1 and 2 are shown in figures 11-4 and 11-5, respectively. All sensors initially cooled very rapidly, and those sensors at depths greater than 0.7 m continued to cool monotonically with time and were still cooling after 300 hr. The thermometers at depths less than 0.7 m responded to the high temperatures of the borestem projecting above the surface. The temperature in the top of the probe 1 borestem, which is projecting above the surface, was 348 K at lunar noon. As shown in figure 11-5, an obstruction prevented heat-flow probe 2 from passing to the bottom of the hole; consequently, the platinum-resistance thermometers in the top section are not on scale. The cooldown duration of probe 2 is longer than that of probe 1, probably because of the lower conductivity material surrounding the borestem and the higher initial heat input that resulted from extended drilling.

Extrapolation to Equilibrium Values

To extrapolate the sensor temperatures to equilibrium values, the first-order approximation of the long-term solution of the cooling-cylinder problem

$$\frac{T - T_\infty}{T_0 - T_\infty} = \left(\frac{4\pi kt}{S_1 + S_2}\right)^{-1} \quad (11\text{-}2)$$

is used, where T is the absolute temperature of the probe in K, T_∞ is the true equilibrium temperature of the probe, and T_0 is the probe initial temperature; and where S_1 and S_2 are the thermal heat capacities per unit length of the inner and outer cylinders, respectively, in W-sec/cm-K, and t is time in seconds. A more complete discussion of the derivation of equation (11-2) and definitions of the variables can be found in the appendix of this section.

If an initial estimate of the equilibrium temperature T'_∞ is made, it can be shown that the error in the estimate δ is given by

$$\delta = \frac{v'_1 - v'_2 \frac{t_2}{t_1}}{\frac{t_2}{t_1} - 1} \qquad (11\text{-}3)$$

where $v_1 = T_1 - T'_\infty$, $v'_2 = T_2 - T'_\infty$ and T_1 and T_2 are two temperatures selected from the long-term cooling history at times t_1 and t_2, respectively. Then the true equilibrium temperature is simply $T'_\infty + \delta$. The equilibrium temperature determined in this way is independent of the initial-temperature estimate. The equilibrium temperatures for all sensors not affected by the diurnal variation are shown in table 11-II. The values also are plotted as functions of depth in figures 11-4 and 11-5. The accuracy of these equilibrium temperatures is ±0.05 K.

At the probe 1 site, the subsurface temperature, which increases regularly with depth, is approximately 252.0 K at a depth of 80 cm. The increase along the lower 60 cm of probe 1 is approximately 1 K. For probe 2, the temperature at a depth of 80 cm is approximately 250.5 K. The two probe 2 sensors that are unaffected by the diurnal variations indicate an increase in temperature with depth at a rate comparable to the rate detected by the probe 1 sensors. This gradient in temperature is a result of the outward flow of heat from the Moon.

Equilibrium Temperature Differences Along the Heat-Flow Probes

The gradient and ring bridges enable measurement of temperature differences between points 47 cm and 28 cm apart on the probe with an accuracy of ±0.001 K, which is far greater than the accuracy of the absolute-temperature measurements. An analysis similar to that used for the equilibration of the individual sensors can be used to extrapolate the temperature differences to equilibrium values. By using the first term of equation (11-13) in the appendix, the equilibrium temperature difference ΔT_∞ between two points on the probes is given by the expression

$$\Delta T_\infty = \frac{\Delta T_1 - \Delta T_2 \frac{t_2}{t_1}}{1 - \frac{t_2}{t_1}} \qquad (11\text{-}4)$$

where ΔT_1 and ΔT_2 are temperature differences measured at times t_1 and t_2, respectively. Equation 11-4 is valid only for very long times ($t > 300$ hr); consequently, only those differential thermometers that have not been affected by the diurnal variations in the first few hundred hours of observation can be used in this analysis.

The only differential thermometers that have not been affected by the diurnal variations are those on the bottom section of probe 1. The calculated equilibrium-temperature difference across the gradient bridge on the bottom section of probe 1 is 0.779 K and, across the ring bridge, 0.483 K. These results can be interpreted in terms of the temperature difference between adjacent points on the borestem wall by taking into account the effect of radiative

TABLE 11-II.—*Lunar Conductivities Determined From Heat-Probe and Cooldown Histories*

Sensor	Depth, cm	Equilibrium temperature, K	Minimum-conductivity case		Maximum-conductivity case	
			Initial temperature, K	Deduced conductivity, W/cm-K	Initial temperature, K	Deduced conductivity, W/cm-K
TG11B	83	251.96	315	0.8×10^{-4}	349	1.5×10^{-4}
TR22A	87	250.53	a323	1.1	--	--
TG12A	91	252.28	317	1.2	349	1.7
TG22B	96	250.70	a323	1.2	--	--
TR12A	100	252.40	317	1.2	349	1.7
TR12B	129	252.87	313	1.9	349	2.9
TG12B	138	253.00	313	2.1	349	3.3

aThe borestem at the probe 2 site was drilled down approximately 1 m during the first extravehicular activity (EVA). During EVA-2, 22 hr later, the borestem was drilled an additional 3 min to a depth of approximately 1.5 m. The initial temperature was estimated by calculating the cooldown from the preemplacement temperature for 22 hr between the EVA periods and adding the additional heat of drilling during EVA-2. This procedure resulted in an estimate of initial temperature (323 K) very close to the value determined by extrapolation of temperature data in the 1st hour after insertion.

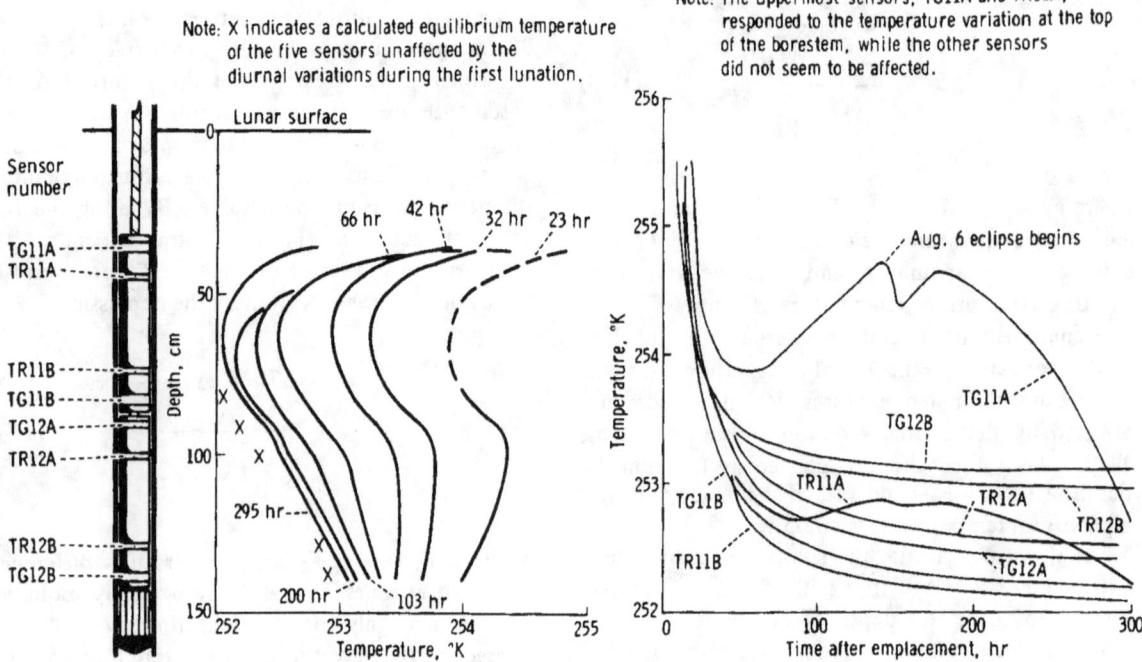

FIGURE 11-4.—Temperature histories of the sensors on heat probe 1 during the first 300 hr after emplacement. The subsurface geometry of the probe and temperature as a function of depth are shown on the left. Temperature as a function of time is shown on the right.

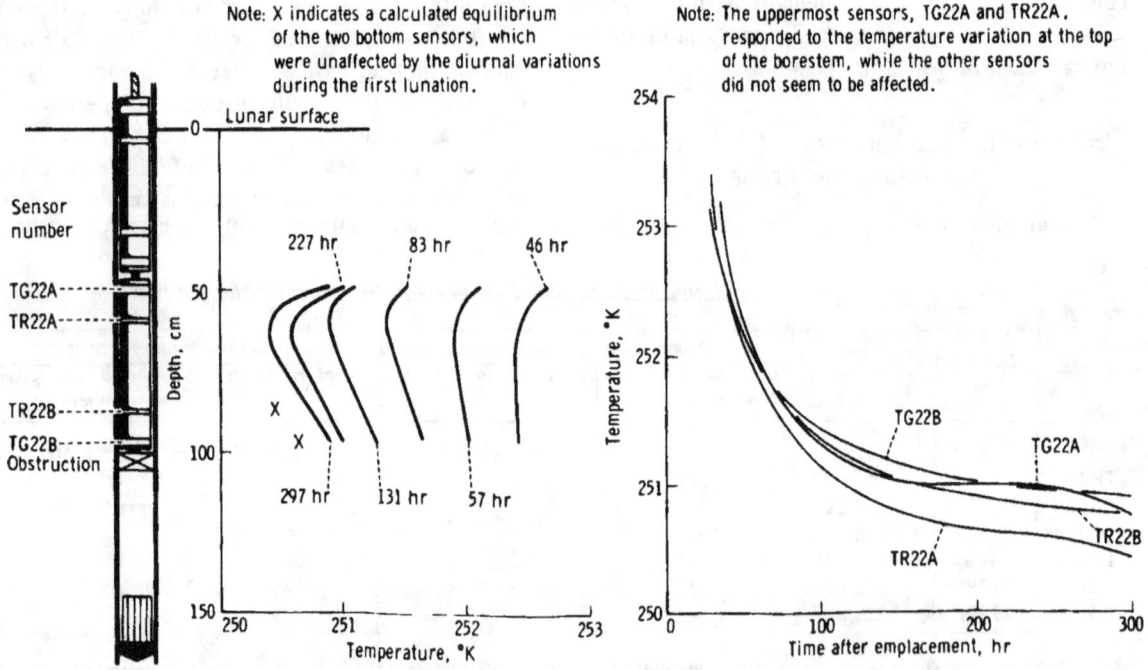

FIGURE 11-5.—Temperature histories of the sensors on heat probe 2 during the first 300 hr after emplacement. The subsurface geometry of the probe and temperature as a function of depth are shown on the left. Temperature as a function of time is shown on the right.

TABLE 11-III. *Summary of Equilibrium-Temperature-Difference Measurements and Subsurface-Temperature Gradients*

Differential thermometer	Equilibrium temperature difference, K		Lunar temperature gradient, K/m
	Probe	Borestem	
Gradient bridge	0.779	0.819	1.74
Ring bridge	.483	.502	1.77

coupling between the walls and the probe and the finite axial conductance of the probe. The temperature difference over the probe is always slightly less than the temperature difference between the adjacent points on the borestem. The ratio of the two, which is called the shorting ratio, was determined experimentally in the laboratory for each section of the probes. The temperature differences at points in the borestem adjacent to the differential thermometers after the shorting ratio has been applied are listed in table 11-III.

The relatively high axial conductance of the borestem results in some axial shunting of the steady-state heat flow; therefore, to determine the undisturbed gradient (i.e., the gradient at large radial distances from the borestem), some correction must be made. The shunting effect can be estimated by modeling the borestem as a prolate spheroid surrounded by a medium with a lower conductivity. By using an effective axial conductivity of 2.3×10^{-3} W/cm-K for the borestem and a lunar conductivity of 1.7×10^{-4} W/cm-K, the model indicates that a plus-1-percent correction should be applied to the borestem temperature gradients. This correction has been applied to the temperature gradients listed in table 11-III.

Diurnal Temperature Variations

Variations in temperature synchronous with the solar phase were observed at depths as great as 70 cm during the first one and a half lunations after emplacement. The temperature variations measured by probes 1 and 2 are shown in figures 11-6 and 11-7, respectively. The peak-to-peak amplitude of the variation at the top of probe 1 is approximately 6 K, a 43-to-1 attenuation of the 260 K temperature excursion measured in the part of the borestem that projects above the lunar surface. The ring sensor TR11A, which is located 9 cm below the top of the probe, measured variations with a 2 K amplitude. The sensors on the lower section of probe 2 that detected variations are somewhat deeper (49 and 58 cm) and recorded correspondingly smaller amplitudes. There are two interesting features of the observed variations. The phase shift of the peaks is extremely small, in view of the large attenuation factors, and a considerable portion of the high-frequency component of the solar radiation penetrates to these depths, as indicated by the rapid rates of temperature change at dawn and sunset. These features suggest that much of the heat transfer to the probe occurs by direct radiative exchange with the upper part of the borestem. The thin, aluminized Mylar disk that is located on top of each probe as a radiation shield apparently does not prevent significant radiative exchange between the top of the probe and the lunar surface.

The heat exchange during a lunation cycle is very complex, because the borestem conducts heat from and toward the lunar surface more efficiently than the regolith material. Thus, the low nighttime temperatures penetrate downward along the borestem, which enlarges the low-temperature area viewed by the top of the probe and, hence, increases the heat loss from the probe to the surface. A similar, but opposite, effect occurs during the day. This phenomenon may, in part, explain the asymmetry of the plots of temperature as a function of time.

Diurnal temperature variations that propagate along the borestem have an important effect on the mean temperature in the borestem. Because the conductivity of the borestem is not as temperature dependent as the adjacent lunar material, heat will be lost more readily along the borestem at night. Consequently, the heat balance over a full lunation will require that the borestem, to depths that diurnal variations penetrate, have a lower mean temperature at a given depth than the regolith. Thus, a net cooling of the borestem in the upper meter can be anticipated, which is an effect that is already apparent in the substantial decrease in peak temperature during the second lunation. Comparisons with the cooling curves of deeper sensors show that this difference cannot be explained by cooling from initial temperatures alone. This cooling effect results in a gradient in mean temperature in the upper meter of the borestem that is unrelated to the heat flow from the interior.

It is essential to note that the mean temperatures and temperature differences in those sections of the

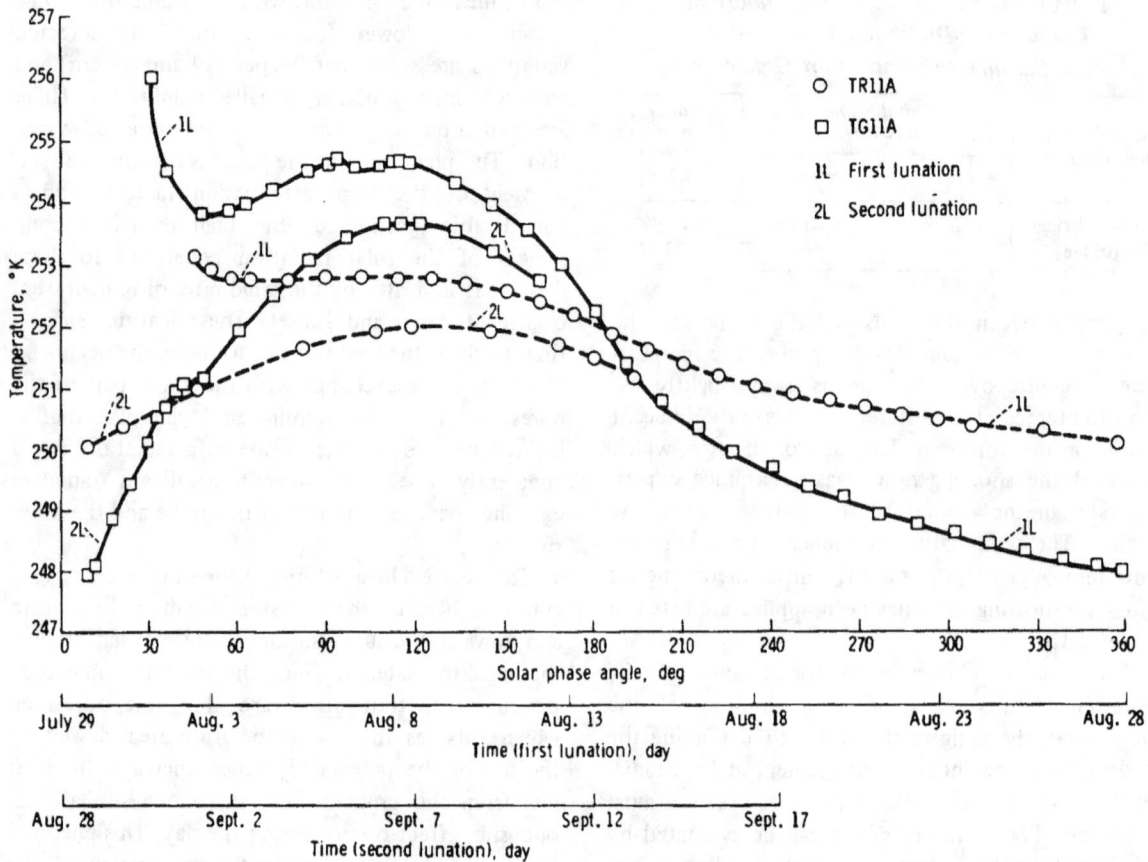

FIGURE 11-6.—Temperature as a function of solar phase angle for probe 1 sensors TR11A and TG11A (the sensors that detect diurnal variations) for the first one and a half lunations after emplacement.

borestem that see diurnal variations cannot be used to determine gradients related to the heat flow from the lunar interior until the effect of temperature-dependent conduction in the borestem and the surrounding lunar material is analyzed and the effect quantitatively determined. Such an analysis is beyond the scope of this preliminary report, but the analysis will be made on future subsurface-temperature data after the upper part of the borestem has equilibrated nearer to the mean periodic steady-state temperature. This analysis will add two independent measurements of heat flow to the result already reported here.

Implications of the Large Mean-Temperature Gradients in the Upper 50 cm at the Hadley Rille Site

By using a finite-difference model to generate daytime lunar-surface temperatures (which depend almost exclusively on the solar flux) and by using the reduced thermocouple temperatures to obtain lunation nighttime surface temperatures, a mean lunar-surface temperature of 217 K (±3 K) was obtained. This result indicates an increase in mean temperature (35 K higher than the mean surface temperature) at depths beyond which the diurnal variation penetrates. This phenomenon can be explained in terms of a strong temperature dependence of the thermal conductivity, which previously has been investigated by Linsky (ref. 11-4) and others. Because of the near lack of an atmosphere on the Moon, radiative transfer of heat between and through particles of the lunar fines can contribute significantly to the effective thermal conductivity. This temperature-dependent conductivity has been found to obey a relation of the form (ref. 11-3)

$$k(T) = k_c + k_r T^3 \qquad (11\text{-}5)$$

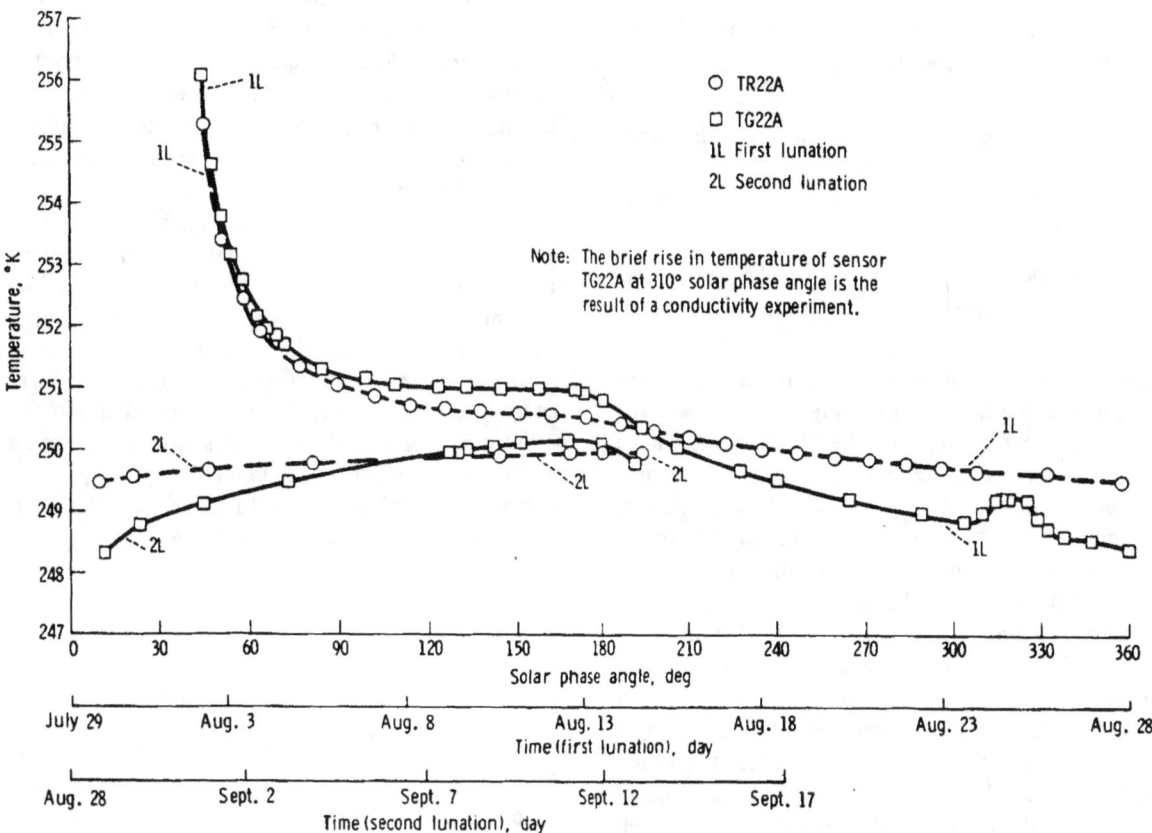

FIGURE 11-7.—Temperature as a function of solar phase angle for probe 2 sensors TR22A and TG22A (the sensors that detect diurnal variations) for the first one and a half lunations after emplacement.

where k_c is the contribution from conduction and $k_r T^3$ represents the radiative exchange between and through particles. Linsky (ref. 11-4) has used computer models of the lunar surface to evaluate this effect in the absence of a steady-state heat flow. By interpolating from these models, the relative contributions of the conductive and radiative terms can be estimated. For a difference of 35 K in mean temperature between the surface and depths at which no significant time variations of temperature exist, the ratio of radiative to conductive terms is approximately 2 at a temperature of 350 K. The relatively small steady-state gradient (1.75 K/m) produced by the measured steady-state heat flow will have only a slight effect on this ratio. Conductivity measurements have been performed for a wide range of temperatures on returned lunar samples from the Apollo 11 and 12 missions (ref. 11-9). The results also indicate the significant temperature dependence of conductivity. For Apollo 11 and 12 samples, the data cited in reference 11-9 indicated that the ratios are 0.5 and 1.5, respectively. The conductivity of the more highly temperature-dependent lunar fines from the Apollo 12 site seem to be more comparable to the upper regolith conductivity at the Hadley Rille site. Further refinement of surface-thermocouple data, combined with a more accurate determination of conductivity as a function of depth and direct measurements of the conductivity of returned Apollo 15 samples, will result in the first directly measured profile of regolith conductivity to a depth of 1.5 m.

CONDUCTIVITY OF THE REGOLITH

Preliminary Deductions from the Heat-Flow-Probe Histories

The rate of equilibration of the probes depends on the thermal diffusivity κ of the surrounding lunar material, the ratio α of heat capacity per unit volume of the lunar material to the heat capacity per unit volume of the heat probe, and the contact conductances H_1 and H_2, where

$\alpha = 2\pi b^2 \rho c / (S_1 + S_2)$

c = heat capacity per unit mass of the surrounding lunar material, W-sec/cm-K

ρ = density of the surrounding lunar material g/cm^3

H_1, H_2 = contact conductances of the inner- and outer-cylinder boundaries, respectively, W/cm^2-K

A more complete discussion can be found in the appendix of this section. By using an estimate of the volumetric heat capacity of the lunar material ρc, a value for α can be determined, because the thermal properties of the probe and the borestem are known. From an analysis of the cooling history of the probes, an estimate of the diffusivity, and, thus, the conductivity, can be made. Measurements of the heat capacity of samples that represent a wide range of lunar rock types result in very uniform values (ref. 11-10). The density of the regolith material is quite variable; preliminary measurements of samples taken by core tubes at the Hadley Rille site result in values that range from 1.35 to 1.91 g/cm^3. At the depth of the probes, the densities are probably near the high end of this range and not so variable. For the analysis described in this section, a density of 1.8 g/cm^3 and a heat capacity of 0.66 W-sec/g-K have been assumed.

It is not possible to determine a value of κ from the ratios of temperatures at various times during the cooldown, because, as the long-term solution indicates, the temperature ratios depend solely on the ratio of the times. Bullard (ref. 11-11) has pointed out this property of cooling cylinders in his discussion of sea-floor heat-flow measurements. To estimate a value for κ, the initial temperature must be known. Estimates of the initial temperature can be made by extrapolating data recorded soon after the probes were inserted to the time the borestem was emplaced. This estimate is considered to be a minimum value, because cooling during the first several minutes is faster as a result of enhanced radiative transfer at high temperatures. Alternately, the assumption can be made that the initial temperature is the temperature of the borestem before emplacement, plus some estimated temperature rise as a result of the heat produced during drilling. Temperatures recorded before emplacement by probe 2, which was stored temporarily in the drill rack between EVA-1 and EVA-2, were used as estimates of the borestem temperature before emplacement. The temperature rise that resulted from drilling is estimated to be 15 K/min, based on estimated torque levels. The initial-temperature estimates based on these assumptions are considered to be maximum estimates.

The cooling histories of all subsurface sensors that are not affected by diurnal variations were analyzed to determine the conductivity of the surrounding lunar material for the two limiting estimates of initial temperature. By using the equilibrium temperature T_∞ for each sensor, the ratio $(T-T_\infty)/(T_0-T_\infty)$ was determined for the first few hundred hours of equilibration. A typical plot of this ratio as a function of time is shown in figure 11-8. The procedure for determining conductivity is to make an initial estimate of the parameters h and A, where

$h = k/bH_2$
$A = S_2 \gamma \kappa / b^2 (S_1 + S_2)$
$\gamma = S_1 / 2\pi a H_1$

a = radius of the inner cylinder (heat-flow probe) in centimeters

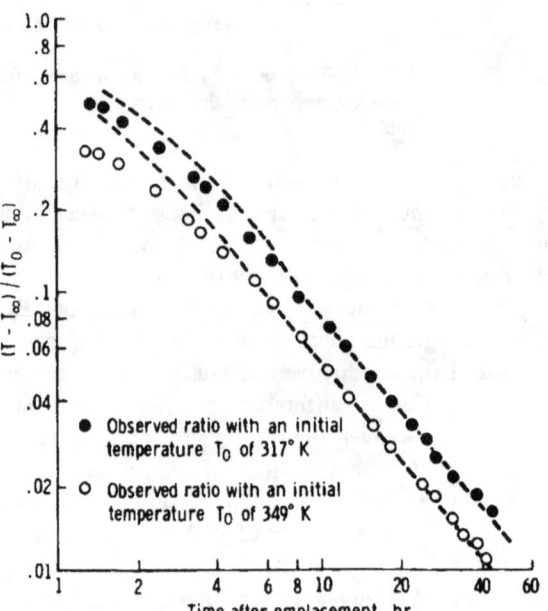

FIGURE 11-8.—Ratio of temperatures measured by sensor TG12A during the initial equilibration of heat probe 1 with the lunar subsurface compared with the theoretical cooldown curves computed from equation (11-12) (dashed lines).

Then, by equating observed temperature ratios for several times in the cooling history with the ratios computed from equation (11-12) in the appendix, a value for the dimensionless parameter τ can be found that corresponds to each time. Once a value for τ is known, k can be determined from the relation $\tau = kt/\rho c b^2$. The conductivity value determined for a large value of τ will be the most accurate, because, at long times, $f(A,h,\alpha,\tau)$ is nearly independent of h and A, where $f(A,h,\alpha,\tau)$ designates the right-hand side of equation (11-12) in the appendix. By using this initial long-time estimate and comparisons of observed and computed ratios at early times, a best value of k and H_2 can be determined with two or three repetitions of the procedure.

Theoretical curves fitted to the data obtained from sensor TG12A by this procedure and ratios for both limiting initial-temperature estimates are shown in figure 11-8. The parameters h and A have been chosen to fit data for times greater than 6 to 8 hr. The theoretical curve at earlier times lies well above the observed data; however, it is not possible to find a value for H_2 to fit data obtained at times earlier than 6 hr without degrading the fit at later times. To obtain the most accurate value of k, the curve must be fit to the data for large values of τ.

In table 11-II, the maximum and minimum conductivity values determined by the procedure are shown arranged in order of increasing depth. The conductivities that were determined for minimum and maximum initial-temperature estimates differ, on the average, by 50 percent. The more accurate conductivity measurements, which were made by using the heaters that surround the gradient sensors, resulted in values that lie within the ranges listed in table 11-III. The deduced conductivity values are considerably higher than the value obtained from measurements on returned lunar fines. The value for the returned lunar fines is approximately 2.2×10^{-5} W/cm-K at 250 K (ref. 11-12). The higher conductivity values that were obtained may be representative of fragmental regolith material in a more dense and compressed state than the surface fines.

In Situ Measurements

Six in situ conductivity measurements in mode 2, which is the low-conductivity mode, were conducted at the end of the first lunar night and during the first half of the second lunar day. The two heaters on the upper section of probe 2 were not turned on because the gradient bridges were off scale. The mode 2 measurements indicated the subsurface conductivity to be in the lower range of measurement and, in addition, showed that a substantial contact resistance exists between the borestem and the lunar material. A decision was made, therefore, not to run the mode 3 (high-conductivity mode) measurements at this time because of the possibility that the gradient sensors might reach temperatures potentially dangerous to the sensor calibration. Mode 3 measurements are planned at some future time after the effects of heater turnon are examined by using the conductivities determined from the mode 2 results.

Three of the conductivity measurements have been analyzed. Two of these measurements were obtained by the use of heaters on the lower section of probe 1, the section across which the best temperature gradient was determined. The third measurement was obtained by the use of the upper heater on the lower section of probe 2.

The interpretation of the response of the temperature-gradient sensor to heater turnon, in terms of the lunar conductivity, is accomplished by using a detailed finite-difference model (ref. 11-13). A simple analytical model of the gradient-sensor long-term ($t > 20$ hr) performance deduced from the experimental data and the finite-difference models will be briefly discussed in the following paragraphs.

The temperature increase as a function of time at a given heater-sensor location upon heater turnon depends on the quantity of heat generated and the rate at which the generated heat can diffuse outward from the heater source. This rate will depend on the thermal properties of the material that surrounds the source. The heat will propagate axially along the probe and radially from the probe to the drill casing, across the contact-resistance layer outside the casing, and into the lunar medium. Both radiative transfer and conductive transfer are involved in the dissipation of heat. Shortly after heater turnon, the rate of temperature increase at the gradient sensor will depend primarily on the thermal properties of the probe and the borestem in the immediate vicinity of the heater and on the resistive gaps between the probe and the borestem and between the borestem and the lunar material. As the near-sensor probe parts and the borestem temperature increases, a temperature drop is established across the resistive gaps. When this temperature difference builds to a relatively large value, heat will flow out from the borestem, across the contact-resistance gap, and into the medium; and the rate of temperature increase at the sensor will level off. At long times (times greater than 1000 min in this experiment), the temperature increase $\Delta v(t)$, measured at the sensor, closely fits a relation of the form

$$\Delta v(t) = C_1 \ln(t) + C_2 \qquad (11\text{-}6)$$

where C_1 and C_2 are constants that depend on the contact conductances H_1 and H_2 and the properties of the lunar material. The finite-difference thermal model of the probe in the lunar material shows the same long-term characteristics. This relationship has the same form as the long-term solution to the problem of a uniformly heated infinite cylinder (ref. 11-14). As in the case of the long-term solution for a cylinder, it has been determined from the finite-difference models that, at long times after heater turnon, the constant C_1 is almost solely a function of the conductivity of the surrounding material and the heat input. Thus, the slope $[\Delta v(t_2) - \Delta(t_1)]/\ln(t_2/t_1)$, for times greater than 1000 min, is a sensitive measure of the conductivity of the surrounding material for a constant heat input. Plots of temperature increase as a function of time for the three conductivity measurements are shown in figure 11-9 and compared with best-fitted theoretical models.

The magnitude of temperature increase at any time greater than 0.5 hr after heater turnon is very sensitive to the magnitude of the contact conductance H_2. The value of the contact conductance, however, has no detectable effect on the slope of the curve of Δv as a function of $\ln(t)$ at long times; therefore, the determination of a value for k can be made independently of H_2 by matching slopes at times greater than 1000 min. By using this value of k, a value for H_2 was determined by varying H_2 in the finite-difference models until the experimental curves of Δv as a function of t were bracketed within a small tolerance. Examples of models that bracket the experimental curves of the three in situ experiments are shown in figure 11-9.

A rather accurate determination of the conductivity of the lunar subsurface material that surrounds each heater location can be made. For example, at heater location H23 (the location of sensor TG22A), the long-time slope data could be bracketed by models of $k = 1.3 \times 10^{-4}$ and $k = 1.4 \times 10^{-4}$ W/cm-K. A linear interpolation between these models resulted in a value $k = 1.37(\pm 0.02) \times 10^{-4}$ W/cm-K. However, the assumption cannot be made that the models represent the physical situation this accurately; a value $k = 1.4(\pm 0.1) \times 10^{-4}$ W/cm-K would be more realistic. Further examination of the effects of the errors introduced by the assumptions about the model parameters (probe properties, heat-transfer linkages, etc.) must be made so the actual precision of the k values can be determined. From previous limited parametric studies, a range of ±10 percent should represent a maximum bound in the error of the k-value determinations.

The best determinations of conductivity values are listed in table 11-IV. As shown in the table, the conductivity determinations from the heater experiments fall within the range of the k predictions of the initial probe cooldown analyses and indicate a significant increase of conductivity with depth.

The contact-conductance values determined from the in situ measurements vary. The contact conductance probably corresponds to a thin zone around the borestem that is filled with lunar fines. If the assumption is made that these fines have a conductivity of 2×10^{-5} W/cm-K, which is similar to the conductivity of the surface fines, the widths of the disturbed zones would be 2.7, 2.0, and 1.3 mm for the locations of sensors TG22A, TG12A, and TG12B, respectively. The larger value of H_2 at the location of sensor TG12B might result from greater compaction of the fines, rather than a thinner zone. The thicker disrupted zone around probe 2 may have resulted from the longer period of drilling.

STEADY-STATE HEAT FLOW FROM THE LUNAR INTERIOR BELOW THE HADLEY RILLE SITE

The conductivity of the regolith is shown by the measurements returned from the experiment to be significantly variable with depth over the lower section of probe 1. To compute the heat flow from temperature differences over a finite depth interval, the thermal resistance must be known. The thermal resistance can be calculated from the equation

$$R_{x_1 - x_2} = \int_{x_1}^{x_2} \frac{dx}{k(x)} \qquad (11\text{-}7)$$

where x_1 and x_2 are the end points of the interval.

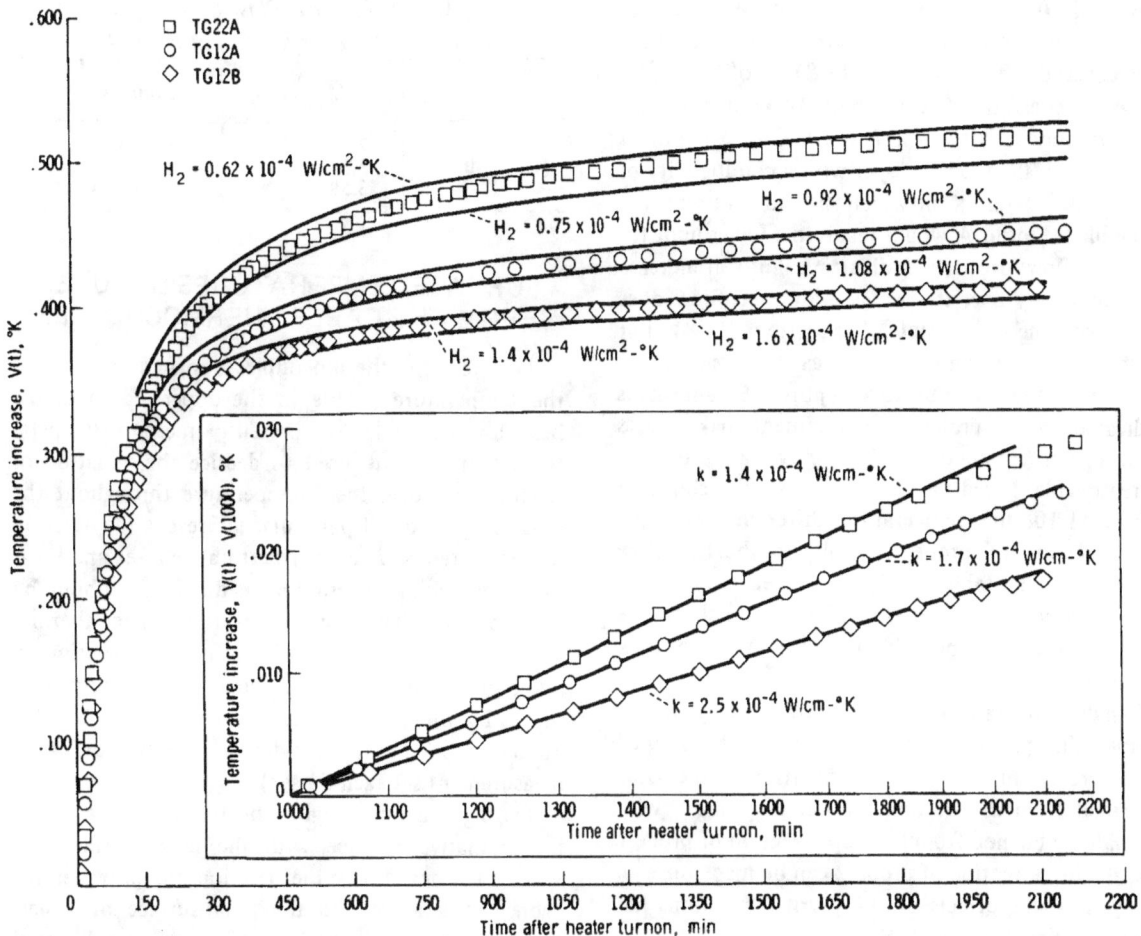

FIGURE 11-9.—Temperature increase as a function of time after heater turnon for heaters located at sensors TG22A, TG12A, and TG12B. Two computed models that closely bracket the data are shown for each of the three heater locations. Temperature increase as a function of time after 1000 min is shown in the inset on an expanded scale. The solid lines in the inset are best fitting computer models.

TABLE 11-IV.—*Conductivity Determinations From in Situ Experiments*

Heater sensor	Depth, cm	Thermal conductivity, k, W/cm-K	Contact conductance of the borestem, H_2, W/cm^2-K
TG22A	49	$1.40 (\pm 0.14) \times 10^{-4}$	0.7×10^{-4}
TG12A	91	$1.70 (\pm 0.17)$	1.0
TG12B	138	$2.50 (\pm 0.25)$	1.5

Thus, for the flux to be determined, k-value variation in the interval between the gradient sensors, which are located at depths of 91 and 138 cm, must be known. Accurate measurements of k were made only at the end points; however, a constraint can be applied on the variation with depth from the ratio of the temperature differences measured by the ring bridge and the gradient bridge. If the heat flow is uniform with depth, the constraint required by the ratio of temperature differences is

$$\frac{\int_{100}^{129} \frac{dx}{k(x)}}{\int_{91}^{138} \frac{dx}{k(x)}} = \frac{\Delta T_{\text{ring}}}{\Delta T_{\text{gradient}}} = 0.613 \quad (11\text{-}8)$$

Three possible conductivity profiles are shown in

figure 11-10. Profile B is based on the trend of conductivities from the cooldown curves and obeys the constraint of equation (11-8). Profile A also obeys the constraint of equation (11-8) but includes a uniform conductivity of 1.7 W/cm·K to a depth of 136 cm and, then, a thin layer with a conductivity of 2.5×10^{-4} W/cm·K in which the bottom sensor is embedded. Profile A would result in a lower limit for the heat flow; profile C indicates a uniform increase in conductivity over the probe section. Profile C does not obey the constraint of equation (11-8), but defines an upper limit for the heat-flow value. The trend of conductivity up to a depth of 50 cm that is indicated by the probe 2 measurement makes cases with higher conductivity than shown in profile C unreasonable. Based on these three profiles shown in figure 11-10, the temperature difference over the lower section of probe 1 results in the heat-flow values listed in table 11-V. The uncertainty of the conductivity measurements (±10 percent) should be considered as error bounds on each of the heat-flow values listed in table 11-V.

Analysis of data obtained during a full year will enable the previous determinations to be refined considerably. In addition, a comparison of the value obtained from the bottom section of probe 1 with the values obtained from the upper section of probe 1 and the lower section of probe 2 can be made once an analysis of the effects of the diurnal variations has been completed.

TABLE 11-V.—*Heat-Flow Data Obtained From the Lower Section of Probe 1*

Profile	Heat flow, W/cm²	Comment
A	2.99×10^{-6}	Lower limit
B	3.31	Best value
C	3.59	Upper limit

SURFACE TEMPERATURES DEDUCED FROM THE CABLE THERMOCOUPLES

Of the eight thermocouples designed to measure the temperature profile in the upper 1.5 m of the heat-flow borehole, six presently measure temperatures that may be used to deduce the variation of lunar-surface brightness temperature throughout the lunation period. Of particular interest is the determination of temperature during total eclipses and lunar nights, which are measurements difficult to obtain by Earth-based telescopic observation. The thermocouples in the cable of the heat-flow experiment, lying on or just above the lunar surface, provide a means by which these measurements can be obtained at a sampling rate previously unattainable (one measurement set each 54 sec).

During the lunar night, the thermocouples come into radiative balance with the lunar surface and space. To determine the relationship between the cable temperature and the lunar-surface brightness temperature, the heat balance for a small cylindrical cable element of radius a and length dl can be considered. The heat balance for such an element arbitrarily oriented above the lunar surface during the lunar night can be represented by

$$F_{C-M} 2\pi a dl \sigma \epsilon_{IR} \epsilon_M \sigma T_M^4 - 2\pi a dl \epsilon_C \sigma T_C^4 - V \rho c \frac{\partial T_C}{\partial t} = 0 \quad (11-9)$$

where the first term is the energy received by the cable element from the Moon per unit time, the second term is the energy lost from the cable element per unit time, and the third term is the energy required to change the temperature of the cable element per unit time; and where

F_{C-M} = view factor of the cable element to the lunar surface

σ = Stefan-Boltzmann constant (5.67×10^{-8} W-sec/m²·K)

ϵ_M = lunar-surface emittance

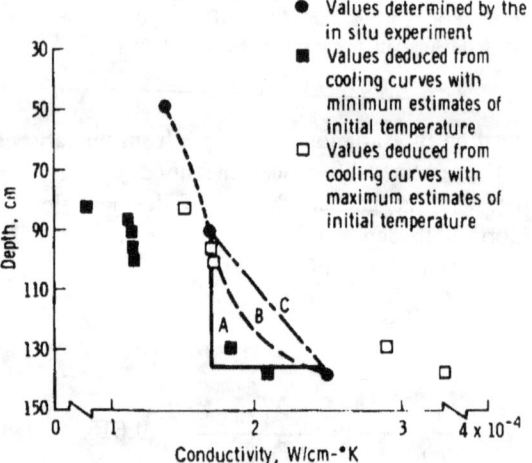

FIGURE 11-10.—Conductivity as a function of depth, with three possible conductivity profiles (A, B, and C) (table 11-V).

ϵ_C = cable-element emittance
α_{CIR} = cable-element infrared absorptance
V = cable-element volume
ρc = volumetric heat capacity of the cable element
T_M = lunar-surface brightness temperature
T_C = cable-element temperature

For $\epsilon_M = 1$ and a flat lunar surface, equation (11-9) reduces to

$$T_M^4 = \frac{1}{\alpha_{CIR}} \left(\frac{a\rho c}{\sigma} \frac{\partial T_C}{\partial t} + 2\epsilon_C T_C^4 \right) \quad (11\text{-}10)$$

The term $\frac{a\rho c}{\sigma} \frac{\partial T_C}{\partial t}$, which accounts for the thermal time constant of the cylinder, is retained only for eclipse calculations, because the constant is on the order of minutes.

The orientations of the thermocouples that are outside the borestems are unknown. The time at which a given thermocouple reaches the maximum temperature and the value of that maximum are strong functions of the orientation of the cable section in which the thermocouple is embedded. This effect, which is shown in figure 11-11, is a result primarily of the variation in the incidence angle of the solar radiation; therefore, lunar-surface brightness temperatures deduced during the lunar day are subject to error that results from the orientation uncertainties.

For the calculation of the lunar-surface brightness temperatures, the assumption was made that the view factors from the cable to space and from the cable to the surface are identical. The irregular horizon formed by the Apennine Mountains increases the effective view factor to the lunar surface, and the view factor to space is reduced correspondingly. Some of the thermocouples may be close enough to the surface so that local topographic irregularities affect the horizon seen by the cable. For a 10-percent increase in the effective view area to the topographic surfaces, the calculated surface brightness temperatures would be reduced by 2.25 percent during the lunar night.

A more serious error in lunar-night calculations results from the uncertainty in the values of the cable absorptance and emittance. For the temperatures given in figure 11-12, the emittance-to-absorptance ratio was assumed to be unity. A 20-percent increase in this ratio would result in an increase of 4.5 percent in the calculated value of the surface brightness temperature. A value of 0.97 was chosen for both the cable infrared emittance and absorptance, where the assumption was made that the sections of cable in which the thermocouples are embedded are covered by a significant amount of lunar-surface material. Photographs show that most of the cable areas are coated with lunar material.

The surface-brightness-temperature history for the first lunar night, as deduced from thermocouple temperatures, is shown in figure 11-12. A very rapid cooling of the surface is indicated for the first 80 hr after sunset; subsequently, the rate of cooling slows significantly. A cooling curve (ref. 11-15) based on astronomical observations at two different latitudes is shown for comparison. Two theoretical curves based on finite-difference calculations of thermal models of the lunar surface and subsurface are also shown. Curve A, which is derived from a model with a linear conductivity increase starting at a depth of 8 cm (inset, fig. 11-12), duplicates the rate of cooling for times greater than 80 hr, whereas curve B, which is based on a temperature-dependent-conductivity model (ref. 11-12), duplicates the earlier part of the observed curve. For both models, the heat capacity was defined by the following equation (from ref. 11-16)

$$c(T) = -0.034 T^{1/2} + 0.008 T - 0.0002 T^{3/2} \quad (11\text{-}11)$$

The assumed densities are 1.2 g/cm^3 for curve A and 1.0 g/cm^3 for curve B. The flattening of the observed curve could, in part, be a result of a significant increase in density in the upper several centimeters.

The temperature dependence of conductivity that is indicated by the cooling curves is in agreement with the large increase in mean temperature described previously. The substantial increase in conductivity and density with depth that is suggested by the flattening of the cooling curves is in agreement with earlier conclusions that were based on astronomical observations (e.g., ref. 11-16). The mechanical properties of the soil measured near the Apollo 15 ALSEP site revealed that the shear strength increased rapidly with depth in the upper 20 cm (sec. 7). This increase in shear strength probably is related to a near-surface density gradient.

Surface brightness temperatures during the umbral stage of the August 6 eclipse were deduced from the

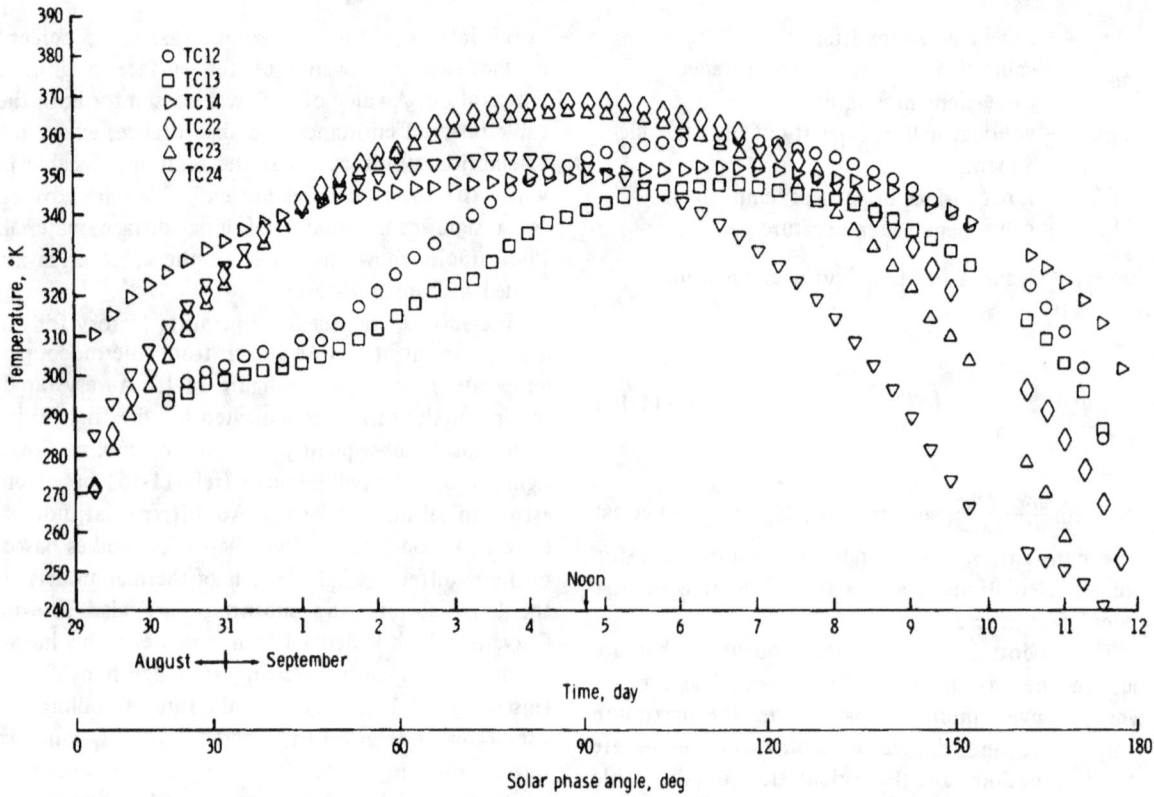

FIGURE 11-11.—Thermocouple temperatures as functions of time for the second lunar day. Thermocouples TC12 and TC13 measure the probe 1 cable temperatures; TC22, TC23, and TC24 measure the probe 2 cable temperatures; and TC14 measures the probe 1 borestem temperature.

thermocouple temperatures. For these calculations, the thermal time constant of the cable must be considered. The thermal mass of the cable per unit length in which the thermocouple is embedded was estimated by summing the properties of the 35 conductors. Uncertainties result from the fact that noise in the thermocouple measurements makes an accurate determination of the slope difficult. A sample of thermocouple data for the later part of the umbra was reduced by determining the rate of change of the cooling curve by graphical techniques. These very preliminary results indicate that the umbral temperatures reached at the lunar surface correspond well with the temperatures predicted by the theoretical curve based on the relationship between conductivity and temperature for lunar fines (ref. 11-12).

In summary, the surface-temperature data for lunar night and the umbral part of the August 6 eclipse support the conclusion that the upper few centimeters of the lunar-surface material have a thermal conductivity-versus-temperature relationship similar to that found for samples of lunar fines measured in the laboratory. The lunar nighttime observations reveal a substantial conductivity gradient with depth that probably results from increasing material density with depth.

DISCUSSION OF HEAT-FLOW-EXPERIMENT RESULTS

Local Topographic Effects

The heat-flow determinations at the Hadley Rille site are susceptible to a number of disturbances. Corrections for some of these disturbances (such as the thermal perturbations caused by visible topographic features) can readily be made with sufficient accuracy, but other disturbances may result from refraction associated with sloping interfaces between

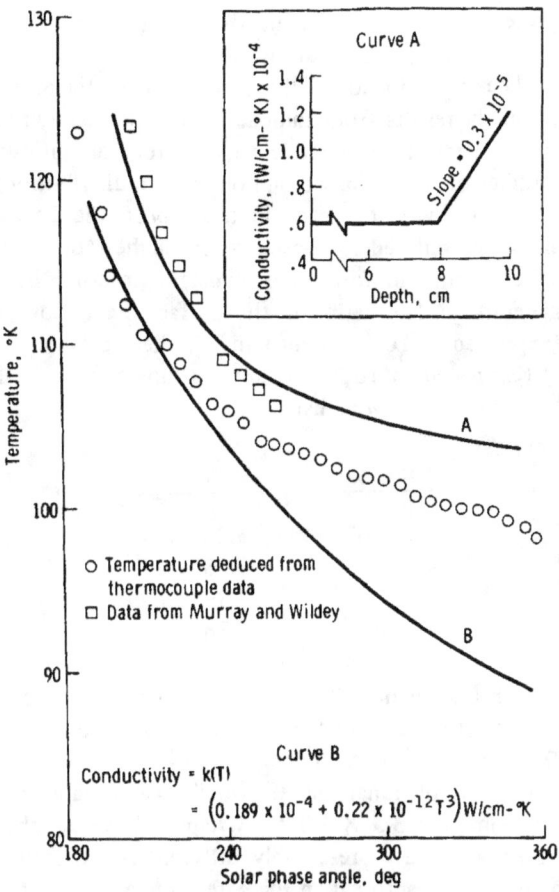

FIGURE 11-12.—Surface cooling as a function of time for the first lunar night, as deduced from thermocouple measurements. Two theoretical curves are given as references. Curve A is the expected curve for a conductivity increase with depth, starting at a depth of 8 cm; curve B is the theoretical curve for a temperature-dependent conductivity. The data from reference 11-16 are included for comparison. Conductivity as a function of depth for the curve A data is shown in the inset.

materials of differing thermal conductivity. Such interfaces, if present at all, are hidden in the lunar interior, and only a qualitative discussion of the effects can be given.

Probe 1, from which data were used for the preliminary determination of heat flow, is located in a small crater that has been almost filled and nearly obliterated by later bombardment. The topography associated with this old crater is so subdued that no correction for the topography is required. However, the possibility remains that the event that produced this crater also locally altered both the thickness and the physical properties of the regolith. The truth of this hypothesis cannot be verified because no observational evidence is available; however, after the data from probe 2 have been analyzed, a comparison of the two heat-flow values may provide more information.

The two most conspicuous and important topographic features near the heat-flow-experiment site are Hadley Rille and the Apennine Front. The topographic effect of the rille was calculated by fitting the rille profile with a two-dimensional Lees-type valley (refs. 11-17 and 11-18). This procedure results in a correction of 4.5 percent, and further allowance for the Elbow, where the rille abruptly changes direction, reduces the correction to 3.5 percent.

A second effect of the topography around the Hadley Rille site is that the surface in the area is shaded during part of the day; consequently, the average surface temperature is lower than flat portions of the lunar surface. This local cold-spot effect has been estimated quantitatively in two ways. The radiation balance for a point halfway down the rille wall (including factors for incoming solar radiation and radiation from the opposite wall), at appropriate times of the lunar day, was calculated and used to derive the mean temperature in the rille. Alternatively, the temperature at the vertex of the rille, which was assumed to have a symmetrical V-shape, was calculated from the solar input alone; for this geometry, the vertex sees neither wall. The temperature was assumed to vary linearly from the vertex to the top of the rille. The radiation-balance method resulted in a correction of 5 to 10 percent, and the vertex method resulted in a correction of 10 to 20 percent. The former value is considered to be the more reliable, mainly because the profile of the rille does not particularly resemble a V-shape. A correction of 10 percent is considered to be reasonably close to the upper limit of the effect that results from the cold rille wall.

The thermal effect of the Apennine Front was estimated in a preliminary way from the two-dimensional-slope method of Lachenbruch (ref. 11-19). By assuming two-dimensional symmetry, the curves given by Lachenbruch indicate that the correction, which is negative in this case, is 10 percent at a maximum. Actually, the heat-flow-experiment site is in an embayment in the Apennine Front, which invalidates the two-dimensional approximation and reduces the correction. A correction of 5 percent would be more realistic.

In summary, the topographic effect of Hadley Rille is approximately canceled by the effect of the Apennine Front, which leaves only the cold-spot effect of the rille as a remaining correction resulting from the visible topography. This latter correction is the largest correction, in any case, and seems most likely to approximate 10 percent of the measured heat flow.

Implications of the Results

In this section, the view was adopted that the heat flow observed at the Hadley Rille site is representative of the moonwide value, in spite of the possibilities of local and regional disturbances and large-scale variations in the lunar heat flow. Thus, the measurement is considered at face value, with full realization that future measurements may produce major changes in the conclusions. Because of the preliminary nature of the results, simple models were used for this report. Only the linear equation of heat conduction was considered, for which the thermal diffusivity was constant.

The value that is considered to be an upper limit to the heat flow resulting from the initial heat is calculated by assuming that, at the end of a convective stage in the early history of the Moon, temperatures throughout the Moon lay along the solidus for lunar basalt (ref. 11-20). After 3×10^9 yr, the heat flow resulting from these very high initial temperatures is in the range of 0.2×10^{-6} to 0.4×10^{-6} W/cm^2. If, as suggested by the greater ages of all known lunar rocks, partial melting throughout the Moon took place earlier, still lower values of heat flow would be associated with the initial heat. It is likely that the assumed initial temperatures are too high; however, by revising the assumed temperatures downward, the flux from the initial heat is further reduced. The initial heat contributes little to the present lunar heat flow, and the contribution can be neglected for present purposes.

The major portion of the heat flow from the Moon probably results from radioactive heat generation in the interior. It is possible to construct an infinite number of models based on nonuniform distributions of radioactivity; however, in this report, the discussion was confined to consideration of the Moon as a sphere with uniform and constant internal heat generation. The ratio of the surface heat flow q (expressed in 10^{-6} W/cm^2) and the heat production Q (expressed in 10^{-13} W/cm^3) is shown for several times in the following list.

Time, yr	q/Q, cm
3×10^9	3.22×10^7
4	3.58
5	3.89
Infinite	5.78

With a lunar heat flow of 3.3×10^{-6} W/cm^2, the value of Q must be in the range of 0.57×10^{-13} to 1.0×10^{-13} W/cm^3. This number is far lower than the heat production of lunar basalt, which has a value of approximately 3.5×10^{-13} W/cm^3. However, the basaltic rocks are presumably differentiates that are far more radioactive than the parent material. On the other hand, ordinary chondrite and type I carbonaceous chondrites generate heat at rates of approximately 0.17×10^{-13} and 0.22×10^{-13} W/cm^3, respectively. The respective average rates of heat generation over the last 4.5×10^9 yr are 0.61×10^{-13} and 0.67×10^{-13} W/cm^3. Even these last figures are barely sufficient to provide the necessary flux. The conclusion is that, if the observed lunar heat flow originates from radioactivity, then the Moon must be more radioactive than the classes of meteorites that have formed the basis of Earth and Moon models in the past.

APPENDIX

Equilibration of an Infinitely Long Cylinder in a Homogeneous Medium by Conduction

In this appendix, the theory of the cooling of a cylinder by conduction is reviewed and applied to the heat-flow probe. Cylindrically shaped probes are commonly used in making geothermal measurements, and, as a consequence, the theory of heat flow in cylindrical coordinates has been thoroughly investigated (refs. 11-11, 11-14, and 11-21). The investigation of the effect of a finite contact resistance between a cylinder and the surrounding medium (ref. 11-14) is particularly applicable to problems of the cooling of the lunar probes. In this appendix, the solution (ref. 11-14) has been extended to a somewhat more complex model that includes a solid cylinder inside a thin-walled concentric cylinder which, in turn, is surrounded by an infinite medium with a conductivity k. The cylinders are assumed to be perfect conductors; that is, each cylinder is isothermal, which is very nearly true in the case of the heat-flow probe. Contact resistances exist at the two cylindrical boundaries. In this model, the inner cylinder is an idealization of the heat-flow probe, which is radiatively coupled to the borestem (the concentric cylinder). The borestem, in turn, loses heat by conduction to the surrounding regolith, of conductivity k, through a thin zone of lunar material that has been disturbed by drilling and, hence, has a different conductivity. The initial temperature of the surrounding infinite medium is zero; and, initially, the two inner cylinders have a temperature v_0.

By defining the dimensionless parameter τ as $\kappa t/b^2$ (where κ is the thermal diffusivity of the surrounding infinite medium, t is time in seconds, and b is the radius of the outermost cylinder), the ratio of the inner-cylinder temperature $v(\tau)$ to the inner-cylinder initial temperature v_0 is given by

where

$\alpha = 2\pi b^2 \rho c/(S_1 + S_2)$
$h = k/bH_2$
$A = S_2 \gamma \kappa/b^2(S_1 + S_2)$
$B = \gamma \kappa/b^2$
$\gamma = S_1/2\pi a H_1$
a = radius of the inner cylinder (heat-flow probe) in centimeters
b = outer radius of the concentric outer cylinder (borestem) in centimeters
S_1, S_2 = thermal heat capacities per unit length of the inner and outer cylinders, respectively, W-sec/cm-K
H_1, H_2 = contact conductances at the inner- and outer-cylinder boundaries, respectively, W/cm²-K
ρ = density of the surrounding material, g/cm³
c = heat capacity per unit mass of the surrounding material, W-sec/g-K
k = thermal conductivity of the surrounding material, W/cm-K
κ = thermal diffusivity ($k/\rho c$) of the surrounding material, cm²/sec

Also, $J_0(u)$, $J_1(u)$, $Y_0(u)$, and $Y_1(u)$ are the zero- and first-order Bessel functions of the first and second kinds.

By following the method described in reference 11-11, the expression on the right-hand side of equation (11-12) can be redesignated $f(A,h,\alpha,\tau)$; therefore, equation (11-12) can be written as $v(\tau)/v_0 = f(A,h,\alpha,\tau)$. A more complete derivation of equation (11-12) and the resulting tabulated values will be published at a later date. Two typical plots of $f(A,h,\alpha,\tau)$ are shown in figure 11-8. The values of k

$$\frac{v(\tau)}{v_0} = \int_0^\infty \frac{(1 - Au^2)\exp(-u^2\tau)\,du}{u\left(\{u(1-Au^2)J_0(u) - [hu^2(1-Au^2) - \alpha(1-Bu^2)]J_1(u)\}^2 - \{u(1-Au^2)Y_0(u) - [hu^2(1-Au^2) - \alpha(1-Bu^2)]Y_1(u)\}^2\right)}$$

(11-12)

and H_2 that were used for computing the theoretical curves are given in table 11-II. Other parameters are $\alpha = 3.33$, $H_1 = 3.25 \times 10^{-4}$ W/cm^2·K, $\gamma = 445$ sec, and a and b are 0.95 and 1.259 cm, respectively.

By using asymptotic values of the Bessel functions for large values of τ (after a method outlined in ref. 11-21), a solution that is valid for long times can be derived as follows.

$$\frac{(v)}{t_0} = \frac{1}{2\alpha\tau} - \frac{1}{\ln a^2\tau^2}\left\{4\pi\gamma k\left[\frac{2S_1 \cdot S_2}{(S_1 \cdot S_2)^2}\right] + \ln 2 + \alpha - 2\ln\left(\frac{b}{a}\right)\right\} \quad (11\text{-}13)$$

where ln $(S) = 0.5772$ (Euler's constant). The long-term solution exhibits some interesting features of the function $f(A,h,\alpha,\tau)$. At long times ($\tau > 20$), the function becomes nearly equal to $(2\alpha\tau)^{-1}$, or $(4k\pi t/S_1 + S_2)^{-1}$, which is independent of ρc. A log-log plot of $f(A,h,\alpha,\tau)$ is shown in figure 11-8; for long times, the function describes a straight line with a slope of very nearly -1. This property of the function enables the temperature histories to be extrapolated to equilibrium values by the procedure described in the text of this section and a value for k to be determined independently of ρc from long-time portions of the equilibration curves. In addition, as shown by equation (11-13), the contact conductances H_1 and H_2, which are contained in the constants γ and h, become less important with time (because they are multiplied by t^{-2}).

REFERENCES

11-1. Wesselink, A.J.: Heat Conductivity and Nature of the Lunar Surface Material. Bull. of the Astron. Inst. Neth., vol. 10, 1948, pp. 351-363.

11-2. Birkebak, Richard C.; Cremers, Clifford J.; and Dawson, J.P.: Thermal Radiation Properties and Thermal Conductivity of Lunar Materials. Science, vol. 167, no. 3918, Jan. 30, 1970, pp. 724-726.

11-3. Watson, K.I.: Thermal Conductivity Measurements of Selected Silicate Powders in Vacuum from 150-350 K, II. An Interpretation of the Moon's Eclipse and Lunation Cooling Curve as Observed Through the Earth's Atmosphere from 8-14 Microns. Thesis, Calif. Inst. Technol., 1964.

11-4. Linsky, Jeffrey L.: Models of the Lunar Surface Including Temperature Dependent Thermal Properties. Icarus, vol. 5, 1966, pp. 606-634.

11-5. Lee, William H.K.; and Uyeda, Seiya: Review of Heat Flow Data. Terrestrial Heat Flow, ch. 6, William H.K. Lee, ed., Am. Geophys. Union of Nat'l. Acad. of Sci.-Nat'l Res. Council (Washington), 1965, pp. 87-190.

11-6. Baldwin, J.E.: Thermal Radiation from the Moon and the Heat Flow Through the Lunar Surface. Royal Astronomical Society Monthly Notice 122, 1961, pp. 513-522.

11-7. Troitskiy, V.S.; and Tikhonova, T.V.: Thermal Radiation from the Moon and the Physical Properties of Its Upper Mantle. NASA TT F-13455, 1971.

11-8. Fricker, Peter E.; Reynolds, Ray T.; and Summers, Audrey L.: On the Thermal History of the Moon. J. Geophys. Res., vol. 72, no. 10, May 15, 1967, pp. 2649-2663.

11-9. Cremers, C.J.; and Birkebak, R.C.: Thermal Conductivity of Fines from Apollo 12. Proceedings of the Second Lunar Science Conference, vol. 3, A.A. Levinson, ed., MIT Press (Cambridge, Mass.), 1971, pp. 211-216.

11-10. Robie, Richard A.; Hemingway, Bruce S.; and Wilson, William H.: Specific Heats of Lunar Surface Materials from 90 to 350 Degrees Kelvin. Science, vol. 167, no. 3918, Jan. 30, 1970, pp. 749-750.

11-11. Bullard, E.C.: The Flow of Heat Through the Floor of the Atlantic Ocean. Proc. Royal Soc. London, A, vol. 222, 1954, pp. 408-429.

11-12. Cremers, C.J.; Birkebak, R.C.; and White, J.E.: Lunar Surface Temperatures at Tranquility Base. AIAA paper 71-79, AIAA Ninth Aerospace Sciences Meeting (New York, N.Y.), Jan. 25-27, 1971.

11-13. Langseth, M.D., Jr.; Drake, E.I.; and Nathanson, D.: Development of an In Situ Thermal Conductivity Measurement for the Lunar Heat Flow Experiment. Lunar Thermal Characteristics, John Luca, ed., AIAA Lunar Progress Series, 1971.

11-14. Jaeger, J.C.: Conduction of Heat in an Infinite Region Bounded Internally by a Circular Cylinder of a Perfect Conductor. Australian J. Phys., vol. 9, 1956, pp. 167-179.

11-15. Murray, Bruce C.; and Wildey, Robert L.: Surface Temperature Variations During the Lunar Nighttime. Astrophys. J., vol. 139, no. 2, Feb. 15, 1964, pp. 734-750.

11-16. Winter, D.F.; and Saari, J.M.: A Particulate Thermophysical Model of the Lunar Soil. Astrophys. J., vol. 156, no. 3, June 1969, pp. 1135-1151.

11-17. Birch, Francis: Flow of Heat in the Front Range, Colorado. Bull. Geol. Soc. Am., vol. 61, no. 6, June 1950, pp. 567-630.

11-18. Jaeger, J.C.; and Sass, J.H.: Lees's Topographic Correction in Heat Flow and the Geothermal Flux in Tasmania. Geofis. Pura Appl., vol. 54, 1962, pp. 53-63.

11-19. Lachenbruch, A.H.: Rapid Estimation of the Topographic Disturbance to Superficial Thermal Gradients. Rev. Geophys., vol. 6, no. 3, Aug. 1968, pp. 365-400.

11-20. Ringwood, A.E.; and Essene, E.: Petrogenesis of Apollo 11 Basalts, Internal Composition and Origin of the Moon. Proceedings of the Apollo 11 Lunar Science Conference, vol. 1, A.A. Levinson, ed., Pergamon Press (New York, 1970, pp. 769-799.

11-21. Blackwell, J.H.: A Transient-Flow Method for Determination of Thermal Constants of Insulating Materials in Bulk. J. Appl. Phys., vol. 25, 1955, pp. 137-144.

ACKNOWLEDGMENTS

The authors wish to express their sincere appreciation to the many individuals whose enthusiasm and efforts have made possible the successful undertaking of this experiment. In particular, we wish to thank the personnel of the five corporations that developed, fabricated, and tested the essential instrumentation: Bendix Corporation of Ann Arbor, Michigan; Arthur D. Little, Inc., of Cambridge, Massachusetts; The Data Systems Division of Gulton Industries, Inc., of Albuquerque, New Mexico; Martin-Marietta Corporation of Denver, Colorado; and Rosemount Engineering Company of Minneapolis, Minnesota. The advice, encouragement, and active support of M. Ewing and the efforts of H. Gibbon, K. Peters, and R. Perry, all of the Lamont-Doherty Geological Observatory, were essential to the success of the program. The support of G. Simmons of the NASA Manned Spacecraft Center throughout the development of the experiment also is appreciated.

12. Suprathermal Ion Detector Experiment (Lunar Ionosphere Detector)

H. Kent Hills,[a] Jürg C. Meister,[a] Richard R. Vondrak,[a] and John W. Freeman, Jr.[a][†]

The suprathermal ion detector experiment (SIDE), part of the Apollo lunar surface experiments package (ALSEP), is designed to provide information on the energy and mass spectra of the positive ions close to the lunar surface (the lunar exosphere). These ions can be classified into two groups: (1) those that result from the ionization of gases generated on the Moon by natural and manmade sources, and (2) those that arrive from sources beyond the near-Moon environment.

The ions generated on the Moon are of intense interest because possible sources of these ions are sporadic outgassing from volcanic or seismic activity, gases from a residual primordial atmosphere of heavy gases, evaporation of solar-wind gases accreted on the lunar surface, and exhaust gases from the lunar module (LM) descent and ascent engines and from the astronauts' portable life-support equipment.

An example of the significant results of this experiment is the recent report[1] of the detection of water vapor in the lunar exosphere by the Apollo 14 SIDE. Water from some unknown depth below the surface is believed to be liberated by seismic activity, after which it vaporizes instantly upon exposure to the vacuum of the lunar atmosphere. The vapor is then dispersed over a wide area; some fraction of it becomes ionized and is subsequently detected by the SIDE.

In addition, evidence for the operation of a prompt ionization and acceleration mechanism operating in the lunar exosphere and a preliminary measurement of the decay time for the heavier components of the Apollo exhaust gases have been presented in reference 12-1, based on data from the Apollo 12 and 14 SIDE instruments. Electric and magnetic fields near the lunar surface can be studied by observing their effects on the motions and energies of the ions after they are generated. The network of three SIDE instruments (Apollo 12, 14, and 15) now operating on the Moon allows more precise determination of the dimensions and motions of ion clouds moving across the lunar surface. In this section, the instruments at these different sites are used to follow the motions of ions apparently resulting from two meteoroid-impact events, from the LM-ascent-stage impact, and from short-duration events occurring during the lunar night.

The ions from distant sources include those from the solar wind and also those which are part of the magnetosphere of the Earth. The magnetic field of the Earth deflects the incoming solar-wind ions so that they do not reach the surface. Instead, these ions undergo complex motions in the magnetosphere, which surrounds the Earth and extends to great distances in the direction away from the Sun (the magnetospheric tail). Because the Moon does not have a strong magnetic field, the solar wind can impinge directly on the lunar surface. This plasma interaction between the solar wind and the solid Moon can, therefore, be studied by means of the instrumentation on the lunar surface. In addition, the motions of ions in the magnetosphere can be investigated during those periods when the Moon passes through the magnetospheric tail of the Earth. In this report, the highly directional flow of energetic ions down the magnetosheath (or boundary region of the magneto-

---

[a]Rice University.
[†]Principal investigator.

[1]J. W. Freeman, Jr., paper presented at the Lunar Geophysics Conference, Houston, Tex., Oct. 18-21, 1971.

sphere) is described by using the three different look directions of the three SIDE instruments. The effects of the interaction of the LM ascent-engine exhaust with the magnetosheath ions observed at the Apollo 15 site are also discussed.

The principal contributions to the SIDE scientific objectives reported in this paper are as follows.

(1) Multiple-site observations of the apparent movement of an ion source presumed to be a neutral gas cloud resulting from a meteoroid impact

(2) Observation of the complicated ion event related to the Apollo 15 LM impact

(3) Observation of a sharp decrease in the magnetosheath fluxes at the time of LM lift-off

(4) Three-directional observations of the 500- to 1000-eV ions streaming down the magnetosheath

(5) Further observations of short-duration energetic-ion events occurring during the lunar night

INSTRUMENT

The Apollo 15 SIDE instrument is basically identical to those flown on the Apollo 12 (ref. 12-2) and 14 (ref. 12-3) missions. The only major difference is in the mass ranges covered by the three instruments. However, the Apollo 15 instrument is completely described herein.

Description

The SIDE consists of two positive-ion detectors. The first of these, the mass analyzer detector, is provided with a Wien velocity filter (crossed electric and magnetic fields) and a curved-plate electrostatic energy-per-unit-charge filter in tandem in the ion flight path. The requirement that the detected ion must pass through both filters allows a determination of the mass per unit charge. The ion sensor itself is a channel electron multiplier operated as an ion counter that yields saturated pulses for each input ion. The second detector, the total ion detector, uses only a curved-plate electrostatic energy-per-unit-charge filter. Again, the ion sensor itself is a channel electron multiplier operated as an ion counter. Both channel electron multipliers are biased with the input ends at -3.5 kV, thereby providing a postanalysis acceleration to boost the positive-ion energies to yield high detection efficiencies. The general detector concept is illustrated in figure 12-1; figure 12-2 is a

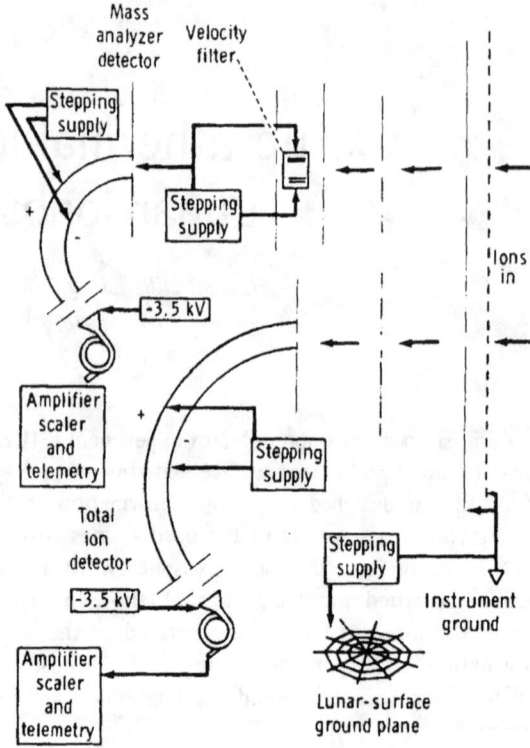

FIGURE 12-1.—Schematic diagram of the SIDE.

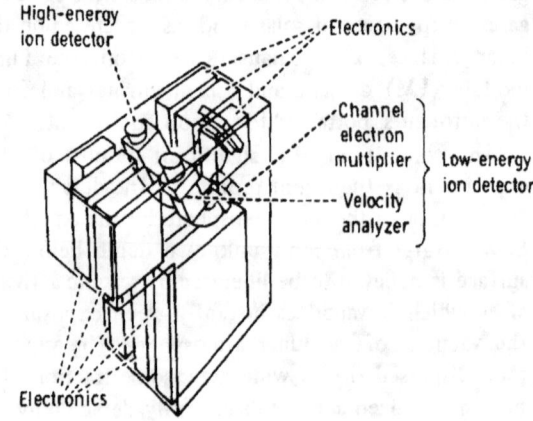

FIGURE 12-2.—Cutaway drawing showing the interior of the SIDE.

cutaway drawing that illustrates the location of the filter elements and the channel electron multipliers.

A primary objective of the experiment is to provide a measurement of the approximate mass-per-unit-charge spectrum of the positive ions near the lunar surface as a function of energy for ions from approximately 50 eV down to near-thermal energies.

Therefore, the mass analyzer detector measures mass spectra at six energy levels: 48.6, 16.2, 5.4, 1.8, 0.6, and 0.2 eV. Only the two highest energy levels of the mass analyzer detector were well calibrated in the laboratory, although data from the other levels are still useful. For the Apollo 15 mass analyzer detector, 20 mass channels span the mass spectrum from 1 to approximately 90 atomic-mass units per charge (amu/Q). The total ion detector measures the differential positive-ion energy spectrum (regardless of mass) from 3500 eV down to 10 eV in 20 energy steps.

To compensate for the possibly large (tens of volts) lunar-surface electric potential, a wire screen is deployed on the lunar surface beneath the SIDE. This screen is connected to one side of a stepped voltage supply, the other side of which is connected to the internal ground of the detector and to a grounded grid mounted immediately above the instrument and in front of the ion entrance apertures (fig. 12-1). The stepped voltage is advanced only after a complete energy and mass scan of the mass analyzer detector (i.e., every 2.58 min). The voltage supply is programed to step through the following voltages: 0, 0.6, 1.2, 1.8, 2.4, 3.6, 5.4, 7.8, 10.2, 16.2, 19.8, 27.6, 0, -0.6, -1.2, -1.8, -2.4, -3.6, -5.4, -7.8, -10.2, -16.2, -19.8, and -27.6. This stepped supply and the ground screen may function in either of two ways. If the lunar-surface potential is large and positive, the stepped supply, when on the appropriate step, may counteract the effect of the lunar-surface potential and thereby allow low-energy ions to reach the instrument with their intrinsic energies. However, if the lunar-surface potential is near zero, then on those voltage steps that match or nearly match the energy level of the mass analyzer detector or the total ion detector (1.2, 5.4, etc.), thermal ions may be accelerated into the SIDE at energies optimum for detection. The success of this method depends on the Debye length and on the extent to which the ground-screen potential approximates that of the lunar surface. It is not yet possible to assess either of these factors; however, the data from the Apollo 12 and 14 ALSEP instruments indicate that the ground-screen voltage often has little influence on the response of the instrument to the incoming ions.

The SIDE is shown deployed on the lunar surface in figure 12-3. The experiment is deployed approximately 15 m northeast of the ALSEP central station. The top surface stands 0.5 m above the lunar

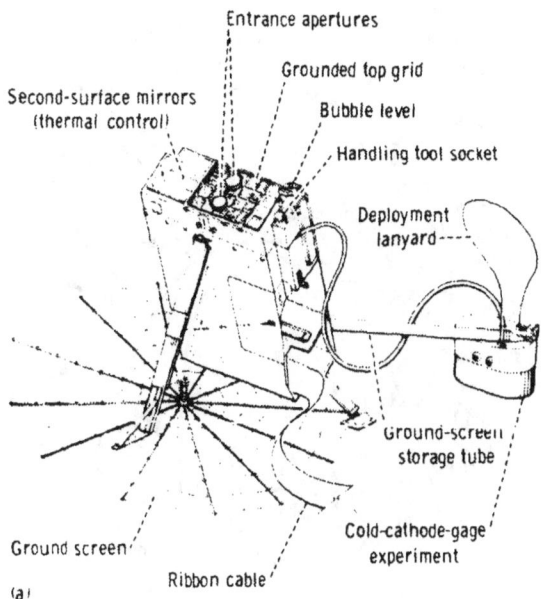

FIGURE 12-3.—The SIDE as deployed on the Moon. (a) External diagram. (b) NASA photograph AS15-86-11596.

surface. The instrument is tilted 26° from vertical toward the south so that the sensor look directions include the ecliptic plane; the look axes are directed 15° to the east. The field of view of each sensor is roughly a square solid angle, 6° on a side. The sensitivities of the total ion detector and of the mass analyzer detector are approximately 5×10^{17} and 10^{17} counts/sec/A of entering ion flux, respectively. The look directions of the Apollo 12, 14, and 15

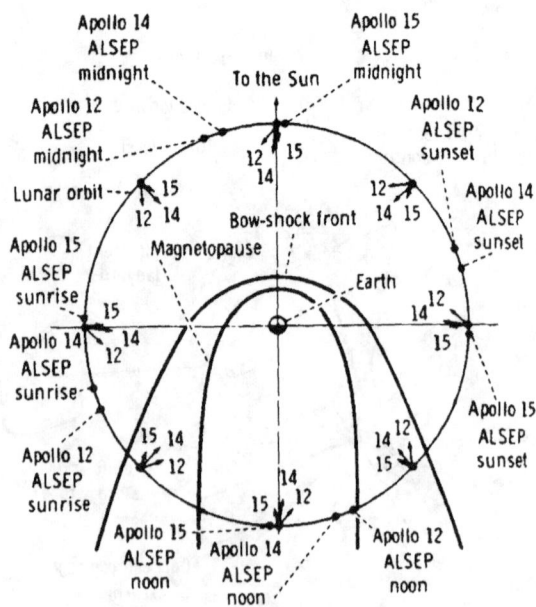

FIGURE 12-4.—The look directions of the Apollo 12, 14, and 15 SIDE instruments at various points along the lunar orbit. (The diameter of the Earth is not drawn to scale.)

instruments are shown in figure 12-4 in an Earth-Sun coordinate system at various points along the lunar orbit. As shown in the figure, the three instruments have look directions that cover a wide range of angles and thus allow study of the directional characteristics of the fluxes of ions streaming down the magnetosheath region into the tail of the magnetosphere of the Earth. Furthermore, it is possible to monitor the ion fluxes on both sides of the magnetospheric tail to check for asymmetry in the flow patterns.

Performance

At the time of preparation of this report, the operation of the SIDE and the associated cold-cathode-gage-experiment (CCGE) electronics continued to be excellent; all temperatures and voltages were nominal. The high voltages within the instrument have been commanded off for the periods of higher instrument temperature (up to ~353 K), centered on local noon, to allow the instrument to outgas without danger to the electronic components. The high voltages were not operated when the instrument temperature was greater than 310 K on the lunar day of deployment. They were operated the following lunar day up to 323 K in the morning and from 328 K down in the afternoon; no problems arose. Present plans are to increase the operating temperature limit by 10 K each successive lunar day until full-time operation is reached. The background counting rates have been quite low, well under 1 count/sec even at an internal temperature of 328 K.

RESULTS

Apollo 15 LM Ascent

The Apollo 15 SIDE measured the differential fluxes of magnetosheath ions during the ascent of the Apollo 15 LM. The expanding cloud of LM exhaust gas caused large changes in these fluxes. Before ascent, the magnetosheath ion energy spectrum was generally peaked near 1 keV, with fluxes at 3.5 keV smaller by a factor of 100 and negligible fluxes of ions with energy below 100 eV. As shown in figure 12-5, after LM ascent, the magnetosheath ion fluxes were abruptly reduced by as much as a factor of 100, and the differential energy spectrum was greatly distorted. The integral fluxes between 17:12 and 17:16 G.m.t. are probably unreliable because the fluxes are changing significantly in the time required to accumulate an integral spectrum measurement (13.3 sec). Consequently, the flux reduction and spectrum distortion are best seen in the differential flux data. Full recovery of the fluxes to preascent intensities was not attained until approximately 8 min after ascent. For approximately 5 min (starting 2.5 min after ascent), the mass analyzer detected fluxes of 48.6-eV ions with mass (amu/Q) in the range that includes helium. Masses heavier than 6 amu/Q were not being sampled at that time because of operation in the high-time-resolution mode for the CCGE.

The reduction in the magnetosheath ion fluxes may be due to the LM exhaust gas stopping the ions by charge exchange and collisional deenergization before they reach the SIDE. Another possibility is that the LM exhaust gas distorts the magnetosheath magnetic field so that the ions are deflected from the SIDE look direction. Preliminary range-energy calculations indicate that the mass of the LM exhaust gas was probably sufficient to stop the ions. The recovery time may thus indicate the time required for the local dispersal of the LM exhaust gas.

Simultaneously with the preceding events, the Apollo 14 SIDE also measured the fluxes of magnetosheath ions. No unusual changes were detected in the time period that included the LM

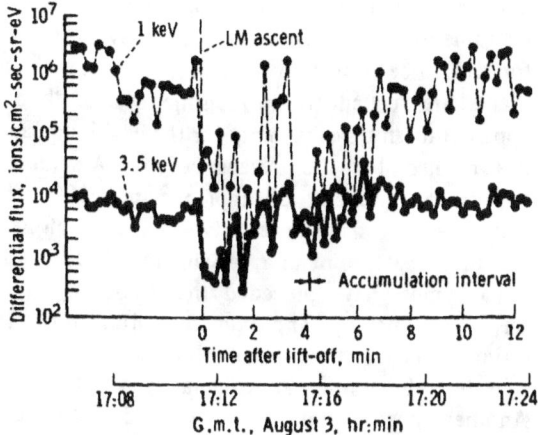

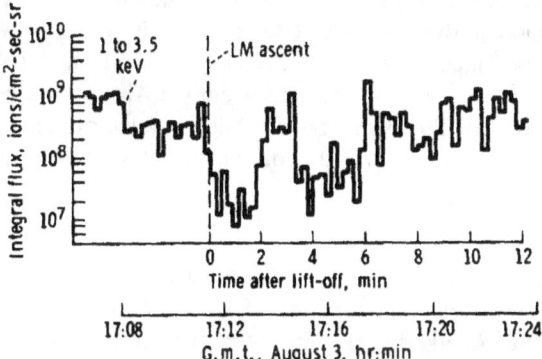

FIGURE 12-5.—Apollo 15 ALSEP differential and integral fluxes of magnetosheath ions during the ascent of the Apollo 15 LM. Greenwich mean time and minutes after ascent are shown. The integral flux is obtained from the 11 energy channels, which cover the range 1 to 3.5 keV. The accumulation interval of the differential flux data points is 1.13 sec (or 1/11 of the integral flux-accumulation interval). The separation between successive differential data points is 13.3 sec.

ascent. This fact confirms the theory that the events at the Apollo 15 site shown in figure 12-5 are local effects produced by the LM exhaust gas and do not result from large-scale temporal changes in the magnetosheath ions. A detailed analysis of this event may provide further insight into the behavior of gas clouds on the lunar surface.

Impact Events

On August 26, 1971, an ion event apparently caused by meteoroid impact was detected by both the Apollo 15 and 14 SIDE instruments. A time delay between the observations indicates an apparent movement of the source. At 22:14 G.m.t., the total ion detector of the Apollo 15 SIDE began detecting ions with energies between 3000 and 3500 eV. The event lasted approximately 130 sec, and no significant counts were recorded before or after this period.

At 22:39 G.m.t., after a delay of nearly 26 min, the total ion detector at the Apollo 14 site began to detect high-energy ions. During a period of 560 sec, ions with energies between 2500 and 3500 eV were detected. The ion observations at both sites are given in figure 12-6. The similar high energies recorded, which are very unusual for this time of the lunar cycle, indicate that the events observed at the two different sites probably had a common source. The SIDE at the Apollo 12 site, 180 km to the west of the one at the Apollo 14 site, did not observe any significant counts during the entire period in question, possibly because of the different look direction. The Apollo 12 instrument is looking 15° to the west of the local vertical, whereas those of Apollo 14 and 15 are looking 15° to the east. This orientation would imply that the ion fluxes observed were directional and generally pointed to the west. It is also possible that the source of the ions moved through a region from which the ions could not reach the Apollo 12 SIDE detector.

The Apollo 14 and 15 passive seismic experiments recorded a seismic event of impact character at approximately 21:00 G.m.t.[2] The impact point was estimated to be within 1000 km of both the Apollo 14 and 15 ALSEP sites.

Despite the long delay of approximately 74 min between the onset of the seismic signal and that of the energetic ions, the two events may be related. At the Apollo 15 site, the time was approximately 31 hr before sunrise; the sunlight terminator was 440 km east of the instrument, and the distance to the assumed average solar wind overhead was 14 km. The seismic data place the impact point on the night hemisphere of the Moon. Therefore, for the impact-generated gas atoms to be ionized, the cloud had to expand into either the solar wind or the sunlight. A more precise determination of the impact point on the basis of seismic data will allow an estimate of the time required for the gas cloud to arrive at the ionization region. Once ionized, the particles would have to travel thousands of kilometers along the interplanetary electric field to attain the observed energies, assuming that the $\mathbf{V} \times \mathbf{B}$ solar-wind-induced

---

[2]G. V. Latham, private communication, Sept. 13, 1971.

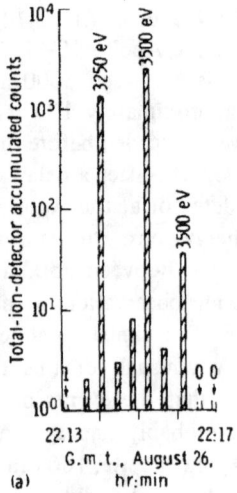

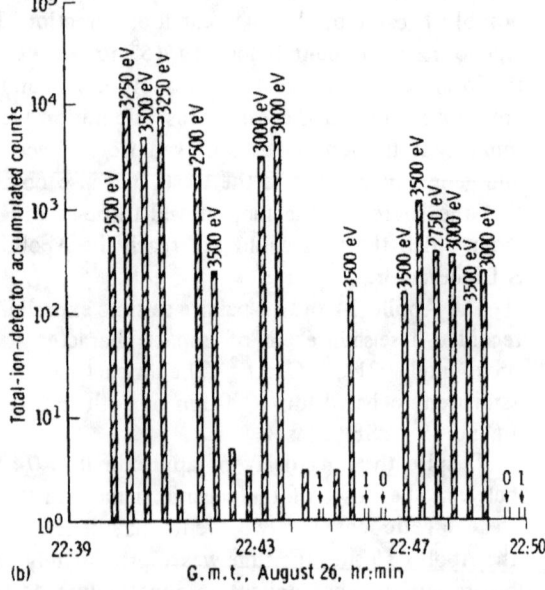

FIGURE 12-6.—Observation of the high-energy-ion event on August 26 by the Apollo 15 and 14 SIDE instruments. All total-ion-detector counts accumulated in the five adjoining 2500- to 3500-eV channels of each 20-channel spectrum are given. Practically all significant counts fell within these five channels. The energy of the peak channel is indicated. This event occurred approximately 75 min after the observations of a seismic event of impact character. The 26-min delay between arrivals at the Apollo 15 and 14 instruments should be noted. (a) Apollo 15 SIDE. (b) Apollo 14 SIDE.

electric field is the acceleration mechanism. Detailed calculations will show if all these factors can account for a delay of more than an hour.

A similar event, with a delay time of 36 min, was detected March 19 by the Apollo 14 and 12 SIDE instruments (ref. 12-3). Both of these events are now thought to be related to the impact events. If the ion events are related to the seismic events, then an apparent mean travel velocity for the gas cloud can be determined for the August event. Assuming a seismic-wave velocity of approximately 100 m/sec (ref. 12-4) and a distance between the seismometer and the impact point of approximately 1000 km, the impact must have preceded the recording of the seismic events by approximately 160 min. This estimated impact time implies that the travel velocity of the cloud was approximately 0.08 km/sec. Another apparent travel velocity could be calculated from the site separation and the time delay between ion arrivals at the two sites. This velocity, approximately 0.7 km/sec, would be valid only if the ion source moved directly from the Apollo 15 site to the Apollo 14 site. Such motion is ruled out here by the seismic information on the impact location.

Lunar Module Impact

At 03:03:36 G.m.t. on August 3, the LM ascent stage impacted the Moon 93 km west-northwest of the Apollo 15 ALSEP. Later, the SIDE detected an ion event involving ions with energies of 10 to 20 eV and dominant masses in the range of 16 to 20 amu/Q. These ions apparently originated from ionization of the gas cloud generated by the impact. At impact time, the total ion detector was recording typical magnetosheath ion spectra with energies peaked at approximately 1 keV, and the mass analyzer detector gave a constant zero reading. The mass-analyzer-detector data at 16.2 eV after the impact are included in figure 12-7(a), which shows the accumulated counts in the mass channels (13 to 16) in which the peak readings occurred. No significant counts were recorded at 48.6 eV, the only other energy channel observed during this event. The total-ion-detector counts accumulated in the 10- and 20-eV channels (those nearest the 16.2-eV energy range of the ions detected by the mass analyzer detector) are given in figure 12-7(b). The accumulated counts of the 500- to 1500-eV channels of the total ion detector are shown in figure 12-7(c). These counts are representative of the magnetosheath ion intensity. The observation of impact-related ions may be divided into two separate time segments. Part 1 covers the first 7 min after impact, and part 2 covers the time between 23 and 28 min.

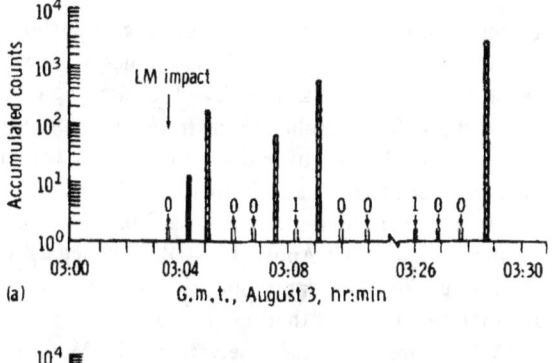
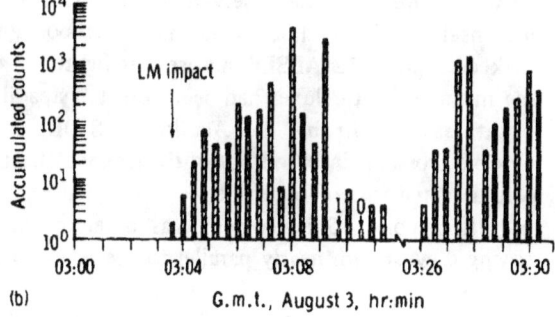
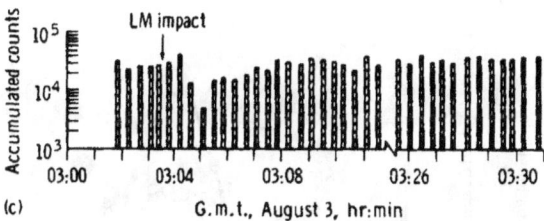
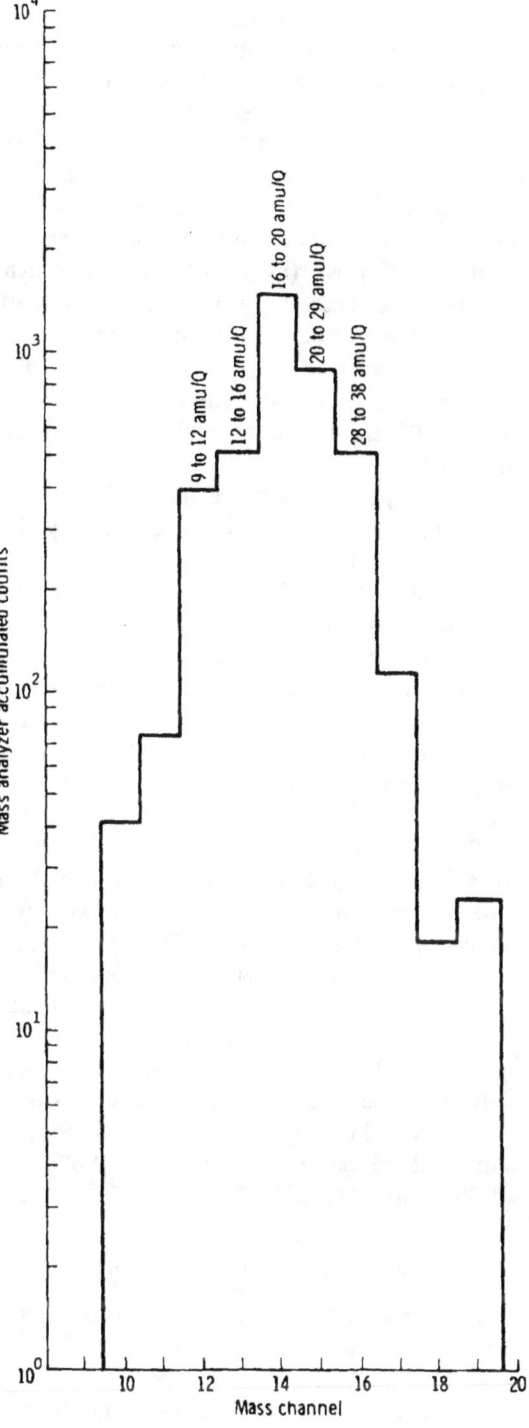

FIGURE 12-7.—Observation of the Apollo 15 LM impact. (a) Mass analyzer detector: sum of the counts accumulated in the mass channels 13 to 16 (masses 12 to 38) at the 16.2-eV energy step. (b) Total ion detector, 10 to 20 eV: counts accumulated in the adjoining 10- and 20-eV energy channels that are nearest the 16.2-eV range of ions detected by the mass analyzer detector. Significant counts were recorded only during the two time periods shown in the figure. Only background counts were recorded at other times, including the interval from 03:12 to 03:25 G.m.t. (c) Total ion detector, 500 to 1500 eV: counts accumulated over an energy range between 500 and 1500 eV. These counts represent typical magnetosheath ions.

The first response of the SIDE was detected (by the mass analyzer detector) approximately 1 min after impact, indicating fairly monoenergetic ions in the mass range between 16 and 20 amu/Q (fig. 12-8). These ions were probably preceded by ions of roughly 1-keV energy that caused a slightly increased counting rate of magnetosheath ions at 03:04:20 G.m.t. (fig. 12-7). However, this increase may have

FIGURE 12-8.—Cumulative mass spectrum observed by the mass analyzer detector, including all counts shown in figure 12-7. The mass ranges (amu/Q) of each channel are indicated. The spectrum was taken at the 16.2-eV energy step.

resulted from the normal fluctuations of the magnetosheath ion intensities. For the following 5 min, both the mass analyzer detector and the total ion detector indicated the presence of ions with energies between 10 and 20 eV and masses in the range 16 to 20 amu/Q. At the same time, the counting rates resulting from magnetosheath ions in the energy range between 500 and 1500 eV showed a significant decrease (by a factor of 8), which is believed to be caused by charge exchange or scattering in a neutral gas cloud (or both). The LM impact presumably generated a gas cloud that moved outward from the impact site. The low-energy ions seen by the SIDE are thought to result from ionization of this cloud.

The apparent travel velocity of the cloud, calculated from the distance between the Apollo 15 ALSEP and the impact point and from the time delay between impact and the first observation of ions, is 1.6 km/sec. If the neutral gas cloud arrived at the ALSEP site at the same time as the first ions, as implied by the magnetosheath ion data, the expansion velocity would correspond to a gas temperature of approximately 2000 K. The mass spectrum and the energy of the ions of part 2 of the event are in very good agreement with those of part 1. This agreement leads to the conclusion that both parts form one single event despite the separation in time. However, in part 2, no evidence exists for an associated neutral gas cloud. During part 2, the maximum flux for the total event was observed as 3×10^7 ions/cm²-sec-sr-eV. If it is assumed that the gas generated by the impact was expanding in a hemispherical shell, the fluxes at the Apollo 14 and 12 sites would be near the limit of detection for the instrument. This may explain the absence of any impact-related ion observations by the Apollo 12 and 14 SIDE instruments.

Magnetosheath of the Earth

The Apollo 12, 14, and 15 SIDE instruments are deployed so that their respective look directions, depicted in figure 12-4 for the case of zero libration, are approximately 38° west, 2° west, and 19° east of the Earth. Thus, a three-point observation can be made of the angular distribution of ions in the magnetosheath. The result of such an observation reveals a highly directional flow of energetic ions moving downstream parallel to the magnetosheath. Simultaneous observations at the three sites were obtained in real-time operations on August 3, shortly before the LM impact. Sample energy spectra recorded by the total ion detectors are shown in figure 12-9. The Apollo 15 instrument recorded a relatively steady flux of magnetosheath ions, with an energy spectrum peak at 750 to 1000 eV. The Apollo 14 detector measured a less intense flux of ions with similar energies. The Apollo 12 SIDE recorded only a small counting rate, approximately background level for that instrument at that temperature.

At the time of these observations, the Moon was just inside the bow-shock front near the position marked "Apollo 12 ALSEP sunrise" in figure 12-4. The magnetosheath fluxes had been detected steadily for at least 15 hr by the Apollo 14 SIDE, and multiple crossings into and out of the magnetosheath were detected earlier. It can be seen from figure 12-4 that the Apollo 15 instrument was detecting ions moving downstream nearly parallel to the bow-shock

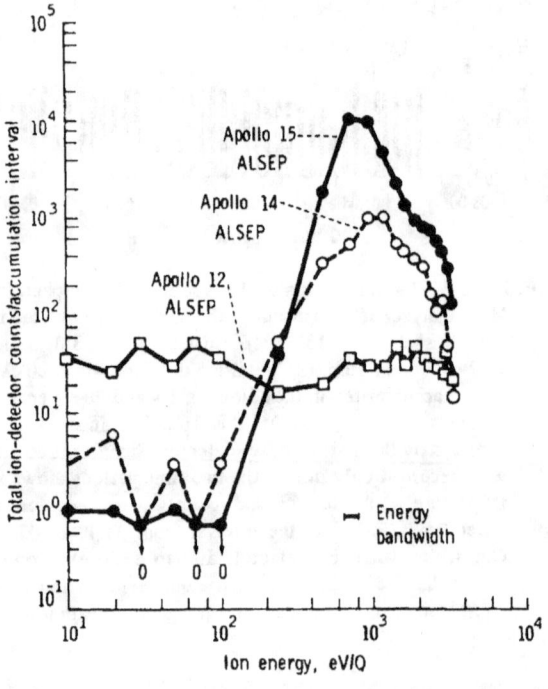

FIGURE 12-9.—Ion-energy spectra detected by the three SIDE instruments in the magnetosheath region on August 3. The Apollo 12 ALSEP spectrum began at 03:02:34 G.m.t., the Apollo 14 ALSEP spectrum at 03:02:33 G.m.t., and the Apollo 15 ALSEP spectrum at 03:02:25 G.m.t. The three instruments look in widely separated directions, as shown in figure 12-4. The accumulation interval for one data point is 1.13 sec of the 1.2 sec/frame.

front, while the Apollo 14 and 12 instruments were detecting ions moving at angles of approximately 21° and 57°, respectively, to the shock front. Thus, the intensity of the ions streaming down the magnetosheath was reduced by a factor of 10 at 21° angle and was further reduced by an additional factor of approximately 40 at 57°. The low counting rates of the Apollo 12 SIDE at this time are typical of the data recorded by that instrument at this place in the magnetosheath over the previous 21 months, although, occasionally, the streaming ions have been observed. On the other side of the magnetospheric tail, the situation is reversed. There, the Apollo 12 and 14 instruments look upstream, while the Apollo 15 detector looks at a large angle to the shock front.

Miscellaneous Observations

The Apollo 15 SIVB impact, which occurred before the deployment of the Apollo 15 ALSEP, was monitored by the SIDE instruments of Apollo 12 and 14. As in previous such events (refs. 12-3 and 12-5), the positive ions produced from the resulting gas cloud were observed. The event was relatively well defined at the Apollo 14 ALSEP site, but the Apollo 12 SIDE recorded only a very brief, low-intensity event. This difference probably results from the fact that the Apollo 12 instrument was looking generally away from the impact, whereas the Apollo 14 instrument was looking toward it. In addition, the Apollo 12 site was farther into the dark side of the Moon at the time and thus was farther removed from areas where ions could be produced by solar ionization of the neutral gas cloud.

Three LM-cabin-depressurization events were observed after deployment of the Apollo 15 ALSEP. The SIDE/CCGE was operated at these times primarily to enable the CCGE to detect the neutral-gas-pressure effects of the cabin venting. Thus, the mode of operation was one that optimized CCGE temporal resolution and that cycled through only the first 11 SIDE mass channels, covering the range up to 6 amu/Q. Although counts were recorded by the mass analyzer detector during these events, no spectra are presented here because of the small mass range covered and the rapid temporal variations of intensity observed during these periods.

Observations during the August 6 total eclipse of the Sun were made with the Apollo 12 and 14 SIDE instruments, but no unusual events attributable to the eclipse were recorded. The Apollo 15 instrument was operating with high voltages commanded off, as planned, while outgassing of the instrument took place.

The Apollo 15 SIDE high voltages were commanded on near sunset of the first lunar day and operated throughout the night and into the next day, as discussed in the paragraph on performance of the instrument. At various times throughout the night, short-duration ion events were observed, similar to those recorded by the two earlier SIDE instruments (refs. 12-1, 12-6, and 12-7). The present three-site network of observing instruments makes it possible to distinguish between events of a local nature and those of global scale. Apparent motions of ion events can also be determined in many cases. Complete analysis must await the receipt of flight data tapes from all three ALSEP units. The available data, mostly from the Apollo 14 and 15 instruments, show some events that are observed at only one site (either 14 or 15) and other events that are observed at both sites. The addition of the simultaneous Apollo 12 data will allow more comprehensive determinations of the character of these events.

CONCLUSIONS

The Apollo 15 SIDE is performing excellently and is returning very useful scientific data. These data are especially valuable in conjunction with the simultaneous data from the Apollo 12 and 14 SIDE instruments, which are still operating. Preliminary analysis of the data yields the following significant observations.

(1) Multiple-site observations of ion events apparently related to a meteoroid impact have led to the determination of apparent motions of the ion source. The source is presumed to be neutral gas that moves outward from the impact and becomes ionized. The apparent travel velocity of the gas cloud is calculated as approximately 0.08 km/sec.

(2) A complicated ion event related to the Apollo 15 LM impact was detected by the Apollo 15 SIDE. A small flux of 10- to 20-eV ions was observed within a minute after the impact. At 26 min after the impact, an intense flux of 10- to 20-eV ions was recorded. These ions exhibited a broad mass spectrum, with a peak in the range 16 to 20 amu/Q. Both the meteoroid event and the LM-impact event produced a second ion-flux increase observed at a time several minutes after the initial increase.

(3) At LM ascent, a strong decrease occurred in the magnetosheath ion fluxes being detected at the time. This decrease, which lasted approximately 8 min, could be attributable to energy loss in the relatively dense exhaust gas, to losses by charge exchange, or to temporary deviations of the magnetosheath ion-flow direction caused by the exhaust gas.

(4) The 500- to 1000-eV ions streaming down the magnetosheath have been observed simultaneously by all three SIDE instruments, located at different sites and looking in different directions. The ion flux is strongly peaked in the downstream direction, decreases by a factor of 10 within a 21° change in direction, and further decreases by an additional factor of approximately 40 within the next 36°.

(5) As with the Apollo 12 and 14 SIDE instruments, short-duration energetic ion events have been observed during the lunar night by the Apollo 15 SIDE. Analysis of these events from simultaneous observations at three sites will lead to a more comprehensive determination of the characteristics and causes of these events.

REFERENCES

12-1. Freeman, J. W., Jr.; Fenner, M. A.; Hills, H. K.; Lindeman, R. A.; et al.: Suprathermal Ions Near the Moon. Presented at the 15th General Assembly, International Union of Geodesy and Geophysics (Moscow), Aug. 1971.

12-2. Freeman, J. W., Jr.; Balsiger, H.; and Hills, H. K.: Suprathermal Ion Detector Experiment (Lunar Ionosphere Detector). Sec. 6 of Apollo 12 Preliminary Science Report. NASA SP-235, 1970.

12-3. Hills, H. K.; and Freeman, J. W., Jr.: Suprathermal Ion Detector Experiment (Lunar Ionosphere Detector). Sec. 8 of Apollo 14 Preliminary Science Report. NASA SP-272, 1971.

12-4. Latham, Gary V.; Ewing, Maurice; Press, Frank; Sutton, George; et al.: Passive Seismic Experiment. Sec. 6 of Apollo 14 Preliminary Science Report. NASA SP-272, 1971.

12-5. Freeman, J. W., Jr.; Hills, H. K.; and Fenner, M.A.: Some Results from the Apollo 12 Suprathermal Ion Detector. Proceedings of the Second Lunar Science Conference, vol. 3, A. A. Levinson, ed., MIT Press (Cambridge, Mass.), 1971, pp. 2093-2102.

12-6. Lindeman, Robert A.: Recurring Ion Clouds at the Lunar Surface. M. S. Thesis, Rice Univ., June 1971.

12-7. Freeman, J. W., Jr.: Energetic Ion Bursts on the Night Side of the Moon. J. Geophys. Res., vol. 76, no. 34, Dec. 1, 1971.

ACKNOWLEDGMENTS

The authors gratefully acknowledge the support of those who contributed to the success of the SIDE. Particular thanks are extended to Wayne Andrew Smith, Paul Bailey, James Ballentyne, and Alex Frosch of Rice University. Thanks are also due to Martha Fenner, Robert Lindeman, Rene Medrano, and John Benson, graduate students in the Space Science Department who assisted with the project. Hans Balsiger, a former European Space Research Organization/National Aeronautics and Space Administration (ESRO/NASA) international fellow, also contributed to the SIDE project. Time Zero Corp. was the subcontractor for the design and fabrication of the instrument. Personnel of the NASA Manned Spacecraft Center and of Bendix Aerospace Systems Division also provided valuable support. This research has been supported by NASA contract NAS 9-5911. J. Meister is an ESRO/NASA fellow.

13. Cold Cathode Gage Experiment (Lunar-Atmosphere Detector)

F. S. Johnson,[a][†] *D. E. Evans,*[b] *and J. M. Carroll*[a]

OBJECTIVES OF THE EXPERIMENT

Although the lunar atmosphere is known to be tenuous, its existence cannot be doubted because the solar wind striking the lunar surface constitutes one source, and there may be other sources as well. The most significant source of lunar atmosphere, if it should prove detectable, is degassing from the interior. Such degassing would constitute useful information on how planetary atmospheres originate.

The gas concentration at the lunar surface must depend on the balance between source and loss mechanisms as well as on properties of diffusion over the lunar surface. The dominant loss mechanisms for lunar gases are thermal escape for particles lighter than neon and escape through interaction with the solar wind after photoionization has occurred for neon and heavier particles. The gas particles lighter than neon have such high thermal velocities that a significant fraction of them can escape from the gravitational field of the Moon owing to their greater-than-escape velocity. The average lifetime on the Moon for helium is approximately 10^4 sec. Heavier particles, with lower thermal velocities, have longer lifetimes; the lifetime for neon is approximately 10^{10} sec, and the lifetime for heavier particles is much longer.

Particles exposed to solar ultraviolet radiation become ionized in approximately 10^7 sec; and, once ionized, the particles are accelerated by the electric field associated with the motion of the solar wind. The initial acceleration is at right angles to the direction of both the solar wind and the embedded magnetic field; then, the direction of motion is deviated by the magnetic field so that the ionized particle acquires an average velocity equal to the solar-wind-velocity component perpendicular to the embedded magnetic field. The time required for this acceleration is approximately the ion gyro period in the embedded magnetic field. The radii of gyration for most ions are comparable to or greater than the lunar radius. As a consequence of this acceleration process, particles in the lunar atmosphere are largely swept away into space within a few hundred seconds (the ion gyro period) after becoming ionized. Thus, the time required for ionization regulates the loss process, which results in lifetimes for particles in the lunar atmosphere on the order of 10^7 sec.

The cold cathode gage experiment (CCGE) was included in the Apollo lunar surface experiments package (ALSEP) to evaluate the amount of gas present on the lunar surface. The CCGE indications can be expressed as concentration of particles per unit volume or as pressure, which depends on the ambient temperature in addition to the concentration. The amount of gas observed can be compared with the expectation associated with the solar-wind source to obtain an indication of whether other sources of gases are present. Contamination from the lunar module (LM) and from the astronaut suits constitutes an additional source, but one that should decrease with time in an identifiable way. In the long run, measurements of actual composition of the lunar atmosphere should be made with a mass spectrometer to examine constituents of particularly great interest geochemically and to identify and discriminate against contaminants from the vehicle system.

INSTRUMENT DESCRIPTION

The essential sensing element of the CCGE consists of a coaxial electrode arrangement, as shown in figure 13-1. The cathode consists of a spool that is

---
[a]The University of Texas at Dallas.
[b]NASA Manned Spacecraft Center.
[†]Principal investigator.

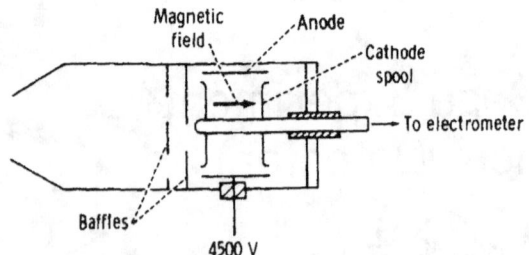

FIGURE 13-1.—Diagrammatic representation of the cold-cathode ionization gage used in the CCGE.

surrounded by a cylindrical anode. A magnetic field of approximately 0.090 T is applied along the axis, and 4500 V are applied to the anode. A self-sustained electrical discharge develops in the gage in which the electrons remain largely trapped in the magnetic field with enough energy to ionize any gas particles that they strike. The current of ions collected at the cathode is a measure of the gas density in the gage.

The response of the CCGE in terms of cathode current as a function of gas concentration is shown in figure 13-2. The CCGE response depends to a rather modest degree on the gas composition; thus, as long as the gas composition remains unknown, a fundamental uncertainty remains in the interpretation of the data. Usually, the results are expressed in terms of equivalent nitrogen response (i.e., the concentration of nitrogen that would produce the observed response). The true concentration varies from this result by a factor that is usually less than 2.

Instrument temperature was monitored by means of a sensor on the CCGE. Because no temperature control exists, the temperature range is approximately 100 to 350 K.

The CCGE was closed with a dust cover that did not constitute a vacuum seal. The cover was removed on command by using a squib motor and was then pulled aside by a spring. Because the CCGE was not evacuated, adsorbed gases produced an elevated level of response when the gage was initially turned on. Baking the CCGE on the lunar surface at 350 K for more than a week during each lunar day drove the adsorbed gases out of the gage.

Electronic Circuitry

A description of each of the major CCGE assemblies follows.

Electrometer amplifier.—An autoranging, autozeroing electrometer amplifier monitors current

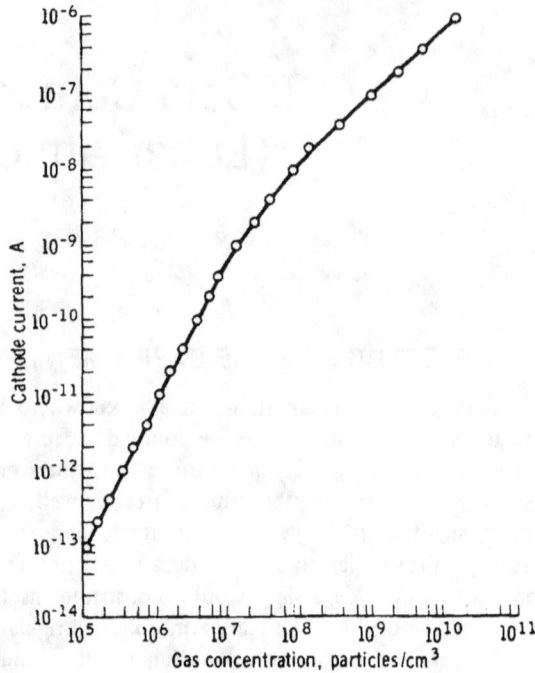

FIGURE 13-2.—Response curve for the CCGE.

outputs from the sensor or from the calibration-current generators in the 10^{-13}- to 10^{-6}-A range. The output ranges from -15 mV to -15 V. The output of the electrometer is routed to an analog-to-digital converter. The electrometer consists of a high-gain, low-leakage differential amplifier with switched high-impedance feedback resistors and an autozeroing network.

The electrometer operates in three automatically selected overlapping ranges: (1) most sensitive, (2) midrange, and (3) least sensitive. Range 1 senses currents from approximately 10^{-13} to 9.3×10^{-11} A; range 2, currents from approximately 3.3×10^{-12} to 3.2×10^{-9} A; and range 3, currents from approximately 10^{-9} to 9.3×10^{-7} A.

Power supply.—The 4500-V power supply consists of a regulator, a converter, a voltage-multiplier network, and the associated feedback network of a low-voltage power supply. The regulator furnishes approximately 24 V for conversion to a 5-kHz square wave, which is applied to the converter transformer. The output of the converter transformer is applied to a voltage-multiplier network (stacked standard doublers), the output of which is filtered and applied to the CCGE anode.

COLD CATHODE GAGE EXPERIMENT

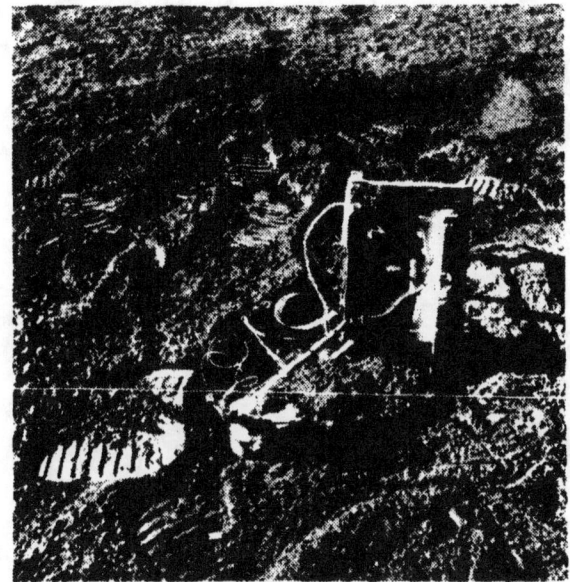

FIGURE 13-3.—The CCGE and SIDE as deployed on the lunar surface. The cold-cathode ionization gage is attached to the lower end of the extended leg of the SIDE (AS15-86-11597).

Deployment

The electronics for the CCGE are contained in the suprathermal ion detector experiment (SIDE), and the command and data-processing systems of the SIDE also serve the CCGE. The CCGE is attached to an extended leg of the SIDE on its northeast face, approximately 33 cm from the SIDE. The experiment was deployed so that the LM descent stage was outside the CCGE field of view, which looked northward. In figure 13-3, the CCGE is shown deployed on the lunar surface.

RESULTS

The CCGE was turned on at approximately 19:34 G.m.t. on July 31, 1971. On initial activation, the gage indicated full scale; but, after approximately 30 min of operation, the output began to drop. The high voltage was then commanded off to allow the instrument to outgas. The gage has not been operated for prolonged periods during the lunar day because of voltage restrictions placed on the high-voltage power supply in the SIDE package, as described in section 12, Suprathermal Ion Detector Experiment.

The experiment was operated four more times for periods of approximately 30 min each to observe the effects of the LM depressurizations for the second and third periods of extravehicular activity (EVA) and for the equipment jettison and to observe the effects of the LM lift-off from the lunar surface. In each of the three LM depressurizations, the output of the experiment was driven to full scale for approximately 30 sec, as indicated in figure 13-4 for the third EVA. The double off-scale peaks separated by approximately 30 sec were caused by the cracking and the closing of the depressurization valve on the LM; and the third peak resulted from opening the hatch. The response during the ascent-stage lift-off is shown in figure 13-5.

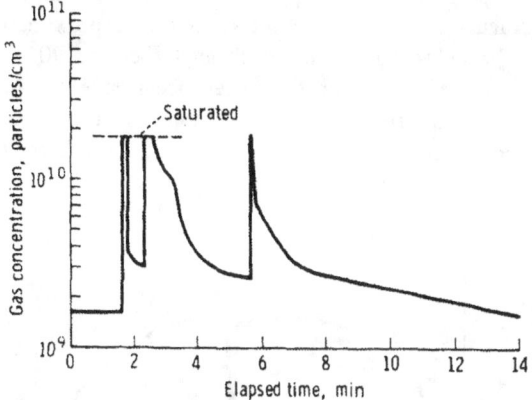

FIGURE 13-4.—Gas concentration detected during depressurization of the LM for the third EVA.

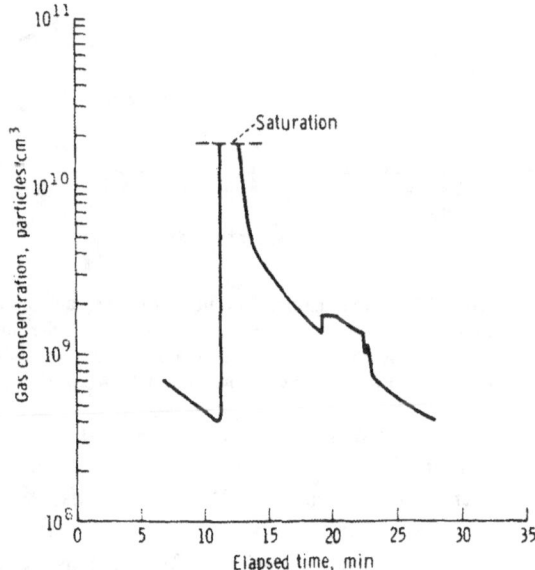

FIGURE 13-5.—Gas concentration detected during ascent-stage lift-off.

The gage was off scale for approximately 90 sec, after which the gas concentration fell rapidly. The cause of the approximate 4-min increase in response at approximately 8 min after lift-off is not known. At 20 min after lift-off, the gas concentration was back to approximately the value that prevailed before lift-off.

The temperature history of the gage during the first month on the lunar surface is shown in figure 13-6. The temperature rose to a maximum of approximately 350 K, near local noon. The sharp dip in the temperature curve near midday was caused by an eclipse of the Sun. A sharp increase in the rate of temperature fall at approximately 09:20 G.m.t. on August 13 indicates sunset at the gage approximately 5 hr before the Sun zenith angle became 90°; the calculated time for the latter occurrence is 14:14 G.m.t. Sunrise occurred at approximately 00:20 G.m.t. on August 29, approximately 17 hr after the time of 90° zenith angle, 07:22 G.m.t. on August 28.

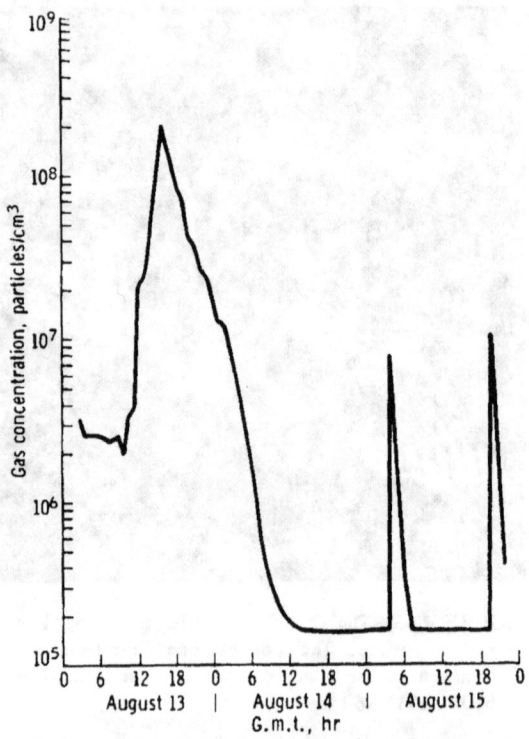

FIGURE 13-7.—Gas concentration detected during the first lunar evening after deployment of the CCGE. Sunset occurred at approximately 09:20 G.m.t. on August 13.

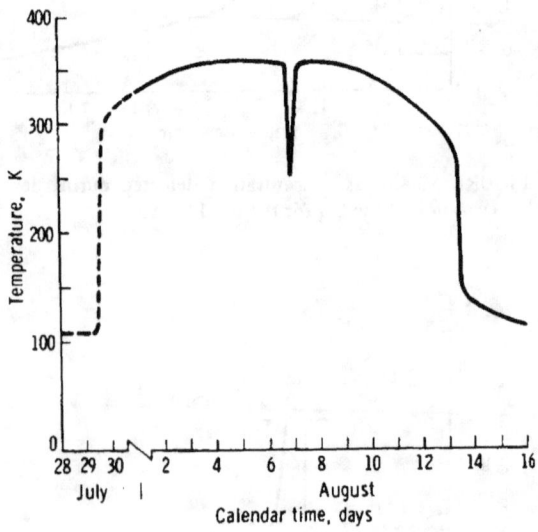

FIGURE 13-6.—The temperature history of the cold-cathode ionization gage during the first month on the lunar surface. The sharp dip in temperature near midday was caused by a lunar eclipse at the Apollo 15 ALSEP site.

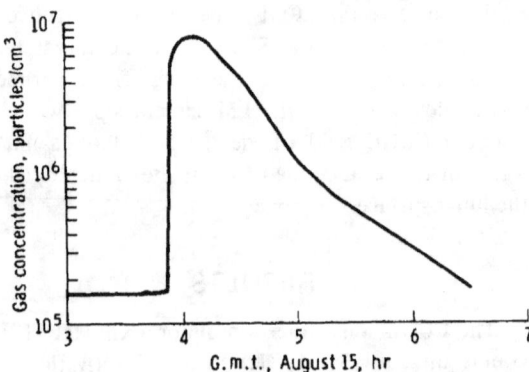

FIGURE 13-8.—Variation in gas concentration from 03:00 to 06:00 G.m.t. on August 15.

The response history during the first sunset is shown in figure 13-7. A large increase in gas concentration occurred just after sunset, which (according to the temperature data) occurred at 09:20 G.m.t. on August 13. The increase lasted approximately a day, after which the response fell to a low value that is characteristic of lunar nighttime conditions. The source of the increase is not known, but it was probably the LM. An increase of lesser magnitude but of longer duration occurred after the first lunar sunset on Apollo 14.

The two shorter duration peaks on August 15, also shown in figure 13-7, are shown in greater detail in figures 13-8 and 13-9. A single peak somewhat similar

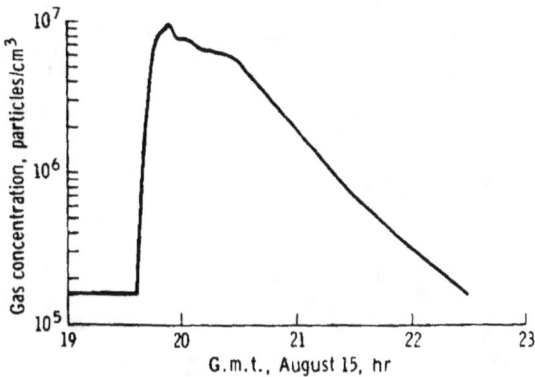

FIGURE 13-9.—Variation in gas concentration from 19:00 to 22:00 G.m.t. on August 15.

to these was seen on Apollo 14, also not long after sunset. No other peaks of this magnitude have been seen on Apollo 14 after the first month of operation, and this circumstance suggests that the peaks shown in figures 13-8 and 13-9 were caused by gas release at the LM.

The gas concentrations observed during sunlit periods appear to be caused by release of adsorbed gases associated with the landing operations. However, at the low nighttime temperatures, the contaminant gases remain adsorbed on the lunar surface; the observed concentrations are believed to be representative of natural ambient conditions. The observed concentrations are lower than might be expected from the solar wind, which should provide a nighttime atmospheric concentration of neon in excess of 10^6 atoms/cm^3 if as much neon is released from the lunar surface as impinges upon it from the solar wind. The fact that the observed concentrations are almost an order of magnitude lower than the expected concentration suggests that the lunar surface is not saturated with neon and that the rate of neon release from the lunar surface is much slower than the rate of neon implantation.

The LM impact was not detected by the Apollo 14 CCGE, and data are not complete enough at present to determine whether the LM impact was detected by the Apollo 15 CCGE. The CCGE was turned off at the times of the solar eclipse on August 6 and the solar flare on September 1. No easily recognizable correlations were found between transient gas events as seen on the CCGE and the response of other ALSEP instrumentation.

The data presented in this report are preliminary and may be changed significantly when data tapes become available.

ACKNOWLEDGMENTS

The CCGE sensor was supplied by the Norton Company, with the assistance of Frank Torney. The electronics were built by Marshall Laboratories (now Time Zero Corporation) under subcontract to Rice University. The authors are grateful to Dr. John Freeman and his associates at Rice University for their assistance in assuming contractual responsibility for development and production of the gage electronics in connection with the SIDE.

14. Laser Ranging Retroreflector

J. E. Faller,[a†] *C. O. Alley,*[b] *P. L. Bender,*[c] *D. G. Currie,*[b] *R. H. Dicke,*[d]
W. M. Kaula,[e] *G. J. F. MacDonald,*[f] *J. D. Mulholland,*[g] *H. H. Plotkin,*[h]
E. C. Silverberg,[i] *and D. T. Wilkinson*[d]

CONCEPT OF THE EXPERIMENT

During the Apollo 15 mission, the third and largest U.S. laser ranging retroreflector (LRRR) was deployed on the lunar surface in the area near Hadley Rille. Ground-based stations can conduct short-pulse laser ranging during both lunar day and lunar night to this Apollo 15 array and the Apollo 11 (Sea of Tranquility area) and Apollo 14 (Fra Mauro area) retroreflector packages. These arrays are deployed at well-separated sites (fig. 1-1, sec. 1). The returned signal from the LRRR has an intensity 10 to 100 times greater than that reflected by the natural surface. The use of the LRRR eliminates the time-stretching of the pulse that results from the light being reflected back from different parts of the lunar surface. An observation program is being actively followed to obtain an extended sequence of high-precision Earth-Moon distance measurements that will, over a number of years, provide the data from which a variety of information about the Earth-Moon system can be derived (refs. 14-1 to 14-8). Preliminary analysis of ranging data from the three retroreflector arrays presently indicates that substantial corrections in their assumed position coordinates will be required. Full utilization of the Apollo arrays, as well as of the French-Russian array carried on Luna 17, will require an observing program lasting decades and using ground stations located around the world.

An obvious and immediate use of these data will be to define more precisely the motion of the Moon in its orbit. Another experimental result will be the measurement of the lunar librations—the irregular motions of the Moon about its center. The three Apollo arrays, which are well separated in longitude and latitude, will permit a completely geometrical separation of the lunar librations.

With two or three regularly observing stations well separated geographically, both components of polar motion as well as universal time can be determined. Periods as brief as 1 day in the rotation and polar motion of the Earth can be found if the data are frequent enough, but a considerably larger number of stations is needed if short-period variations are to be monitored regularly. The laser-ranging method, with its expected ± cm or better range accuracy, is capable of achieving an accuracy of a few centimeters for polar motion and crustal movements and of 100 μsec for universal time. Present accuracies, as determined by conventional astronomical observations, are 1 to 2 m for polar motion and approximately 5 msec for universal time (UT 1).

Accurate measurements of terrestial global plate motions by means of laser ranging may test whether the present rates are the same as the average past rates that have been deduced from observed displacements of geological features and remanent magnetic records.

Observations of the changes in the rotation rate of the Earth should provide clues into the nature of the core-mantle coupling and, hence, of the properties of the core and lower mantle. In addition, possible changes in the total angular momentum of the atmosphere of the Earth, which are believed to cause the annual and semiannual terms in the rotation rate

---

[a]Wesleyan University.
[b]University of Maryland.
[c]Joint Institute for Laboratory Astrophysics.
[d]Princeton University.
[e]University of California at Los Angeles.
[f]Council on Environmental Quality.
[g]University of Texas at Austin.
[h]NASA Goddard Space Flight Center.
[i]University of Texas, McDonald Observatory.
[†]Principal investigator.

of the Earth, may be sufficient to cause observable changes in the rotation rate of the Earth even for periods as short as a few days.

To begin checking present astronomical information concerning polar motion and Earth rotation, the major factor required is the improvement of the basic lunar ephemeris. The initial range uncertainties for the Apollo 11 and 14 retroreflectors were approximately 300 m. So far, using Apollo 11 LRRR data through July 1970, it has been possible to improve the range-prediction accuracy substantially. With the much greater frequency of data from the Apollo 11 LRRR that has been obtained since October 1970; data from the Apollo 14 LRRR that have been obtained since February 1971; and, now, data from the Apollo 15 LRRR, it should be possible to fit the lunar motion accurately as soon as the necessary analytical work has been done.

Finally, the sensitivity afforded by the presence of these reflecting arrays on the lunar surface will make it possible to use the Moon as a testing ground for gravitational theories. Many observers are interested in discovering whether the tensor theory of gravity is sufficient or if a scalar component is necessary as has been suggested. A definitive test of the hypotheses may be obtained by monitoring the motion of the Moon. Additionally, the possibility exists of seeing some very small but important effects in the motion of the Moon that are predicted by the general theory of relativity.

PROPERTIES OF THE LRRR ARRAYS

Each of the three arrays is a wholly passive device containing small, fused-silica corner cubes with front-face diameters of 3.8 cm. The Apollo 11 (ref. 14-9) and 14 (ref. 14-10) arrays are almost identical; each array contains 100 corner cubes. The Apollo 15 LRRR (fig. 14-1) contains 300 small, fused-silica corner cubes. Each corner cube in the array has the property of reflecting light parallel to the incident direction; that is, a light beam incident on a corner cube is internally reflected in sequence from the three back faces and then returned along a path parallel to the incident beam (fig. 14-2). This parallelism between the reflected and incident beams ensures that the reflected laser pulse will return to the vicinity of origin on the Earth.

The temperature gradients in the individual corner cubes are minimized by recessing each reflector by

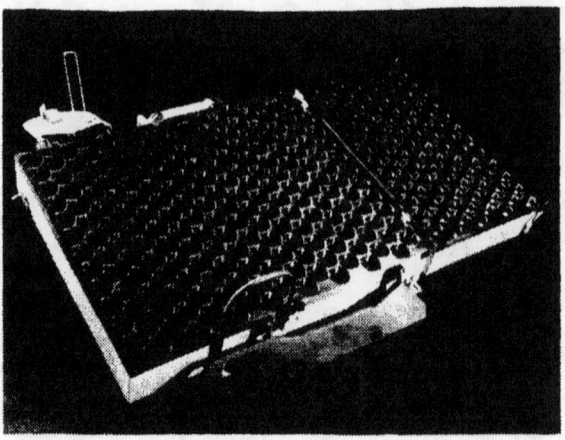

FIGURE 14-1. - Apollo 15 LRRR in deployed configuration.

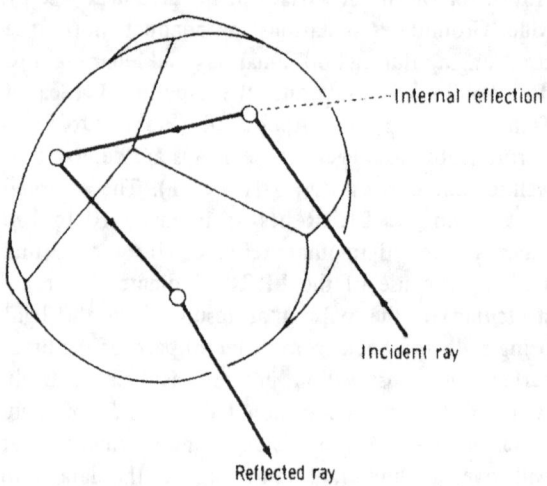

FIGURE 14-2. - Corner cube, showing how a light beam is reflected.

half its diameter in a circular socket. Each individual reflector is tab-mounted between two Teflon rings to afford the maximum thermal isolation (fig. 14-3). The mechanical mounting structure also is used to provide passive thermal control by means of its surface properties. A comparison of the calculated thermal performance expected from the Apollo 11, 14, and 15 arrays is shown in figure 14-4. During storage, transportation, handling, and flight, a transparent polyester cover assembly protects the arrays from dust and other contamination.

Mechanically, the Apollo 15 array consists of a hinged two-panel assembly (one panel containing 204 reflectors and the other containing 96 reflectors) mounted on a deployment-leg assembly. This leg was

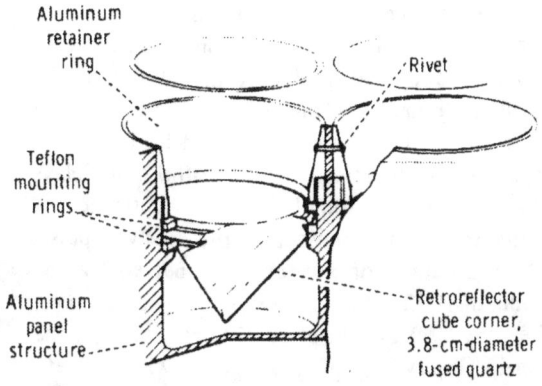

FIGURE 14-3.—Cutaway drawing of corner-cube mounting.

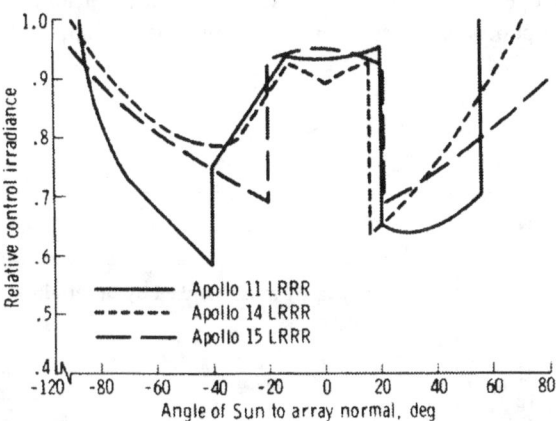

FIGURE 14-4.—Comparison of calculated thermal performance expected from Apollo 11, 14, and 15 LRRR arrays.

FIGURE 14-5.—Apollo 15 LRRR array deployed on lunar surface (AS15-85-11469).

extended in deployment to support the retroreflector array at an elevation of approximately 26° to the lunar surface (fig. 14-5). In both panels, the cubes are arranged in a close-packed configuration to minimize the weight and overall size of the array. A comparison of these parameters for the three Apollo arrays is given in table 14-I. A Sun-compass assembly attached to the larger panel provides azimuthal alinement of the arrays with respect to the Sun, and a bubble level provides alinement with the lunar horizontal.

The Apollo 15 LRRR was deployed during the first period of extravehicular activity approximately 43 m southwest of the Apollo lunar surface experiments package central station (that is, approximately 140 m west of the lunar module). Leveling and alinement, to point the array toward the center of the Earth libration pattern, were accomplished with no difficulty. As a result of contingencies during the lunar-surface phase of the mission, photographic documentation was insufficient to determine deployment accuracy. However, both the astronauts' voice record and subsequent debriefing indicate that the array was properly deployed on the lunar surface.

Successful range measurements to the Apollo 15 array were first made from the McDonald Observatory of the University of Texas on August 3, 1971. In fact, a few returns had been received the preceding day, but these returns were not recognized until later

TABLE 14-I.—*Apollo LRRR Array Particulars*

Parameter	Apollo 11	Apollo 14	Apollo 15
Size, cm			
Height (stowed)	29.2	30.0	30.0
Width (stowed)	68.6	63.8	69.5
Width (deployed)	68.6	63.8	105.2
Length	66.0	64.8	64.8
Weight, kg	23.59	20.41	36.20
Number of retroreflectors	100	100	300
Retroreflector size, cm (front-face diameter)	3.8	3.8	3.8

because of heavy noise blanking that resulted from the initial range uncertainty. Experience thus far indicates that no serious degradation occurred during lunar module ascent-stage firing. Visual guiding of the telescope on the Apollo 15 site is facilitated by nearby lunar landmarks, which should aid other stations in their acquisition of this retroreflector array. A firing record for the Apollo 15 LRRR is given in table 14-II.

GROUND-STATION OPERATION

At present, range measurements to all three retroreflector packages at nearly all lunar phases are being made at the McDonald Observatory with NASA support. A line drawing of the laser-ranging station at the McDonald Observatory is shown in figure 14-6.

The present accuracy of ±30 cm for the lunar-distance measurements at the McDonald Observatory is limited mainly by problems in calibrating the electronic time delays in the system. The installation of a new calibration system is planned for late 1971; this system will, in effect, eliminate the time delays by using the same photomultiplier and electronics for both the transmitted pulse and the received pulse. Thus, an accuracy of ±15 cm is expected by the beginning of 1972.

The ruby-laser system presently being used at the McDonald Observatory gives 3-J pulses with a repetition rate of one every 3 sec. The total pulse length between the 10-percent-intensity points is 4 nsec. The root-mean-square variation in the observed transit time, caused by the laser-pulse length and the jitter in the photomultiplier receiving the returned signal, is 2

TABLE 14-II.—*Record of Firings for the Apollo 15 LRRR*

[McDonald Observatory, September 13, 1971]

Date, 1971	Time, c.d.t	Number of laser firings	Number of returns	Comments
Aug. 2	8:00 to 10:00 p.m.	490	4	Returns heavily noise-blanked by uncertainty in range
Aug. 3	8:30 to 10:30 p.m.	300	19	
Aug. 4	10:30 to 12:20 p.m.	400	32	
Aug. 5 to 6	11:00 p.m. to 1:00 a.m.	700	18	
Aug. 7	12:00 p.m. to 2:00 a.m.	150	24	
Aug. 8	2:00 to 4:00 a.m.	80	14	
Aug. 12	5:30 to 8:30 a.m.	150	21	
	10:10 to 10:30 a.m.	127	6	
Aug. 14	6:15 to 7:30 a.m.	50	0	Stopped by clouds
	8:00 to 8:45 a.m.	50	6	
Aug. 26	8:00 to 9:00 p.m.	100	0	Partly cloudy, computer-guided
Aug. 28	7:30 to 9:30 p.m.	200	12	
Aug. 29	7:30 to 9:30 p.m.	200	19	
Aug. 30	7:30 to 9:30 p.m.	150	13	
Aug. 31	10:00 to 11:00 p.m.	200	15	Extremely good return for the conditions; i.e., 3 arc-sec seeing
	12:30 to 1:20 a.m.	50	7	
Sept. 10	6:30 to 7:45 a.m.	100	25	Returns on 6 successive shots
Sept. 11	5:30 to 7:00 a.m.	100	30	Best signal so far for an extended run on any corner
Sept. 12	6:00 to 8:00 a.m.	180	13	Computer-guided
Sept. 13	6:25 to 7:00 a.m.	30	6	Computer-guided
Totals[a]		3807	284	[b]0.075 return/shot

[a]Comparative signals on the other two corner reflectors for the same period: Apollo 11 LRRR, 4130 shots, 150 returns, 0.037 return/shot; Apollo 14 LRRR, 5045 shots, 243 returns, 0.048 return/shot.

[b]The Apollo 15 returns will appear depressed because of the inability of the McDonald Observatory electronics to detect the multiple returns on days such as September 10. The smaller corner reflectors left by Apollo 11 and Apollo 14 will only rarely produce multiple-photoelectron returns. Thus, the improvement in signal brought about by the larger reflector is underestimated when calculated in this manner.

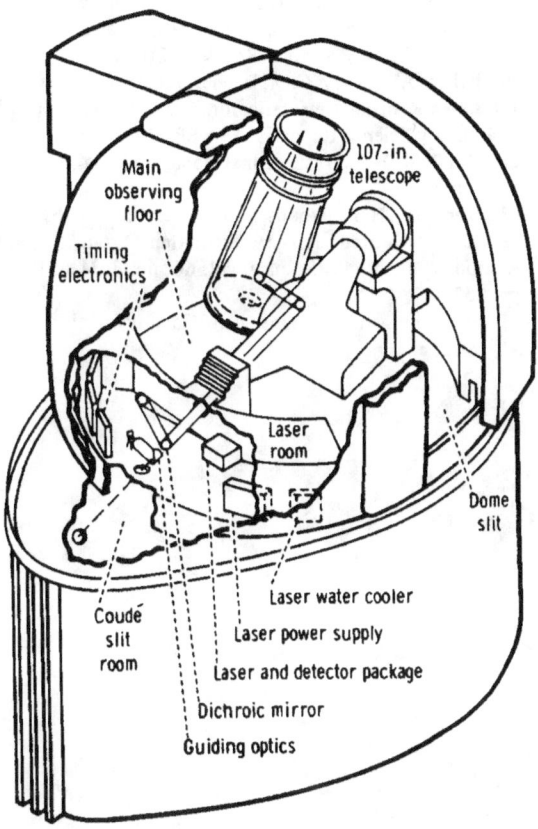

FIGURE 14-6.—Laser-ranging station at McDonald Observatory.

nsec. This variation results in the present ±30-cm statistical uncertainty for a single shot. Improvement to less than 1-nsec (±15-cm) statistical uncertainty is expected by averaging the range residuals over a period of a few minutes.

The uncertainty in the range correction for the effect of the atmosphere has been shown to be less than 6 cm up to zenith angles of 70° (ref. 14-2). This result was based on using the surface value of the atmospheric refractive index as a predictor for the correction, as is often done in radio work. Recently, it has been pointed out that very much better corrections for the optical case can be obtained by using the surface pressure as the predictor (refs. 14-11, 14-12, and 14-13). It now seems likely that the total error in the range correction for zenith angles of up to 70° will be less than 1 cm for normal atmospheric conditions.

Present indications are that lasers will be available shortly with reduced beam divergence, much shorter pulse lengths, and sufficient power to permit lunar ranging. The use of subnanosecond laser pulses will permit significantly greater measurement accuracy. The laser system proposed for the Lunar Ranging Experiment Team ranging station planned in Hawaii will have a 0.2-nsec pulse length, and it is expected that a short pulse length will be tried very soon at the McDonald Observatory. With care, an accuracy of 0.1 nsec seems achievable for the timing electronics. The accuracy of range data obtainable using a 0.2-nsec-pulse-length laser should be 3 cm or better, including an allowance of approximately 1 cm for the uncertainty in the atmospheric corrections at 70° from the zenith.

SUMMARY

With the Apollo 11 and 14 arrays, the placing of the Apollo 15 retroreflector array completes a three-array network. The larger signals obtainable with this array provide for a greater frequency of returns and will allow laser ranging to be carried out with telescopes of smaller aperture. This fact should encourage participation by a number of ground stations in other countries in monitoring the variations in the lunar distance by using these arrays, which give every indication of providing primary benchmarks on the lunar surface for years to come.

REFERENCES

14-1. Alley, C.O.; Bender, P.L.; Dicke, R.H.; Faller, J.E.; et al.: Optical Radar Using a Corner Reflector on the Moon. J. Geophys. Res., vol. 70, no. 9, May 1, 1965, pp. 2267-2269.

14-2. Alley, C.O.; and Bender, P.L.: Information Obtainable From Laser Range Measurements to a Laser Corner Reflector, Continental Drift, Secular Motion of the Pole, and Rotation of the Earth. Symp. no. 32 IAU, William Markowitz and B. Guinot, eds., D. Reidel Pub. Co. (Dordrecht, Holland), 1968, pp. 86-90.

14-3. Alley, C.O.; Bender, P.L.; Currie, D.G.; Dicke, R.H.; and Faller, J.E.: Some Implications for Physics and Geophysics of Laser Range Measurements From Earth to a Lunar Retroreflector. The Application of Modern Physics to the Earth and Planetary Interiors. Proceedings of the NATO Advanced Study Institute, S.K. Runcorn, ed., Wiley-Interscience (London and New York), 1969, pp. 523-530.

14-4. Faller, James; Winer, Irvin; Carrion, Walter; Johnson, Thomas S.; et al.: Laser Beam Directed at the Lunar Retroreflector Array: Observations of the First Returns. Science, vol. 166, no. 3901, Oct. 3, 1969, pp. 99-102.

14-5. Alley, C.O.; Chang, R.F.; Currie, D.G.; Mullendore, J.;

et al.: Apollo 11 Laser Ranging Retroreflector: Initial Measurements From the McDonald Observatory. Science, vol. 167, no. 3917, Jan. 23, 1970, pp. 368-370.

14-6. Alley, C.O.; Chang, R.F.; Currie, D.G.; Poultney, S.K.; et al.: Laser Ranging Retroreflector: Continuing Measurements and Expected Results. Science, vol. 167, no. 3918, Jan. 30, 1970, pp. 458-460.

14-7. Faller, James E.; and Wampler, E. Joseph: The Lunar Laser Reflector. Sci. Amer., vol. 222, no. 3, Mar. 1970, pp. 38-49.

14-8. Eckhardt, Donald H.; and Dieter, Kenneth: A Nonlinear Analysis of the Moon's Physical Libration in Longitude. The Moon, vol. 2, no. 3, Feb. 1971, pp. 309-319.

14-9. Alley, C.O.; Bender, P.L.; Chang, R.F.; Currie, D.G.; et al.: Laser Ranging Retroreflector. Sec. 7 of Apollo 11 Preliminary Science Report. NASA SP-214, 1969.

14-10. Faller, J.E.; Alley, C.O.; Bender, P.L.; Currie, D.G.; et al.: Laser Ranging Retroreflector. Sec. 11 of Apollo 14 Preliminary Science Report. NASA SP-272, 1971.

14-11. Hopfield, H.S.: Trans. Am. Geophys. Union. Vol. 51, 1970, p. 266.

14-12. Hopfield, H.S.: Tropospheric Effect on Electromagnetically Measured Range: Prediction From Surface Weather Data. Radio Science, vol. 66, no. 3, Mar. 1971, pp. 357-367.

14-13. Saastamoinen, J.: Trans. Am. Geophys. Union. Vol. 51, 1970, p. 266.

15. Solar-Wind Composition Experiment

J. Geiss,[a][†] F. Buehler,[a] H. Cerutti,[a] and P. Eberhardt[a]

The solar wind carries solar material into interplanetary space and thus provides the possibility of direct studies of solar matter. By measuring the relative abundances of ions in the solar wind, elemental and—in particular—isotopic abundances in solar matter can be determined. Time variations in the abundances of ions in the solar wind have to be studied to gain an understanding of the dynamics and the fractionation processes occurring in the photosphere-corona transition zone and in the acceleration region of the solar wind. Only when these processes are sufficiently understood is it possible to interpret solar-wind-abundance data correctly in terms of solar abundances.

The study of elemental and isotopic abundances in the Sun is of fundamental importance. By comparing them with abundances in planetary objects (i.e., planets, satellites, and meteorites), unique information on the origin and evolution of these objects can be obtained. Furthermore, solar-wind-abundance data are essential for a detailed interpretation of the trapped gases in meteorites and lunar material. Based on these investigations, the evolution of the lunar surface and the possibility of a transient lunar atmosphere can be studied. Noble-gas studies of the solar wind may also help in tracing the evolution of the terrestrial atmosphere.

It is well established that the helium/hydrogen ratio in the solar wind is highly variable and ranges from less than 0.01 to 0.25, with an average of approximately 0.04 (refs. 15-1 to 15-5). During periods of low solar-wind-ion temperature, the elements oxygen, silicon, and iron have been measured by means of the high-resolution electrostatic analyzers on board the Vela satellites, and, in some cases, even helium-3 (^{3}He) has been detected (refs. 15-6 and 15-7). In the Apollo Program, a different technique is used for studying elemental and isotopic abundances in the solar wind.

During the Apollo 11, 12, 14, and 15 missions, aluminum foils were deployed on the lunar surface and used as targets for collecting solar-wind ions. The foils were returned to Earth, and the implanted solar-wind particles were analyzed in the laboratory. From the Apollo 11, 12, and 14 solar-wind composition (SWC) experiments, the absolute fluxes and relative abundances in the solar wind of ^{4}He, ^{3}He, neon-20 (^{20}Ne), and ^{22}Ne have been obtained so far (refs. 15-8 to 15-10). In the case of the Apollo 12 and 14 experiments, ^{21}Ne was detected also, and the abundance in the solar wind of argon-36 (^{36}Ar) during the Apollo 14 foil-exposure time was determined (ref. 15-10).

The exposure times of the SWC foils were increased in successive Apollo missions. The Apollo 11 foil captured the solar-wind particles for only 77 min; the exposure times during the Apollo 12 and 14 missions were approximately 19 and 21 hr, respectively. During the Apollo 15 lunar stay, an exposure time of 41 hr was obtained. This increase in exposure time probably will lead to more detailed and more accurate solar-wind-abundance determinations. In this report, preliminary results of the first analyses on relatively small sections of the Apollo 15 foil are presented.

Particle experiments and magnetometers flown on Explorer 35 or placed on the lunar surface during the Apollo 12 and 14 missions have been used to establish that the Moon behaves in principle like a passive obstacle to the solar wind, and no evidence of a lunar bow shock has been found (refs. 15-11 to 15-15). Thus, during the normal lunar day, the solar-wind particles strike the lunar surface with essentially unchanged direction and energy, except perhaps in a

---

[a]Physikalisches Institut, University of Bern.
[†]Principal investigator.

few places where local magnetic fields are unusually high (ref. 15-16). In fact, it was shown from the Apollo 11 and 12 SWC experiment data that helium reaches the lunar surface in an undisturbed, highly directional flow (refs. 15-8 and 15-9). The Apollo 12 solar-wind spectrometer has recorded the plasma flow that arrives at the Apollo 12 lunar surface experiments package site as a function of the phase of the lunar day. The result is that, most of the time, the plasma flux is not affected by the proximity of the surface of the Moon (ref. 15-15). Thus, experiments deployed on the lunar surface should yield solar-wind-abundance data that are valid for the undisturbed solar wind.

PRINCIPLE OF THE EXPERIMENT

A piece of aluminum foil 30 cm wide and approximately 140 cm long was exposed to the solar wind on the lunar surface by the Apollo 15 crew on July 31, 1971, at 19:36 G.m.t. The foil was positioned perpendicular to the solar rays in the azimuthal direction (fig. 15-1), exposed for 41 hr 8 min, and returned to Earth. Laboratory experiments have determined that solar-wind ions arriving with an energy of approximately 1 keV/nucleon penetrate approximately 10^{-5} cm into the foil (ref. 15-17), and a large and calibrated fraction is firmly trapped (refs. 15-18 and 15-19). In the laboratory, the returned foil is analyzed for implanted solar-wind noble-gas atoms. Parts of the foil are melted in ultra-high-vacuum systems, and the noble-gas atoms of solar-wind origin thus released are analyzed with mass spectrometers for elemental abundance and isotopic composition. Further details of the principle and the procedures of this experiment have been discussed elsewhere (refs. 15-18, 15-20, and 15-21).

INSTRUMENTATION AND LUNAR-SURFACE OPERATION

The experiment hardware was similar to that used on the Apollo 11, 12, and 14 missions (ref. 15-21). The experiment consisted of a metallic telescopic pole approximately 4 cm in diameter and 38 cm in length when collapsed. In the stowed position, the foil was enclosed in the tubing and rolled up on a spring-driven reel. The instrument weighed 430 g. When extended on the lunar surface, the pole was approximately 1.5 m long, and a 30- by 130-cm foil area was exposed. Only the foil assembly was recovered at the end of the lunar exposure; it was rolled on the spring-driven reel and returned to Earth. The instrument is shown deployed on the lunar surface at the Apollo 15 landing site in figure 15-1. The reel handle was color coded to give the exact angular position of the reel and the portion of foil rolled around it. Detailed analyses of this portion of the foil are expected to yield the angular distribution of the arriving solar-wind ions. After examination of a number of Apollo 15 photographs, it was concluded that the foil was standing vertically (within a few degrees) on the lunar surface. After retrieval, the return unit was placed in a special Teflon bag and returned to Earth.

PRELIMINARY RESULTS

For the initial analyses, four small pieces from the upper part of the foil (samples 3-1, 3-2a, 3-2b, and 3-3) were cleaned of possible lunar-dust contamination by means of the ultrasonic treatments that had proved effective during the previous Apollo SWC experiment analyses. The results of these first measurements are presented in table 15-I. Along with these measurements on the flight foil, several pieces that had been cut from the Apollo 15 foil before flight for the purpose of noble-gas blank measurements were analyzed. The blanks that had been deter-

FIGURE 15-1.—Apollo 15 SWC experiment deployed on the lunar surface (AS15-87-11781).

TABLE 15-I.—*First Preliminary Results From the Analyses of the Foil From the Apollo 15 SWC Experiment*

Sample no.	Area, cm^2	^{4}He concentration, $\times 10^{10}$ atoms/cm^2	^{4}He/^{3}He	^{4}He/^{20}Ne	^{20}Ne/^{22}Ne	^{22}Ne/^{21}Ne
^{a}3-1	9.0	157	2350	490	13.7	30
3-2a	4.1	169	2320	440	13.5	33
3-2b	5.9	158	2360	460	13.7	31
^{a}3-3	8.7	160	2360	490	13.6	30

^{a}Oxide layer on back side of foil removed, and replicating film technique applied for cleaning.

mined in this way were subtracted from the noble-gas concentrations measured in the pieces of the flight foil, and areal concentrations of the solar-wind particles were obtained (table 15-I). The foil blanks for helium and neon were 0.02 and 0.8 percent, respectively, relative to the solar-wind-particle content. In addition, one piece of the flight foil that had been covered during exposure and thus deliberately protected from the solar wind was analyzed. In fact, no solar-wind gases were found. The upper limits for the helium and neon concentrations were 0.05 and 0.3 percent, respectively, relative to the exposed foil. Samples 3-1 and 3-3 were further cleaned by applying a replicating film technique to both surfaces. Moreover, the oxide layer was removed from the back side of these two pieces. These procedures are known to reduce significantly a possible residual-dust contamination. For the helium and neon isotopes, the data from the four foil pieces are in good agreement. This agreement is further evidence that the helium and neon data given in table 15-I are not appreciably affected by a residual-dust contamination.

Solar-wind argon was detected in samples 3-1 and 3-3 of the Apollo 15 foil. A preliminary ^{20}Ne/^{36}Ar value greater than 20 has been obtained. There is an indication that the argon abundance relative to neon for Apollo 15 is higher than for Apollo 14; a value of 37^{+10}_{-5} was obtained for the ^{20}Ne/^{36}Ar ratio in the solar wind during the Apollo 14 foil exposure. Analyses on larger pieces of the Apollo 15 foil will be performed to determine this ratio with sufficient accuracy.

The average ^{4}He flux during the Apollo 14 exposure period can be calculated by using the data given in table 15-I. The trapping probabilities of the foil for noble-gas ions depend only slightly on energy in the general solar-wind-velocity region. For helium with a velocity of approximately 300 km/sec, the trapping probability is 89±2 percent for normal incidence and approximately 17 percent less for an incidence angle of 56°.

The angular distribution and the average angle of incidence on the Apollo 15 foil have not yet been determined. Thus, for the purpose of this report, the average angle of incidence is estimated. The average solar elevation during the foil exposure was 31°. By taking into account the effects of aberration and corotation, an angle of incidence on the foil of 56° is obtained for the solar wind. With this assumption, the ^{4}He flux during the Apollo 15 SWC foil exposure can be calculated (table 15-II), together with the ^{4}He fluxes previously determined for the Apollo 11, 12, and 14 exposure periods (refs. 15-8 and 15-9).

From the experimental results given in table 15-I, preliminary values for the relative abundances of ions in the solar wind during the Apollo 15 foil-exposure period have been calculated (table 15-III). Weighted averages of the measured ratios were taken and corrected for differences in trapping probabilities. Thus, the ^{4}He/^{20}Ne ratio was corrected by 14 percent (ref. 15-19), and the trapping probability for ^{3}He was assumed to be 2 percent lower than the probability determined for ^{4}He. The isotopic ratios of neon were not corrected. Because only relatively small fractions of the respective foils have been analyzed so far, the data and limits of error given in table 15-III for Apollo 14 and 15 must be considered as preliminary.

DISCUSSION

The neon flux during the Apollo 15 mission was very high and the exposure time was twice as long as that of previous missions, and, therefore, a relatively good value for the ^{21}Ne abundance has been

TABLE 15-II — ^{4}He *Flux Averages During the Times of Foil Exposure*

[Data for Apollo 14 and 15 are preliminary. Data for Apollo 11, 12, and 14 are from refs. 15-9 and 15-10.]

Mission	Exposure initiation date and G.m.t., hr:min	Exposure duration, hr:min	Average solar-wind ^{4}He flux, $\times 10^6$ cm^{-2} sec^{-1}
Apollo 11	July 21, 1969, 03:35	01:17	6.2 ± 1.2
Apollo 12	Nov. 19, 1969, 12:35	18:42	8.1 ± 1.0
Apollo 14	Feb. 5, 1971, 15:16	21:00	4.2 ± 0.8
Apollo 15	July 31, 1971, 19:36	41:08	17.7 ± 2.5

TABLE III.—*Abundance Ratios of Solar-Wind Ions Determined For the Foil-Exposure Periods of the Apollo Missions*

[Data for Apollo 14 and 15 are preliminary. Data for Apollo 11 and 12 are from ref. 15-9.]

Condition	^{4}He/^{3}He	^{4}He/^{20}Ne	^{20}Ne/^{22}Ne	^{22}Ne/^{21}Ne
Apollo 11	1860 ± 140	430 ± 90	13.5 ± 1.0	--
Apollo 12	2450 ± 100	620 ± 70	13.1 ± 0.6	26 ± 12
Apollo 14	2230 ± 140	550 ± 70	13.65 ± 0.50	--
Apollo 15	2310 ± 120	550 ± 50	13.65 ± 0.30	31 ± 4
Earth atmosphere	7 × 10^5	.3	9.8	34.5

obtained even from the small foil pieces analyzed so far. The isotopic composition of neon in the solar wind is very different from that in the terrestrial atmosphere (ref. 15-9). The relation between ^{20}Ne, ^{21}Ne, and ^{22}Ne in solar-wind gases trapped in lunar-surface material is such that this difference can be explained by a mass fractionation process in the atmosphere (ref. 15-22). It is important to confirm this conclusion by direct solar-wind observation, because the isotopic composition of trapped lunar neon appears to have been altered by diffusion or other secondary processes. Because of its low abundance, ^{21}Ne is a very sensitive indicator of particle irradiation (ref. 15-23). An analysis of larger pieces of the Apollo 15 foil should yield the ^{21}Ne abundance with sufficient accuracy to determine whether the difference between terrestrial and solar neon is entirely caused by fractionation processes in terrestrial neon or whether there exists, either in the Sun or in Earth atmosphere, an excess of ^{21}Ne produced by nuclear reactions.

Large differences exist between the helium fluxes measured during the four Apollo missions (table 15-II). In particular, the Apollo 14 and 15 fluxes differ by a factor of approximately 4; however, the relative abundances of helium and neon and also the isotopic abundances are very similar (table 15-III). Comparison with proton flux measurements obtained from unmanned spacecraft will show whether a similar difference existed in the fluxes of hydrogen during the Apollo 14 and 15 missions or whether the abundances of the helium and neon isotopes changed relative to hydrogen by a large and rather uniform factor.

The variations in the ^{4}He/^{3}He and ^{4}He/^{20}Ne ratios in the table 15-III are small but significant. Figure 15-2 shows that a correlation exists between the ^{4}He/^{3}He and ^{4}He/^{20}Ne ratios obtained from the four Apollos missions. An anticorrelation between these two ion-abundance ratios would be expected if electromagnetic-separation effects were operative either in the corona (ref. 15-24) or near the Moon (ref. 15-9). Conversely, theoretical considerations (refs. 15-24 and 15-25) on the acceleration of ions in the corona predict larger fluctuations in the ^{4}He abundance than in the ^{3}He and ^{20}Ne abundances; and, in fact, the ^{20}Ne/^{3}He ratio is the same for the four SWC experiments, within the limits of errors.

The data obtained from electrostatic analyzers on the Vela 3A and 3B satellites suggest a dependence of the helium/hydrogen ratio on solar activity (ref. 15-5). Explorer 34 results have shown an association

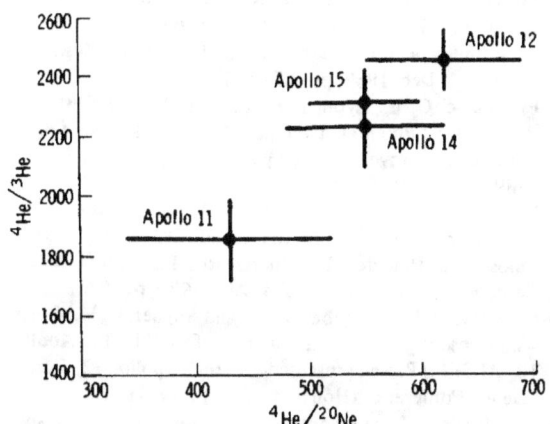

FIGURE 15-2.—Correlation between the ^{4}He/^{3}He and ^{4}He/^{20}Ne solar-wind-abundance ratios as determined for the four Apollo foil-exposure times.

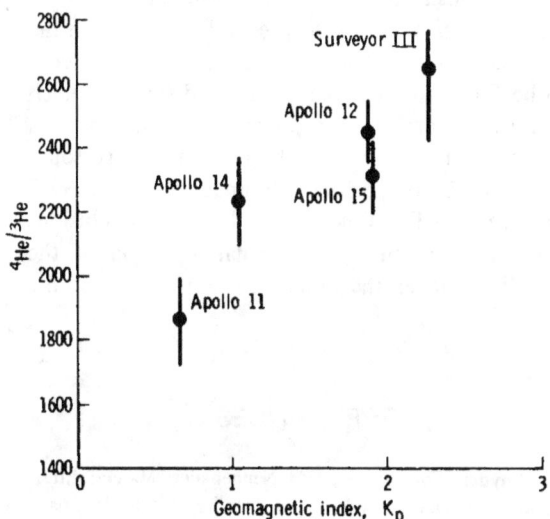

FIGURE 15-3.—Correlation between the solar-wind ^{4}He/^{3}He abundance ratio and the level of disturbance in the solar wind as indicated by the geomagnetic index K_p. The K_p values were averaged with a time shift of 8 hr relative to the solar-wind collecting periods. The results obtained from the analysis of returned Surveyor III material are described in reference 15-9.

of high helium/hydrogen ratios with geomagnetic storms (ref. 15-26). It has been observed that interplanetary shocks are followed—with a delay of approximately 8 hr—by a helium-rich plasma interpreted as the driver gas that produces the shock (refs. 15-27 and 15-28). Thus, it is somewhat well established that the helium-hydrogen ratio in the solar wind is linked to solar activity and the level of disturbances in the solar wind. To gain a better understanding of the dynamic processes underlying this observation, it is important to study the behavior of other ion species. In figure 15-3, the ^{4}He/^{3}He abundance ratio is plotted as a function of the geomagnetic index K_p. Included is the ^{4}He/^{3}He ratio obtained from a Surveyor III aluminum tube returned from the Moon by the Apollo 12 crewmen. After a correction of 2 percent for the difference in the trapping probabilities of the two helium isotopes, the measured value (ref. 15-29) represents the average ^{4}He/^{3}He ratio in the solar wind during the exposure period of this material (April 20, 1967 to November 20, 1969). The lower limit of error in this case includes an estimated possible change of the ^{4}He/^{3}He ratio as a result of helium diffusion from the aluminum. To allow for a delay between disturbance in the solar wind and change in composition as suggested by the delayed arrival of the helium-rich plasma after a shock (ref. 15-28), the K_p values were averaged with a time shift of 8 hr relative to the solar-wind collecting periods. In the case of the Surveyor material, the occurrence of two periods of solar-wind exposure during each lunar day (ref. 15-29) was correctly taken into account. In the case of the Apollo 11 material, the K_p indices were low and rather uniform during the 9 hr preceding and including the foil exposure.

CONCLUSIONS

A positive correlation is quite clearly indicated in figure 15-3. Because the general level of solar activity is expected to decline further in 1972, it is hoped that experiments on the remaining Apollo flights will record solar-wind abundances at low geomagnetic activity and firmly establish the correlation indicated in figure 15-3. This correlation may actually be expected on the basis of models of solar-wind-ion acceleration (ref. 15-24). This type of correlation will help in estimating the long-time average of the ^{4}He/^{3}He ratio in the solar wind. Only after this present average ^{4}He/^{3}He ratio is established will it be possible to investigate conclusively the question of a secular variation of this ratio in the outer convective zone of the Sun. Solar-wind helium implanted in lunar dust and breccias actually gives higher ^{4}He/^{3}He ratios than those obtained from the SWC experiments. However, so far, the observed differences are relatively small and could have resulted at least in part from diffusion or other secondary effects in the lunar material. A possibility exists that lunar material will be found that contains solar-wind samples much older than those studied previously. A comparison between such samples and the present average solar-wind helium composition obtained from the SWC experiments will allow more definite conclusions about the existence of a secular variation of the $^4\text{He}/^3\text{He}$ ratio in the outer convective zone of the Sun.

REFERENCES

15-1. Snyder, Conway W.; and Neugebauer, Marcia: Interplanetary Solar-Wind Measurements by Mariner II. Space Res., vol. 4, 1964, pp. 89-113.

15-2. Wolfe, J. H.; Silva, R. W.; McKibbin, D. D.; and Mason, R. H.: The Compositional, Anisotropic, and Nonradial Flow Characteristics of the Solar Wind. J. Geophys. Res., vol. 71, no. 13, July 1966, pp. 3329-3335.

15-3. Hundhausen, A. J.; Asbridge, J. R.; Bame, S. J.; Gilbert, H. E.; and Strong, I. B.: Vela 3 Satellite Observations of Solar Wind Ions: A Preliminary Report. J. Geophys. Res., vol. 72, no. 1, Jan. 1967, pp. 87-100.

15-4. Ogilvie, K. W.; Burlaga, L. F.; and Wilkerson, T. D.: Plasma Observations on Explorer 34. J. Geophys. Res., vol. 73, no. 21, Nov. 1968, pp. 6809-6824.

15-5. Robbins, D. E.; Hundhausen, A. J.; and Bame, S. J.: Helium in the Solar Wind. J. Geophys. Res., vol. 75, no. 7, Mar. 1970, pp. 1178-1187.

15-6. Bame, S. J.; Hundhausen, A. J.; Asbridge, J. R.; and Strong, I. B.: Solar Wind Ion Composition. Phys. Rev. Lett., vol. 20, no. 8, Feb. 19, 1968, pp. 393-395.

15-7. Bame, S. J.; Asbridge, J. R.; Hundhausen, A. J.; and Montgomery, Michael D.: Solar Wind Ions: $^{56}\text{Fe}^{+8}$ to $^{56}\text{Fe}^{+12}$, $^{28}\text{Si}^{+7}$, $^{28}\text{Si}^{+8}$, $^{28}\text{Si}^{+9}$, and $^{16}\text{O}^{+6}$. J. Geophys. Res., vol. 75, no. 31, Nov. 1970, pp. 6360-6365.

15-8. Bühler, F.; Eberhardt, P.; Geiss, J.; Meister, J.; and Signer, P.: Apollo 11 Solar Wind Composition Experiment: First Results. Science, vol. 166, no. 3912, Dec. 19, 1969, pp. 1502-1503.

15-9. Geiss, J.; Eberhardt, P.; Bühler, F.; Meister, J.; and Signer, P.: Apollo 11 and 12 Solar Wind Composition Experiments: Fluxes of He and Ne Isotopes. J. Geophys. Res., vol. 75, no. 31, Nov. 1970, pp. 5972-5979.

15-10. Geiss, J.; Bühler, F.; Cerutti, H.; Eberhardt, P.; and Meister, J.: The Solar-Wind Composition Experiment. Sec. 12 of Apollo 14 Preliminary Science Report. NASA SP-272, 1971.

15-11. Lyon, E. F.; Bridge, H. S.; and Binsack, J. H.: Explorer 35 Plasma Measurements in the Vicinity of the Moon. J. Geophys. Res., vol. 72, no. 29, Dec. 1967, pp. 6113-6117.

15-12. Ness, N. F.; Behannon, K. W.; Scearce, C. S.; and Cantarano, S. C.: Early Results from the Magnetic Field Experiment on Lunar Explorer 35. J. Geophys. Res., vol. 72, no. 23, Dec. 1967, pp. 5769-5778.

15-13. Siscoe, G. L.; Lyon, E. F.; Binsack, J. H.; and Bridge, H. S.: Experimental Evidence for a Detached Lunar Compression Wave. J. Geophys. Res., vol. 74, no. 1, Jan. 1969, pp. 59-69.

15-14. Freeman, J. W., Jr.; Hills, H. K.; and Balsiger, H.: Preliminary Results from the Apollo 12 ALSEP Lunar Ionosphere Detector I. General Results. Trans. Amer. Geophys. Union, vol. 51, no. 4, Apr. 1970, p. 407.

15-15. Clay, D. R.; Neugebauer, M.; and Snyder, C. W.: Solar Wind Observations on the Lunar Surface With the Apollo 12 ALSEP. Paper presented at the Apollo 12 Lunar Science Conference (Houston), Jan. 11-14, 1971.

15-16. Mihalov, J. D.; Sonett, C. P.; Binsack, J. H.; and Moutsoulas, M. D.: Possible Fossil Lunar Magnetism Inferred From Satellite Data. Science, vol. 171, no. 3974, Mar. 5, 1971, pp. 892-895.

15-17. Davies, J. A.; Brown, F.; and McCargo, M.: Range of Xe^{133} and Ar^{41} Ions of Kiloelectron Volt Energies in Aluminum. Can. J. Phys., vol. 41, no. 6, June 1963, pp. 829-843.

15-18. Bühler, F.; Geiss, J.; Meister, J.; Eberhardt, P.; et al.: Trapping of the Solar Wind in Solids. Earth Planet. Sci. Lett., vol. 1, 1966, pp. 249-255.

15-19. Meister, J.: Ein Experiment zur Bestimmung der Zusammensetzung und der Isotopenverhältnisse des Sonnenwindes: Einfangverhalten von Aluminium für niederenergetische Edelgasionen. Ph. D. thesis, Univ. of Bern, 1969.

15-20. Signer, Peter; Eberhardt, Peter; and Geiss, Johannes: Possible Determination of the Solar Wind Composition. J. Geophys. Res., vol. 70, no. 9, May 1965, pp. 2243-2244.

15-21. Geiss, J.; Eberhardt, P.; Signer, P.; Bühler, F.; and Meister, J.: The Solar-Wind Composition Experiment. Sec. 8 of Apollo 11 Preliminary Science Report. NASA SP-214, 1969.

15-22. Eberhardt, P.; Geiss, J.; Graf, H.; Grögler, N.; et al.: Trapped Solar Wind Noble Gases, Exposure Age and K/Ar Age in Apollo 11 Lunar Fine Material. Proceedings of the Apollo 11 Lunar Science Conference, Supp. 1, vol. 2, A. A. Levinson, ed., Pergamon Press (New York), 1970, pp. 1037-1070.

15-23. Wetherill, G. W.: Variations in the Isotopic Abundances of Neon and Argon Extracted from Radioactive Minerals. Phys. Rev., vol. 96, 1954, pp. 679-683.

15-24. Geiss, J.: On Elemental and Isotopic Composition of the Solar Wind. Paper presented at the Asilomar Conference on Solar Wind, Mar. 1971.

15-25. Geiss, Johannes; Hirt, Peter; and Leutwyler, Heinrich: On Acceleration and Motion of Ions in Corona and Solar Wind. Solar Phys., vol. 12, 1970, pp. 458-483.

15-26. Ogilvie, K. W.; and Wilkerson, T. D.: Helium Abundance in the Solar Wind. Solar Phys., vol. 8, 1969, p. 435.

15-27. Hirshberg, J. A.: Solar Wind Helium Enhancements Following Major Solar Flares. Paper presented at the Asilomar Conference on Solar Wind, Mar. 1971.

15-28. Hirshberg, J.A.; Alksne, A.; Colburn, D. S.; Bame, S. J.; and Hundhausen, A. J.: Observation of a Solar Flare Induced Interplanetary Shock and Helium-Enriched Driver Gas. J. Geophys. Res., vol. 75, Jan. 1970, p. 1.

15-29. Bühler, F.; Eberhardt, P.; Geiss, J.; and Schwarzmüller, J.: Trapped Solar Wind Helium and Neon in Surveyor 3 Material. Earth Planet. Sci. Lett., vol. 10, no. 3, Feb. 1971, pp. 297-306.

ACKNOWLEDGMENTS

The authors are indebted to J. Meister, presently at Rice University, for his part in the development and preparation of the experiment. The hardware construction and foil analyses were supported by the University of Bern and the Swiss National Science Foundation.

16. Gamma-Ray Spectrometer Experiment

James R. Arnold,[a,†] *Laurence E. Peterson,*[a] *Albert E. Metzger,*[b]
and Jack I. Trombka[c]

The Apollo 15 gamma-ray spectrometer experiment is one of a group of three orbiting geochemical experiments flown for the first time on this mission. The broad objective of geochemical mapping of the lunar surface is common to all three experiments. Because the landing sites for which chemical information is available are limited in number and not necessarily representative, it is important for understanding the origin and evolution of the Moon to have a source of data concerning chemical composition over as wide an area of the Moon as possible. The very detailed studies of returned lunar samples have permitted some progress in identifying major components of the soil, some components indigenous to the sites explored and some external to them. One task of the orbital geochemical experiments is to verify the occurrence of these suggested components or of others. More importantly, the expectation is that regions of the Moon in which each distinct material is the dominant constituent will be mapped. During transearth coast, the gamma-ray spectrometer measured the diffuse cosmic gamma-ray flux.

BASIC THEORY

Gamma rays are absorbed or scattered by passing through lunar soil or rock on the order of tens of centimeters thick. The gamma-ray experiment, therefore, can sample the composition of the Moon to a depth of this order. This layer is generally well within the regolith and can be reasonably assumed to be well mixed.

The chemical information in a gamma-ray spectrum is carried by discrete lines with energies that are characteristic of individual elements. Two broad classes of such lines exist. The first class, which traditionally is called natural radioactivity, results from the decay of potassium-40 (^{40}K), and the radioactive daughters of thorium (Th), and uranium (U). Important examples are the 1.46-MeV line emitted in the decay of ^{40}K to argon-40 and the 2.62-MeV line of thallium-208 (^{208}Tl), a daughter of ^{232}Th. The second class is composed of the lines that result from the bombardment of the lunar surface by high-energy charged particles, the cosmic rays. These particles interact with the lunar surface to produce secondary particles and excited nuclei.

The galactic cosmic rays (GCR) are responsible for nearly all the emitted gamma rays of this type. A typical GCR particle is a proton with a kinetic energy comparable to the rest mass, approximately 10^9 eV. The GCR particle interacts near the lunar surface to produce a cascade of lower energy particles, of which the most important are neutrons. These neutrons in turn give rise to excited nuclei that are capable of emitting line radiation in three ways. First, they may scatter inelastically, leaving the target nucleus in an excited state. This process is very important for neutron energies of a few million electron volts. Each major element produces such lines. An example is the 0.84-MeV line of iron (Fe). The second major process is neutron capture. Neutrons lose energy by successive collisions until they either escape from the surface or are captured. The binding energy of the added neutron, typically approximately 8 MeV, is emitted from the product nucleus in a complex decay scheme, which sometimes contains a few dominant lines. The line emitted by Fe at 7.64 MeV is an important case. Finally, gamma rays are emitted by radioactive nuclides produced by nuclear reactions, such as aluminum-26 (^{26}Al) from Al and silicon (Si).

---

[a]University of California, San Diego.
[b]Jet Propulsion Laboratory.
[c]NASA Goddard Space Flight Center.
[†]Principal investigator.

These reactions are generally less important, but not negligible.

The Sun emits high-energy particles during major solar flares, the so-called solar cosmic rays (SCR). The important energy region for these particles is in the range of 10 to 100 MeV. The particles lose energy mainly by ionization, but sometimes nuclear reactions occur. Except during the occurrence of a solar flare, the rapid processes of capture and scattering discussed previously cannot be observed. The SCR-induced radioactivity is found to be a small component in the Apollo 15 data.

The expected intensities of the spectrum lines as functions of chemical composition can be calculated from a knowledge of the physical processes involved. In the case of the natural radioactivities, this calculation is simple and unambiguous. For the lines induced by high-energy bombardment, the required fluxes and cross sections are known only approximately. The availability of "ground truth" for areas such as Mare Tranquillitatis and the Apollo 15 landing site, overflown by the Apollo 15 command-service module, is of great value. In the absence of the main data tapes for the experiment, only a limited amount of data processing has been possible to date. The main results available from the quick-look data show clear evidence of compositional differences among areas of the maria and highlands.

INSTRUMENT DESCRIPTION

The gamma-ray spectrometer is shown in figure 16-1, and a generalized block diagram is shown in figure 16-2. Within the cylindrical thermal shield are the detection, amplification, encoding, and data-processing systems that identify and characterize as functions of time and energy the incident gamma rays. The sensing element of the detector is a 2.75- by 2.75-in. cylindrical scintillation crystal of sodium iodide (Tl activated) (NaI(Tl)). The energy lost by gamma rays in traversing the crystal is converted, by means of ionization processes, into light that is sensed and transformed into a proportionate charge output by a 3-in.-diameter photomultiplier tube (PMT), which is optically coupled to the NaI(Tl) crystal. After amplification and shaping, the output signal passes to an analog-to-digital converter (ADC), which is controlled by a crystal-clock pulse generator that transforms the detector signal into an accurately measured pulse train. This pulse train is

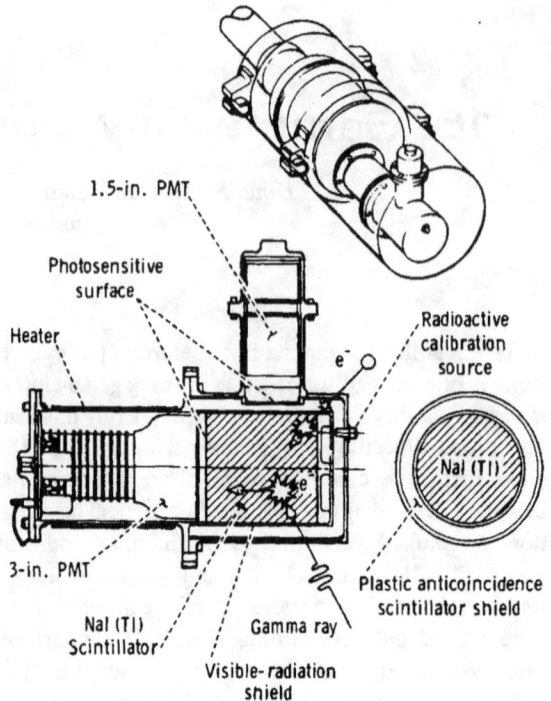

FIGURE 16-1.—Sketch of the gamma-ray spectrometer.

counted in an accumulator and results in a number proportional to the charge output of the detector for that particular event. The data are sent on an event-by-event basis either to the spacecraft telemetry system for direct transmission or to a tape recorder for intermediate storage when the spacecraft is behind the Moon.

The NaI(Tl) scintillator responds to charged particles as well as to gamma rays. To eliminate charged-particle events, a plastic-scintillator shield surrounds the NaI(Tl). The plastic scintillator detects all charged particles above a minimum energy, but has a low probability of interacting with gamma rays. Events in the plastic scintillator produce a signal in the 1.5-in. PMT that is transmitted to a gate ahead of the ADC to inhibit (veto) the acceptance of a coincident pulse from the NaI(Tl) PMT.

In addition to the accumulators that process the primary-data pulse train, separate accumulators are provided for counting the number of events in the plastic-scintillator shield, the number of events coincident in both the NaI(Tl) and plastic scintillators, and a live-time pulse train. The live-time pulse train provides the factor to derive the rate at which gamma

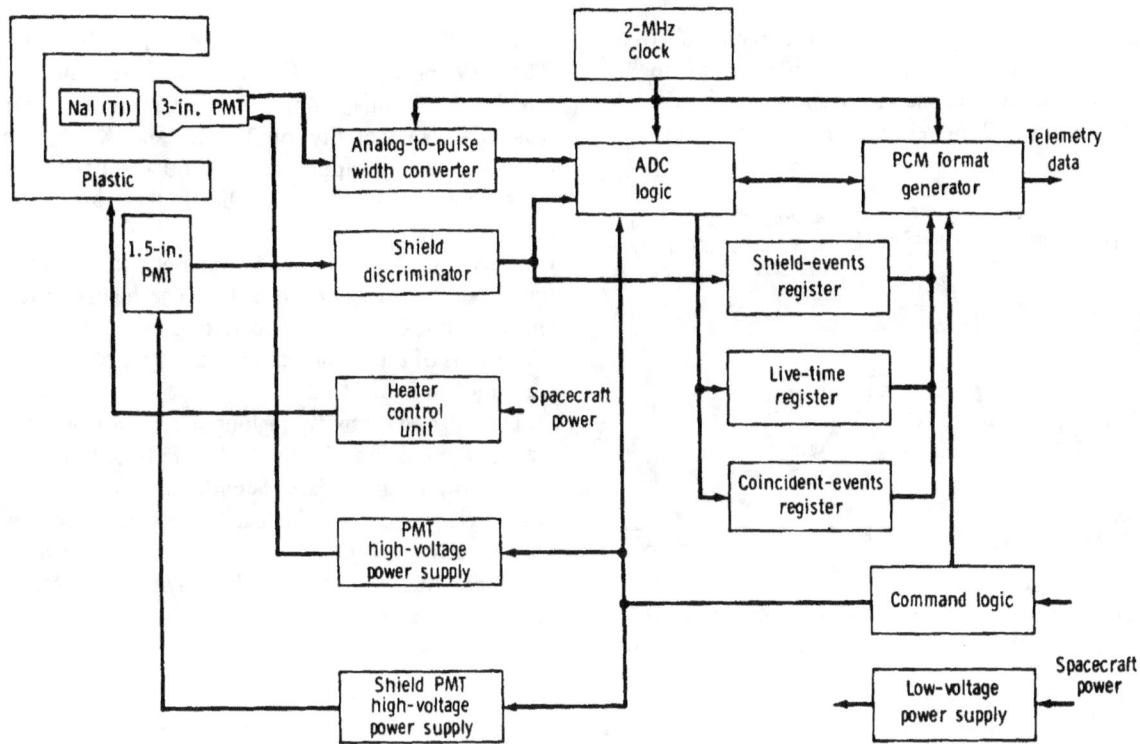

FIGURE 16-2.—Block diagram of the gamma-ray spectrometer.

rays are entering the NaI(Tl) crystal from the number that are analyzed. The command capability of the instrument allows the veto function to be enabled and disabled, the high-voltage bias on the 3-in. PMT to be varied in steps amounting to a total range in gain of a factor of approximately 3, and the power to be turned on and off.

Besides the passive thermal control provided by the striped paint pattern on the thermal shield, a thermal-control circuit supplies power to a heating blanket around the central detector when a control sensor indicates that the temperature in the immediate vicinity of the central detector has dropped below 15° C. The purpose of this capability is to minimize thermally induced variations in gain that, if rapid and continuous, would degrade the energy resolution of the instrument.

In normal operation, the instrument is deployed on a 7.6-m boom that extends normal to the scientific instrument module bay surface. The purpose of the boom is to decrease the response of the instrument to cosmic-ray interactions and to radioactive sources in the spacecraft. The importance of these effects can be seen from the observation that the count rates observed when the boom was retracted during transearth coast were close to those in lunar orbit with the boom extended; that is, the instrument response to the spacecraft was approximately the same as that to the Moon. In the extended position, the instrument response to the spacecraft is reduced to a few percent of the total count rate.

RESULTS

The experiment operated successfully in lunar orbit and during transearth coast. The total time during which data were taken in lunar orbit was close to nominal. During transearth coast, the mapping-camera door was open, so corrections for the Th contained in the camera lens as well as in the structural material of the guidance and navigation system must be made. Two instrument problems were encountered during the mission: a drift in gain that gradually stabilized and a discontinuous zero offset that occurred for a period during the transearth coast. The causes of both problems now seem to be understood; the data analysis will not be affected.

The most striking results seen so far are shown, in part, in figure 16-3. The energy region above the 0.51-MeV positron peak, up to and including the 2.62-MeV Th line, contains a major contribution from the radioactivity of Th, U, and K. A high counting rate of approximately 60 counts/sec occurred in this broad energy band; therefore, good statistical precision can be achieved. Thus, the accumulations over 5 min, or approximately 15° of longitude, can be compared directly. The high contrast and reproducibility of this counting rate over different regions of the Moon on two successive revolutions are shown in figure 16-3.

The highest activity regions are in the western maria, Oceanus Procellarum and Mare Imbrium. Mare Tranquillitatis and Mare Serenitatis have a much lower activity, and the highlands are still lower. The lowest activity seen is in the eastern far-side highlands. The highest levels of radioactivity seen are well below those seen in the Apollo 14 soil samples and

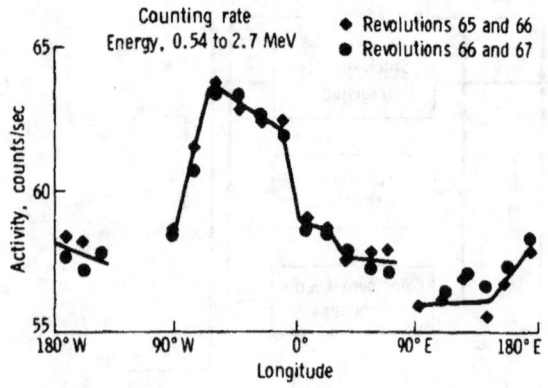

FIGURE 16-3.—Data counts for lunar revolutions 65 to 67.

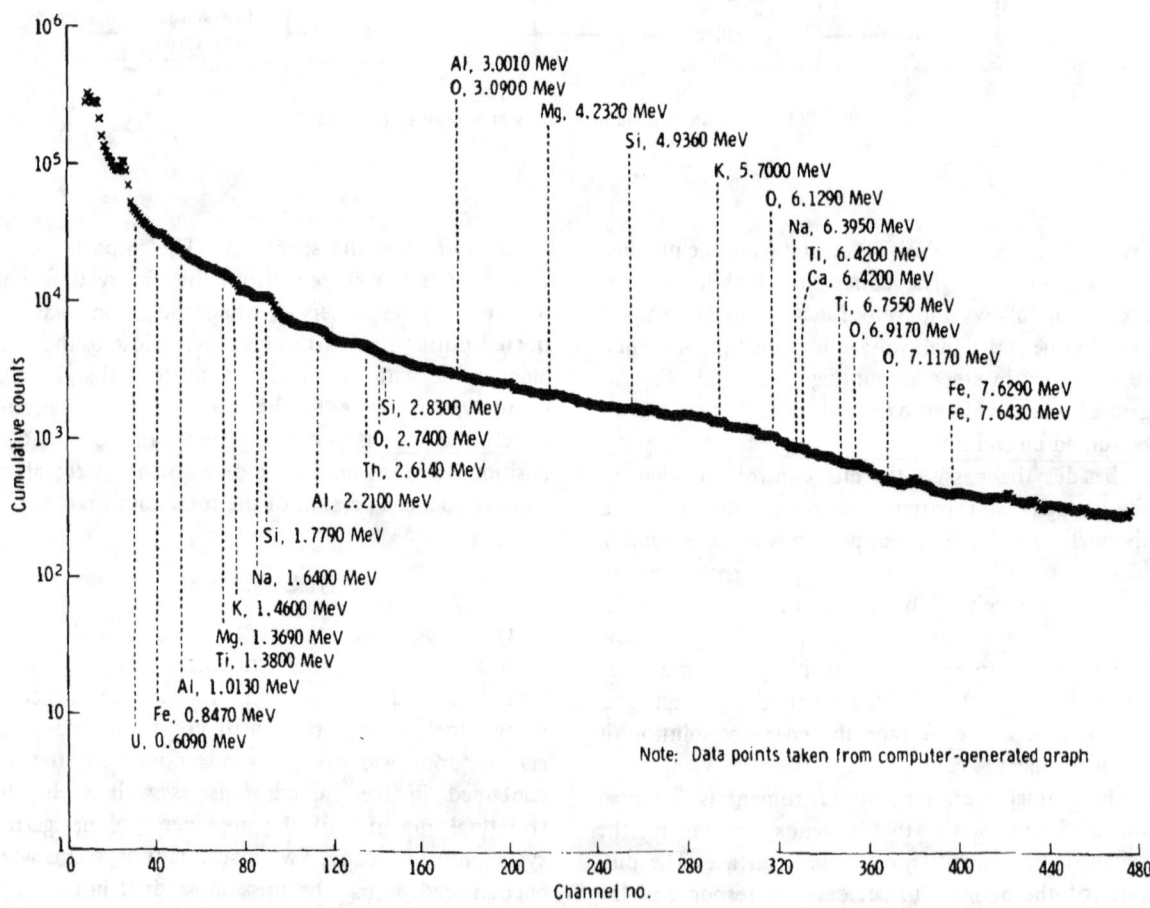

FIGURE 16-4.—Measured pulse-height spectrum integrated over approximately 7 hr.

are in the general range of the radioactivity observed in the Apollo 12 soil samples. This analysis also has been conducted for the earlier revolutions 39 to 41. Data are not as complete but are broadly similar. A sampling of data from other revolutions shows the same tendencies. In other energy bands, and especially for individual lines, data for single passes have less statistical significance. In the energy band from 3 to 9 MeV, a definite low count is seen in the region of Mare Imbrium, but the interpretation is not yet clear.

The measured gamma-ray spectra can be considered to be made up of two components—a featureless continuum and a discrete spectrum that contains the information from which inferences about surface composition can be made. A pulse-height spectrum integrated over a number of revolutions around the Moon is shown in figure 16-4. Positions of significant discrete gamma-ray lines are noted. By determining and removing the continuum component, the discrete line spectrum becomes more obvious. Figure 16-5 shows the result of removing the continuum component from the spectrum up to 4 MeV (fig. 16-4). By knowing the characteristic energies and intensities of the various lines in the discrete spectrum, elemental abundances for the lunar surface can be determined.

Preliminary examination of transearth-coast data permits a number of interesting conclusions to be made. Spectra with the boom extended provide the first measurements of the total cosmic gamma-ray background of energies up to 27 MeV. This spectrum (fig. 16-6) confirms previous measurements of energies up to approximately 1 MeV, but falls below the singular ERS-18 (military satellite) data points in the 2- to 6-MeV energy range. The spectrum obtained from the Apollo 15 mission is a steep continuum,

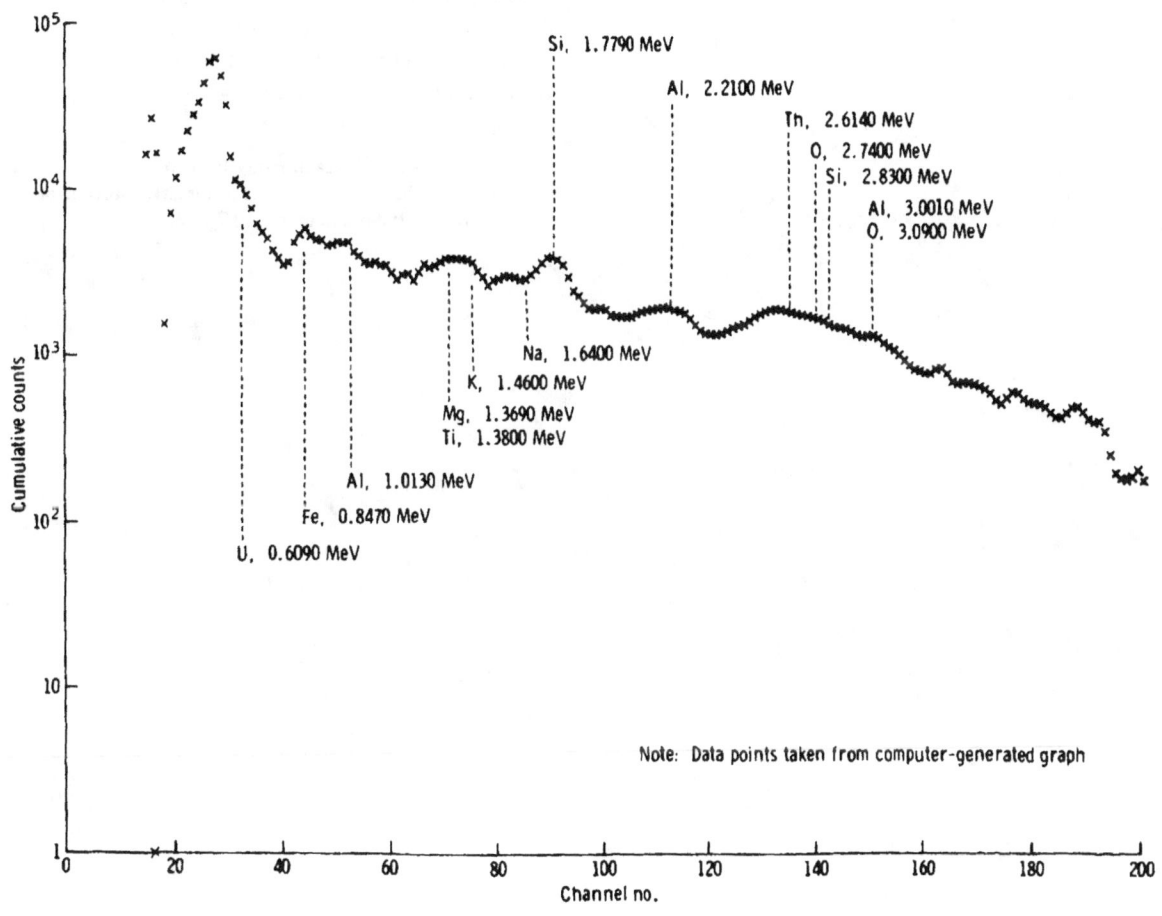

FIGURE 16-5.—Pulse-height spectrum of the discrete lines after the continuous spectrum has been deleted.

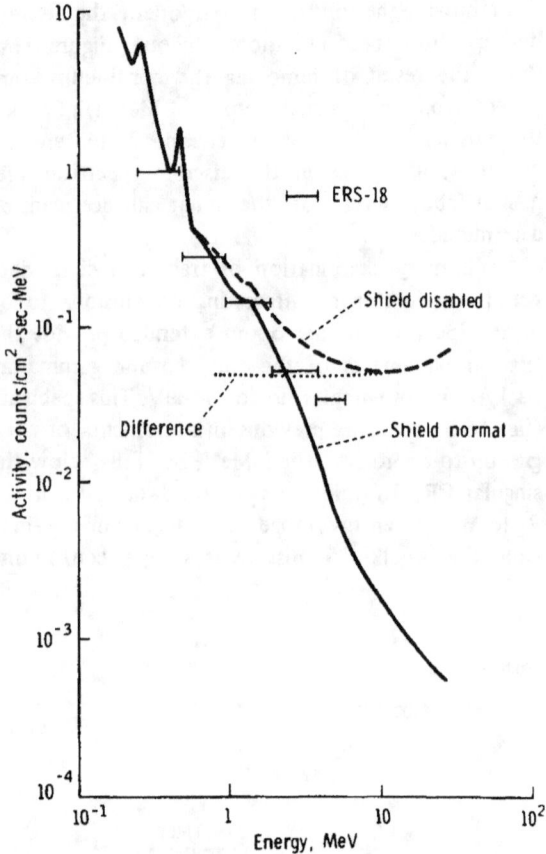

with very little structure and some indication of a change in slope for energies greater than 20 MeV. Further analysis is required to determine whether this effect is instrumental, involves the detector response, or represents a real feature of cosmic gamma rays. The most extreme models of cosmic-ray acceleration during early epochs of the expansion of the universe seem to be ruled out by the present results.

The line at 0.51 MeV, which is apparently a result of positron annihilation, is more clearly visible in the data obtained during transearth coast. Whether the line is of cosmic origin or is associated with materials near the detector may be difficult to determine. Most components of the spectrum measured in interplanetary space are factors of 3 to 7 below the components measured for the Moon and for the spacecraft with the boom retracted. Therefore, spacecraft production is not a dominant background effect for the lunar measurements. Based on the rather small intensity change with the boom extensions at 2.4, 4.6, and 7.6 m, the conclusion can be tentatively made that spacecraft production is also not important for the gamma-ray astronomy data.

FIGURE 16-6.—Cosmic gamma-ray background spectra. The data were obtained during the transearth portion of the mission with the boom extended 7.6 m.

17. X-Ray Fluorescence Experiment

*I. Adler,[a][†] J. Trombka,[a] J. Gerard,[a][b] R. Schmadebeck,[a]
P. Lowman,[a] H. Blodgett,[a] L. Yin,[a] E. Eller,[a] R. Lamothe,[a]
P. Gorenstein,[c] P. Bjorkholm,[c] B. Harris,[c] and H. Gursky[c]*

The X-ray fluorescence spectrometer, carried in the scientific instrument module (SIM) bay of the command-service module (CSM), was used principally for orbital mapping of the lunar-surface composition and, secondarily, for X-ray astronomical observations during transearth coast. The lunar-surface measurements involved observations of the intensity and characteristic energy distribution of the secondary or fluorescent X-rays produced by the interaction of solar X-rays with the lunar surface. The astronomical observations consisted of relatively long periods of X-ray measurement of preselected galactic sources such as Cygnus (Cyg X-1) and Scorpius (Sco X-1) and of the galactic poles.

COMPOSITIONAL MAPPING OF THE LUNAR SURFACE

Theoretical Basis

The production of X-rays characteristic of an element can be understood in terms of the simple Bohr theory of the atom. The electrons surround the nucleus in an orderly fashion in a series of shells — K, L, M, and so forth — from the nucleus out to the valence shell. Within any given atom, the binding energy decreases from the inner electrons, where the binding energy is greatest, to the outer shells.

Because characteristic X-ray spectra result from the filling of the vacancies produced by the ejection of these inner-shell electrons, greater input energy is required to excite the K spectrum than is required to excite the L spectrum, and even less energy is required to excite the M spectrum. The situation is summarized in figure 17-1. Tabulated within figure 17-1 are the absorption-edge energies E_k, which are required to ionize the atoms in the K shell, and $E(K_\alpha)$, the energies of the resulting X-ray lines. Therefore, to produce the characteristic X-rays, an incident energy in excess of the binding energy of the electron is necessary.

In the fluorescence experiment described in this section, the production of the characteristic X-rays follows the interaction of solar X-rays with the lunar surface. The result of numerous calculations indicates that the typical solar X-ray spectrum is energetically

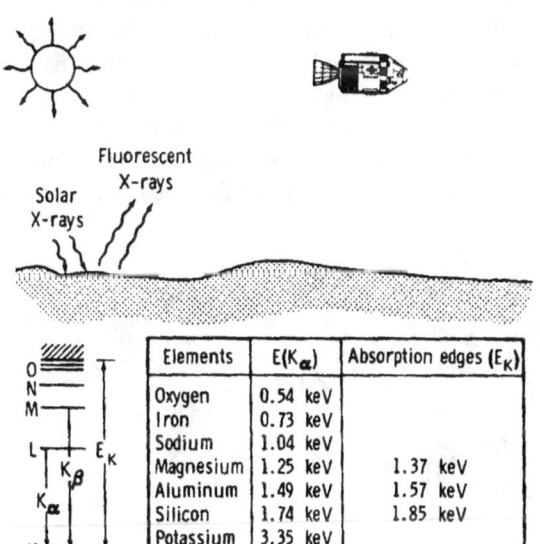

FIGURE 17-1.—The X-ray fluorescence at the lunar surface. On the left are inner electron transitions resulting in characteristic K X-ray spectra. Absorption-edge energies and the energies of the resulting X-ray lines are tabulated.

---
[a]NASA Goddard Space Flight Center.
[b]National Academy of Sciences.
[c]American Science and Engineering.
[†]Principal investigator.

capable of producing measurable amounts of characteristic X-rays from all the abundant elements with atomic numbers of approximately 14 (silicon (Si)) or less. During brief periods of more intense solar activity, observation of radiation from elements of higher atomic number also should be possible.

In the final analysis of the data, however, some features of the X-ray production of the Sun must be considered in greater detail. The solar X-ray flux varies greatly on time scales of minutes to hours. In addition, systematic changes occur that are associated with the 11-yr solar cycle.

This long-term trend is shown in figure 17-2 (ref. 17-1). A curve of the smoothed sunspot number is shown on the same axis as the observed X-ray intensity. A definite correlation exists between the sunspot number and the X-ray intensity, so the extrapolated smoothed sunspot number is an indication of the expected solar intensity for the next few years.

The solar X-ray flux, with low-resolution instruments such as proportional counters, decreases with increasing energy. If a strictly thermal mechanism of production is assumed, variable coronal temperatures are found somewhere between 10^6 to 10^7 K. Such variations in temperature produce changes in both flux and spectral composition. Thus, changes must be expected not only in fluorescent intensities but in the relative intensities from the various elements being observed. For example, if the solar spectrum hardens (larger fluxes of higher energies) or if there is an increase in characteristic line intensities on the high-energy side of the absorption edge of the heavier element, an enhancement of the intensities from the heavier elements relative to the lighter ones would be observed.

An X-ray monitor was used to follow the possible variation in solar X-ray intensity and spectral shape. In addition, detailed simultaneous measurements of the solar X-ray spectrum were obtained in flight during the mission from the various Explorer satellites that measure solar radiation.

An estimate of the emitted solar X-ray flux for an assumed temperature of 4×10^6 K and a gray-body emitter is shown in figure 17-3. Superimposed on this

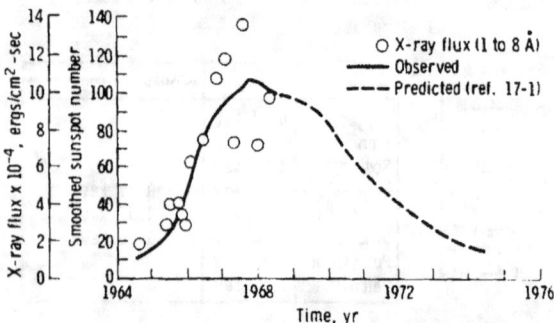

FIGURE 17-2.—Smoothed sunspot number and observed X-ray intensity as a function of time during the solar cycle.

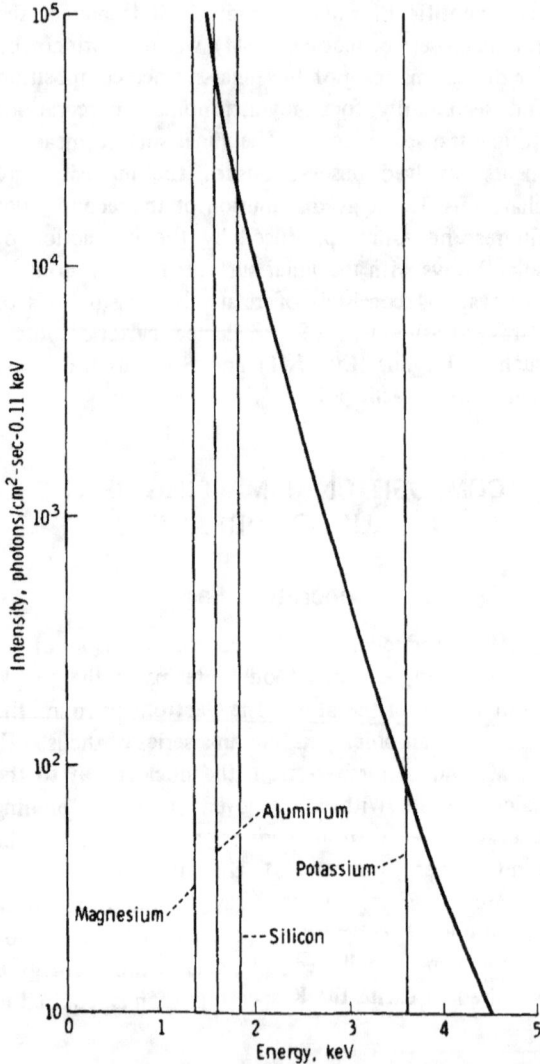

FIGURE 17-3.—The solar spectrum calculated for a coronal temperature of 4×10^6 K. Lines showing the positions of the absorption edges for Mg, Al, Si, and K are superimposed.

curve along the energy axis are the K-shell absorption edges for magnesium (Mg), aluminum (Al), Si, and potassium (K). Only the solar X-rays with energies on the high side of the absorption edges are capable of exciting these elements and to a degree depending on the incident flux and ionization cross section. Therefore, under quiet-Sun conditions, the solar flux is most suitable for exciting the light elements, including the major rock-forming elements Si, Al, and Mg.

Equipment is used to measure both X-ray intensities and energies because the Sun will produce secondary X-rays characteristic of such rock-forming elements as Mg, Al, and Si. A Bragg spectrometer cannot be used for precise wavelength selection because the amounts of secondary X-rays produced are relatively small. Instead, low-resolution but high-sensitivity techniques are required, such as proportional counters and pulse-height analysis (ref. 17-2). Selected X-ray filters are also used to provide additional energy discrimination.

Description of the Instrument

The X-ray fluorescence and alpha-particle experiment is shown in figure 17-4, and a functional configuration of the spectrometer is shown in figure 17-5. The spectrometer consists of three main subsystems.

(1) Three large-area proportional counters that have state-of-the-art energy resolution and 0.0025-cm-thick beryllium (Be) windows

(2) A set of large-area filters for energy discrimination among the characteristic X-rays of Al, Si, and Mg

(3) A data-handling system for count accumulation, sorting into eight pulse-height channels, and, finally, for relaying the data to the spacecraft telemetry

The X-ray detector assembly consists of three proportional-counter detectors, two X-ray filters, mechanical collimators, an inflight calibration device, temperature monitors, and associated electronics. The detector assembly senses X-rays emitted from the lunar surface and converts the X-rays to voltage pulses that are processed in the X-ray processor assembly. Provisions for inflight calibration are made by means of programed calibration sources, which, upon internal command, assume a position in front of the three detectors for calibration of gain, resolution, and efficiency. Thermistors located at strategic points sense the temperature of the detector assembly for telemetry monitoring and temperature control of the detectors by means of heaters located near the proportional counter windows.

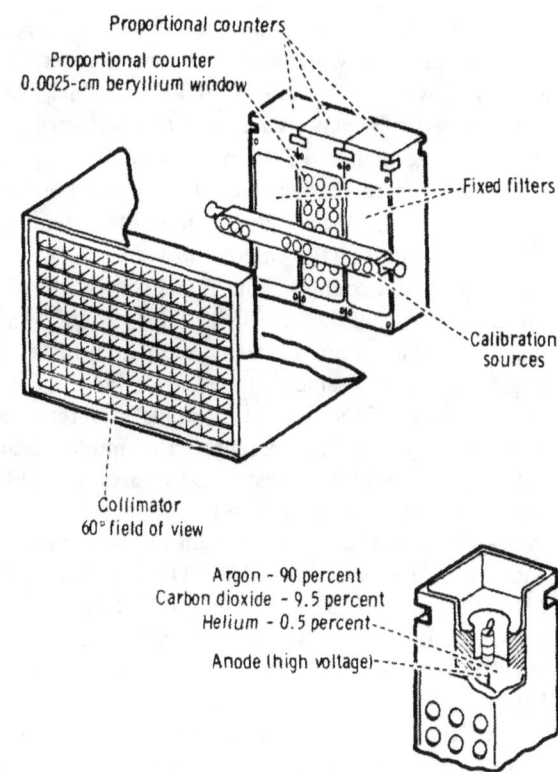

FIGURE 17-5.— Functional configuration of the X-ray spectrometer.

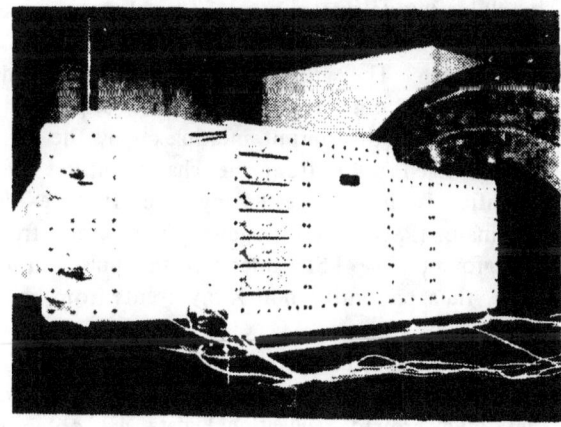

FIGURE 17-4.—The X-ray fluorescence and alpha-particle experiment. The 10 alpha-particle detectors are shown to the left of the proportional counter collimator.

The three proportional counters are identical, each having an effective window area of approximately 25 cm^2. The window consists of 0.0025-cm-thick Be. The proportional counters are filled to a pressure of 1 atm with the standard P-10 mixture of 90 percent argon, 9.5 percent carbon dioxide, and 0.5 percent helium. To change the wavelength response, filters are mounted across the Be window aperture on two of the proportional counters. The filters consist of a Mg foil and an Al foil 5.08×10^{-4} to 1.27×10^{-3} cm thick. The third counter does not contain a filter. A single collimator assembly is used to define the field of view (FOV) of the three proportional counters as a single unit. The collimator consists of multicellular baffles that combine a large sensitive area and high resolution but are restricted in FOV. The FOV determines the total flux recorded from the lunar surface and the spatial resolution. The FOV is specified as ±30° full width, half maximum (FWHM), in two perpendicular directions. The FWHM is the total angular width at which the collimator drops to one-half of its peak response.

The X-ray photons passing through a proportional-counter Be window ionize the gas inside by an amount proportional to the X-ray photon energy. A very stable high-voltage power supply provides a bias voltage for the operation of the proportional counters. This high voltage across the counter produces an electrical-field gradient and, hence, a multiplication effect that results in a charge output proportional to the incident X-ray energy. A charge-sensitive preamplifier that converts the input charge to an output pulse by storing it on an integrating capacitor is mounted on each proportional counter. This pulse has a fast rise time, determined primarily by the response of the preamplifier; a slow decay, determined by the integrator decay time; and an amplitude proportional to the X-ray energy. The preamplifier gain is set for an output scale factor of approximately 0.2 V/keV. Each of the preamplifier outputs is applied to the X-ray processor assembly that sorts the outputs according to the peak amplitude level.

The inflight calibration device consists of a calibration rod with radioactive sources (Mg and manganese-K radiation) that normally face away from the proportional counters. Upon internal command from the X-ray processor assembly, the rod is rotated 180° by a solenoid driver, thereby positioning the sources to face the proportional counters. Magnetically sensitive reed relays provide feedback signals indicating when the rod is fully in the calibration mode or fully in the noncalibration mode. These feedback signals are flag bits in the data telemetry output. The calibration command signal is generated in the X-ray processor assembly. The calibration cycle repeats every 16 min and continues 64 sec.

The X-ray processor assembly processes X-ray data received from the detector assembly and from the solar monitor. The lunar X-ray data and the solar X-ray data from the solar monitor are sorted, counted, stored, and sent to telemetry. Processing of the data from one detector is shown in a functional block diagram in figure 17-6. The pulse received from the charge-sensitive preamplifier is amplified and operates as many as eight voltage discriminators, depending on the voltage level. The discriminator outputs are processed logically in the pulse-routing logic to obtain an output pulse in one of eight data channels, depending on the highest-level discriminator operated. Thus, eight channels of differential pulse-height spectra are obtained. The first seven channels are equal in width, and the eighth channel contains the integral number of events with energies greater than channel 7. The pulses from each data channel are counted by the counters in the counter-shift-register logic. Every 8 sec, the contents of the counters are transferred to the shift registers, and the counters are reset. The data are then sequentially shifted out of the shift registers to telemetry at a 10-word/sec rate. Each telemetry word consists of eight bits. Each counter is 16 bits long, thereby supplying two telemetry words. The telemetry-word-output sequence is divided into four groups of 20 words each. The groups are obtained from the 20-word-long shift registers that are sequentially gated through the output multiplexer by the main timing. Each pulse from the charge-sensitive preamplifier is also processed by a pulse-shape discriminator (PSD) that distinguishes X-ray events from background. The PSD gates off the pulse-routing logic, thus preventing non X-ray events from being counted.

Data transmitted from the CSM are recorded at ground stations on magnetic tape. As raw spectral data and ground navigational data are obtained, ephemeris data are derived. The X-rays observed by each of the three lunar detectors are sorted into seven equal-interval energy channels. The bare detector is

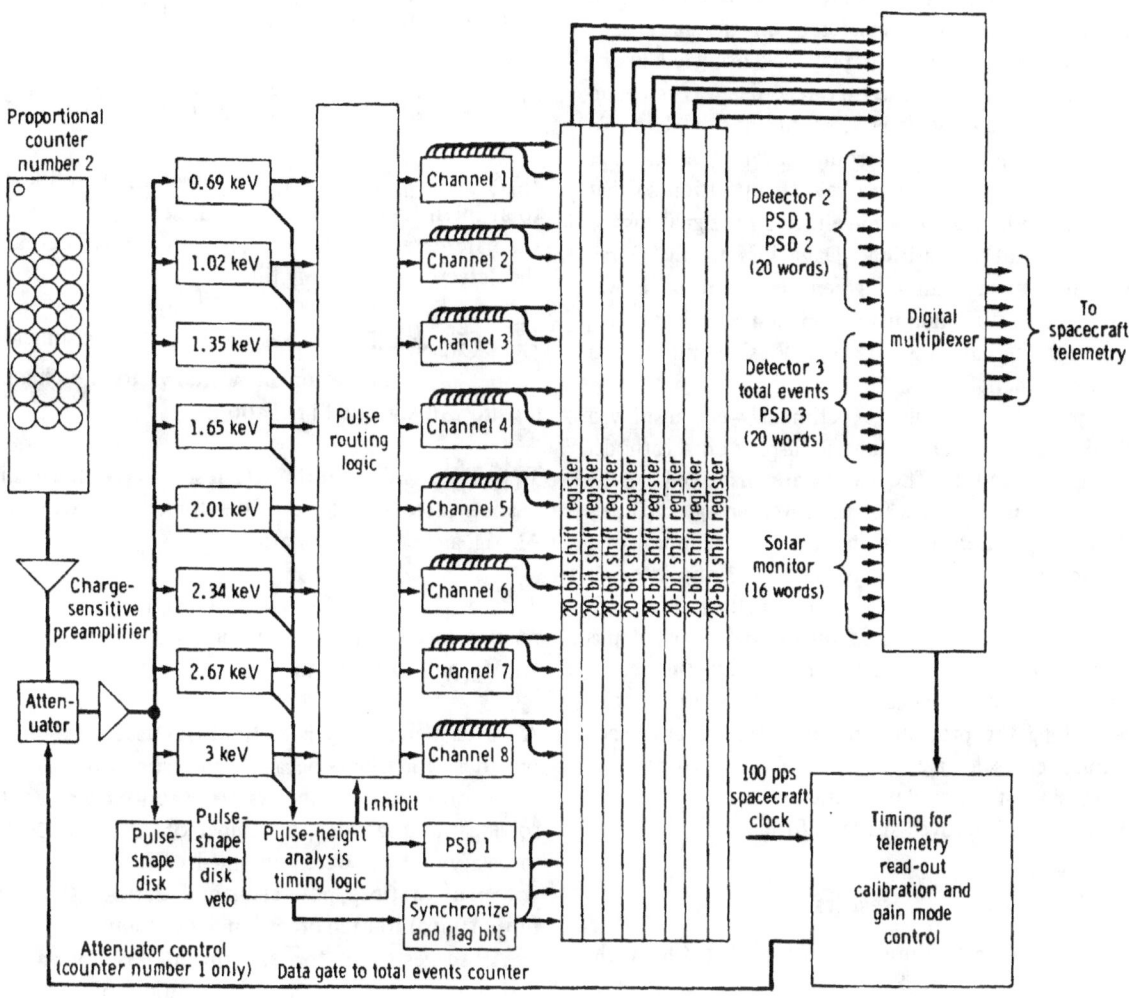

FIGURE 17-6.—Block diagram showing the data flow in the X-ray spectrometer.

operated in two modes: the high-gain (normal) mode and the attenuate mode. After the initial turnon, the bare detector operates 9.87 sec in the attenuate mode, covering a range of approximately 1.5 to 5.5 keV. At the end of this cycle, the bare detector automatically goes into the normal mode, which covers the range 0.75 to 2.75 keV. During flight, the bare detector automatically operates for 6 hr in the normal mode and 2 hr in the attenuate mode. The filtered surface detectors continuously cover the spectral range of 0.75 to 2.75 keV. The solar monitor also covers a range of approximately 1 to 3 keV.

The printout indicates the number of pulses of a given energy that were observed for each detector in a given energy channel for a given measurement interval. Histograms that are subsequently reduced by alternate techniques to line intensities and finally to chemical concentrations can be plotted from this printout.

Operation of the Experiment

The X-ray experiment began to function 84 hr into the flight, during the third revolution around the Moon. From 84 to 102 hr ground elapsed time (GET), the orbit was approximately 8 by 60 n. mi. After 102 hr, the orbit was circularized and maintained at approximately 60 n. mi. until transearth coast. During the orbital period, more than 100 hr of surface measurements were made. The solar-monitor detector was used for simultaneous monitoring of solar X-ray flux. Lunar-far-side data were recorded on magnetic tape and telemetered on the near side. Because both record and playback occurred in the same tape direction, the tape recorder was rewound each time, causing some loss of far-side data. To maximize the amount of ground coverage, an attempt was made to alternate the rewind operation so that the operation would occur either after signal loss or before signal acquisition. Data loss in the X-ray experiment occurred only when the tape recorder was rewound on the illuminated portion of the far side; however, the data loss has not seriously compromised the experiment.

The data from the experiment were displayed almost real time as a numeric display on a cathode-ray-tube monitor. The data were displayed in the form of running sums for the seven energy channels for each of the four detectors. The data, supplied as hard-copy printout, were updated at 1-min intervals and at regular intervals of approximately 4 hr. This preliminary report is based on the reduction of this quick-look data. The results presented in this report are, of necessity, degraded in terms of spatial resolution along the projected ground track because each minute represents a 3° longitudinal displacement. The processing of prime data, obtained at 8-sec intervals, should yield improved spatial information.

Results

Alternate procedures for processing the flight information are available, depending on the degree of sophistication warranted by the data. The data in this report have been treated in a very simple fashion based on the energy discrimination afforded by the selected X-ray filters. The following assumptions have been made.

(1) All three proportional counters have identical characteristics

(2) The detectors are 100-percent efficient for the radiation transmitted through the detector window

(3) Background corrections can be made using the measured far-side fluxes

(4) The effect of X-ray scattering on the Al/Si intensity ratios, although authentic, can be ignored in these first estimates

Based on these assumptions, three simultaneous expressions of the following form were written.

$$I_{Bare} = T^{Be}_{Al} I_{Al} + T^{Be}_{Mg} I_{Mg} + T^{Be}_{Si} I_{Si} \quad (17\text{-}1)$$

$$I_{Al\ filter} = T^{Be}_{Al}\left(I_{Al} T^{Al}_{Al}\right) + T^{Be}_{Mg}\left(I_{Mg} T^{Al}_{Mg}\right) + T^{Be}_{Si}\left(I_{Si} T^{Al}_{Si}\right) \quad (17\text{-}2)$$

$$I_{Mg\ filter} = T^{Be}_{Al}\left(I_{Al} T^{Mg}_{Al}\right) + T^{Be}_{Mg}\left(I_{Mg} T^{Mg}_{Mg}\right) + T^{Be}_{Si}\left(I_{Si} T^{Mg}_{Si}\right) \quad (17\text{-}3)$$

I_{Bare}, $I_{Al\ filter}$, and $I_{Mg\ filter}$ are, respectively, the total intensities summed over all seven channels for the bare detector, the detector with the Al filter, and the detector with the Mg filter.

T^{Be}_{Al}, T^{Be}_{Mg} and T^{Be}_{Si} are, respectively, the transmission factors of the detector Be windows to the characteristics Al, Mg, and Si radiation.

T^{Al}_{Al}, T^{Al}_{Mg}, and so forth are, respectively, the transmission factors of the Al filter for the characteristic Al, Mg, and so forth radiation.

T^{Mg}_{Al}, T^{Mg}_{Mg}, and so forth are, respectively, the transmission factors of the Mg filter for the characteristic Al, Mg, and so forth radiation.

A least-squares analysis method was used to solve the previous equations because an estimation of the statistical validity of the results is easily obtained. In solving the equations, a number of Al/Si and Mg/Si intensity ratios were obtained. These are X-ray intensity ratios, not elemental ratios, that are proportional to each other but are not equal.

As suggested in the experiment description, the flux intensity and energy distribution of the solar X-rays were expected to have a significant effect on the nature of the fluorescent X-rays. The purpose of the solar monitor was to observe the intensity and energy distribution of the solar X-rays as a function of time, simultaneously with the surface measurements. Figure 17-7 is a plot of the integrated intensities registered by the solar monitor for the period corresponding to the surface measurements. These values were observed at the subsolar point. A plot is also shown for comparing the surface intensities as observed with the solar detector. With the exception of revolutions 67 and 73, the solar flux was fairly stable, varying less than ±30 percent of the mean value. This stability was equally reflected in the surface data, which indicate a stable incident flux as well as a stable spectral distribution.

During revolutions 67 and 73, a marked increase in flux and, apparently, a hardening of the energy

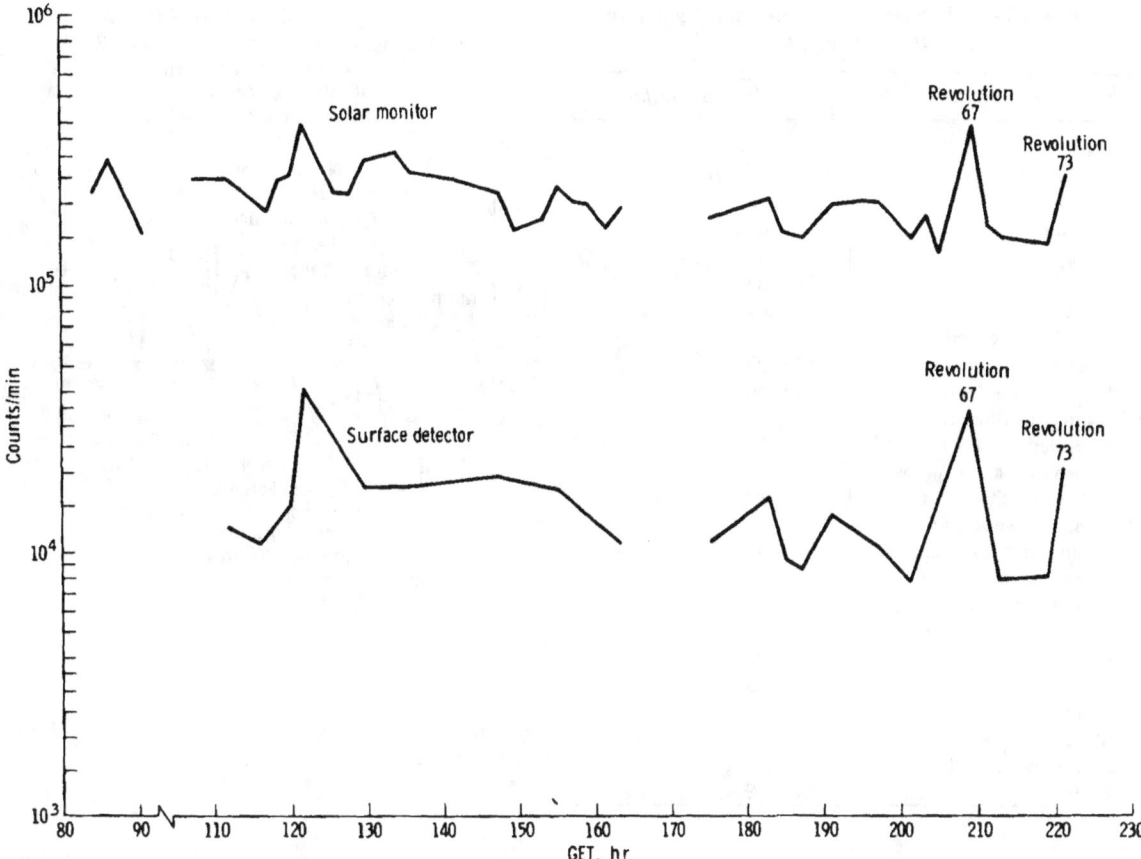

FIGURE 17-7.—Comparison of solar X-ray flux and surface flux.

spectrum occurred. This phenomenon was reflected in reduced Al/Si intensity ratios, which were probably caused by enhanced Si excitation. On subsequent revolutions, the solar-flux intensity and the Al/Si intensity ratios returned to more nearly normal values.

Some of the Al/Si intensity ratios were averaged for corresponding revolutions because of their close overlap. The agreement between values on the close ground tracks was approximately ±10 percent on the average. (The scatter is based on a 1-σ variation calculated by taking repetitive values for a given area.) An examination of the ground tracks from east to west indicates that the spacecraft passed over such features as the craters Gagarin and Tsiolkovsky; the far-side and eastern-limb highlands; the mare areas Mare Smythii, Mare Crisium, Mare Fecunditatis, Mare Tranquillitatis, Mare Serenitatis, Mare Imbrium, and Oceanus Procellarum; the Haemus Mountains; and the Apennine Mountains. The Al/Si intensity ratios varied by more than a factor of 2 between the eastern-limb highlands and the mare areas, where the ratios were the lowest.

These values are detailed in table 17-I and are plotted synoptically in figures 17-8 and 17-9. Figure 17-8 represents a northerly trajectory (revolutions 16, 26 to 28, and 34), and figure 17-9 represents a more southerly course (revolutions 63, 64, and 72). The following observations can be made.

(1) The Al/Si intensity ratios are lowest over the mare areas and highest over the nonmare areas (an average of 0.67 as opposed to 1.13). The extremes vary from 0.58 to 1.37, by a factor of more than 2.3.

(2) The value for the Apennine Mountain region is 0.88, and the lower average value of the Haemus Mountains is 0.83. On the other side of the Apennine Mountains in the Archimedes Rille area, the observed value is 0.64. On either side of the Apennine

TABLE 17-I.—*Averaged Al/Si Intensity Ratios for Various Lunar Features*

Location	Intensity ratios
Mare areas	
Serenitatis	0.58 ± 0.06 (rim, 0.71 ± 0.05)
Imbrium	.59 ± .04
Crisium	.71 ± .02 (rim, .80 ± .09)
Tranquillitatis	.71 ± .05 (rim, .84 ± .04)
Fecunditatis	.73 ± .07 (rim, .94 ± .14)
Smythii	.73 ± .07 (rim, 1.00 ± .11)
Archimedes Rille area	.64 ± .03
Apennine Mountains	.88 ± .03
Haemus Mountains	.83 ± .10
Highlands east of Mare Serenitatis to 40° E	.80 ± .08
Highlands west and southwest of Mare Crisium	1.08 ± .13
Highlands between Mare Crisium and Mare Smythii	1.07 ± .11
Highlands between Mare Smythii and Tsiolkovsky	1.16 ± .11
Highlands east of Mare Fecunditatis	1.29 ± .23
Highlands east of Tsiolkovsky	1.37 ± .25
Area north of Schröter's Valley	.73 ± .05
Area northeast of Schröter's Valley	.64 ± .09

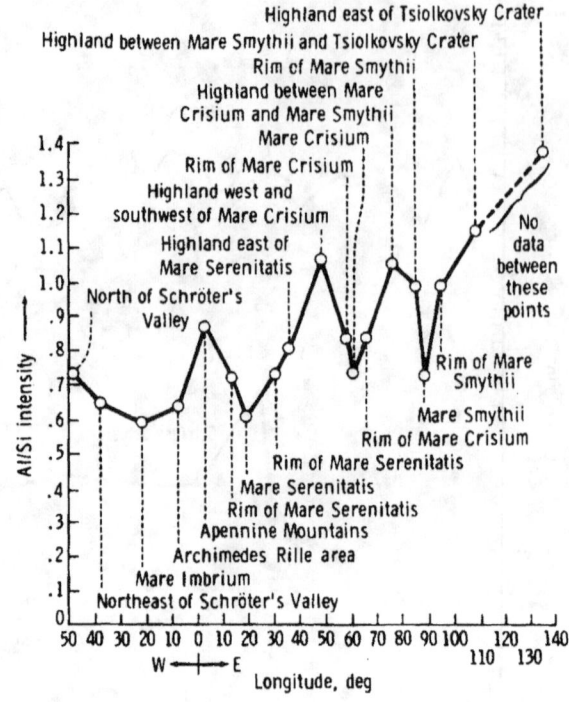

FIGURE 17-8.—Plot of Al/Si intensity ratios along a northerly ground track.

Mountains, the values for Mare Imbrium and Mare Serenitatis are 0.59 and 0.58. The Apennine Mountains have values between those of the maria and the eastern-limb highlands.

(3) An examination of the Al/Si coordinate plot reveals that the values tended to increase from the western mare areas to the eastern-limb highlands.

(4) As predicted, the rim areas are intermediately between the mare areas and the surrounding highlands.

Another interesting correlation between Al/Si ratios and albedo values along selected ground tracks is shown in figures 17-10 (ref. 17-3) and 17-11. Generally, higher Al/Si ratios correspond to higher albedo values. There are occasional deviations from these relationships caused by surface features; for example, the Copernican-type crater identified in the plot in figures 17-10 and 17-11. The composition of areas with different albedos can be inferred by using the X-ray data. It is possible, for example, to state whether the albedo variations are related to chemical differences or to the nature and, perhaps, age of a given feature.

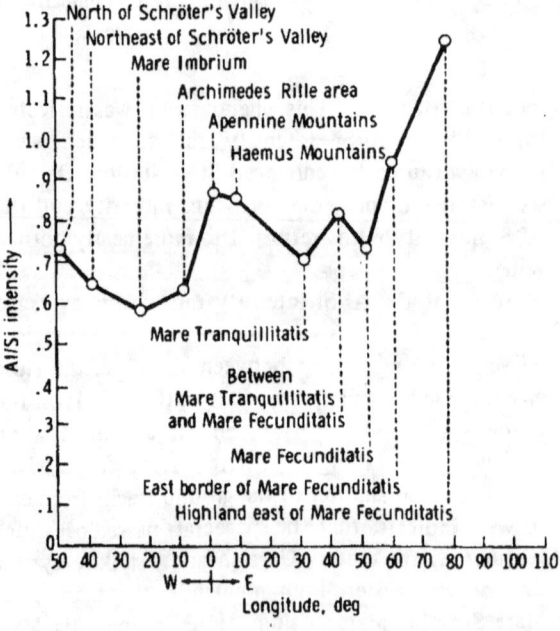

FIGURE 17-9.—Plot of Al/Si intensity ratios along a southerly ground track.

Other interesting correlations exist, but the comments on these are necessarily of a very preliminary nature. An attempt has been made to plot Al/Si intensity ratios along a gravitational profile. The observed correlation indicates that the Al/Si intensity ratios vary inversely with the gravitational values. For example, the lowest Al/Si intensity ratios are found in the regions of greatest positive gravitational anomalies.

Finally, an attempt has been made to arrive at actual concentration ratios for Al/Si. These concentrations are determined by an approach that is both theoretical and empirical. The theoretical calculations are based on the assumption of a quiet Sun and a coronal temperature of 4×10^6 K. These conditions give an X-ray energy distribution, consisting of both a continuum and characteristic lines, that is consistent with the solar-monitor observations shown normalized in figure 17-12. The X-ray energy distribution and various soil compositions, as determined from the analysis of lunar samples, have been used to calculate a relationship between Al/Si intensity ratios as a function of chemical ratios.

The empirical approach involves the assumption that the soil values from the Apollo 11 site at Tranquillity Base and the Luna 16 soil values from Mare Fecunditatis are ground-truth values. With these values and the theoretically calculated slopes, the

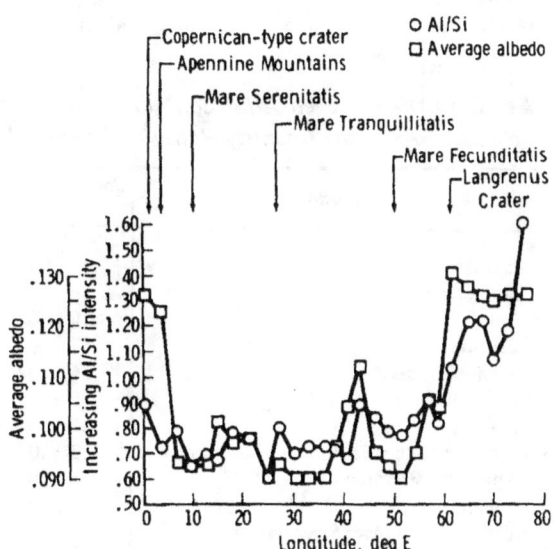

FIGURE 17-10.—Plot of Al/Si intensity ratios as a function of normal albedo values. Albedo values are for points corresponding to average Al/Si intensities for read-out intervals shown on map, for revolutions 63 and 64 (figs. 17-13 and 17-14).

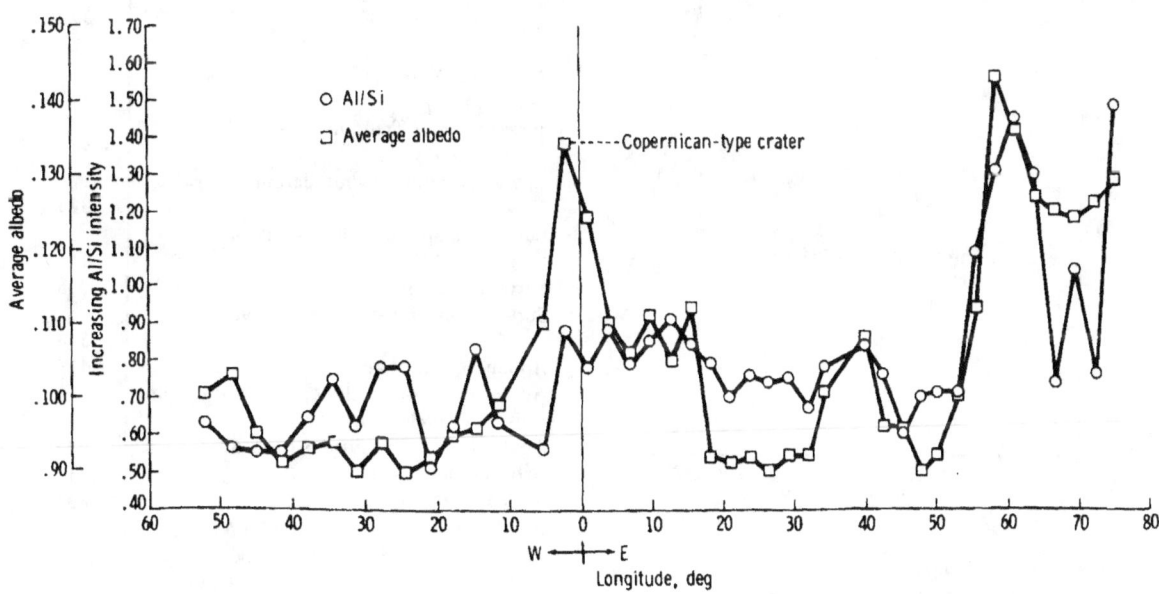

FIGURE 17-11.—Comparison of albedos and Al/Si intensity ratios for revolution 72.

values of Al/Si concentrations shown in table 17-II have been determined for various parts of the Moon along the ground track. Some typical values for various lunar rocks, as determined by the chemical analysis of lunar materials, are included in this table.

The results of the X-ray fluorescence experiment are still in the preliminary stages of analysis. Never-

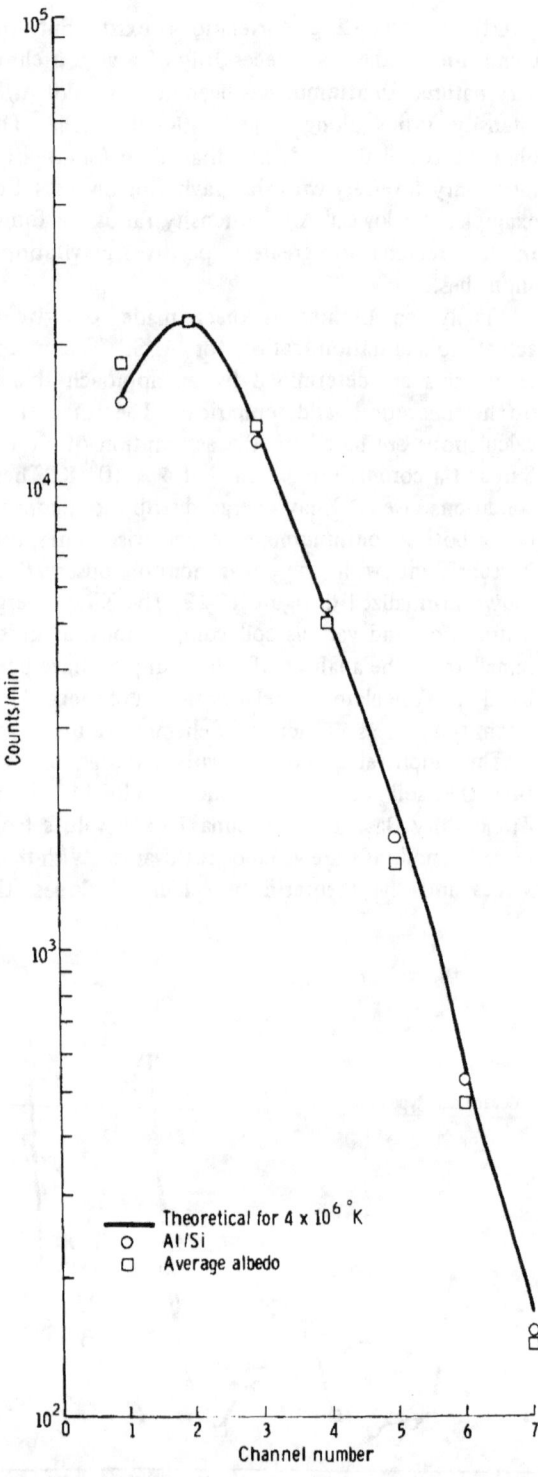

FIGURE 17-12.—Comparison of theoretical spectral distribution of detector (radiation 36° from normal) with solar monitor data taken ±12 min from subsolar point.

TABLE 17-II.—*Averaged Al/Si Concentration Ratios Calculated From the Intensity Ratios of Table 17-I*

Location	Concentration ratios
Mare areas	
Serenitatis	0.29 ± 0.02
Imbrium	.30 ± .03
Crisium	.37 ± .01
Tranquillitatis	.36 ± .02
Fecunditatis	.38 ± .03
Smythii	.37 ± .04
Archimedes Rille area (to Palus Putredinis)	.33 ± .01
Apennine Mountains	.46 ± .01
Haemus Mountains	.43 ± .05
Highlands east of Mare Serenitatis to 40° east	.42 ± .04
Highlands west and southeast of Mare Crisium	.56 ± .05
Highlands between Mare Crisium and Mare Smythii	.56 ± .06
Highlands between Mare Smythii and Tsiolkovsky	.60 ± .07
Highlands east of Mare Fecunditatis	.66 ± .11
Highlands east of Tsiolkovsky	.69 ± .12
Area north of Schröter's Valley	.37 ± .03
Area northeast of Schröter's Valley	.32 ± .04
Apollo 11, Mare Tranquillitatis bulk soil (ref. 17-8)	.37
Surveyor, Mare Tranquillitatis regolith (ref. 17-9)	.35
Apollo 12, Oceanus Procellarum average of soils (ref. 17-10)	.325
Luna 16, Mare Fecunditatis bulk soil (ref. 17-11)	.416
Surveyor VI, Sinus Medii regolith (ref. 17-9)	.338
Surveyor VII, rim-of-Tycho regolith (ref. 17-9)	.546
Apollo 14, Fra Mauro soils (ref. 17-12)	.413
Apollo 15, Hadley Rille-Apennine Mountains, soils (3 preliminary)	.379
Apollo 11 and 12, anorthositic gabbros	.637
Gabbroic anorthosites	.819
Anorthositic fragments	.885
Apollo 12, noritic material (ref. 17-7)	.416
Apollo 12, KREEP (refs. 17-13 and 17-14)	.394
Apollo 12, dark 12013 (refs. 17-13 and 17-14)	.327

theless, some tentative interpretations are possible, especially when the data are viewed in the context of previous Apollo missions. These interpretations are based on comparisons of the relative Al/Si intensity ratios with albedo, regional geology, and analyses of returned samples; however, interpretation is subject to several limitations at this stage of analysis. Firstly, the Al/Si intensity ratios refer to large areas because the plotted data were read out at 1-min intervals and because of the 60° FOV of the X-ray spectrometer. Secondly, the X-ray fluorescence experiment is inherently a measurement of surface composition and provides no information on the subsurface composition below depths of approximately 0.1 mm except for the mixing effect of gardening.

Subject to these limitations, the following general conclusions have been formulated from the X-ray experiment about the geology and evolution of the Moon.

(1) The Al/Si ratios shown in tables 17-I and 17-II confirm that the highlands and maria do indeed have different chemical and mineralogical compositions. This conclusion is stated explicitly because, to date, there have been only two sample-return missions to the highlands (Apollo 14 and 15). Furthermore, this conclusion confirms that the albedo difference between the highlands and maria (figs. 17-13 and 17-14) is, at least in part, the expression of chemical differences. With the evidence from returned samples, the X-ray experiment indicates that two major types of materials are exposed. The high Al/Si ratios of the highlands indicate that the highlands are related to the plagioclase-rich fractions of the returned samples, whereas the low Al content of the maria is consistent with the compositional analysis of numerous mare basalts (table 17-II).

The distinct chemical and mineralogical compositional differences between the maria and the highlands limit the effectiveness of horizontal transport of material. If such a mechanism were effective for long-distance transport, the spectrometer would probably not detect compositional differences in the highlands

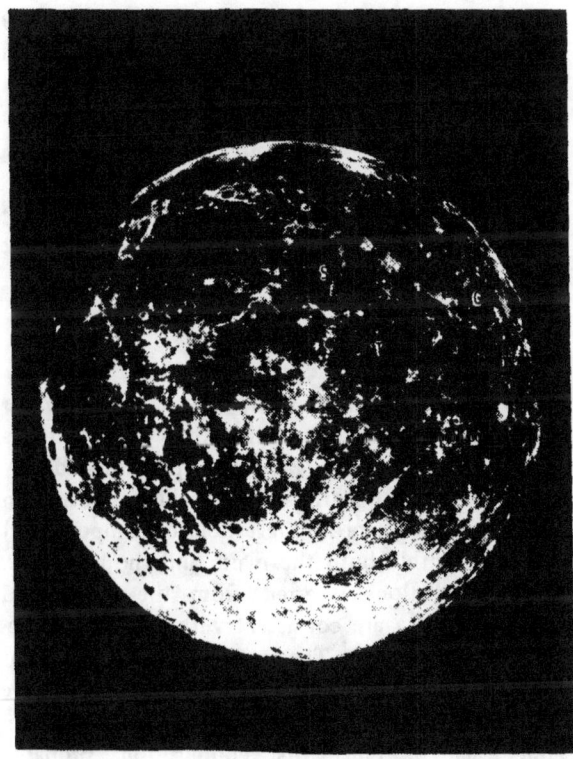

FIGURE 17-13.—Full-Moon telescopic photograph emphasizing albedo differences of the near side. Labeled areas are Mare Serenitatis (S), Mare Crisium (C), and Mare Tranquillitatis (T).

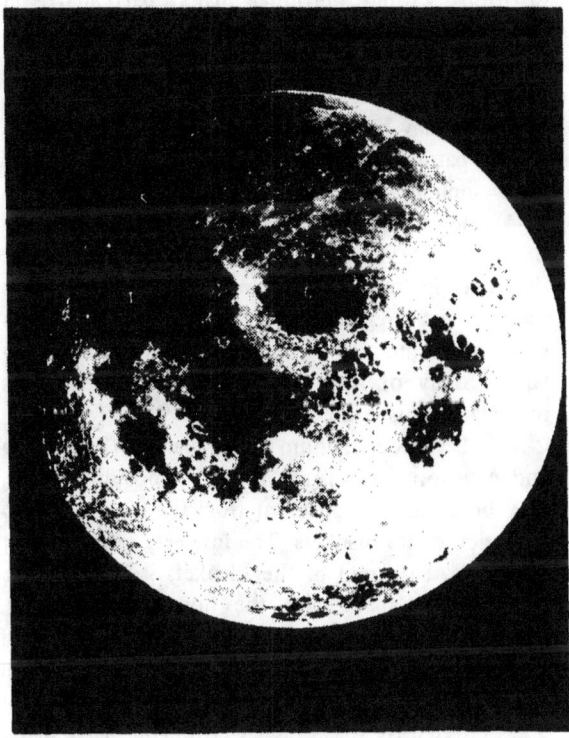

FIGURE 17-14.—High-Sun-angle photograph of east limb and part of far side, emphasizing albedo differences. Labeled areas are Mare Serenitatis (S), Mare Crisium (C), and Mare Tranquillitatis (T) (AS11-44-6667).

and maria because the instrument detects only the fine-grained surface material that would be widely distributed by electrostatic suspension. Some blurring of the mare-highland contacts probably has been caused by the cumulative effect of minor impacts. Horizontal transport, by this mechanism (minor impacts), has not carried material farther than approximately one-tenth the diameter of, for example, Mare Serenitatis.

(2) As shown in table 17-II, some indication exists of chemical differences among maria; the raw data indicate that the circular maria have less Al than the irregular maria. If these differences are authentic, they may provide information about the origin of the mare basins; the circular maria are characterized by mascons and apparent ejecta blankets whereas the irregular maria are not. The X-ray spectrometer data require more refinement for this conclusion to be confirmed; however, these data agree with the recent work of Soderblom (ref. 17-4), who found that the circular maria (red) could be distinguished spectrally from the irregular maria (blue).

(3) There appear to be systematic chemical variations within individual maria, particularly in Mare Crisium and Mare Serenitatis. The raw data indicate that the edges of these maria have a greater Al content than the centers. The possibility that these variations may be real is supported by the systematic albedo patterns in Mare Serenitatis (ref. 17-5) and more indirectly by the work of Soderblom (ref. 17-4). However, two factors require clarification. The first is the FOV, because intensity readings near mare rims may include the highland areas. The second is that the mare regolith near the borders may contain substantial amounts of ejecta from the nearby highlands. Study of X-ray data taken at 8-sec intervals may clarify this result.

(4) The Mare Imbrium ejecta blanket (Fra Mauro and Alps formations) (ref. 17-6), and possibly others, may be chemically different from the highlands outside such ejecta blankets. The intensity ratios suggest that the Al content of the blankets is intermediate between the maria and the nonejecta highlands. Assuming this difference is real, the mare-basin ejecta blankets represent more basic material derived from some depth in the Moon. However, the ejecta-blanket areas covered by the X-ray spectrometer, particularly the Haemus Mountains south of Mare Serenitatis, included substantial areas of highland-mare material (shown as Eratosthenian mare by Wilhelms, ref. 17-6). Therefore, definite conclusions concerning the composition of the mare ejecta cannot be formulated at this time.

(5) The X-ray experiment supports the belief that the first major geochemical event of the geological evolution of the Moon after its formation was the development of a global, differentiated, Al-rich crust. Despite the preliminary nature of these data, the highland areas can be pinpointed that have Al/Si ratios corresponding to analyzed lunar materials of anorthositic affinity (ref. 17-7). As shown in table 17-II (refs. 17-7 to 17-14), the anorthositic gabbros have an Al/Si ratio of 0.64; gabbroic anorthosites, 0.82; and anorthositic fragments, 0.89. Based on the X-ray fluorescence concentration ratios shown in table 17-II, the anorthositic material could be derived from any of several distinct highland areas on or near the east limb of the Moon. This assumption does not exclude other highland surfaces as the actual source of these materials but emphasizes the possibility that large definite areas exist from which they could be derived.

Negative evidence from the Apollo 15 gamma-ray experiment and the data from Luna 10 (ref. 17-15) indicate that the highlands do not have large amounts of granitic material that would otherwise be a possible explanation for the higher Al content of the highlands. Also, the scarcity of granitic compositions in samples returned to date from the Apollo 11, 12, and 14 missions and from Luna 16 is negative evidence.

Summary

Preliminary analysis of only part of the X-ray fluorescence experiment data, that pertaining to the relative Al/Si ratios, shows that the experiment has been highly successful. Major compositional differences between the maria and the highlands have been confirmed, and there are suggestions of compositional variations within both the maria and the highlands. The compositions inferred from the X-ray and gamma-ray data are consistent with the study of returned lunar samples. The main preliminary results of the X-ray fluorescence experiment are a tentative confirmation that the Moon has a differentiated highland crust of feldspathic composition, probably similar to the various materials of anorthositic affinity found in samples extracted during the Apollo 11, 12, 14, and 15 missions.

ASTRONOMICAL X-RAY OBSERVATIONS

The objective of the galactic X-ray observations is to study in detail the temporal behavior of pulsating X-ray sources. Approximately 100 sources of X-ray emission have been discovered beyond the solar system. These sources include a large variety of unusual objects, such as supernova remnants, energetic external galaxies, quasars, and a large number of sources in our galaxy that cannot be associated with any previously known class of objects. Several of these sources are detectable as emitters of radio waves or can be seen as faint stars, but emission from these sources occurs predominantly at X-ray frequencies. One of the prime objectives of astrophysics is to understand the nature of these X-ray sources. The first X-ray astronomy satellite launched by NASA, Explorer 42 (UHURU), has recently discovered rapid variability or pulsations in the output from several sources (refs. 17-16 and 17-17). The rapid variability occurs on a time scale of minutes, seconds, or less, implying that the emitting regions are very small, much smaller than the Sun, although the regions are emitting approximately a thousand times more power. Because this is such an unusual phenomenon, the rapid variability may provide the clue that is needed to understand the mechanisms that drive these sources. The objective of the Apollo observation is continuously to record the emission from several objects for a period of approximately 1 hr. The Apollo Program capability for observing time variations is unique, because the spacecraft can be pointed at the source for the entire time, whereas, UHURU can observe for only approximately 1 or 2 min per sighting. Consequently, the Apollo Program provides the capability of determining whether periodicities exist in the range of 10^1 to 10^3 sec.

Two powerful X-ray sources observed by Apollo 15, Sco X-1 and Cyg X-1, can also be seen on the Earth in other regions of the electromagnetic spectrum. Sco X-1 is detectable in both visible light and radio emission, but Cyg X-1 is detectable only in radio emission. It is known that the visible light and radio emissions are also variable, but how the light or radio variability correlates with the X-ray variability is still unknown. Specific predictions concerning the ranging of X-ray and other variability from no correlation to complete correlation are made on the basis of data obtained from special models of these X-ray objects. Consequently, to broaden the scope of the investigation, arrangements were made for ground-based observatories to monitor visible light and radio emissions simultaneously with Apollo 15 observations. The Apollo observations were made during Houston daylight hours, which is an unfavorable situation for observatories in North America. Fortunately, observatories located at more easterly points in nighttime hours were able to acquire Sco X-1 and Cyg X-1 data simultaneously with Apollo 15 observations. The optical flux from Sco X-1 was observed by the Crimean Astrophysical Observatory in the U.S.S.R. and by the Wise Observatory in Israel with time resolutions of 20 sec and 4 min, respectively. Radio emissions from Sco X-1 and Cyg X-1 were observed by the Westerbork Observatory in the Netherlands. The specific objective of the first phase of the analysis is to apply Fourier analysis to all these data.

Theoretical Basis

The observation of rapid variations in several galactic X-ray sources is rather conclusive evidence that the sources are highly compact objects. Significant intensity variations within a few seconds or less indicate that the diameter of the X-ray emitting region is comparable to the emitting region of the Earth or smaller, although energy is being emitted at a rate of a thousand Suns. Because the behavior of the pulsating X-ray sources is rather phenomenal and not easily explained, the simplest approach to a valid explanation is to attempt an association of these sources with strange, previously known objects. The small size of these sources indicates that the X-ray source is possibly a white dwarf, a neutron star, or the hypothetical black hole. However, associating the pulsating X-ray source with any of these three objects or explaining the profuse radiation of energy presents rather formidable problems. The most common of the three sources are the white dwarfs; however, no white dwarf has been observed to radiate energy so copiously, to emit X-rays, or to vary so rapidly. If the pulsating X-ray sources are white dwarfs, the sources represent a rather unique subgroup. No nebulosity surrounds the X-ray source (which would suggest that a supernova explosion has taken place), so the source cannot be explained as a neutron star (the superdense core remnant of a supernova explosion). The X-ray source is possibly an old neutron-star remnant in which the surrounding nebulosity has dissipated. However, the older, known neutron stars do not emit

X-rays, and their radio emissions show a rather stable, well-defined period. Black holes are the hypothetical, infinitely dense remnants of supernova explosions of massive stars. Although the intense gravitational field of a black hole does not allow radiation to be emitted, the suggestion has been made that the accretion of matter from a nearby neighbor onto the black hole would result in an irregularly varying X-ray source. In fact, according to some viewpoints, this process of accretion is the only way in which a black hole can be detected.

One important objective is to determine the mechanism that causes such a small object to radiate energy so copiously. The X-rays presumably come from either the thermalization of high-velocity matter accreting onto a small object or from the dissipation of rotational or vibrational mechanical energy contained in the object itself. Possibly, a study of the time variability will answer this question. Each mechanism would predict a frequency spectrum. An explanation based on the dissipation of rotational energy may result in a rather regular variability, whereas vibrational energy or matter accretion as the source would indicate a different kind of time variability. The time variability can be studied as a function of X-ray energy. The X-ray detectors are sensitive to the range 1 to 3 keV. The presence of filters over two of the three detectors permits the effective subdivision of this range into three parts for the analysis.

The bright X-ray source, Sco X-1, is an especially interesting candidate for the Apollo studies because this source is detectable as a 12th-magnitude blue star in visible light, is observable in infrared radiation, and is a radio source. Thus, the observational power of a broad range of astronomical disciplines can be used to study this object. An understanding of the mechanisms driving Sco X-1 may provide a basis for understanding many galactic sources. No rapid variability has been detected from Sco X-1. However, in the Apollo Program, it is possible to observe Sco X-1 in a new time regime and to look for variability on the order of a few minutes.

Equipment

The same instrumentation is used in the Apollo astronomical X-ray observations as in the X-ray fluorescence studies of lunar-surface chemical composition. A detailed description of the instrument has been given in the first part of this section.

Description of the Instrument

The FOV of the detectors is limited to an angular acceptance of 30° by 30° (FWHM) by mechanical slat collimators. Energy from a cosmic source reaches the counters as a perfectly parallel beam and is seen at maximum intensity when the direction is normal to the plane of the counters. The intensity is reduced off axis by the cosine and shadowing effects of the collimators. When the angle of the source exceeds 60° off axis, no counts are detected; however, variation of the angle will cause an apparent variation in the intensity of an otherwise steady source.

The nature of the X-ray emission from two of the Apollo 15 targets, Sco X-1 and Cyg X-1, and from most cosmic X-ray sources is such that the unfiltered detector counts more strongly than the two that are filtered. Most of the useful data on cosmic sources will come from the unfiltered detector, although the combination of the three detectors will provide important information regarding the time variability as a function of energy. The time resolution of the Apollo 15 instrument is 8 sec, the minimum integration time. This time permits periodicities down to approximately 5 sec and random variability of approximately 10 sec to be detected.

Deployment and Operation of the Instrument

The instrument, hard-mounted in the SIM bay of the CSM, is used for the galactic X-ray astronomical observations during the transearth-coast phase of the mission by pointing the instrument at a cosmic X-ray source and continuously counting for a sustained period. The spacecraft is required to hold the pointing position accurately to within 1° for approximately 1 hr of observation. The integrated count output for each of the three counters is recorded successively in 8-sec periods. Signals are divided into eight channels of pulse amplitude. A summary of the observing schedule is given in table 17-III. Two of the objects, Sco X-1 and Cyg X-1, were observed simultaneously in radio waves by the Westerbork Observatory, and Sco X-1 was observed simultaneously in visible light by the Crimean Astrophysical Observatory and the Wise Observatory. The X-ray spectrom-

TABLE 17-III.—*Apollo 15 Astronomical X-Ray Schedule*

Source	Start, GET, hr:min	Duration of observation, min:sec	Date	G.m.t., hr:min
Centaurus	226:28	1:05	August 5	00:02
Midgalactic latitude	237:15	0:50	August 5	10:49
Scorpius	245:45	0:35	August 5	19:19
Cygnus	246:35	0:55	August 5	20:09
South galactic pole	261:50	0:55	August 6	11:24
North galactic pole	274:15	0:30	August 6	23:49
Galactic plane anticenter	275:00	1:00	August 7	00:34

eter can be operated either in a normal-gain mode, in which the pulse height or energy channels are concentrated on the region 0.7 to 3 keV, or in an attenuate mode, in which all energies are doubled. Except for the first pointing position, the instrument was in the normal-gain mode. The X-ray source observed in the first pointing position is known to be deficient in photons below 3 keV. Consequently, the instrument was placed in the attenuate mode for that position.

The FOV of the instrument is 30° by 30°, or approximately 1 sr of solid angle. This FOV is rather large—so large that, if the instrument were pointed directly at Sco X-1 and Cyg X-1, other neighboring sources would be in the FOV at the same time. To avoid these potentially confusing difficulties, pointing positions were intentionally chosen to be several degrees off the source. As a result, the full intensity of the sources was not observed.

The origin of a rather strong component of diffuse cosmic X-rays is still unknown. The flux contained 1 sr that is actually greater than that of almost all the discrete sources. One of the outstanding questions of X-ray astronomy is whether this flux is truly isotropic or whether there is increased emission from relatively nearby clusters of galaxies. If the flux is isotropic to a high degree, the flux originates from sources at very large distances, possibly at the edge of the universe. The diffuse flux should not show any time variability. Several of the pointing positions, numbers 2, 5, 6, and 7, did not include discrete sources, only diffuse X-ray background. There were two reasons for devoting some of the time to observing the diffuse flux: (1) to determine the degree of isotropy and (2) to provide a control of the ability to determine time variability. If variability is observed in these positions, there is an indication of extraneous systematic effects in the instrument, in the data transmission, or in the method of analysis.

Results

Preliminary results from the observations of Sco X-1 and Cyg X-1 have been obtained from analysis of the quick-look or thrift data. There are indications of variability in the count rate; however, the apparent intensity of a source is a function of the angle from which the source is viewed. A change of 1° in the angle will cause an apparent intensity change of 1.5 percent. Consequently, spurious variations in the intensity of a source are induced by spacecraft motion. Because records of the spacecraft attitude during the pointing positions were not available for this section, no definite conclusions can be made concerning the relationship of observed variability effects to the source.

The count rate as a function of time for Sco X-1 is shown in figure 17-15. Data are not included for several time intervals because of the appearance of calibration sources. In the final data analysis, the contribution of the calibration sources will be subtracted, and the missing intervals will be supplied. The X-ray data are shown for two intervals of energy, 1 to 3 keV and >3 keV. The former is the sum of all the events in channels 1 to 7, and the latter is the interval of energy for channel 8. Two features are seen qualitatively in the data. A weak quasi-cyclical variation in intensity exists for three periods of approximately 5-min duration, and a very pronounced increase in the X-ray counting rate occurred at the end of the Apollo observations. Both features appear in both energy intervals. These effects are statistically significant, and if the effects are not caused by

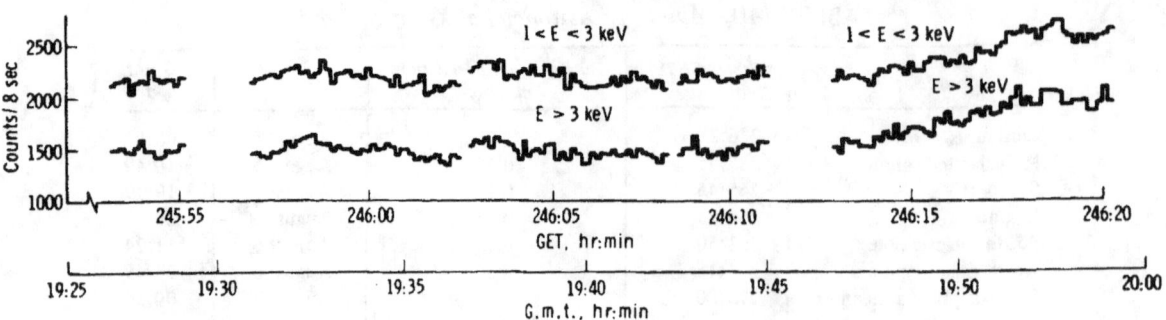

FIGURE 17-15.—The X-ray counting rate of Sco X-1 in two energy (E) bands as a function of time as observed during the Apollo 15 mission. Data have been omitted during periods when a calibration source appears.

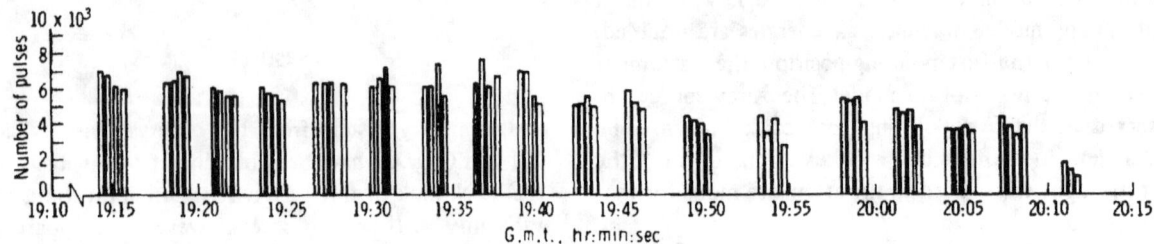

FIGURE 17-16.—Optical flux of Sco X-1 at the time of the Apollo 15 X-ray observation.

changes in spacecraft attitude, the statistics indicate that Sco X-1 is varying rapidly.

Data from simultaneous optical observations by the Crimean Astrophysical Observatory are shown in figure 17-16. Sky background has been subtracted. The statistical precision of these data is approximately 1 to 2 percent smaller than the apparent variations in optical density; however, systematic effects that could lead to variations have not been investigated. Discounting possible corrections for systematic errors, the X-ray intensity increases simultaneously as the optical intensity decreases. Several minutes of data from Cyg X-1 observation are shown in figure 17-17. The large peak is the appearance of a calibration source. The 8-sec time resolution of the instrument is insufficient to reveal the very rapid time variations that are described in reference 17-16. However, a statistically significant rise and fall in intensity begins at approximately 246 hr 57 min GET and ends at approximately 247 hr 03 min GET. In this case also, the possible factor of spacecraft motion is still in question.

Summary

The Apollo 15 X-ray detector was successfully pointed at three pulsating X-ray sources and at four locations dominated by the diffuse X-ray flux. During the observation period, the count rate from Sco X-1 and Cyg X-1 showed statistically significant changes of approximately 10 percent in intensity during a period of several minutes; however, possible changes in spacecraft attitude might account for the variations. Simultaneous optical observations of Sco X-1 by the Crimean Astrophysical Observatory tentatively indicate an anticorrelation between changes in X-ray and optical intensities during the final minutes of the Apollo observations. If the increase in flux of Sco X-1 observed during the Apollo 15 mission is not a result of changes in spacecraft attitude, the change in the X-ray intensity of Sco X-1 will be the most rapid ever observed.

Final analysis of the Apollo data will necessarily involve a comparison with UHURU results that cover different time regimes. The fast time resolution of the

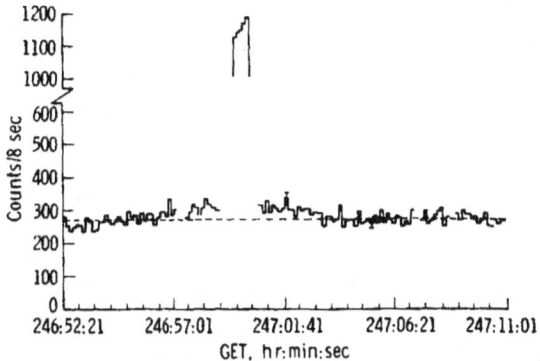

FIGURE 17-17.—The X-ray flux of Cyg X-1 during 13 min of 1-hr observation during the Apollo 15 mission. A dashed horizontal line representing a constant intensity is shown for reference. The large peak is the effect of a calibration source.

spacecraft is only 8 sec, whereas the superior time of the UHURU is 0.1 sec. Because UHURU is an orbiting satellite, short-duration observations can be made during several days, but the ability in the Apollo missions to monitor variations continuously for approximately 1 hr will fill an important gap in the observation of the pulsating sources.

REFERENCES

17-1. Adler, I.; Gorenstein, P.; Gursky, H.; and Trombka, J.: Advances in X-Ray Analysis. vol. 13. Plenum Press, 1970.

17-2. Adler, Isidore: X-Ray Emission Spectrography in Geology. Elsevier (New York), 1966.

17-3. Pohn, H. A.; and Wildey, R. L.: A Photoelectric-Photographic Study of the Normal Albedo of the Moon. Part A, Lunar and Planetary Investigation. Astrogeologic Studies, Dec. 1966, pp. 211-234.

17-4. Soderblom, L. A.: The Distribution and Relative Ages of Regional Lithologies in the Lunar Maria. Abstracts with Programs, vol. 2, no. 70, Geol. Soc. Am. 1970 Annual Meeting, pp. 690-691.

17-5. Carr, M. H.: Geologic Map of the Mare Serenitatis Region of the Moon. U.S. Geological Survey, Misc. Geol. Inv. Map I, p. 489.

17-6. Wilhelms, D. E.: Summary of Telescopic Lunar Stratigraphy. Part A, Lunar and Planetary Investigation. Astrogeologic Studies Annual Progress Report, July 1, 1965, to July 1, 1966, pp. 235-298.

17-7. Marvin, U. B.; Wood, J. A.; Taylor, G. J.; Reid, J. B.; et al.: Relative Proportions and Probable Sources of Rock Fragments in the Apollo 12 Soil Samples. Proceedings of the Second Lunar Science Conference, vol. 1, A. A. Levinson, ed., MIT Press (Cambridge, Mass.), 1971, pp. 679-700.

17-8. Levinson, A.A., ed.: Proceedings of the Apollo 11 Lunar Science Conference. Pergamon Press (New York), 1970.

17-9. Mason, Brian; and Melson, William G.: The Lunar Rocks. John Wiley & Sons, Inc., 1970.

17-10. Levinson, A. A., ed.: Proceedings of the Second Lunar Science Conference. MIT Press (Cambridge, Mass.), 1971.

17-11. Vinogradov, A. P.: Preliminary Data on Lunar Ground Brought to Earth by Automatic Probe Luna 16. Proceedings of the Second Lunar Science Conference, vol. 1, A. A. Levinson, ed., MIT Press (Cambridge, Mass.), 1971, pp. 1-16.

17-12. Lunar Sample Preliminary Examination Team: Preliminary Examination of Lunar Samples from Apollo 14. Science, vol. 173, no. 3998, Aug. 20, 1971, pp. 681-693.

17-13. McKay, D. S.; Morrison, D. A.; Clanton, U. S.; Ladle, G. H.; et al.: Apollo 12 Soil and Breccia. Proceedings of the Second Lunar Science Conference, vol. 1, A. A. Levinson, ed., MIT Press (Cambridge, Mass.), 1971, pp. 755-774.

17-14. Meyer, C.; Brett, R.; Hubbard, N. J.; Morrison, D. A.; et al.: Mineralogy, Chemistry, and Origin of the KREEP Component in Soil Samples from the Ocean of Storms. Proceedings of the Second Lunar Science Conference, vol. 1, A. A. Levinson, ed., MIT Press (Cambridge, Mass.), 1971, pp. 393-412.

17-15. Chernov, G. M.; Kirnozov, F. F.; Surkov, I. A.; and Vinogradov, A. P.: Measurements of γ Radiation of the Moon's Surface by the Cosmic Station Luna 10. Geochemistry (USSR), vol. 8, 1966, p. 891.

17-16. Oda, M.; Gorenstein, P.; Gursky, H.; Kellogg, E.; et al.: X-ray Pulsations from Cygnus X-1 Observed from UHURU. Ap. J. (Letters), vol. 166, no. 1, May 1971, p. L1.

17-17. Giacconi, R.; Gursky, H.; Kellogg, E.; Schreier, E.; et al.: Discovery of Periodic X-ray Pulsations in Centaurus X-3 from UHURU. Ap. J. (Letters), vol. 167, no. 2, July 1971, p. L67.

18. Alpha-Particle Spectrometer Experiment

Paul Gorenstein[a,†] *and P. Bjorkholm*[a]

INTRODUCTION

Analysis of returned lunar samples has revealed significant concentrations of uranium and thorium in lunar-surface material. Both elements are radioactive and are the first members of two distinct, highly complex decay series that terminate in stable isotopes of lead. Unstable isotopes of radon gas are produced as intermediate products in these series. Uranium produces radon-222 (^{222}Rn), and thorium produces ^{220}Rn. Radon is a rather special component of the decay series because of its gaseous nature. Radon may diffuse above the lunar surface where it would remain gravitationally trapped to form an exceedingly rare atmosphere. As a result, the radioactive decay of the radon isotopes and their daughter products would have the effect of enhancing the levels of radioactivity upon the surface of the Moon in an important way. Local differences in uranium and thorium concentrations and uneven effects that promote or inhibit diffusion of gases would lead to inhomogeneities in the distribution of radon across the surface of the Moon.

The level of radon emanation is extremely small (probably less than 10^{-2} atoms/cm^2-sec, based on previous measurements) but, hopefully, detectable. The objective of the Apollo alpha-particle spectrometer experiment is to map the radon emanation of the Moon and to find potential areas of high activity. Detection of radon isotopes and their daughter products is based on detecting the decay alpha particles and measuring their unique energy with detectors having good energy resolution.

The alpha-particle spectrometer, mounted in the scientific instrument module, detects individual alpha particles and measures their energy. The presence of radon and its daughter products appears as a distinct set of peaks in an energy spectrum. There is a background of cosmic-ray effects. Spatial resolution is achieved by restricting the field of view of the Apollo alpha-particle spectrometer to 45° by 45° (full width at half maximum). In practice, the achievable spatial resolution is determined by the counting-rate level of the experiment and can be significantly smaller than the experiment field of view if the counting rate is of sufficient strength.

The primary objective of the Apollo observation is to construct a radon map of the Moon for the region overflown by the Apollo spacecraft and within the instrument field of view. Hopefully, spatial differences related to differences in uranium and thorium concentrations would be detected. If any common volatiles such as water vapor or carbon dioxide outgas locally on the Moon, then trace quantities of radon quite possibly will be included. In that case, the ability to detect a very small number of decaying radon atoms would provide a very sensitive means of identifying and locating lunar areas of outgassing.

BASIC THEORY

The migration of radon gas from natural concentrations of uranium and thorium in the ground is a well-known terrestrial phenomenon. Various effects promote this diffusion of radon into the atmosphere. Across the surface of the Earth, gross differences exist in the amount of radon emanation; these differences reflect local differences in the concentrations of uranium and thorium and the ability of radon to diffuse through the soil. Generally, a high degree of atmospheric radon activity exists in an area of high concentration of uranium and thorium. Volcanic activity and the evolution of volatiles from the ground are usually accompanied by radon emanation. Hence, a radon-emanation map of the Earth would be exceedingly nonuniform.

---

[a]American Science and Engineering, Inc.
[†]Principal investigator.

It is not unreasonable to expect that analogous effects occur across the surface of the Moon. Significant concentrations of uranium and thorium comparable to terrestrial values have been found in the analysis of returned lunar samples. However, conditions on the Moon are quite different from those on the Earth. The most important difference is the absence of an appreciable lunar atmosphere and the very high vacuum conditions in the lunar soil. Because the mean free path in the lunar atmosphere is large, radon atoms from the lunar surface move in ballistic trajectories. Emitted at thermal velocities of approximately 0.15 km/sec, corresponding to 300 K, radon atoms are acted upon by the gravitational pull of the Moon. Typically, these atoms reach a maximum altitude of approximately 10 km and fall back upon the surface. Essentially, no radon atoms have sufficient velocity to escape. The half-life of ^{222}Rn (3.8 days), the daughter product of uranium, is greatly different from that of ^{220}Rn (55 sec), the daughter product of thorium. Since ^{222}Rn has a much longer half-life than ^{220}Rn, the ^{222}Rn has time to diffuse through more material and can originate from greater depths in the Moon. Because the concentration of uranium is about a fourth that of thorium, the concentration of ^{222}Rn in the lunar atmosphere would be expected to be much greater than that of ^{220}Rn.

The ^{222}Rn atoms have sufficient time after initial diffusion to travel a ballistic trajectory and return to the lunar surface to be absorbed and eventually re-emitted. This process can be repeated many times. Conversely, ^{220}Rn has barely enough time to follow a single trajectory before decay. Thus, the activity of ^{220}Rn will be much more localized that that of ^{222}Rn. It is extremely difficult to determine a priori how the very high vacuum conditions in the soil affect the diffusion of radon. Previous measurements of lunar radon activity (to be discussed subsequently) lead to the conclusion that these conditions retard diffusion in general. However, the presence of crevices or fissures in local regions, which increase the amount of exposed surface, might be expected to enhance the quantity of radon evolved into the atmosphere. Volcanic or thermal sources of ordinary volatiles such as water vapor or carbon dioxide, should they exist on the Moon, may also be sources of radon, as they are on the Earth. The movement of these common gases through rocks and material that contain uranium and thorium would very likely sweep radon to the surface. Because the uncertainties in this process are so large, even a qualitative estimate of the amount of radon reaching the surface is not possible.

In the Apollo 15 experiment, detection of radon is based on the fact that alpha particles are produced in radon decay. The kinetic energies of the alpha particles emitted by the radon isotopes of the uranium and thorium series and the energies of the alpha particles from their subsequent daughter products are given in tables 18-I (uranium) and 18-II (thorium). Alpha particles from the decay of radon above the lunar surface will be seen at their full energy because no significant slowing down can occur in the lunar atmosphere. When radon decay occurs,

TABLE 18-I.— *Uranium Decay Series*[a]

Isotope	Half-life	α-energy, MeV	Relative intensity[b]
^{222}Rn	3.823 days	5.490	100
^{218}Po	3.05 min	6.002	50
^{214}Pb	26.8 min	β-	--
Bismuth-214 (^{214}Bi)	19.7 min	β-	--
^{214}Po	164 × 10^{-6} sec	7.687	50
^{210}Pb	21 yr	β-	--
^{210}Bi	5.01 days	β-	--
^{210}Po	138.4 days	5.305	50
^{206}Pb	Stable	--	--

[a]Starting with ^{222}Rn. The decay series to this point is $^{238}_{92}$U $\xrightarrow{\alpha}$ $^{234}_{90}$Th $\xrightarrow{\beta-}$ $^{234}_{91}$Pa $\xrightarrow{\beta-}$ $^{234}_{92}$U $\xrightarrow{\alpha}$ $^{230}_{90}$Th $\xrightarrow{\alpha}$ $^{226}_{88}$Ra $\xrightarrow{\alpha}$ $^{222}_{86}$Rn, where Pa is protactinium and Ra is radium.

[b]Relative intensities normalized to 100 decays of ^{222}Rn above the lunar surface.

TABLE 18-II.— *Thorium Decay Series*[a]

Isotope	Half-life	α-energy, MeV	Relative intensity[b]
^{220}Rn	55 sec	6.287	14
^{216}Po	0.158 sec	6.777	7
^{212}Pb	10.64 hr	β-	--
^{212}Bi	60.0	{6.090, 6.051}	{0.7, 1.8}
^{212}Po	.304 × 10^{-6} sec	8.785	4.5
Thallium-208	3.10 min	β-	--
^{208}Pb	Stable	--	--

[a]Starting with ^{220}Rn. The decay series to this point is $^{232}_{90}$Th $\xrightarrow{\alpha}$ $^{228}_{88}$Ra $\xrightarrow{\beta-}$ $^{228}_{89}$Ac $\xrightarrow{\beta-}$ $^{228}_{90}$Th $\xrightarrow{\alpha}$ $^{224}_{88}$Ra $\xrightarrow{\alpha}$ $^{220}_{86}$Rn, where Ac is actinium.

[b]Relative intensities normalized to 100 decays of ^{222}Rn above the lunar surface and assuming a 7-to-1 ratio for the ^{222}Rn/^{220}Rn, as reported in reference 18-3.

the heavy residual nucleus recoils with a kinetic energy of approximately 100 keV. Half of the recoiled nuclei are directed downward toward the lunar surface. The distance in which the heavy-recoil nucleus is brought to rest is very much smaller than the range of the alpha particles that it will emit subsequently. Hence, when the radon descendants decay, the upward-directed alpha particles will not be slowed significantly by the lunar material.

The conditions expected at the lunar surface are shown in figure 18-1. Radon atoms moving in ballistic trajectories above the Moon decay, resulting in the emission of monoenergetic alpha particles and the creation of an active deposit on the lunar surface. The active deposit is itself a source of several groups of monoenergetic alpha particles. Conversely, no alpha particles will reach the surface from radon that decays at a depth exceeding the alpha-particle range. Typically, this depth is approximately 10 μm. Thus, alpha particles emitted between 0 and 10 μm below the surface of the Moon are degraded in energy. Hence, the intensity of monoenergetic alpha-particle emission is highly dependent on the effectiveness of the diffusion process.

In an early estimate (ref. 18-1), terrestrial conditions were assumed for the diffusion coefficient and concentrations. When the actual concentrations of uranium and thorium are used (as determined from lunar samples), this model results in a prediction of a rate of two disintegrations/sec/cm^2 for ^{222}Rn and approximately 10^{-2} disintegrations/sec/cm^2 for ^{220}Rn. Observations of alpha emission from the Moon indicate that, if radon is present, the activity levels are considerably smaller than this value. A measurement from the lunar-orbiting Explorer 35 (ref. 18-2) indicated no alpha-particle emission and set an upper limit approximately one-tenth the value predicted in reference 18-1. Turkevich et al. (ref. 18-3), reporting on background data obtained in the Surveyor V, VI, and VII alpha-backscatter experiments, cited evidence for a radioactive deposit at Mare Tranquillitatis (Surveyor V) with an intensity of 0.09 ± 0.03 alpha particles/sec/cm^2. The instrument was deployed on the surface; the radon itself was well above it in high trajectories. Thus, if radioactive equilibrium existed between radon and its daughter products, the total alpha emission observed at high altitudes would be several times greater than that found in these experiments. At the other two sites, Sinus Medii (Surveyor VI) and the rim of Tycho Crater (Surveyor VII), only upper limits to the alpha activity are reported; these are lower than the Mare Tranquillitatis values by a factor of approximately 2 or 3.

Two other indirect measurements of alpha activity involve looking for the active deposit on returned samples that have been exposed to lunar radon. Lindstrom, Evans, Finkel, and Arnold (ref. 18-4) looked for an excess of the radon daughter polonium-210 (^{210}Po) in Apollo 11 samples. They found that "the expected and actual ^{210}Po activity is the same to within the 3.2-percent limit set by the counting statistics limits," which implies that the effect of the active deposit is less than the 10^{-4} value predicted in reference 18-1. This is the most pessimistic of all the experiments. However, all or nearly all the active deposit that resides entirely in the first micrometer of surface material possibly could have been blown away by the action of the lunar module descent engine. A similar measurement was made on the Surveyor III camera visor (ref. 18-5), which was returned to Earth from Oceanus Procellarum by the Apollo 12 astronauts. No evidence of the deposit was found, and an upper limit can be set that is approximately six times smaller than the value reported in reference 18-3 for Mare Tranquillitatis. However, the slightest amount of abrasion or erosion could remove most of the active deposit. The net result from these pre-Apollo orbital measurements is that the active deposit on the lunar surface is probably several hundred times smaller than would be predicted from terrestrial diffusion rates. However, the expected ultimate sensitivity of the Apollo alpha-particle spectrometer is still sufficient to detect lunar alpha particles even if the

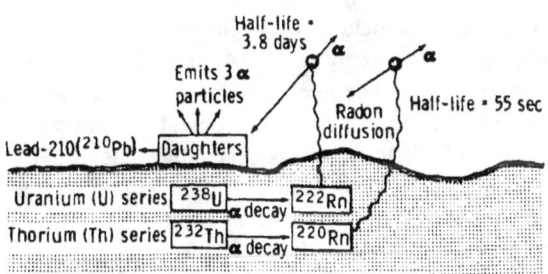

FIGURE 18-1.—Schematic representation of diffusion of radon through lunar material to gravitationally trapped condition above surface. Radon, which is traveling in a ballistic trajectory, decays and ejects an alpha particle. The nucleus is recoiled to the lunar surface, where it will eventually emit three additional alpha particles.

average rate of activity is several times lower than the lower limit found previously (ref. 18-4).

An interesting question concerns the degree to which radon remains localized. Evidently, most of the shorter lived isotope ^{220}Rn (half-life = 55 sec) decays on its first ballistic trajectory. The alpha emission from ^{220}Rn and its daughter is confined to a region with a radius of 10 km around the point of emanation, thus preserving the localization to a very high degree. On the other hand, ^{222}Rn (half-life = 3.8 days) has sufficient time to migrate a considerable distance before decay. The largest uncertainty is the accommodation time, or the elapsed time between the return to the surface of a freely falling radon atom and its re-emission on a new trajectory. If a thermal velocity of 0.15 km/sec, an emission angle of 45°, and an accommodation time of zero are assumed for the average ^{222}Rn atom, the migration of the nucleus is approximately 1000 km. For either nonzero accommodation time or lower temperatures, a smaller spread of the activity will occur. In any case, some degree of localization may be preserved. A theoretical model for the displacement of ^{222}Rn, in which a pileup of ^{222}Rn is predicted at the sunrise terminator of the Moon, is described in reference 18-6.

EQUIPMENT

The function of the alpha spectrometer is to detect and measure the energies of alpha particles emitted by the radon isotopes and their daughter products. The sensing elements are 10 totally depleted silicon surface-barrier detectors, each approximately 100 μm thick, having 3 cm^2 of active area and a 90° field of view, and operating at a −50-V bias. Additional gold, aluminum, and nickel layers are used at the contacts to ensure that exposure to visible light will not degrade the experiment performance. The thickness of the detectors was chosen so that any background protons (deuterons or tritons) would produce an output pulse of less than that for a 5-MeV alpha particle, while the output for alpha particles up to 12 MeV would be linearly proportional to energy. This design precludes the necessity for discriminating against other particles in any other way.

The 10 detector preamplifier outputs are merged in a single summing amplifier and processed by a single analog-to-digital converter (ADC). Although the use of one ADC minimizes the complexity of the hardware, the noise from all 10 preamplifiers is summed, resulting in a resolution degradation of a factor of approximately 3. To circumvent this problem, each preamplifier has a bias offset corresponding to 350 keV. This offset effectively removes the noise, except from the preamplifiers that are giving a signal pulse, and allows the use of a single ADC without significant degradation in resolution.

The ADC converts the energy pulse into a 9-bit digital signal. If the most significant bit is a 1, the ADC is disabled and the digital signal held until the next telemetry read-out (every 100 Msec). If the most significant bit is zero, the ADC is reset, and the next pulse is processed. This design allows the instrument to digitize to a 9-bit accuracy and transmit only 8 bits. Therefore, only the upper half of the digitized energy range is telemetered. Physically, this is reasonable because the alpha energies of interest range from 5.3 to 8.8 MeV; also, the usage of telemetry time by any low-energy background is prevented. The actual telemetered energy range of the instrument was 4.7 to 9.1 MeV. Parallel circuitry generates an analog signal of 0.25 to 4.75 V in steps of 0.5 V; this signal identifies the detector that originated any given pulse.

Because the digital telemetry is limited to 80 bits/sec (10 counts/sec), an additional circuit is used that generates an analog signal proportional to the time from the end of one telemetry read cycle to the sensing of the first pulse with energy greater than 4.7 MeV. This signal allows the dead-time correction of the data if the count rate exceeds approximately 20 counts/sec. Exclusive of housekeeping, the output consists of an 8-bit energy word, an analog voltage identifying the detector, and an analog voltage exponentially proportional to the count rate.

Five of the detectors had energy-calibration sources in their field of view. The sources were ^{208}Po with an alpha energy of 5.114 MeV. The count rate of these sources was approximately 0.1 count/sec. An additional energy calibration originated from a small amount of ^{210}Po that was accidentally deposited on the detector surface during testing. This contamination was on all 10 detectors in varying amounts. The worst case was approximately 0.047 count/sec, and the best was 0.000 counts/sec at launch.

The spectrometer was turned on at 15:47 G.m.t. on July 29, 1971, and remained on until 12:43 G.m.t. on August 7, except for short periods during major burns and during water and urine dumps. The spectrometer functioned as expected during this period, except for occasional bursts of noise in two detectors (6 and 8). The cause of this noise has not been determined.

RESULTS

The major results to be discussed in this subsection have been deduced mainly from the quick-look data. The final data tapes have not yet been received. The quick-look data have serious limitations. The major disadvantage of the quick-look data is that the energy spectra from all 10 detectors have been summed, and the individual spectra cannot be separated. During the mission, a slight differential gain shift probably occurred between the various detector channels. This gain shift and the intermittent noise on two detectors degrade the resolution in the quick-look data.

Also, because the observed count rates were very low during a single 5-min accumulation, no statistically significant energy spectra could be accumulated in the energy range above the onboard sources. The low count rate and the resolution degradation of the quick-look data prevent using the energy information at this time. Only energy-integrated count rates can be used as an indicator of the presence of radon or its daughter products.

Detectors 1, 4, 5, 9, and 10 do not have onboard calibration sources and have the least amount of ^{210}Po contamination. Therefore, the count rate of these detectors was examined as a function of time. Detector 1 had to be dropped from the analysis, however, because of a programing error in the quick-look display program. This error caused a count to be registered for detector 1 if there were a real count in detector 1 or if there were no counts in any detector.

Because a scan of the data showed no obvious localized region of alpha emission, the data were examined for variations on a larger scale. The available data between 12:53 G.m.t. on August 1 and 17:34 G.m.t. on August 4 were divided into 5-min intervals, neglecting data taken when the spacecraft was out of lunar-attitude orientation and those taken with the reaction control system plume-protection door closed. The data were then arranged so that all data taken during the same portion of the orbits were overlayed (i.e., all passes over the sunset terminator were alined and added together). However, because of the arbitrary starting time of the quick-look processing, the times are overlayed to an accuracy of only 2.5 min.

Approximately 22 hr of data were treated in this fashion. The results are shown in figure 18-2. The starting point is the sunset terminator, and the approximate times of the subsolar point and the sunrise terminator are shown. The magnitudes of the error bars vary because of the nonuniform amount of data available from the quick-look analysis. That is,

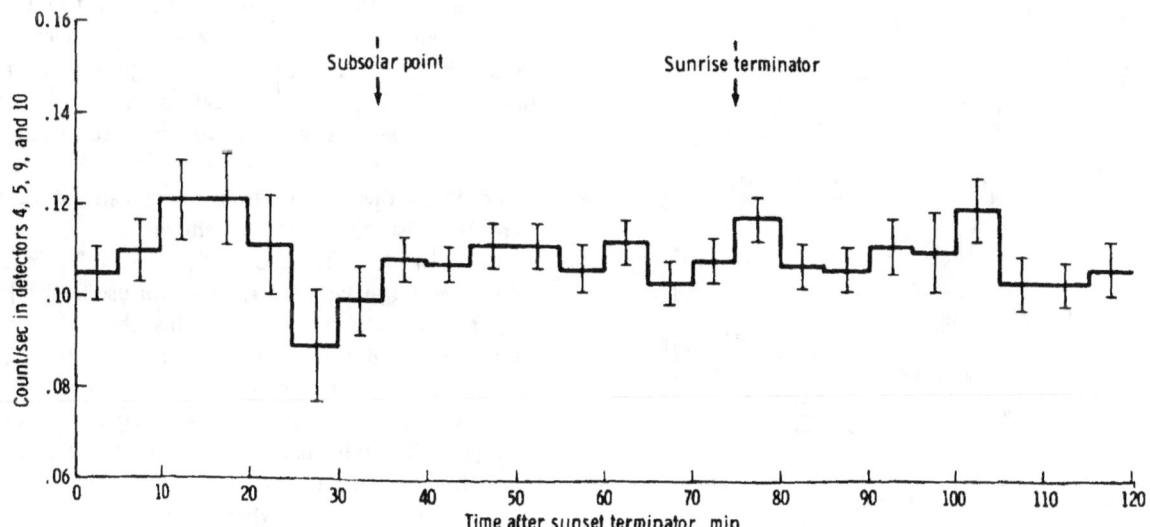

FIGURE 18-2.—Summed count rate of detectors 4, 5, 9, and 10 as a function of time from sunset terminator. These data are derived from 21 hr 21 min of data between 12:53 G.m.t., August 1, and 17:34 G.m.t., August 4.

more data were available from real-time telemetry than from that processed via the data-storage equipment. Although possible spatial variations exist in these data, they are at most two standard deviations. In particular, a small dip may occur just before the subsolar point, and a small increase in count rate may occur just after the sunrise terminator. However, these variations are of very marginal statistical significance. To improve the statistics, all data taken on the sunlit portion of the Moon were added together and, similarly, data taken on the dark side were added. The average count rates on the sunlit and dark sides were 0.108 ± 0.001 and 0.110 ± 0.002 count/sec, respectively. These rates apply to an effective instrument area of approximately 14 cm^2 (five detectors without calibration sources and a field of view of approximately 2 sr). This count rate is 0.016 times that expected if radon diffusion were consistent with the terrestrial model of reference 18-1 and if the

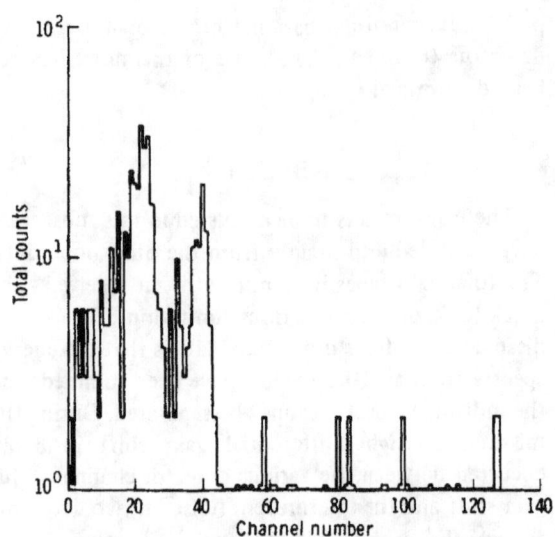

FIGURE 18-4.—An energy spectrum derived from 15 min of telemetry data. Channels above 140 have been left out for clarity. The count scale is logarithmic. Channels containing single counts are shown as small boxes slightly above the 10^0 base line. This representation exaggerates the apparent significance of channels containing two or more counts.

uranium and thorium concentrations are as found in the Apollo 11 and 12 samples, which is a factor of approximately 3 smaller than the concentrations assumed in reference 18-1.

The signal-to-noise ratio can be dramatically improved by examining the energy spectra of each individual detector and observing the count rate only in those energy channels where alpha particles from radon decay are expected. An energy spectrum as displayed by the quick-look data processor is shown in figure 18-3. The onboard calibration source (^{208}Po) and the contamination (^{210}Po) are indicated. Using these points as an energy calibration, the expected position of ^{222}Rn is shown.

In addition to these quick-look data, 15 min of telemetry data have been received for use in checking the data-reduction programs. This check has been performed, and the resulting energy spectrum (excluding detectors 6 and 8) is shown in figure 18-4. The energy spectrum shows the same structure as that in figure 18-3, but the resolution is better. The better resolution is partially because the quick-look spectrum has 128 channels, whereas the spectrum derived from the telemetry data has 256 channels.

No statistically significant indication of any alpha emission can be deduced from these 15 min of data.

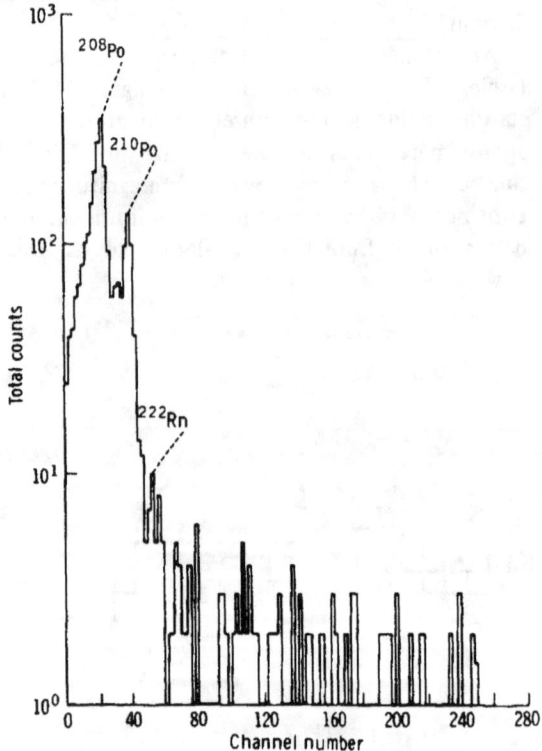

FIGURE 18-3.—An energy spectrum as displayed by the quick-look data system. This spectrum includes 47 min 23 sec of data beginning at 04:40 G.m.t. on July 30. The expected position of alpha particles from ^{222}Rn is indicated. This position has been determined by extrapolation from the ^{208}Po and ^{210}Po peaks.

The sensitivity of this instrument can be demonstrated in figure 18-4. If the measured activity of the daughter products of ^{222}Rn is assumed to be as reported in reference 18-3, then approximately 126 ± 42 counts should be in the ^{210}Po peak channel (above the existing contamination in approximately channel 40), 68 ±54 counts in the ^{218}Po peak channel (approximately channel 80), and 57 ±42 counts in the ^{214}Po peak channel (approximately channel 175). (Although this region of the spectrum is not shown, there are only two counts between channels 165 and 185.) The count rate observed for ^{222}Rn is less than 10^{-4} counts/cm^2/sec/sr. If radioactive equilibrium is assumed between ^{222}Rn and ^{210}Po, the implied rate of ^{210}Po disintegration is less than 6×10^{-4} particles/cm^2/sec. This is a factor of 50 less than that reported in reference 18-3. This result is not necessarily inconsistent with that report, because radioactive equilibrium may not exist and because Apollo 15 did not overfly the Surveyor V site. However, it can be concluded that the levels of ^{210}Po reported for the Surveyor V site are not likely to be typical for the Moon at the time of the Apollo 15 flight.

REFERENCES

18-1. Kraner, Hobart W.; Schroeder, Gerald L.; Davidson, Gilbert; and Carpenter, Jack W.: Radioactivity of the Lunar Surface. Science, vol. 152, no. 3726, May 27, 1966, pp. 1235-1236.

18-2. Yeh, Richard S.; and Van Allen, James A.: Alpha-Particle Emissivity of the Moon: An Observed Upper Limit. Science, vol. 166, no. 3903, Oct. 17, 1969, pp. 370-372.

18-3. Turkevich, Anthony L.; Patterson, James H.; Franzgrote, Ernest J.; Sowinski, Kenneth P.; and Economou, Thanasis E.: Alpha Radioactivity of the Lunar Surface at the Landing Sites of Surveyors 5, 6, and 7. Science, vol. 167, no. 3926, Mar. 27, 1970, pp. 1722-1724.

18-4. Lindstrom, Richard M.; Evans, John C., Jr.; Finkel, Robert C.; and Arnold, James R.: Radon Emanation from the Lunar Surface. Earth Planet. Sci. Lett., vol. 11, July 1971, pp. 254-256.

18-5. Economou, T.E.; and Turkevich, A.L.: Examination of Returned Surveyor III Camera Visor for Alpha Radioactivity. Proceedings of the Second Lunar Science Conference, vol. 3, A.A. Levinson, ed., MIT Press (Cambridge, Mass.), 1971, pp. 2699-2704.

18-6. Heymann, D.; and Yaniv, A.: Distribution of Radon-222 on the Surface of the Moon. Nature Phys. Sci., vol. 233, no. 37, Sept. 23, 1971, pp. 37-38.

19. Lunar Orbital Mass Spectrometer Experiment

J.H. Hoffman,[a][†] R.R. Hodges,[a] and D.E. Evans[b]

The scientific objective of the orbital mass spectrometer experiment is to measure the composition and distribution of the ambient lunar atmosphere. Data from the experiment are applicable to several areas of lunar studies. One such field is the understanding of the origin of the lunar atmosphere. Light gases, such as hydrogen, helium, and neon, probably originate from neutralization of solar-wind ions at the surface of the Moon, while Ar^{40} is most likely due to radioactive decay of K^{40}, and Ar^{36} and Ar^{38} may be expected as spallation products of cosmic ray interactions with surface materials. Molecular gases, such as carbon dioxide, carbon monoxide, hydrogen sulfide, ammonia, sulphur dioxide, and water vapor, may be produced by lunar volcanism.

Another field of application of spectrometer data is related to transport processes in planetary exospheres. The exosphere of the Earth (like that of almost any other planet) is bounded by a dense atmosphere in which hydrodynamic wind systems complicate the problem of specifying appropriate boundary conditions for exospheric transport. This contrasts sharply with the situation in the lunar atmosphere, which is entirely a classical exosphere, with its base the surface of the Moon. The lunar exosphere should be amenable to accurate, analytical study, and experimental determination of the global distributions of lunar gases can provide a reasonable check on theory, giving confidence to the application of theoretical techniques to transport problems in the terrestrial exosphere.[1]

Some of the gases thought to be dominant in the lunar atmosphere are hydrogen, helium, neon, and argon, with the abundance of neon exceeding the others by about an order of magnitude (ref. 19-1).

Hodges and Johnson (ref. 19-2) have recently shown that light gases with negligible production and loss rates tend to be distributed at the lunar surface as the inverse 5/2 power of temperature, while heavier gases are influenced by the rotation of the Moon. Neon falls into the former category, and its concentration on the antisolar side should be about 32 times that on the sunlit side. Its scale height on the dark side is about one-fourth that on the sunlit side, and, thus, at a satellite altitude of 100 km, the diurnal fluctuation of neon concentration should be less than a factor of 2. Argon, being a heavier gas, is expected to be noticeably influenced by the rotation of the Moon. It has a slightly less diurnal variation than neon and a longitudinal shift of its maximum toward sunrise, resulting in a concentration at sunrise that is approximately twice that at sunset.

Water vapor and other condensable gases probably exist in the lunar atmosphere, but not on the dark side, or near the poles, where the surface temperature is below 100 K and adsorption removes every particle that comes in contact with the surface. Gases adsorbed in continuously shadowed regions near the poles are unlikely to reenter the atmosphere, but at lower latitudes the rotation of the Moon transports adsorbed gases into sunlight where they are released into the atmosphere. Since surface heating occurs rapidly, this release probably occurs entirely within a few degrees longitude from the sunrise terminator, creating a pocket of gas.

The firing of the ascent rocket of the lunar module and the impact of the jettisoned lunar module ascent stage with the surface are good examples of known point sources of gas on the lunar surface. If the rate of spreading of these gas clouds around the Moon could be detected by the mass spectrometer in orbit, the diffusion rates for the various gases could be

---

[a]The University of Texas at Dallas.
[b]NASA Manned Spacecraft Center.
[†]Principal investigator.
[1] Hodges, R.R.: Applicability of a Diffusion Model to Lateral Transport in the Terrestrial and Lunar Exosphere. To be published in Planetary and Space Science.

calculated. Also, the escape rates of gases of various molecular weights could be determined. Because the gases from the rocket will be adsorbed on the lunar surface materials, the measurement of outgassing rates of these adsorbed gases should be useful. From this information, the amount of contamination of the lunar atmosphere due to the firing of rocket motors, both past and future, near the surface could be estimated.

INSTRUMENT

A sector-field, dual-collector, single-focusing mass spectrometer, with its electronics packaged in a controlled thermal environment, is mounted on a 7.3-m bistem boom that is extended from the service module of the Apollo spacecraft. Control of the experiment functions, as well as the boom extension and retraction, is provided by a set of five switches in the command module (CM) operated by a crew member according to the mission time line or by request from the Mission Control Center. The dimensions of the instrument are approximately 30 by 32 by 23 cm, and the weight is 11 kg. Figure 19-1 is a photograph of the mass spectrometer. The scoop mounted on top of the package is a gas inlet plenum that is oriented along the spacecraft velocity vector (the ram direction) when the command-service module (CSM) is flown backwards in its −X direction.

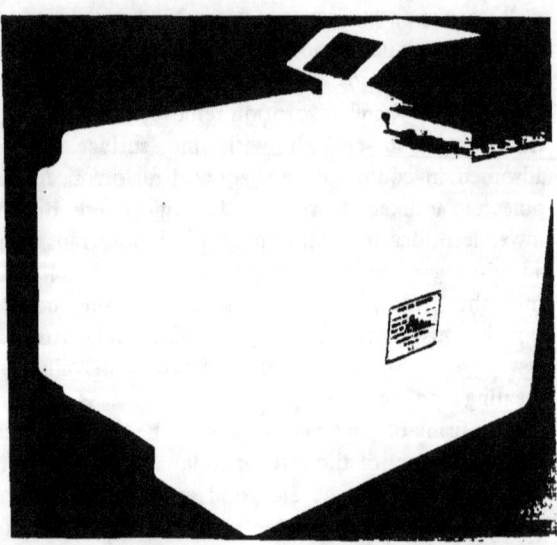

FIGURE 19-1.—Photograph of the lunar orbital mass spectrometer. The plenum containing the ion source is at top of photograph.

The plenum is on the outboard side of the instrument with respect to the CSM, such that the distance to a plane passing through the entrance aperture from any point of the CSM is 5.5 m minimum. In order that a gas molecule emanating from the CSM be detected, it must either undergo a collision within the field of view of the plenum aperture and be deflected into the aperture, or orbit the Moon so as to intersect the path of the plenum aperture. This system was shown to be a very effective discrimination mechanism against direct CSM outgassing by comparative tests in lunar orbit and in transearth coast (TEC), where there were no returning, orbiting particles.

The plenum contains the mass spectrometer ion source (a Nier type) employing redundant tungsten (with 1 percent rhenium) filaments mounted on either side of the ionization chamber. An emission-control circuit activated by the ion source switch (ON position) in the command module powers the filaments. Two small heaters, consisting of ceramic blocks with imbedded resistors, are mounted on the sides of the ionization chamber. In order to outgas the ion source during flight, these heaters are activated by the ion-source switch (STANDBY position). The ion-source temperature reaches 573 K in 15 min. Several outgassing periods during the flight maintained the ion source in a reasonably outgassed state.

When the CSM is oriented to orbit the Moon with the −X-axis forward (flying backward), the instrument plenum angle of attack is near zero and the native gases of the lunar atmosphere are scooped into the plenum. To determine the background spectra from the CSM and the instrument outgassing, the +X spacecraft axis is pointed forward with the plenum aperture in the wake, preventing native gases from entering the plenum.

The mass analyzer is a single-focusing permanent magnet with second-order angle focusing achieved by circular-exit field boundaries, giving a mass resolution of better than 1 percent valley at mass 40 amu. Two collector systems permit simultaneous scanning of two mass ranges, 12 to 28 amu and 28 to 66 amu. Figure 19-2 is a schematic drawing of the analyzer.

Voltage scan is employed utilizing a stepping high-voltage power supply. The ion-accelerating voltage (sweep voltage) is varied in a stepwise manner (590 steps are required to scan the spectra), with a dwell time on each step of 0.1 sec. An enable pulse from the data handling system steps the ion-accelerating voltage. The voltage step number, which de-

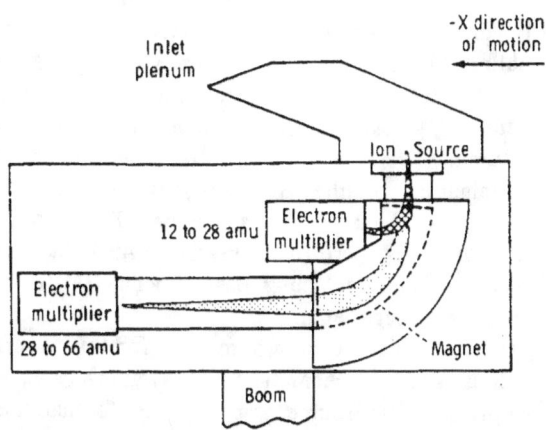

FIGURE 19-2.—Schematic drawing of the magnetic analyzer, showing two ion-beam trajectories.

termines the mass number of the ion being detected, is found by counting from step one, which is indicated by a sweep start flag (Data C). The ion accelerating voltage sweep is generated by varying this voltage in a series of 590 steps from 620 to 1560 V. Between each sweep 30 additional steps at zero V are used to determine background counting rate, and to apply an internal calibration frequency. The sweep start flag (Data C) indicates data or background, and serves as a marker for the start of each sweep. The minimum number of steps between adjacent mass peaks below mass 54 is 12.

The detector systems, employing electron multipliers, preamplifiers, and discriminators, count the number of ions that pass through each collector slit on each of the sweep voltage steps. The ion-count number is stored in a 21-bit accumulator (one for each channel) until sampled by the scientific data system, a 64-kilobit/sec telemetry link to Earth. Just prior to sampling, each data word is compressed pseudologarithmically into a 10-bit word consisting of a 6-bit mantissa and a 4-bit multiplier. This system maintains 7-bit accuracy throughout the 21-bit range of data counts.

Two switches in the CM control the data-counting system. Gain of the electron multipliers is adjustable by controlling the high voltage applied to the multiplier by the MULT switch. The normal operation mode is LOW, but the HIGH position may be used if the multiplier gain should decrease during flight. Similarly, in the discriminator circuit of the pulse-counting system, a sensitivity-level change is selectable by the DISC switch. Normal mode is discriminator HIGH. During the flight the discriminator LOW setting was used to reduce an occasionally high background counting rate due to scattered sunlight in the analyzer. With sufficient multiplier gain, the counting rate ratio at multiplier LOW and discriminator LOW to multiplier HIGH and discriminator HIGH should be greater than 0.95. This condition was met during the flight of this instrument, indicating the ion-counting efficiency was very high.

A housekeeping circuit monitors 15 functions within the instrument that are sampled serially, one per second, by the data system. Parameters, such as certain internal voltages, electron emission in the ion source, filament currents (to determine which filament was operating), multiplier voltages, sweep voltages, temperatures, multiplier and discriminator settings, and instrument current, are monitored.

The mass spectrometer analyzer, magnet, ion source, and detectors are mounted to a baseplate which bisects the instrument package. A conetic housing surrounds these components as is seen in the front portion of the picture of figure 19-1. On the opposite side is a passively controlled thermal environment containing the electronics. Attached to the baseplate is a flange that mates to a similar one on the boom. Opposite to the flange is the plenum that serves as the gas-entrance system to the mass spectrometer.

Thermal control of the electronics is accomplished mainly by a passively controlled environment. An active heating element, a 5-W heater, which is operated by a thermal control circuit, is attached to the interior of the electronics housing. It is powered at temperatures below 273 K and is turned off at 279 K. During the Apollo 15 mission, the heater was inoperative as the electronics temperature remained between 278 and 313 K, indicating that the passive thermal system was adequate to protect the electronics from the lunar and interplanetary environments.

CALIBRATION

Initial calibration of the mass spectrometers, performed in a high-vacuum chamber at The University of Texas at Dallas, verified that the proper mass ranges were scanned, and tested the resolution, linearity, mass discrimination, and dynamic range of the analyzer. Neon was introduced into the vacuum chamber with isotopic partial pressures ranging from

10^{-11} to 10^{-7} torr. As is shown in figure 19-3, the instrument response was linear up to 1×10^{-8} torr where the onset of saturation of the data-counting system occurred. The sensitivity of the instrument was verified to be greater than 3×10^{-5} A/torr, enabling the instrument to measure partial pressures down to 10^{-13} torr. The uncertainties in the introduction of gases into the chamber, in the pressure measurement, and in the wall effects precluded the determination of the absolute sensitivity in The University of Texas at Dallas chamber. The absolute calibration was performed in the NASA Langley Research Center Molecular Beam Facility (MBF) as reported by Yeager et al.[2]

The molecular beam system used for this experiment is shown schematically in figure 19-4. A high-pressure gas source was required to maintain inlet pressures from 0.1 to 10^4 torr, at a constant known temperature between 295 and 301 K. The high source pressure was then reduced by passing the gas through a porous silicate-glass plug into a molecular furnace. The conductance of the plug C_p was experimentally determined (in situ) for all test gases. Gas molecules, upon entering the molecular furnace, equilibrated to the known wall temperature, and effused through a precision aperture into a beam with an angular distribution determined by the Knudsen

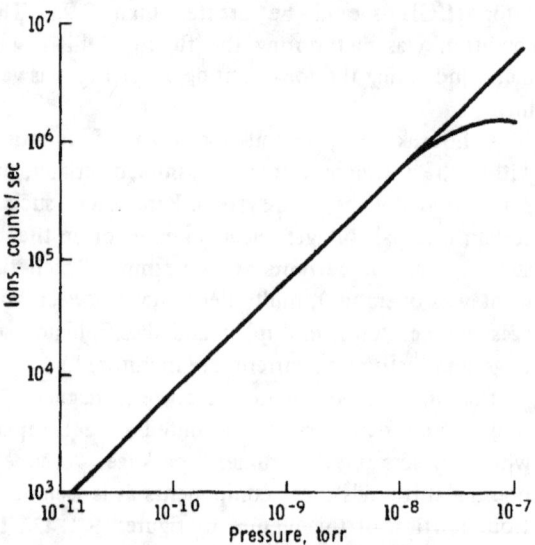

FIGURE 19-3.—Dynamic range of the mass spectrometer for neon, including neon-21 and neon-22 peaks to extend the pressure range.

[2] Yeager, P.; Smith, A.; Jackson, J.J.; and Hoffman, J.H.: Absolute Calibration of Apollo Lunar Orbital Mass Spectrometer. To be published in Journal of Applied Physics.

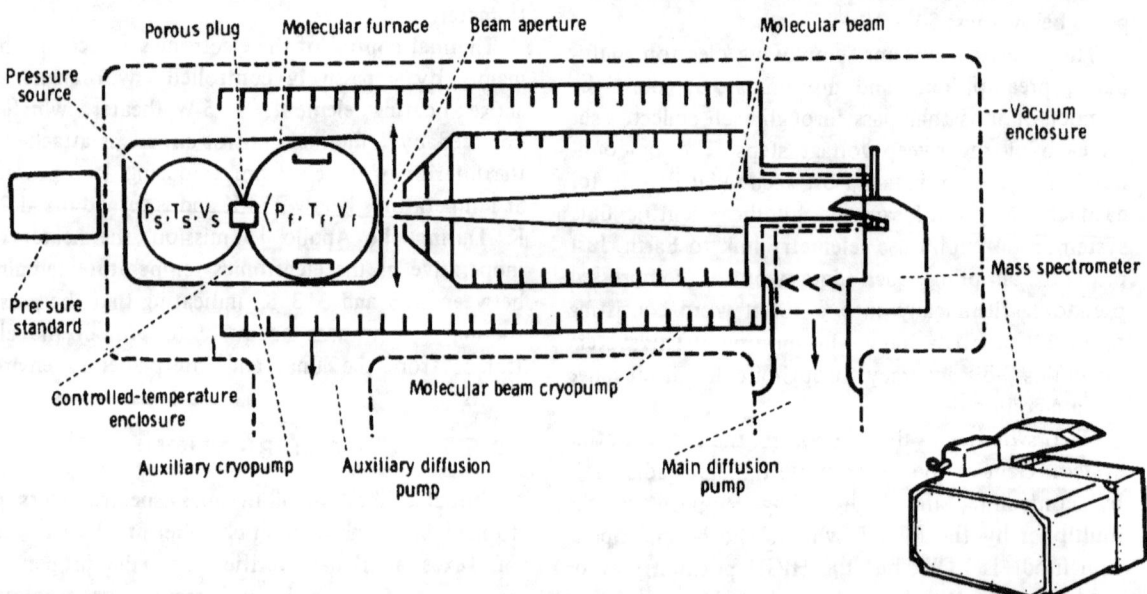

FIGURE 19-4.—Sketch of the Molecular Beam Facility at the NASA Langley Research Center. Pressure, temperature, and volume (P, T, and V) are indicated by subscript as source S and furnace f.

cosine law. The beam passed through a set of cryo baffles at 4.2 K and through a copper tube, also at 4.2 K, into the mass spectrometer plenum, the axis of which was alined with the beam axis.

The mass spectrometer was mounted in the MBF with the electronic package in the guard vacuum of the system and the plenum located as described previously. An externally controlled mechanical linkage allowed the plenum-beam angle to be varied from 0° to 40° with reference to a horizontal axis perpendicular to the beam axis (the spacecraft yaw direction). Pitch angles of −5°, 0°, and +5° (with reference to a vertical axis perpendicular to the beam axis) could be set manually with the system open. Separate tests were conducted with a combination of the three pitch angles and various yaw angles from 0° to 40°. The mass spectrometer inlet was completely enclosed by a 4.2 K extension copper tube so that the back scattering of molecules into the inlet is essentially eliminated. The 4.2 K extension tube was enclosed by a 77 K wall of the guard system.

A large amount of data was generated from these tests using three flight instruments and one qualification model. Figure 19-5 shows a typical set of curves of the output counting rate for neon and argon as a function of MBF beam flux; 10^{10} molecules/cm²-sec is equivalent to 6.5×10^{-12} torr of argon in the plenum. It can be seen that pressures below 10^{-13} torr in the plenum are readily measurable. The data show a linear response well within the molecular beam accuracy (6 percent).

Figure 19-6 shows the variations in output for the flight units with change in yaw angle. Up to about 20° the output falls off according to the cosine law, as would be expected. However, between 20° and 40° the fall-off is close to a $\cos^2$ function. This seems to be due to incoming molecules not reaching equilibrium with the walls before they are measured or escape from the plenum. Up to 20° yaw, the input beam hits the rear wall of the plenum only. Beyond 20°, it begins to strike the side walls from which gas molecules may more readily escape without being measured. The qualification model was also tested at +5° and −5° pitch angles. The output at negative pitch angles is higher than for positive pitch angles. This is due to the shape of the inlet and the resulting in and out gas flux. The inlet opening is not perpendicular to the direction of the beam, but becomes more nearly so at negative pitch angles. This results in the effective beam opening being smaller than the actual exit size for an incoming beam, but

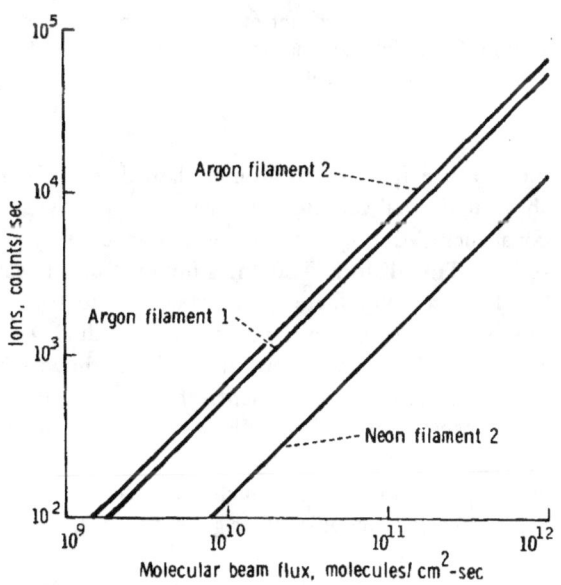

FIGURE 19-5.—Calibration of mass spectrometer output as a function of molecular beam flux for neon and argon. Combining these results with figure 19-3 gives a dynamic range of nearly 6 decades.

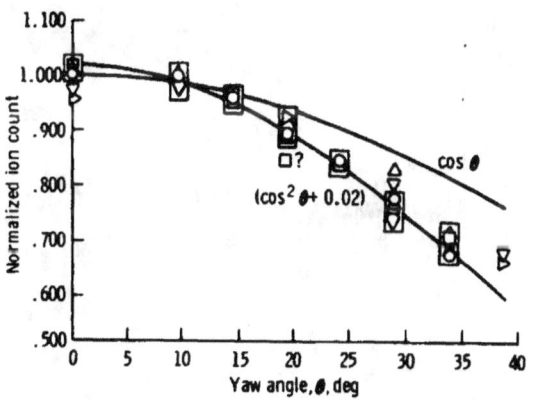

Symbol	Gas	Unit	Pressure, N/m²	Pitch angle, deg
○	Argon	Flight 1	2.06×10^5	0
□	Neon	Flight 1	2.06	0
◇	Argon	Flight 2	2.06	−5
△	Neon	Flight 2	2.06	−5
▽	Argon	Qualification model	2.06	+5
▷	Neon	Qualification model	6.88×10^4	+5

FIGURE 19-6.—Angular response of the mass spectrometer to off-axis beams. Flight Unit 3 data are within the bounded region shown for Flight Units 1 and 2.

remaining the geometrical opening for the outgoing flux. The measured pitch angle response follows a cosine law.

During flight, the instrument is mounted on a long boom which is susceptible to thermal twisting, equivalent to a spacecraft yaw maneuver, and bending, equivalent to a pitch motion. Preliminary results from the flight indicate the boom did twist as a function of the Sun angle by an amount very close to that predicted by models of boom twist, 35° to 40°. The bending was very slight.

RESULTS

The mass spectrometer experiment produced about 40 hours of data in lunar orbit and 50 additional hours during TEC. Instrument performance was quite satisfactory. Preliminary quick-look-type data showed large numbers of peaks in the mass spectra of relatively large amplitude. Figure 19-7 shows typical spectra from lunar orbit. The peak amplitudes (counts per second) are plotted as a function of sweep-voltage step number. The upper spectrum shows peaks from 66 amu to 27 amu; the lower shows peaks from 28 amu to 12 amu. The mass 18 peak (water vapor) saturated the counting system.

In figure 19-8 the counting rates for three gases are

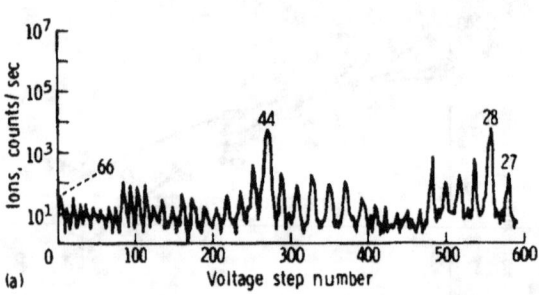
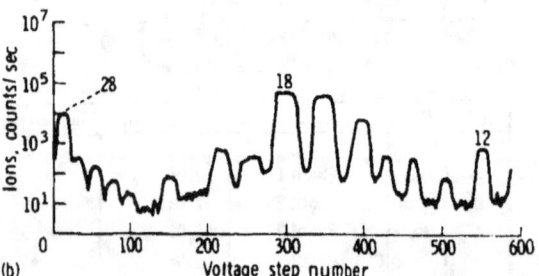

FIGURE 19-7.—Mass spectra for lunar orbit. (a) High mass range from 66 to 27 amu. (b) Low mass range from 28 to 12 amu.

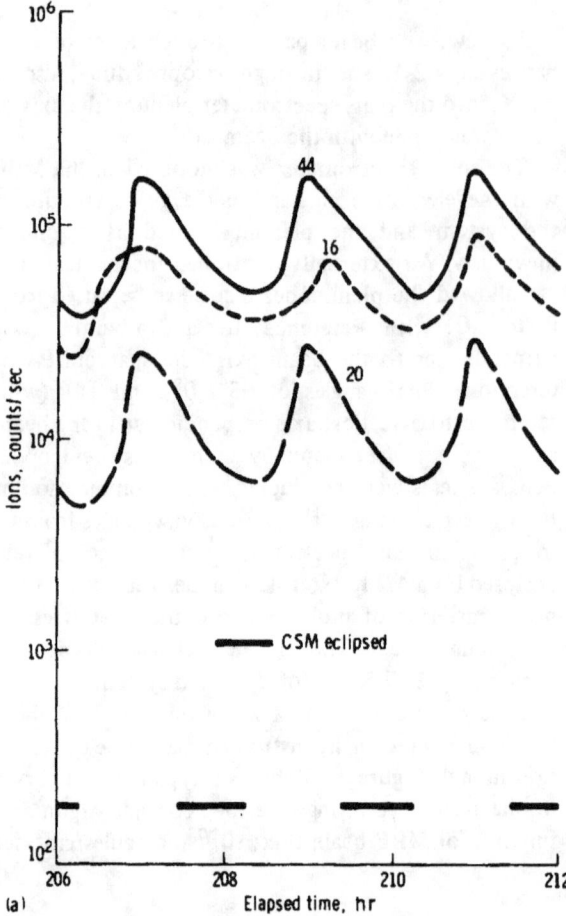

FIGURE 19-8.—Mass spectrometer data for three gas species. (a) Data recorded during lunar orbit. (b) Data recorded in passive-thermal-control attitude of TEC.

plotted as a function of ground elapsed time (GET), showing diurnal variations in lunar orbit and the same constituents during the passive-thermal-control attitude of TEC. During TEC the amplitude of all peaks has decreased by a factor in the range of 5 to 10. Also during TEC, a boom-retraction test, reducing the boom length in four steps to 1.25 m, showed no increase in any gas constituent. This implies that the mass spectrometer plenum is very effective in preventing molecules originating at the CSM from entering the ion source. Such outgassing molecules form essentially a collisionless gas within 1.25 m from the CSM surface.

The major gases observed in lunar orbit do not appear to have a significant velocity with respect to the spacecraft as the observed densities are not a

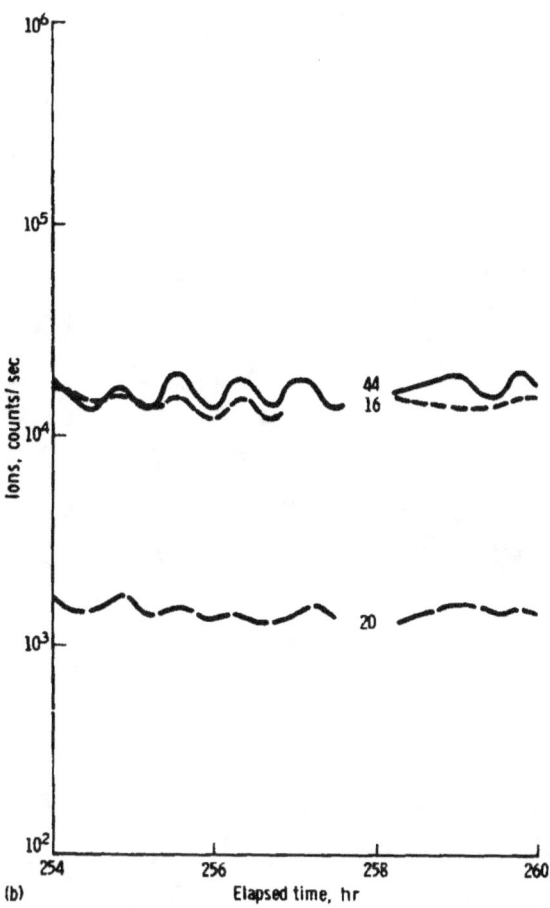
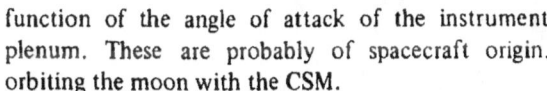

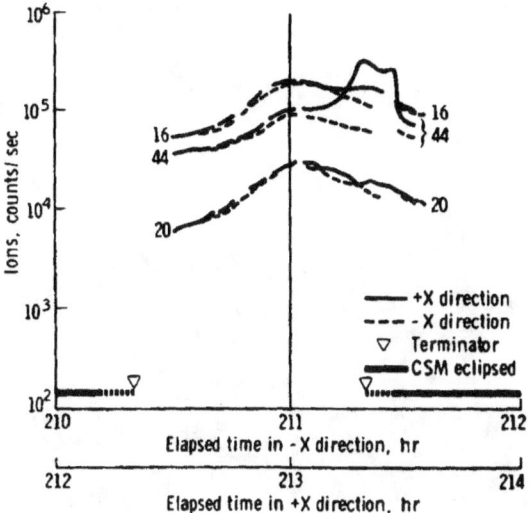

FIGURE 19-9.—Data from lunar orbit for −X (ram) and +X (wake) orientations for selected mass peaks.

function of the angle of attack of the instrument plenum. These are probably of spacecraft origin, orbiting the moon with the CSM.

Figure 19-9 is a plot of several gas species counting rates for two successive revolutions about the Moon, the first in −X forward CSM orientation, the second in +X. Except for the region near the terminator where the mass 44 peak (probably carbon dioxide) exhibits a large increase, the amplitudes (gas concentrations) of all peaks are essentially independent of plenum orientation.

CONCLUSION

The lunar orbital mass spectrometer experiment, from reduction of a small amount of quick-look data, appears to have operated successfully, observing a large number of gas molecules in the vicinity of the spacecraft in lunar orbit. Many of these molecules are most likely of spacecraft origin, orbiting the Moon with the CSM. Interesting variations of this gas cloud are observed as a function of orbital parameters, but extensive analysis of the complete set of data soon to become available from tapes recorded during the mission will be required to understand the significance of these results.

REFERENCES

19-1. Johnson, Francis S.: Lunar Atmosphere. Rev. of Geophys. and Space Phys., vol. 9, no. 3, Aug. 1971, pp. 813-823.

19-2. Hodges, R.R., Jr.; and Johnson, F.S.: Lateral Transport in Planetary Exospheres. J. Geophys. Res., vol. 73, no. 23, Dec. 1, 1968, pp. 7307-7317.

ACKNOWLEDGMENTS

The authors wish to express their appreciation to the large number of people who contributed to the success of this experiment. Many personnel of The University of Dallas made major contributions to the successful design, fabrication, testing, and flight of the mass spectrometer.

The NASA Langley Research Center Molecular Beam Facility calibrations were made possible through the efforts of Paul Yeager and Al Smith.

20. S-Band Transponder Experiment

W. L. Sjogren,[a†] *P. Gottlieb,*[a] *P. M. Muller,*[a] *and W. R. Wollenhaupt*[b]

The S-band transponder experiment derives data from three lunar-orbiting objects—the command-service module (CSM), the lunar module (LM), and the subsatellite. Each object provides detailed information on the near-side lunar gravitational field. In this section, the primary emphasis is on the low-altitude (20 km) CSM data. The LM data cover a very short time span and are somewhat redundant with the CSM data. The resolution of the high-altitude (100 km) CSM data is not as great as that of the low-altitude data and will be examined later. The low-altitude CSM and LM data coverage and the complementary coverage obtained during the Apollo 14 mission are shown in figure 20-1. The experiment uses the same technique of gravity determination employed on the Lunar Orbiter, in the data of which the large anomalies called mascons (refs. 20-1 to 20-3) were first observed. No special instruments are required on the CSM and LM other than the existing real-time navigational system. The data consist of variations in the spacecraft speed as measured by the Earth-based radio tracking system.

TECHNIQUES

The schematic drawing in figure 20-2 shows the basic measuring system. A very stable frequency of 2115 MHz obtained from a cesium reference is transmitted to the orbiting spacecraft. The transponder in the spacecraft multiplies the received frequency by the constant 240/221 (to avoid self-lockup) and transmits the signal to the Earth. (The transmitted and received frequencies are within the S-band region.) At the Earth receiver, the initial transmitted frequency, multiplied by the same constant, is subtracted, and the resulting cycle-count differences are accumulated in a counter along with the precise time at which differencing occurred. These cycle-count differences are the doppler shift in frequency f_d caused by radial component V_r of the spacecraft velocity or $2V_r/c \times 2300$ MHz where c is the speed of light. At times of high resolution, the counter is read every second, whereas at low resolution, it is read once a minute. Not only is the cycle-count difference recorded, but the fractional part of the cycle is measured. This process allows a resolution in the measurements of approximately 0.01 Hz or 0.65 mm/sec. This observable speed (range rate) is often referred to as doppler or line-of-sight velocity.

The raw data represent or contain many components of motion, and these must be removed before gravity analysis can proceed. Factors that must be accounted for include the tracking-station rotation about the Earth spin axis; the spacecraft motion perturbed by point-mass accelerations from the Sun, Earth, Moon, and planets; and atmospherics and signal-transit times. All these quantities are known a priori and are determined to accuracies well beyond those required to evaluate local gravitational effects.

Two approaches are possible for reducing the resulting velocity data. The first approach directly differentiates the velocity observations, and gravitational accelerations are immediately determined. The second approach estimates a surface-mass distribution from a dynamic fit to the observations.

The reduction procedure for the first approach was performed using the Jet Propulsion Laboratory orbit-determination computer program (ref. 20-4), which contains the theoretical model with all the dynamic constraints and parameters previously mentioned. Each revolution of data (approximately 65 min) was evaluated independently, with the doppler observations least-squares-fitted assuming a spherical Moon and solving for (adjusting) only the six state parameters of initial position and velocity.

---
[a]Jet Propulsion Laboratory.
[b]NASA Manned Spacecraft Center.
[†]Principal investigator.

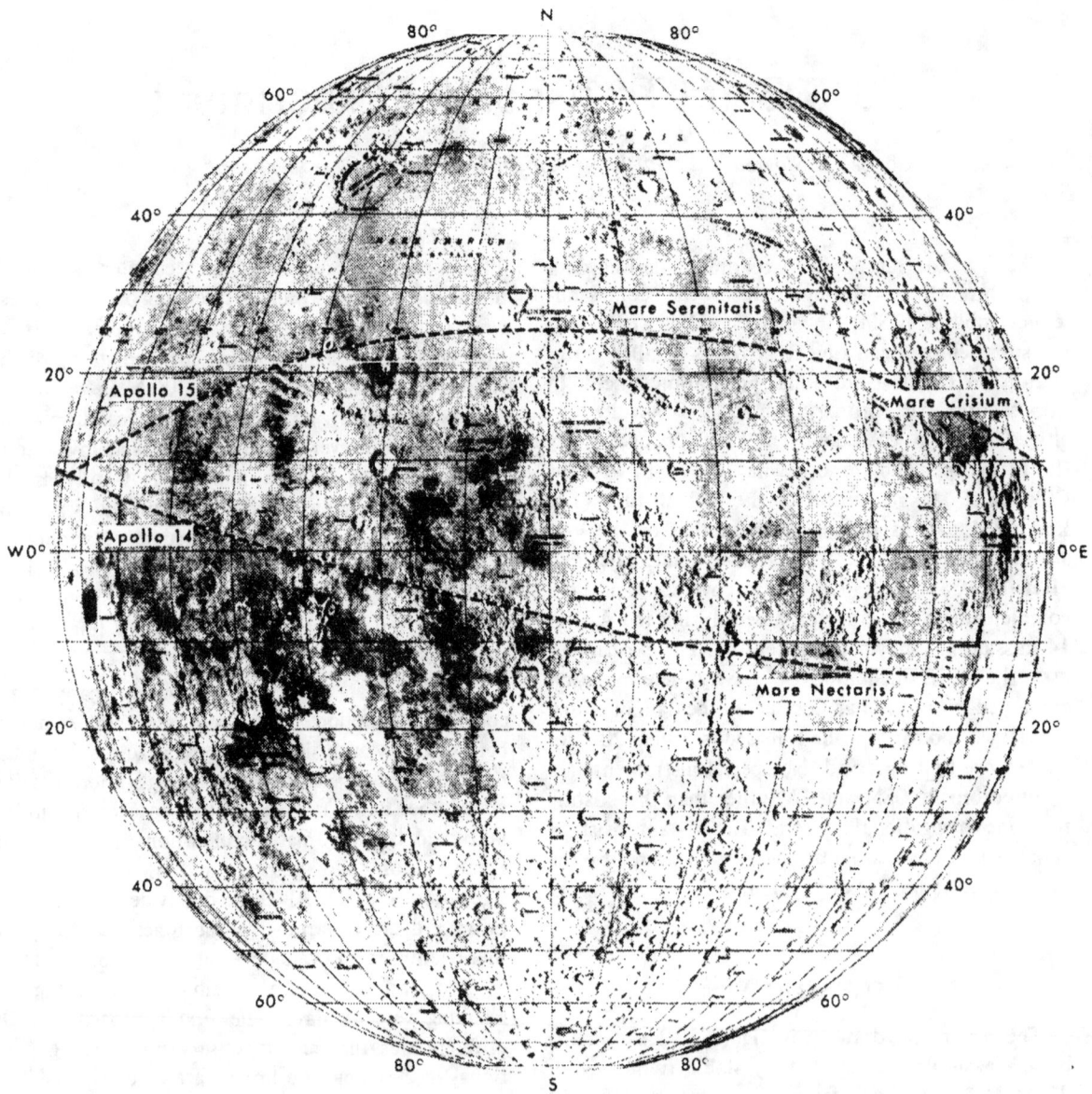

FIGURE 20-1.—The CSM and LM low-altitude gravity data coverage during the Apollo 14 and 15 missions.

The resulting systematic residuals (i.e., real observations minus theoretically calculated observations) are then attributed to lunar gravitational effects. The velocity residuals from revolution 4 are shown in figure 20-3. The signature is very definite and far above the noise level of 0.01 Hz. The rapid change in velocity is clearly evident over Mare Crisium and Mare Serenitatis.

Once these residuals are checked for consistency between adjacent revolutions and all obviously erroneous points are removed, analytic patched cubic splines with continuous second derivative are least-squares-fitted to the residuals. These functions are then differentiated, and the line-of-sight accelerations ("gravity") are analyzed and correlated with the subspacecraft lunar track and existing topography. At present, two of the 10 low-altitude CSM revolutions have been screened; when all are completed, the accelerations as a function of spacecraft lunar latitude and longitude will be plotted on a 1:1 000 000

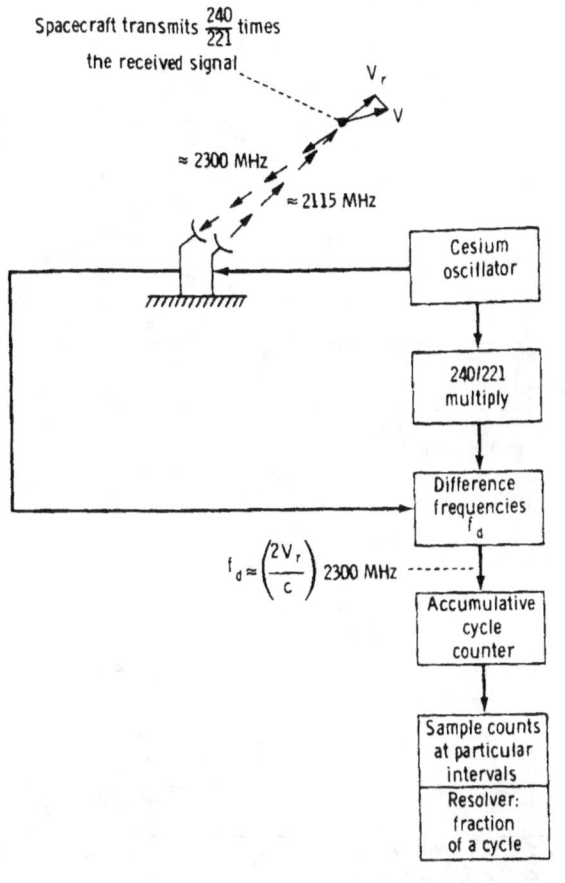

FIGURE 20-2.—Simplified schematic drawing of doppler transponder system.

Mercator projection through ±70° of longitude. Because these are line-of-sight (rather than vertical) gravity components, a geometric effect exists that shifts gravity-feature locations toward the limbs and reduces the amplitude slightly. The shift is a little more than 1°, and the amplitude reduction is about 30 percent for an object at longitude 50° E or 50° W. Altitudes over the 100-km band vary from 17 km at periapsis (approximately longitude 26° E) to 40 km at longitude 40° W. No normalization factor has been applied to bring the accelerations to a constant-altitude surface. A plot of the accelerations as a function of longitude from revolution 4 data is shown in figure 20-4. The validity of the structure in this profile was verified by the independent reduction of the adjacent revolution, which produced the same signature.

This same approach will be used on the subsatellite data and, hopefully, complete near-side coverage will be obtained between latitude 30° N and 30° S at several relatively constant altitudes (i.e., 100 km, 60 km, 30 km). There should be periods when altitudes of 30 km or less will provide high resolution over new regions similar to the low-altitude CSM resolution obtained during the Apollo 15 mission. At the time of preparation of this report, the subsatellite was functioning well. Data collection is proceeding, and the lifetime of the subsatellite is expected to be greater than a year.

The second approach is to estimate a dense surface

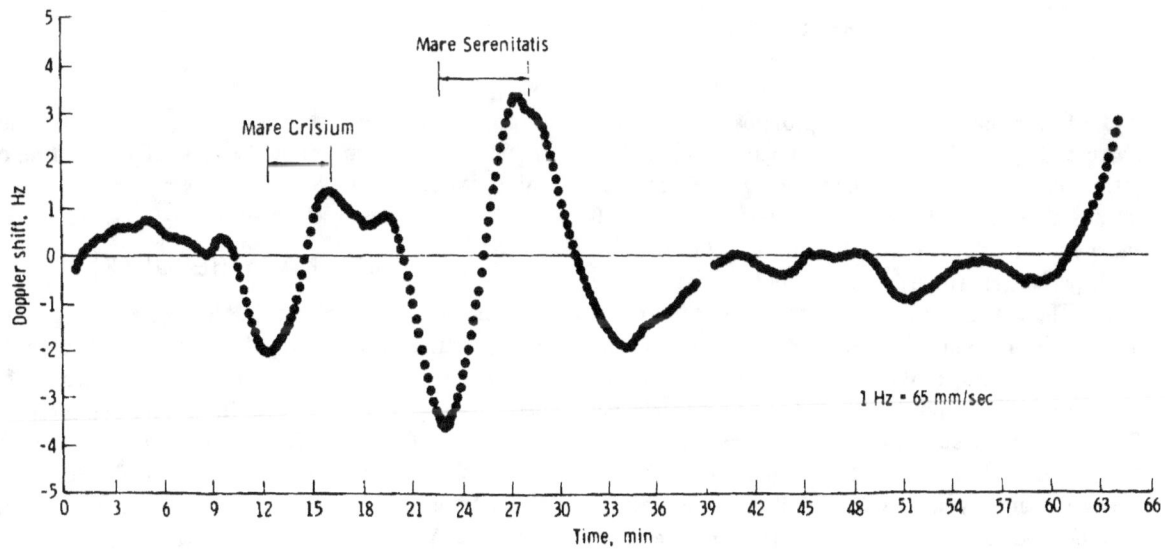

FIGURE 20-3.—Doppler residuals from revolution 4 on July 30, 1971, 02:43 G.m.t.; Goldstone tracking data.

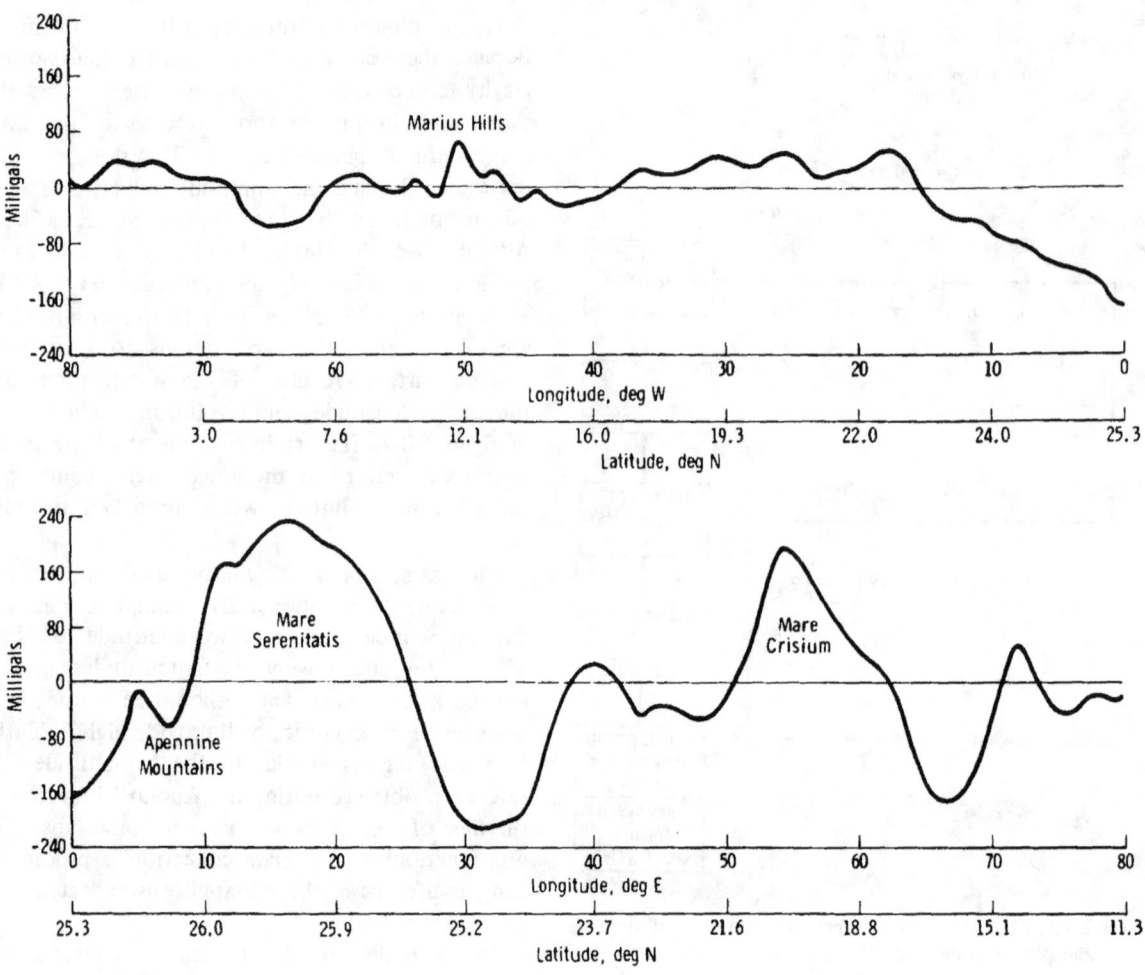

FIGURE 20-4.—Line-of-sight gravity component derived from revolution 4.

grid of disk masses spaced approximately 1° to 2° apart and lying along the orbital surface track. This estimation will involve determining 200 to 300 masses along with the state parameters from some 20 independent orbits—10 at low altitude (17 km (9.3. n. mi.), periapsis). These data are from the CSM and LM only. There will also be data from the subsatellite and, with the data, a 5° grid of surface disks will be estimated (approximately 400 disks between longitude 95° E and 95° W and latitude 25° N and 25° S). These results will be more quantitative than those derived from the first approach because all the geometric and dynamic effects will be accounted for and no spurious effects will result from the least-squares operations (which reduce the absolute amplitude of the residuals and sometimes introduce erroneous negative accelerations (ref. 20-5)). This reduction is in process at the time of the writing of this section.

PRELIMINARY RESULTS

The most striking features in figure 20-4 are the large positive-gravity anomalies over Mare Serenitatis and Mare Crisium. The shapes of the two differ because of geometric effects (i.e., Mare Serenitatis at longitude 19° E and Mare Crisium at longitude 59° E), as shown somewhat obviously by the following simulations. As an initial attempt to compare results with the Apollo 14 reductions (refs. 20-6 and 20-7), two simulations were made. The first was to generate tracking data with a theoretical model of the Moon having surface-disk features at the Mare Serenitatis and Mare Crisium location. Then, the same fitting process used on the real data was applied to the simulated data, and acceleration profiles were obtained. The second simulation was the same as the first except that the theoretical model of the Moon had spherical bodies 100 km deep rather than the surface disks. Comparisons of each of these simulations with the real data are shown in figures 20-5 and 20-6. The surface disk matches remarkably well both the Mare Serenitatis and Mare Crisium mascon profiles, whereas the deeply buried spherical body matches poorly. This was precisely the same result noted for the Mare Nectaris mascon from Apollo 14 data reduction (ref. 20-7). Moreover, the mass per unit area in these simulations was held at the Mare Nectaris mascon value of 500 kg/cm^2, which strongly suggests that the mass distribution per unit area is about the same for all three mascons (i.e., simulated-data acceleration amplitude agrees well with the real-data acceleration amplitude) (ref. 20-8). The broadness of the profile depends on the radius of the surface disk, which was 245 km for the Mare Serenitatis mascon (centered at longitude 19° E and latitude 26° N) and 210 km for the Mare Crisium mascon (centered at longitude 59° E and latitude 17.5° N). Both mascons for the spherical-body case were assumed to be a mass of 9×10^{-6} of the lunar mass. The difference in shape of the acceleration profile between Mare Crisium and Mare Serenitatis (figs. 20-5 and 20-6) is due to geometrical effects, and the simulations illustrate the same effect (figs. 20-5 and 20-6). The spherical bodies have peak amplitudes closer to the feature center but are still shifted, primarily because of the 100-km depth. In the preceding subsection on techniques, the comment about geometrical effects of 1° or so in the higher longitudes is for relatively small surface features and does not apply to the mascons.

Another interesting feature is the definite shoulder in the Mare Serenitatis gravity anomaly at longitude 12° E. (There is also a hint of similar shouldering at longitude 20° E.) Again, compensating for geometrical effects and possible surface-feature correlation, these locations are shifted closer to longitude 13.5° E and longitude 21.5° E, respectively. This shouldering is an indication of an abrupt excess mass change, implying a change in surface or subsurface structure. Possibly, it is due to subsurface ring structure usually associated with circular maria but ill defined for Mare Serenitatis. This arc structure is visible in Whitaker's work (ref. 20-9) about halfway between the Apennine Mountains and the Tycho Crater ray.

The Apennine Mountains show up quite clearly as a local high with an approximately 85-milligal positive anomaly (obtained by smoothing the Mare Serenitatis curve into the low at longitude 0°). If a 4-km height, a 100-km width, and a 200-km length are assumed for a nonisostatic feature having a density of 3.0 g/cm^3, the anomaly should be 300 milligals, which implies that there has been partial isostatic compensation.

The Marius Hills at longitude 50° W exhibit a 60-milligal positive gravity. This is confirmation of the Lunar Orbiter data, which also reveal a positive anomaly in this region. It seems odd that the highlands

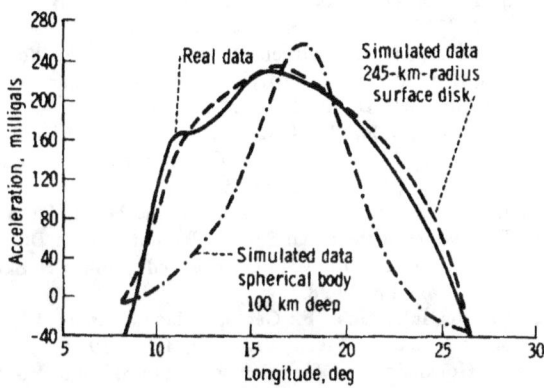

FIGURE 20-5.—Simulated gravity compared to real gravity for Mare Serenitatis.

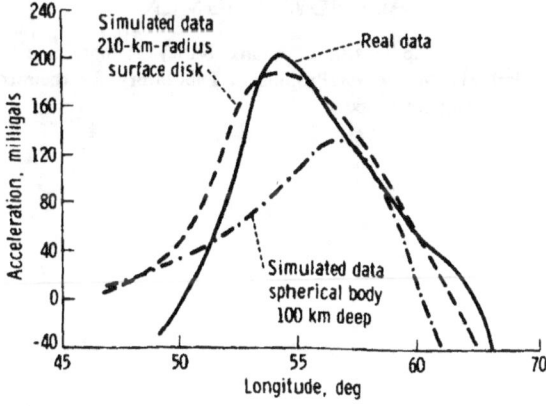

FIGURE 20-6.—Simulated gravity compared to real gravity for Mare Crisium.

terrain south of Archimedes Crater at longitude 5° W does not reveal any distinct signature although it is apparently of the same extent as the Marius Hills. These two areas are, however, different geologic blocks (as shown on U.S. Geological Survey maps (refs. 20-10 and 20-11)).

The large negative-gravity anomalies east of Mare Crisium and Mare Serenitatis are possibly part of a negative ring structure similar to that determined from the Lunar Orbiter data for Mare Orientale. This assumption will be confirmed or denied once the solid coverage is obtained from the subsatellite.

Other areas of interest, such as the low at longitude 65° E, the low-amplitude but slightly positive gravity for Oceanus Procellarum, and the low at longitude 0°, are generally consistent with Lunar Orbiter data results. Mare Imbrium does not appear as a significant high because the trajectory path is outside the inner ring. However, at longitude 18° W, there is a small positive-gravity tongue (ref. 20-12) that extends out of the Mare Imbrium basin, and this shows up mildly as 50 milligals.

CONCLUSIONS

It appears quite evident that the mascons are indeed near-surface features with a mass distribution of approximately 500 kg/cm^2. Definite structural features are visible in the Mare Serenitatis gravity signature. These features will be further investigated. The Apennine Mountains are a local gravity high and have partial isostatic compensation.

The substatellite is operating well, and new information is being accumulated that should greatly enhance understanding of the detailed mass distribution of the Moon. Reduction of these data began in December 1971 when a solid block of data was obtained.

REFERENCES

20-1. Muller, P. M.; and Sjogren, W. L.: Mascons: Lunar Mass Concentrations. Science, vol. 161, no. 3843, Aug. 16, 1968, pp. 680-684.

20-2. Muller, P. M.; and Sjogren, W. L.: Lunar Gravimetrics. Proceedings of Open Meeting of Working Groups at the 12th Plenary Meeting of COSPAR, Space Research X, North Holland Pub. Co. (Amsterdam), 1970, pp. 975-983.

20-3. Wong, L.; Duetchler, G.; Downs, W.; Sjogren, W. L.; et al.: A Surface-Layer Representation of the Lunar Gravitational Field. J. Geophys. Res., vol. 76, no. 26, Sept. 10, 1971, pp. 6220-6236.

20-4. Warner, M. R.; Flynn, J. A.; and Hilt, D. E.: Jet Propulsion Laboratory Engineering Planning Doc. 426-8, May 2, 1969.

20-5. Gottlieb, P.: Estimation of Local Lunar Gravity Features. Radio Science, vol. 5, 1970, pp. 303-312.

20-6. Sjogren, W. L.; Gottlieb, P.; Muller, P. M.; and Wollenhaupt, W. R.: S-Band Transponder Experiment. Sec. 16 of Apollo 14 Preliminary Science Report, NASA SP-272. 1971.

20-7. Sjogren, W. L.; Gottlieb, P.; Muller, P. M.; and Wollenhaupt, W. R.: Lunar Gravity via Apollo 14 Doppler Radio Tracking. Science, vol. 175, 1972.

20-8. Booker, John R.; Kovach, Robert L.; and Lu, Lee: Mascons and Lunar Gravity. J. Geophys. Res., vol. 75, no. 32, Nov. 10, 1970, pp. 6558-6574.

20-9. Whitaker, E. A.: The Surface of the Moon. Ch. 3 of The Nature of the Lunar Surface. Wilmot N. Hess, Donald H. Menzel, and John A. O'Keefe, eds., Johns Hopkins Press, 1965, pp. 79-98.

20-10. McCauley, John F.: Geologic Map of Herelius Region of the Moon. U.S. Geol. Survey Map I-491, 1967.

20-11. Hockman, R. G.: Geologic Map of the Montes Apenninus Region of the Moon. U.S. Geol. Survey Map I-463. 1966.

20-12. Sjogren, W. L.; et al.: Lunar Surface Mass Distribution from Dynamical Point-Mass Solution. The Moon, vol. 2, no. 3, Feb. 1971, pp. 338-353.

ACKNOWLEDGMENTS

The authors wish to thank Nancy Hamata and Ray Wimberly of the Jet Propulsion Laboratory for their computer-program support.

21. Subsatellite Measurements of Plasmas and Solar Particles

K. A. Anderson,[a†] *L. M. Chase,*[a] *R. P. Lin,*[a] *J. E. McCoy,*[b]
and R. E. McGuire[a]

On August 4, 1971, at 21 hr 00 min 30.81 sec G.m.t., the Apollo 15 astronauts launched a small scientific spacecraft into an orbit around the Moon. Within 20 min, information about the magnetic fields, plasmas, and energetic particles in the vicinity of the Moon was transmitted to Earth. The Apollo 15 particles and fields subsatellite has already provided several weeks of nearly continuous data coverage and is designed to operate for at least several months.

Immediately following the subsatellite launch, the astronauts observed and photographed the subsatellite in space, which was the first time this had been achieved. A photograph of the subsatellite several seconds after launch is shown in figure 21-1.

The particles and fields subsatellite is instrumented to make the following measurements.

(1) Plasma and energetic-particle fluxes (sec. 21)
(2) Vector magnetic fields (sec. 22)
(3) Velocity of the subsatellite to a high precision for the purpose of determining lunar gravitational anomalies (sec. 20)

DESCRIPTION OF THE PARTICLES AND FIELDS SUBSATELLITE

The small scientific subsatellite has a mass of approximately 38 kg and a length of 78 cm. The subsatellite cross section is hexagonal, and the distance between opposite corners is approximately 36 cm. The subsatellite has three deployable booms hinged from one of the end platforms. To one of the booms is attached a two-axis fluxgate-magnetometer sensor, and tip masses are attached to the other two booms to provide balance and a proper ratio of moments of inertia to avoid precession.

The subsatellite has a short cylindrical section attached to one of the end platforms. This cylinder fits into a barrel that is attached to the scientific instrument module of the command-service module (CSM). A compression spring pushes the subsatellite away from the CSM and, at the same time, imparts a spin. Precessional and rotational motion imparted to the subsatellite by the launch and by the boom deployment was removed by a wobble damper. The spin axis of the subsatellite was initially pointed normal to the ecliptic plane; very precise pointing of the CSM by the astronauts resulted in an error of less than 1°. The spin period is 5 sec. Each of the six sides of the subsatellite forms a solar panel. The total power output of the arrays is approximately 24 W; averaged over an orbit around the Moon, the total power is 14 W. The power subsystem also includes a battery pack of 11 silver-cadmium cells.

A basic scientific requirement placed on the subsatellite was that it must obtain particles and fields data everywhere on the orbit around the Moon; this requirement necessitated the inclusion of a data-storage capability in the subsatellite. The magnetic-core memory unit used for this purpose provides a capacity of 49 152 bits. Data can be read into the memory at a rate of 8 bits/sec, which allows coverage of nearly the entire revolution (2-hr period). Data can also be read in at 16 bits/sec, if a need exists to obtain better time resolution in the measurements, at the expense of covering only approximately one-half the orbit. Real-time data at the rate of 128 bits/sec also can be acquired from the experiments, but the

---
[a]University of California at Berkeley.
[b]NASA Manned Spacecraft Center.
[†]Principal investigator.

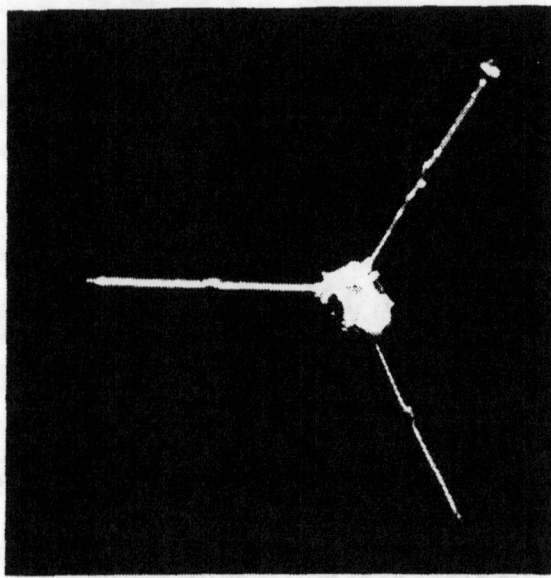

FIGURE 21-1.—The particles and fields subsatellite a few seconds after ejection from the scientific instrument module. The photograph is from a frame of the motion-picture film taken by the astronauts.

trade-off is that battery power, as well as solar-cell power, must be used beyond a certain point. During normal operation, the transmitter is commanded to turn on after the subsatellite appears from behind the Moon. Real-time housekeeping data and scientific data are transmitted for a short time to ensure that the receiving stations are locked on to the signal. The data in the memory unit then are dumped in 512 sec at a rate of 128 bits/sec. The transmitter then is turned off, and accumulation of data in the storage unit begins again.

The perilune altitude of the subsatellite on the first revolution around the Moon was 102 km, and the apolune altitude was 139 km. The orbital period is approximately 120 min, and the orbital inclination, with respect to the lunar equator, is 28.5°. The sense of revolution around the Moon is clockwise, as viewed from the north. Perturbations on the orbit affect the perilune; after approximately 50 revolutions, the orbit had become nearly circular at an altitude of 120 km. The perilune then decreased to 100 km on the 120th revolution, after which it again began to increase. The inclination of the orbit is not appreciably changed by the perturbations. A long-term variation of the lowest perilune reached in the shorter term cycle also will occur. The expectation is that, after some months, the subsatellite may reach altitudes as low as 30 km, before again increasing.

The subsatellite scientific instrumentation includes several fixed-voltage electric-field analyzers for electron-energy measurement over the range of 500 eV to 15 keV. Because of power, weight, and volume limitations, no proton measurements are made in this range. The decision in favor of emphasizing electrons rather than protons was a result of the fact that electron shadowing by the Moon gives more physical information; however, protons, as well as electrons, are measured in the two solid-state telescopes over the energy range of 20 keV to 4 MeV. The main parameters of the detectors are listed in table 21-I.

OBJECTIVES AND RESULTS OF THE PLASMA AND ENERGETIC-PARTICLE EXPERIMENT

Results from the magnetic-field and gravitational-field experiments are discussed in sections 22 and 20, respectively, of this report. In this section, the plasma and energetic-particle detectors and some of the results obtained to date from the detectors are briefly discussed.

The main objectives of the plasma and energetic-particle experiment are as follows: (1) to describe the various plasma regimes in which the Moon moves, (2) to determine how the Moon interacts with the plasmas and magnetic fields in the environment, and (3) to determine certain features of the structure and dynamics of the Earth magnetosphere.

The primary results obtained to date are as follows.

(1) A wide variety of particle shadows has been measured; the shadow shapes agree well with the theory that has been developed.

(2) The cavity made by the Moon in the solar wind is highly structured and variable. The observations show that the interaction of the Moon with the solar wind is complex and probably involves features that cannot be explained by a fluid-dynamic treatment of the interaction.

(3) The energy spectrum of solar-flare electrons has been determined over a wider energy range than had been possible in previous space experiments.

(4) Detailed observations have been made of particle populations in the magnetosheath and in the Earth bow shock as the Moon moves through these regions.

TABLE 21-I.—*Summary of Detector Characteristics*

Detector designation	Detector type	Energy range		Geometric factor, cm^2-sr	Angular aperture	Angle to spin axis, deg	Approximate value of minimum detectable flux, particles/cm^2-sr-sec
		Protons, MeV	Electrons, keV				
SA_{1-6}	Open solid-state detector, with anticoincidence detector in back (six-channel pulse-height analyzer)	0.05 to 6	20 to 300	0.45	40°, cone	0	0.01
SB_{1-6}	Same as SA_{1-6}, except with a 500-$\mu g/cm^2$ foil over the detector	.3 to 6	25 to 300	.45	40°, cone	0	.01
C_1	Channel electron multiplier in a hemispherical-plate electrostatic analyzer	No response	.53 to .68	3.2×10^{-4}	20° by 60°; full width, half maximum	90	10^4
C_2	Channel electron multiplier in a hemispherical-plate electrostatic analyzer	No response	1.75 to 2.25	3.2×10^{-4}	20° by 60°; full width, half maximum	90	10^4
C_3	Channel electron multiplier in a hemispherical-plate electrostatic analyzer	No response	5.8 to 6.5	1.4×10^{-3}	15° by 60°; full width, half maximum	90	10^4
C_4	Funnel-mouthed channel electron multiplier in a hemispherical-plate electrostatic analyzer	No response	5.5 to 6.5	.27	18° by 60°; full width, half maximum	90	.1
C_5	Funnel-mouthed channel electron multiplier in a hemispherical-plate electrostatic analyzer	No response	13.5 to 15.0	63	13° by 60°; full width, half maximum	90	.1

DESCRIPTION OF THE PLASMA AND ENERGETIC-PARTICLE DETECTORS

In the following paragraphs, the plasma and energetic-particle detectors are described in some detail.

Energetic-Particle Telescopes

Absolute fluxes and energy spectra of electrons and protons in the energy range of 20 to approximately 700 keV are obtained from two telescopes in which solid-state particle detectors are used (fig. 21-2). Each telescope contains a 25-mm^2, Si(Li), surface-barrier detector. In terms of particle kinetic energies, the detector has a thickness that stops electrons with energies less than 320 keV and protons with energies less than 4 MeV. Behind this detector is a second detector with a 50-mm^2 area. The output of the back detector is placed in anticoincidence with the front detector. This arrangement rejects undesired events, such as the following.

(1) Particles that enter at the acceptance angle of the telescope and emerge from the front detector are rejected, because the energy loss in the front detector does not give the total energy of the particle.

(2) Particles that enter the detectors from directions other than the directions defined by the collimators are rejected. An energy loss of 20 keV or greater in the rear detector is required for the anticoincidence logic to operate.

The collimator is passive and consists of 11 aluminum rings with knife edges that define a 15°-half-angle entrance cone. The finite area of the front detector allows some particles that make an angle as large as 20° with respect to the detector axis to be counted. All internal collimator surfaces are anodized to a flat black to reduce illumination by sunlight on the light-sensitive surface-barrier detectors.

The front detector is a fully depleted surface-barrier detector that is mounted with the active surface-barrier side away from the collimator. Thus, the positively biased aluminum-coated surface is the particle-entrance surface. This orientation minimizes radiation damage and light-sensitivity effects, while providing a thin (40 µg/cm^2) entrance window. The opposite surface of the front detector is the surface barrier covered by evaporated gold with a thickness of 40 µg/cm^2. The rear detector is oriented in the opposite direction so that the surface barriers of the two detectors directly face each other.

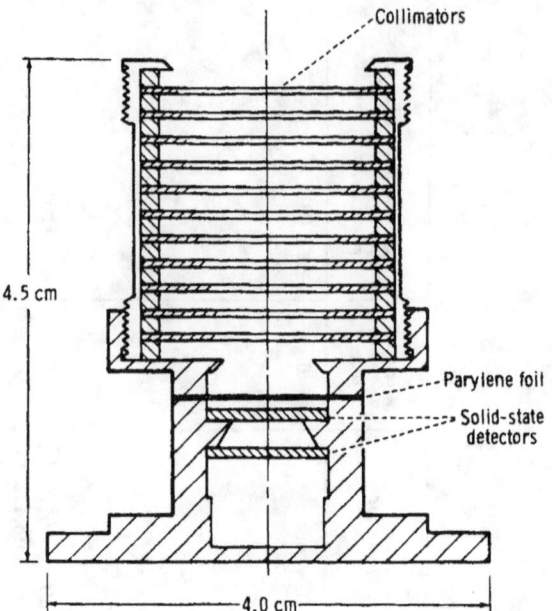

FIGURE 21-2.—One of the two solid-state telescopes used on the subsatellite to detect electrons and protons in the energy range of 20 keV to 4 MeV. The other telescope is identical to this one, except it has no foil.

One of the telescopes has an organic foil (with a thickness of 375 µg/cm^2 ± 10 percent) in front of the front solid-state detector. This foil stops incident protons with energies as large as 310 keV, but reduces the energy of a 26-keV electron by only 5 keV. Thus, except for a small energy shift, a flux of electrons with energies in the 20- to 320-keV range would cause both telescopes to count at the same rate; however, when protons are incident on the telescopes, the counting rates will show large differences. In addition to this means of particle discrimination, use also can be made of the fact that protons and electrons of the same energy are shadowed by the Moon quite differently.

Each of the two telescopes has its own analog-signal processor. When a particle stops in the front detector (no veto from the rear detector), the analog output passes into a stacked-discriminator pulse-height analyzer. The analyzer is switched from one telescope to the other.

Detector pulses are analyzed into eight energy channels: the nominal energy thresholds of six of these channels are listed in table 21-II. The upper two channels are transmitted in calibration mode only and are substituted for low-energy proton data, as shown in table 21-II. The electron thresholds are switched when the analyzer is switched from one telescope to the other so that the channel edges correspond to the same incident electron energy to compensate for the loss in the foil (approximately 25 percent at the lowest threshold). The foil and the 320- and 520-keV thresholds are adjusted so that (1) the 40- to 340-keV protons detected by the open telescope are degraded below the lowest threshold of the shielded telescope and (2) the 340- to 520-keV protons detected by the open telescope deposit 20 to 320 keV in the shielded telescope. When the 340- to 520-keV-proton fluxes detected by the telescope covered by the foil are less than 10 times the electron fluxes, these constraints allow a subtraction of the proton and electron spectra.

TABLE 21-II.—*Solid-State-Telescope Energy Channels*

Channel number	Normal-mode energy range, keV	Calibration-mode energy range, keV
1	[a]20 to 320	[a]20 to 320
2	40 to 80	40 to 80
3	80 to 160	80 to 160
4	160 to 320	160 to 320
5	320 to 520	2000 to 4000
6	520 to 670	>4000

[a]The lowest threshold is adjusted to provide a specified noise counting rate. This threshold can be increased approximately 5 keV by ground command.

A weak radioactive source (plutonium 239) is placed near the front detector in each telescope. Alpha particles from these sources provide well-defined and known energy deposits as a check on detector and electronics stability.

The low-energy thresholds of the telescopes may be varied by ground command. This feature was included because of the desire to operate the telescope as near the thermal-noise levels as possible. The two threshold settings that are available on the subsatellite are 18 and 25 keV. Because of somewhat higher temperatures than anticipated, the threshold was raised to 25 keV during the first week of operation in orbit. Potentiometer adjustments are provided to enable variation of the lowest thresholds and adjustment of the higher energy channels to match proton edges. Other adjustments are provided to allow for energy loss in the absorber foil.

Electrostatic Analyzer

The electrostatic-analyzer assembly consists of four electrostatic analyzers, analog electronics, high-voltage power supplies, and logic circuits in the programing and data-handling subassembly. Each electrostatic analyzer consists of two concentric sections of spherical copper plates. The outer plate in each pair is grounded, while the inner plate is raised to a positive potential. The plates are shaped to provide a 180° by 90° angular segments for the electron trajectories.

The force experienced by an electron entering the analyzer is directed toward the common center of the plate pair. If the angle that the velocity vector of the incoming electron makes with the normal to the aperture is small enough, and if the energy of the electron lies within an interval determined by the plate radii and the bias on the inner plate, then the electron will traverse the entire 180° path from the entrance aperture to the exit aperture. The maximum elevation angle accepted is determined by the radii of the plates. Thus, each pair of plates operates to measure the electron flux from within a certain solid angle and energy band, over an area determined by the active area of the detector. An electron detector that consists of one or several channel multipliers or funnel-mouthed channel multipliers is placed at the exit aperture. These multipliers are connected to preamplifiers and discriminators.

Analyzers C_1 and C_2 are geometrically identical and use one channel multiplier (without a funnel) to detect intense fluxes of low-energy electrons. Analyzers C_1 and C_2 differ only in the plate voltage and, hence, in the mean detected energy. For analyzers C_3 and C_4, the same set of plates is used, but the output of analyzer C_4 is derived from two funnel-mouthed multipliers in parallel. The output of analyzer C_3 is derived from one small-aperture multiplier. This arrangement permits a wider dynamic range to be measured. The funnel multipliers are surrounded by a plastic scintillator that is viewed by a photomultiplier, which is connected in anticoincidence to eliminate charged-particle counts induced by cosmic rays.

An important design requirement is the rejection of ultraviolet light. This requirement is met in a number of ways. A 180° analyzer transmits a photon only after several reflections. The probability of this occurrence is minimized by serration and gold blacking of the outer-plate inner surface. The serration promotes the backward reflection of light, and the gold blacking promotes its absorption. The combined procedure reduces the incoming photon flux by a factor greater than 10^{12}. The entire assembly must be mounted so that no light can enter except through the entrance aperture. The electrostatic analyzers are oriented perpendicular to the subsatellite spin axis. To avoid spin biasing of the data as a result of directional anisotropy of particle flux, data are accumulated for integral spin periods only.

In addition, the output of analyzer C_5 (fig. 21-3) is time-division multiplexed so that particle intensities from various sectors of the subsatellite rotation can be obtained. The sectoring is made with respect to the magnetic-field vector, as sensed by the transverse magnetometer. The sectors are defined as follows:

Sector I: $-45°$ to $+45°$ of the magnetic-field vector ($\pm 5°$)

Sector II: $+45°$ to $+90°$ and $+270°$ to $+315°$ of the magnetic-field vector ($\pm 5°$)

Sector III: $+90°$ to $+135°$ and $+225°$ to $+270°$ of the magnetic-field vector ($\pm 5°$)

Sector IV: $+135°$ to $+225°$ of the magnetic-field vector ($\pm 5°$)

This technique of sectoring based on the magnetometer output enables a particle-pitch-angle distribution to be obtained directly. A functional diagram of the subsatellite systems is shown in figure 21-4.

The spiral continuous-channel electron multipliers are operated in the saturated mode with a gain of approximately 4×10^8. Detection of a single particle produces an output charge of 6×10^{-11} C, or approximately 600 times the threshold signal. This signal triggers the operation of the overlap loop, which enables the amplifier to recover in a fraction of a microsecond. A second pulse (following within a microsecond) will be detected even though the continuous-channel electron multiplier (CCEM) has recovered only a small fraction of its gain. This system enables accurate counting with CCEM's at very high rates (in excess of 100 kHz).

The same amplifier and discriminator are used to detect veto events in the plastic scintillator. The low-threshold sensitivity enables the photomultiplier to be run at relatively low gain. A signal charge of approximately 50 photoelectrons is expected, which corresponds to a photocathode input of approximately 8×10^{-18} C. For a signal-to-threshold ratio of 10,

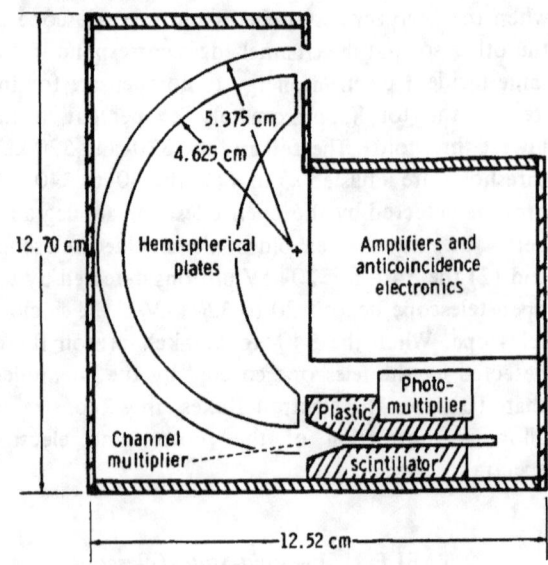

FIGURE 21-3.—Analyzer C_5, which is the largest electrostatic analyzer flown on the subsatellite. Analyzer C_5 measures electrons in the energy interval of 13.5 to 15 keV, with high sensitivity. Analyzers C_1 to C_4 are similar, but measure electrons at lower energies.

the photomultiplier gain need be only approximately 1.25×10^5; a substantial margin is thus available. Overload signals from protons stopping in the scintillator may correspond to energy inputs as large as 100 MeV. This input is approximately 600 times the threshold, which is a dynamic range easily accommodated by the overload circuit; therefore, vetoes will not be missed as a result of overloads.

The fixed deadtime, which is nominally 6 μsec for the CCEM discriminators, is used to keep the spacecraft accumulators from missing pulses. The leading edge of the discriminator pulse starts a digital output that triggers the accumulator. The deadtime of the phototube discriminator is set at 3 μsec. After a 3 ± 5 μsec delay from the leading edge of a CCEM discriminator pulse, an output pulse is transmitted by the logic, if a veto has not been sensed during the delay interval. This timing ensures that the accumulators have sufficient time to recover between pulses.

THEORY OF PARTICLE-SHADOW FORMATION BY THE MOON

A large particle-absorbing object, such as the Moon, when placed in a flux of charged particles, will

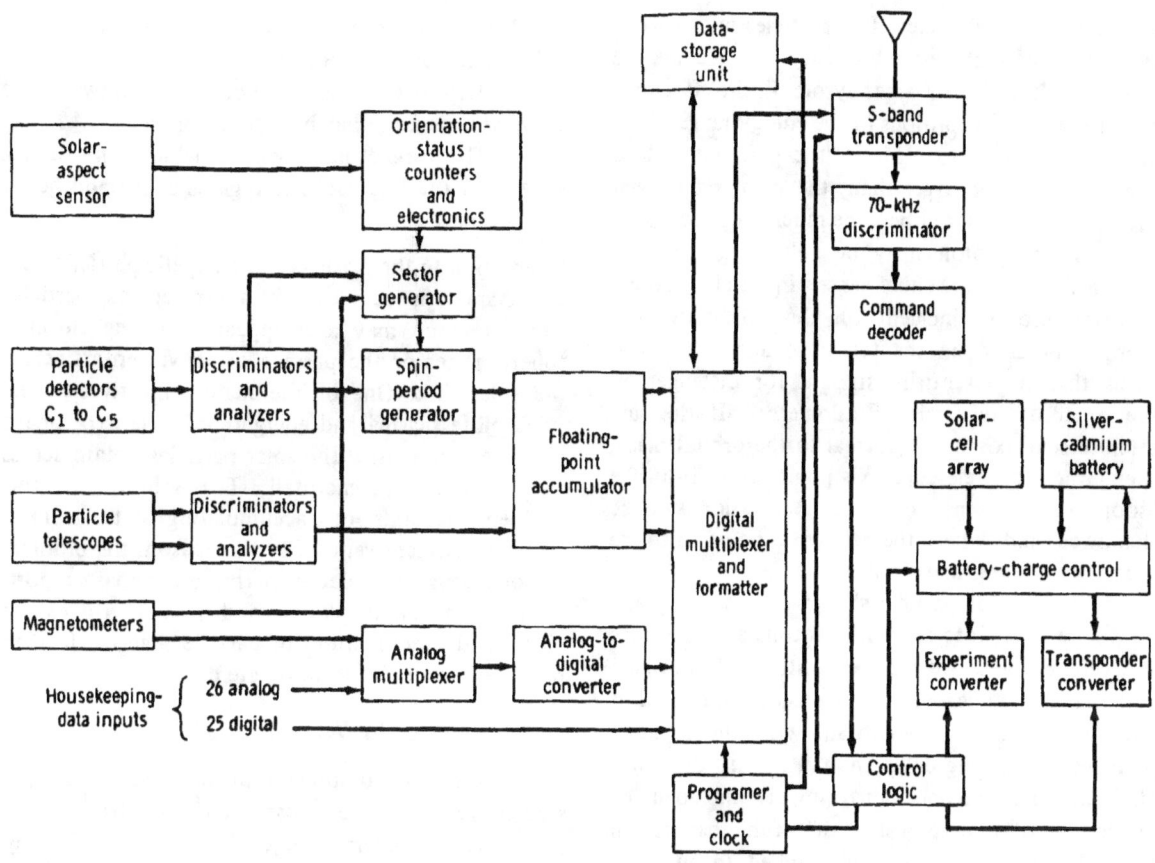

FIGURE 21-4.—Functional diagram of the subsatellite systems.

create a cavity or shadow region in that flux. The process is roughly analogous to the optical shadow created by an opaque object placed in a beam of light; however, the magnetic fields present in space will cause particle gyromotion around the field lines so that the particle trajectories will be helical. A further difference from the optical case is introduced by the motion of the magnetic-field lines and the particle-drift effects. A satellite in a low lunar orbit will be particularly sensitive to gyromotion and drift effects. Much physical information can be derived from the detailed structure of the particle shadows. It is, therefore, important to have a quantitative theory of shadow formation. One of the authors (R. E. McGuire) has developed such a theory that enables both particle-drift effects and detector-response effects to be quantitatively predicted. In this section, the shadow theory has been applied to the particle telescopes carried by the particles and fields subsatellite.

The three most significant factors in the determination of specific shadow shapes are particle gyroradius, the sense of particle rotation, and the angle between the detector axis and the orbital plane. To illustrate the importance of the sense of particle gyration, consider a 50-keV electron in a 10-γ magnetic field; the gyroradius of this electron, at a 70° pitch angle, is approximately 73 km. If the subsatellite is assumed to be above the lunar magnetic terminator, which is defined as the great circle that is tangent to the magnetic-field lines, and if an observer looks along the magnetic field toward the incident particles in such a way that the orbital plane of the subsatellite is horizontal and the Moon is on his left, then the subsatellite spin axis is nearly vertical. Electrons will rotate counterclockwise as they approach the subsatellite and can enter the telescopes without being blocked by the Moon; electrons rotating clockwise, however, will be blocked. On the opposite side of the orbit, only clockwise-rotating electrons can be detected. For particles with smaller gyroradii, this east-west asymmetry will become less pronounced; while, for larger gyroradii, the effect will become larger. For protons in the energy range of the detectors (25 keV to 4 MeV), the gyroradii will be comparable to, or larger than, the lunar radius, and the shadow effects will become much more complex. Some of the protons may be able to go out and around the Moon for either sense of particle rotation; however, protons incident on the solid-state telescopes will arrive in the telescope apertures along paths that are essentially straight for distances on the order of hundreds of kilometers. If the subsatellite spin axis is not normal to the orbital plane, then, where the telescope axis points away from the Moon, neither sense of particle rotation will be shadowed, and, where the telescope points into the Moon, both senses may be shadowed out.

The theoretical particle shadows, as seen by the solid-state telescopes for both protons and electrons at several energies, are shown quantitatively in figure 21-5. The specific case considered is of magnetic-field lines in the Earth magnetotail, with the incident particles propagating down the tail toward the Earth (fig. 21-5(a)). Particle mirroring is included by making use of an empirical model of the magnetotail field. The results also can be applied to an interplanetary flux of particles with scattering or mirroring of the particles behind the Moon. In figures 21-5(b) and 21-5(c), the assumption has been made that the subsatellite spin-axis orientation is normal to the magnetic-field vector and the orbital plane; while, in figure 21-5(d), the assumption has been made that the orientation is normal to the magnetic-field vector, but 60° from the orbital plane. The dashed curve in figure 21-5(d) is derived for a magnetic-field direction opposite the direction shown (fig. 21-5(a)). The marked terminator regions are where the subsatellite is above the magnetic terminator. A magnetic-field strength of 10 γ and a circular orbit at 110 km also have been assumed.

INITIAL RESULTS FROM THE PLASMA AND ENERGETIC-PARTICLE EXPERIMENT

As the Moon revolves around the Earth, it encounters several distinct regions of magnetized plasmas.

(1) The solar wind

(2) The bow shock, which stands on the sunward side of the Earth magnetosphere

(3) The magnetosheath, which lies between the bow shock and the Earth magnetosphere

(4) The magnetotail, which includes the plasma sheet and the high-latitude regions above and below the plasma sheet

In addition to the plasmas and energetic particles that are characteristic of each of these regions, particles from the Sun may also appear following chromospheric flares or the passage of active centers across the solar disk. One of the major objectives of the subsatellite plasma and energetic-particle experiment is to determine how the solar particles obtain access to the Earth magnetotail. The solution to this problem depends on accumulating data from a number of solar events. For this reason, the problem is not discussed in detail in this preliminary report. Instead, measured plasma and particle fluxes are given, and some features of particle shadows formed by the presence of the Moon are described.

Solar-Wind Electrons

The electron component of the solar wind was studied by Montgomery, Bame, and Hundhausen (ref. 21-1) over the energy range of 20 to 700 eV. They found that, at low energies (up to approximately 70 eV), the electron population could be described by a Maxwellian distribution. At energies greater than approximately 100 eV, a non-Maxwellian tail became dominant. One of the electrostatic analyzers, C_1, responds to electron energies between 530 and 680 eV, which is a range that overlaps the energy range measured by the Vela satellite experiment. Solar-wind electrons clearly have been detected by analyzer C_1; solar-wind electrons also have been detected by analyzer C_2 at energies between 1750 and 2250 eV. A velocity distribution function could not be determined with the information obtained; however, fluxes could be measured with considerable precision. At a mean energy of 600 eV, the electron flux is variable, but it usually lies between 1.5×10^5 and 5×10^5 electrons/cm^2-sr-sec-keV. At a mean energy of approximately 2 keV, the electron flux is approximately 150 electrons/cm^2-sr-sec-keV at a time when the flux at 600 eV was 4.5×10^5 electrons/cm^2-sr-sec-keV. The form of the distribution is unknown; however, as an indication of the spectral shape, by fitting a power law to the two measured points, an exponent of -5.2 for the differential

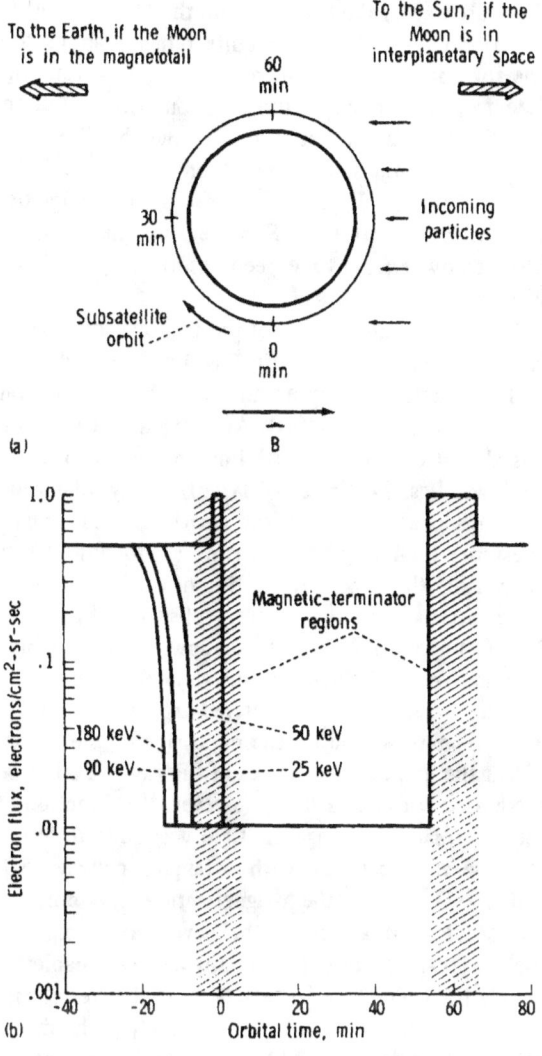
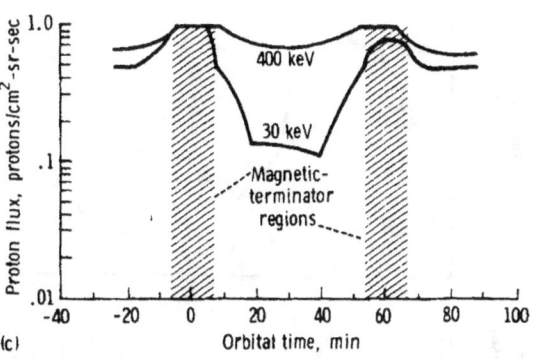
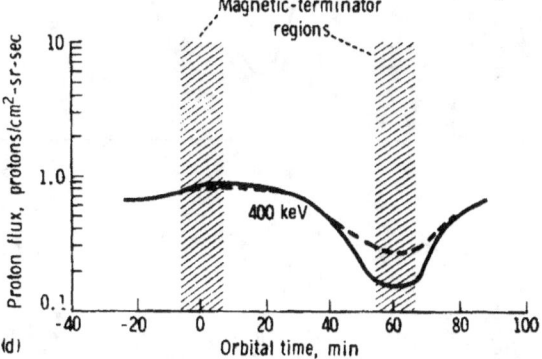

FIGURE 21-5.—Predicted particle shadows at various particle energies measured by the solid-state telescopes. (a) The Moon relative to the Earth, the Sun, the magnetic field, and the direction of the incoming particles. If the Moon is in interplanetary space, the direction of the magnetic field can be different from the direction shown. (b) Electron flux as a function of subsatellite orbital time (fig. 21-5(a)) for several electron energies when the subsatellite spin axis was oriented normal to the magnetic-field vector and the orbital plane. (c) Proton flux as a function of subsatellite orbital time (fig. 21-5(a)) for 30- and 400-keV protons when the subsatellite spin axis was oriented normal to the magnetic-field vector and the orbital plane. (d) Proton flux as a function of subsatellite orbital time (fig. 21-5(a)) for 400-keV protons when the subsatellite spin axis was oriented normal to the magnetic-field vector and 60° to the orbital plane. The dashed curve is derived for a magnetic-field direction opposite the direction shown in figure 21-5(a).

energy spectrum was obtained. The thermal speeds of the electrons that were measured in this experiment are much higher than the solar-wind bulk speed. For 600-eV electrons, the thermal speed is 14 000 km/sec. In deriving these flux values, the assumption was made that the solar-wind electrons at these high energies are essentially isotropic.

When the Moon is in the free-streaming solar wind, it is well known that the solar-wind plasma is absorbed on the sunward side and that a cavity, with reduced plasma density, is formed on the antisolar side (refs 21-2 and 21-3). The flux of solar-wind electrons with energies of 600 eV for three revolutions of the subsatellite around the Moon is shown in figure 21-6. At the times of the measurements, the Moon was situated approximately 25° west of the Sun-Earth line, well away from the Earth bow shock, and the interplanetary magnetic field was approximately $\phi = 180°$.[1] Two significant facts were observed about the shadows produced by the Moon in the flux of fast solar-wind electrons.

(1) At the times of the measurements, the fast

---
[1] Private communication with P. J. Coleman.

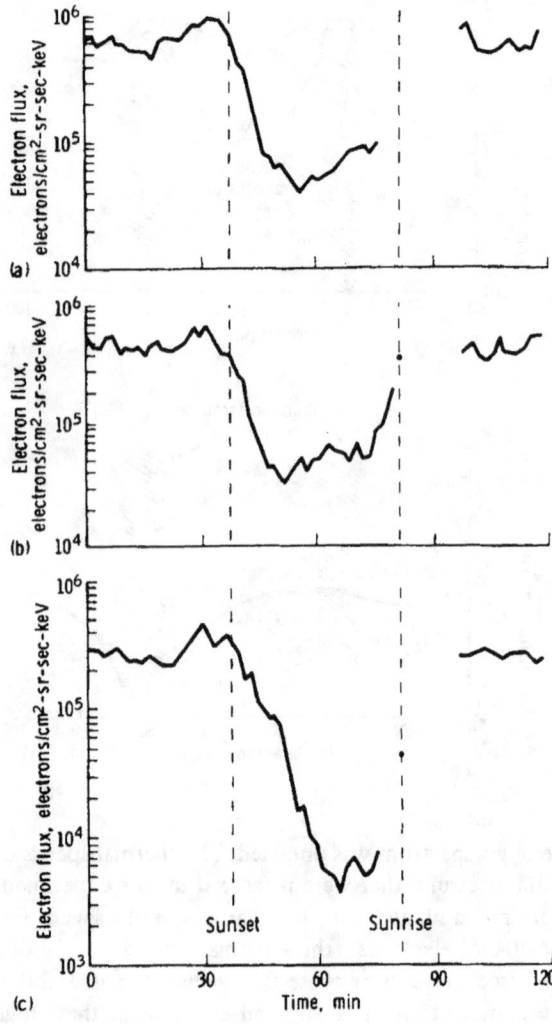

FIGURE 21-6.—Subsatellite measurements of the shadow produced by the Moon in the solar wind (as indicated by the flux of 600-eV electrons), where the Moon was located approximately 25° west of the Sun-Earth line and the orientation of the interplanetary magnetic field was approximately $\phi = 180°$. (a) August 19, 09:22 to 11:22 G.m.t. (b) August 19, 19:23 to 21:23 G.m.t. (c) August 20, 01:22 to 03:22 G.m.t.

solar-wind electrons are mostly excluded from the region directly downstream from the Moon. The edges of the cavity region correspond closely to the sunset and sunrise on the subsatellite, as would be expected; however, a small fraction of the fast electrons penetrate some distance into the cavity.

(2) Increases in electron intensity occur immediately before sunset on the subsatellite. These increases are from 50 to 100 percent. At these times, the spacecraft is above a line from the Sun, tangent to the Moon limb. The subsatellite remains above this line for approximately 14 min. During this time, the detectors receive solar-wind electrons from all directions. When the subsatellite is below this line as it crosses the sunlit side of the Moon, the detectors receive only those electrons that have thermal velocities directed away from the Sun, because the electrons further downstream have been absorbed by the lunar surface.

Thus, the main features of the shadows in a flux of fast solar-wind electrons (fig. 21-6) can be explained.

The situation was quite different about a day later (figs. 21-7(a) and 21-7(b)). At that time, the Moon was almost on the Sun-Earth line. In the case of these shadows (figs. 21-7(a) and 21-7(b)), many solar-wind electrons penetrated into the cavity near the sunset (western) boundary. Also, strong shadowing effects far outside the expected location of the boundary on the sunrise (eastern) side were observed. In figures 21-7(c) to 21-7(f), the shadows have become comparatively shallow. Approximately 20 to 30 percent of the flux encountered in the free stream was still experienced on the night side of the Moon.

All the anomalous features of the cavity cannot be accounted for at this time; however, the direction of the interplanetary magnetic field was quite different at these various times (with ϕ ranging between 80° and 105°).[2] Thus, the magnetic field was pointing almost at right angles to the cavity axis, and this magnetic-field orientation apparently enables a sizable fraction of the fast electrons in the free stream to move into the cavity immediately behind the Moon at speeds comparable to the bulk speed of the solar wind.

Energetic-Electron Fluxes in Interplanetary Space

For a period of several days in August, a weak flux of electrons in the energy range of 25 to 300 keV was observed while the Moon moved from near the Sun-Earth line to 60° east of the line. The flux of electrons was approximately 20 electrons/cm^2-sr-sec over this energy range, and the flux remained rather constant. The spectral slope is not yet well known, but dJ/dE is proportional to $E^{-\gamma}$, where J is the electron flux, E is the electron energy, and γ is between 3 and 4. These electrons are shadowed by

---

[2] Ibid.

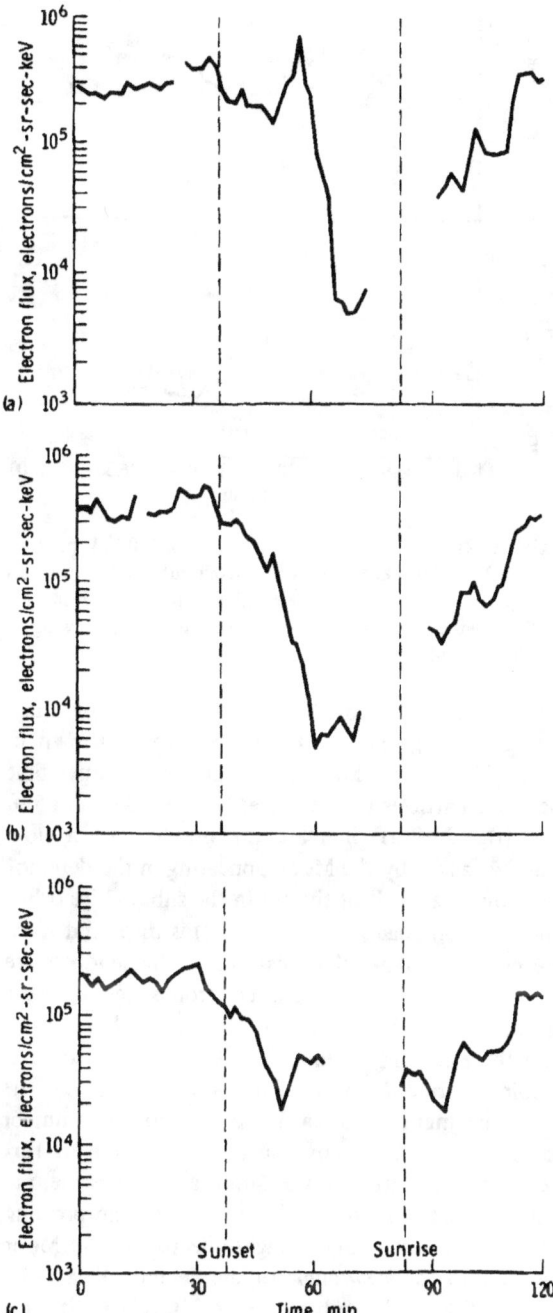
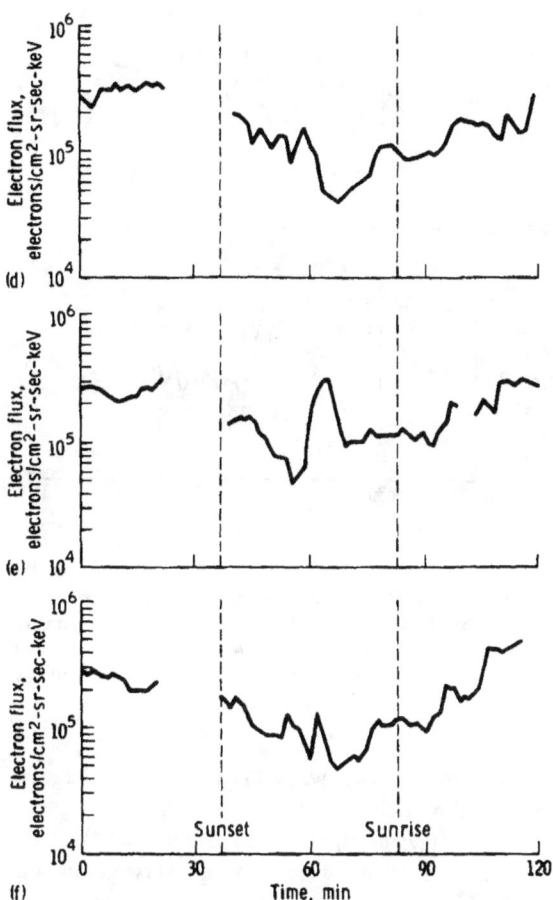

FIGURE 21-7.—Subsatellite measurements of the shadow produced by the Moon in the solar wind (as indicated by the flux of 600-eV electrons), where the Moon was located near the Sun-Earth line and the orientation of the interplanetary magnetic field varied from $\phi = 80°$ to $\phi = 105°$. (a) August 20 and 21, 23:17 to 01:17 G.m.t. (b) August 21, 01:17 to 03:17 G.m.t. (c) August 21, 11:17 to 13:17 G.m.t. (d) August 22, 07:12 to 09:12 G.m.t. (e) August 22, 09:12 to 11:12 G.m.t. (f) August 22, 11:12 to 13:12 G.m.t.

the Moon, as shown in figure 21-8(a). Although the foil discrimination technique indicates clearly that the particles were electrons, the nature of the shadow confirmed this identification. The axes of the shadows observed during this interval of several days in August lay close to the direction of the interplanetary magnetic field, as expected; however, the shadows occurred on the sunward side of the Moon (fig. 21-8(b)). This observation indicates that, for a period of several days, a flux of energetic electrons moved predominantly in a direction generally toward the Sun. For a day in the middle of this period, a solar event occurred, and the shadows moved to the anti-sunward side of the Moon, which indicates that the particles were moving away from the Sun guided by the interplanetary magnetic field. After the impulsive solar particles decayed away, the shadows again moved to the sunward side of the Moon.

A steady energetic-particle flux directed predominantly toward the Sun evidently may be ac-

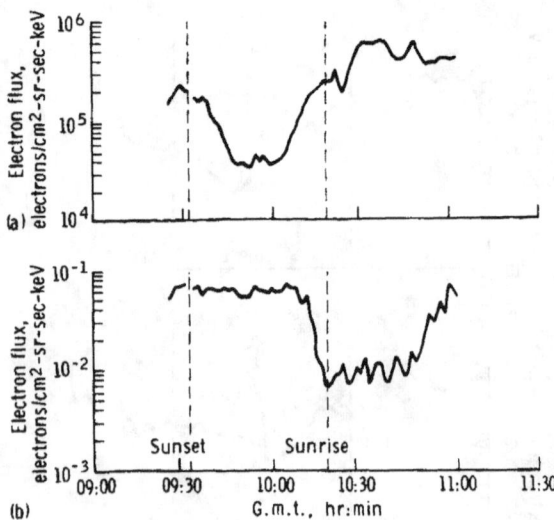

FIGURE 21-8.—The shadowing of the solar wind on the anti-sunward side of the Moon on August 26 as compared with the shadowing of 25- to 300-keV electrons on the sunward side of the Moon on August 26. (a) Shadowing of the solar wind. (b) Shadowing of 25- to 300-keV electrons.

counted for in two ways (1) a steady leakage of terrestrial particles from the bow shock, magnetosheath, or magnetosphere may occur, or (2) at the time the observations were made, a reverse gradient in the background solar cosmic-ray fluxes may have existed. Further study incorporating detailed magnetic-field information will enable a determination between these two possibilities.

Energetic-Solar-Particle Event on September 1 to 5

The subsatellite was turned off for the last 12 hr of September 1 in order to bring the battery up to full charge for the upcoming magnetotail pass. During this time, a large energetic-solar-particle event began, apparently triggered by a solar flare in an active region beyond the west limb of the Sun. This event produced both energetic electrons and protons that were observed by the subsatellite for at least 4 or 5 days during its second magnetotail passage.

The maximum fluxes observed were on the order of 10^2 to 10^3 times the background electron and proton fluxes. The high-intensity fluxes and long duration of the event permitted detailed studies to be made of the dynamics of the magnetotail over many days.

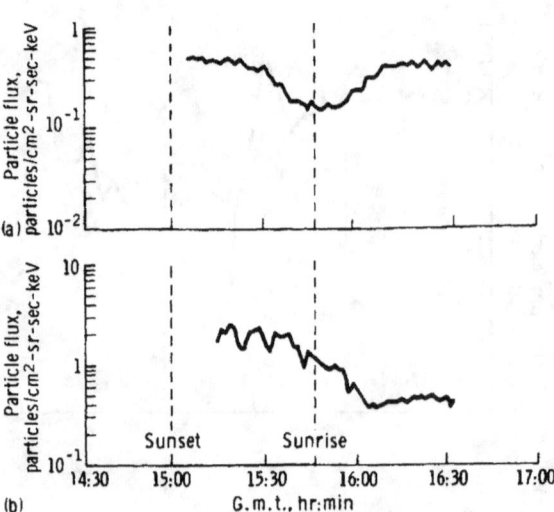

FIGURE 21-9.—The shadowing of 320- to 520-keV protons and 20- to 300-keV electrons for one subsatellite orbit on September 4 during the September 1 to 5 solar event. (a) Shadowing of 320- to 520-keV protons. (b) Shadowing of 20- to 300-keV electrons.

A single orbit of the subsatellite during magnetotail passage late in the solar event is shown in figure 21-9. The data for the proton channel that measures particles with energies between 320 and 520 keV (fig. 21-9(a)) indicate the presence of a slow shadow caused by the Moon appearing in the detector aperture as a result of the tilt in the subsatellite orbit. The electron shadow (fig. 21-9(b)) is displaced from the proton shadow; the locations of the shadows are shown in figure 21-10. The electron shadow occurs on the sunward side of the Moon, which indicates that the electrons do indeed enter the magnetotail at a point beyond the orbit of the Moon (ref. 21-4). The strong magnetic field near the Earth provides a mirror to reflect the electrons caught in the tube of flux between the Earth and the Moon and causes them to impact the lunar surface. In the case of the protons, which have essentially straight-line paths, the Moon moves into the view angle of the detector as a result of the tilt in the subsatellite orbit relative to the detector axis.

Analyzers C_4 and C_5, which have large geometric factors, are capable of extending measurements of the solar-flare-electron spectrum to energies as low as approximately 6 keV. The electron spectrum that was measured early in the event is shown in figure 21-11. The time chosen was at the minimum of proton flux; therefore, contamination by protons was negligible. The spectrum is extremely hard (dJ/dE is

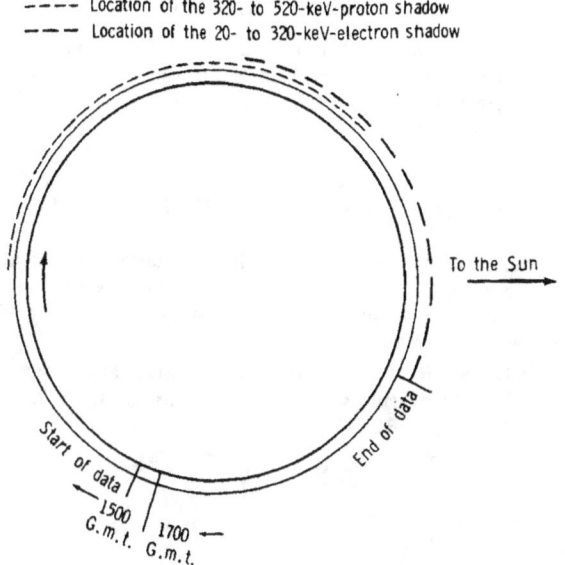

FIGURE 21-10.—Projections of the locations of the 320- to 520-keV-proton and 20- to 300-keV-electron shadows on the selenocentric ecliptic plane for the data shown in figure 21-9.

proportional to $E^{-1.5}$) and was taken approximately 6.5 hr after the flare. The solar flare on September 1 occurred behind the western limb of the Sun; therefore, the case may be that the higher energy particles were preferentially transported to the Earth, at least at these early times.

Two Black Brant IIIB sounding rockets, which were instrumented in much the same way as the particles and fields subsatellite, were launched into the September 1 solar-particle event from Resolute Bay, Northwest Territories, Canada. Data from these sounding-rocket experiments will make it possible to compare solar-particle fluxes near the Moon with intensities over the Earth polar cap. The rockets were launched at 13:52 G.m.t., September 4, and 13:44 G.m.t., September 5. Real-time data acquisition from the subsatellite and subsatellite orbit predictions was used to determine the exact times of launch for the sounding rockets. Two more launches will be conducted in early May 1972.

CONCLUSIONS

Analysis of initial data from the plasma and energetic-particle experiment of the particles and fields subsatellite has led to the following conclusions.

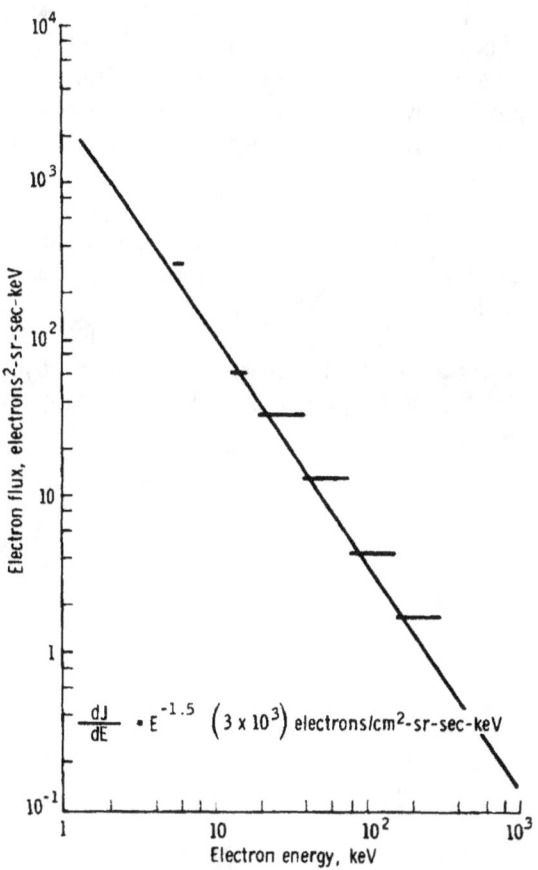

FIGURE 21-11.—Energy spectrum of solar-flare electrons with energies from approximately 6 to 300 keV. The data were collected on September 2 at 01:58 G.m.t. The lowest two energy data points were obtained from electrostatic analyzers C_4 and C_5; the highest four energy data points were obtained from the shielded solid-state telescopes.

(1) The cavity formed in the solar wind by the Moon has been observed in the fast-electron component of the solar wind. When the interplanetary magnetic field is alined approximately along the solar-wind flow, the electrons are almost completely excluded from the cavity. When the magnetic field is more nearly alined perpendicular to the solar-wind flow, the shadow structure, as defined by the fast-electron component, of the solar wind becomes extremely complex. The shadow structure becomes much broader than the lunar diameter and may become very shallow. (That is, a considerable fraction of the electrons is able to enter the cavity.)

(2) A weak flux of electrons in the energy range of 25 to 300 keV was found to move predominantly

in a sunward direction for a period of several days while the Moon was upstream from the Earth in interplanetary space. The intensity of this flux was approximately 20 electrons/cm^2-sr-sec for electrons in the 25- to 300-keV energy range. Whether these particles have a solar or terrestrial origin has not yet been determined.

(3) Following an important solar flare on September 1, a flux of solar electrons was measured at the subsatellite. The electron spectrum was determined for the energy range of 6 to 300 keV. A power-law fit to this spectrum resulted in $dJ/dE = E^{-1.5} (3 \times 10^3)$ electrons/cm^2-sr-sec-keV 6 hr after the flare.

REFERENCES

21-1. Montgomery, Michael D.; Bame, S. J.; and Hundhausen, A. J.: Solar Wind Electrons: Vela 4 Measurements. J. Geophys. Res., vol. 73, no. 15, Aug. 1, 1968, pp. 4999-5003.

21-2. Lyon, E. F.; Bridge, H. S.; and Binsack, J. H.: Explorer 35 Plasma Measurements in the Vicinity of the Moon. J. Geophys. Res., vol. 72, no. 23, Dec. 1, 1967, pp. 6113-6117.

21-3. Colburn, D. S.; Currie, R. G.; Mihalov, J. D.; and Sonett, C. P.: Diamagnetic Solar-Wind Cavity Discovered Behind Moon. Science, vol. 158, no. 3804, Nov. 24, 1967, pp. 1040-1042.

21-4. Lin, R. P.: Observations of Lunar Shadowing of Energetic Particles. J. Geophys. Res., vol. 73, no. 9, May 1, 1968, pp. 3066-3071.

22. The Particles and Fields Subsatellite Magnetometer Experiment

Paul J. Coleman, Jr.,[a†] *G. Schubert,*[a] *C. T. Russell,*[a] *and L. R. Sharp*[a]

INTRODUCTION

Magnetic-field measurements obtained with the Apollo 12 lunar-surface magnetometer and the Apollo 14 handheld magnetometer have established the existence of a significant lunar remanent magnetization, have provided estimates of the distribution of electrical conductivity of the interior of the Moon, and have established an upper limit on the average magnetic permeability of the Moon. The basic objectives of the lunar subsatellite magnetometer experiment are to extend the measurements of the lunar magnetic field (the permanent as well as the induced components) and to study the interaction of the Moon with the field and charged particles of its environment. Specific objectives are to map the remanent magnetic field of the Moon, to map the electrical conductivity of the lunar interior, and to study the properties of the plasmas in cislunar space by measuring the magnetic effects of the interactions of these plasmas with the Moon. Included in this last objective is a search for magnetic disturbances caused by material that leaves the Moon by one process or another and is ionized while still nearby. Finally, the magnetometer is part of the instrumentation for the lunar-particle-shadowing experiment (sec. 21).

The multiplicity of the objectives of the magnetic-field study is made possible by the geometry of the orbit of the Moon, which passes through three very different regions in near-Earth space (fig. 22-1): the region in which the solar-wind flow is essentially undisturbed by the presence of the Earth; the magnetosheath in which the flow is drastically modified by the obstacle presented by the Earth magnetic field; and the geomagnetic cavity, which consists of the space threaded by the magnetic field from the Earth.

As shown in figure 22-1, this cavity extends to approximately 10 Earth radii R_E in the direction of the Sun. The exact extent of the cavity in the opposite direction (the length of the geomagnetic tail) is unknown; however, the extent is more than 100 Earth radii.

Information concerning the electromagnetic properties of the Moon and the interactions of the Moon with the lunar plasma environment has been ascertained from Explorer 35 (lunar-orbiting satellite). Two orbits of this satellite (for times separated by 6 months) are shown in figure 22-2. The area traversed by Explorer 35 during a 12-month period is indicated by bounding circles marking the loci of perilune and apolune; the orbit of the subsatellite is also shown.

When the Moon is in the solar wind, the region directly behind the Moon (downstream from the Moon) is essentially devoid of solar-wind plasma. This downstream cavity and the rarefaction waves on the boundary of the cavity (fig. 22-2) were discovered by Colburn et al. (ref. 22-1) with the Explorer 35 magnetometer. The essential difference between the interaction of the solar wind with the Moon and the interaction of the solar wind with the Earth is shown in figures 22-1 and 22-2. Specifically, most of the solar wind that interacts with the Moon simply hits the Moon and stops (fig. 22-2), whereas most of the solar wind that interacts with the Earth is diverted around the entire geomagnetic cavity and forms the magnetosheath (fig. 22-1).

THE MAGNETOMETER

The Apollo 15 subsatellite magnetometer system consists of two fluxgate sensors mounted orthogonally at the end of a 1.83-m boom and an electronics unit housed in the spacecraft. A block diagram and a photograph of the magnetometer are shown in figures

---
[a]University of California at Los Angeles.
[†]Principal investigator.

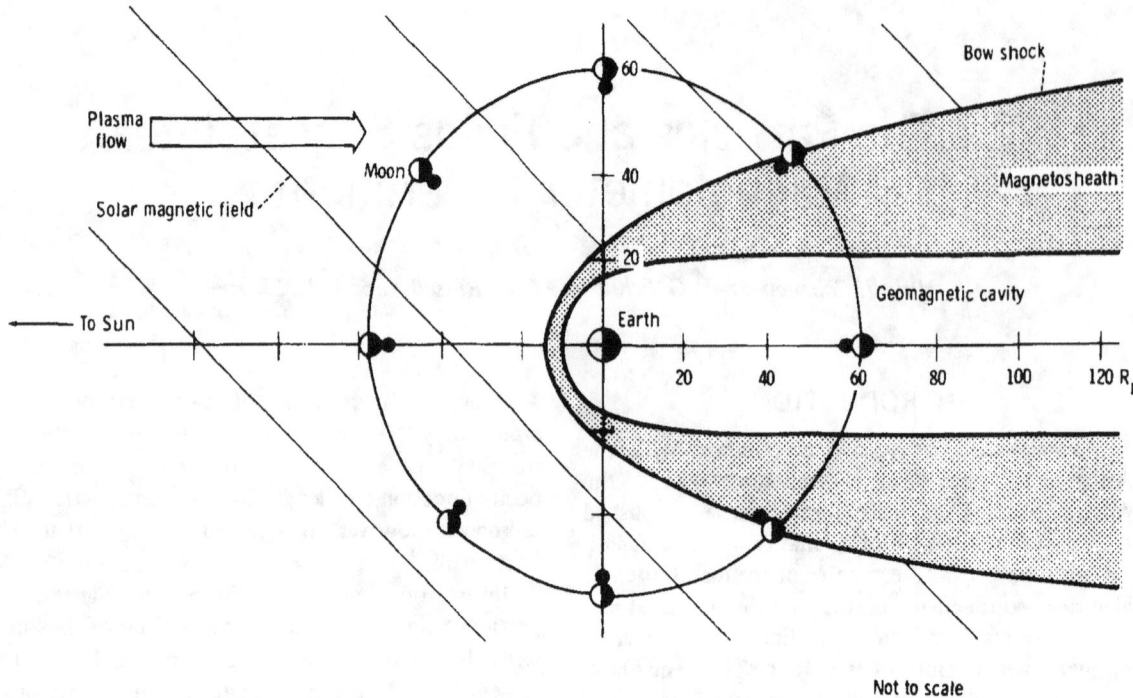

FIGURE 22-1.—The three regions of near-Earth space traversed by the Moon. The plane of the figure is essentially the ecliptic plane; the large dot adjacent to the lunar surface marks the approximate location of the Apollo 12 lunar-surface magnetometer.

22-3 and 22-4, respectively. The specifications of the instrument are listed in table 22-I.

The two magnetometer sensors are oriented parallel B_P and transverse B_T to the spin axis of the spacecraft. The measured quantities that are used to define the vector field are (1) the magnitudes of the parallel component, (2) the absolute value of the transverse component, and (3) the angle between the transverse component and the component of the Sun-spacecraft vector transverse to the spin axis. The subsatellite data storage unit records field measurements on the far side of the Moon for playback when the subsatellite is in view from the Earth.

REMANENT MAGNETIZATION

When geomagnetic activity is at a relatively low level, the magnetic field in the geomagnetic tail is quite constant. Thus, the best possibilities for the detection of lunar remanent magnetism are provided by the subsatellite magnetometer data that are recorded when the Moon is in the geomagnetic tail during quiet intervals. During the preliminary analysis of the quick-look data that were recorded during the first traversal of the geomagnetic tail, the existence of measurable levels of remanent magnetism over much of the subsatellite orbit was revealed. The average values of B_P and B_T computed for nine successive revolutions during which the Moon was in the geomagnetic tail are shown in figure 22-5.

The major features of the structure in the traces are associated with large craters lying within 10° of the band defined by the ground tracks of the nine revolutions. The most obvious feature is that apparently associated with Van de Graaff crater, which produces a 1-γ variation in the field as the satellite sweeps past it. Van de Graaff is approximately 9° across, and the center of the crater is located approximately 8° from the satellite ground track. Other prominent features of the data are associated with craters Hertzsprung, Korolev, Gagarin, and Milne, and the Mare Smythii. The location of the subsatellite ground track relative to the aforementioned major craters is shown in figure 22-6.

The identification of the craters associated with each of these anomalies should be viewed with some caution. Since the process of averaging over 9 revolutions effectively filters out variations over latitude and longitude ranges of approximately 5° as well as

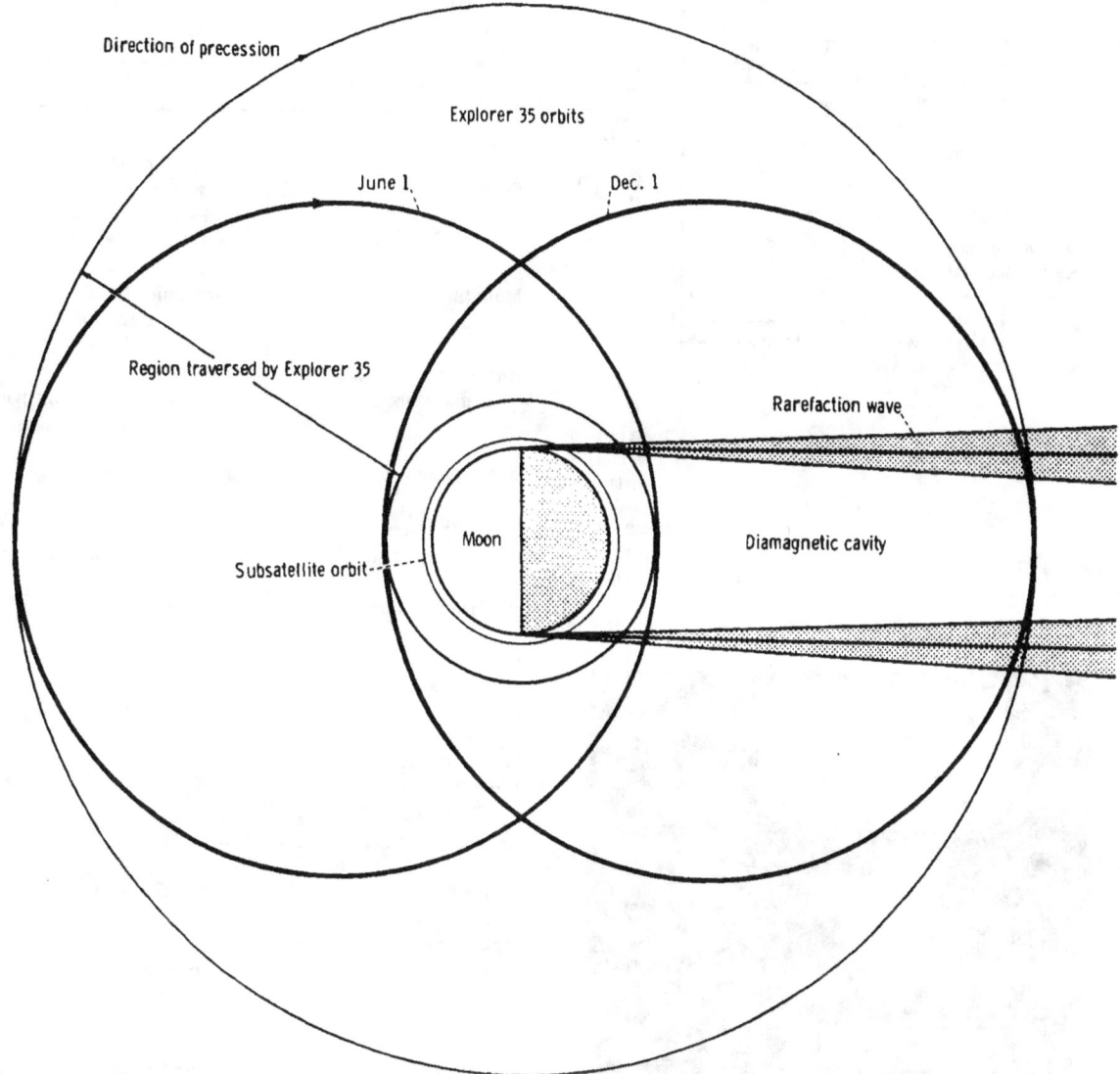

FIGURE 22-2.—The interaction of the solar wind with the Moon and the revolutions of the lunar subsatellite and Explorer 35. The actual inclination of the subsatellite orbital plane to the ecliptic plane is approximately 25° to 30°; the Explorer 35 orbital plane is inclined by approximately 15°.

temporal variations in the background field of the magnetotail of the Earth, the features have been identified with the largest nearby craters. However, some of these features may be associated with smaller craters. For example, the positive anomaly occurring as the subsatellite passes by Gagarin Crater may instead be due to negative anomalies associated with the craters Thomson and Pavlov. Similarly, the anomaly that has been associated with Mare Smythii may be due to some small crater closer to the orbital track.

The magnetic-field measurements used in this preliminary analysis do not include the final preflight calibrations. Thus, although the measured variations are accurate, the absolute values may be in error by a few gammas. Also, performing the data processing entirely by hand has precluded determining the orientation of the component perpendicular to the satellite spin axis. However, the preliminary results show that detailed mapping of the lunar remanent magnetization will be obtained from the subsatellite orbital data.

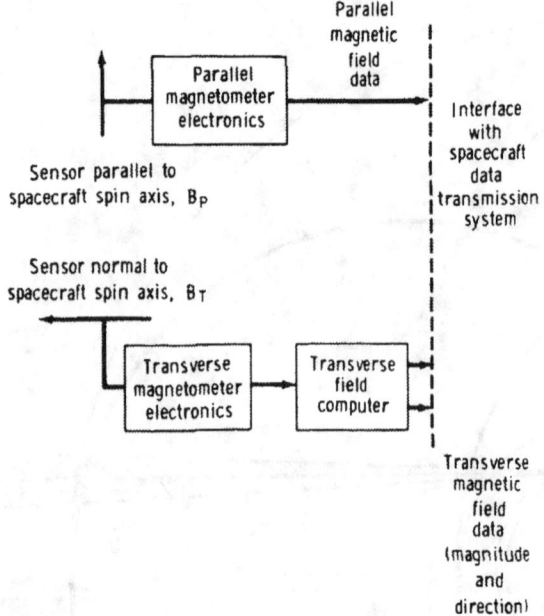

FIGURE 22-3.—Block diagram of the Apollo 15 subsatellite magnetometer.

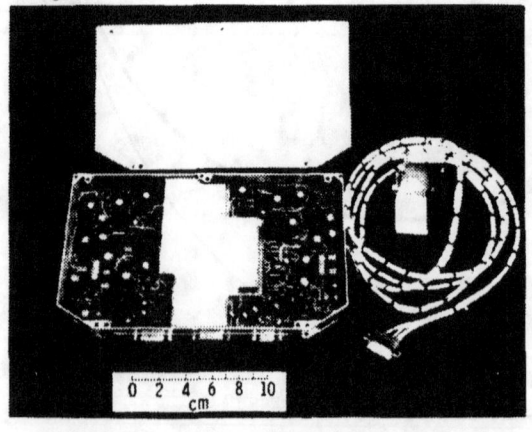

FIGURE 22-4.—Apollo 15 subsatellite magnetometer.

TABLE 22-I.—*Apollo Subsatellite Magnetometer Specifications*

Characteristic	Specification
Type	Second-harmonic, saturable core fluxgate
Sensor configuration	Two sensors, one sensor parallel B_P and one perpendicular B_T to the satellite-spin axis
Mounting	Sensor unit at end of 1.83-m boom; electronics unit in spacecraft body
Automatically selected dynamic ranges, γ	0 to ±50 at higher sensitivity, 0 to ±200 at lower sensitivity
Resolutions, γ	0.4 and 1.6, depending on range
Sampling rates:	
Real time	B_P every 2 sec, B_T every sec
High-rate storage	B_P and B_T magnitude and B_T phase once every 12 sec
Low-rate storage	B_P and B_T magnitude and B_T phase once every 24 sec
Power, W	0.70
Weight:	
Electronics unit, kg	≈ 0.8
Sensor unit, kg	≈ 0.2
Size:	
Electronics unit, cm	27.9 by 15.9 by 3.8
Sensor unit, cm	1.5 (diameter) by 7.6
Operating temperature range, K	344 to 172

The plot of B_T in the nine-revolution average (fig. 22-5) suggests that the remanent field is weaker on the near side than on the far side and that most of the major craters, with the possible exception of Gagarin, are associated with local minima in B_T. This near-side/far-side asymmetry evokes the speculation that the remanent field observed is attributable to irregularities in a magnetized crust. This crust has been grossly disturbed over a broad region of the near side, possibly by the infall of the bodies that created the ringed maria. On the far side, the crust has been disturbed primarily by more localized crater formation.

Analysis of the samples returned from the Apollo 11 and 12 sites revealed remanent magnetization as great as 10^{-2} emu/cm^3. If this value is used as an upper limit on the magnetization of lunar material, then the minimum scale size is approximately 10 km for a spherical body magnetized at this level and producing a 1-γ variation at the subsatellite orbit. The field at the surface of such a region and, therefore, the maximum field that could be produced by such a region on the surface of the Moon is roughly 1000 γ. Such a volume would have a magnetic dipole moment of approximately 10^{16} G-cm^3. For a more typical remanent magnetization of 10^{-5} emu/cm^3, the scale size would be 100 km, and the surface field would be approximately 10 γ for this dipole moment. The data shown in figure 22-5 also indicate that any lunar-

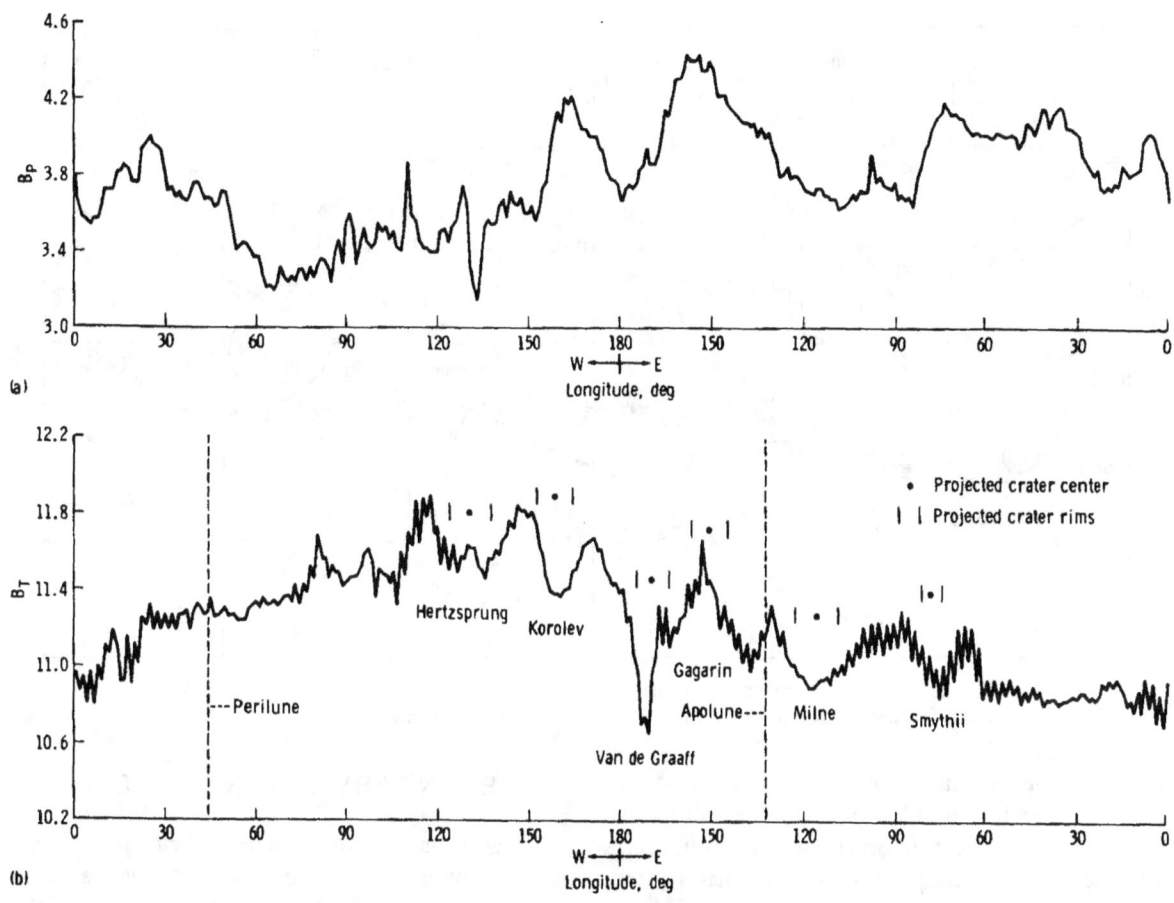

FIGURE 22-5.—Average measurements of B_P and B_T obtained during nine successive revolutions while the Moon was in the geomagnetic tail. The saw-tooth nature of the plots in certain regions is the result of statistical noise and the high resolution used for the plot.

centered magnetic dipole must have a magnetic moment of less than 4×10^{19} G-cm³ corresponding to a surface field strength in the 1.5- to 3-γ range.

A permanent magnetic field of 38 ± 3 γ was detected at the Apollo 12 site with the lunar-surface magnetometer (ref. 22-2). Permanent fields of 103 ± 5 γ and 43 ± 6 γ were detected at sites separated by 1.12 km in the Fra Mauro region that was explored by the Apollo 14 astronauts (ref. 22-3). The lunar subsatellite has passed directly over both sites, but no significant field variation was observed over either. Thus, the surface fields observed to date must be of relatively small-scale size, as indicated by the field gradient measured at the Apollo 14 site.

As the altitude of perilune of the subsatellite decreases, more information will be obtained on the fossil fields at the Apollo 12 and 14 sites. As pointed out by Sonett et al. (ref. 22-4), if both were produced by the same magnetizing field, a field in excess of 10^3 γ must have existed some billion years after the formation of the Moon, or 3.4 billion years ago; further, the field must have existed for 300 million years, because the magnetized rocks from the Apollo 11 and 12 sites are 3.4 and 3.7 billion years old, respectively. A detailed map of the permanent magnetization on the Moon will provide additional information on the ancient magnetizing field and the history of the magnetized material subsequent to magnetization.

ELECTRICAL CONDUCTIVITY

Information concerning the electrical conductivity of the interior of the Moon has been obtained through an analysis of simultaneous magnetic-field measurements at the Apollo 12 site and from the

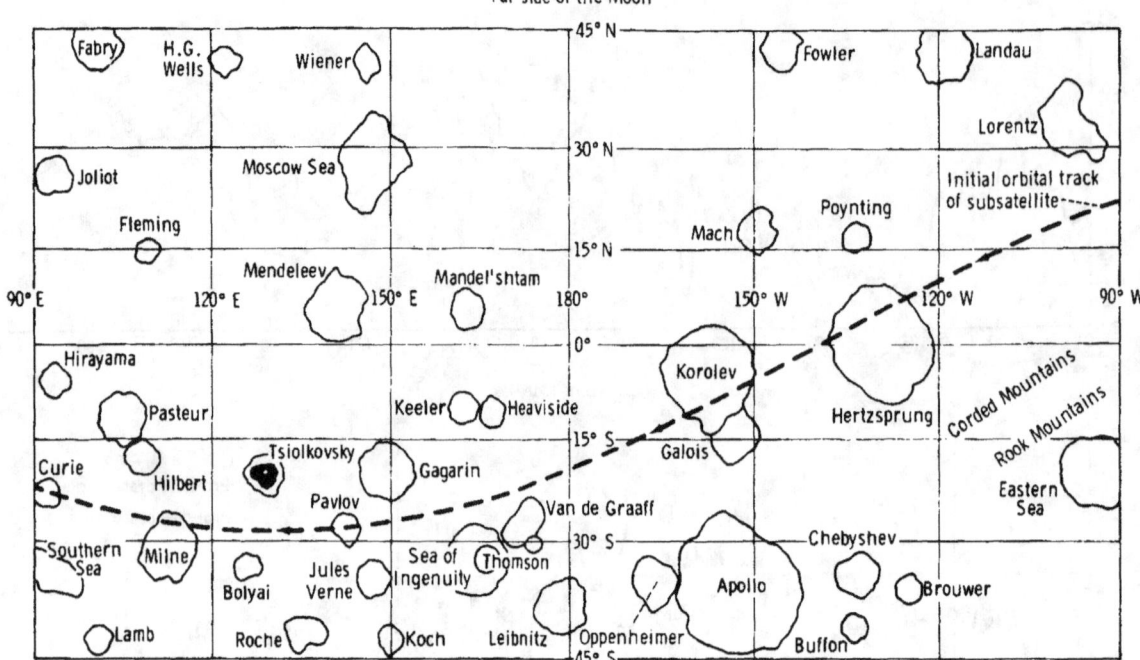

FIGURE 22-6.—Mercator projection of the far side of the Moon showing the ground track of the satellite for the fifth revolution of the nine-revolution sequence shown in figure 22-5.

lunar-orbiting satellite, Explorer 35. The previously obtained results include the radial conductivity profile that provides information on mantle-core stratification, the temperature of the mantle, the near-surface thermal gradient and heat flux, and the composition of the interior (refs. 22-4, 22-5, and 22-6). The experimental technique employed in these studies is essentially a measurement of the response of the Moon to changes in the solar-wind magnetic field.

Data recorded during five successive revolutions of the lunar subsatellite when the Moon was in the solar wind are plotted in figure 22-7. From the standpoint of the conductivity studies, the important feature of these plots is the greater variability of the magnetic field on the day side, or upstream side, of the Moon. The magnetic field measured at the subsatellite when the Moon is in the solar wind includes a component attributable to lunar induction. The presence of this component indicates that data from the subsatellite magnetometer, along with simultaneous data from the lunar-surface magnetometers and Explorer 35 magnetometer, can be used to produce a detailed, three-dimensional model of the interior conductivity. The conductivity studies are just beginning, and further results are not yet available.

BOUNDARY-LAYER STUDIES

Observations of the magnetic field and plasma obtained with Explorer 35 have revealed a fairly consistent picture of the large-scale interaction of the solar wind with the Moon. As shown schematically in figure 22-2, the absence of a lunar bow shock allows most of the solar-wind plasma to reach the lunar surface, where it is absorbed. Consequently, when the Moon is in the solar wind, a so-called diamagnetic cavity exists behind the Moon, or downstream from the Moon. The essential magnetic feature of this cavity is an interior magnetic field approximately 1.5 γ stronger, on the average, than the exterior field. At the boundary of this cavity, there is a sharply localized decrease in the field magnitude that is approximately coincident with the boundary of the optical shadow of the Moon.

The preliminary analysis of the data from the subsatellite magnetometer indicates that a diamagnetic increase appears also at the lower altitude of the subsatellite. A 12-revolution average of the measurements was recorded while the Moon was in the solar wind (fig. 22-8). The field enhancement between satellite sunset and satellite sunrise is readily distinguishable and is approximately 1 γ.

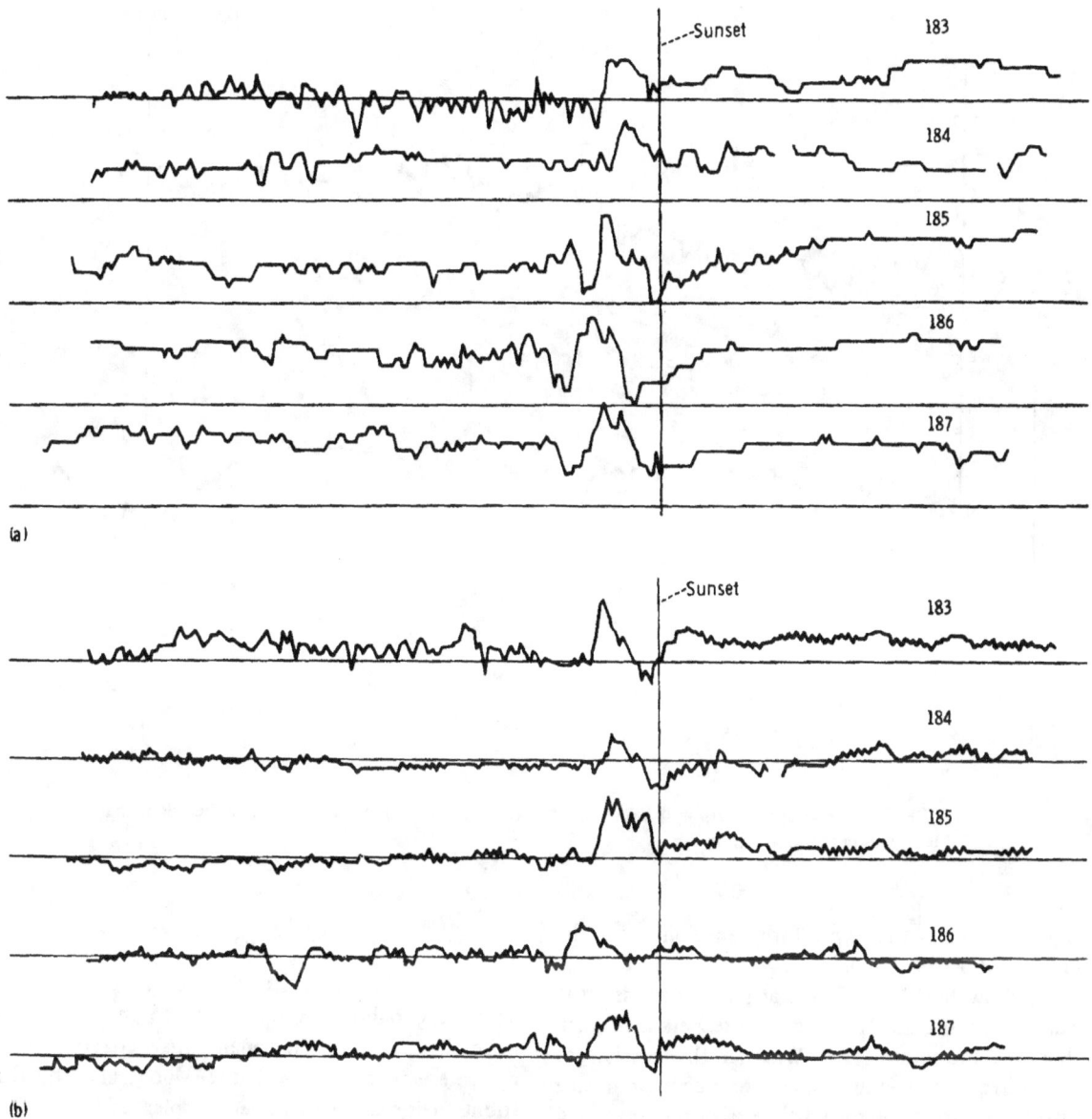

FIGURE 22-7.—Plots of B_P and B_T for five consecutive revolutions of the subsatellite when the Moon was in the solar wind. (a) Plots of B_P for revolutions 183 to 187. (b) Plots of B_T for revolutions 183 to 187.

The existence of sporadic field disturbances adjacent to the rarefaction wave at the boundary of the diamagnetic cavity (fig. 22-2) is evident from a study of Explorer 35 data. Mihalov et al. (ref. 22-7) have shown that these disturbances in the solar-wind flow occur when certain regions of the lunar surface are at the limbs. The greatest concentration of disturbance sources was found in a 15° square near Gagarin Crater.

Preliminary analysis of the subsatellite magnetometer data indicates that strong disturbances, or limb effects, are present most of the time. These disturbances produce more or less characteristic variations in the subsatellite magnetometer record. Record sections from revolutions 183 to 187 are shown in figure 22-7. The data analyzed to date suggest that the disturbances, such as those apparent near the sunset line, occur when near-side regions are

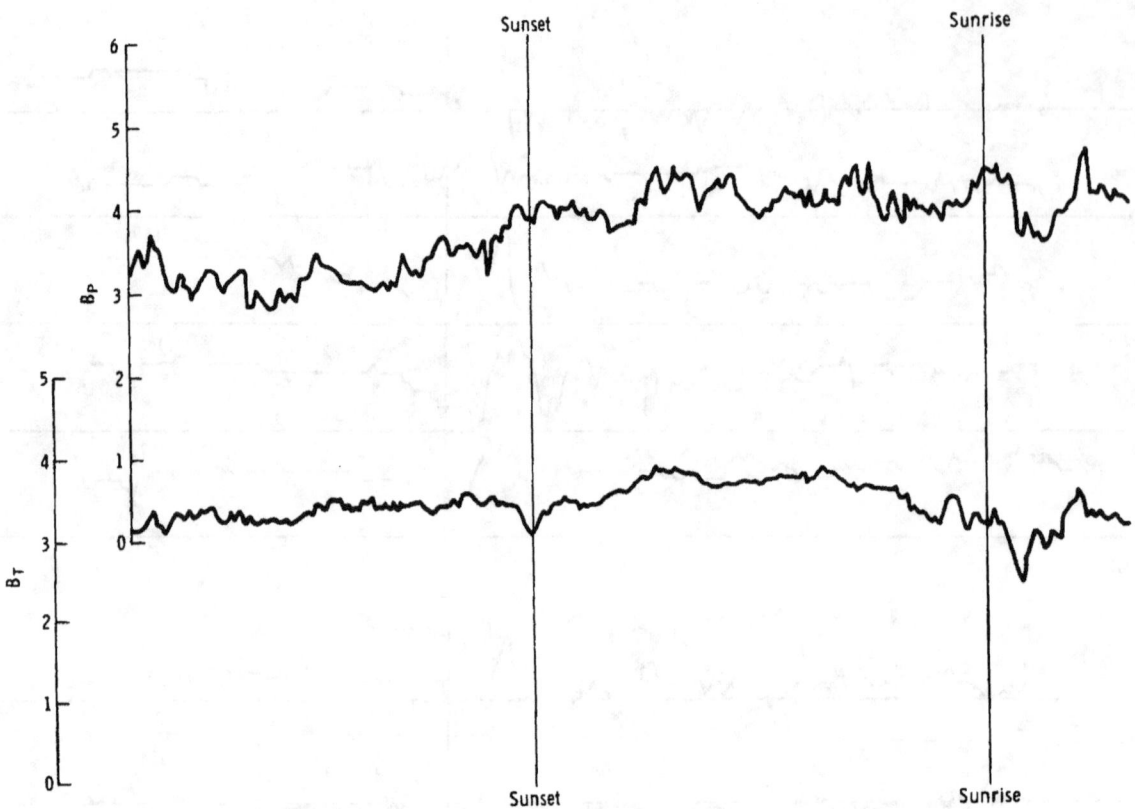

FIGURE 22-8.—Averages of B_P and B_T computed from 12 revolutions of the subsatellite when the Moon was in the solar wind.

at the limbs as well as when far-side regions are at the limbs. The 12-revolution averages of B_P and B_T (fig. 22-8) show that limb effects at satellite sunrise (the sunset limb on the Moon) are more persistent than those at the other limb. Whether this persistence results from the effectiveness of the area in producing disturbances or from some property of the solar wind (such as the orientation of the magnetic field) is yet unknown. The detection of relatively strong remanent fields in the vicinity of Gagarin Crater may be consistent with the suggestion of Mihalov et al. (ref. 22-7) that the limb effects detected from Explorer 35 are caused by localized regions of enhanced magnetic fields. However, the preliminary analysis of the subsatellite data is indicative that limb disturbances are present more often than not and are just as great when many other regions are at the limb. Further study is required to establish the causes of these disturbances in the solar-wind flow.

SUMMARY

A preliminary analysis of the data from the magnetometer onboard the Apollo 15 subsatellite is indicative that (1) remanent magnetization is a characteristic property of the Moon, and the distribution is such that a rather complex pattern or fine structure is produced; (2) a detailed mapping of the distribution is feasible with the present experiment; (3) lunar induction fields produced by transients in the interplanetary magnetic field are detectable at the satellite orbit; (4) the magnetometer data will provide estimates of the latitude and longitude dependences in the distribution of interior conductivity; (5) the plasma void, or diamagnetic cavity, that forms behind the Moon when the Moon is in the solar wind extends to some altitude below the satellite orbit and probably to the lunar surface; and (6) the flow of the solar wind near the limbs is usually rather strongly disturbed. These conclusions are tentative because verification depends upon more detailed analysis, improvements in statistical accuracy, and comparisons of the data with data from the lunar-surface magnetometers and the Explorer 35 magnetometer.

REFERENCES

22-1. Colburn, D. S.; Currie, R. G.; Mihalov, J. D.; and Sonett, C. P.: Diamagnetic Solar-Wind Cavity Discovered Behind Moon. Science, vol. 158, no. 3804, Nov. 24, 1967, pp. 1040-1042.

22-2. Dyal, P.; Parkin, C. W.; and Sonett, C. P.: Apollo 12 Magnetometer: Measurements of a Steady Magnetic Field on the Surface of the Moon. Science, vol. 169, Aug. 21, 1970, pp. 762-764.

22-3. Dyal, P.; Parkin, C. W.; Sonett, C. P.; DuBois, R. L.; and Simmons, G.: Lunar Portable Magnetometer Experiment. Sec. 13 of Apollo 14 Preliminary Science Report, NASA SP-272, 1971.

22-4. Sonett, C. P.; Schubert, G.; Smith, B. F.; Schwartz, K.; and Colburn, D. S.: Lunar Electrical Conductivity from Apollo 12 Magnetometer Measurements: Compositional and Thermal Inferences. Proceedings of the Second Lunar Science Conference, Vol. 3, A. A. Levinson, ed., The MIT Press (Cambridge, Mass.), 1971, pp. 2415-2431.

22-5. Dyal, Palmer; and Parkin, Curtis W.: The Apollo 12 Magnetometer Experiment: Internal Lunar Properties from Transient and Steady Magnetic Field Measurements. Proceedings of the Second Lunar Science Conference, Vol. 3, A. A. Levinson, ed., The MIT Press (Cambridge, Mass.), 1971, pp. 2391-2413.

22-6. Sonett, C. P.; Colburn, D. S.; Dyal, P.; Parkin, C. W.; et al.: Lunar Electrical Conductivity Profile. Nature, vol. 230, no. 5293, Apr. 9, 1971, pp. 359-362.

22-7. Mihalov, J. D.; Sonett, C. P.; Binsack, J. H.; and Moutsoulas, M. D.: Possible Fossil Lunar Magnetism Inferred from Satellite Data. Science, vol. 171, no. 3974, Mar. 5, 1971, pp. 892-895.

ACKNOWLEDGMENTS

The authors express appreciation to G. Takahashi and his staff at Time Zero Corporation for their efforts in the design and fabrication of the subsatellite magnetometer; to T. Pederson and R. Brown and their staff at TRW Systems Group for their work in the design, fabrication, and testing of the subsatellite and the integration of the magnetometer; and, particularly, to C. Thorpe who controlled the magnetic fields of the subsatellite.

The University of California at Los Angeles engineering team was led by R. C. Snare. Preliminary circuit designs were done by the late R. F. Klein. The testing and calibration of the magnetometer was supervised by F. R. George.

23. Bistatic-Radar Investigation

H. T. Howard[a][†] and G. L. Tyler[a]

Observation of the lunar surface by Earth-based optical, radio, and radar techniques has produced a large body of knowledge about the Moon. However, because of severe limitations of distance and geometry, tentative conclusions as to the physical nature of the lunar surface could be obtained only through tedious analysis, usually involving a hypothetical, parametric model. Direct scientific exploration of the surface has placed the interpretation of remote sensing experiments on a much firmer foundation. Since the known geological surface properties severely constrain the electromagnetic models, particular parameters of interest may now be determined remotely with confidence.

The Apollo 15 bistatic-radar experiment uses the orbiting command-service module (CSM) as one terminal of a two-station or bistatic radar. Signals from the spacecraft transmitters are directed toward that portion of the lunar surface that produces the strongest echo on the Earth. Because of the changing geometry, the CSM must be maneuvered to maintain this condition. For each near-side pass, echoes are received from a roughly 10-km-diameter spot that moves across the lunar surface with the CSM. The bistatic investigation is, thus, a scanning experiment. The size of this spot and its precise location is determined by the scattering laws for the lunar surface. In particular, the quasi-specular mechanism dominates the scattering process. The signals received on the Earth are processed by methods that preserve the frequency, phase, polarization, and amplitude information contained in them. These properties are presented as a function of location on the lunar surface. The characteristics of the echoes are compared with the known characteristics of the transmitted signals; the differences between the characteristics are used in conjunction with a well-developed (ref. 23-1) and generally accepted scattering theory to derive quantitative inferences about the Moon. Lunar crustal properties such as dielectric constant, average slope and slope probability, density, small-scale surface roughness, and imbedded rocks to a depth of 20 m may be determined. These properties are of interest to several disciplines concerned with the problems of lunar origin, evolution, and history. The results are proving to be most useful in understanding the processes that have produced and modified the crust and in distinguishing between adjacent and subjacent geological units. The experimental observations are also of intrinsic interest to those involved in the study of electromagnetic scattering.

Radar techniques and the associated interpretive theories have developed rapidly in recent years. The Apollo 14 bistatic-radar experiment provided a basis of comparison between the radar results and results of other techniques. Comparisons have been made with both interpretive geologic maps and quantitative topographic work using primarily photogrammetric techniques. To date, the agreement between these diverse methods has been excellent.

Two such comparisons are shown in figures 23-1 and 23-2. Values of lunar root-mean-square (rms) slopes derived from simultaneous very-high-frequency (VHF) and S-band observations have been plotted as a function of lunar longitude. Portions of the geologic maps (ref. 23-2) made from Earth-based observations are included. The rms S-band slope is generally 5° to 6° over most of the longitude 10° W to 50° W range shown. At approximately 41°, this slope is reduced by a factor of two. This change occurs as the radar target zone crosses a previously mapped scarp and enters Oceanus Procellarum. The decrease in slope at 36.5° is indicative of a previously unmapped change in surface characteristics. Examination of the Lunar Orbiter photography shows that this particular area is characterized by a total lack of small craters, while the surrounding areas are covered with them. The very large slopes associated with the crater Lansberg

---

[a]Stanford University.
[†]Principal investigator.

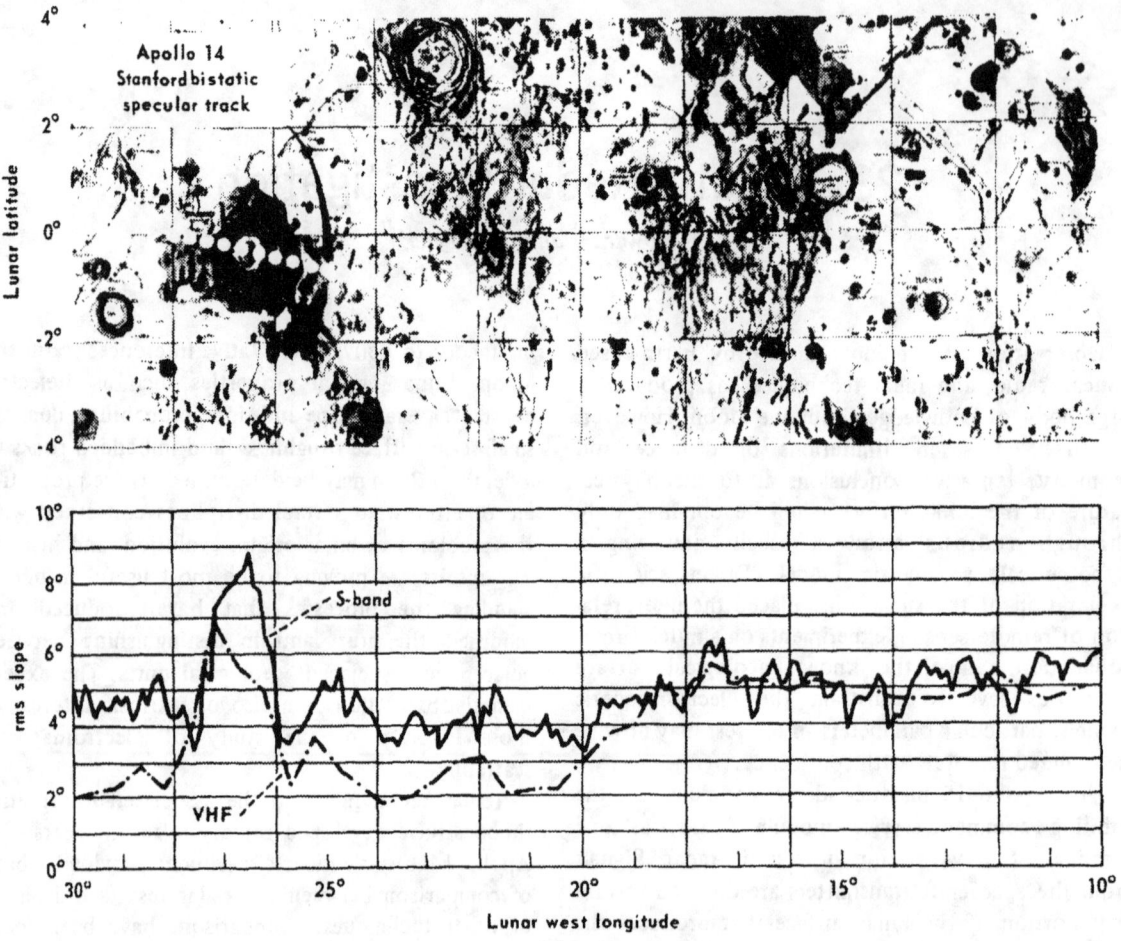

FIGURE 23-1.—A comparison of bistatic-radar slopes with a geologic map of the Copernicus and Riphaeus Mountains regions (ref. 23-2). The series of dots crossing the geologic map represents successive data points from the Apollo 14 bistatic-radar experiment. The dots are approximately the size of the experimental resolution cell. The plots, given below the geologic maps, are in registration with the longitude scale. The rms slopes, as inferred from simultaneous observations at 13 cm (S-band) and 1.2 m (VHF), are given. These slopes apply to a set of scales roughly 20 times the wavelength.

mean little quantitatively, because the region containing the crater is clearly inhomogeneous and violates the assumptions of the scattering theory at that point. The slopes deduced from VHF returns behave quite differently. Initially, the VHF slopes agree closely with the S-band results in the rough highlands areas. Once the reflecting area moves into mare, the VHF slopes are much smaller—typically less than half the S-band result. Simultaneously, the VHF changes character in another way—a large portion of the received signal becomes depolarized—suggesting reflection from wavelength-size (and smaller) material located below the surface. This differential change between the VHF and S-band results is attributed to a change in the probability distribution of the slopes on the scale of the VHF and S-band wavelengths (1.15 and 0.13 m, respectively). It will soon be possible to extend the techniques used here to investigations of the Earth and other planets—the Earth, for resource explorations; Mars, for similar goals as were attempted for the Moon; and Venus, shrouded in dense clouds, for the only information man may ever gain about the construction of the surface.

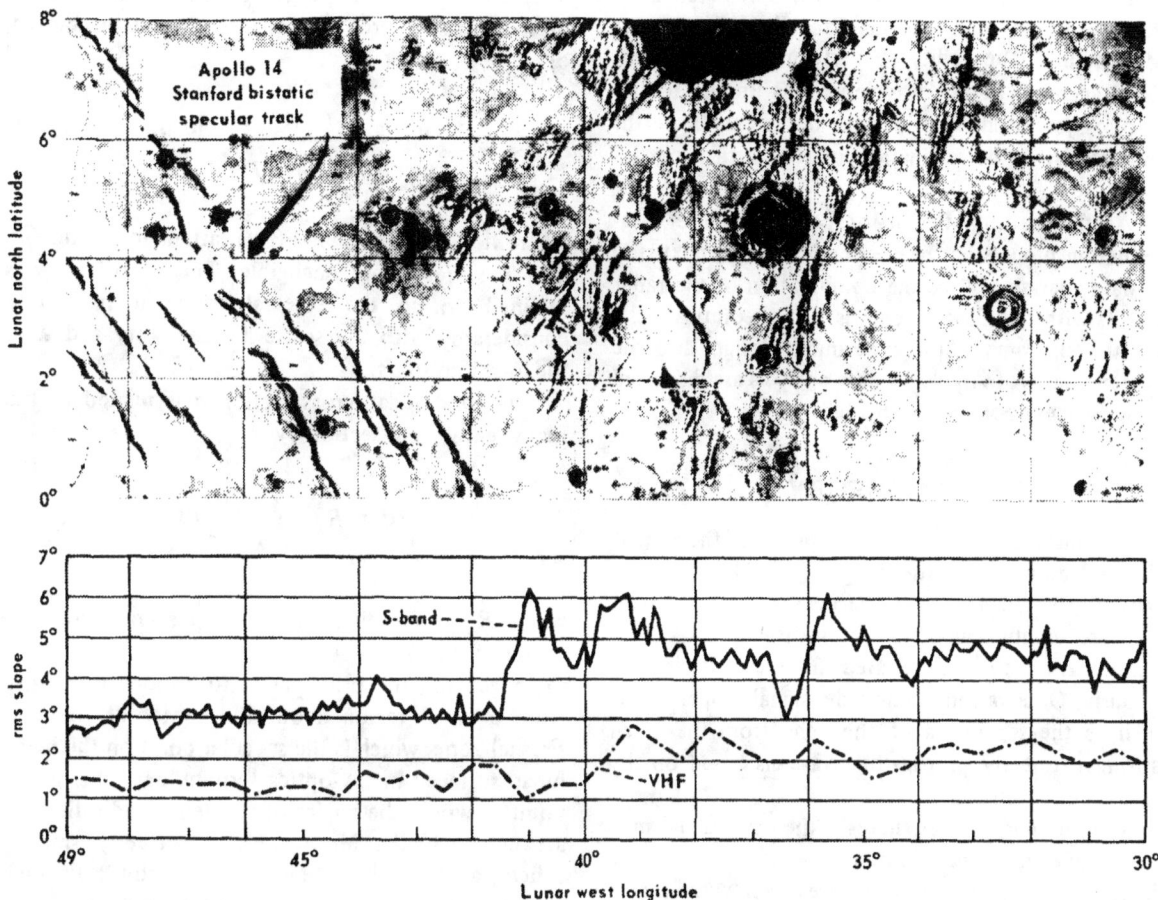

FIGURE 23-2.—Bistatic-radar, geological map comparison for the Kepler region of the Moon (ref. 23-2).

The Apollo 15 bistatic-radar experiment discussed in this report used continuous-wave transmissions from both the S-band telemetry system (the spacecraft-to-Earth communications link) and the VHF communications system (normally used for voice transmissions and ranging between the CSM and lunar module in lunar orbit). The overall techniques employed are similar to those that had been used previously with the Lunar Orbiters 1 and 3 and Explorer 35 spacecraft (refs. 23-3 and 23-4) and were identical to those used with Apollo 14 (ref. 23-5). The one departure from the Apollo 14 configuration was in the use of the S-band high-gain antenna rather than the S-band omnidirectional antenna system. This change, through additional flexibility, permitted significant improvement in the quality of both the VHF data and the S-band data. Simultaneous S-band and VHF observations were successfully conducted for one complete near-side pass while the CSM was maneuvered to maintain a constant orientation of the spacecraft antenna toward the quasi-specular reflection area. There were four complete passes of VHF data alone. The VHF data were obtained with the antenna pointed directly at the surface below the CSM. Echoes of the S-band transmissions were received with the NASA 64-m-diameter antenna located at Goldstone, Calif.; VHF echoes were received with the 46-m-diameter antenna of the Stanford Center for Radar Astronomy located on the Stanford University campus.

Excellent data were received during these five observation periods. These data represent nearly an order-of-magnitude improvement in the signal-to-noise ratio over the signal-to-noise ratio during the Apollo 14 experiment. Also, the Apollo 15 experiment provides the first bistatic-radar data from a number of significant lunar features, Mare Serenitatis, the Apennines, mid-Oceanus Procellarum, and the Marius Hills. The ability to perform the experiment simultaneously at two wavelengths is unique to the Apollo system and provides the observer a powerful tool to use in determination of the vertical crustal structure.

At this writing, the major portion of the data-reduction process has been completed. The next step in data analysis will be the combination of the radar results with the CSM ephemeris. The theoretical basis for the experiment, the experiment design, and the current status of Apollo 15 data analysis are discussed in subsequent portions of this section.

BASIC THEORY

The bistatic-radar echo is composed of the sum of the reflections from the area of the Moon that is mutually visible from the spacecraft and the Earth. Because continuous-wave transmissions are used, echoes from this entire area are received simultaneously. Observation of the echo-signal properties, as well as the separation of the echo from the often stronger, directly propagating wave, is based on the frequency spectrum.

For the purposes of analysis, the echo signal may be considered as a composite of the reflections from a large number of elemental surfaces, each with an area ds. Then, from the radar equation, the power received from a particular area ds is

$$dP_R = \frac{P_T G_T}{4\pi r_1^2} \sigma_0 \, ds \, \frac{1}{4\pi r_2^2} A \quad (23\text{-}1)$$

where
- dP_R = power received from the elemental area ds
- P_T = transmitted power
- G_T = gain of the transmitting antenna
- r_1 = distance from the transmitter to ds
- σ_0 = incremental radar cross section at ds
- ds = elemental area on the lunar surface
- r_2 = distance from ds to the receiving antenna
- A = effective aperture of the receiving antenna

The geometry of the experiment is illustrated in figure 23-3. The total received power is obtained by integration over the surface S (the surface that is mutually visible from the spacecraft and the Earth), or

$$P_R = \frac{P_T G_T A}{(4\pi)^2 r_2^2} \int_S \frac{\sigma_0}{r_1^2} ds \quad (23\text{-}2)$$

where it is assumed that the variations in r_2 and G_T over the area S are negligible. The quantity σ_0 is retained within the integral, since it may vary considerably with the scattering geometry and with the locations of ds.

The total radar cross section σ is related to the incremental cross section by

$$\sigma = R_p^2 \int_S \frac{\sigma_0}{r_1^2} ds \quad (23\text{-}3)$$

where R_p is the distance of the spacecraft from the center of the Moon.

In general, the principal contributions to σ arise from a small region about the center of the first Fresnel zone, which is the specular point on the mean lunar surface. In ray-optics terminology, this is the point at which the angles of incidence and reflection are equal. If the Moon were a perfectly smooth sphere, all the echo would originate from a Fresnel-zone-size spot surrounding this point. By roughening the surface through the introduction of large-scale (with respect to a wavelength) topographic undulations, this spot is caused to break up into a number of glints. The location of these glints will correspond to specular reflection from local surface undulations. Echoes that result from this type of surface are designated quasi-specular.

Quasi-specular reflection constitutes the principal scattering mechanism in this experiment. To the first order, the effects of surface material composition and shape are separable; that is, if $\hat{\sigma}$ is the radar cross section of a perfectly conducting surface of a particular shape, then the radar cross section of a dielectric surface of precisely the same shape and with reflectivity ρ is

$$\sigma = \rho \hat{\sigma} \quad (23\text{-}4)$$

Substituting equations (23-3) and (23-4) into equation (23-2) and solving for ρ yields

$$\rho = \frac{(4\pi)^2 r_2{}^2 R_p{}^2 P_R}{P_T G_T A \hat{\sigma}} \qquad (23\text{-}5)$$

Thus, if $\hat{\sigma}$ can be found, then ρ may be determined.

The quantity $\hat{\sigma}$ may be computed on the basis of statistical surface models. The results of one such computation for a gently undulating surface with a gaussian height distribution and autocorrelation function are given for bistatic geometry in figure 23-3. The radar cross section is plotted as a function of the angle formed by the spacecraft, the center of the Moon, and the Earth. Radius is measured from the center of the Moon (ref. 23-4). The dashed curve corresponds to a perfectly smooth sphere, and the solid curve corresponds to a surface with rms slopes of $10°$. The effect of the surface slopes on the cross section is second order. The reflectivity of the Moon is inferred from the data by measuring the total echo powers received and by normalizing the results by the use of the theoretical reflectivity values (plotted in fig. 23-3 from the data given in ref. 23-6) for a perfectly conducting sphere. The inferred reflectivity is then compared with the reflectivity of a dielectric surface under oblique geometry. The effective dielectric constant of the surface may be determined directly from the observation of the Brewster angle and indirectly by a quantitative comparison of the reflectivity values.

The bandwidth of the echo depends directly on the surface slope. For the surface model considered previously, the bandwidth is given by

$$\Delta f = 4.9 \left(\frac{v_s}{\lambda}\right) \cos \gamma_s \tan \beta_0 \qquad (23\text{-}6)$$

where Δf = one-half the power bandwidth of the echo spectrum
v_s = velocity of the specular point with respect to the lunar surface
λ = wavelength
γ_s = angle of incidence on the mean surface at the specular point
β_0 = rms surface slope

On the basis of this model, surface slopes may be obtained directly from the orbital parameters and the width of the echo spectrum.

Physically, the spectrum is broadened according to the probability-density function of the surface slopes and the doppler shift. In figure 23-4, reflection from

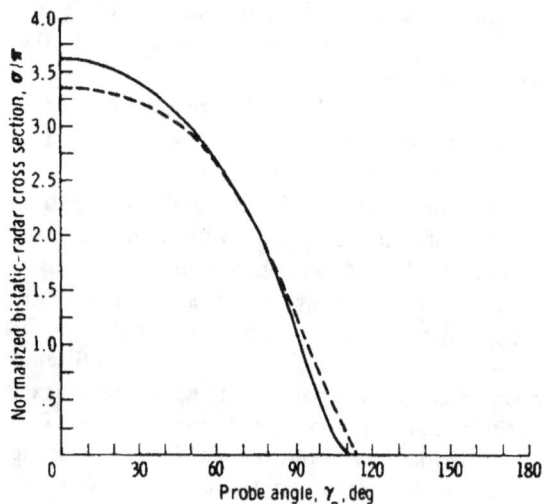

FIGURE 23-3.—Normalized bistatic-radar cross section σ/π as a function of spacecraft-Moon-Earth angle γ_p for a perfectly reflecting sphere with a gently undulating surface.

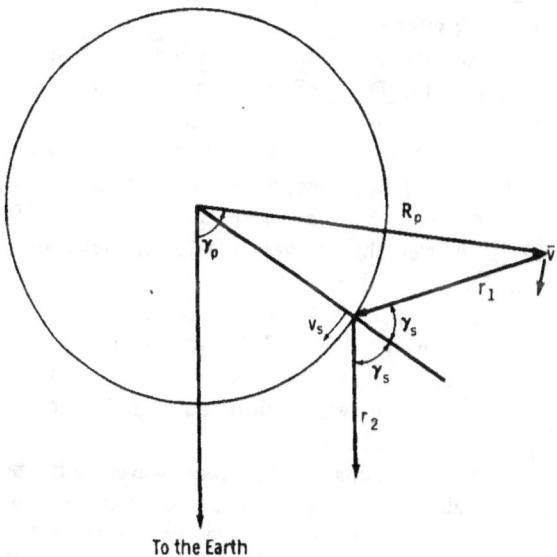

FIGURE 23-4.—Bistatic-radar geometry.

a point ahead of the mean reflecting point will have a greater doppler shift (determined by the angle between $\bar{r}_1$ and $\bar{v}$) than those points behind the mean reflecting point. At every point (on the basis of the quasi-specular model), the probability of obtaining a reflection depends on the probability of finding a local surface undulation with the proper slope. Slopes, then, also determine the surface resolution of the experiment. Reflections are obtained from an area with a radius of 5 to 10 km (which is approximately equal to the rms slope multiplied by the spacecraft altitude).

Inferences based on this model are valid on a set of scales related to the wavelength of the probing wave. Quantitative comparisons of slope distributions inferred from Explorer 35 data and from photogrammetry have been conducted with good results for a limited number of locations on the lunar surface (ref. 23-7). Under fairly broad assumptions, the quasi-specular scattering may be considered to occur at a fictitious surface that is a low-pass-filtered version of the actual surface. Although the bounds on this filter cut-off have been only approximated, it is known that they scale with the length of the probing wave. Typically, slopes on the order of 10 wavelengths (or longer) are expected to be important in the scattering process. Thus, for quasi-specular scattering, bistatic S-band data are sensitive to surface structure on the order of 1.3 m (and larger); and for VHF, 12 m is the lower bound. A more complete discussion of these theoretical concepts and results is available in reference 23-6.

Surface-reflectivity measurements are also sensitive to wavelength. Dry geological materials with approximately the density of the lunar regolith exhibit loss tangents that are independent of the radio frequency for frequencies greater than approximately 10 MHz (ref. 23-8). Penetration depths between 10 and 20 wavelengths are typical. Thus, the reflection coefficient inferred from S-band data will be sensitive to vertical structure within the lunar crust to a depth of 1 to 2 m and, from VHF data, to a depth of approximately 10 to 20 m. Such penetration effects have been observed with Explorer 35 data obtained at a wavelength of 2.2 m (ref. 23-9).

Diffuse scattering arises from wavelength-size (and smaller) surface structures and from second-order effects of the gently undulating surface. In the lunar case, the diffuse component of the reflected radiation is normally associated with the presence of large numbers of wavelength-size (or smaller) rock or rock fragments. Very small rocks will be in the Rayleigh regime and will not contribute individually to the echo. Some attempts to provide quantitative descriptions of diffuse scattering in terms of rock distributions from the Moon have been made (ref. 23-10). However, in terms of surface structure, the diffuse scattering is not understood nearly so well as the quasi-specular scattering.

Experimentally, quasi-specular and diffuse scattering can be distinguished by polarization and coherence properties and by the scattering law (ref. 23-11). Quasi-specular scattering, which by definition originates from those portions of the surface that produce mirrorlike reflections, is deterministically polarized and is the predominant scattering mechanism. Although the echo polarization will change with variations in the polarization of the illuminating wave and the geometry, the polarization is the same as would be produced by a smooth surface of the same material. Diffusely scattered waves are not expected to exhibit this behavior. To the extent that the diffuse component arises from randomly oriented structures on or within the surface, it will be unpolarized. A decomposition of the echo spectrum into the polarized and unpolarized components provides a mechanism for separating the scattering from large-scale (wavelength) surface and small-scale, randomly oriented features or roughness.

EQUIPMENT DESCRIPTION

Schematic block diagrams of the receiving and data-processing systems are shown in figure 23-5. The NASA Deep Space Network 64-m-diameter parabolic antenna at Goldstone, Calif., was used to receive the S-band signals. Both the open- and closed-loop receivers, which were installed for the Mariner spacecraft orbited about Mars in late 1971, were used for the bistatic-radar experiment. Normal Apollo mission operations are conducted with the regular ground-station receivers. A signal-conditioning unit processes the Mariner receiver 10-MHz intermediate-frequency output. This unit determines the overall system bandwidth for bistatic echoes, provides signal-level control, and produces an audiofrequency for magnetic-tape recording.

The use of the open- and closed-loop systems provides redundancy and additional operational flexibility. Because of the cost of the digital data reduction is directly proportional to the signal bandwidth, it is desirable to keep this bandwidth as small as possible. However, the possibility existed that the CSM-transmitted frequency would change by several times the echo bandwidth during the experiment.

Eighty-kHz filters are used in conjunction with the open-loop receivers. This bandwidth was selected to insure that the echo would be within the passband at all times during observation with only a single, predetermined frequency setting. The closed-loop

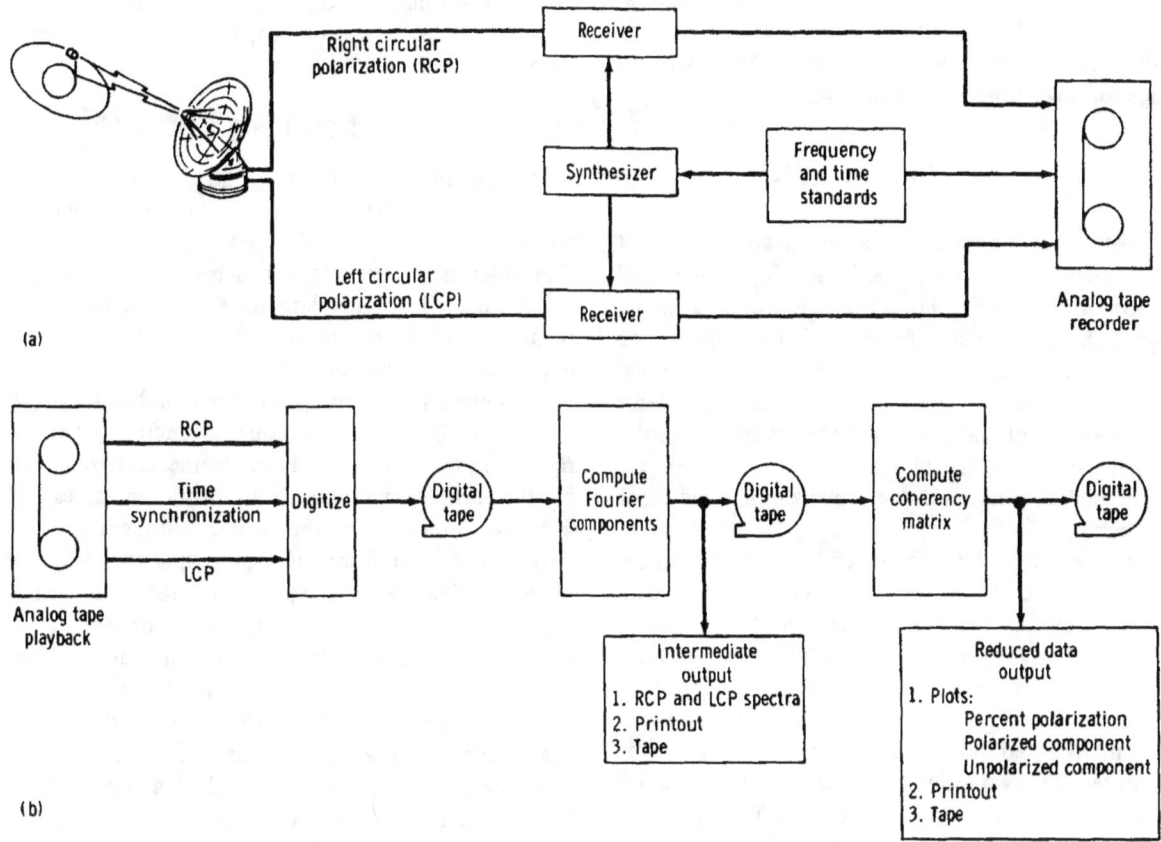

FIGURE 23-5.—System schematics. (a) Signal receiving system. (b) Data-processing system.

bandwidth is 20 kHz. This bandwidth is sufficiently wide to insure that the echo is within the passband as long as the receiver is locked on the direct signal from the CSM. The disadvantage of using the closed-loop system alone is that brief periods exist when the direct signal fades below the threshold required for receiver lock. Wideband data can be processed for those periods when lock is lost. On the Apollo 15 mission, there was only one period of several minutes when the direct signal was below the closed-loop threshold.

Both left and right circular polarizations are received. All receivers are driven from a single frequency source, which enables the relative phase between polarizations to be preserved. System noise temperature in this configuration is near 30 K when the antenna is aimed at the sky alone and 192 K when the Moon fills the beam of the antenna.

The VHF receiving facility was the 46-m-diameter parabolic antenna at the Stanford Center for Radar Astronomy. A complete, open-loop receiver was constructed, which consists of solid-stage preamplifiers for 259.7 MHz with frequency conversions to 50 MHz, 10 MHz, and baseband. The data bandwidth is determined at 10 MHz by 3.5-kHz-wide multipole crystal filters. This bandwidth is narrower than the bandwidth used at S-band, by the ratio of the transmitted frequencies. Only open-loop channels were used. As was the case for the S-band receiving system, left and right circularly polarized signals were received, and the system is coherent. The system noise temperature is approximately 700 K.

Open-loop operation is similar at S-band and VHF. Based on a doppler ephemeris calculated from elements supplied by Manned Spacecraft Center personnel, the receivers are tuned so that the direct signal will be centered in the passband at the time the CSM crosses the Earth-Moon line. The closed-loop receiver is initially tuned according to the operational-frequency predictions for the CSM. Once lock is achieved, the receiver automatically compensates for doppler effects.

Two magnetic-tape recorders are used simultaneously for data recording. Tapes are started at different times so that overlapping records (with no gaps for tape changes) are available.

DATA REDUCTION

Data reduction consists of a three-step process that is independent of either the S-band (open or closed loop) or the VHF. This process is outlined in the signal channel shown in figure 23-5. First, the analog tapes are replayed and digitally sampled. The sampled data are converted to weighted Fourier coefficients and spectral estimates. Finally, the weighted Fourier coefficients from the two polarization channels are combined to determine the polarization spectra of the echo.

As the analog data are played back, the signals are low-pass filtered to avoid aliasing of the high-frequency tape-recorder noise in the sampling process. Sampling is synchronized with the original recording time by the use of a NASA 36-bit time code and a synchronizing waveform, both of which are multiplexed onto the data tracks of the tape recorder. The two receiver channels, for right and left circular polarization, are sampled simultaneously so that the coherence between channels is preserved. Calibration signals are recorded, sampled, and processed in the same manner as the data.

Weighted Fourier coefficients are computed using fast Fourier transform techniques. Groups of either 1024 or 2048 data samples from each channel are multiplicatively weighted with a sine-squared data window, and the Fourier coefficients are computed. Because the three analog-data sources are each sampled at different rates, the corresponding frequency and time resolutions of the spectral estimates are not uniform. When the effects of the data window are considered, spectral resolution of approximately 40 Hz is achieved with the closed-loop S-band data. A spectral resolution of approximately 5 Hz is obtained with the VHF data.

The Fourier coefficients are easily manipulated to provide a variety of data presentations. For example, sums of the squares of successive Fourier coefficient magnitudes yield spectral estimates of the received signals. The time resolution and stability of these estimates may be varied simply by changing the number of terms included in the time average. Spectral estimates for signals in two orthogonal polarizations may be combined with the cross spectra to obtain the polarization properties of the echo (ref. 23-5).

RESULTS

At the time of this writing, all the data obtained have been sampled, the Fourier coefficients computed, and the preliminary spectra examined. The polarimetry processing has been carried out on the S-band data. The data have not yet been correlated with position of the reflecting region except in a most general way.

A summary of several minutes of S-band data is shown in figure 23-6. Frequency increases to the right, and the ordinate is linear in the polarized part of the echo power spectrum. Time increases by 2.5-sec intervals in the vertical direction. The frequency resolution is approximately 40 Hz. Numerous features, that move from right to left with time, are apparent. These features correspond to reflections from relatively discrete areas. The variation in doppler shift as the spacecraft approaches and then passes one of these areas produces the characteristic S-shaped signature. Spectra such as these are typical of both the Apollo 14 and 15 data; however, on the whole, the Apollo 15 spectra do appear to represent echoes from much smoother, more homogeneous terrain than echoes from Apollo 14 data.

CONCLUSIONS

Any conclusions drawn at this stage of the analysis must of necessity remain somewhat tentative. On the basis of the S-band data, the area surveyed during the Apollo 15 mission is largely homogeneous (on a scale of a few centimeters) and very similar to those regions sampled at lower latitudes during the Apollo 14 mission. While there are distinct variations in centimeter-to-meter-length slopes, the vertical structure of the surface itself appears to be remarkably uniform. The unpolarized data suggest that there are exceptions to this rule in the form of regions that must contain an unusually heavy population of centimeter-size fragments or rocks. However, this aspect of the data must be examined further.

The Apollo 15 VHF data are of significantly (about the order of magnitude) higher signal-to-noise ratio than any previously obtained. The effects of the bulk-surface electrical properties and variations in

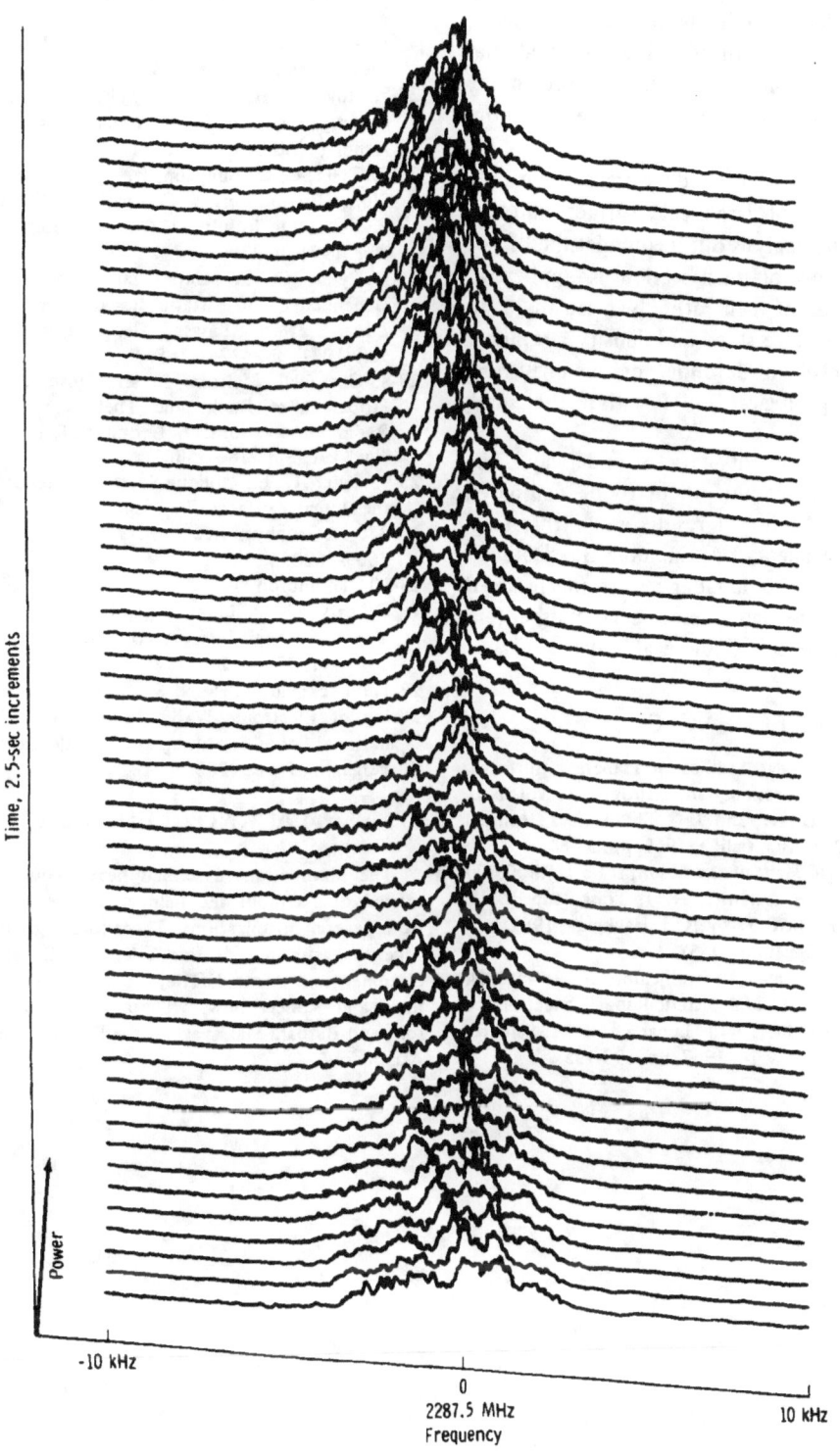

FIGURE 23-6.—Apollo 15 bistatic-radar spectra, S-band polarized component. This series of spectra is typical of the Apollo 15 S-band data over the entire data pass.

slope statistics are clearly present in the data. The Apollo 14 VHF data exhibit marked surface penetration effects and depolarization. It is anticipated that the advance in signal-to-noise ratio achieved in the present observations will permit the study of more subtle effects of this type.

In general, the Apollo 15 data are substantially free of the effects of large-scale surface inhomogeneities; thus, the analysis outlined under the "Basic Theory" subsection of this section is applicable. These data, because of their rather high sensitivity to small changes in surface slope and density, are likely to be most useful in defining, or distinguishing among, various geological units within the mare basins sampled.

Preliminary analysis of data is underway. The nature of the data permits them to be compared directly with the Apollo 14 results and with results from earlier experiments on Explorer 35. Detailed comparisons of the bistatic-radar results and reduced Orbiter photography are being accomplished in conjunction with the U.S. Geological Survey.

REFERENCES

23-1. Evans, J. V.: Radar Studies of Planetary Surfaces. Annual Review of Astronomy and Astrophysics, vol. 7, L. Goldberg, David Layzer, and J. G. Phillips, eds., Annual Reviews, Inc. (Palo Alto, Calif.), 1969, pp. 39-66.

23-2. USGS Geologic Maps of the Moon, scale 1:1,000,000 (LAC 58 by H. H. Schmitt, N. J. Trask, and E. M. Shoemaker, 1967; LAC 57 by R. J. Hackman, 1962; and LAC 76 by R. E. Eggleton, 1965).

23-3. Tyler, G. L.; Eshleman, V. R.; Fjeldbo, G.; Howard, H. T.; and Peterson, A. M.: Bistatic-Radar Detection of Lunar Scattering Centers with Lunar Orbiter I. Science, vol. 157, no. 3785, July 14, 1967, pp. 193-195.

23-4. Tyler, G. L.; Ingalls, D. H. H.; and Simpson, R. A.: Stanford Telemetry Monitoring Experiment on Lunar Explorer 35. Final Rept. SU-SEL-69-066, Stanford Electronics Lab., Oct. 1969.

23-5. Howard, H. T.; and Tyler, G. L.: Bistatic-Radar Investigation. Sec. 17 of Apollo 14 Preliminary Science Report, NASA SP-272, 1971.

23-6. Tyler, G. L.; Simpson, R. A.; and Moore, H. J.: Lunar Slope Distributions: A Comparison of Bistatic Radar and Photographic Results. J. Geophys. Res., vol. 76, no. 11, Apr. 1971, p. 2790.

23-7. Tyler, G. L.; and Ingalls, D. H. H.: Functional Dependences of Bistatic Radar Frequency Spectra on Lunar Scattering Laws. J. Geophys. Res., vol. 76, no. 20, July 1971, pp. 4775-4785.

23-8. Campbell, Malcolm J.; and Ulrichs, Juris: Electrical Properties of Rocks and Their Significance for Lunar Radar Observations. J. Geophys. Res., vol. 74, no. 25, Nov. 1969, pp. 5867-5881.

23-9. Tyler, G. L.: Oblique-Scattering Radar Reflectivity of the Lunar Surface: Preliminary Results From Explorer 35. J. Geophys. Res., vol. 73, no. 24, Dec. 1968, pp. 7609-7620.

23-10. Thompson, T. W.; Pollack, J. B.; Campbell, M. J.; and O'Leary, B. T.: Radar Maps of the Moon at 70-cm Wavelength and Their Interpretation. Radio Science, vol. 5, no. 2, Feb. 1970, pp. 253-262.

23-11. Beckman, Petr; and Spizzichino, André: The Scattering of Electronic Waves From Rough Surfaces. International Series of Monographs on Electromagnetic Waves, Pergamon Press, 1963.

ACKNOWLEDGMENTS

The observations described herein could not have been carried out without the help of a large number of people from several organizations. The authors gratefully acknowledge the assistance of Jim Raleigh of Bellcom, Inc.; Allan Chapman and Booth Hartley of the Jet Propulsion Laboratory; and Robert Dow, William Faulkerson, John Williamson, and Barbara Warsavage of the Stanford Center for Radar Astronomy.

24. Apollo Window Meteoroid Experiment

Burton G. Cour-Palais,[a†] *Robert E. Flaherty,*[a] *and Milton L. Brown*[b]

The purpose of the Apollo window meteoroid experiment is to use the Apollo command module (CM) heat-shield-window surfaces to obtain additional information about the flux of meteoroids with masses of 10^{-9} g and less, to examine the residue and the morphology of the craters produced by these meteoroids to obtain information regarding the dynamic and physical properties of the meteoroids, and to discover possible correlations with the lunar-rock-crater studies.

In addition to information regarding meteoroid flux, this experiment could yield information on the mass density and, possibly, on the composition of meteoroids. To determine the mass density would require the assumption that the velocity distribution, as determined from optical and radar observations of much larger meteoroids, is applicable to the smaller meteoroids.

Laboratory test data are currently being generated. Glass targets, identical to the CM heat-shield windows, are being impacted by particles of different sizes, mass densities, and velocities using electrostatic accelerators for correlation with the observed crater characteristics.

When the effects of entry heating, subsequent immersion in salt water, and all the other contaminating sources are accounted for, the composition of the meteoroid residue in the crater or in the fused glass will be determined by use of a scanning-electron-microscope (SEM) nondispersive X-ray detector. The significance of the use of this detector is that compositions and mass densities obtained in space can be compared with those obtained from the lunar rock samples without the long-term exposure to the environmental effects of space. Knowledge of the mass density is also important for designing meteoroid shielding. So far, mass density can only be inferred from observations of meteoroid breakup in the atmosphere. Controversy exists as to whether the typical meteoroid is a low-density (1.0 g/cm^3 or less) dust or ice ball or a stony object that froths during atmospheric entry and breaks off in chunks. If it is the latter, the mass density could be 2.5 to 3.0 g/cm^3.

With the exception of the Apollo 11 mission, all the CM heat-shield windows have been examined for meteoroid impacts 50 μm and larger in diameter. So far, a total of 10 possible impacts has been observed. The flux represented by the number of impacts observed and the area-time of exposure by the Apollo windows is compatible with the flux estimates obtained from the results of penetration sensors mounted on the Pegasus 1, 2, and 3 satellites, by the Explorer 16 and 23 satellites, and by the Surveyor III shroud. The Apollo 15 windows had not been received for study at the time of publication of this report.

BASIC THEORY

Meteoroids are solid particles moving in interplanetary space that originate from both cometary and asteroidal sources. They are classified as sporadics when the orbits are random and as streams when many have nearly identical orbits. A meteor is the light phenomenon associated with the interaction of a meteoroid with the atmosphere of the Earth. The portion that survives interaction with the atmosphere and is found on the surface of the Earth is a meteorite. It is generally accepted that most meteorites are of asteroidal origin (ref. 24-1) and that the typical meteoroid originates from a cometary nucleus, is frangible, and does not reach the surface of the Earth. Thus, very little is known about the composition and mass density of meteoroids, whereas meteorites have been collected and examined very thoroughly (ref.

---

[a]NASA Manned Spacecraft Center.
[b]Lockheed Electronics Corporation.
[†]Principal investigator.

24-2). The typical meteoroid has been described (ref. 24-3) as a conglomerate of dust particles bound together by frozen gases or "ices"; another author (ref. 24-4) postulates that meteoroids are "dust balls." The mass density of these conglomerates is assumed to be no greater than 1 g/cm^3 because of the evidence of breakup high in the atmosphere. Recent experiments with carbonaceous chondrites in arc jets (ref. 24-5) have shown that a sufficient amount of water is present to cause frothing during the entry heating and that this frothy material breaks off along the path because of aerodynamic pressure. Thus, laboratory evidence has shown that the breakup of meteoroids in the atmosphere of the Earth is not necessarily indicative of a low-density conglomerate.

The near-Earth flux of meteoroids entering the atmosphere has been determined from photographic observation of meteors, radar echoes from the ionized column produced by meteoroids, and direct measurements by satellite detectors. The results of these observations have been combined in the plot of cumulative number/m^2-sec for each size (fig. 24-1). Details of the observation techniques, direct measurements, and conversion of the observed data to mass may be found in references 24-6 to 24-8. Detectors flown on spacecraft have furnished information on the meteoroid flux in the mass range of 10^{-13} to 10^{-6} g (refs. 24-8 to 24-11). Fluxes for masses 10^{-7} g and less have been detected primarily by acoustic-impact (microphone) sensors, whereas fluxes of 10^{-9} to 10^{-6} g have been determined by the detection of perforations in thin metallic-sheet sensors. The acoustic-impact-sensor measurements (ref. 24-8) indicate a much higher particle flux than do the penetration sensors for the same mass range (refs. 24-9 to 24-11). The examination of the Gemini spacecraft windows for meteoroid impacts confirmed the lower flux estimates obtained by the penetration sensors (ref. 24-12) for masses of 10^{-7} g and less. The cumulative flux plot of figure 24-1 reflects the low flux estimate of the penetration sensors and the Gemini window examination. Because the proponents of the higher flux measurements obtained from the acoustic sensors do not completely accept the penetration-sensor results, it is believed that the Apollo window meteoroid experiment, because of the much greater total time-area exposure to the environment than the Gemini windows, could settle the controversy.

DESCRIPTION OF THE INSTRUMENT

The Apollo window meteoroid experiment is passive in that it uses approximately 0.4 m^2 of the external surfaces of the Apollo CM windows as meteoroid-impact detectors. The windows are made of 99-percent-pure fused silica and are mounted more or less flush with the external surface of the heat-shield contour. The rendezvous windows were originally included in the total area of glass to be scanned but have not been used since the Apollo 10 mission for the following reasons.

(1) The rendezvous windows are inset into the heat shield and are fairly well shielded from the environment.

(2) In all missions involving lunar module (LM) rendezvous, the surfaces of the rendezvous windows are subject to impacts by particles emanating from the thrusters of the LM reaction control system (RCS). During the postflight examination, the Apollo

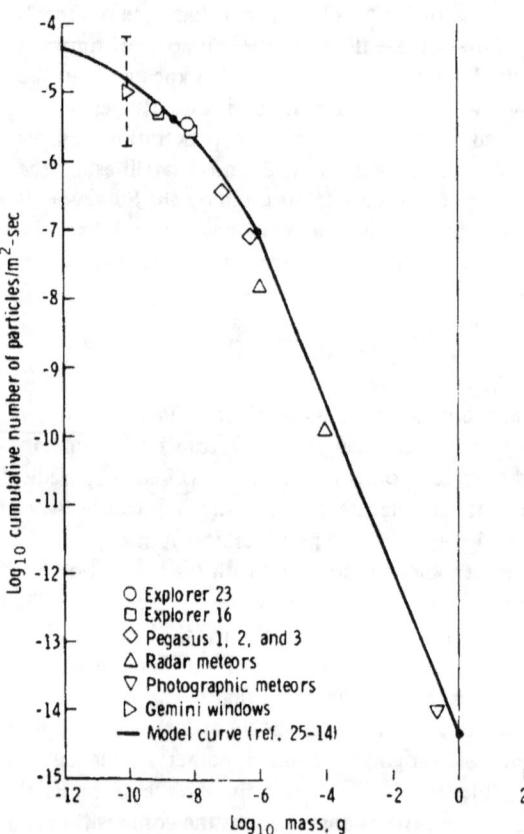

FIGURE 24-1.—Comparison of cumulative meteoroid flux-mass data with the adopted model.

9 rendezvous windows showed a significant increase in pitting over that on previous missions with no LM rendezvous.

The window is an ideal detector because the surface spallation area can be 25 to 100 times the diameter of the impacting meteoroid. This ratio allows a 20X optical scan to detect a 50-μm crater (which would be caused by a 0.5- to 2-μm meteoroid). Hence, the time necessary to scan the large areas involved is considered reduced, and the detection threshold is small enough to include cosmic dust particles.

During entry, the heat-shield windows are subjected to a surface temperature of almost 1175 K for the side windows; the hatch window is subjected to a somewhat lower temperature. These temperatures are well below the annealing and softening temperatures of the glass; therefore, the crater morphology should not be affected. However, any residue in the craters will be affected. It is hoped that both the temperature effect and the contamination caused by ablative products from the heat shield and the subsequent immersion in sea water can be accounted for in the analyses.

EXAMINATION PROCEDURE

The heat shield and hatch windows from the Apollo 7 to 14 command modules were scanned before flight at 20X magnification to determine the general condition of the external surfaces. All chips, scratches, and other features that could be confused with meteoroid impact craters were noted on a surface map. As a result of the stringent quality control and optical requirements, the windows have been generally free of such defects, and this mapping practice has been discontinued. The windows are received with a fairly thick coating of surface contamination, which is removed before the optical scan by careful scraping with a razor blade and by washing with water and isopropyl alcohol.

The windows are next scanned at 20X magnification with a stereozoom microscope by two different observers, and suspicious surface effects are marked on a postflight map. This map is used to relocate the possible craters that remain after comparison with the preflight map for individual examination at much higher magnifications. From experience with hypervelocity impacts in fused silica, it is possible to separate the meteoroid impacts from other surface effects (ref. 24-13). All laboratory hypervelocity impacts, when viewed with bottom lighting, have less than 10 percent of the total damaged area blacked out because of total reflection; the central area consists of pulverized glass that is generally 4 to 6 times the diameter of the impacting particle. This central zone, often raised above the surrounding chipped area, is dome shaped and has a shallow depression at the point of impact for speeds of 5 to 7 km/sec. For very high velocities (10 km/sec and faster), the center of this dome shows signs of fusion, and a lip often extends above the original surface of the glass. The central pulverized zone is surrounded by an inner ring of a generally rough, chipped-out appearance and an outer ring of a generally smoothed, conchoidal, terraced look. The periphery of the spalled surface is very rarely symmetrical because of the inhomogeneous fracture strength of the glass.

Every suspected meteoroid crater is subjected to the following procedure.

(1) The crater is photographed in detail with top and bottom lighting.

(2) The crater depth and diameter are measured.

(3) A section of the window containing the crater is removed by coring or sectioning as close as possible to the crater.

(4) The window section is prepared for residue analysis with the SEM nondispersive X-ray detector by applying a thin carbon coating.

(5) Residue analysis is performed, and all constituents are recorded.

(6) The same window section is cleaned, and a thin gold coating is applied for crater photography at high magnifications.

The data obtained from these examinations will be compared with the laboratory test data mentioned previously, and the following meteoroid characteristics will be determined.

(1) The probable meteoroid mass, using assumed velocity and impact angle

(2) The flux of particles of this probable mass and larger, using exposure time and area

(3) Probable meteoroid composition, determined from elemental constituents and by comparison with lunar-crater studies, meteorites, and comets

(4) Impact shock pressure, estimated by correlating actual crater characteristics with test data

(5) Probable mass density, estimated by using Hugoniot equations for the Apollo window glass and

from the results of the impact-shock-pressure determination

RESULTS

The Apollo 14 windows were received with the outer surfaces contaminated with the hard deposit that has been present on all the Apollo windows examined to date. The results of a chemical analysis of samples taken from the Apollo 9 windows and possible sources of the contaminants are listed in table 24-I. The sources chosen were the Mylar protective coating on the heat-shield surface, scrapings from an RCS thruster nozzle, and a piece of charred heat-shield material. It is apparent that high to very high concentrations of sodium were found in the thruster-nozzle crust and in the heat-shield "char," but only a trace of sodium was found in the Mylar. Because the Mylar ablates before CM splashdown, the origin of the sodium is clear. The concentration of magnesium found in the thruster material was greater than 10 percent, whereas it was less than 1 percent in the heat-shield char and the Mylar, indicating another possible source. Also, the Mylar, rich in titanium and silicon, is identified as the source of these two elements on the window surfaces. Because the CM entry and landing conditions are similar for all missions, it is assumed that the Apollo 9 results are typical of all other spacecraft.

After scanning approximately 3 m² of Apollo heat-shield windows at a general level of 20X magnification, 10 meteoroid impacts have been tentatively identified. Of these, the five craters shown in figure 24-2 were found on the Apollo 7 window. One crater was found on the Apollo 8 window, one on the Apollo 9 window, one on the Apollo 13 window, and two on the Apollo 14 window. The Apollo 13 crater is shown in figure 24-3. The surface spallation diameters of these craters range from 50 to 500 μm.

A number of surface effects from low-velocity particles have been found after many of the flights. Typical examples have a very different appearance from the probable meteoroid craters (fig. 24-4). Inasmuch as the origins of these effects are not likely to be meteoroids, they are excluded from consideration.

A list of the craters is given in table 24-II, which includes a preliminary estimate of the exposure (product of window area and time) to the meteoroid environment for each mission, the resultant flux, the upper and lower 95 percent confidence limits, and an

TABLE 24-I.–Apollo 9 Window Surface Contaminants and Possible Sources

Sample area	Al	B	Ca	Cr	Cu	Fe	Pb	Mg	Mn	Mo	Ni	P	K	Si	Ag	Na	Sn	Ti	W	Zn
Hatch window	Tr	–	Tr	Tr	Tr	<1	–	>10	Tr	–	–	–	–	1 to 10	–	>10	Tr	<1	–	Tr
Side windows	<1	Tr	<1	Tr	<1	<1	Tr	>10	Tr	Tr	–	–	–	1 to 10	Tr	>>10	Tr	1 to 10	–	Tr
Rendezvous windows	<1	Tr	<1	<0.1	<1	<1	Tr	>10	Tr	Tr	–	–	–	>10	Tr	>>10	Tr	1 to 10	–	Tr
CM Mylar layer	<1	Tr	Tr	Tr	Tr	Tr	Tr	<1	Tr	–	Tr	–	–	>10	–	Tr	Tr	>>10	–	<0.1
Thruster-nozzle residue	Tr	–	Tr	–	Tr	<1	–	>10	Tr	Tr	–	–	–	1 to 10	–	>>10	–	<1	–	–
CM heat-shield char	<1	<0.1	Tr	–	Tr	<1	–	<1	–	–	–	–	–	1 to 10	–	>10	–	1 to 10	–	–

ᵃTr = trace (in range of ppm).

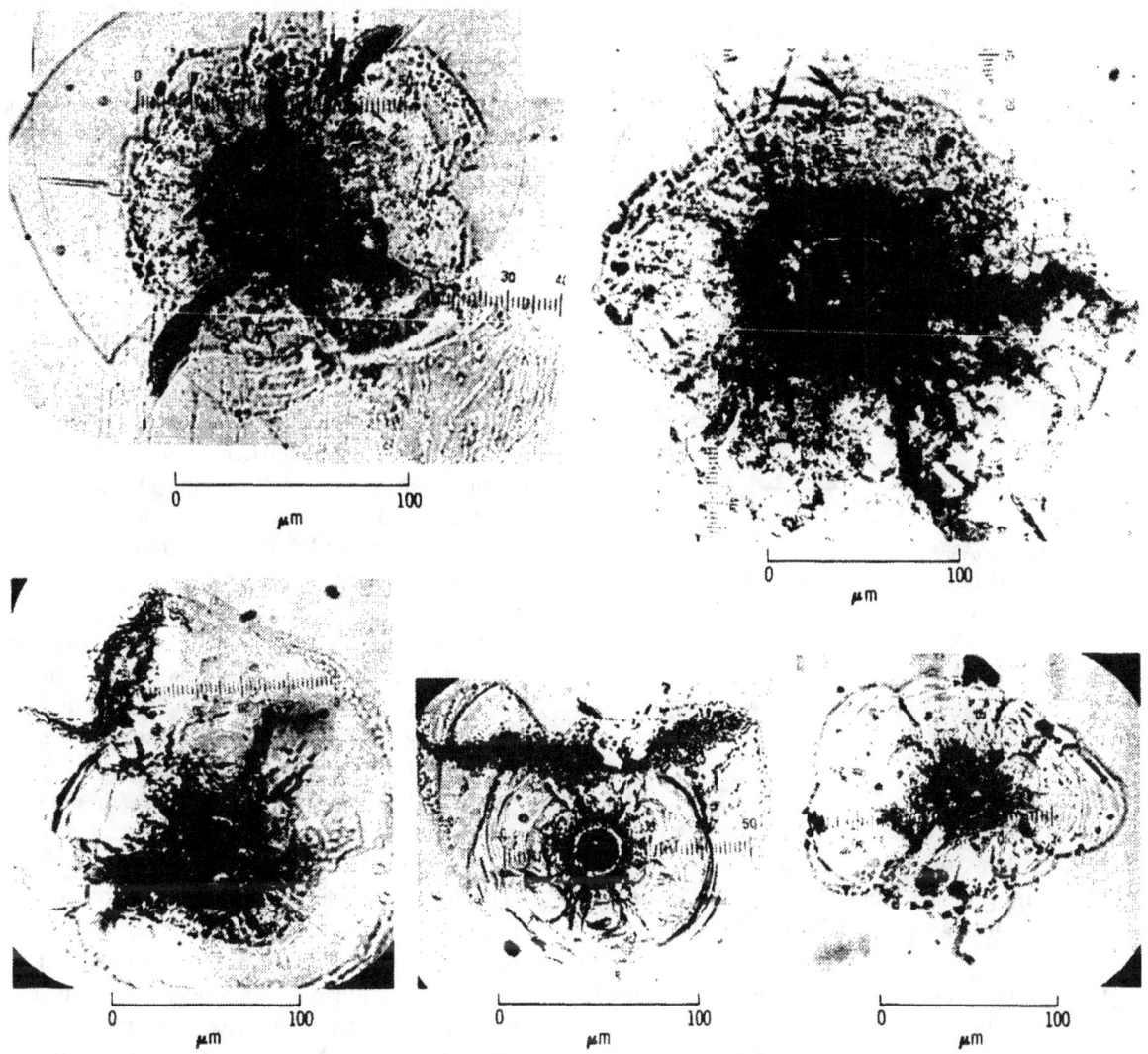

FIGURE 24-2.—Suspected meteoroid impact craters found on the Apollo 7 window surfaces.

estimate of the impacting meteoroid mass. The mass determination is based on a spall-diameter-to-meteoroid-diameter ratio of 100 to 1 and a mass density of 2 g/cm^3. The diameter ratio used was obtained from a series of hypervelocity tests conducted by the authors at the NASA Manned Spacecraft Center by simulating the expected meteoroid-impact shock characteristics. The flux estimate includes the window area, mission time, planetary shielding during the Earth- and lunar-orbital periods, an allowance for the additional shielding provided by the LM during Earth-Moon transit, and a factor to account for the focusing effect of the gravitational field of the Earth (ref. 24-14) on the meteoroid population. The resulting data points are shown in figure 24-5 compared with the near-Earth meteoroid-flux model of reference 24-14 and the Surveyor III shroud data (ref. 24-15). It is evident that the Apollo and the Surveyor III results agree fairly well and that the data lie within the reference environment model.

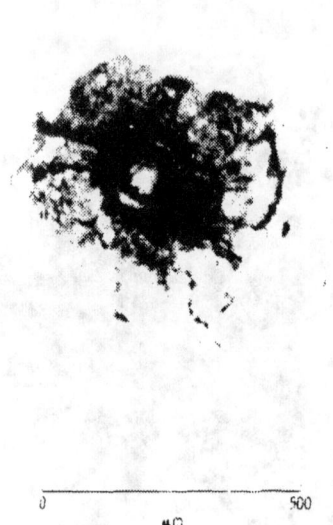

FIGURE 24-3.—Suspected meteoroid impact crater found on the Apollo 13 right-hand side-window surface.

No data on the detailed SEM analyses of the crater characteristics and residue were available at the time of this report.

DISCUSSION

The preliminary estimates of the flux of particles obtained from the examination of the Apollo windows are generally below expectations. One or more of the following reasons may be the cause of this low flux.

(1) The flux model (ref. 24-14) used for comparison could be high.

(2) A number of meteoroid impacts are possibly being missed during the 20X optical scan.

(3) The product of the effective surface area of the windows and the total exposure time may be overestimated, possibly because of incorrect estimates of either or both of these factors.

(4) Micrometeoroids cannot penetrate the deposit on the window surfaces with sufficient energy to create a typical impact crater.

The fourth reason can probably be eliminated because the contaminants found on the window surfaces appear to be associated with the final events of each mission (except for contamination caused by the thrusters). Any thruster deposit on the windows during the mission must have been minimal because the astronauts have not commented about thruster contamination. Any film too light to be noticeable would have had little effect on the formation of a typical meteoroid crater.

In calculating the product of the window area and exposure time, it has been assumed that, during lunar orbit, the windows are completely shielded from the environment by the Moon. This assumption is based on the fact that the hatch and side windows are oriented toward the lunar surface for photography and landmark observation for most of this phase. A detailed examination of the Apollo 15 spacecraft-attitude history during lunar orbit has not yet been obtained.

In addition to the planetary shielding, the LM is assumed to shield a quarter of the available solid angle for the side and hatch windows during the translunar portion of the mission. This assumption can be checked experimentally by determining the shadow cast by the LM on each window. This determination will be made if it proves to be of importance.

The possibility of not detecting all the meteoroid impacts during the optical scan has been minimized. Each window is scanned by two independent observers who have had considerable experience in the appearances of craters and surface defects in glass. It is thought that the discrepancy between predicted and measured flux values is not the result of inadequate inspection.

The final possibility for the low flux recorded by the Apollo window meteoroid experiment is the reference flux model used for comparison. As mentioned previously, the model environment is based on the penetration-detector results and is already considered to be a low estimate by some meteoroid experimenters. The Apollo and Surveyor III data have been enhanced by the gravitational focusing factor to make them comparable to the reference model environment based on near-Earth measurements. The Apollo 7 (Earth orbital without an LM) and Apollo 13 (circumlunar abort with an LM) missions yielded results very close to the predicted values. Therefore, the assumption regarding LM shielding of the windows could be fairly accurate, and the Earth-orbital shielding for the Apollo 7 mission may have been fortuitously close.

The mass estimate for the meteoroids that caused the Apollo window craters appears to be reasonably correct. If the ratio of the measured spallation diam-

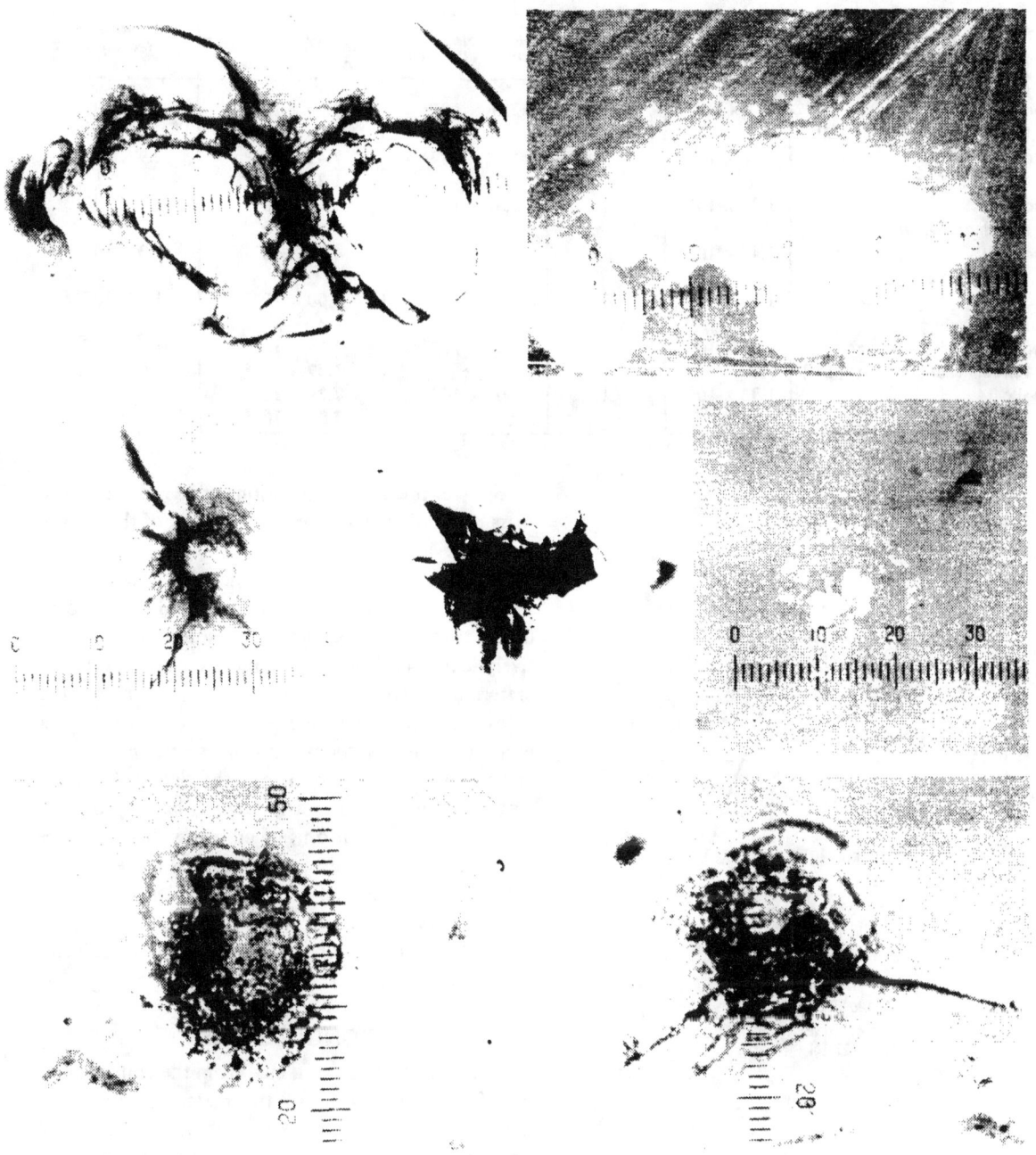

FIGURE 24-4.—Typical low-velocity surface effects.

eter to meteoroid diameter were 50 instead of 100, the value used, the mass values would be approximately an order of magnitude larger. This ratio would not have any great effect on the correlation of the data with the reference model, except for the Apollo 13 mission. If the Apollo 13 result were adjusted to conform to the reference environment model, this would mean that the exposed window area would have to be larger, that is, the LM shielding would be less than the 25 percent assumed. Because the effect

TABLE 24-II.—*Meteoroid Craters and Related Information*

Mission	Window exposure, m^2-sec	Number of impacts	Meteoroid flux, number/m^2-sec	95 percent confidence limits, number/m^2-sec	Minimum meteoroid mass, g
Apollo 7 (Earth orbital without LM)	2.21×10^5	5	2.26×10^{-5}	5.29×10^{-5} 7.23×10^{-6}	2.1×10^{-12}
Apollo 8 (lunar orbital without LM)	1.46×10^5	1	1.32×10^{-5}	7.41×10^{-5} 1.32×10^{-6}	2.9×10^{-13}
Apollo 9 (Earth orbital with LM)	1.78×10^5	1	5.57×10^{-6}	3.12×10^{-5} 5.57×10^{-7}	8.4×10^{-12}
Apollo 10 (lunar orbital with LM)	1.49×10^5	0	--	4.86×10^{-5} --	1.3×10^{-13}
Apollo 12 (lunar landing)	1.79×10^5	0	--	4.00×10^{-5} --	1.3×10^{-13}
Apollo 13 (circumlunar abort with LM)	1.42×10^5	1	1.37×10^{-5}	7.64×10^{-5} 1.37×10^{-6}	1.3×10^{-10}
Apollo 14 (lunar landing)	1.93×10^5	2	2.00×10^{-5}	7.24×10^{-5} 2.00×10^{-6}	1.3×10^{-13}

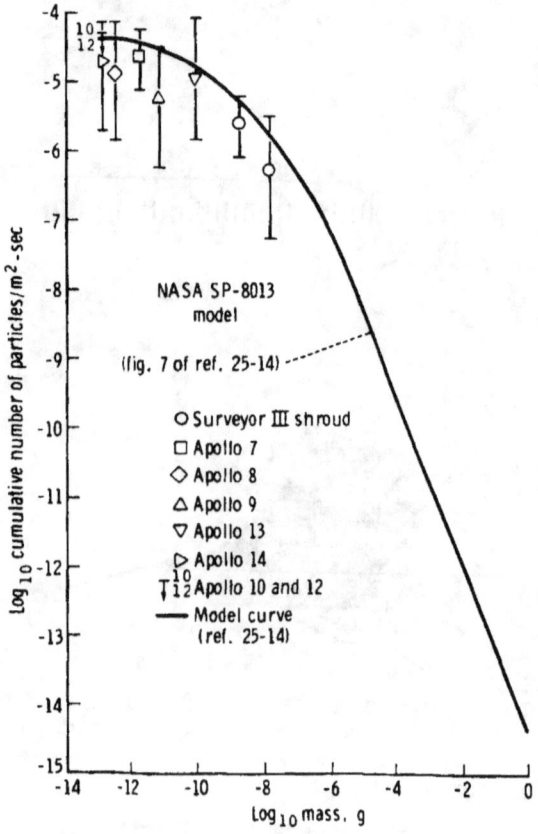

FIGURE 24-5.—Comparison of Apollo window experiment data with the reference model environment.

of reducing the LM shielding for the Apollo 13 would be to lower the other lunar-mission data points, this solution is not readily acceptable.

For the present, the questions of LM shielding and mass calibration must await the results of the experiments previously mentioned.

The extent of the window contamination described in the previous subsection could cause some doubt that the meteoroid composition can be positively identified from the residue associated with each crater. Certainly, the uncombined surface meteoroid debris will be contaminated. However, the impact event generates a shock pressure sufficient to melt and vaporize the meteoroid and the glass. Laboratory impact tests with glass targets have produced "mounds" in the bottoms of craters, which indicates the presence of the projectile material in association with the host material (ref. 24-16). Hence, it is extremely likely that the meteoroid elements can be identified within similar mounds in the 99-percent-pure fused silica.

CONCLUSIONS

Useful data can be obtained from the crater counts and from the analysis of the meteoroid residue in combination with the fused glass in each crater. A preliminary estimate of the flux resulting from seven Apollo spacecraft is in agreement with the Surveyor III data but is lower than the model environment (ref. 24-14).

REFERENCES

24-1. Whipple, Fred L.: On Maintaining the Meteoritic Complex. Studies in Interplanetary Particles, Fred L. Whipple, Richard B. Southworth, and Carl S. Nilsson, eds., Special

Rep. 239, Smithsonian Astrophysical Observatory (Cambridge, Mass.), 1967, pp. 1-46.

24-2. Wood, J. A.: Physics and Chemistry of Meteorites. The Moon, Meteorites and Comets. Vol. IV of The Solar System, Barbara M. Middlehurst and Gerard P. Kuiper, eds., University of Chicago Press, 1963, pp. 337-401.

24-3. Whipple, Fred L.: The Meteoritic Risk to Space Vehicles. Vistas in Astronautics, Morton Alperin, ed., Pergamon Press (Los Angeles), 1958.

24-4. Opik, Ernst Julius: Physics of Meteor Flight in the Atmosphere. Interscience Pub., Inc. (New York), 1958.

24-5. Allen, H. Julian; and Baldwin, S., Jr.: Frothing as an Explanation of the Acceleration Anomalies of Cometary Meteors. J. Geophys. Res., vol. 72, no. 13, July 1, 1967, pp. 3483-3496.

24-6. Lovell, Alfred Charles Bernard: Meteor Astronomy. Clarendon Press (Oxford), 1954.

24-7. McKinley, Donald William Robert: Meteor Science and Engineering. McGraw-Hill Book Co., Inc., 1961.

24-8. Alexander, W. M.; McCracken, C. W.; Secretan, L.; and Berg, O.: Review of Direct Measurements of Interplanetary Dust from Satellites and Probes. Proceedings of the Third International Space Science Symposium. Wolfgang Priester, ed., Interscience Pub., Inc. (New York), 1963, pp. 891-917.

24-9. Hastings, Earl C., Jr.: The Explorer 16 Micrometeoroid Satellite. Supplement III, Preliminary Results for the Period May 27, 1963, Through July 22, 1963. NASA TM X-949, 1964.

24-10. O'Neal, R. L.: The Explorer 23 Micrometeoroid Satellite. Description and Results for the Period Nov. 6, 1964, Through Nov. 5, 1965. NASA TN D-4284, 1968.

24-11. Clifton, Stuart; and Naumann, Robert: Pegasus Satellite Measurements of Meteoroid Penetration (Feb. 16, 1965, Through Dec. 31, 1965). NASA TM X-1316, 1966.

24-12. Zook, Herbert A.; Flaherty, Robert E.; and Kessler, Donald J.: Meteoroid Impacts on the Gemini Windows. Planet. Space Sci., vol. 18, no. 7, July 1970, pp. 953-964.

24-13. Flaherty, Robert E.: Impact Characteristics in Fused Silica for Various Projectile Velocities. J. Spacecraft and Rockets, Vol. 7, no. 3, Mar. 1970, pp. 319-324.

24-14. Cour-Palais, B. G.; Whipple, Fred L.; D'Aiutolo, C. T.; Dalton, C. C.; et al.: Meteoroid Environment Model—1969 (Near Earth to Lunar Surface). NASA SP-8013, 1969.

24-15. Cour-Palais, B. G.; Flaherty, R. E.; High, R. W.; Kessler, D. J.; et al.: Results of the Surveyor III Sample Impact Examination Conducted at the Manned Spacecraft Center. Proceedings of the Second Lunar Science Conference, vol. 3. A. A. Levinson, ed., MIT Press (Cambridge, Mass.), 1971, pp. 2767-2780.

24-16. Carter, J. L.; and McKay, D. S.: Influence of Target Temperature on Crater Morphology and Implications on the Origin of Craters on Lunar Glass Spheres. Proceedings of the Second Lunar Science Conference, vol. 3. A. A. Levinson, ed., MIT Press (Cambridge, Mass.), 1971, pp. 2653-2670.

25. Orbital-Science Investigations

This section contains a preliminary analysis of orbital observations and photography. Part A is a summary of geological observations from lunar orbit, a mission objective for the first time on the Apollo 15 mission. These visual observations are closely linked with the orbital photography because of the complementary nature of the data.

Parts B to T pertain to the scientific results of the orbital photography. The metric and panoramic cameras, which were carried for the first time on the Apollo 15 mission, produced an impressive number of high-quality lunar photographs. Examples of the quick-look photogrammetric results based on the capabilities inherent in both photographic systems are given in Parts B to D.

Parts E to N include geologic interpretations of both metric- and panoramic-camera photographs. These parts may be subdivided as follows: (1) discussions of the Apollo 15 landing site or some aspects of the landing-site photography (Parts E, F, and G), (2) preliminary geological analyses of two regions that are being considered (among other locations) as candidate Apollo 17 landing sites (Taurus Mountains, Parts H and I; and Proclus region, Parts J and K), and (3) discussions of selected features in the far-side highlands and in the near-side maria (Parts L to N).

The latter parts of this section deal with some results of the command module (CM) photography. Parts O to S include preliminary analyses of lunar-surface photographs taken at varying Sun illumination angles and in earthshine. The final part (Part T) deals with CM photography of astronomical phenomena.

The discussions found in this section should be considered only as illustrations of what could be done by detailed study of the orbital-science data from the Apollo 15 mission. Comprehensive reports on the geological significance of visual observations from lunar orbit will be published elsewhere. Detailed studies of more than 6000 lunar-surface photographs obtained during the Apollo 15 mission presently are underway. Tectonic and geologic maps of both regional and local extent will be prepared when base-map products become available. Additional reports also will become available as unusual and new lunar-surface features are detected for the first time in the Apollo 15 photographs.

PART A

VISUAL OBSERVATIONS FROM LUNAR ORBIT

Farouk El-Baz[a] *and Alfred M. Worden*[b]

The objective of visual observations from lunar orbit was fulfilled with extraordinary success on the Apollo 15 mission. The purpose of the observations was to complement photographic and other remotely sensed data from lunar orbit. Geologic descriptions of the regional or local settings of particular lunar-surface features and processes were obtained. These descriptions provided solutions to geologic problems that were difficult to solve by any other means, as will be explained subsequently. These observations constitute a significant complement to instrument-gathered data and can be accomplished only by man.

---

[a]Bellcomm, Inc.
[b]NASA Manned Spacecraft Center.

Concept and Background

The use of the capabilities of the human eye in making observations from lunar orbit is an integral part of manned lunar exploration. This is so because of (1) the fact that the extensive dynamic range and color sensitivities of the human eye cannot be matched by any one film type or sensing instrument and (2) the need, in special cases, for on-the-scene interpretation of observed features or phenomena.

The resolving power of the unaided human eye is 3×10^{-4} rad (approximately one-sixtieth of a degree), corresponding to a resolution of approximately 30 m from a 60-n. mi. circular orbit—adequate for problems related to regional distributions and tectonic trends. Another advantage of visual observations is that of stereoscopic viewing of large areas.

In regard to color sensitivity, the unaided human eye, under good viewing conditions, can distinguish 1×10^6 color surfaces—a precision 2 to 3 times better than the most accurate photoelectric spectrophotometers. This capability is best used in detecting subtle color differences among lunar-surface units at varying Sun angles and viewing directions. The dynamic range of the eye also makes visibility possible within what appear in photographs as hard shadows and washout regions. The contrast in a scene does not affect the eye as much as it does photographic film.

Most of these human characteristics and unique abilities have been tested and demonstrated on previous manned flights. The lunar-orbital periods of the Apollo missions provide an excellent opportunity to use human visual capability to supplement photographic, geochemical, and geophysical data gathered from lunar orbit. In addition to recording unexpected phenomena, visual observations aid in the regional mapping and characterization of major units as well as in the understanding of certain small-scale features such as the following.

Color Tone of Surface Units

It has long been established by special photographic and photoelectric methods that subtle color differences exist in the lunar maria. Ground-truth data indicate that these tonal differences reflect compositional variations. These data are available for small areas in the near-side maria. Visual observations may extend the comparison to larger portions on the near side as well as to the otherwise-inaccessible far-side maria. Subtle color differences in the highlands may also characterize these lunar-surface materials.

Global Tectonic Trends

Extensive fracture systems, which delineate some of the major tectonic trends, are detectable in photographs. However, global tectonic trends with subtle surface expressions appear to occur. The delineation of such trends requires repetitive observations at varying Sun angles and viewing directions.

Small-Scale Features

A number of problems related to localized features of significant importance to man's understanding of the Moon could best be solved by visual observation from lunar orbit. These problems include the nature and extent of bright swirls of unknown origin recognized so far in three mare regions; dark, wall-like features such as those in King Crater; banding and possible layering on numerous fault scarps; and the ray-excluded zones around a number of probable impact craters.

Data Acquisition

To make scientifically valuable observations from lunar orbit, the crewman should be familiar with (1) the orbital groundtracks (i.e., the geology of overflown surface units and the location of key features on the Moon), (2) the critical characteristics of the observed features and their relationship to the problems to be solved, and (3) the method whereby one may describe the observed phenomena to a geologically oriented audience.

The crewmembers in general, and the command module pilot (CMP) in particular, were briefed before the flight on all these points. In the course of training, the CMP was also able to conduct simulations of the activity by studying aerial photographs of geologically complex regions in the United States and by flying over these areas and adding information to the photographs by making and recording visual observations. Debriefings held after these flights emphasized applications to lunar geologic problems.

Of the many interesting features overflown by the Apollo 15 crew, the authors selected 15 targets for detailed study. Segments of time in the flight plan were allocated to the task, and the astronauts were supplied with photographs of the features and a list of the questions to be answered. Examples of these photographs and questions will be presented later in this part. A complete list of the 15 visual-observation targets is given in table 25-I.

Observations were made from the command module windows without disturbing the operations of the scientific instrument module. Data were acquired on all visual-observation targets, including two not scheduled in the flight plan. On far-side passes, observations were recorded on the onboard tape recorder; and, on near-side passes, observations were recorded by real-time voice communications with the Mission Control Center. Where appropriate, observations were recorded by marking onboard graphics and charts. After the mission, a debriefing was held during which the authors discussed the studied features and the geologic significance of the observations. Excerpts from both the real-time comments and the debriefing statements (edited for clarity) will be given in the following discussion.

TABLE 25-I.—*Targets for Visual Observation*

No.	Target designation	Longitude, deg[a]	
		Start	Stop
1	Northwest Tsiolkovsky Crater	135 E	125 E
2	Picard Crater	60 E	54 E
3	Proclus Crater	53 E	45 E
4	Cauchy Crater	51 E	40 E
5	Littrow Crater	35 E	27 E
6	Dawes Crater	31 E	26 E
7	Sulpicius Gallus Crater	18 E	9 E
8	Hadley Rille	10 E	2 E
9	Imbrian flows	15 W	24 W
10	Harbinger Mountains	35 W	44 W
11	Aristarchus Plateau (in earthshine)	46 W	52 W
11	Aristarchus Plateau	44 W	54 W
12 and 13	Post-transearth-injection areas	(b)	(b)
[c]14	Mare Ingenii	170 E	157 E
[c]15	Ibn Yunus area	96 E	80 E

[a]Spacecraft-nadir coordinates.
[b]Not applicable.
[c]Targets not scheduled in the flight plan.

Significant Results

A brief summary of the significant results of the visual observations on Apollo 15 has recently been published (ref. 25-1). An attempt will be made in the following discussion to detail some of these results and to point out their geological significance.

Because of the limitations of space, only a selection of the visual-observation targets will be discussed. (The references contain more complete information.) Eight target areas will be discussed in the order of location relative to the mission groundtracks, from east to west. The locations of these targets are shown in figure 25-1.

Tsiolkovsky Crater

Tsiolkovsky is a 200-km-diameter crater that dominates a large area on the lunar far side. It is a relatively fresh crater, probably of Eratosthenian age. The ejecta blanket is bright at high Sun-illumination angles; however, under these lighting conditions, the rays are not visible. The crater displays a relatively flat, smooth floor, which appears in all photographs to be extremely dark. On the floor is a large central peak consisting of segments arranged in the shape of a "W."

The continuous ejecta blanket of Tsiolkovsky Crater can be traced about one crater diameter out from the crest of the crater rim. Secondary craters and crater chains originating from Tsiolkovsky are superposed on older surface units and formations for greater distances. On the northwestern rim of the crater is a lineated unit that is thought to be a flow of some sort.

Most of the aforementioned features of Tsiolkovsky Crater and its environs were the subject of visual-observation target 1 (figs. 25-2 and 25-3). Excerpts from real-time CMP descriptions arranged in the sequence in which they appear in the air-to-ground transcript follow. The lunar revolution (LR) and ground-elapsed time (GET) in days, hours, minutes, and seconds will be provided for each entry.

LR33/05:22:26:09 GET: Tsiolkovsky is big enough that you do get some, at least, optical impressions with the central peak being higher than the ridges. But I think it's just because the basin is big enough, is far enough across, that, as you're looking from one rim to the other, the curvature kind of gets to you and makes the central peak appear higher than it is. But it is a very, very high central peak; it's a very large mass. And, as matter of fact, on the last couple of revolutions, I've been watching the central peak, and I'm pretty sure that I can see some layering in the central peak. There should be some pictures of it. I got some pictures looking down on it. But it looks like a big slab that's been stuck up on edge.

LR34/05:23:36:31 GET: The central peak of Tsiolkovsky is very large; spur peaks on the south and east sides, getting blocky on the north side. What appears to be some layering is visible on the south and west exposed scarps of the peak, dipping to the north at about 30°.

The lineated segment of the northwestern rim of

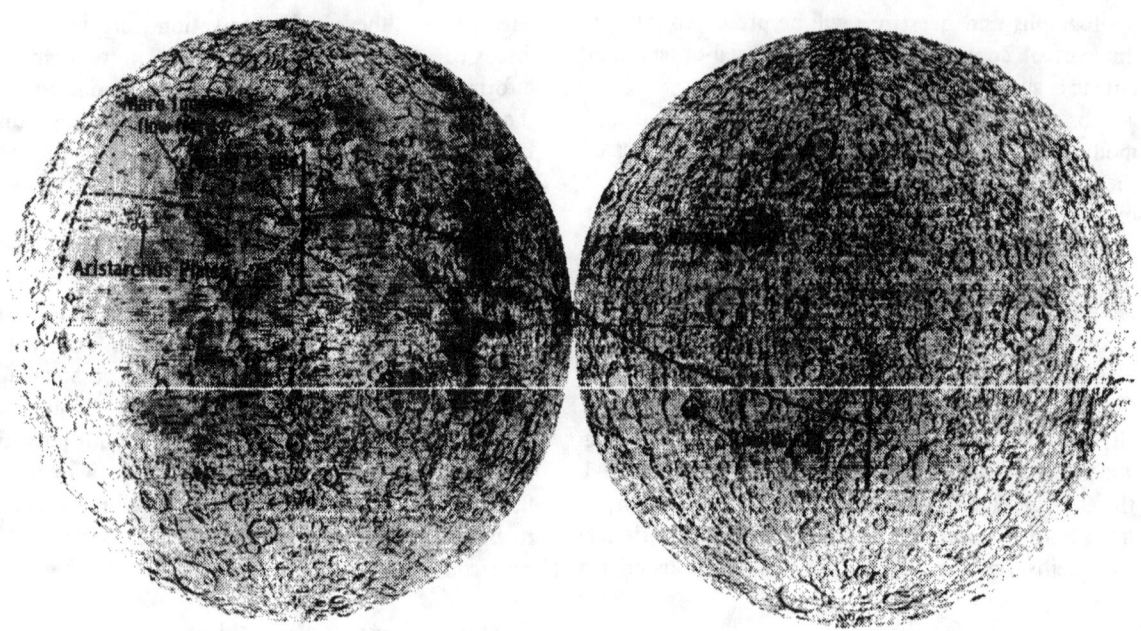

FIGURE 25-1.—Lunar groundtracks of the Apollo 15 mission on the lunar far side (right) and near side (left). The solid line represents the groundtrack of the first lunar revolution, and the dashed line represents the groundtrack of the last (75th) revolution. The terminators (nearly vertical lines) bound the sunlit portions of the lunar surface to the farthest extent of visibility from the command module windows; visibility extends approximately 20° north and south of the subspacecraft point from a 110-km altitude.

Tsiolkovsky, which has an area of approximately 80 km² was interpreted during the flight as a landslide or rock glacier.

LR33/05:22:26:28 GET: I look at it every time I go by, and there's just no question in my mind at all that it is a rock avalanche. It does have some interesting qualities about it, though. And it's a little bit hard for me to decipher right now, but it seems like the density of crater impacts in that slide is greater than in the surrounding terrain, even though the slide had to be emplaced on top of the surrounding terrain. Maybe it's just that the craters are fresher looking in that particular material, but there's no question about the lineaments being parallel to the direction of the travel of the flow; all the characteristics that I've seen of a rock avalanche.

LR34/05:23:37:30 GET: On the west side, the rim is a very large, clean scarp; and, when I say clean, it goes almost from the basin floor to the rim crest in one large chunk. That scarp appears to define the limits of a couple of fault zones that go through the rim of Tsiolkovsky. I couldn't trace the fault zones beyond Tsiolkovsky very well from the vantage point I had. But they're very distinct in the wall itself. And one fault zone coincides with or occurs in the same location as the southernmost edge of what appears to be a rock glacier extending northwest into Fermi. Now, that rock glacier has all the flow banding and the lobing toes characteristic of what we consider a rockslide. However, one feature about that slide that I mentioned before is that it has what looks like fairly fresh craters; in other words, a higher density of craters than on the surrounding floor of Fermi, although the Fermi floor looks much older; it's much smoother and more like the Cayley Formation. [The Cayley Formation is a highland basin fill on the near side of the Moon.] Looking more to the south, I see no evidence of another rockslide. The pictures might indicate or might hint at some kind of a rockslide there, but it appears that it's more like ejecta, now. The picture doesn't clearly show the ejecta from Tsiolkovsky, but the ejecta pattern and the flow lines are at least observable around most of Tsiolkovsky, and the ones that we see on the south and west side of Tsiolkovsky seem to be more ejecta than anything else. I couldn't see any distinct unit there that could have been a flow, such as the one in the northwest corner. And it appears that what lineaments there are in that particular part of the rim are merely ejecta patterns. Looking into Waterman, there is a small flow that goes into Waterman, but it doesn't come from Tsiolkovsky itself; and I couldn't locate the source of the flow, but it seemed to have come down the side of Waterman wall and out into the crater. I do have some pictures of it; maybe we can tell from the pictures. But as you're looking at the picture in the landmark book [fig. 25-2], it looks like there's maybe a breach in the wall of Tsiolkovsky, between Tsiolkovsky and

V-1A TSIOLKOVSKY REGION (128.5°E, 20°S)

Describe pertinent details relative to:
1. Structures and possible layering on the central peaks of Tsiolkovsky.
2. Nature of light-colored floor material and relationship to surrounding units.
3. Variations in texture and structure along segments of the wall of Tsiolkovsky.
4. Rim deposits due south of the crater and possible volcanic fill of the crater Waterman.
5. Origin and inter-relationship of crater pair due north of Tsiolkovsky.

FIGURE 25-2.—Visual-observation target V-1A, Tsiolkovsky Crater region.

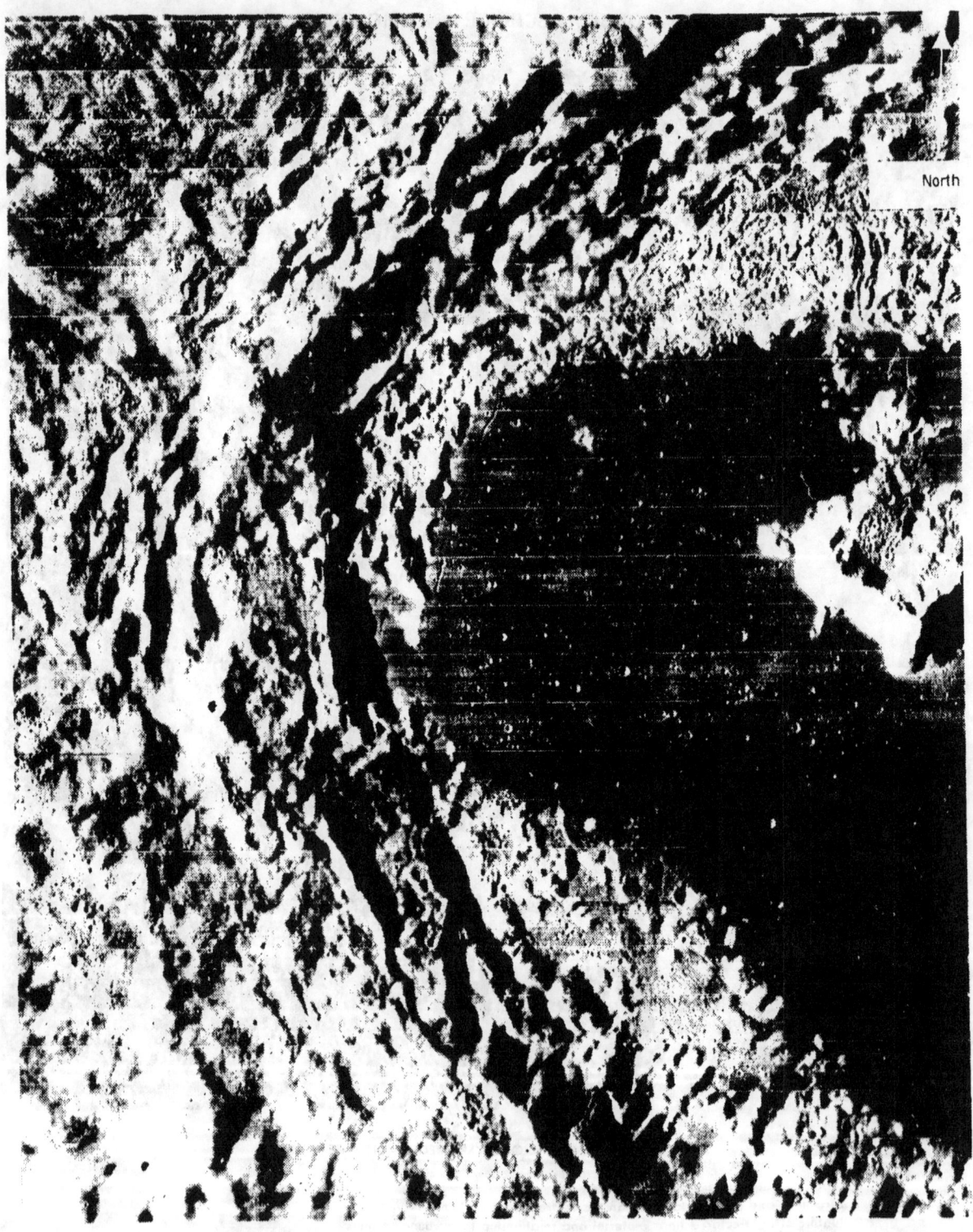

V-1B NW RIM OF TSIOLKOVSKY (126.5°E, 20.5°S)

Determine the nature of the flow units on the northwest rim of Tsiolkovsky in terms of:
1. Structural control.
2. Relationship to the crater Fermi.

Waterman, possibly allowing some flow into Waterman. Well, visually, that particular breach in the wall doesn't show up. The terrain there is much more level than it would appear in the photo. The flow does come from the direction on the north side of Waterman. My impression would be that the flow came down the side of Waterman, possibly out of some fracture or fault related to Tsiolkovsky, but outside the basic rim itself.

Item 5 in figure 25-2 is related to a crater pair, with intricate relationships, due north of Tsiolkovsky Crater. Relative ages were confirmed by visual observations and by scientifically interesting descriptions.

LR34/05:03:42:51 GET: One other comment on the crater pair just north of Tsiolkovsky. The smaller crater on the east side apparently was the original crater, with an impact occurring alongside it. That appears to have induced mass wasting or some kind of a rock avalanche into the smaller of the two craters. The larger crater to the west has a fairly intact rim, being faulted into a couple of places where it crosses the rim of another crater to the west. But that rim is fairly intact. The rim that was apparently moved or obliterated by the most recent impact was the rim of the smaller crater, and that's where all the rock debris has fallen.

During the postmission visual-observation debriefing, one important aspect of Tsiolkovsky Crater was stressed; namely, the apparent color of the crater floor. "The floor is a gun-metal-gray color. It's about the color of a metal cabinet. Everything was that same gray color; so the only apparent aspects were the differences in texture. In fact, Tsiolkovsky is lighter than the other mare areas. It's a gun-metal-gray color."

Orbital photography of Tsiolkovsky Crater, obtained both by the unmanned Lunar Orbiters and by the Apollo crews, shows the floor of Tsiolkovsky to be extremely dark. Therefore, this mare unit was thought to be one of the youngest on the Moon. However, the visual impressions indicate that the mare material on the floor of Tsiolkovsky is no darker than other mare regions of Eratosthenian age. A postmission crater-density comparison between the floor of Tsiolkovsky and Eratosthenian mare units in Oceanus Procellarum indicates a similarity in age.

The reason for the apparent darkness of the floor material in Tsiolkovsky Crater is probably that photographic exposures depend on average scene brightness. Because Tsiolkovsky is surrounded by very bright highlands, photographs taken of the crater and its environs will expose mostly for the surrounding highlands and, therefore, underexpose the mare fill in the floor. The human eye is at an advantage, however. It responds to the scene as a whole and makes the relative brightness levels of each and every unit distinctly separable. This is one illustration of the capabilities of the human eye in making observations from lunar orbit.

Also discussed were what appeared to be lava flows on the northeast rim of Tsiolkovsky Crater. These flows are reminiscent of the flows to the south within Waterman Crater (fig. 25-2). The flow on the northeastern rim (fig. 25-4) appeared to be "really viscous; the lava was piled up! I could see down into some of the craters. The lava was piled up on one side, and a cross section of the area where the lava had been piled up along the rim of the crater would show this. When the Sun angle was low, I could see a shadow all the way around. The unit appeared to have flowed down into the crater, up the far side, and then curled back on itself. There was a great big lobe of lava at the end."

The differences between the aforementioned flow and the mare material that fills the Tsiolkovsky Crater floor were obvious. The flow on the northeastern corner (fig. 25-4), as well as the flow within Waterman Crater, originated at a point higher than the Tsiolkovsky floor. These flows very distinctly have come down the side from the top of ridges on the Tsiolkovsky rim. These flows probably originated from the extrusion of viscous lava, whereas the mare-type flows filled the crater floor.

This type of flow unit is not restricted to the immediate vicinity of Tsiolkovsky Crater. Several of these units were identified in other craters; one example is the floor fill in Litke Crater, approximately 100 km northwest of Tsiolkovsky (fig. 25-2). In this area, it

looks like some of the flows came straight up, almost like a gusher, and then went up and stopped. In some areas, it looks almost like a beret turned upside down and stuck in the bottom of the crater. They are very round, and they appear to be isolated units. They're very flat, too, but they must have a high flow front. They must be fairly high, because there is a very, very distinct rim around the inside. So there was a hump or a topographic prominence at the terminus of the flow. Yes, they're almost curled over on themselves. I had the distinct impression that this was what happened to some of them; except, instead of being right out on a level, they just stopped right along the crater wall. That's why all the material was piled up against the crater wall, and it appeared

FIGURE 25-3.—Visual-observation target V-IB, northwest rim of Tsiolkovsky Crater.

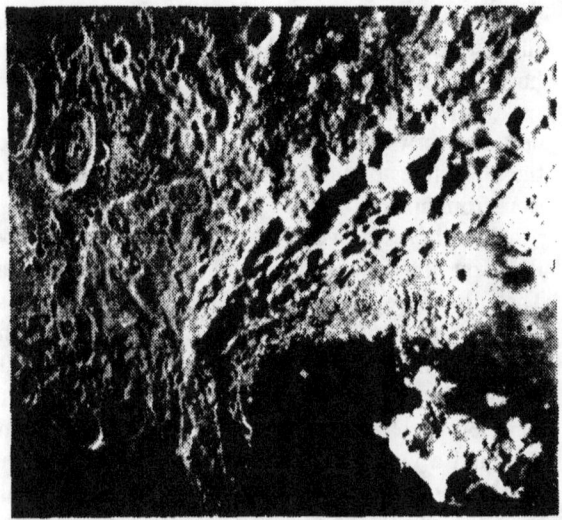

FIGURE 25-4.—Metric camera photograph of part of Tsiolkovsky Crater and environs. This forward oblique view was obtained during a photographic pass on which the spacecraft was pointed 25° off the vertical. The craters in the upper left portion are Luetke and Delporte. The lineated unit that starts at the northwestern rim of Tsiolkovsky and extends onto the floor of the large crater Fermi is noteworthy.

FIGURE 25-5.—Detail of the terminal portion of a lineated landslide or rock glacier on the northwest rim of Tsiolkovsky Crater (fig. 25-4). The relatively high population of fresh-appearing small craters on the lineated unit at the right as compared with the older floor materials of Fermi Crater at left (AS15-94-12812).

as if it had curled back down underneath itself, but still inside the crater. It looks like something that just has been dropped, as a whole unit, right there—something very sticky and viscous that appears to have draped itself over the crater floor.

The mention during the flight of the higher frequency of craters on the rock glacier or landslide on the northwestern rim of Tsiolkovsky Crater was the one item related to this visual-observation target that could best be checked by studying the returned photographs. A photograph of the terminus of the landslide where the lineaments are radial to Tsiolkovsky and parallel to the flow direction is shown in figure 25-5, which clearly shows that more sharp and small craters exist on the relatively younger floor unit than on the relatively older floor fill of the Fermi Crater basin to the west.

The discrepancy in crater density is restricted only to small craters. The excessive population of these craters on the younger of the two units may be due to one of the following reasons.

(1) The presence of drainage craters, which may have developed by the seismic shaking of the surface and by the drainage of the material in the void spaces initially sealed over by the flow layers

(2) The absence of a thick regolith on the rock glacier because of the relative youth. (Small impacts would tend to appear fresher and to "live longer" on such a unit, whereas the floor of Fermi Crater exhibits the characteristics of a very thick regolith.)

Swirls in Mare Marginis

Unusual, light-colored surface markings were observed in Mare Marginis (refs. 25-2 and 25-3). Identical markings were distinguished in Mare Ingenii on the lunar far side and were associated with those in Mare Marginis on the eastern limb of the Moon (ref. 25-4). Although not scheduled in the flight plan (table 25-1), visual observations of the two regions were made during the Apollo 15 mission.

LR1/03:07:13:57 GET: One of the interesting things we've noted is the variation in albedo from white to dark gray, with many variations of gray in between. Many times, this albedo change appears without any significant change in topography. There are many variations in the albedo all over the surface. I guess our general consensus is, it's gray. We haven't noticed any brown yet. Another interesting fact that we've all noticed is that it looks like a great desert across

FIGURE 25-6.—Sinuous, light-colored markings in the area north of Mare Marginis on the eastern limb of the Moon.

which we've had a number of duststorms. And, in many places, you can see the tracks or the swirls across the surface, which looks like a great duststorm has been blowing across the surface, primarily indicated by the albedo change.

It was confirmed, therefore, that the swirls in Mare Marginis (fig. 25-6) have no associated topographic expressions as is also the case with Mare Ingenii (fig. 25-7).

LR2/03:09:15:41 GET: I took a look at the light-colored swirls in the bottom of the mare. No elevation is associated with those light-colored swirls, and they're very distinct when you look at them at this angle.

The two observations were made of similar features under different Sun angles and viewing directions. Mare Ingenii was observed with a Sun angle of approximately 15°, looking obliquely south of the spacecraft groundtrack; Mare Marginis was observed with a Sun angle of approximately 85°, looking forward and north of the spacecraft groundtrack. In both cases, however, they were "very, very distinct, both with no topographic expression that I could see. But they are just as plain as they are in the picture."

The search for topographic expressions was planned and conducted to test a theory of the origin of these features (ref. 25-4). Positive topographic expressions would indicate that these markings have either an exogenetic origin (such as ejecta or irregular ray material of some sort, or even as cometary tail deposits) or an endogenetic origin (such as extrusive volcanic materials). A complete lack of topographic expression would indicate an endogenetic source without the actual extrusion of materials.

The positioning of these two regions on the Moon is striking: the swirls of Mare Marginis lie diametrically opposed to the center of Mare Orientale, and the swirls of Mare Ingenii lie on the opposite side of the

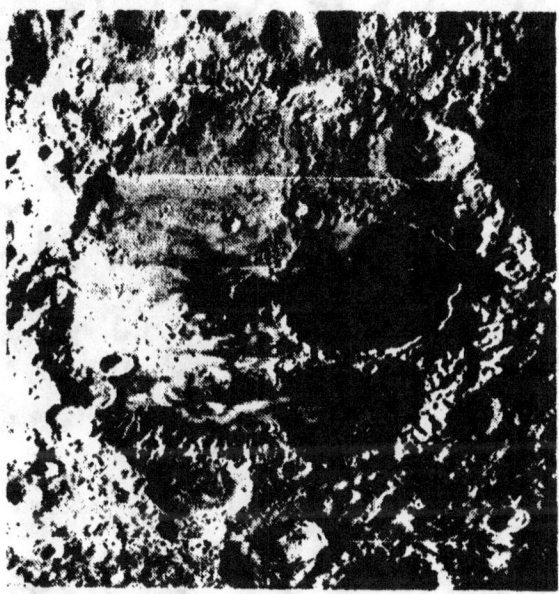

FIGURE 25-7.—Swirls of light-colored markings in the dark basin fill of Mare Ingenii, which is centered at approximately latitude 33° S, longitude 164° E.

center of Mare Imbrium. This observation suggests that these markings are perhaps due to chemical alteration of the surface material; the alteration may have been caused by the release of gases from the lunar interior. The release of gases, in turn, may have been triggered by seismic-wave attenuations at the antipodal areas of the impacts that created Mare Imbrium and Mare Orientale. This subject will be treated in greater detail at a later time.

Picard Crater Region

Picard Crater is located in the western part of Mare

Crisium. The interest in the crater itself, in the surrounding mare material, and in associated features is evident in figure 25-8. Following is a selection of the comments made during the flight.

LR26/05:08:03:45 GET: I'm just coming over Picard at the present time and wanted to make a comment that it looks like there are several ring structures inside the basin itself. They're all concentric, and I don't see a great deal of relief on those that look like they're in the bottom of the basin. But, looking at the scarps around the outer ring, Picard looks like it's just almost a caldera type. They look almost like fault plains along the outside. And I can see in the outer wall very distinct layering. For instance, right on the top is a

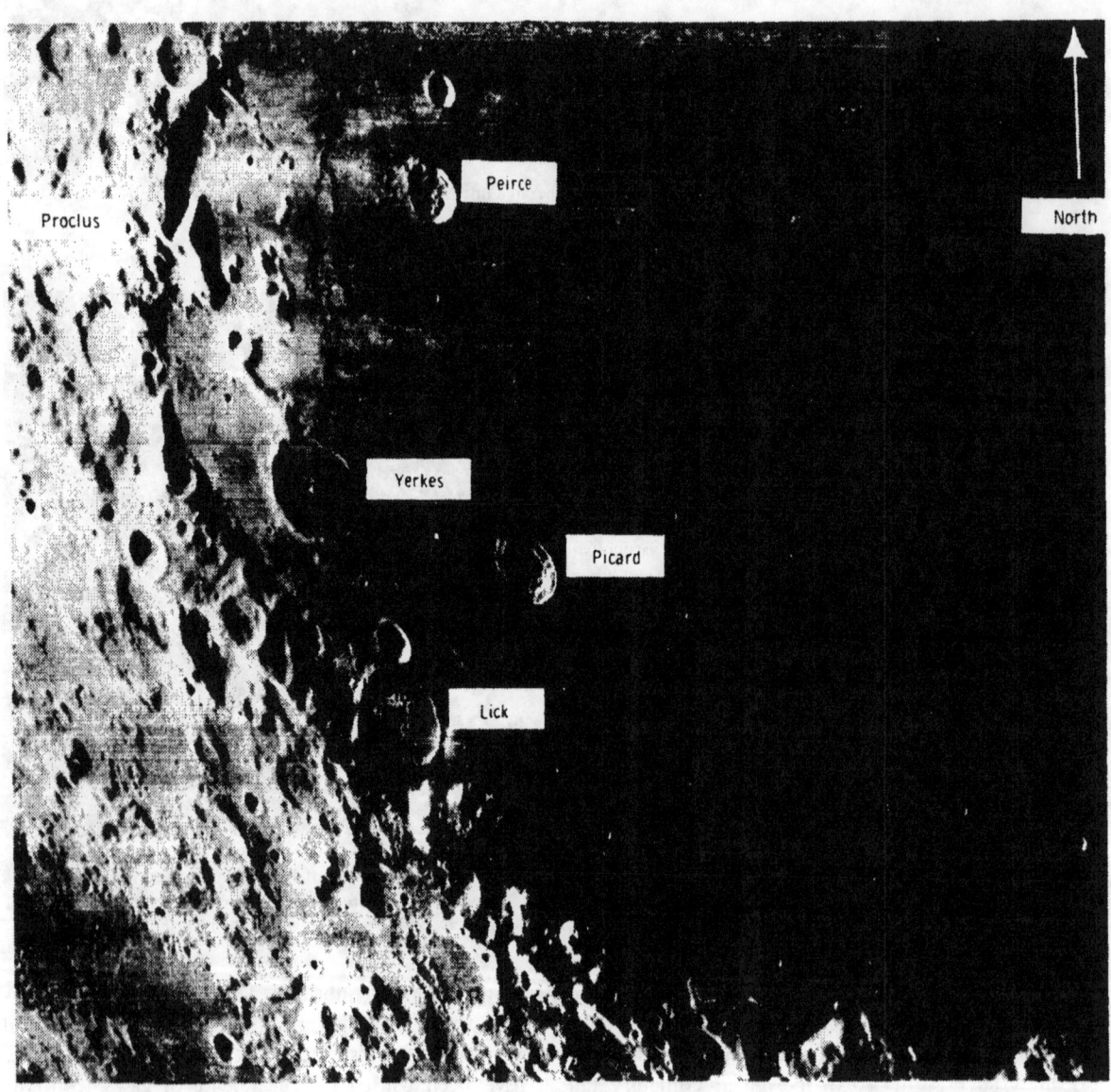

V-2 PICARD REGION (54.8°E, 14.7°N)

1. Describe subtle color-tones within western Mare Crisium, if any
2. Is there a difference in the color of the crater Picard and the surrounding mare material?
3. Describe the similarities and/or differences between the craters Picard, Lick, Yerkes and Peirce; compare to the multi-ringed craters of Mare Smythii.

FIGURE 25-8.—Visual-observation target V-2, Picard Crater region.

very thin dark layer that runs all the way around. And there's a light-colored layer. And then there are alternating dark and light layers all within about the same distance from the top of the crater, all the way around.

LR34/05:23:49:24 GET: Endeavour is coming up over Picard. Talking about the color variations, Picard is a slightly different color than the rest of the mare basin. I would consider Crisium to be a light brownish gray. Picard itself is more of a brown tone and has a darker halo around it. I can see some of the brown material just on the outside of the rim, and outside of that is some darker material that gradually turns into the gray of Mare Crisium. Inside Picard, I can see six distinct rings that go around the inside of Picard. And the walls of Picard are very shallow. It looks like a very shallow dishlike basin. And I can see some definite layering, particularly in the upper boundary of the rim.

The freshness of detail in Picard Crater and in its layering (fig. 25-9) was compared to the situation in other nearby craters.

LR36/06:01:48:01 GET: While we're going over Crisium here, looking down at Picard, at Peirce, and at Lick Delta, they all look alike. They all have the same ring structure; all have the same low rims. The rims look very shallow compared to the rims on the other craters I've seen around. Also, they all have a slightly darker halo effect around the entire crater. But the color difference is very subtle.

This description contrasts with the other large craters in the same region, namely Lick and Yerkes, which display different characteristics.

LR34/05:23:49:24 GET: Now, Lick looks like it's almost completely obscured. It looks very much like a collapse. All I can see is a little bit of a ring, a color variation, with some positive relief. And then, inside the crater looks very much like outside it, as far as the color and the texture are concerned. However, it does appear to slope gently in towards the center. Lick looks to me like a very large collapse feature, with the same kind of material both inside and outside the basin. And I would make the same comment about Yerkes.

Proclus Crater

The classical example of ray-excluded zones around impact craters is that of Proclus Crater on the western rim of Mare Crisium (ref. 25-5). Rays from Proclus extend in all directions except for a segment on the west (fig. 25-10). The rays could be traced to the east over much of the surface of Mare Crisium. Following are descriptions of the regional and local settings made by the Apollo 15 crew.

LR1/03:07:18:37 GET: The rays extending from Proclus are very light in color for about 240° to 260° around, and then there's a region of dark albedo. And our orientation presently with the spacecraft is such that we're having a tough time figuring out north and south. But the inner walls of Proclus are very light in color, almost white. The outer ring has a somewhat light-gray appearance, and the difference in the rays is really between a light and a dark gray, as distinguished from the inner walls, which are quite white.

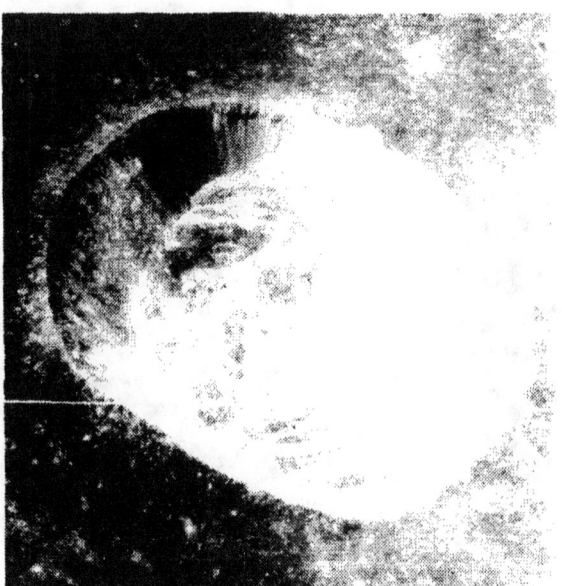

FIGURE 25-9.—Panoramic camera view of Dawes Crater showing numerous layers along the crater wall, especially in the upper portions. Layers on craters walls were also observed in Picard and Peirce Craters in Mare Crisium (fig. 25-8).

The walls exhibit some debris on the upper slopes, maybe the upper 30 percent. I can see, on one side of the crater, some large blocks. On another side, I can see what appears to be a large slump block or a large slumping of the wall that goes about halfway down and takes approximately 15° of the rim of the crater with it. The floor is very irregular and rough, almost constant medium-gray color, somewhat darker than the light-gray rays and somewhat lighter than the dark gray on the surface, which does not seem to be covered with a ray pattern. There are a few ridges on the floor, arcuate ridges, and some domes which are quite prominent.

LR2/03:09:22:54 GET: Proclus, coming up from the east, is really spectacular. You can very distinctly see the difference in color of the albedo in the excluded zone of Proclus; and, as you're coming up across Crisium with Proclus ahead, you can see the ray pattern very distinctly, extending out across Crisium, and follow the ray patterns almost as far as you like. And the excluded zone in the ray pattern is just very distinct at this point. And, from this angle looking at Proclus, about a crater diameter out to maybe a diameter and a half or so, you can see many small, bright, fresh craters, which appear to be in the general direction of a ray, like part of the ejecta blanket. They occur within a diameter to a diameter and a half of Proclus, and they're about the same brightness as the inner walls of Proclus, and they're small, just small craters. I do see one which you might call a loop, which would suggest secondaries. They just seem to lie in the general direction of the rays of the ejecta from Proclus.

During the debriefing, the CMP commented,

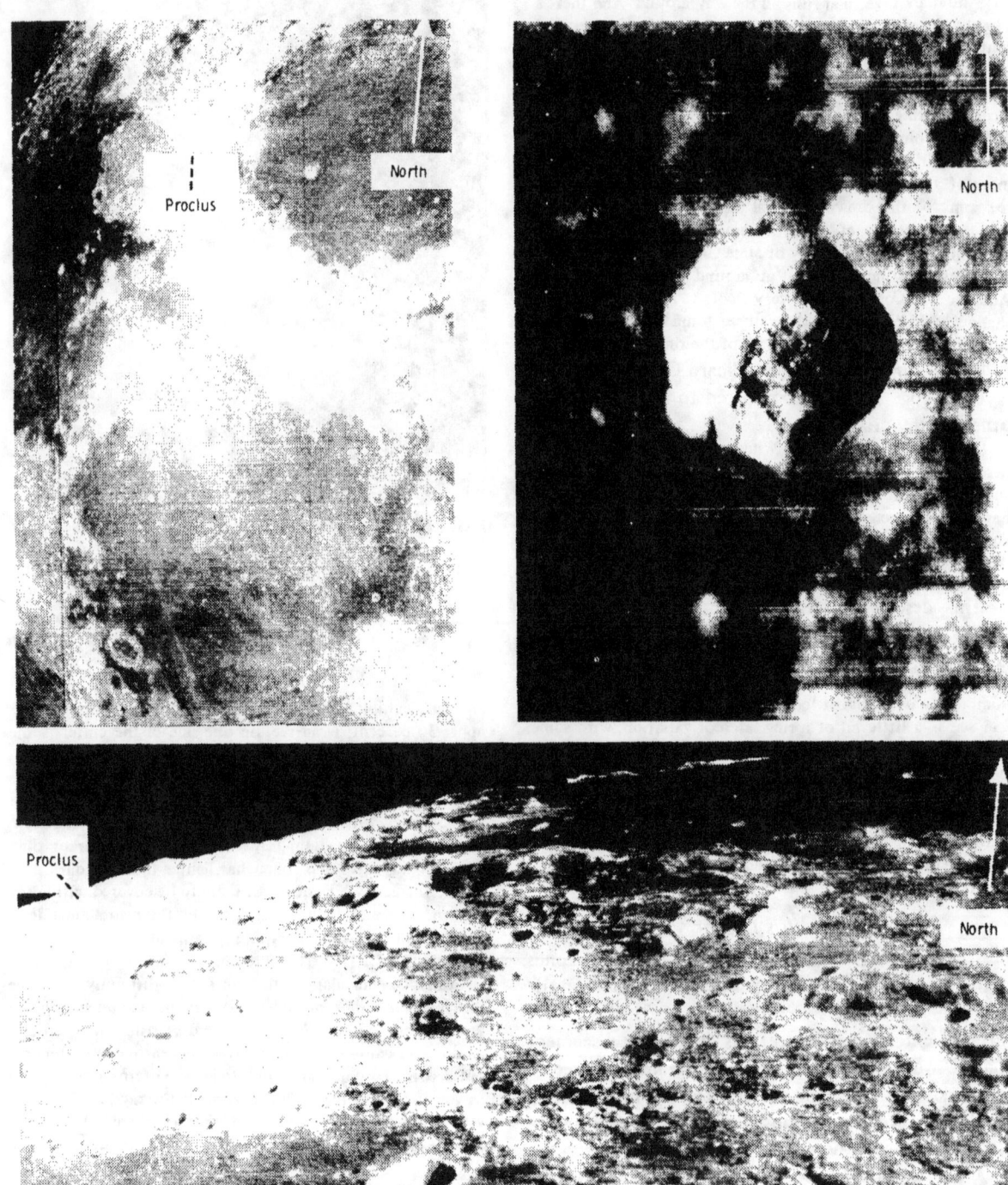

V-3 PROCLUS (47° E, 16.4° N)

Observe the crater Proclus and its environs both vertically and obliquely to determine:
1. What are the major differences between the surrounding units?
2. Does the crater Proclus straddle two different surface units?
3. Are there any topographic features which would indicate that shadowing was responsible for ray-asymmetry?

FIGURE 25-10.—Visual-observation target V-3, Proclus Crater.

It's very strange the way the ejecta from, particularly, Proclus crosses Crisium. It's almost like flying above a haze layer and looking down through the haze layer at the surface. Ejecta from that crater doesn't look like it's resting on the top of Crisium. It looks like it's suspended over it. It gives a very filmy, very gauzy appearance to the whole thing. It must be very thin. And I guess the reason it looks like it's just draped or suspended over it is that almost any way you look at it, if it goes through a crater or it goes through a wrinkle ridge or it goes through any topographic feature, it doesn't make any difference from what angle you view it; those lines of the ejecta pattern are straight. When the ray goes through a topographic prominence or a negative depression of some sort, you still see the ray through that.

The areas where you don't see the ray seem to be independent of topographic features. There would be some areas where you wouldn't see any rays from Proclus. I guess the reason for that is that they might have been covered up by material from another crater that was a different color. So you get rather vague, obscure patterns on the ground where the ray patterns cross, and you don't see the ray from Proclus. But it didn't seem to be tied into topographic features. It seemed to be more a function of just overlapping rays or ray patterns built on top of each other. But that's about all you can see in Crisium. It's like looking at a star field and trying to pick out some familiar stars. At first, all you see are stars; then, you have to orient yourself by picking out a star that you know; and then you pick out the other stars around it. This is almost the same thing. When you first look at Crisium, all you see is the ray pattern. You don't see any features at all; you just see the ray pattern. And then you start picking out some craters or some features that you know. From there, of course, you can see through the ray pattern, and it doesn't distract you anymore.

The visual observation of Proclus was planned to determine the probable origin or cause of the ray-excluded zone. These zones occur around a number of lunar impact craters (ref. 25-5), and the cause for such zones may be one or more of the following.

(1) The obliquity of the projectile would cause the ejecta to be distributed all around the crater except for a segment directly below the path of the projectile. In most of these cases, a "rooster tail" pattern develops by the ejection of material in a direction opposite to that of the impact (fig. 25-11).

(2) Topographic "shadowing" by a positive prominence would shield a segment around a crater against deposition of ejected ray material.

(3) Where differences exist in the materials in which the crater is situated, the two types of materials may respond differently to the impact pressure or appear different after being ejected from the impact site.

Observations of Proclus from lunar orbit suggest a fourth class of reasons for ray exclusion; namely,

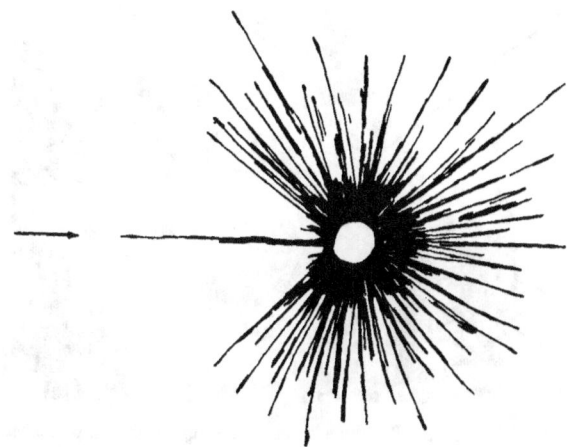

FIGURE 25-11.—Schematic illustration of ejecta distribution around impact craters produced by low-angle projectiles. A continuous ejecta blanket extends about one crater diameter from the rim crest, beyond which rays extend for a greater distance. A ray-excluded zone develops in the direction of the impacting projectile (arrow), where one thin ray develops in the form of a "rooster tail" pattern.

structural control (fig. 25-12). In the case of Proclus, a fault zone appears to have predated the crater. When the impact occurred, the fault plane formed part of the wall of the crater, and a broken-off segment of material may have been uplifted during the impact to inhibit the rays beyond it, later collapsing westward from the fault plane. This interpretation is based on the following comments.

LR26/04:08:07:19 GET: I'm up over Proclus now, one of the visual targets. And a couple of comments about Proclus, which weren't too obvious from the pictures we have seen before. The edge or segment of the crater, which is in the excluded zone of the ray pattern (the ray pattern, by the way, is very distinct even from directly overhead), that little segment of the crater wall seems to be discontinuous with the rest of the crater. In other words, the crater, if you made a circular ring and you showed that as the crater, then this little chunk in that quarter where the excluded zone is, lies outside of what you would describe as a circle for the crater itself. It's like a little dimple in the crater itself. And I can't see anything in particular there close to the rim that would account for any physical shadowing of the ray pattern. But I can see a diagonal fault zone that runs down into that little dimple that I just described a minute ago, and runs into that dimple from the east side. I couldn't pick one out on the west side, but it's very distinct on the east side. And, in addition to that, I didn't see a great deal of difference in the terrain or in the structure of the terrain across the excluded zone.

LR34/05:23:52:18 GET: I'm looking at Proclus now.

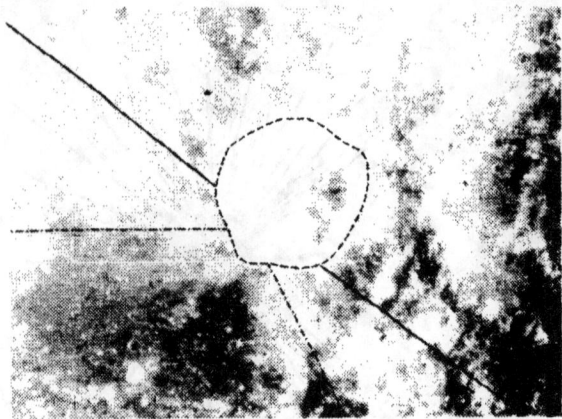

FIGURE 25-12.—Proclus Crater. The circular line emphasizes the shape of the rim crest; the dashed lines enclose the ray-excluded zone southwest of the crater; and the solid lines depict a probable fault the plane of which forms the western part of the crater wall. As discussed in the text, ray exclusion in this case appears to have been due to the structural setting of the impact site (met AS15-0959).

And remember, yesterday we were talking about variations in the crater wall? Well, I don't exactly know how to describe it, but there is a tremendous variation in the wall, which does line up with the ejecta pattern. There's almost a straight wall on the side of Proclus that shows a ray-excluded zone. And then there is some breakthrough directly in the middle of that wall, which makes Proclus look like it's almost a circular crater. However, the truth is that Proclus looks like an elongate crater with one wall dipping quite steeply into the crater. And that wall is oriented perpendicular to a line bisecting the excluded zone, dipping into the crater. And then right in the middle of that portion, it looks like a small piece of that wall was also ejected, but it was only at the top part of the fault scarp. And so, if you look at it from the right angle, you can see almost a flat plate, which looks like it's cut right into Proclus; and to the north and west of that flat plane is the Crater Proclus, and to the south and east is a small chunk out of the top of it that coincides with the central part of the excluded zone.

During the debriefing, the CMP commented, "I had the distinct impression of an inclined fault zone that ran down into Proclus, dipping towards the crater. To the west of the fault zone, a portion of the rim looks like a half cone that has been blasted out in the direction of the ray-excluded zone—ray shadowing by structural control."

Littrow Crater Area

The Apollo 15 crew was asked to describe the geologic setting of the Littrow Crater area (fig. 25-13).

LR1/03:07:23:11 GET: We're coming up to Serenitatis, and it really looks like an ocean. The landforms, as we approach, are very rugged, very highly cratered, rounded, and we get to the shoreline and we see a few wrinkle ridges that have been smoothed out. And we can see on the far side of the horizon the mountains which pick up again on the western side of Serenitatis. We're coming up over Serenitatis now. We're almost over le Monnier at the present time, and we can see the Littrow area just out in front of us. And it is, in fact, about three different shades (fig. 25-14). You can see in the upland area and particularly what looks like down in the valleys a darker color, and it does look like it's a light powdering or dusting over the entire area. And then, as you get out further into Mare Serenitatis, there's another layering which is a little bit lighter in color. And then, out at the last edge of the wrinkle ridge, out beyond that is the last layer, and the rest of Serenitatis looks fairly light in color. So I'd say that the central part of Serenitatis is light, out beyond the first wrinkle ridge is a darker layering, and we're not up close enough to see what it is yet; then, as you get up into the highlands around le Monnier and Littrow area itself, there's what appears to be a light dusting of dark material, and it certainly looks volcanic from here. Off to the left of that, to the west, we can pick up Sulpicius Gallus pretty clearly right now. Looking down into the Sulpicius Gallus area, looking at some of the wrinkle ridges and some of the rilles—the arcuate rilles down there—I can make out some distinct color patterns that seem to run parallel to the arcuate rilles and along the wrinkle ridges. And there is a very subtle darker color, again almost as if it was some kind of cinder fallout along the ridges and along some of the rilles.

During the television transmission from lunar orbit, some of the local setting was also described.

LR9/03:22:59:14 GET: You can see we are up over one of the Littrow rilles now, and you can see some of the wrinkle ridges. The rilles are graben-type arcuate rilles, and the ridges, in places, look like they could be nothing more than a flow that stopped there, a flow front. In other places, they look like buckled material from underneath, folded to give it some elevation.

LR23/05:02:18:17 GET: Okay. I'm directly over Littrow at the present time. And I can see all the way around to the Apennine Front, encompassing all of Serenitatis between here and there except to the north over by Posidonius. So I got a very good view of Sulpicius Gallus. And the observation I wanted to make, in particular, pertains to the distinct way that the rilles do follow the old mare basin. And the fact that the second color band that we discussed in Littrow seems to be continuous right on across the basin into Tranquillitatis and on around—almost a shelf, a continental shelf appearance—into the Sulpicius Gallus area. There is a darker coloring in the uplands in Littrow and closer to the front or closer to the basin scarp. But the second band seems to go all the way around Sulpicius Gallus. And then, as you follow Sulpicius Gallus on around a little bit more to the west, that color banding is still there. I can see a distinct boundary between it and the Serenitatis basin itself, the inner basin, but it turned into a little more brownish color from the gray color that we saw before.

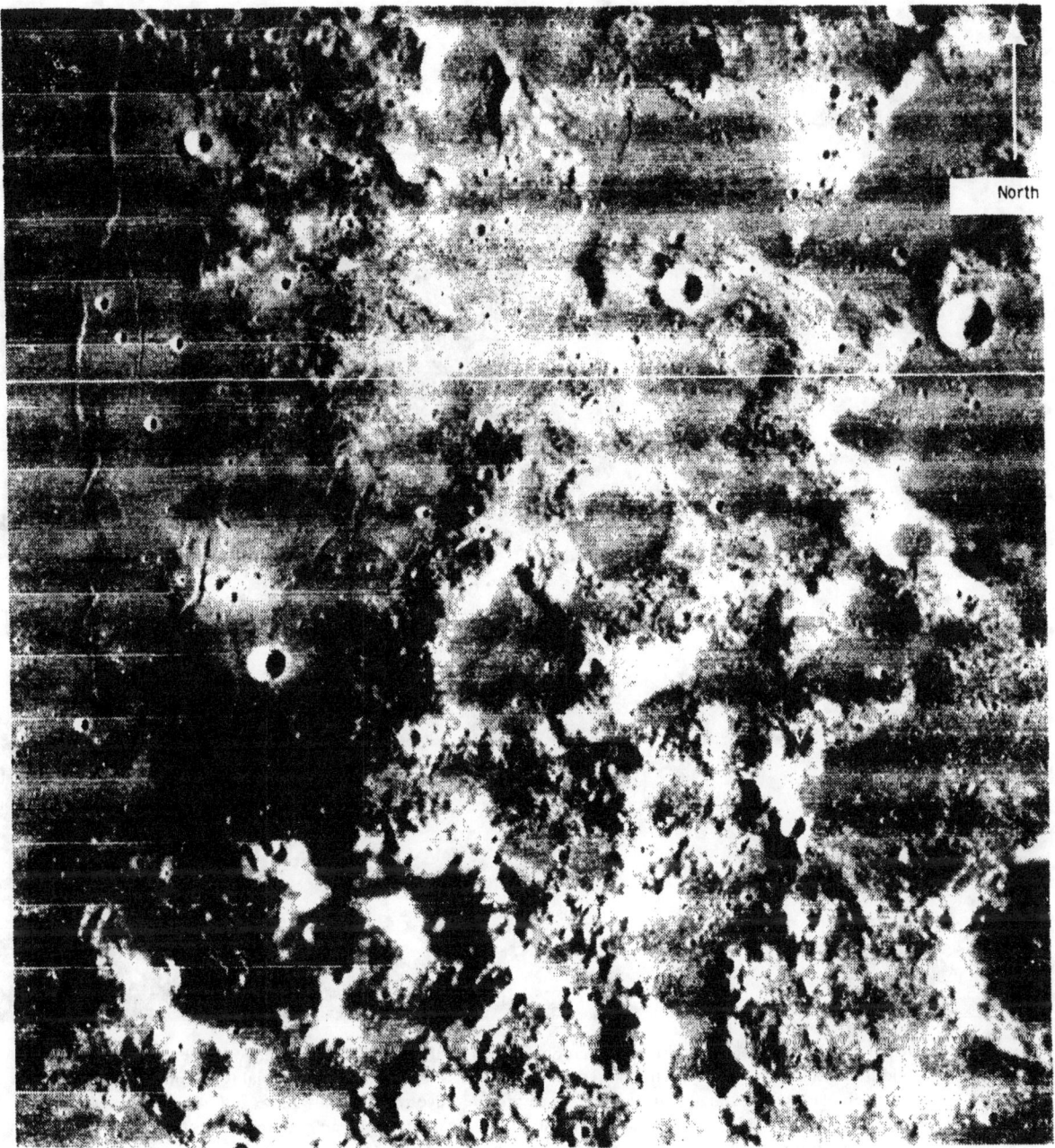

V-5 LITTROW AREA (31.3°E, 21.8°N)

Use photograph to locate and describe the following:
1. The highland units on the eastern rim of Mare Serenitatis.
2. Three mare units which may be distinguished by relative albedo.
3. Sources of the darkest mare unit which mantles older highland and mare materials.
4. Floor fill and wall structure of Rimae Littrow.

FIGURE 25-13.—Visual-observation target V-5, Littrow Crater area.

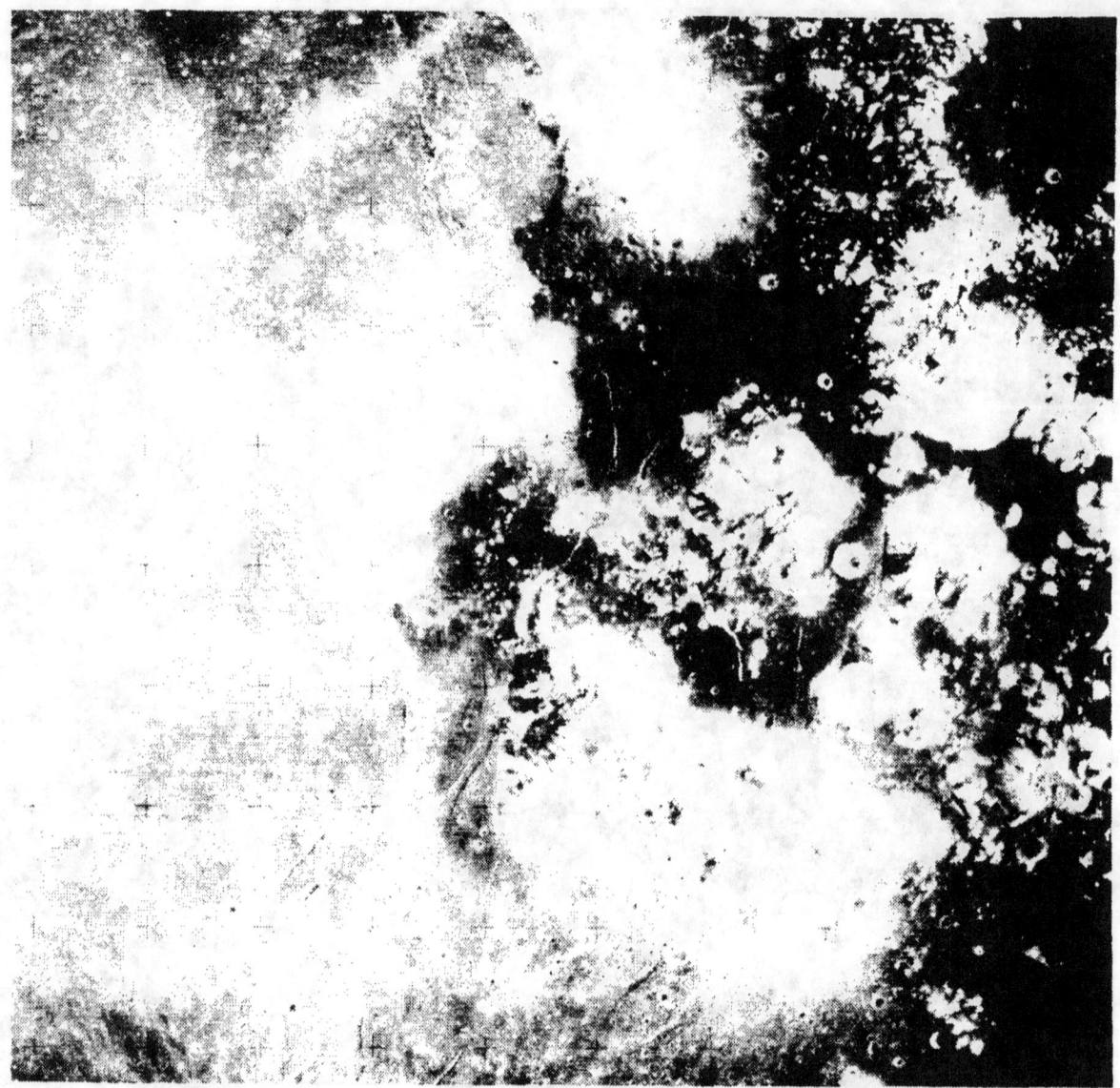

FIGURE 25-14.—Southeastern portion of Mare Serenitatis. To the left is a representative segment of the middle portion of relatively light-colored mare material. Surrounding this material is an annulus of relatively dark-colored mare, which includes an arcuate ridge system. The valleys and low areas between the Taurus Mountains on the southeast corner are occupied by an extremely dark unit with smoother textural characteristics than those of the mare materials (met AS15-1113).

LR25/05:06:17:03 GET: If I had to give you the opinion right now, I'd say the dark area in Littrow was all some kind of ash. I'm not sure it's flow. But it certainly looks like a deposit over the entire surface. You can see it mostly in the upland areas, some in the mare areas, but mostly in valleys and in depressions. This stuff seems to have collected almost like there was some mass wasting down the hills, making the valleys darker in color and maybe a little thicker with the kind of material. But there are still at least three different distinctive colorations in the Littrow area, going from dark gray to a sort of brownish color. And it was the dark gray that looked like it was an ash fall to me.

During the following revolution came the most important single observation of the entire flight.

LR26/05:08:12:46 GET: Okay. I'm looking right down on Littrow now, and a very interesting thing. I see the whole area around Littrow, particularly in the area of Littrow where we've noticed the darker deposits; there is a whole series of small, almost irregular-shaped cones, and they have a very distinct dark mantling just around those cones. It looks like a whole field of small cinder cones down there. And I say cinder cones because they're somewhat irregular in shape. They're not all round, they're positive features, and they have a very dark halo, which is mostly symmetric, but not always, around them individually.

LR27/05:10:12:14 GET: Just going over the Littrow area again. I described what, at least some, looked like fumarolic vents. They look like small cinder cones to me. And every time I look at them, that firms up my impression more and more that they're volcanic cinder cones.

Later during the mission, asked whether the cinder cones were fairly evenly distributed or whether they concentrated in spots on the darkest unit, the CMP responded as follows.

LR37/06:05:58:59 GET: They're concentrated in spots on the darkest unit, and they seem to be concentrated in localized areas. Also, within the darker units, there'd be a relatively high density of these small cones, and then a few scattered ones in the outlying areas. But I would say they were concentrated within the darker areas, more on the lowland side, in the valleys and in what looks like the lower areas. And within concentrations of cinder cones, there seems to be one locus of major activity, one locus of the greatest number of cones, and then they thin out beyond that.

Discovery of the cinder cones in this area (which was later confirmed by the panoramic camera photographs shown in fig. 25-15) and delineation of the dark deposit as an ash fall suggest late volcanism that postdates the major episodes of mare-material extrusion. The dark deposit is relatively younger than Eratosthenian-age mare materials; that is, younger than 3.2×10^9 yr. The unit and the source vents appear much younger than any formation in the surrounding mare material. The cinder cones, as observed during the flight,

were very sharp, very distinct, and quite small. And there were neat symmetrical things, too. They weren't broken down any way that I could see. They didn't look like they'd been warped much since they first got there. I didn't see any of them partially obscured by an impact. Some of the aprons around them are symmetrical, some unsymmetrical. Around the cones themselves, there's certainly a blanket that was laid down that is even darker than the surrounding material. The black spots are what drew my eye to it to begin with. The aprons are even darker than the material that covers the whole area here.

The geology of this area and the significance of the cinder cones are discussed in detail in part I of this section.

Landing-Site Area

To extrapolate the ground-truth data (mainly collected samples and observations and study of the surface features and their local geologic setting) to the regional geologic setting of the Apollo 15 Hadley-Apennine landing site, visual observations from lunar orbit were planned (figs. 25-16 and 25-17). Following are some comments that pertain to the questions raised before the flight.

LR1/03:07:29:00 GET: We're coming up here on the terminator, and the area I guess we called Crackled Hills really looks like crackled hills. If you distinguish between the mountains, which are very prominent and smooth, the surface between the first small mountain range and what is now the terminator is relatively flat with a very rough texture—very irregular, lower, crackled hills. Jim calls it a gun-metal gray, and that's a very good term, I think, for the color that we're seeing now. As we approach the terminator, of course, the relief stands out even more. The shadows are getting much longer, and the peaks of the mountains, as they're silhouetted against the Crackled Hills, seem to have a diffuse shadow at the top. The shadow, as it goes from the base of the mountain to its peak, is very sharp. And around the top of the mountain, it becomes more diffuse, not quite as sharp, and begins to blend in with the surface on which it's being cast. There's quite a bit of shadow now, but we have Aristillus and Autolycus very clearly. With a low Sun angle, the surface between those two large craters and the rim of Imbrium—the eastern rim of Imbrium—is very rough, quite a bit of debris, and it looks like it probably came out of the two craters. I believe we can see Hadley C, just barely on the shadows. Aristillus and Autolycus both have their eastern rims exposed to the sunlight, and we get a pretty good look at the elevation on the rim.

During the television transmission from lunar orbit, some relevant remarks were also made.

LR9/03:23:03:51 GET: And here we come up on the ridges on the west side of Serenitatis, just at the foothills of the Hadley-Apennines. As you can see, there is some relief as we look back to the south there. There is some very distinct relief in the shore of Serenitatis, with some wrinkle ridges that follow the contour. And some look like fairly distinct arcuate rilles that also follow the contour. I think that when we get up very close here, you can see in the field of view a lineament that looks like it might be some sort of a collapsed lava tube. And you can also see down in here that the mare material looks like it's pooling in the foothills of the mountains; and, in some places, you can see what appears to be a frozen shoreline, so to speak. At the landing site, you can see the edge of the rille there as we go beyond it. And you can see the blocky features inside the rille. And now we're out across the plain on the other side.

The type of detail that was required necessitated the use of an aid to the eye for better resolution. The monocular carried on the spacecraft has a 10X power and is not well suited for detailed study from orbit. However, some results were obtained by its planned use.

LR23/05:02:26:53 GET: I got the monocular out. I can vouch for the rocks and the blocks in the bedrock in the rille.

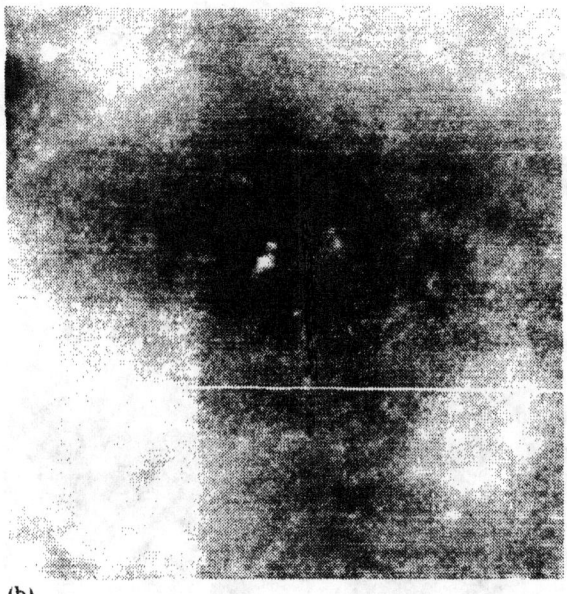

(b)

FIGURE 25-15.—View of southwestern Taurus Mountains in the Littrow Crater area. (a) Terra massif units surrounded by dark, smooth deposit (pan AS15-9554). (b) Enlargement of same photograph showing what appears to be a volcanic cone surrounded by a smooth ejecta blanket that probably consists of pyroclastic deposits.

I can see it from here with the monocular. Where I am now, just south of Archimedes, this hummocky, hilly terrain south of Archimedes is, in fact, quite full of rilles, although they're very subdued rilles. They don't have much definition to them, even in this low Sun angle. But they're combinations of linear rilles, which seem to run northwest-southeast, and sinuous rilles, which have no particular direction. Then, I noticed that a couple of rilles have—in fact, I'm looking at one right now—a small crater pair just to the left of Archimedes and a real light feature running to the east out of it. And that rille feature has a series—a whole succession—of craters running right down the rille. It certainly looks like a volcanic chain. I don't see any rim deposits associated with them. In fact, the craters that I'm looking at are irregular in shape, elongate in direction of the rille, and they look distinctly like collapsed features in a lava tube.

The CMP helped locate the lunar module (LM) landing point during the flight. The process was not easy because of the lack of a good optical system to increase the resolving power of the eye adequately (e.g., 20X to 30X gyroscopically mounted binoculars). The panoramic camera photographs show an interesting situation that might facilitate LM location by the CMP on future flights; namely, a bright halo approximately 150 m across appears to surround the LM almost symmetrically (figs. 25-18 and 25-19).

This bright halo is most probably caused by the disturbance of the uppermost surface layer by LM descent-engine exhaust. The disturbance in this case is manifested by the high albedo of the disturbed region. This subject is separately discussed in greater detail in part E.

At the postmission visual-observation debriefing, some points were clarified concerning the South Cluster and the flow unit east of the landing site (fig. 25-16) at the foot of the Apennine scarp.

I can't say anything about that crater cluster. I looked at it, but it was impossible for me to trace that back anywhere [to either Aristillus or Autolycus]. The only other thing was the flow front, which looks like a front where an area slopes very slightly and then comes up. But you see this flow, and it has a pancake appearance, and it does come up against the front. And you can see the flow front on both sides; almost as if the flow had gone down the valley there and left the front on both sides. The color was darker. I didn't get a really good look at that flow. For one reason or other, I was always looking at something else when I crossed the landing site.

The questions related to the general and local geologic settings of specific features of the landing site were not completely answered. This incompleteness was in part due to the lack of an optical aid and in part to the limited time that was allocated for studying the landing site from orbit. Plans are underway to remedy this situation in future Apollo flights. Data gathered from orbit on the setting of landing-site features are of significance in understanding the locally derived surface data.

Flow Fronts in Mare Imbrium

Discrete flow fronts are known to occur in the lunar maria. The flow fronts of Mare Imbrium are especially distinct. These fronts were found to correspond with small differences in tint or color tone (ref. 25-6) and to relative age differences between the mare units (ref. 25-7). Visual observations of some of these flow fronts revealed that they are not a single front but a complex array of numerous flow fronts.

In one case, the flows seem to have originated from a broad ridge (fig. 25-20).

LR33/05:22:16:28 GET: I am over the spot in Imbrium where some lava flows come out of where a wrinkle ridge is; and, at this low Sun angle, I can very clearly see some lava flows coming out of what appears to be a ridge, extending in both directions from the ridge. During the debriefing, the CMP commented, "On the western edge of Imbrium and in Procellarum, you can see the flow fronts by the difference in the color. You can see the color boundaries before you see the flow scarp. This is because the color is distinct and

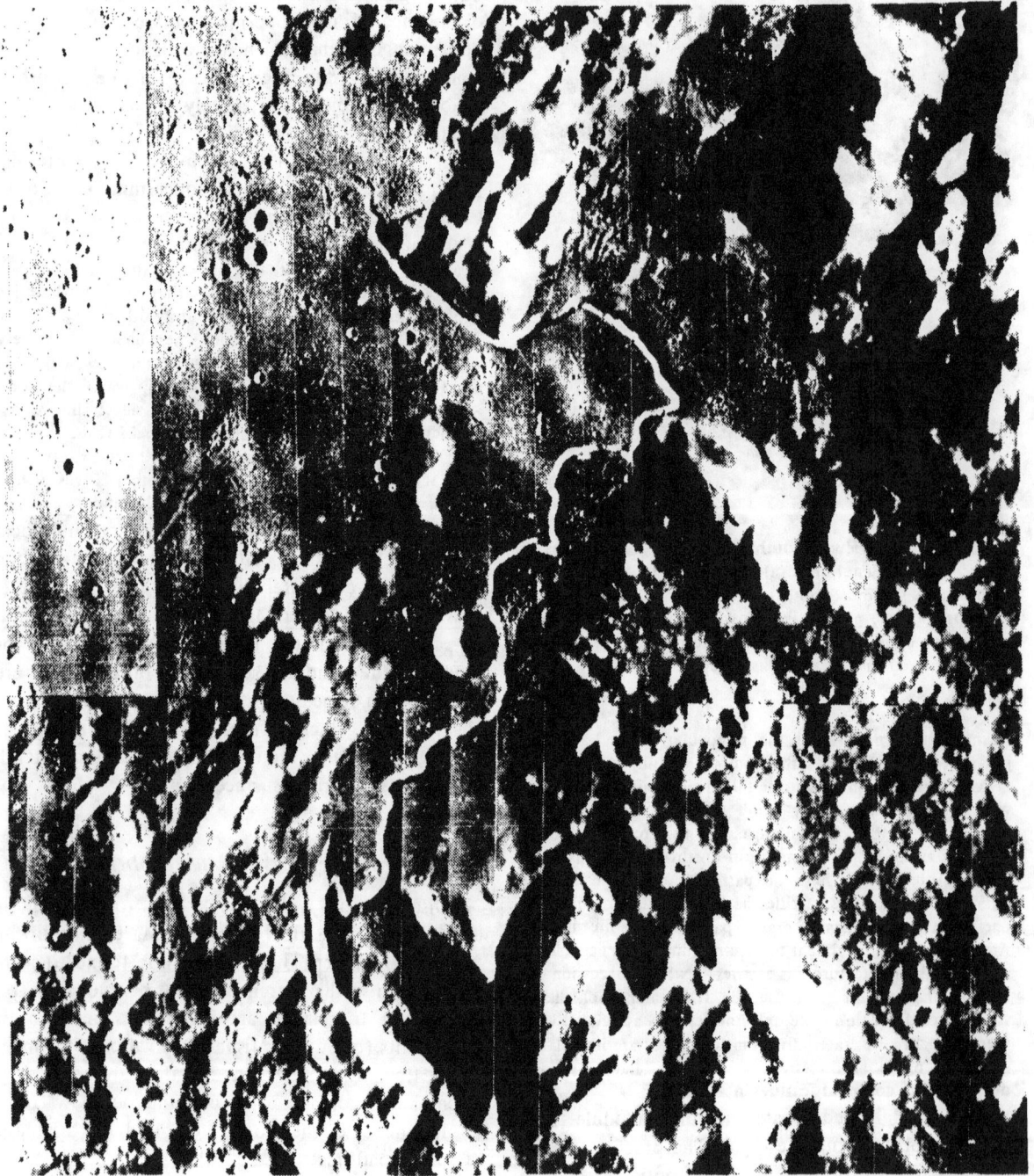

FIGURE 25-16.—Visual-observation target V-8A, Hadley-Apennine region.

V-8B LANDING SITE HADLEY RILLE (3.65° E, 26.1° N)

Use photograph to locate the LM and make observations relative to the local geology of the site.

FIGURE 25-17.—Visual-observation target V-8B, landing site—Hadley Rille.

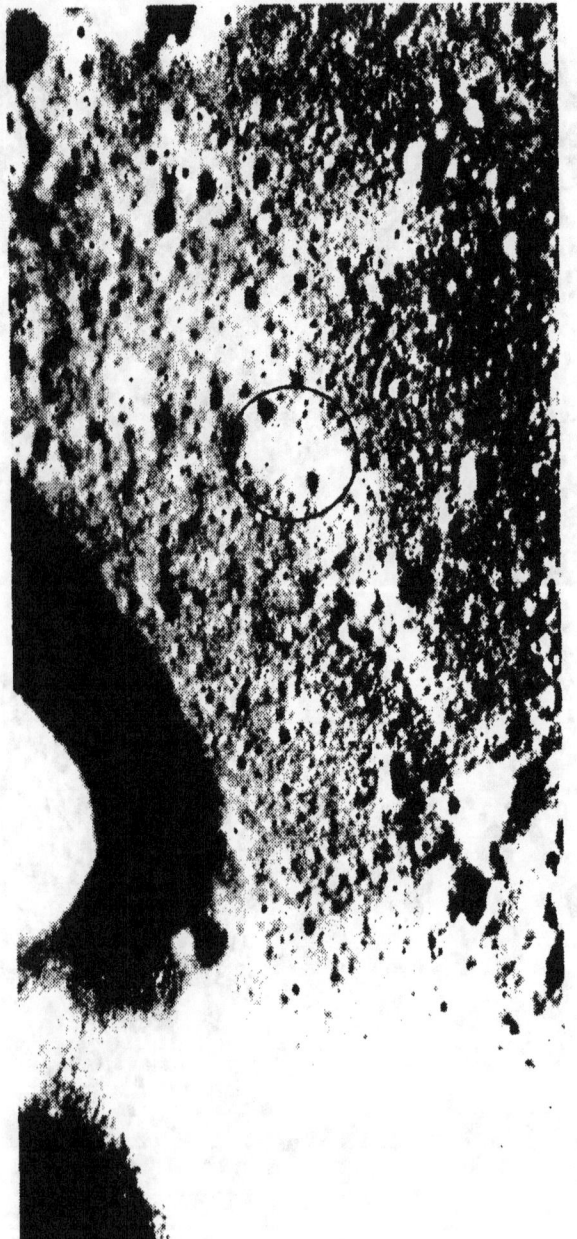

FIGURE 25-18.— Postlanding lunar revolution 16 photograph taken when the Sun-illumination angle was 14° at the landing site. The circle encloses an area of surface brightness most probably caused by surface disturbance from the descent-engine exhaust. Recognition of surface brightening may be used as one tool for easier location of the landed lunar module from the command-service module (pan AS15-9430).

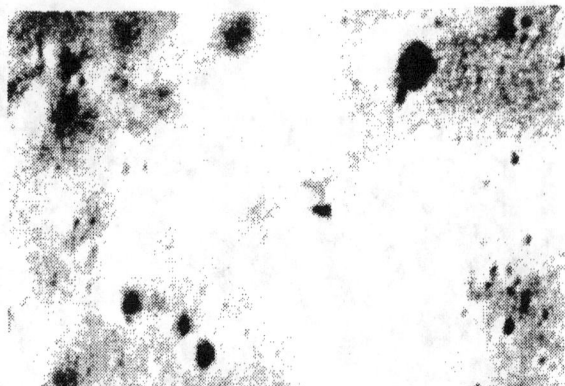

FIGURE 25-19.— This 50X enlargement of a panoramic camera photograph shows the lunar module and its prominent shadow, a small but very bright reflection from the scientific experiments deployment site, and a darkening of the surface between the two sites. The surface darkness is attributed to Rover tracks. The Rover shows up as a small speck very close to and somewhat north of the lunar module.

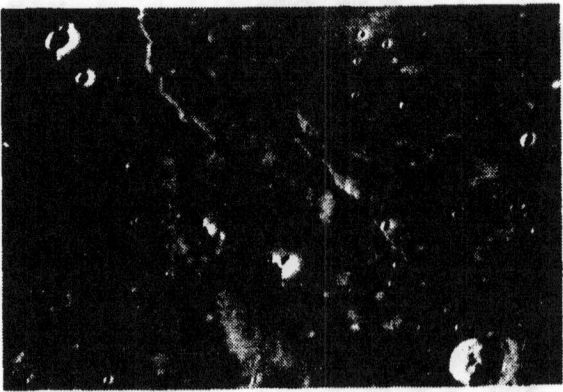

FIGURE 25-20.— Telescopic view of the western part of Mare Imbrium showing a mare ridge with mare flows extending east and west from the ridge.

different from the flow that was underneath it. After you look at one area and you see 10 or 15 flows that are all overlapping and joining in that area, the picture gets a little bit confused. But that's why I say that I get the impression that there are just hundreds of flows that filled up the mare basin. They all look like, for example, if you'd take a pail of water and sluice it out into a skating rink and let it freeze in place; then, if you do that 15 times around the same area, you would get this overlapping mixed-up ice. This was the case in Procellarum.

"In Mare Imbrium, there was a lot more than two flows. In fact, I don't recall seeing any place where it looked like there was just one big flow. All the flows were very thin and appeared as if they came out and froze in place. Even in the case where we had that particular visual-observation target, where there was supposed to be a single flow, particularly in an area like this. You can see the one flow in the photograph because the difference in the color is distinct enough so that you pick it up photographically. But when you get close to it and you look at it with the eye, associated with the major front are flows that are basically the same color, but you begin to notice all kinds of flows in there."

Flow fronts with multiple flows were observed in the metric camera photography (fig. 25-20) and also in near-terminator Hasselblad photographs. The utility of the latter photographs has been established (ref. 25-8), and their significance in photogeologic interpretations has also been stressed (ref. 25-9). An example of this photography that is especially suited for delineating small topographic variations, particularly in the maria at or near flow fronts and wrinkle ridges, is shown in figure 25-21.

Aristarchus Plateau

The Aristarchus Plateau and the surrounding mare materials of Oceanus Procellarum display a multitude of features the characteristics of which are not well understood (fig. 25-22). The area was viewed both in earthshine and in sunlight.

FIGURE 25-21.- Near-terminator view of the mare ridges and flow fronts in Oceanus Procellarum (AS15-98-13356).

Referring to Aristarchus in the debriefing, the CMP said,

In earthshine, that thing really shows up. That's like a really luminous blanket around it. That is really bright, even in earthshine. The ejecta are also bright. That's an impression you get when you're coming up on it, because the only thing you see is the ejecta pattern. And that comes right straight at you out of the darkness as you're going over it. And you see the ejecta pattern, with that one long piece of it that extends out from the crater. Then, as you get up on the crater, the inside of it is fairly bright, too. But the thing that really catches your eye is this ejecta pattern that's out around it, with a big long tail that sticks out to the southwest.

In the latter part of the mission, observations were made of Aristarchus Plateau.

LR70/08:23:19:21 GET: We've all been sitting here looking at Aristarchus a little bit in awe. It looks like probably the most volcanic area that I've seen anywhere on the surface. And certainly it's just very covered with rilles, very deep rilles. Schroter's Valley, for instance, is a magnificent big rille, which looks like it's been worked twice; of course, the large rille and then a smaller rille inside [fig. 25-23].

In reference to the Cobra Head at Schroter's Valley, the CMP said in the debriefing,

I guess I told you before, but the Cobra Head looks like a sheer cliff, a very steep, very high wall. I couldn't even begin to guess how deep it looks, but I bet it's 1000 feet at least. It's like a natural amphitheater where this thing dug down. I never did get a very good look at the bottom of it because it was always in shadow. It was deeper than you could actually see. It was just a big steep wall sitting there. You do get some suggestion of it in the photographs [fig. 25-23]. You could see a portion that looked like something had cut right back into the hill there, like a sandpit. But it cut right back into the hill and just left this great big wall around it where it had cut out part of the hill. Schroter's Valley was really not distinctive other than what you see in the pictures. The rille inside is much more meandering than the basic rille, and I don't recall seeing any layering there.

Speaking of the probable volcanic origin of Schroter's Valley and its source, the CMP made the following comment during the flight.

LR70/08:23:20:08 GET: That whole area in there looks volcanic. And Schroter's Valley certainly comes from up on the plateau. I would guess that the Cobra Head is the source of Schroter's Valley. The elevation is a little subtle from this vantage point, but that would be my guess. And I think we've found something interesting anyway about the ends of the rilles, particularly around the Schroter's Valley area. It looks quite distinctly that the mouths of the rilles or the deltas—have been covered with rising mare materials from the lower elevation, almost as if there was a rille with a deltalike deposit at one time, but then the elevation of the mare came up into the rille far enough to cover all that delta area. So all you see now is what would look like a river into a lake whose elevation has increased.

The observation related to the terminal portions of

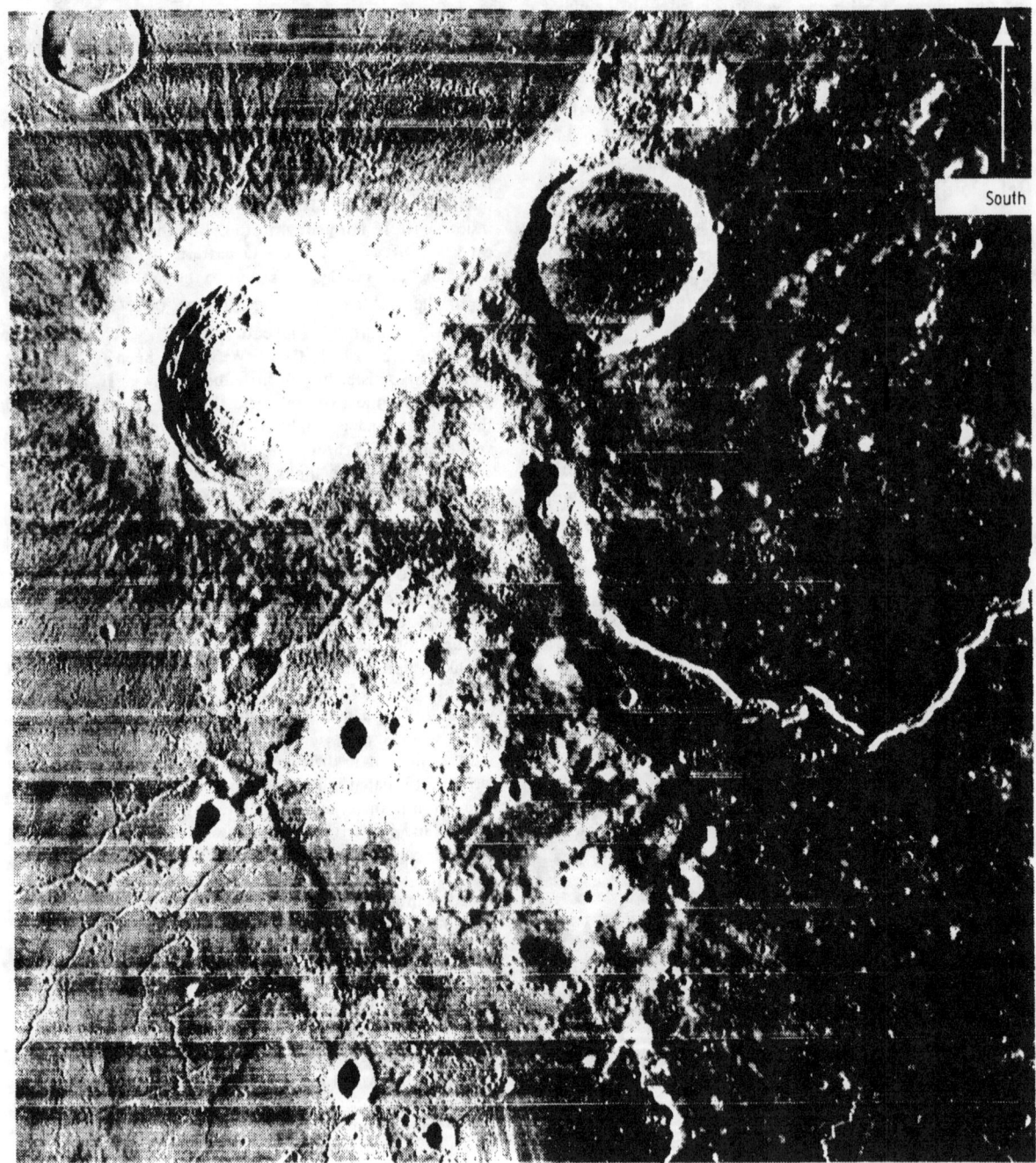

V-11 ARISTARCHUS PLATEAU (48.5°W, 25°N)

Describe the major units of the Aristarchus Plateau (as viewed in earthshine and in sunlight). Note particularly the following:
1. The domical structures and dark deposits.
2. The Cobra Head and its vicinity.
3. Schroter's Valley and its wall structures.
4. The Aristarchus rim and related volcanics.

FIGURE 25-22. Visual-observation target V-11, Aristarchus Plateau.

FIGURE 25-23.—Panoramic photomosaic of Schroter's Valley. The steep interior walls of Cobra Head at left and the sinuous channel that meanders within the valley and crosscuts the wall near the terminus at right are notable (north is toward the bottom, AS15-97-13256 to 13262).

the rilles and what appears to be their flooding with younger mare flows is confirmed by the photography. An example along the northern edge of the Aristarchus Plateau is shown in figures 25-24 and 25-25.

Additional discussion of these features and related flows in Oceanus Procellarum was conducted during the postmission visual-observation debriefing, from which the following is excerpted.

On the west side of Aristarchus, not only do you have the color to help you pick out the flows but you also have the surface texture. The surface texture of the top flow in the lower elevation flows is as shiny as a glass surface. It gives you a very patent-leather appearance—that flow in particular, inside the big rille. But, of course, that's all pockmarked with craters. Where there's not a crater, it looks like there's original surface; it's just as smooth and shiny as a piece of leather, and it's brown. Is it very different from the mare outside? It's a dirty brown color; inside the rille, it looks like a muddy river. And all the mare surface below the rille has been covered, too. It looks like a very viscous, muddy river that has somehow stopped in place. Those flow fronts come back up into the rilles and chop off the terminus. The basic mare around this area is gray. It's not brown; it's gray. You kind of leave the gray of Imbrium as you go farther and farther west, and the next thing you come up over is Aristarchus Plateau. When you get on the other side of Aristarchus Plateau, suddenly the mare is brown; it's not gray any more.

In effect, this description is a characterization of the older Imbrian mare material in Mare Imbrium and farther east as grayish in color. The younger Eratosthenian mare material of Oceanus Procellarum, especially west of the Aristarchus Plateau, is brownish in color. "Imbrium was distinctly gray; Procellarum is distinctly brown."

Conclusions

Visual observations of particular lunar-surface features and processes were made on all Apollo lunar missions. Reports have been published that deal with these observations on Apollo 8 (ref. 25-10), Apollo 10 (ref. 25-11), and Apollo 14 (refs. 25-9 and 25-12).

Before the Apollo 15 mission, however, observations from lunar orbit did not constitute an objective of the mission and, therefore, received only cursory interest. On the Apollo 15 mission, visual observa-

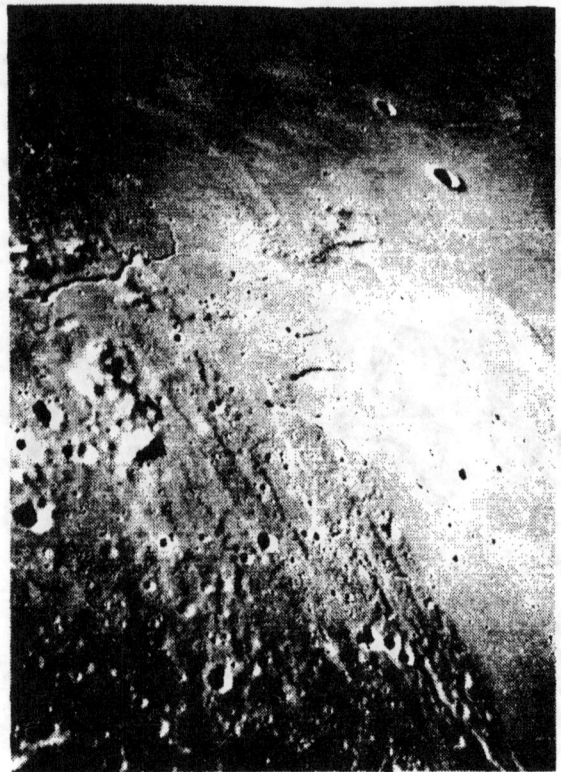

FIGURE 25-24.—Western contact between the Aristarchus Plateau and Oceanus Procellarum (north is toward the bottom of the photograph: AS15-93-12630).

FIGURE 25-25.—Near-terminator view of the western part of the Aristarchus Plateau. It should be noted that the floors of the depressions are filled with the mare material of Oceanus Procellarum (AS15-98-13345).

tions from lunar orbit were treated as a mission objective. The task was conducted systematically, and targets were studied thoroughly. The extraordinary success of this undertaking proved the outstanding capabilities of man and his use in space flight. The unusual sensitivities of the human eye, when combined with the interpretive capabilities of the brain, constitute a combination that cannot be matched by one photographic system. However, this is not a competitive effort. Visual observations were made only to supplement the onboard photographic systems and are considered to complement other remotely sensed data.

Visual observations from lunar orbit proved to be most useful in determining the following.

(1) Subtle color-tone differences between surface units both in the maria and in the highlands may reflect chemical or age differences (or both) among the observed units.

(2) Major tectonic trends and regional settings of lunar-surface units will naturally lead to an integrated picture of the overflown areas and, ultimately, the whole Moon, by further study of the photographs and extrapolation of the knowledge.

(3) The importance and local relationships of small-scale features can be studied. Particular lunar-surface features and processes require more than the stereographic photographs to allow an intelligent deduction of their probable origin and of their role in the formation or modification (or both) of the lunar surface. The orbital periods of the Apollo flights allow the study of these features under varying illuminative conditions and viewing angles.

The single most important and timely observation of the mission was the discovery of cinder cones in the Littrow area. As the probable source of the dark deposit in this area, they provide evidence of extensive volcanism that definitely postdates the late episode of the flooding of the basins by the volcanic lava (approximately 3.2×10^9 yr).

Other observations are also significant (e.g., the distinction of the nature of the landslide or rock glacier on the northwest rim of Tsiolkovsky Crater and the recognition of flooding of the terminal portions of many sinuous rilles by younger mare material). Recognition of layering along crater walls (as opposed to terracing by faults and mass wasting by downward movement of materials along the walls) was achieved for the first time. This recognition gives a new dimension to thinking relative to the nature of the upper layers of the lunar crust.

A trained observer can contribute significantly to achieving the goals of lunar exploration by making and recording visual observations of particular lunar-surface features and processes from lunar orbit.

PART B

PHOTOGRAMMETRIC ANALYSIS OF APOLLO 15 RECORDS

Frederick J. Doyle[a]

The three cameras—stellar, mapping, and panoramic—together with the laser altimeter, all included in the scientific instrument module (SIM) bay, represent an integrated photogrammetric system with extraordinary potential for extending knowledge of the lunar figure, surface configuration, and geological structure.

Data-Analysis Objectives

The eventual objectives of the analysis were defined by the 1965 Falmouth Conference of Lunar Science and Exploration. These objectives are as follows.

(1) To establish a selenodetic coordinate system related unambiguously to the right-ascension and declination system

(2) To derive a reference figure with respect to a point that is representative of the lunar center of mass

(3) To establish a three-dimensional geodetic-control-point system on the lunar surface in terms of latitude, longitude, and height above the chosen reference surface

(4) To provide data for a comprehensive series of medium-scale (1:250 000) topographic and geologic maps

(5) To provide data for large-scale (1:20 000) topographic and geologic maps of landing sites and other areas of high scientific interest

(6) To describe in sufficient detail the gravitational field of the Moon

The instrument package was assembled with these objectives in mind and with a clearly defined data-analysis scheme that would lead to these results. Quite obviously, the extent to which the results can be achieved depends upon the total area photographed, with complete coverage of the lunar surface being the optimum. In any event, the final results will not be obtained until all Apollo missions have been completed and the data from all flights have been included in a single solution.

Data-Analysis Procedure

The basic data-analysis scheme is as follows.

(1) Identification of star image on the stellar photographs, measurement of their image coordinates, and computation of the orientation of the camera in the right-ascension and declination Earth-centered stellar coordinate system

(2) Computation of the orientation of the mapping camera in the stellar coordinate system by using the precalibrated relationship between the stellar and mapping cameras

(3) Transfer of the orientation of the mapping camera from the Earth-centered stellar coordinate system to the Moon-centered geographic system. This transformation involves the ephemeris and physical librations of the Moon

---
[a]U.S. Geological Survey.

(4) Selection of lunar-surface points (approximately 30 per frame) on the mapping camera photographs and identification of them on all overlapping frames on which they appear

(5) Measurement of the coordinates of the selected points and correction of the points for film deformation, lens distortion, and displacement of the focal-plane reseau

(6) Triangulation of groups of photographs. This computation includes the measured image coordinates, the attitudes obtained with the stellar camera, the laser altimeter data, a state vector obtained from the tracking data, and a gravity model

(7) Assembly of the triangulated groups into a single adjusted network. The output of this computer program will be a geometrically homogeneous set of coordinate values for all selected points on the lunar surface, plus the position and orientation of each photograph in the same Moon-centered coordinate system

(8) Preparation of small-scale maps from the mapping camera photographs

(9) Transformation of the panoramic photographs into equivalent vertical photographs for interpretation and mapping

(10) Preparation of large-scale maps from the panoramic photographs

The predicted absolute accuracy of the control-point network is 10 to 15 m. Inherent in the solution is the possibility of improving the lunar ephemeris, libration theory, and gravity model to values commensurate with the 10 to 15 m in point coordinates. In most cases, this accuracy is nearly an order of magnitude improvement over presently known values. Most of the software and hardware for the individual steps have been completed, although a major analytical task still must be completed for the final simultaneous adjustment of all photographs.

The proposed analysis is a long-term project. Efforts to date have been concentrated on the application of isolated steps in the data-analysis scheme to small samples of the photographs to evaluate the quality of the records and to verify the accuracies of the results obtainable.

By agreement between NASA and the Defense Intelligence Agency personnel, all data reduction related to mapping will be performed by the defense mapping organizations—the Army Topographic Command (TOPOCOM) and the Air Force Aeronautical Chart and Information Center (ACIC).

Evaluation of Stellar Photography

The initial evaluation of the Apollo 15 stellar photographs was performed by TOPOCOM personnel. For each of four exposures, the attitude of the stellar camera was determined in a selenocentric inertial coordinate system of 1950.0, along with estimates of the uncertainty in the results obtained.

Two exposures from revolution 4 and one exposure from revolution 15 were selected for measurement. These exposures were arbitrarily given the frame identifications 401, 402, and 1502. One additional frame was later measured and numbered 403. The stellar field imaged on the photographs was identified on the SKALNATE PLESO star charts. Five to 10 reference stars were selected on each of the exposures. These stars were identified in the Boss Catalogue from the right ascensions and declinations read from the charts.

The photographs then were measured on Mann comparators; frame 401 was measured by two different comparator operators. As a result of time limitations, each image (including reseaus, reference stars, and other apparent star images) was measured only once. The number of images measured on each exposure (less the 25 reseau crosses) ranged from 70 to 115. During the computer reductions, considerably fewer (a maximum of 33) were accepted as true star images. The measured data were then entered into the stellar-reduction computer program.

The stellar-reduction computer program operates in two independent phases—a preprocessing phase and a main processing phase. The preprocessor transforms the measured data into the camera system defined by the calibrated values of the reseau coordinates and corrects for systematic film distortion and radial and tangential lens distortion. The main processing portion of the computer program determines the attitude of the camera. This determination is accomplished by first estimating the attitude of the star camera in the inertial system from the preidentified control stars. Next, the approximate right ascension α and declination δ of all measured images are computed. The Boss Catalogue, which is recorded on tape, is then automatically searched for all corresponding similar values of α and δ. The refined attitude of the star camera is then computed from the true α and δ of these identified stars. Finally, the attitude of the terrain camera is computed from the calibrated relative orientation between the two cameras, and the attitudes are transformed to the lunar reference system using the lunar ephemeris and physical libration marks.

The computer program incorporates rigorous error-propagation routines that provide estimates of accuracy for all computed values. However, for the purposes of this test, some of the a priori accuracy estimates were arbitrarily specified. Therefore, only the standard deviations of the stellar camera in the inertial 1950.0 reference system can be considered reliable.

The results of the test are summarized in table 25-II. The table shows that typical standard deviations in roll ω, pitch ϕ, and yaw κ that can be expected are 2.9, 2.2, and 21.2 arc-sec, respectively. These values represent the standard errors in the attitude of the stellar camera resulting from measuring errors of $\sigma\bar{x} = 4.2$ μm and $\sigma\bar{x} = 3.8$ μm.

For this sample, an average of 27 stars was used in the computation of the camera attitude. These stars were identified on the Boss Catalogue tape from an average of 83 apparent star images. Further substantiating these findings is the fact that the values obtained from exposure 401A (suffix A indicates independent measurement made by a second comparator operator) are also typical of the average values. It may reasonably be concluded from these results that, provided the film sample furnished TOPOCOM by NASA is typical of the quality of the remaining exposures, the stellar photographs obtained during the Apollo 15 mission are adequate for determining the attitude of the stellar and terrain cameras.

Evaluation of Mapping Camera Photographs

Geometric integrity of the measurements performed on the mapping camera photographs is the limiting factor in the eventual accuracy of the information to be obtained from the data-analysis procedures. Because the original film is a priceless national asset, it is contemplated that all measurements will be performed on later generation copies; thus, a knowledge of the degradation introduced at each step in the procedure is important.

The first investigation, which was performed by the Mapping Sciences Branch at the NASA Manned Spacecraft Center (MSC), was to evaluate the metric degradation introduced by developing the original negative in the Fultron processor at the MSC. A 22.9- by 22.9-cm glass plate with a 20-mm-square grid was contact printed on a 457-m-long roll of Ektachrome 3400 film in six evenly separated groups of three exposures each. For each exposure, 66 grid intersections were within the 12.7-cm width of the film. The intersections were measured by using the Mann 1210 comparator and then were used to control the mathematical transformations that were applied to each set of the measured image coordinates.

Four geometric-transformation models were applied to each of the 18 sets of image coordinates by the mathematical technique of least squares. Each progressive transformation offered more in the way of corrective capability, and the adaptability to the situation can be judged by the pattern and magnitude of the observational residuals. The transformation models are referred to as the three-, five-, six-, and eight-parameter models (table 25-III). In each case, all the available data points were used in the solution to minimize the effects of any local distortions and to prevent the propagation of random error as systematic error. After the completion of each transformation, the root-mean-square (rms) error of the residuals was computed. In table 25-IV, these results are listed by exposure and applied transformation.

A representative array of residual plots is given in

TABLE 25-II.—*Results of Measurements on Stellar Photographs*

Frame no.	Star images read, no.	Stars found on Boss tape, no.	Stars used in attitude computations, no.	Root-mean-square x/y, μm	σ_ω, arc-sec	σ_ϕ, arc-sec	σ_κ, arc-sec
401	70	34	26	5/4	3.3	2.6	24.6
402	78	38	27	4/4	3.0	2.3	19.1
403	114	36	33	3/2	1.8	1.3	13.4
1502	74	28	22	5/5	3.8	2.9	27.8
401A	79	37	26	4/4	2.8	2.1	21.1
Average	83	35	27	4.2/3.8	2.9	2.2	21.2

TABLE 25-III.—*Three-, Five-, Six-, and Eight-Parameter Transformations*

Transformed values	Transformation			
	3 parameters (α, h, and k)	5 parameters (α, h, k, S_x, and S_y)	6 parameters (α, β, h, k, S_x, and S_y)	8 parameters (A, B, C, A', B', C', D, and E)
x'	$x \cos \alpha + y \sin \alpha + h$	$S_x(x \cos \alpha + y \sin \alpha + h)$	$S_x(x \cos \alpha + y \sin \alpha + h)$	$\dfrac{Ax + By + C}{Dx + Ey + 1}$
y'	$y \cos \alpha - x \sin \alpha + k$	$S_y(y \cos \alpha - x \sin \alpha + k)$	$S_y(y \cos \beta - x \sin \beta + k)$	$\dfrac{A'x + B'y + C'}{Dx + Ey + 1}$

TABLE 25-IV.—*Root-Mean-Square Errors After Application of Transformations*

Exposure identification		Root-mean-square error, µm						
Group no.	Exposure no.	3-parameter transformation		5-parameter transformation		6-parameter transformation		8-parameter transformation	
		X	Y	X	Y	X	Y	X	Y
[a]1	1	--	--	8	7	8	7	3	8
1	2	46	11	7	6	6	6	3	6
1	3	43	5	3	3	3	3	2	3
2	1	36	9	4	5	3	5	3	5
[a]2	2	38	9	6	6	5	6	2	6
2	3	37	7	3	5	3	5	2	5
3	1	46	8	5	4	4	4	3	4
3	2	45	8	4	3	3	3	3	3
3	3	45	7	3	3	2	3	2	3
4	1	45	7	3	3	3	3	3	3
4	2	43	8	3	4	3	4	3	4
4	3	38	6	3	3	3	3	3	3
5	1	47	9	3	4	3	4	2	4
5	2	49	7	3	2	3	2	2	3
5	3	44	4	2	3	2	3	2	3
6	1	52	10	4	4	4	3	2	4
6	2	53	9	5	4	4	4	2	4
6	3	56	11	4	3	3	3	3	3

[a]Incomplete exposure data.

figures 25-26 to 25-29, which show the residuals after the application of each transformation model to one of the 18 exposures. Examination of figure 25-26 shows a clearly defined shrinkage pattern from the original glass-grid measurements. In figures 25-27 to 25-29, distinctly systematic errors in the remaining residuals are evident; however, the magnitudes are small (table 25-IV).

Three exposures were used to estimate the measurement repeatability that could be expected from the use of the film images. One data set at the beginning of the strip, one near the middle, and one near the end were remeasured an additional three times for coordinate comparisons. It was found that these images could be repeatedly measured to approximately 5 µm. The rms residual values listed in table 25-IV for the five-, six-, and eight-parameter transformations approach this limit as a noise level. The film handling was insufficiently controlled to justify a conclusion that all the film deformation was caused by the processing. However, the results indicate that the proposed correction routines are capable of accommodating deformations down to approximately the noise level of the measurements.

Deformation of Second-Generation Negatives

Responsibility for the mensuration and triangulation of the Apollo 15 mapping camera photographs

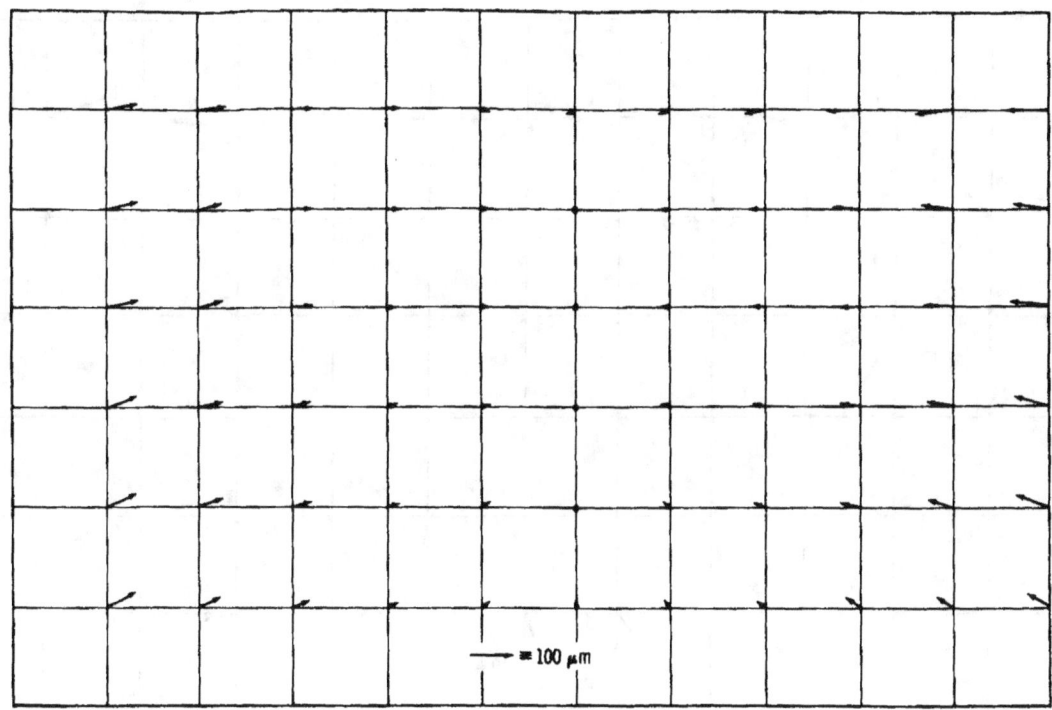

FIGURE 25-26.—Residual after a three-parameter transformation for group 1, exposure 2.

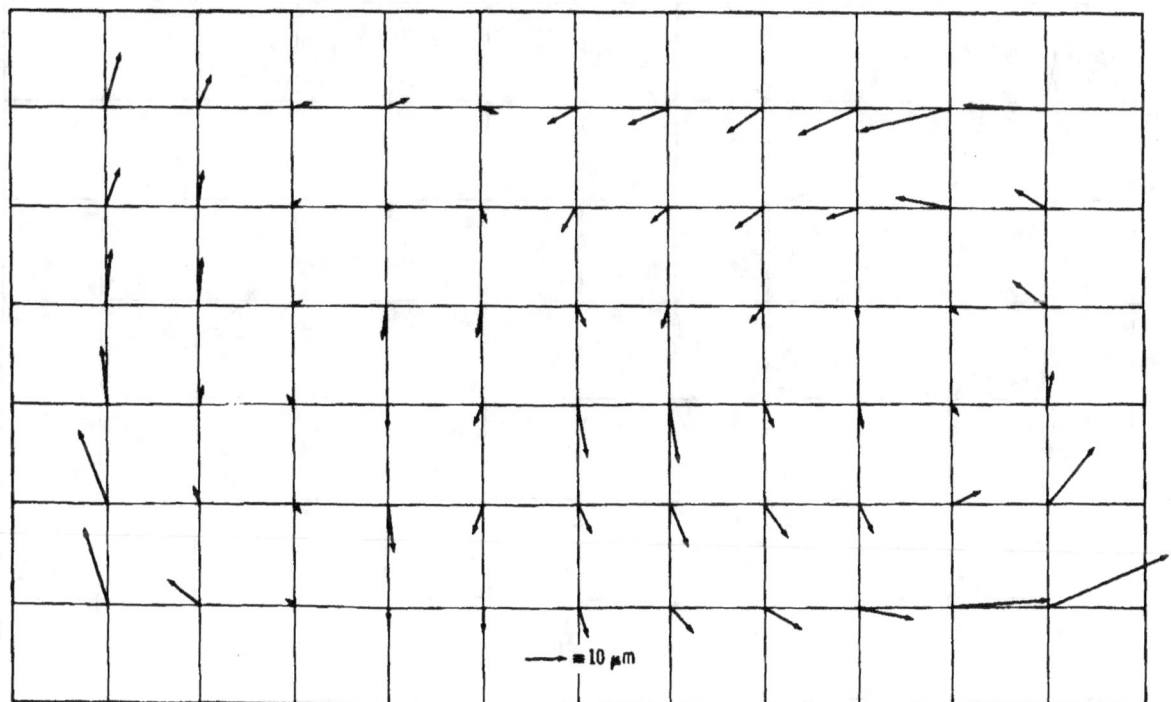

FIGURE 25-27.—Residuals after a five-parameter transformation for group 1, exposure 2.

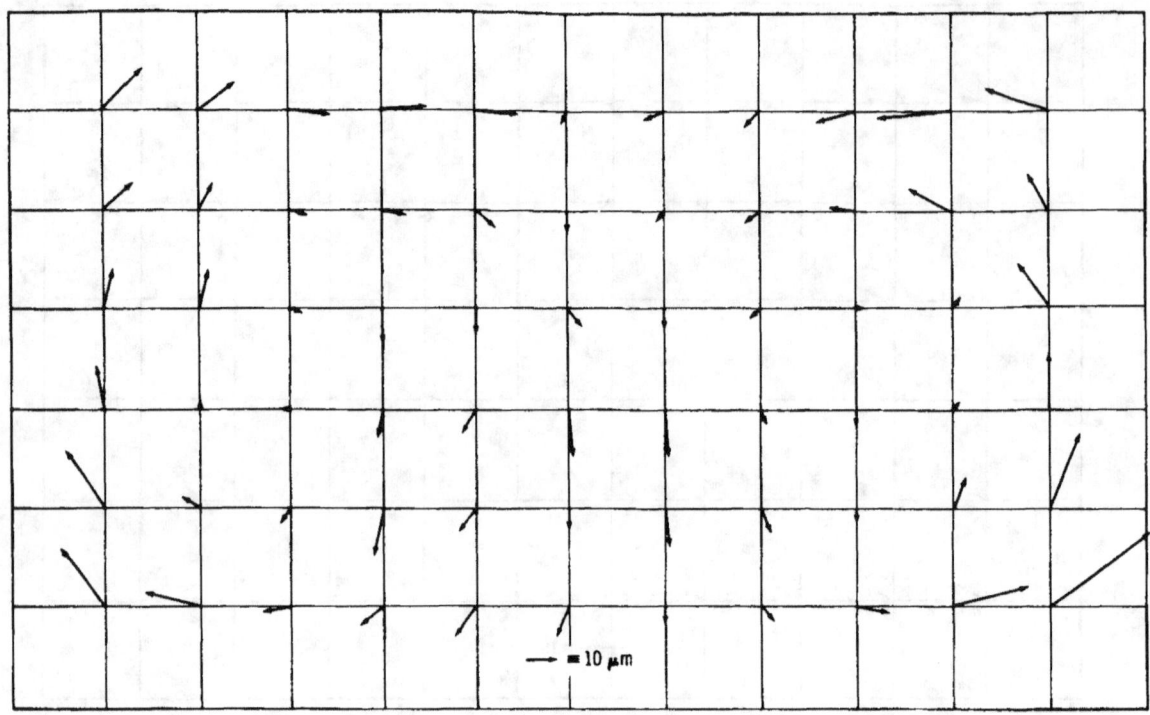

FIGURE 25-28.—Residuals after a six-parameter transformation for group 1, exposure 2.

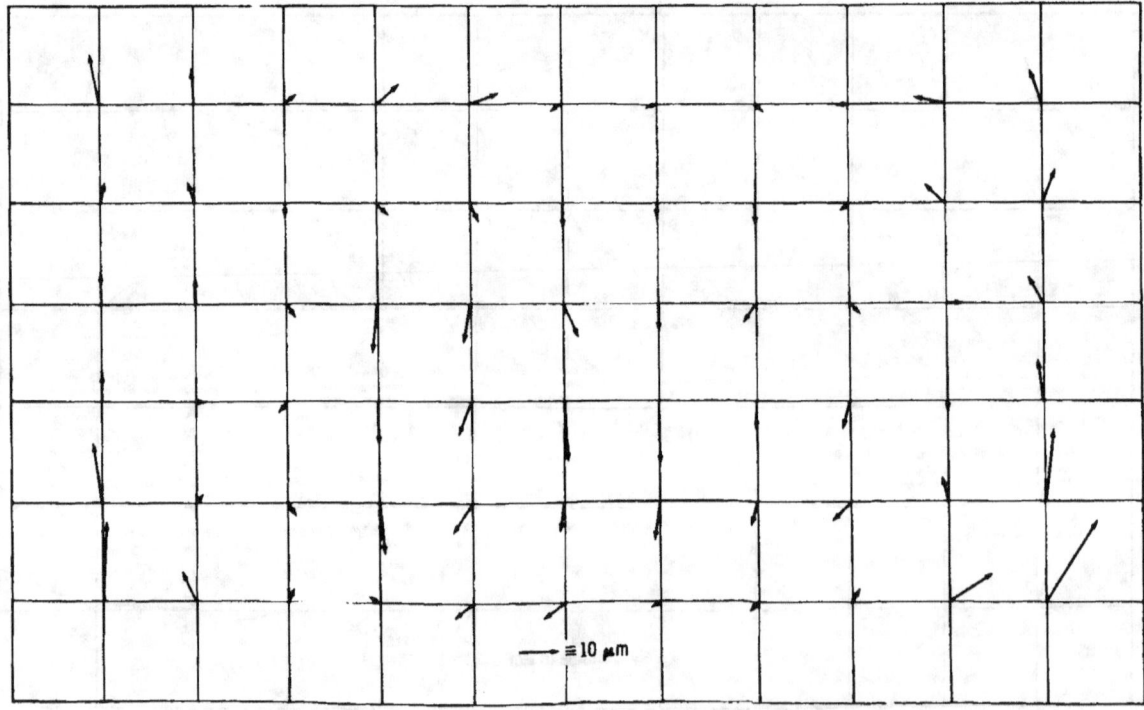

FIGURE 25-29.—Residuals after an eight-parameter transformation for group 1, exposure 2.

has been assigned to the ACIC. A second-generation negative was produced by contact printing on the Niagara printer at the MSC and supplied to the ACIC. Before beginning the point-selection and measuring operation, ACIC personnel investigated the dimensional accuracy of the second-generation negative by measuring 25 reseau crosses on three frames over the Apollo 15 landing site. The standard error of the reseau-coordinate residuals after a linear conformal adjustment of measured values to calibrated values was approximately 8 μm. This error is considered to be a reasonable value, because the predicted accuracy of triangulated control points is based on 10-μm coordinate residuals.

Production of Rectified Prints From Panoramic Camera Photographs

The lunar transforming printer was developed at TOPOCOM to transform the convergent panoramic camera photographs into equivalent vertical photographs. The printer rectifies the original film negative by duplicating the taking system. The light slit sweeps across the cylindrical-negative platen and projects the image through a lens to a folding mirror, then to printing material mounted on the easel. Positive or negative film or paper prints can be produced.

The Apollo photography is convergent 12.5° forward and aft and contains the following distortions, which are corrected by the transforming printer.

(1) *Panoramic distortion.*—the displacement of images that results from the geometry of the focal plane and the scanning action of the camera lens

(2) *Convergent tip distortion.*—the displacement of images that results from the 12.5° inclination of the camera optical axis in the line of flight

(3) *Roll distortion.*—the displacement of images that results from roll of the camera about an axis parallel to the line of flight

The easel may be tilted with respect to the panoramic film to simulate the convergence angle. The easel, which is analogous to the lunar surface, also may be curved into a cylindrical shape with a radius that can be varied to simulate the apparent change in lunar-surface curvature as a function of altitude. The isometric magnification of the output material is 1.9 times the original.

Tests were conducted to determine the resolution and the geometric quality of the rectified prints. The averages of 15 radial- and tangential-resolution readings in line pairs per millimeter (lp/mm), measured at the input scale, are listed in table 25-V. For determining the geometric quality, the rectified print was divided into two segments, one being the center half of the print and the other being the two outer quarters. The circular probable errors of point positions after rectification are listed in table 25-VI. Both resolution and position errors are well within the tolerances specified for the instrument. A section of panoramic camera frame at original contact scale is shown in figure 25-30. The Apollo 15 landing site is located approximately 27° scan angle from the center of the frame. The same area, after rectification in the transforming printer, is shown in figure 25-31.

TABLE 25-V.—*Resolution of Rectified Panoramic Photographs*

Altitude, km	Tilt angle, deg	Ends, lp/mm	Center, lp/mm	Overall, lp/mm
100	0	87	122	103
135	0	83	115	98
21	12.5	91	113	101
90	12.5	92	104	97
100	12.5	95	115	105
110	12.5	95	114	103
120	12.5	89	108	98
135	12.5	86	112	98

TABLE 25-VI.—*Positional Errors in Rectified Panoramic Photographs*

Altitude, km	Tilt angle, deg	Circular probable error, μm		
		Center	Ends	Total
100	0	292	688	529
100	12.5	324	415	372
120	12.5	224	325	279
135	12.5	246	362	310

Production of Enlarged Prints From the Mapping Camera Photographs

The mapping camera configuration was originally selected in anticipation of providing a photographic image base for maps with a scale of 1:250 000. However, recent discussions with lunar geologists

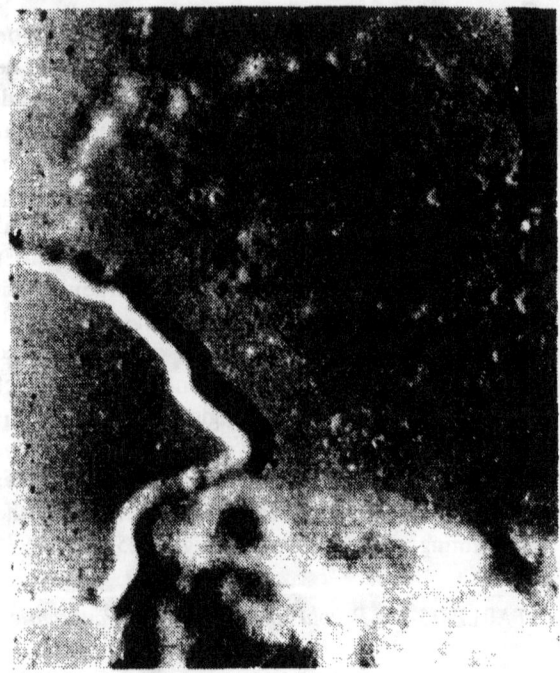

FIGURE 25-30.—Portion of the original photograph of the Hadley Rille site (pan AS15-9793).

have resulted in a recommendation for 1:100 000 photographic-image maps. This scale is considered to be too large for the mapping camera but too small to be made economically from the panorama photographs. Enlargements of the mapping camera photographs were made at TOPOCOM to evaluate the quality that might be expected for each of these scales.

A mapping camera frame enlarged 5.3 times to approximately a scale of 1:270 000 is shown in figure 25-32; the resolution of the photograph is approximately 11 lp/mm. The same photograph enlarged 16.3 times to approximately a scale of 1:90 000 with a resolution of approximately 4 lp/mm is shown in figure 25-33. The quality of these photographs, as reproduced, is representative of the quality of the maps that can be expected for the two scales.

Heighting Accuracy of the Panoramic Photographs

Several surface experiments and geologic interpretations will require a precise knowledge of topographic elevations. An evaluation of the heighting accuracy from the panoramic photography was performed by contractor personnel for the Mapping Sciences Branch at MSC. Control points for use in orienting the panoramic photographs were established from a stereoscopic model of the Apollo 15 metric frame photography. Nine points were then used as common control points for orienting three panoramic stereoscopic models in the AS-11-A1 analytical plotter. Model 1 was set from panoramic photographs AS15-9370 and 9375 of revolution 16. In this model, the Hadley area was located in the approximate center of the photograph. These photographs were taken at a time of low Sun angle, which provided good contrast for stereoviewing. Model 2 was set from panoramic photographs AS15-9793 and 9798 of revolution 38. In this model, the Hadley area was located approximately halfway between the center and the outer edge of the photograph, or where the scan angle was approximately 27° from the midscan line. The higher Sun angle in this model resulted in a washed-out effect for purposes of stereoviewing. Model 3 was set from a combination of photographs from the previous models; photograph AS15-9370 of revolution 16 and photograph AS15-9798 of revolution 38 were used.

The fit of the model coordinates to the control coordinates for the three models is described in table 25-VII. After each model was set, 36 points along a pseudotraverse were selected and an elevation was recorded for each point. The AS-11-A1 model-deformation computer program then was used to fit the model to control points 1, 12, 28, and 33. The elevations of the 36 traverse points again were recorded. This procedure was used for each of the three models.

The mean elevation for each point was determined from the three readings from the models without model deformation. Deviation from the mean was found for each point, and a standard deviation for all points was computed. The resulting standard deviation was ±6.3 m. This value is interpreted as an indication of how precisely panoramic stereoscopic models may be set to control points established by triangulation of the mapping camera photographs.

The same procedure was repeated for the three stereoscopic models after the model-deformation computer program was applied. The resulting standard deviation was ±3.4 m. This value is interpreted as an indication of the repeatability, or internal consistency, of elevations that can be read from panoramic stereoscopic models. It is most interesting that this

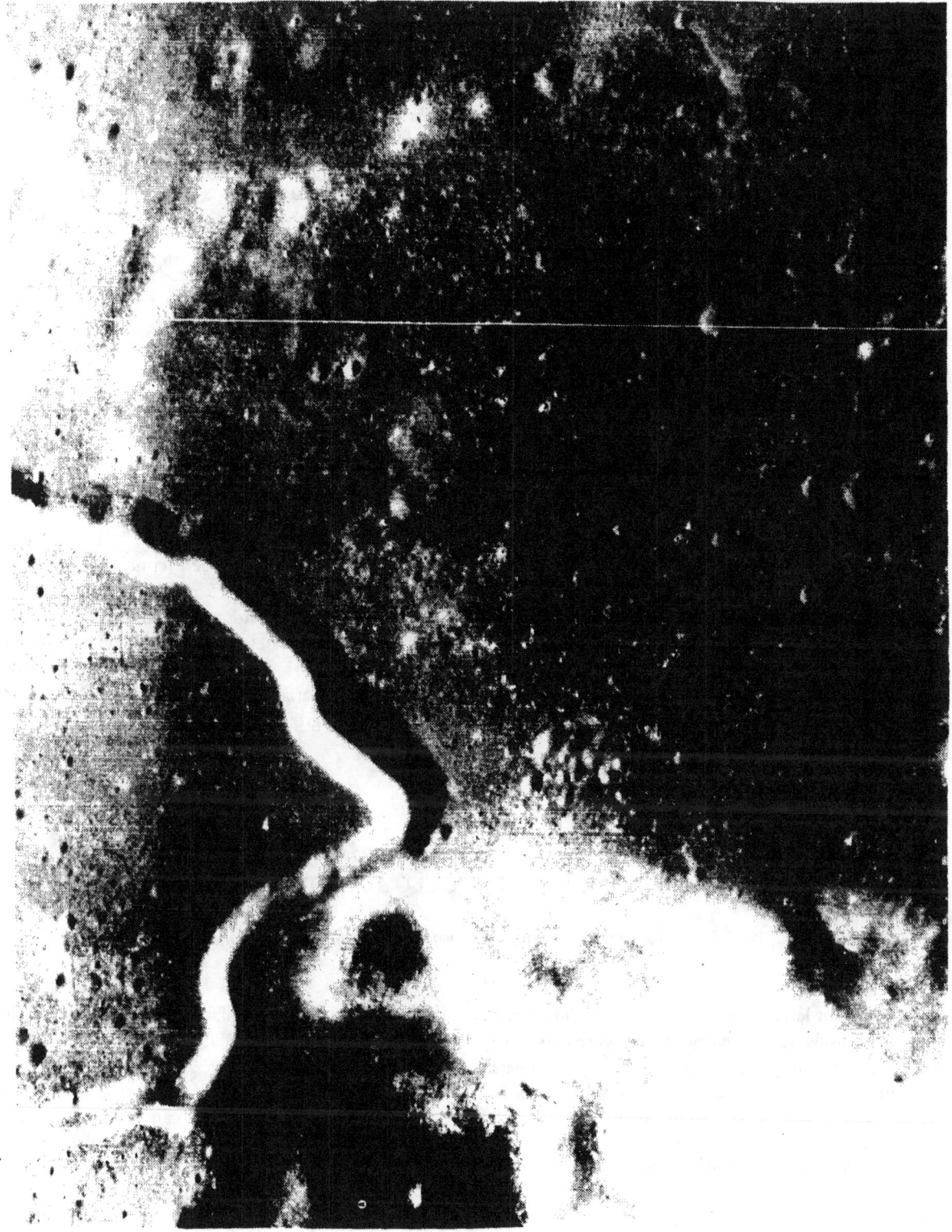

FIGURE 25-31.—Portion of the rectified photograph (fig. 25-30) of the Hadley Rille site.

FIGURE 25-32.—Portion of a mapping camera frame enlarged to a scale of 1:270 000 with a resolution of 11 line pairs/mm.

TABLE 25-VII.—*Model-to-Control-Point Fit*

Point no.	Model 1 (a)			Model 2 (a)			Model 3 (a)		
	ΔX	ΔY	ΔE	ΔX	ΔY	ΔE	ΔX	ΔY	ΔE
1	+6	−4	+13	+7	+7	+5	+10	+9	+4
4	−4	−16	−2	−7	−8	0	−1	−17	−3
6	−1	+3	+14	−3	+5	−5	−3	+4	0
7	−10	−3	+11	−2	0	−1	−9	+3	+4
9	+3	−5	+5	−5	−4	+4	+10	−10	+4
12	+2	−9	−6	+6	−7	−6	+5	−9	−3
28	+7	+5	+1	−3	+6	0	0	−1	−5
31	+11	−19	0	+1	−6	+4	+11	−16	−12
33	−6	−10	+13	0	−11	+1	−4	+2	+2

[a]All values are in meters; difference between the model and the control coordinates, where X is the X coordinate, Y is the Y coordinate, and E is the elevation.

latter value is precisely the number predicted before any photography was obtained.

Conclusions

The tests that have been conducted to date indicate that the Apollo 15 photographs will satisfy all the data-analysis and map-product requirements that were foreseen before the SIM bay instruments were assembled. However, the task of producing the anticipated materials will require the dedicated efforts of a large group of people for many months.

PART C

PHOTOGRAMMETRY OF APOLLO 15 PHOTOGRAPHY

Sherman S.C. Wu,[a] Francis J. Schafer,[a] Raymond Jordan,[a] Gary M. Nakata,[a] and James L. Derick[a]

Mapping of large areas of the Moon by photogrammetric methods was not seriously considered until the Apollo 15 mission. In this mission, a mapping camera system and a 61-cm optical-bar high-resolution panoramic camera, as well as a laser altimeter, were used. The mapping camera system comprises a 7.6-cm metric terrain camera and a 7.6-cm stellar camera mounted in a fixed angular relationship (an angle of 96° between the two camera axes). The metric camera has a glass focal-plane plate with reseau grids. The ground-resolution capability from an altitude of 110 km is approximately 20 m. Because of the auxiliary stellar camera and the laser altimeter, the resulting metric photography can be used not only for medium- and small-scale cartographic or topographic maps, but it also can provide a basis for establishing a lunar geodetic network. The optical-bar panoramic camera has a 135- to 180-line resolution, which is approximately 1 to 2 m of ground resolution from an altitude of 110 km. Very large scale

---
[a]U.S. Geological Survey.

FIGURE 25-33.—Portion of a mapping camera frame (fig. 25-32) enlarged to a scale of 1:90 000 with a resolution of 4 line pairs/mm.

specialized topographic maps for supporting geologic studies of lunar-surface features can be produced from the stereoscopic coverage provided by this camera.

For a preliminary scientific evaluation of the photogrammetric and geologic applications of the Apollo 15 photographs, three models of the metric photography and three models of the panoramic photography were set up in the AP/C analytical plotter in Flagstaff, Arizona, and in the AS-11-A1 analytical plotter at the NASA Manned Spacecraft Center. The areas selected for model coverage were of potential landing sites or contained features of special geologic interest. The models were also selected with consideration given to different Sun angles at the time the photography was taken.

Two contour maps and three profiles were plotted from the models on the AP/C plotter, and three contour maps and nine profiles were plotted from the models on the AS-11-A1 plotter.

For this preliminary evaluation, second-generation positive transparencies from the metric photography and fourth-generation positive transparencies from the panoramic photography were used. Input data to the plotters followed the camera calibration information. Curvature correction was applied to the models and, because no control data were available from the mapping system, scaling information had to be obtained from the postmission data of the unmanned Lunar Orbiter photographs.

Photogrammetric Evaluation

Unlike a frame camera, the panoramic camera has several sources of distortion such as the cylindrical shape of the negative film surface, the scanning action of the lens, the image-motion compensator, and the motion of the spacecraft. Therefore, film products must be processed on a specially designed analytical plotter such as the AS-11-A1.

Before the Apollo 15 mission, within the limits of the capability of the AP/C plotter in handling the panoramic photography, tests were made on five models of the 61-cm panoramic photographs taken in Arizona at an altitude of 18 300 m and at a scale of 1:30 000. The average standard errors of repeated observations of horizontal and vertical ground coordinates from four operators are ±18.3 and ±33.5 cm in the center part and ±0.6 and ±0.8 m in the outer parts. From this test, it was predicted that the standard error of repeatability of horizontal and vertical ground measurements from the Apollo panoramic photography would be in the range of from ±1.1 to ±4.0 m near the center and ±2.2 to ±5.0 m in the outer parts of the photographs. It was also predicted that the contour interval might be as small as 10 or even 6 m.

From this preliminary evaluation, the operator who constructed the models of Apennine-Hadley and Proclus areas (figs. 25-34 to 25-36) using the panoramic photography on the AS-11-A1 plotter obtained a ±1-m repeatability on a well-defined image point and ±3 m on a less well defined image point. Thus, it is possible to obtain a 5-m contour interval with a pair of good-quality photographs. Another operator who worked on the model of Vitruvius (fig. 25-37) on the AS-11-A1 plotter obtained a standard error of repeatability of ±5 m. In this instance, the area mapped was near the edge of the photographs.

Standard errors of horizontal and vertical repeatability from the metric photography on the AP/C plotter are ±7.8 and ±13.9 m, respectively, from a model of Tsiolkovsky Crater (fig. 25-38) and are ±7.4 and ±16.4 m from a model of the landslide area on the northwest side of Tsiolkovsky (figs. 25-39 and 25-40).

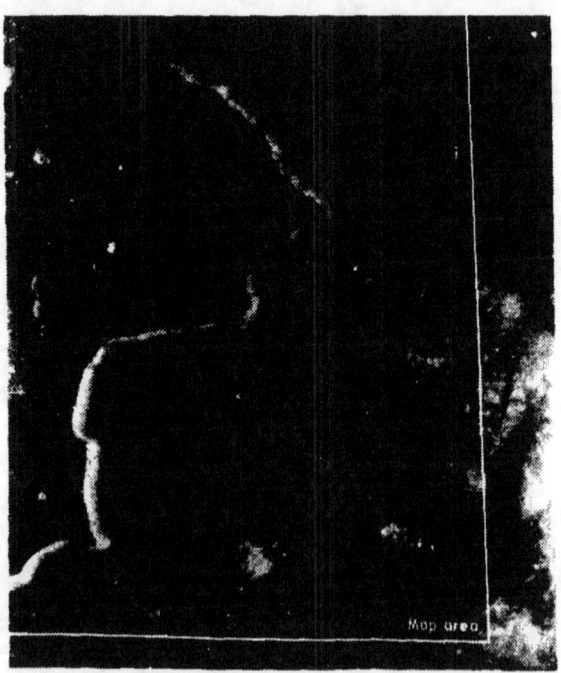

FIGURE 25-34.—Photograph used in a model of Hadley-Apennine landing site (pan AS15-9809).

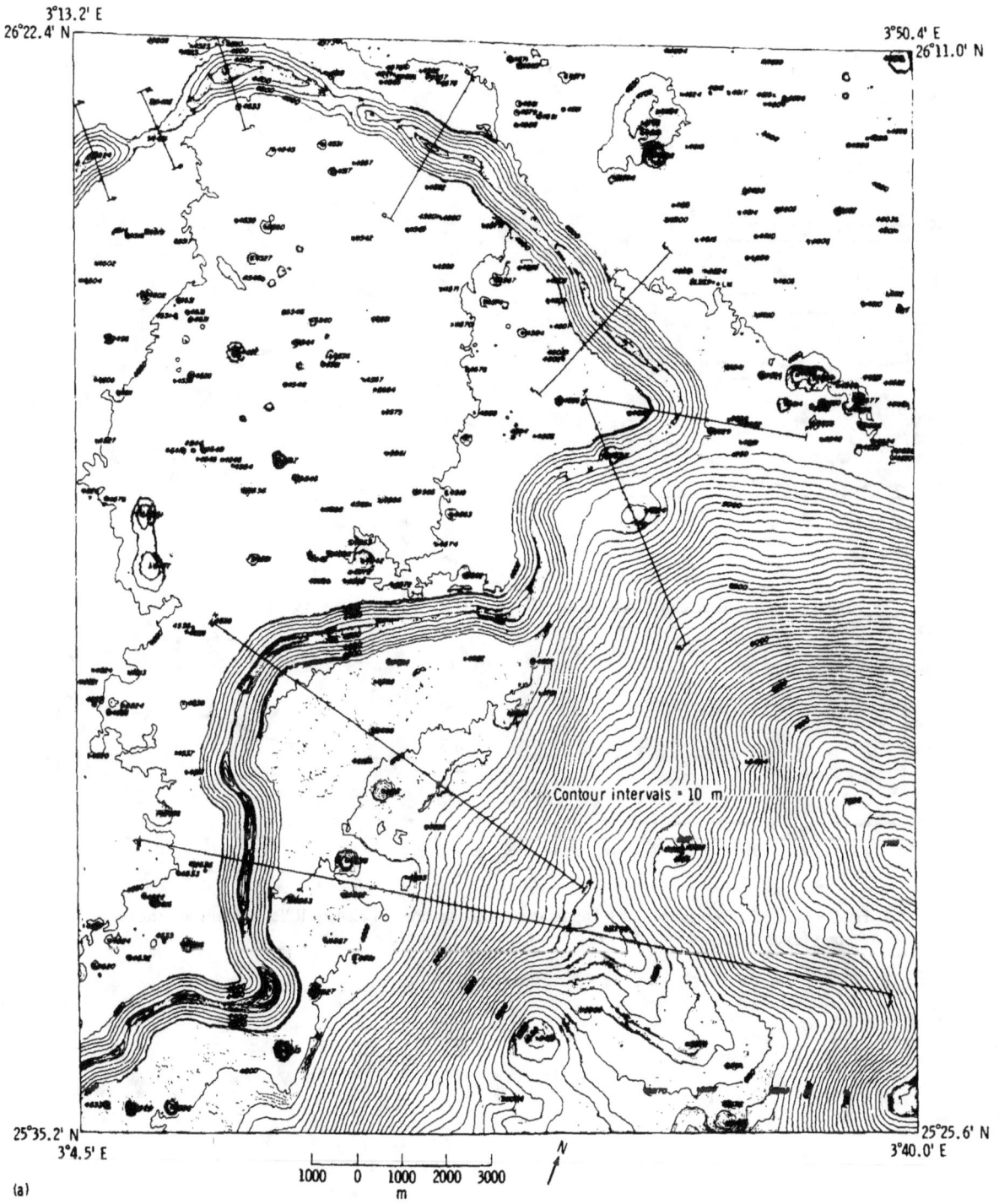

FIGURE 25-35.—Maps of the Hadley-Apennine landing site. (a) Contour map constructed from panoramic photographs AS15-9809 and 9814.

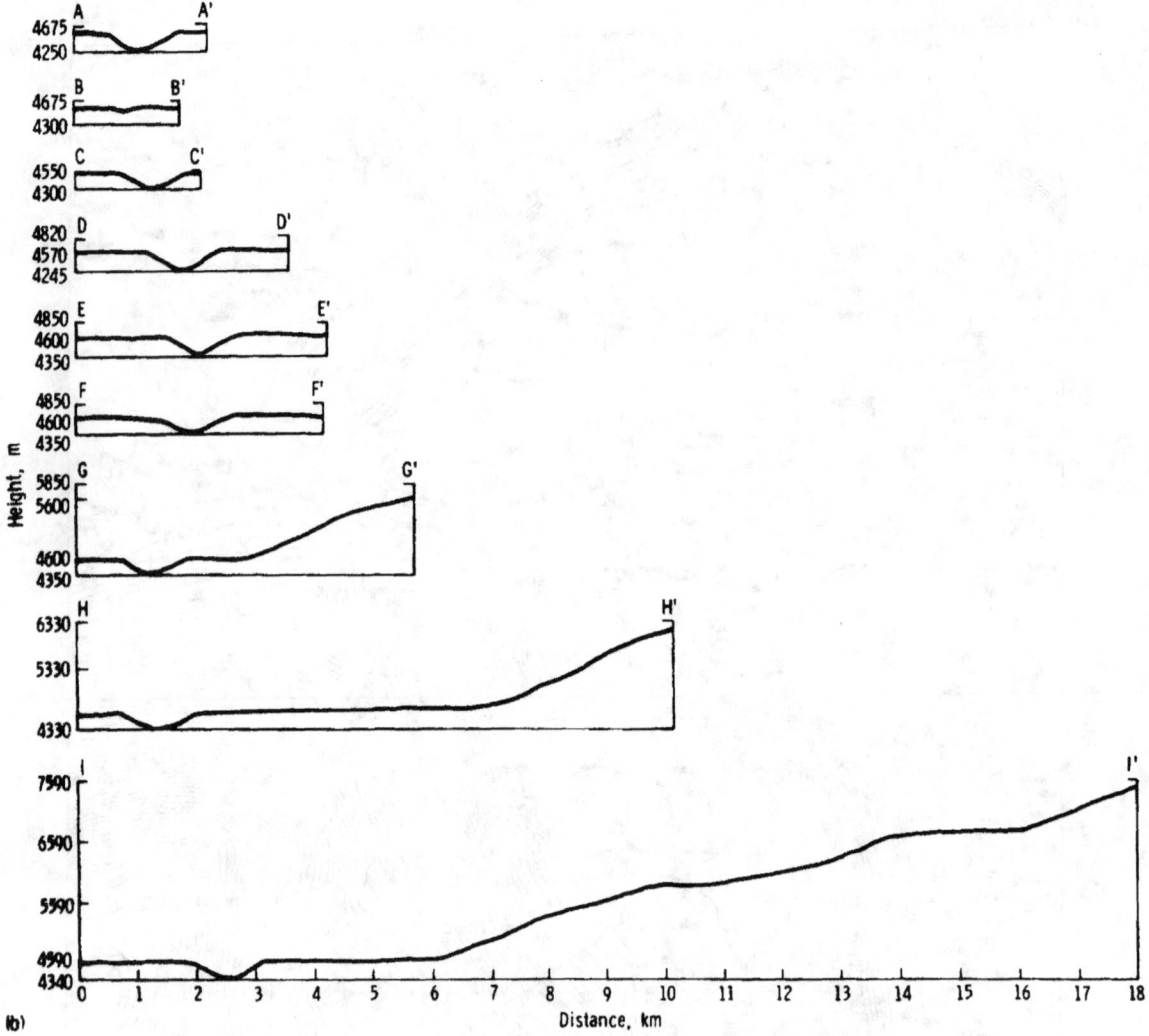

FIGURE 25-35.—Concluded.—(b) Topographic cross sections of Hadley Rille; sections are shown in part (a) of this figure.

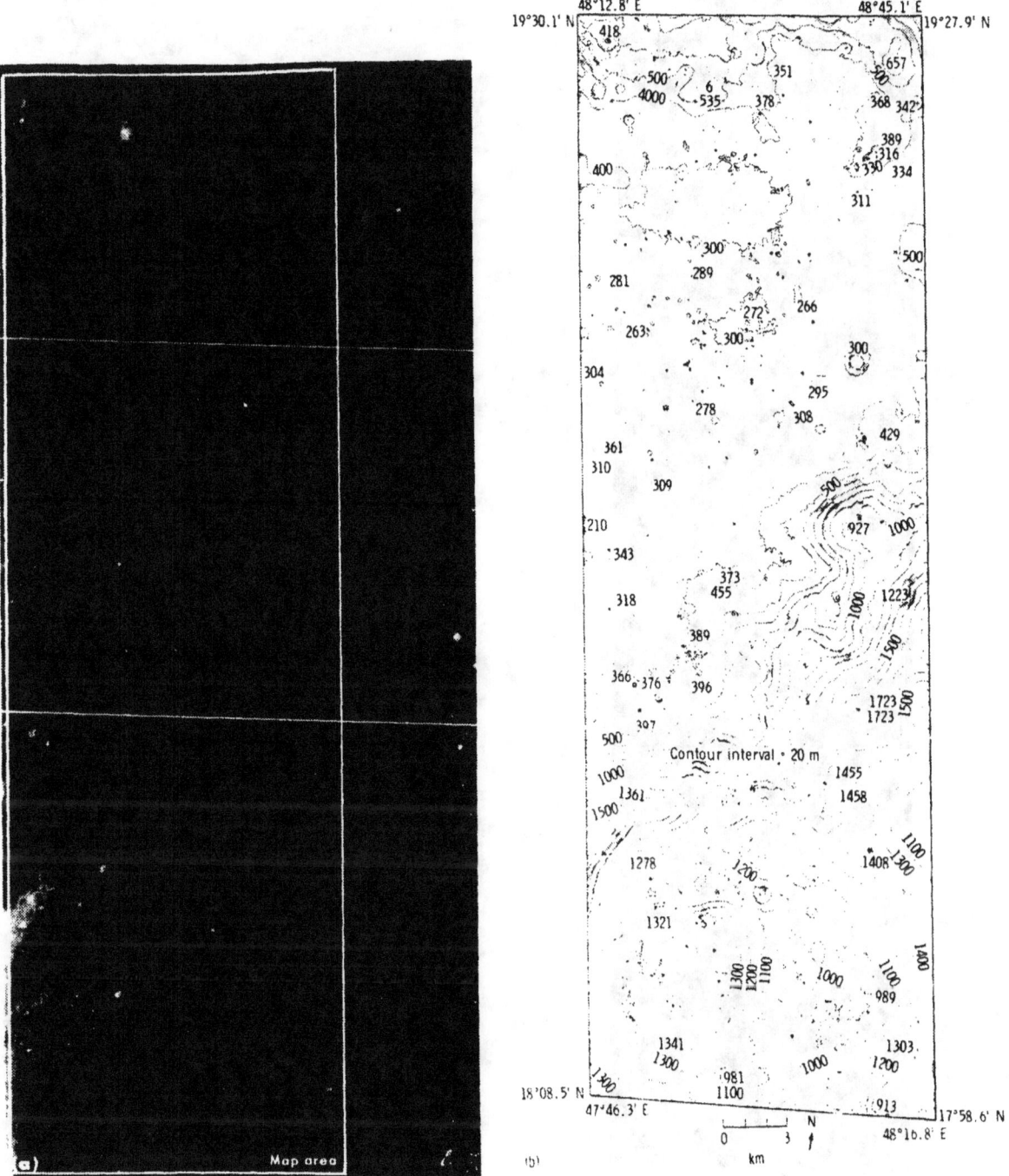

FIGURE 25-36.—Proclus Crater region. (a) Photograph used in model (pan AS15-9243). (b) Contour map constructed from panoramic photographs AS15-9238 and 9243.

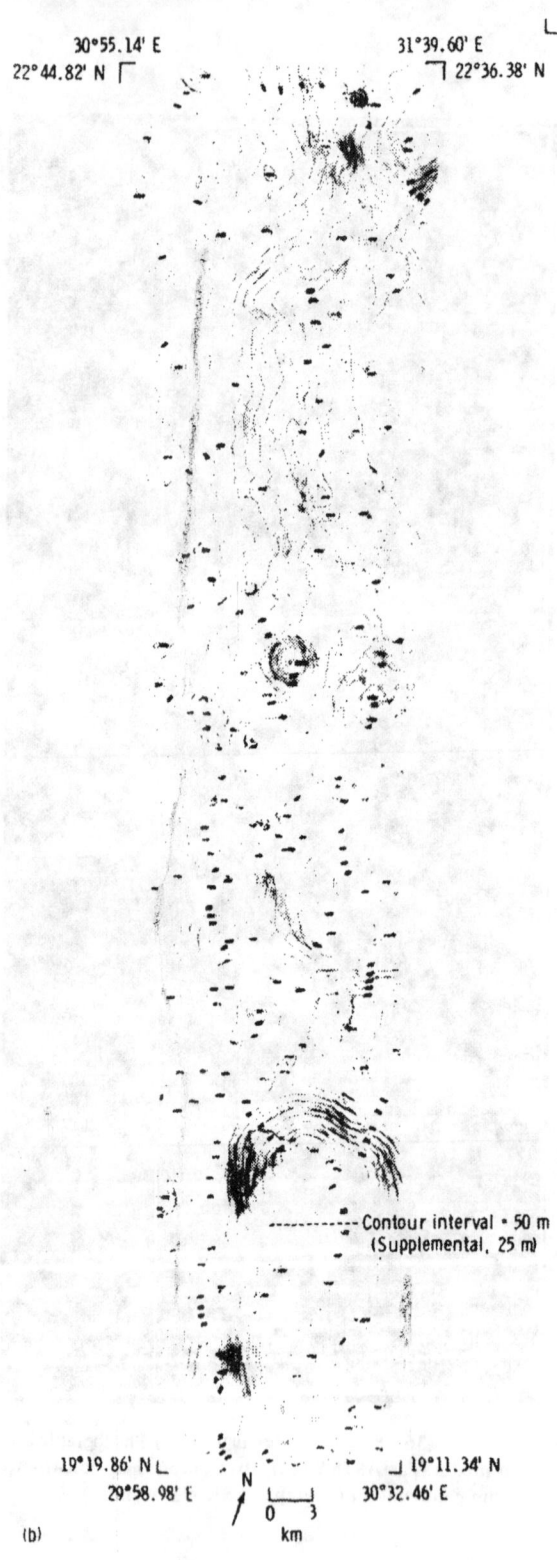

FIGURE 25-37.—Vitruvius Crater region. (a) Photograph used in model (pan AS15-9559). (b) Contour map (scale, 1:50 000) constructed from panoramic photographs AS15-9554 and 9559.

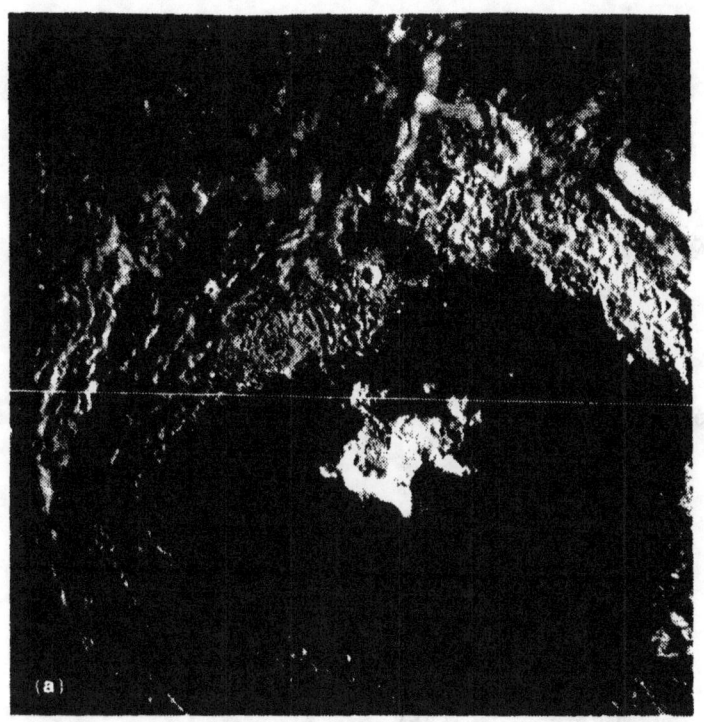
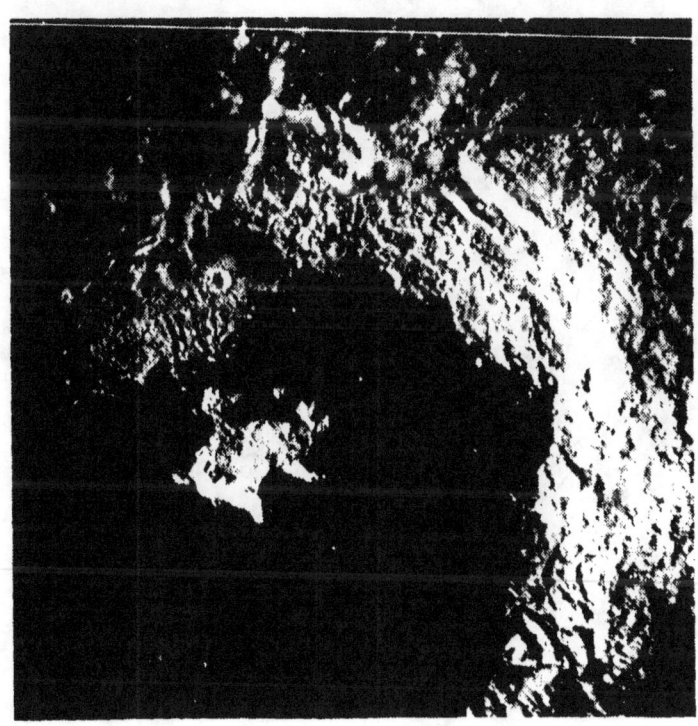

FIGURE 25-38.—Tsiolkovsky Crater region. (a) Photographs used in model (left, met AS15-0889; right, met AS15-0888).

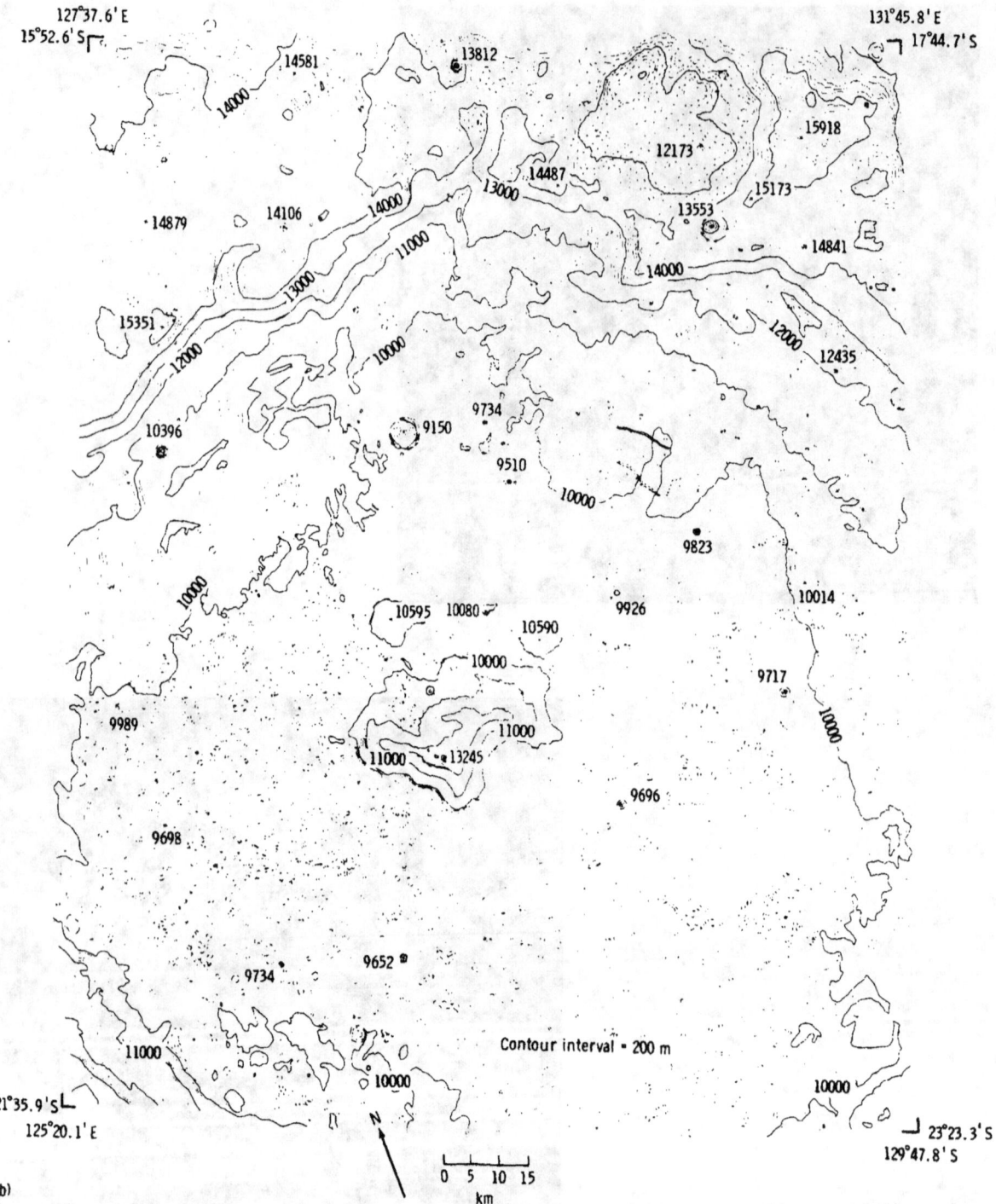

FIGURE 25-38.—Continued—(b) Contour map constructed from same two metric photographs.

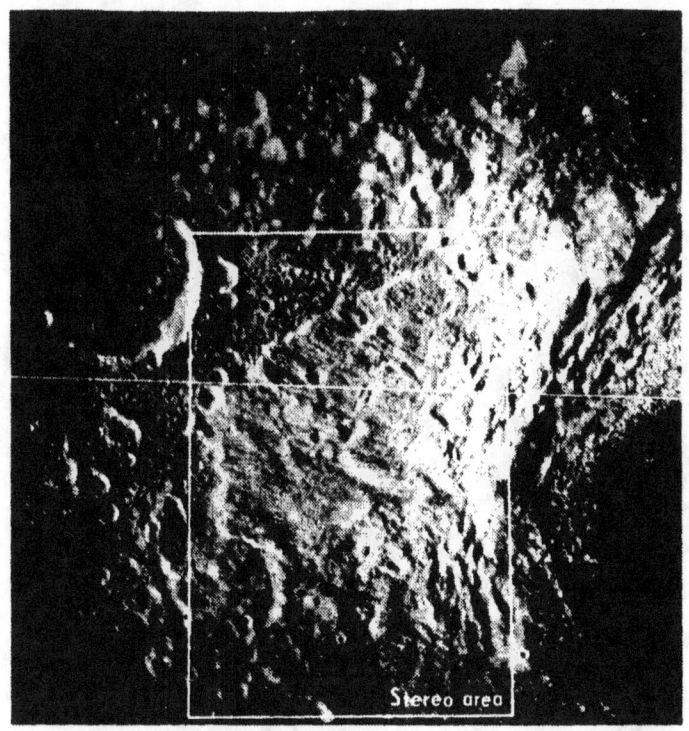
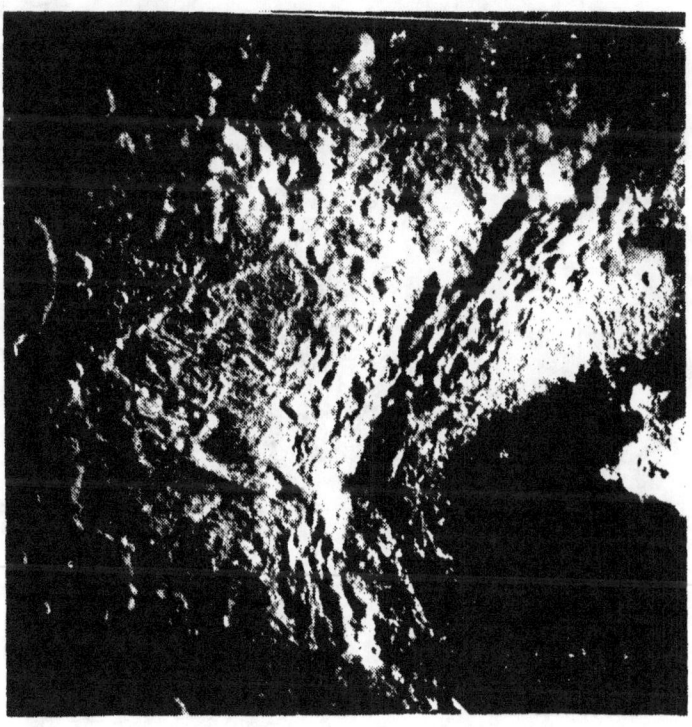

FIGURE 25-39.—Photographs of landslide area northwest of Tsiolkovsky Crater (left, met AS15-0892; right, met AS15-0891).

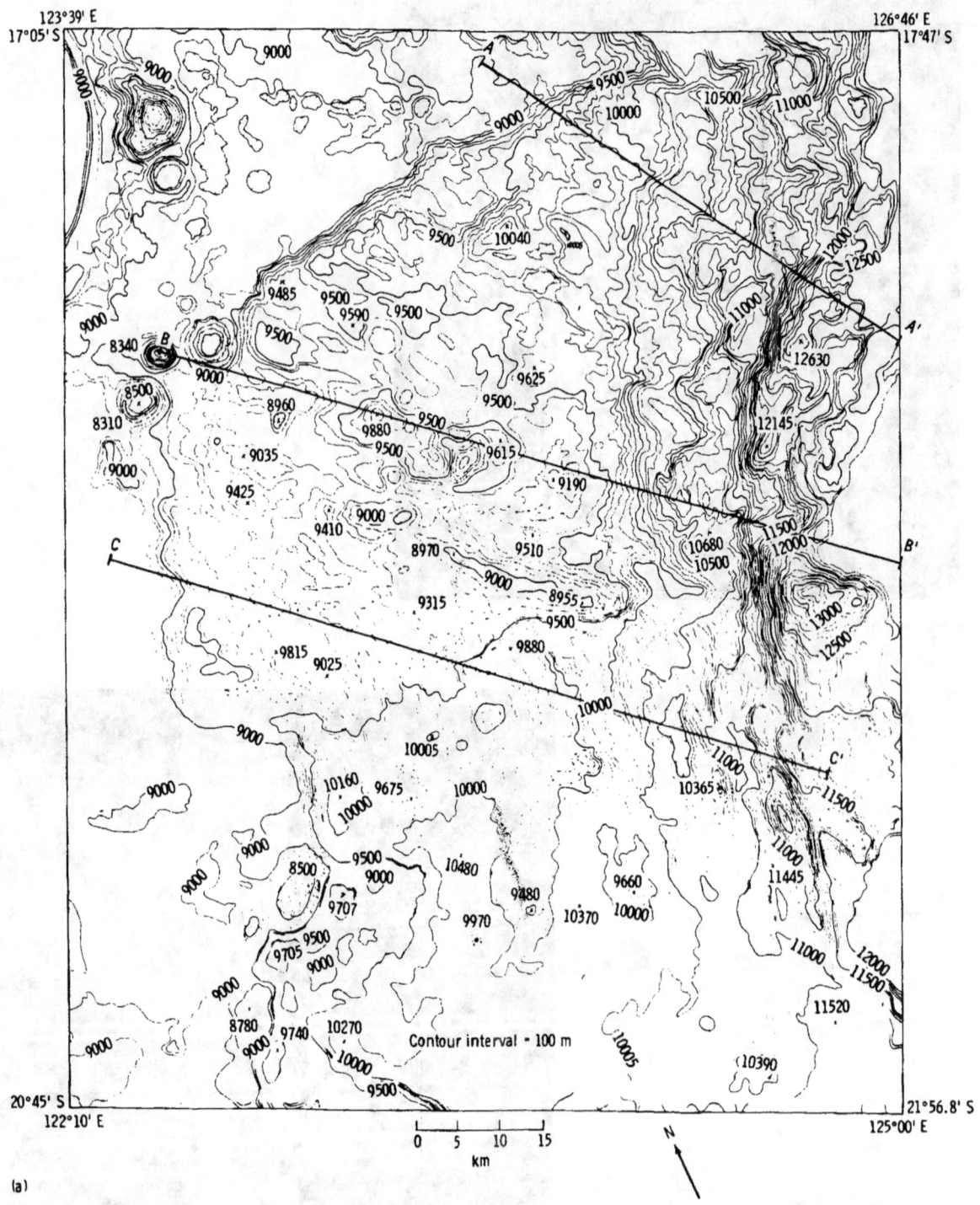

FIGURE 25-40.—Maps of landslide area northwest of Tsiolkovsky Crater. (a) Contour map.

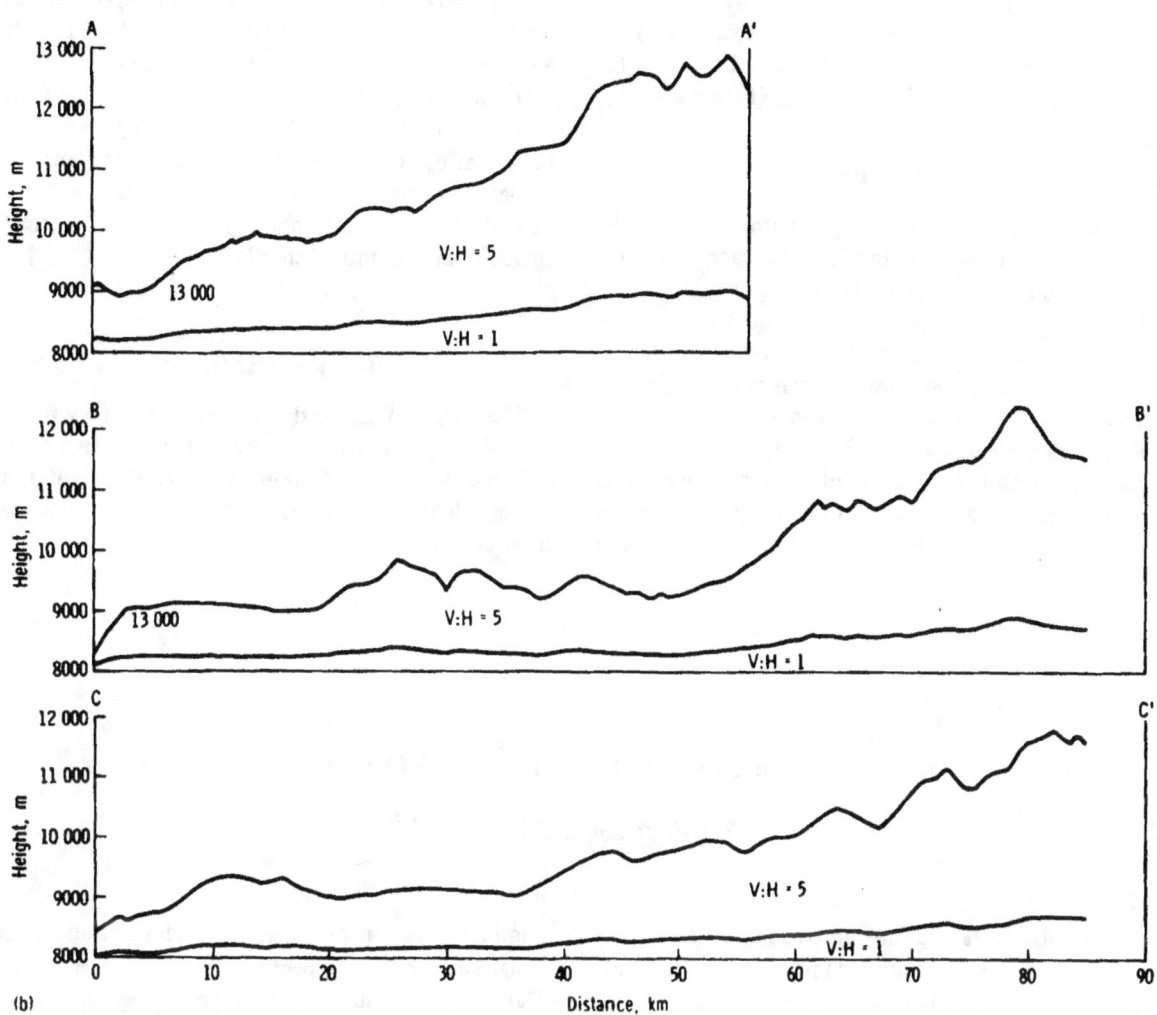

FIGURE 25-40.—Concluded.—(b) Topographic cross sections of area; sections are shown in part (a) of this figure.

TABLE 25-VIII.—*Models Evaluated to Date*

Model	Photographic source	Plotter	Map scale	Contour interval, m	Sun angle, deg
Apennine-Hadley	[a]AS15-9809 and 9814	AS-11-A1	1:25,000	10	49
Vitruvius	[a]AS15-9554 and 9559	AS-11-A1	1:50,000	25 and 50	58
Proclus	[a]AS15-9238 and 9243	AS-11-A1	1:50,000	20	60
Tsiolkovsky	[b]AS15-0888 and 0889	AP/C	1:250,000	100	28 to ~29
Landslide area northwest of Tsiolkovsky	[b]AS15-0891 and 0892	AP/C	1:250,000	100	32 to ~33

[a]Panoramic camera frames.
[b]Metric camera frames.

The models listed in table 25-VIII were evaluated for this preliminary report; all the areas are being mapped geologically to assist in the evaluation of the Apollo 15 landing site and of possible future sites.

Conclusions

The Apollo 15 metric and panoramic photography is the first imagery of the lunar surface that completely fulfills photogrammetric mapping standards. However, except for extending selenodetic control by triangulation, standard photogrammetric equipment cannot be satisfactorily used in compiling maps from this type of photography. The metric photography would have to undergo a photographic enlargement process before it could be used for stereocompilation in a Kelsh, ER-55, or even the Universal autograph such as the A-5 and A-7. To use the panoramic photography for mapping purposes, specially designed analytical plotters such as the AS-11-A1 or the AS-11-B series must be used to correct for the panoramic distortions and to compensate for the variation in scale in the photograph. The AP/C analytical plotter can, however, be used for compiling topographic maps of a very small area in the center part of the panoramic photographs without correcting for the panoramic distortion.

Acknowledgments

The authors wish to thank Harold Masursky for his valuable suggestions; J.W. Van Divier, R.E. Sabala, and the entire Illustrations Unit at the Center of Astrogeology for the preparation of the maps and cross sections.

PART D

APOLLO 15 LASER ALTIMETER

F. I. Roberson[a] and W. M. Kaula[b]

Designed to take an altitude reading at the time of exposure for each metric camera frame, the laser altimeter that was flown on the Apollo 15 mission is part of the metric camera system. In this mode of operation, the altitude, the time exposure, the simultaneous spacecraft attitude derived from stellar photography, and the spacecraft orbit determined from tracking data form a group of parameters that constrain the photogrammetric fit of the overlapping metric camera terrain photographs. An iterative process of triangulation least squares fits all the parameters, including measured coordinates of surface points in overlapping photographs, to an optimum solution. This solution provides geodetic control for cartographic products.

The laser altimeter can also operate independently of the metric camera. This mode of operation, combined with orbital data, provides altitude of the lunar surface with respect to the lunar center of mass. Data acquired about topographic relief around the Moon are extremely useful in concluding facts about the structural basis of mass concentrations.

The Instrument

The laser altimeter consists of a Q-switched ruby laser, transmitting optics, counting timer, receiving optics, and a photomultiplier. When the laser is triggered and fires during operation, a photodiode detects a portion of the output flash, starting a counter that counts increments of 6.67 nsec supplied by a crystal oscillator. When the signal reflected from the lunar surface is detected by the photomultiplier tube through the receiver optics, a stop signal is supplied to the counter. Each 6.67-nsec increment is equivalent to 1-m round-trip traveltime.

Laser output consists of 200-mJ pulses at 6943×10^{-10} m; pulse duration is 10 nsec. The transmitted

---

[a]Jet Propulsion Laboratory.
[b]University of California at Los Angeles.

pulse beam has a 300-μrad beam width, and the receiver optics have a 200-μrad field of view.

Mission Operation

During the Apollo 15 mission, the laser altimeter operated normally through revolution 16. During revolutions 22 and 23, telemetry indications of anomalous performance began to appear. Performance gradually degraded in revolutions 26, 27, and 33; in revolution 38, the altimeter stopped recording valid altitude data. Altimeter data have been reduced only for revolutions 15 and 16 (fig. 25-41). The data are shown as elevations with respect to a sphere of radius 1738 km (centered at the lunar center of mass). The value of 1738 km is used because it is the mean radius of the lunar limb (ref. 25-13).

Interpretation of Results

A harmonic analysis of the data pertaining to east longitude λ obtains a value for elevation h (in km)

$$\begin{aligned}h &= -1.0 - 1.7 \cos \lambda - 1.3 \sin \lambda - 0.7 \cos 2\lambda \\ &\quad + 0.1 \sin 2\lambda \\ &= -1.0 + 2.1 \cos(\lambda - 217°) + 0.7 \cos 2(\lambda - 86°)\end{aligned} \quad (25\text{-}1)$$

The first-degree harmonic is equivalent to a displacement of 2.1 km of the center of mass from the center of volume in a direction almost midway between Mare Serenitatis and Mare Crisium, the two largest near-side positive-gravity anomalies on this track. A displacement as large as 2.1 km implies an inhomogeneity much larger in extent than the mascons, however. This displacement suggests a large-scale

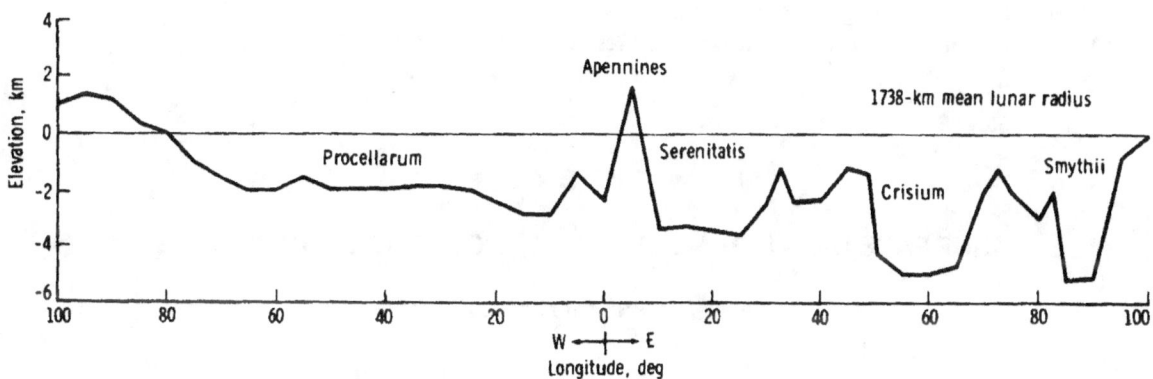

(a)

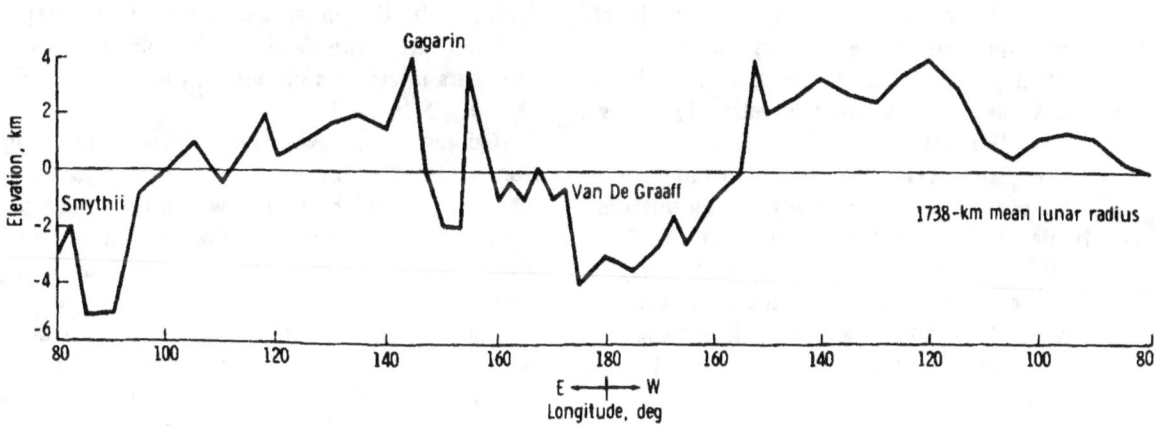

(b)

FIGURE 25-41.—Apollo 15 laser altimeter data. (a) Near side. (b) Far side.

differentiation in the course of the lunar formation (or soon after), the first-degree harmonic having persisted because it does not cause any stresses in the lunar interior. If the differentiation causing the displacement were a varying thickness of an anorthositic crust, this crust would have a maximum thickness of at least $4/3 \times 2.1 \times 3.3/(3.3 - 2.9) = 23$ km on the far side.

The second-degree harmonic is equivalent to a flattening in the direction of the Earth of approximately 0.7 km, bringing the limited sample constituted by these data within 0.3 km of the limb radius determined from visual observations. This variation has an amplitude of approximately an order of magnitude larger than the topographic equivalent of the corresponding term in the gravity field (0.05 km) and is almost exactly out of phase. Because a second-degree harmonic entails stresses, these large differences indicate that almost complete isostatic compensation of the large-scale density differentiations has occurred.

Concerning the best fitting circle (1737-km radius) about the center of mass, the most extensive features are the two far-side highland areas at longitudes 80° to 150° W and 100° to 150° E. These areas also contain the maximum elevations — 5 km. The low features are more numerous but not so extensive; the two lowest elevations (−4 km) are in ringed maria (Crisium and Smythii).

With respect to the best fitting circle about the center of area, displaced 2.1 km from the center of mass toward longitude 143° W, the depression on the far side centered at approximately 175° W becomes the largest feature. This depression is in a region of typical heavily cratered far-side terrain. The depression lies at approximately latitude 27° S, about 500 km from the southern rim of the far-side unfilled mare (ref. 25-14).

Near-side features all have the expected character. With respect to the 1737-km circle, the ringed maria (Serenitatis, Crisium, and Smythii) are all pronounced basins 2 to 4 km deep. Oceanus Procellarum is a rather smooth area depressed by approximately 1 km, and the Apennine Mountains are sharp highlands 2.5 km high.

PART E

SURFACE DISTURBANCES AT THE APOLLO 15 LANDING SITE

N.W. Hinners[a] and Farouk El-Baz[a]

The lunar module (LM) landed at approximately latitude 36° N, longitude 3.5° E in mare material of Palus Putredinis on the eastern edge of Mare Imbrium. Photogeologic maps of the area (ref. 25-15) show that the site is situated in relatively young, ray-mantled, Eratosthenian mare material.

The abundantly cratered mare material in the area undulates gently, sloping generally northward from Mt. Hadley Delta to the south and eastward from Hadley Rille to the west. The ray material, averaging approximately 300 m in width, trends in a northwesterly direction and passes through the landing site. The ray originated either at Autolycus or at Aristillus, probably the former, 150 km to the northwest. At the landing site, the ray material is lighter in color than the surrounding mare material, a fact that was confirmed by the presence of a significant percentage of white fragments and a fair number of glassy fragments in samples collected in the vicinity of the LM (ref. 25-16).

High-resolution panoramic photographs taken from a 110-km orbit of the command-service module (CSM) show, for the first time, the LM structure on the Moon as evidenced by reflected light and by the shadow. This sighting can be most convincingly demonstrated by before and after photography of the landing site (fig. 25-42). The photograph on the left was taken from the LM before landing and at a Sun elevation of approximately 11°; that on the right was taken from the CSM after the LM landing and at a Sun elevation of approximately 14°. The postlanding photograph also shows a bright halo, approximately

---
[a] Bellcomm, Inc.

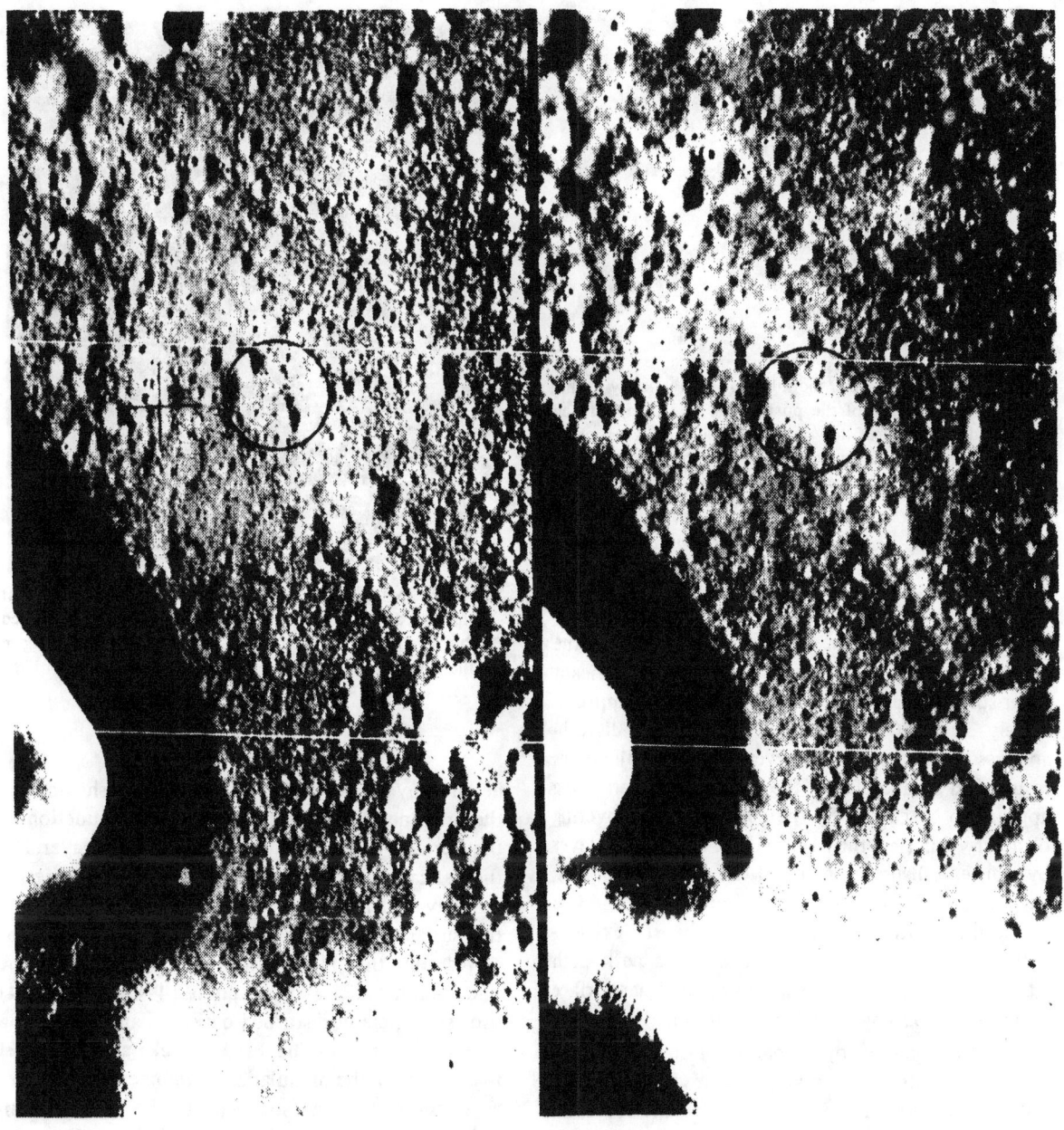

FIGURE 25-42.—Photographs of Apollo 15 landing site showing difference in surface before and after the landing. At left, oblique view as taken from the lunar module on descent approach; at right, view taken from command-service module two revolutions after the landing. The circle encloses the landing site in both views, the latter having a bright halo (left, AS15-87-11719; right, pan AS15-9430).

150 m in diameter, roughly centered on the LM. The halo is attributed to an increase in mare-surface brightness caused by the landing. The symmetry of the halo precludes reflected sunlight as a cause because reflected light should be observed largely only to the east. A relatively bright area southeast of the LM is also visible. This area, in contrast to the halo, is also relatively bright in the prelanding photograph, and this brightness is attributed to the eastward-sloping wall of an old, subdued crater;

FIGURE 25-43.—The lunar module shows very clearly in this 100× enlargement of the postlanding photograph taken after deployment of the Apollo lunar surface experiments package. The dark markings between the lunar module and the experiments site are noteworthy.

however, some enhancement of the brightness as a result of the LM landing is not ruled out. That a surface alteration occurred during landing is not surprising, because, in each Apollo mission, the descent-engine exhaust plume has caused significant lunar-surface erosion at LM altitudes below approximately 30 to 50 m. However, most surface disturbances seen at both the Surveyor and Apollo sites resulted in a darkening of the surface. This darkening appears to be caused by the destruction or covering of a very thin (less than 1 mm) high-albedo skin layer by darker subsurface material (ref. 25-17). In fact, such a disturbance can be seen in figure 25-43, a panoramic photograph taken during the first extravehicular activity. The photograph shows a dark path leading from the LM to the Apollo lunar surface experiments package (ALSEP) deployment site. This darkening is caused by a coating of subsurface soil kicked up by the lunar roving vehicle (Rover) wheels and by the astronauts walking next to the Rover en route to the ALSEP site and by the Rover wheels alone on the return trip. (As is explained, none of the darkening is caused by the Rover wheel tracks themselves.) Considering the above, one would predict that exhaust-induced erosion would destroy the thin, high-albedo skin, leaving a dark halo surrounding the landing point, in direct contrast to what is observed.

The probable answer to the problem lies in a consideration of the lunar photometric function, as

FIGURE 25-44.—Comparative photograph showing bright tracks from the modularized equipment transporter used during the Apollo 14 mission. The surface darkening in the immediate vicinity of and farther to the right of the tracks should be noted (AS14-66-9337).

detailed by Hapke (ref. 25-18), who has shown that the brightness of the lunar soil is a direct function of the porosity or bulk density of the optical layer. His theoretical function predicts that a decrease in porosity will result in a photometric brightening of the surface. This prediction has, in fact, been observed on the Moon and in the laboratory. A photograph taken at the Apollo 14 site (fig. 25-44) shows a generally disturbed darkened area containing two bright tracks. The bright tracks are the wheel marks left by the modularized equipment transporter as it compacted the soil beneath the wheels. Similarly, in laboratory studies (ref. 25-19), it has been shown that compaction of the lunar soil results in an increased brightness. Therefore, it is speculated that the bright halo surrounding the LM is a photometric effect caused by the compaction of the lunar soil under the influence of the dynamic pressure of the descent-engine exhaust gases. Preliminary calculations indicate that such pressures approach the 6.89×10^3 N/m^2 level, quite sufficient to decrease the porosity of the photometric layer greatly. This porosity may initially be as high as 80 to 90 percent, according to Hapke's model. Quantitative calculations on the actual differences in lunar brightness and gas pressures remain to be performed. Porosities of 40 percent (which correspond to bulk soil densities of ~ 1.8 g/cm^3 (ref. 25-19) and would not be changed much by the dynamic gas pressure) do not apply to the photometric layer.

PART F

REGIONAL GEOLOGY OF HADLEY RILLE

Keith A. Howard[a] *and James W. Head*[b]

Study of the sinuous Hadley Rille (fig. 25-45) was a primary goal of the Apollo 15 mission. Local geology of the rille near the landing site is described in section 5 of this report. Preliminary study of orbital photography from the Hasselblad, metric, and panoramic cameras makes possible a description of some regional relationships of the rille. Considerable use is also made of a preliminary topographic map (10-m contour interval) of part of the rille (part C of this section). Contours in the mare area generalized from the map (fig. 25-35(a)) are shown in figure 25-46.

Hadley Rille lies at the base of the Apennine Mountains, which form the southeast boundary of the large, multiringed Imbrium basin. The mountains have prominent fault patterns trending northeast and northwest, respectively concentric and radial to Imbrium (refs. 25-15 and 25-20). For much of its course, Hadley Rille follows a mare-filled graben valley, trending northeast between two high mountain massifs (fig. 25-47). Most of the rille is incised in mare material, but locally the rille cuts into the premare mountains. To the northwest, the mare plain extends through a gap in the mountains to join the main part of Palus Putredinus (the Marsh of Decay), and the rille reaches the wider mare plain through this gap. Continuing north, the rille becomes shallower and indistinct, ultimately intersecting a segment of the linear Fresnel II Rille.

The south end of the rille adjoins an elongate cleftlike depression that cuts into old highlands covered by a dark mantle. This cleft is commonly compared to the source craters of terrestrial lava channels (refs. 25-15 and 25-21). The southeast part of the cleft appears to be controlled by fractures related to the Imbrium basin (ref. 25-15), and the Apollo 15 crew emphasized that the cleft merges with a fracture that cuts far into the highlands to the southeast. The cleft is curved, and the northern part

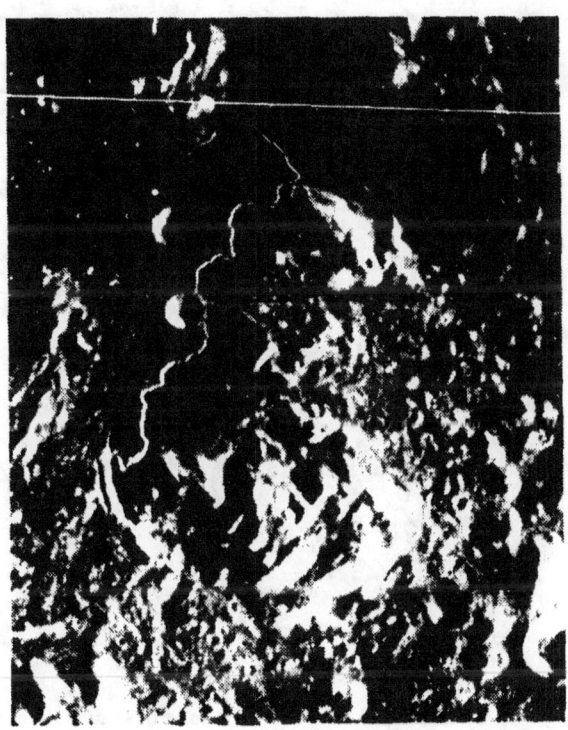

FIGURE 25-45.–Photograph of Hadley Rille (met AS15-M3-1677).

[a]U.S. Geological Survey.
[b]Bellcomm, Inc.

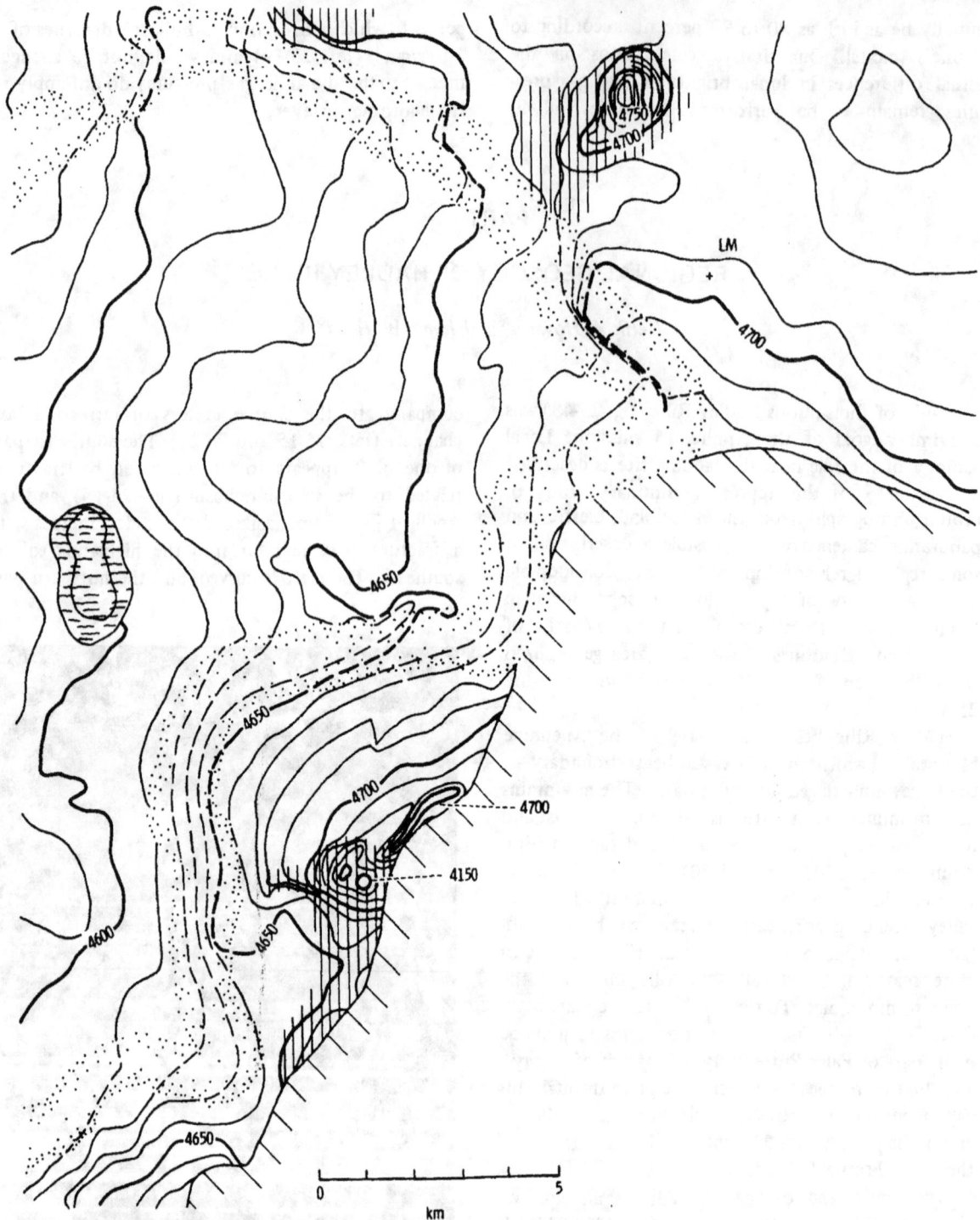

FIGURE 25-46.—Map of part of Hadley Rille (stippled area) near the Apollo 15 landing site. The area is outlined in figure 25-47. In the mare area, contours (in meters) are generalized from a 1:25 000-scale map (fig. 25-35(a)). The contours are extrapolated across the rille as dashed lines to contrast elevations and slopes on either side of the rille. Vertically ruled areas are dark, low hilly regions, which were shown on premission geologic maps (ref. 25-15) as hilly material coated by a dark mantle. Slant hachures indicate Apennine massif. Wavy horizontal pattern indicates materials of a crater doublet of probable volcanic origin.

and slopes northeast instead of west. Elsewhere, the break is smaller or indeterminate; but, in every case where a break exists, the west side of the rille is lower.

Shape of the Rille

The V-shaped profile that characterizes Hadley Rille appears to be a consequence of recession of the walls by mass wasting, so that the talus aprons of the two sides coalesce. Between the landing site and Fresnel Ridge, the rille is discontinuous and consists of a series of coalescing bowl-shaped depressions that are clearly the result of collapse (refs. 25-15 and 25-21). Some parts are shallow and relatively narrow. It is shown in the topographic map (fig. 25-35(a)) that the marginal lip of the rille is lower and less distinct in this area than elsewhere, partly because the immediate mare surface slopes down toward the bowl-shaped depressions. This observation suggests that the mare rocks above the lip are slumped toward the rille.

To test whether individual segments of the sinuous rille may parallel fault structures in the surrounding highlands, azimuth-frequency diagrams were prepared showing cumulative lengths compared to azimuths of the rille (fig. 25-48). It is shown in figure 25-48(a) that, for the southern half of the rille between the cleft and the Apollo 15 landing site, the dominant trends are north and east-northeast. These directions are at substantial angles to the trend of the circum-Imbrium basin graben, which the rille follows, and are also oblique to Imbrian radial structures. This circumstance suggests that individual segments of this part of the rille are not controlled by structures similarly oriented to those in the adjacent highlands. The northerly trend of rille segments parallels part of the cleftlike depression.

In the northern half of the rille northwest of the Apollo 15 site, individual segments tend to follow the main course of the rille—northwest and roughly parallel to Imbrian radial structures (fig. 25-48(b)). Consequently, this part of the rille is less meandering and appears more closely related to structures in the highlands. The impression of structural control is enhanced by faultlike features that intersect this section of the rille (fig. 25-47): a complex of intersecting troughs including Fresnel II Rille at the northwest end of Hadley Rille and four small troughs near Fresnel Ridge. Alternative explanations for the

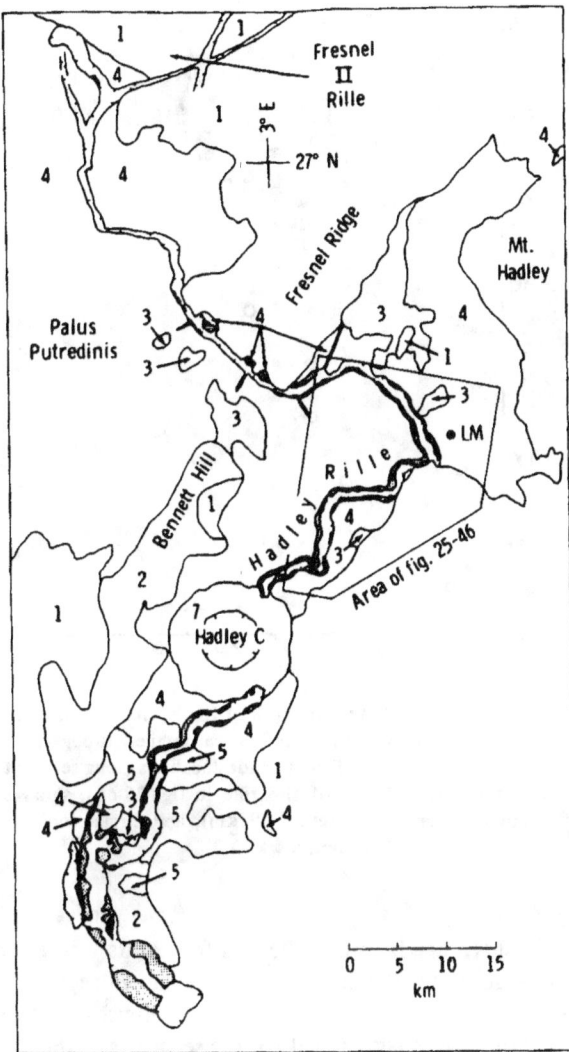

FIGURE 25-47.—Geologic map of Hadley Rille modified from reference 25-15. Numerals indicate (1) materials of premare highlands, (2) very dark mantling highlands materials, (3) material of low hills with the same albedo as adjacent mare material, (4) dark mare basalt, (5) mare basalt with slightly higher albedo, (6) rille or depression (outcrop shown black; bright rocky talus shown shaded), and (7) crater material.

near the mare has a northerly trend that does not reflect fracture trends recognizable in the highlands.

In some areas, the mare surface differs in elevation or in slope on opposite sides of the rille (fig. 25-46). These differences are emphasized by extrapolating contours of the mare surface over the rille. Near the Apollo 15 landing site, the contours break sharply at the rille, because the mare on the east side is higher

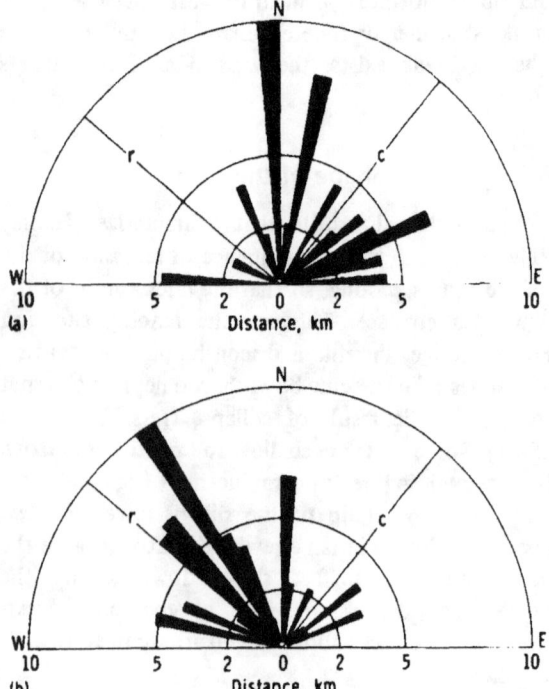

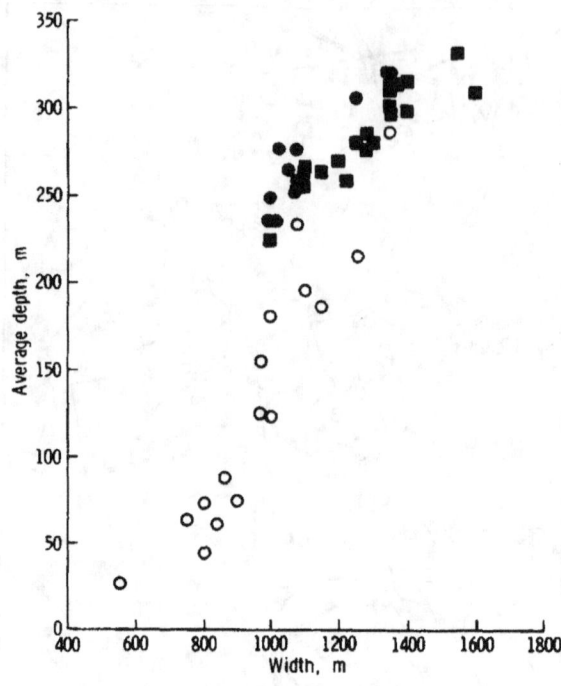

FIGURE 25-48.—Azimuth-frequency diagrams of Hadley Rille. The letter C indicates the trend of structures concentric to Imbrium basin, R indicates the trend of structures radial to Imbrium basin. (a) Southern part of rille, between cleft-shaped depression and Apollo 15 landing site. (b) Northern part of rille, between Apollo 15 site and the intersection with Fresnel II Rille.

FIGURE 25-49.—Width and depth of Hadley Rille in the area of figure 25-46, measured from a detailed topographic map (fig. 25-35(a)). Dots include localities along segment 1 (southern 8.6 km of the rille in fig. 25-46), squares indicate segment 2 (next 24.9 km), and circles indicate segment 3 (northwestern 8.4 km).

four troughs are that they are distributary (ref. 25-21) or tributary (ref. 25-15) lava channels.

Analysis of detailed topography of the part of the rille shown in figure 25-46 indicates an intriguing relationship between depth and width (fig. 25-49) in that the rille is deepest at the widest point. This relationship differs from that shown by river channels, in which depth and width vary inversely, so that the cross-sectional area remains approximately constant. Possibly Hadley Rille was formed by incomplete collapse of a buried lava tube, with more extensive foundering at the widest points.

The depth-to-width ratios are highest in the southern segment and lowest in the northwestern segment (fig. 25-47). These differences indicate that the slopes of the rille walls steepen toward the south, where photography shows that outcrops are thickest and most continuous. Depth-to-width relationships for sharp bends in the rille and for points where the rille abuts the Apennine massif are shown separately in figure 25-50. Both localities have lower depth-to-width ratios than normal for the segments of the rille in which they occur. An explanation for the decreased depth near massifs is that fine debris from the massif partly fills the rille. This relationship near Elbow Crater is described (section 5 of this report) and can also be observed in an area northwest of that shown in figure 25-46 in Orbiter V photograph H-106.

Stratigraphy

Dark mare materials may have been derived, in part at least, from the cleft at the south end of Hadley Rille. In support of this contention, two small dark mare patches are found on either side of the north end of the cleft (fig. 25-47). Possibly related to the emplacement of mare material is a darkening of some of the adjacent older hills (ref. 25-15). The

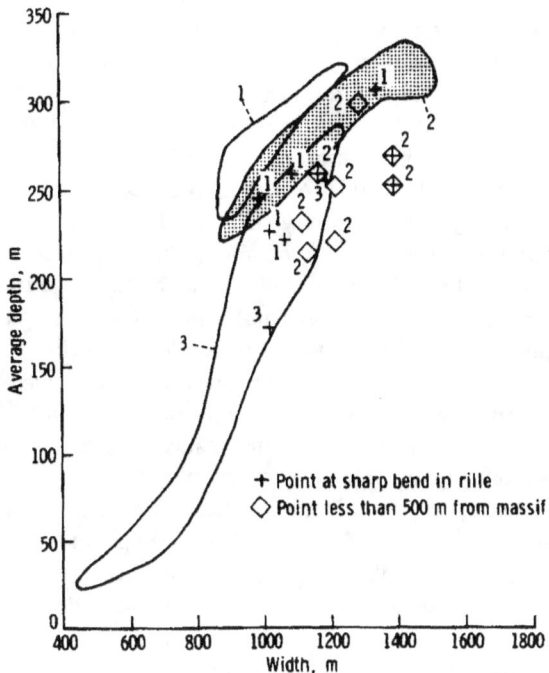

FIGURE 25-50.—Width and depth of points along rille (fig. 25-46) that are located at sharp bends or near massifs. Numbers indicate segments of the rille. Trends shown in figure 25-49 for the three segments are outlined. It should be noted that the numbered points for the three rille segments do not coincide with the trends as determined in figure 25-49.

darkened highlands west of the rille—Bennett Hill and the ridge south of it—are shown by the Hasselblad and metric camera photographs to be muted and smoothed. These characteristics suggest that a thin mantle of material, such as pyroclastics, is responsible for the darkening. It is shown in one Hasselblad color photograph (AS15-97-13232) that the ridge is darker than the mare and that the abrupt albedo boundary coincides with the topographic break. This change in albedo implies that the mare material overlaps the dark mantling material and is younger. A further indication of this superposition is observable at a bright east-facing scarp on Bennett Hill. Bright highland material originally positioned beneath the dark mantle appears to be exposed by mass wasting. The downslope products of the mass wasting are not visible and are presumably buried under mare material.

Near the landing site are several low dark hills (such as North Complex) that have the same albedo as surrounding maria. These hills may be coated with mare lava. One of these hills, 13 km southwest of the landing site, forms the apex of a fan-shaped surface on the mare, descending toward the west (fig. 25-46). The rille curves around the base of this fan. If the mare-lava surface subsided soon after it congealed (ref. 25-15 and section 5 of this report), then possibly the hill is coated by lava and the fan represents a differentially lowered part of the lava surface. The fan is unlikely to be younger than the mare because the northern part is cut by a 1-km-across collapse area that resembles an original lava-surface feature (ref. 25-15). The southern edge of the hill drops abruptly 100 m to the mare surface, so the lava subsidence suggested must amount to 100 m. This subsidence is the same value as suggested for North Complex and the base of Mt. Hadley in section 5 of this report. Differential subsidence of the mare lava may also explain the unusually high elevation of the mare just south of the lunar module, adjacent to Hadley Delta (fig. 25-46).

On the left side of figure 25-46 is a crater doublet of presumed volcanic origin (ref. 25-15). The regional topography of the mare is not modified adjacent to this doublet; apparently, the craters were not an important source of mare lava.

Discontinuous outcrops of mare basalt form a ledge in the upper walls of the rille. The distribution of outcrops recognized from several sources is shown in figure 25-47. Lines of outcrops identified from the ground appear as bright streaks in orbital photography of brightly lit walls and so can be extrapolated away from the landing site. On the side of the rille photographed with grazing sunlight, the outcrop ledge can be identified as a ledge just below the rim. Some thick outcrops are clearly recognizable in Lunar Orbiter V and Apollo 15 Hasselblad and panoramic camera photographs. Outcrops are most continuous in the south (fig. 25-51), and the average thickness of the outcropping ledge appears to diminish northward. Northwest beyond Fresnel Ridge, the rille depth decreases and slopes are evidently not steep enough to expose bedrock.

The thickness of soft material above the top of the major outcropping ledge also decreases toward the north. At the landing site, the vertical distance from the rim to the outcrop ledge is slightly more than 5 m, approximately the regolith thickness in the mare as estimated from crater shapes. South of Hadley C,

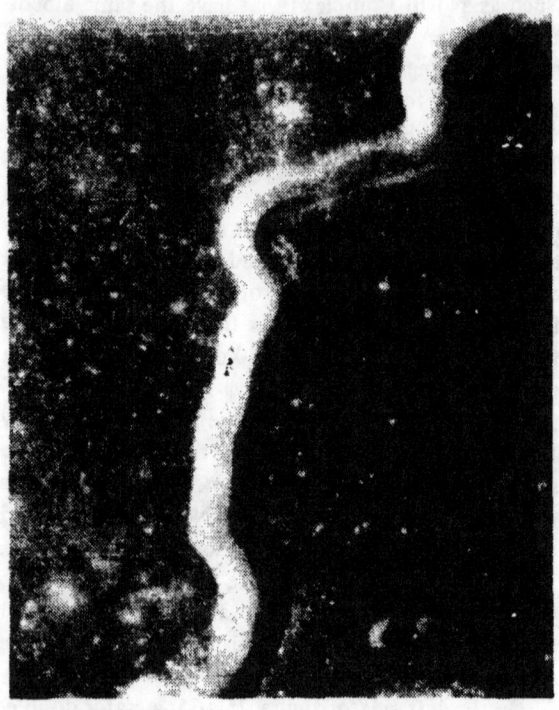

FIGURE 25-51.–Southern part of Hadley Rille, looking northward. The top of the outcropping ledge lies well below the mare surface (AS15-98-10897).

the thickness above massive outcrops is estimated as 15 to 25 m, and rocky protuberances in this interval suggest poorly outcropping strata in addition to the regolith. Orbital photography shows the material above the main outcropping ledge is also thick at the south end of the rille. A regolith produced by impact gardening is present everywhere on the mare surface, but this thickened interval above the outcrop ledge apparently includes, in addition to normal regolith, thin flow units and fragmental material (or both) that thin northward toward the Apollo 15 landing site.

No outcrops occur where the rille abuts premare massifs. This relationship, documented in section 5 of this report for the landing site area, is also recorded in orbital photographs for other rille localities. Apparently, only a thin veneer of hard mare basalt exists at these sites, and the rille cuts into subjacent friable highland material. An analogous situation is shown in Hadley C Crater. This fresh-appearing 6-km-diameter crater postdates Hadley Rille, but remote sensing at various wavelengths shows the crater to be noticeably deficient in blocks, even down to meter and decimeter size (ref. 25-22). Apparently, the hard mare basalt is so thin that Hadley C penetrated deep below it and into more friable basement material.

PART G

LINEAMENTS THAT ARE ARTIFACTS OF LIGHTING

Keith A. Howard[a] and Bradley R. Larsen[a]

Many Apollo 15 orbital photographs, particularly those taken at low Sun-elevation angles, reveal grid patterns of lineaments. In some circumstances, the grid pattern is present in areas where structural control seems unlikely. For example, in an oblique view (fig. 25-52), the ejecta blankets of two fresh impact craters seem to have two intersecting sets of lineaments. Because previous studies of impact craters indicate that concentric and radial trends are commonly present, this pattern is unexpected. A crater-saturated surface on which a faint grid of linear markings can be discerned in shown in figure 25-53. Again, this pattern is unexpected for a surface that

[a]U.S. Geological Survey.

has been exposed to random impacts. In both situations, the azimuths of the main lineaments are approximately symmetrical to the direction of the Sun.

Linear features of debated origin are also evident in landing-site photographs of Silver Spur, Mt. Hadley, and Hadley Delta Mountain (sec. 5 of this report). A structural or stratigraphic origin may be likely for the first occurrence, but this origin is not likely for the other two occurrences because of the presence of a thick regolith and the absence of outcrops or of prominent topographic expression to the lineaments.

Grid patterns of lineaments in scales ranging from kilometers to millimeters have been observed previ-

FIGURE 25-52.—Ejecta blankets of Naumann and Naumann B Craters seem to have grid-lineament patterns conjugate to the Sun direction in this oblique low-Sun photograph (AS15-98-13356).

FIGURE 25-53.—Mare surface (between Schiaparelli and Herodotus D Craters) saturated by small craters shows, at this very low Sun angle, two faint systems of small lineaments that are conjugate to the Sun direction. Main lineament directions are indicated (AS15-98-13348).

ously in many areas of the Moon. Generally, the lineaments are most easily detected when the scene is viewed at low Sun angle. Commonly, the strongest lineament patterns trend northeast and northwest; that is, the patterns tend to be symmetrical about the Sun direction. Recognition of this pattern across most of the near side of the Moon has given rise to the concept of a lunar grid of structural features (presumably fractures) that is globally controlled.

Preliminary results of experiments demonstrate that spurious lineaments and grid patterns can be produced and that the directions are dependent in part upon the position of the light source. The lineations cluster around two maxima that are symmetrical with regard to the direction of the light source. These results will be helpful in interpreting the Apollo 15 photographs, as the reality of the so-called lunar grid is questioned further.

The experiments were designed to duplicate the effect of bright sunlight reflecting from a hummocky surface with little or no diffuse light in the shadowed areas. This effect was accomplished by using a dark room and a light beam with an angular width of 0.5°. For target materials, a finely ground powder of Cu_2O_3 was used first to simulate the lunar photometric function. Gypsum powder, which is more reflective, yielded similar results. Randomly hummocky models were constructed by sprinkling powder through sieves onto a black horizontal surface. Sun elevation, or the angle of incident light, was controlled by tilting the model with respect to the horizontal light beam.

Low oblique lighting caused the appearance of definite conjugate lineament patterns in a grid symmetrical about the lighting direction (fig. 25-54). As a model was rotated, the angle between the light direction and the lineaments remained approximately constant. Many of the lineaments were narrow, discontinuous, bright, or shadowed streaks. Some of the most prominent streaks appeared to be actual linear features formed by the fortuitous arrangement of hummocks in a row. At first glance, many of the patterns appeared quite continuous, but when lines were critically drawn to represent each discernible linear element, the patterns were found to be discontinuous.

Optimum development of the spurious lineaments depends on several factors such as the following.

(1) Contrast between highlighted areas and shadows must be large. If the model is too brightly illuminated, however, the patterns become less obvious. Shadows should be black. Photographs generally enhance the contrast.

(2) The size of individual topographic irregularities (hummocks) must be near the limit of visual resolution at the distance from which the model is viewed. The most useful models were made by sifting uniformly fine target material through progressively smaller mesh sizes down to a minimum size of 350 mesh. This sifting had the effect of superimposing smaller lumps of powder on bigger ones so that hummocks of optimum size existed at any practical viewing distance (fig. 25-55).

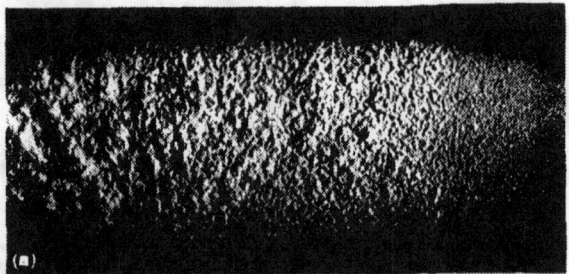

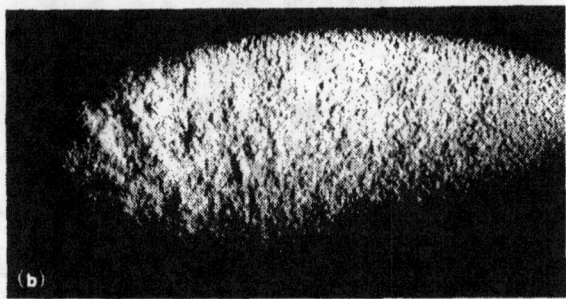

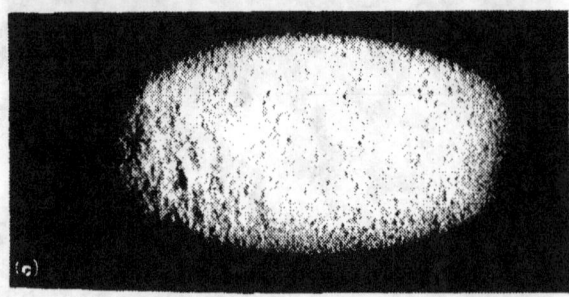

FIGURE 25-55.—Gypsum model, 15 cm wide, viewed normal to the surface, showing lineament effects on three types of terrain. Individual grains and lumps of powder resting on black cardboard are shown on the right, approximately uniform hummocks are shown in the center, and large hummocks on which progressively smaller ones are superposed are shown on the left. Lineament maxima on all models are symmetric about the light direction (from the right). (a) Light elevation, 12°. (b) Light elevation, 22°. (c) Light elevation, 30°.

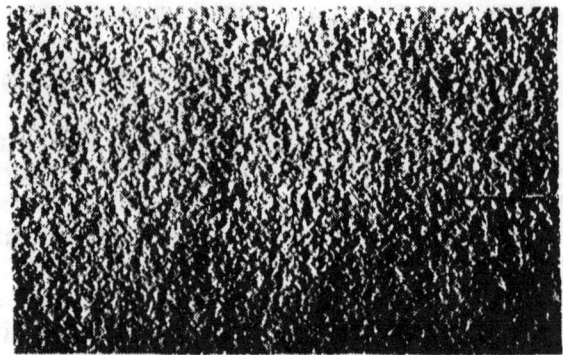

FIGURE 25-54.—Photograph of model showing grid pattern composed largely of spurious lineaments caused by low lighting. Two conjugate lineament sets are 20° to 30° from the axis of the light beam. Source of light is from right (arrow) at an incident angle of 3°. Model is made of gypsum powder sprinkled through 100-mesh sieve onto tabletop.

(3) Most observable lineaments occur at certain favored angles relative to the incident light. The angles are dependent upon the light elevation and also vary according to the topography of each model. The acute angle measured between the azimuthal direction to the light source and the maximum development of lineaments in the models varied from 15° to 55°.

(4) Grazing Sun (fig. 25-54) optimizes the abundance and detectability of lineaments. On the models, faint linear patterns at a Sun elevation as high as 45°

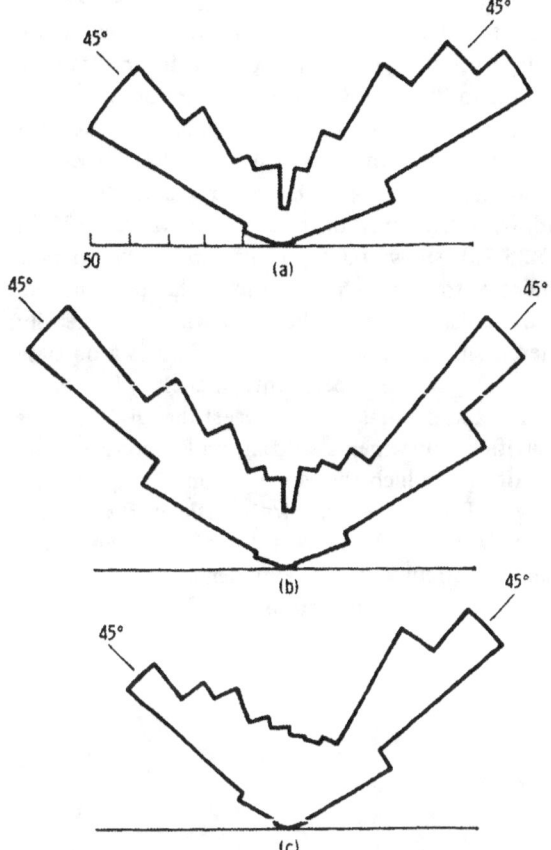

FIGURE 25-56.—Azimuth-frequency diagrams of measured lineaments mapped from the three photographs shown in figure 25-55 and plotted with light source from the right (arrow) to facilitate comparison with plots of lunar lineaments where the Sun is in the east. Lineament trends tend to spread away from light source as the elevation of light increases. (a) Light elevation, 30°. (b) Light elevation, 22°. (c) Light elevation, 12°.

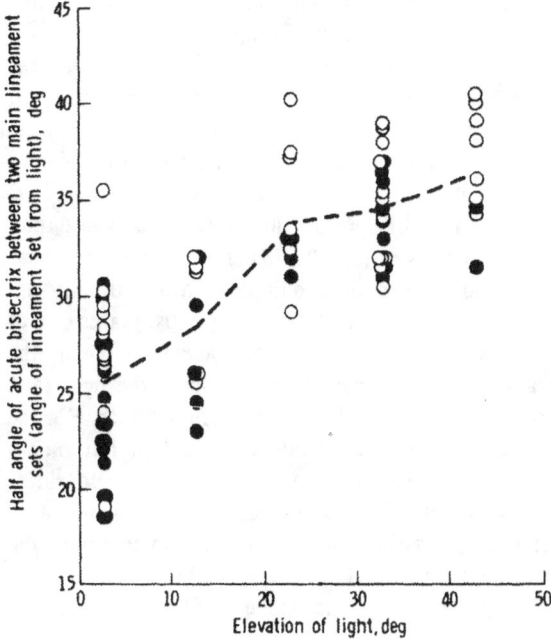

FIGURE 25-57.—Plot showing progressive increase in lineament azimuth away from light azimuth as light elevation is increased. Data are from model shown in figure 25-54. Both authors (black and white dots) measured the angle between the two most prominent appearing lineament sets several times. The half angle is the average angle to the light direction. Dashed line shows average of mean values obtained by each author.

could still be recognized. Increasing the angle of incidence causes the lineament grid to spread out so that acute angles from the direction to the light source increase (figs. 25-56 and 25-57). A few particularly prominent lineaments can be recognized over a 10° to 20° range of Sun angles.

(5) Viewing the model obliquely enhances some or all lineaments because of foreshortening of the scene. This distortion emphasizes features in the direction perpendicular to the line of sight (fig. 25-58). Such distortion is undoubtedly responsible for strengthening the grid patterns seen on the crater rims in figure 25-52. In photographs viewed ob-

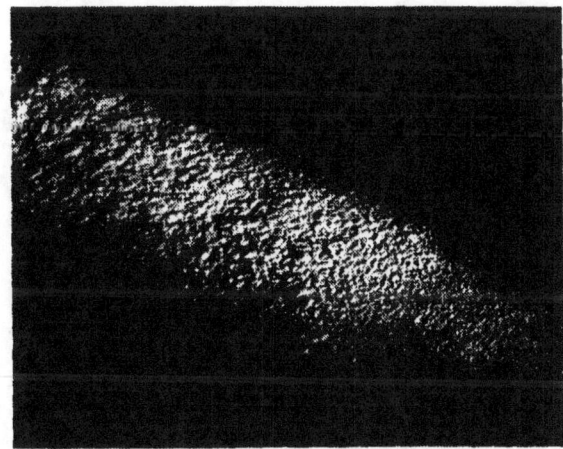

FIGURE 25-58.—Oblique view of model shown in figure 25-55, illustrating effects of foreshortening. Light is from the east (lower right); view is toward the northwest at 30° from the horizontal. Lineaments that trend northeast (horizontally) are enhanced by foreshortening.

liquely, linear patterns in the line of sight also become enhanced in some cases. Oblique viewing of a model can cause a slight azimuthal shift in the apparent maximum direction as opposed to that recognized on the same model when viewed normal to the surface, although some of the individual lineaments may be recognizable from both vantage points.

The cause of spurious linear patterns is not fully understood. Wise (ref. 25-23) showed that, for linear V-shaped valleys, low oblique lighting visually enhances those valleys where the light just grazes one of the valley walls. Accentuated valleys in which the walls are steeper than the elevation of the light tend to occur at acute angles to the light azimuth. For the powder models, fortuitous alinement of hummocks or grains produces discontinuous valleys and ridges that trend in all directions. The oblique light selects or maximizes those segments that coincide with the optimum acute angle to the light. This angle, in turn, is dependent on the slopes of the hummocks. Inasmuch as the frequency distribution of slopes will affect the azimuth distribution of spurious lineaments, azimuthal plots of apparent lineaments may be a clue to the slope-frequency distribution.

A randomly cratered surface apparently behaves similarly to the hummocky surfaces. Close inspection of photographs of Mare Exemplum, a model crater field with randomly distributed craters (refs. 25-24 and 25-25), shows faint grid patterns of lineaments conjugate to the light direction; the patterns are greatly enhanced by oblique viewing. The heavily cratered surface shown in figure 25-53 is a possible lunar analog of this experimental model.

The experimental results suggest that grid patterns of artificial lineaments appear under precisely the conditions in which the Moon is commonly viewed—strong low-angle lighting with little diffusion into shadowed areas. Because of these conditions, considerable difficulty may be experienced in distinguishing real structures from spurious artifacts.

PART H

SKETCH MAP OF THE REGION AROUND CANDIDATE LITTROW APOLLO LANDING SITES

M.H. Carr[a]

The photograph in figure 25-59 and the corresponding map (fig. 25-60) show the geology of part of the lunar surface just east of the Littrow rilles at the eastern edge of Mare Serenitatis. The most striking feature of the region is the extremely low albedo of the area mapped as Eld in the western half of the map. The low albedo is believed to be caused by a thin layer of pyroclastic volcanic material at the surface. Another notable feature is the fresh-appearing ridges that cross the mare and the adjacent terra. The fine-braided texture of these ridges contrasts markedly with the rounded, subdued topography more common to such features, an indication that the ridges here may be unusually young. Also evident is a well-exposed succession of marelike plains units, which probably represent different stages in the filling of the Serenitatis basin. Several sets of rilles are present; most are roughly tangential to the basin and terminate against the different plains units according to the relative ages. In the northwest corner of the map is a relatively fresh volcanic crater chain (Cch) from which material appears to have been ejected over the surrounding terrain, forming rays of volcanic ejecta. The area thus includes an unusually wide variety of lunar features.

The earliest recognizable event in the area is the formation of the Serenitatis basin, probably by impact very early in the history of the Moon. Although Serenitatis ejecta must have been deposited all around the basin, evidence of such material has since disappeared. The pre-Imbrian terra materials (plm, plth, and pltc) presumably include Serenitatis ejecta, but the origin of these units is likely to be complex. Primitive crustal material, volcanic material, and ejecta from other sources may be included. After the formation of the basin, a period of cratering, basin filling, and structural readjustment of the uplands occurred. Littrow and Littrow A Craters were formed at this time. Isostatic adjustment and filling by volcanic and impact ejecta (unit IpIp) have combined to eliminate much of the original relief. Several large blocks of pre-Imbrian terrain now have considerable relief and are surrounded by talus slopes. The relief suggests that the blocks were uplifted long after the formation of the Serenitatis basin, possibly when the Imbrium basin was formed.

Several types of plains-forming units can be recognized, but the relationships between them and the mode of formation are not fully understood. The filling of the upland basins by unit IpIp has already been mentioned; a similar but less cratered unit (Ip) occurs along the western edge of the terra, where it is largely overlain by Eld. North of the map area, however, unit Ip is well exposed and resembles the Apennine Bench material of the Imbrium basin. The unit is crossed by several rilles that abut abruptly against the mare, indicating that Ip is older than the mare unit Im and that the Littrow rilles are of intermediate age. The nature of the unit Ip is obscure,

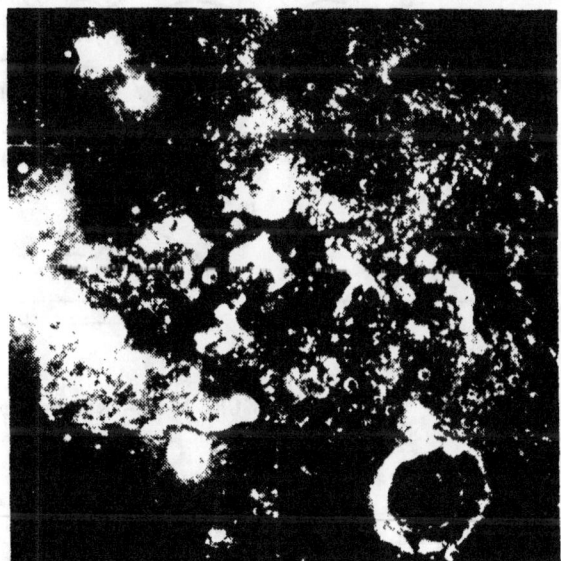

FIGURE 25-59.—Photograph of the Littrow Crater area, which serves as a base for the accompanying map. The large crater filled with dark material midway along the upper edge is Littrow. The bright, medium-sized crater in the lower left is Vitruvius E (met AS15-M3-1113).

---

[a]U.S. Geological Survey.

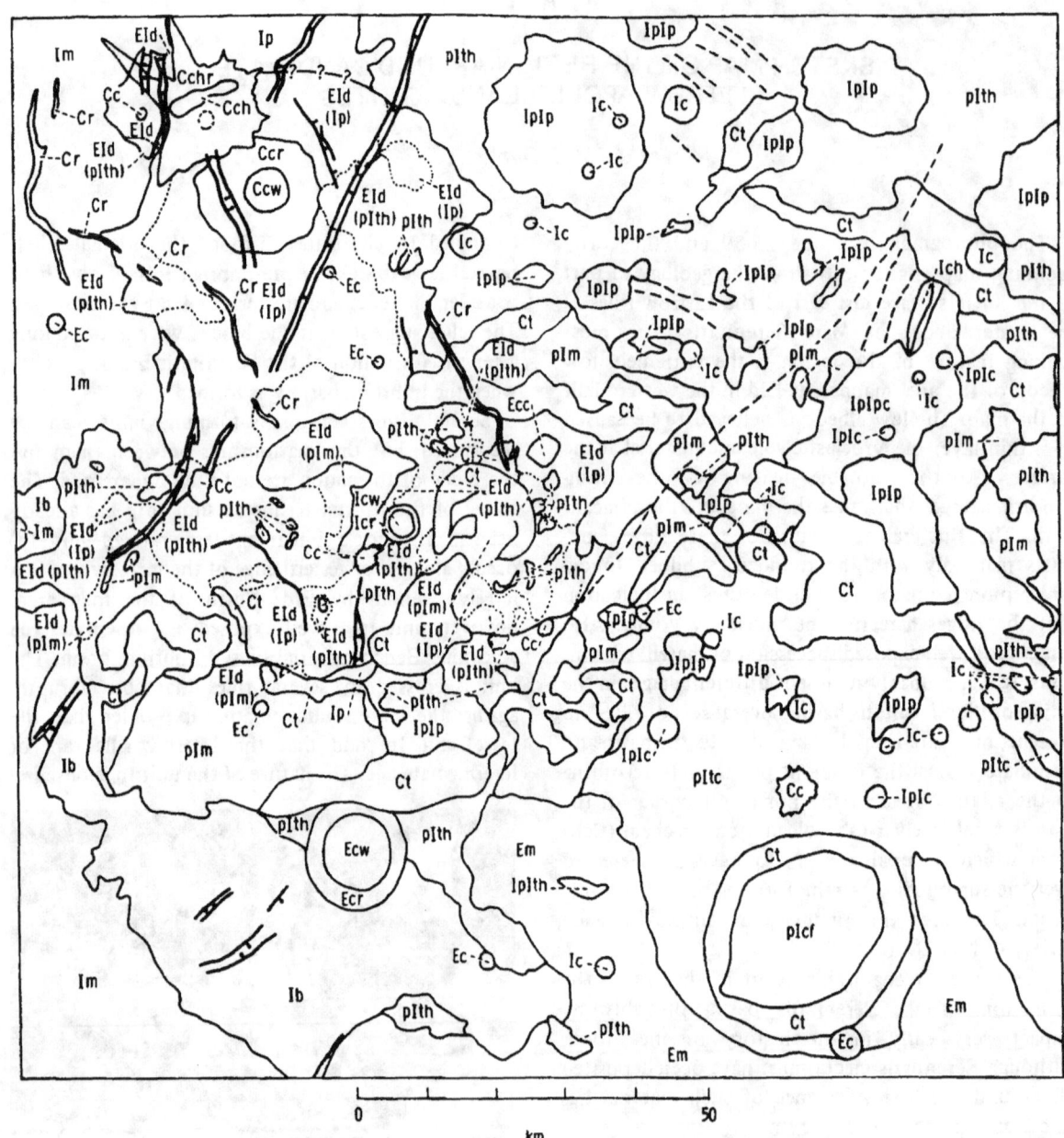

FIGURE 25-60.—Geologic map of the region around Littrow Crater.

but it may be composed of old basaltic flows. A similar plains unit (Ib) occurs south of Vitruvius E Crater. It has an albedo lower than that of the mare unit to the west (Im) but higher than the albedo of the mare unit to the east (Em). Unit Ib is separated on the map from unit Ip, but further study may prove that these units are indistinguishable.

Deposited on top of the two plains units just mentioned and also covering parts of units pIm and pIth is the dark, mantling material EId. The age relationships of the unit are ambiguous, but an Imbrian age is probable; however, the unit appears to cover Im at some locations, and the age is consequently designated Imbrian and Eratosthenian (EI).

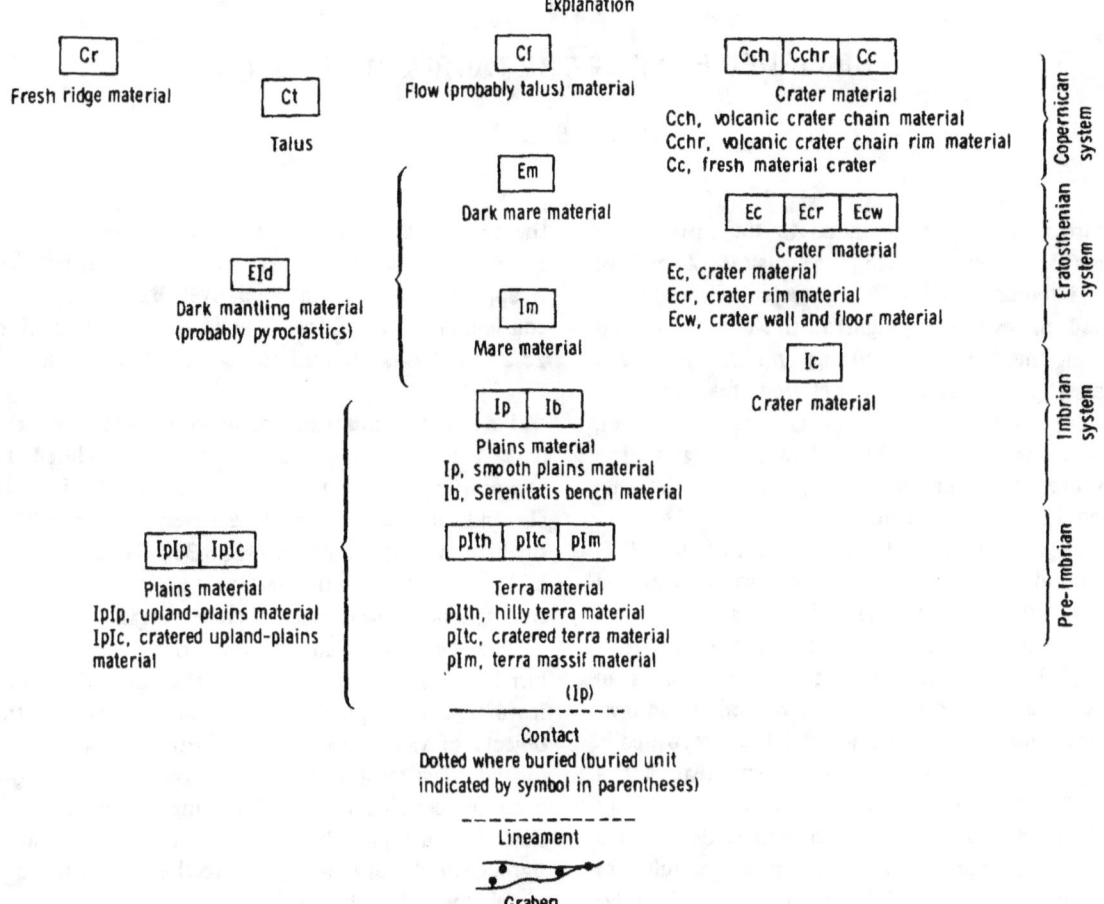

The mantling unit is clearly thin almost everywhere. In hilly terrain, the albedo of the underlying units shows through at the top of small hills where the mantling material has sloughed off, and the rim crests of older craters also protrude through the cover. The deposit has little intrinsic topography, although it clearly subdues the subjacent terrain to a slight degree. Furthermore, the boundaries of the unit appear largely independent of the underlying topography. All these characteristics are suggestive of a thin cover deposited from above, such as would be the case with pyroclastic volcanics. In the center of the map, unit EId appears to be overlain by a flow with a high albedo. This flow is probably a talus deposit derived from the massif to the south.

At the western edge of the map, mare material (Im) appears to embay Ip and pIth, both of which are locally covered with EId. The relationship suggests that in this area, at least, the mare is younger than the dark mantling material.

PART I

THE CINDER FIELD OF THE TAURUS MOUNTAINS

Farouk El-Baz[a]

Mare Serenitatis is bounded to the east by the Taurus Mountains, which are clusters of massive peaks interconnected with old crater rims and light-colored, plains-forming highland fill. The relationship between the Taurus Mountains and the Serenitatis basin is much like that between the Apennine Mountains and the Imbrium basin. Mare Imbrium, however, is younger than Mare Serenitatis. The Apennine mountain peaks are, therefore, less degraded than the Taurus Mountains.

The geomorphological characteristics of Mare Serenitatis and surrounding materials, together with the fact that the area displays a high gravity anomaly second in magnitude only to that of Mare Imbrium (ref. 25-26), indicate that Mare Serenitatis is an impact basin that was filled with volcanic lava flows. The mare materials appear to have been deposited in two major episodes: an early event that left a light-colored mare surface in the middle portion of the basin and a later occurrence that deposited a dark-colored mare unit in an outer annulus or concentric zone (fig. 25-61). A concentric ridge system corresponds closely with this division of the mare materials of Serenitatis. The ridges may have been formed by relatively viscous, extrusive volcanic materials along buried concentric fractures. This hypothesis is supported by the fact that the outer zone of the basin displays numerous fractures and faults in the form of arcuate rilles; for example, the Sulpicius Gallus rilles on the west rim, the Menelaus and Plinius rilles to the south, and the Littrow and Chacornac rilles to the east.

Within the southwestern corner of the Taurus Mountains (the southeastern corner of the Serenitatis rim) a still darker unit occurs (fig. 25-62). In fact, this unit constitutes the darkest surface material on the lunar near side (ref. 25-27) and, probably, on the whole Moon. The unit is centrally located between Littrow Crater to the northeast, Vitruvius Crater to the southeast, and Mt. Argaeus to the southwest. It appears to mantle the highland materials as well as the mare materials. The unit was mapped as the Sulpicius Gallus formation (ref. 25-28); and, because it apparently was formed relatively recently, it was considered a candidate Apollo landing site (ref. 25-29) and was referred to as the Littrow area (fig. 25-63).

Earth-based multispectral data indicate that the dark color of this unit is probably related to compositional characteristics (ref. 25-30). Radar data also indicate that the unit is relatively smooth, with a small number of blocks (ref. 25-31). The recentness of the formation of the dark unit is also indicated by the extremely low density of craters on its surface.

The geologic characteristics of this unit, the intricate relationship between it and the surrounding features, and the probable source of the unit were the objects of visual observations during the Apollo 15 mission. The results of these observations made from orbit (as detailed in part A of this section) indicate that the unit probably consists of volcanic ash or pyroclastic deposits that came to the surface through a multitude of cinder cones.

The regional geologic setting of the area is discussed in section 5 of this report. The local setting and the relationship of the dark deposit to other nearby geologic units will be discussed in this part.

As shown in figure 25-62, the dark material appears to embay the highland massif units. A panoramic camera view of part of the unit is shown in figure 25-64, in which the dark material is visible both in the lowlands between the mountains and, occasionally, on top of massif units that display nearly level or depressed surfaces. Thus, the dark material was probably deposited from above; that is, by settling after ejection from beneath the surface. A different process occurs in the filling of basins by mare materials, which is apparently the result of flowage on the surface after extrusion.

In addition, remnants of the dark deposit can be distinguished on some of the massif slopes. This observation also tends to support deposition from above, followed by mass wasting by the downward movement of the material along steep slopes. The

---
[a]Bellcomm, Inc.

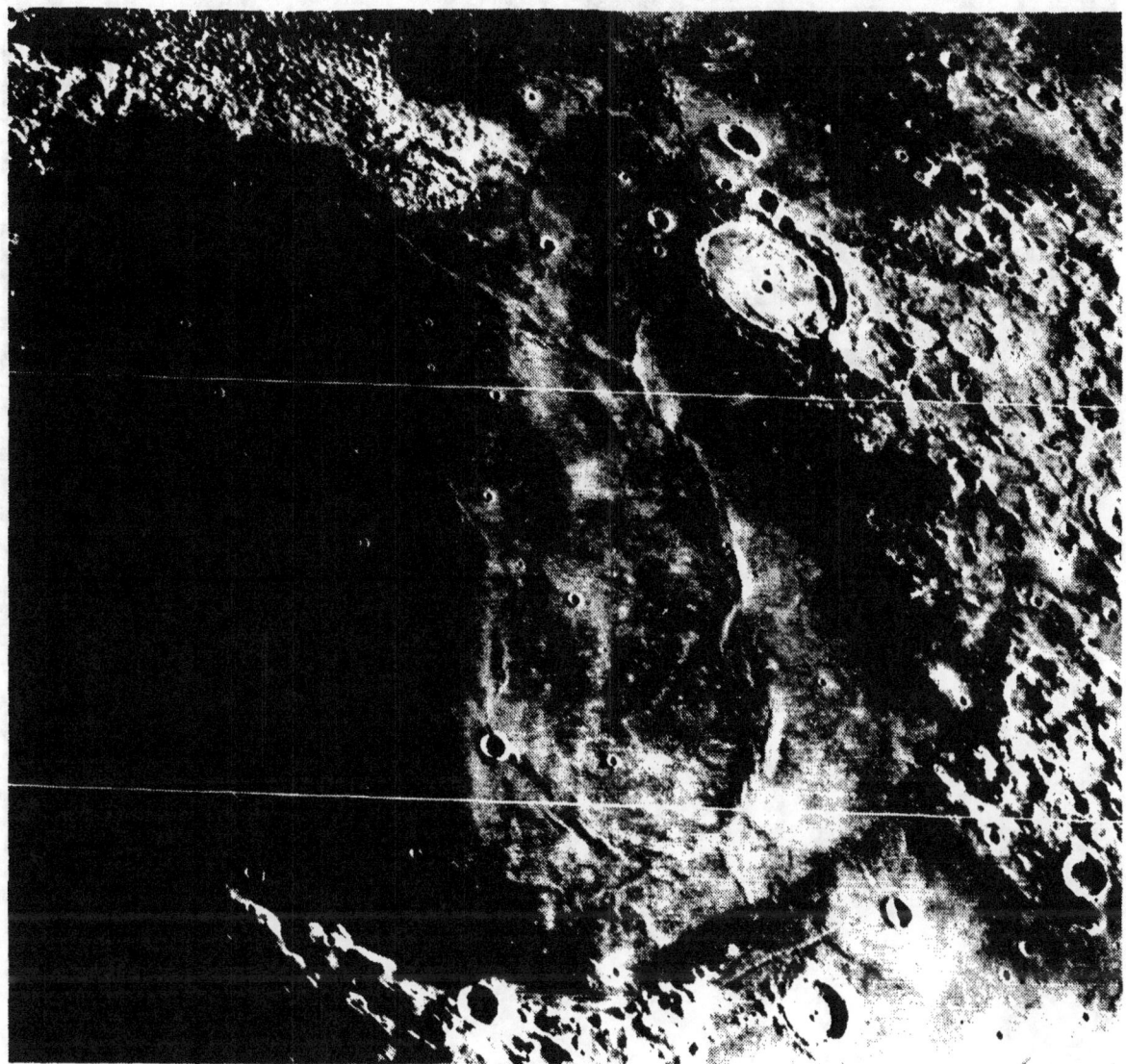

FIGURE 25-61.—Telescopic photograph of Mare Serenitatis showing the light-colored middle portion surrounded with a dark annulus. The unit on the southeast corner is the darkest unit of regional extent on the lunar surface.

dark material, therefore, would be mixed with the light-colored, fine-grained highland material, as evidenced by the fact that contacts between the dark lowland fill and the mountain fronts appear to be gradual.

Deposition of the dark material by gravitational settling after upward ejection does not preclude the formation of the flow scarps. In some cases, flow scarps abut against the foothills of the mountains, leaving V-shaped depressions between the two surfaces. Flow scarps are especially visible on the western borders of the dark unit, where the latter mantles both the mare material of the Serenitatis outer ring and a fresh wrinkle ridge in the mare, as shown in figure 25-63.

Apollo 15 panoramic camera photographs, with a resolution of approximately 2 m, show in great detail some of the cited characteristics and relationships. A

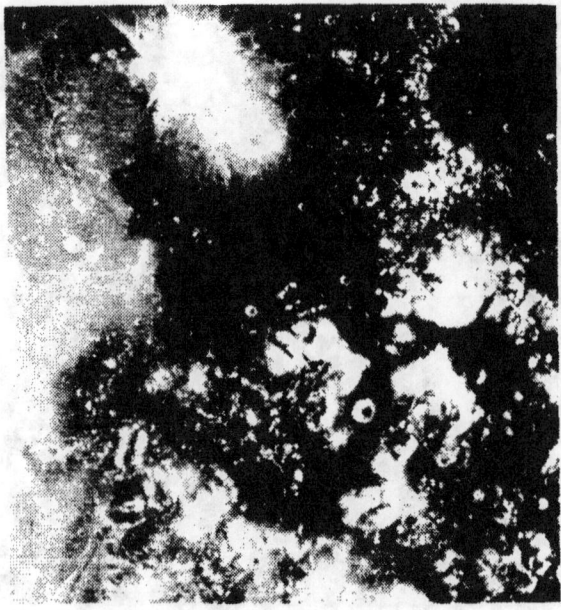

FIGURE 25-62.—Portion of photograph showing the southeastern part of Mare Serenitatis and the southwestern part of the Taurus Mountains (met AS15-1115).

FIGURE 25-63.—Lunar Orbiter V photograph showing a dark and apparently thin deposit that mantles the Serenitatis bench material, the mare material of southeastern Serenitatis, and a fresh-appearing wrinkle ridge in the mare (frame G9-m; framelet width, 4 km).

segment of a panoramic camera frame centered at approximately latitude 20.3° N, longitude 30.5° E is shown in figure 25-64, and a geologic sketch map of part of the same frame is shown in figure 25-65. From these two figures and also from the previous discussion, the geologic history of the area may be summarized as follows.

(1) The Serenitatis basin was formed by a giant impact, which formed a 700-km-wide depression and at least two major systems of fractures, one concentric with the basin and the other radial to it. Massif units of the Taurus Mountains came to the surface by uplifts along the concentric fault system and were segmented by both fracture systems.

(2) At some later time (perhaps 0.5×10^9 yr, calculated by extrapolation of data acquired at Apollo landing sites), a major episode of basin filling by volcanic basalts occurred (Imbrian age). Another episode of filling followed, perhaps another 0.5×10^9 yr later (Eratosthenian age), during which the darker annulus surrounding the older mare material was formed. Following this period, the mare material was faulted along the concentric fracture system and numerous arcuate grabens were formed. One of the latest manifestations of mare-material extrusion was the formation of the wrinkle ridges now visible on the surface.

(3) After the completion of basin filling by mare materials, volcanic eruptions began in the southwestern corner of the Taurus Mountains. Cinder

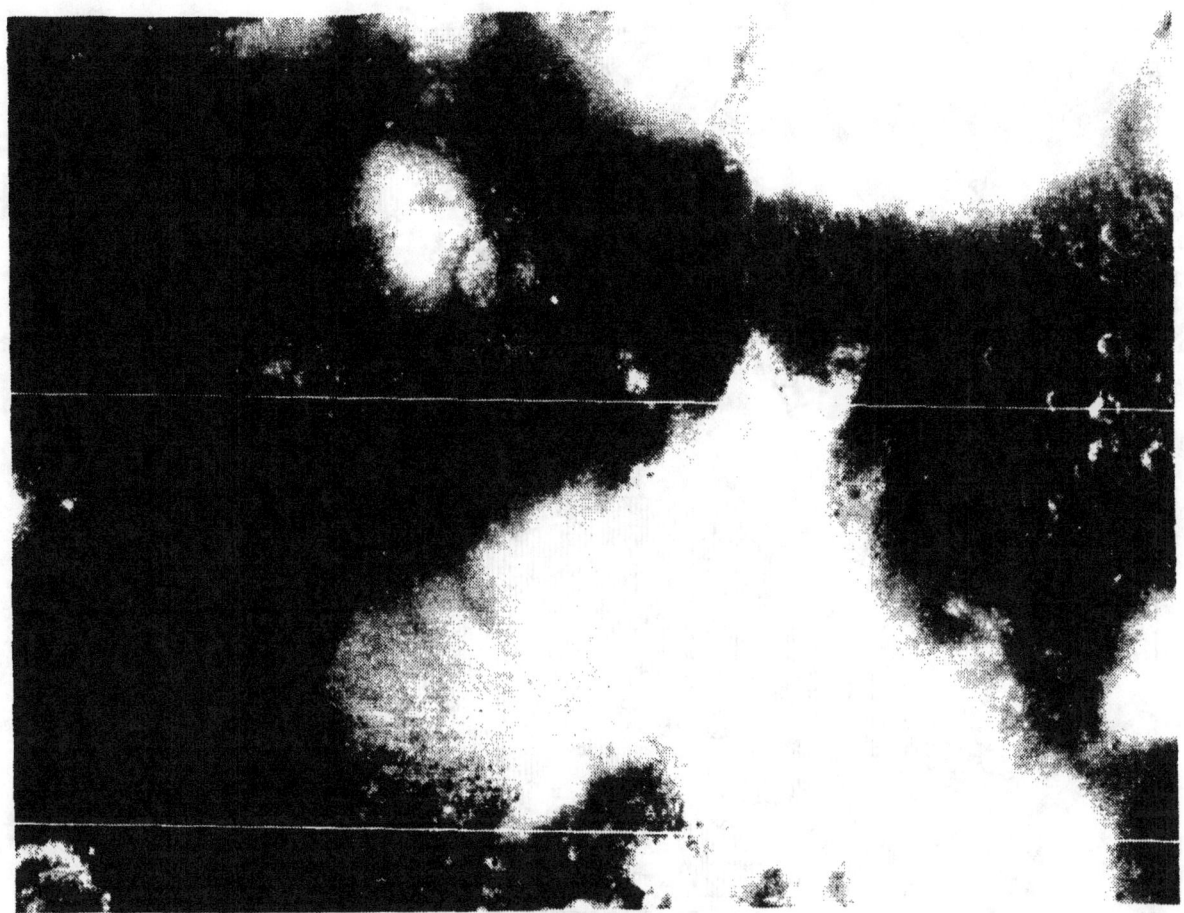

FIGURE 25-64.—Portion of photograph showing the dark and smooth deposit in valleys and level areas between Taurus Mountain massif units (pan AS15-9554).

cones (fig. 25-66), located mainly in the lowlands between the massif units, deposited a dark blanket that mantles the highland materials, the mare materials, and the grabens and wrinkle ridges in the southeastern corner of Mare Serenitatis.

(4) From the time of emplacement, scarps of massif units of the Taurus Mountains were subjected to mass wasting, and this activity may be continuing at the present. (Blocks on the foothills and their tracks on the slopes are shown in fig. 25-64.) One of the manifestations of this process is a unit lineated from north to south, which is interpreted as a landslide that originated from the southern massif unit (fig. 25-65). This lineated unit is probably older than the dark unit because (1) there appear to be similar lineations that were mantled by the dark deposit but still show through it and (2) the asymmetrical dark halo crater within this unit is probably an isolated cinder cone. An alternate interpretation, however, is that the dark halo crater is an impact crater that has exposed an underlying dark deposit.

(5) A fault scarp was formed by the relative upward movement of the materials west of the fault line or by the downward movement of the materials east of it, or perhaps by both movements (fig. 25-65). The fault is the most recent tectonic feature in the area; it bisects the oldest material (Taurus Mountains massif units) as well as the youngest formation (the dark, ashlike mantling material).

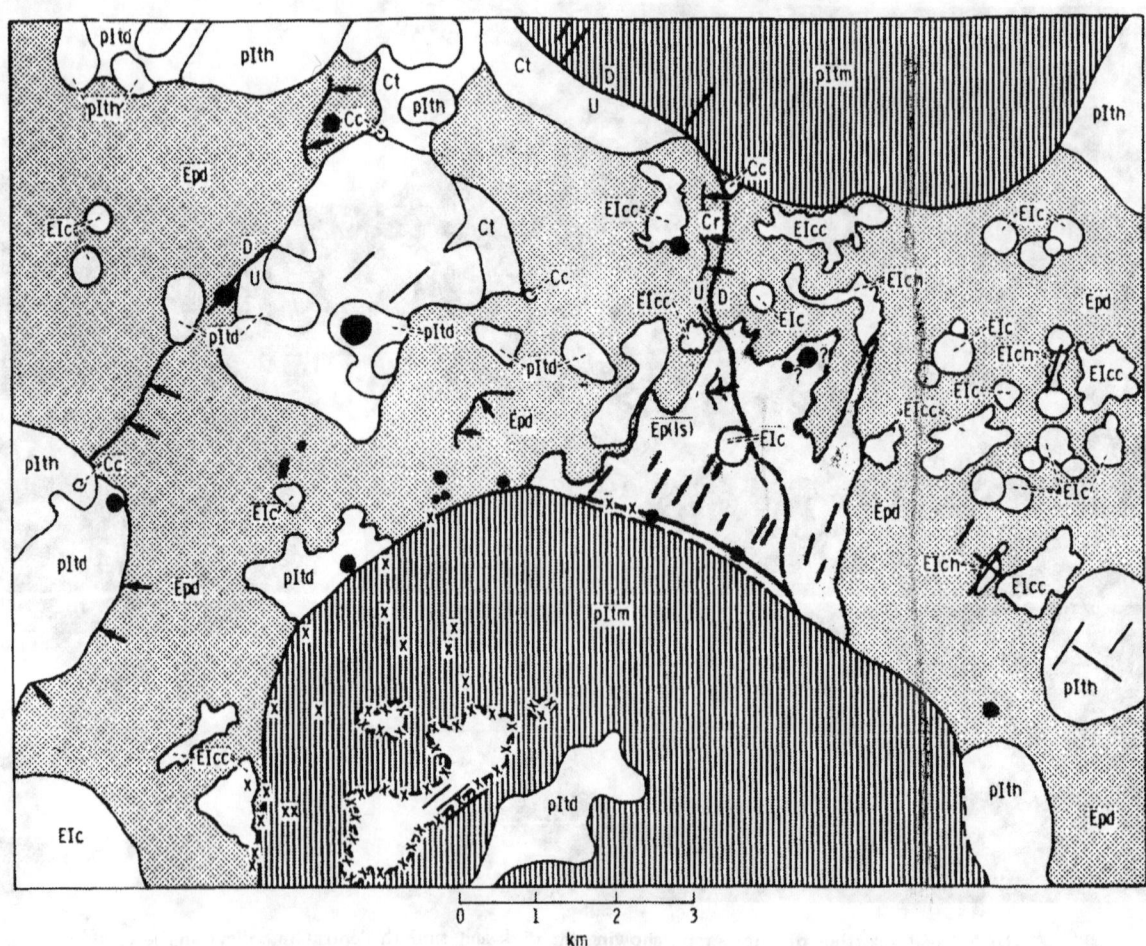

FIGURE 25-65.—Geologic sketch map of the area shown in figure 25-64.

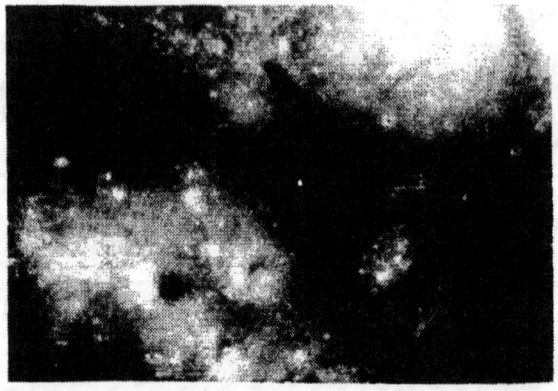

FIGURE 25-66.—Enlargement of photograph showing a probable volcanic cone. A somewhat-dark, smooth-textured ejecta blanket suggests an ash or pyroclastic-type deposit symmetrically surrounding the cone (pan AS15-9559).

Discovery of the cinder cones by Apollo 15 visual observations from lunar orbit and recognition of the nature and variety of geologic features photographed using the panoramic camera have revived interest in the Taurus Mountains region as a possible candidate landing site for the Apollo 17 mission. The area has attracted attention largely because a lunar-landing mission to this site would make possible access to the Taurus Mountains (highland) massif units, which were probably emplaced in pre-Imbrian times, and to the dark mantling deposit, which may be among the youngest lunar-surface units. Association of the dark deposit with what appear to be cinder cones indicates a volcanic origin of the mantle. Examination of this type of region may yield significant clues to the interior composition and the thermal history of the Moon.

Explanation

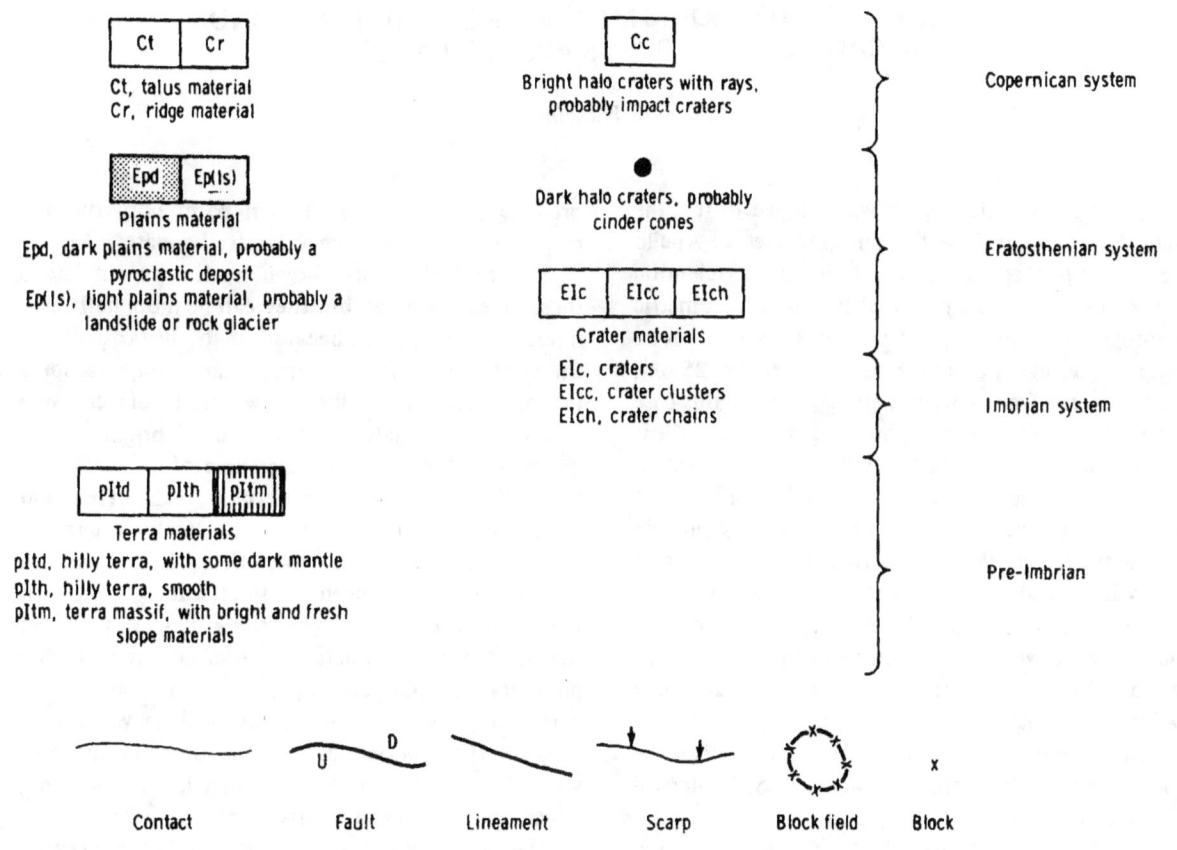

Ct, talus material	
Cr, ridge material	

Bright halo craters with rays, probably impact craters — Copernican system

Plains material
Epd, dark plains material, probably a pyroclastic deposit
Ep(ls), light plains material, probably a landslide or rock glacier

Dark halo craters, probably cinder cones

Crater materials
Elc, craters
Elcc, crater clusters
Elch, crater chains

— Eratosthenian system

— Imbrian system

Terra materials
pItd, hilly terra, with some dark mantle
pIth, hilly terra, smooth
pItm, terra massif, with bright and fresh slope materials

— Pre-Imbrian

Contact   Fault   Lineament   Scarp   Block field   Block

PART J

PRELIMINARY GEOLOGIC MAP OF THE REGION AROUND THE CANDIDATE PROCLUS APOLLO LANDING SITE

Don E. Wilhelms[a]

The Proclus Crater region was mapped to test the value, for photogeologic mapping purposes, of Apollo 15 metric photographs and to estimate the scientific value of the area as a potential landing site. A metric photographic frame (fig. 25-67) serves as a base for a map of the region around Proclus Crater (fig. 25-68), and adjacent frames were overlapped with the base frame to provide stereographic images. The excellent stereocoverage allows easy simultaneous observation of topography and albedo. The large forward overlap and the extensive areal photographic coverage provide the best photogeologic data available to date. Brief study has already refined earlier interpretations of the area (refs. 25-7 and 25-32). Although volcanic units have been shown to be extensive in this region, mass wasting apparently has been more important than volcanism in shaping terra landforms.

The effects of lunar mass wasting, inferred to be present almost from the start of the U.S. Geological Survey lunar geologic mapping effort and since observed on many Lunar Orbiter and Apollo orbital photographs (e.g., ref. 25-33), stand out with unequaled clarity on the Apollo 15 metric photographs in nearly all parts of the terra. Bright, steep slopes, which must consist of fresh talus, are clearly resolved into stripes running downslope; the steeper and younger the slope, the more abundant the talus. The walls of the Copernican bright-rayed Proclus Crater (unit Ccw) have the most talus; the walls of other craters have proportionally less; the old but steep peaks and scarps of the rugged and hummocky terra materials also have much bright talus. Conversely, the lower lying, flat or undulatory areas at the base of the peaks and scarps and between the bright hills are less bright. This material, obviously more stabilized, commonly fills or overlaps craters at the base of scarps, forms aprons that extend up valleys, and gives abundant other evidence of formation by mass wastage from adjacent slopes. Much of the mare-terra contact is gradational; material from the very steep premare scarps of the Crisium basin bordering the mare apparently has slumped onto the mare.

The geologic units shown herein are similar to those of earlier maps, but they can be more definitely interpreted and dated because of the high quality of the photographs used. Four units are intergradational in topography and albedo—two units of relatively low-lying dark materials and two of bright rugged materials. The relationships heretofore described show that the darker materials (plains (Ip) and nondistinctive undulatory terra (It)) originated largely if not entirely by mass wasting. Previously, a volcanic origin had been considered likely, but even the plains contacts are gradational here rather than sharp, lavalike contacts as inferred from earlier photographs. (Of course, plains units elsewhere, particularly extensive ones, may well be volcanic in origin.) The Imbrian age of the units in the area studied is confirmed by the metric photography, which shows a typical Imbrian crater population.

The hummocky unit (pIh) was divided into several units on earlier maps (refs. 25-7 and 25-32). The present study suggests that these units differ only in the thickness of the debris of which they are composed. In figure 25-68, four gradations in a complex pattern of source-debris interrelationships, from relatively well exposed to relatively deeply buried, are represented by the symbols pIh, It (pIh), It, and Ip.

The origin of the hummocky unit is still uncertain. It is gradational in character with the rugged unit and could be lithologically identical. The new photographs tend to confirm the earlier interpretation of the rugged unit as composed of bedrock uplifted when the Crisium basin formed, analogously to the Apennine Mountain peaks, such as Mons Hadley and Hadley Delta of the Imbrium basin (refs. 25-15 and 25-32). Alternatively, the hummocky material in the Proclus region and in the extensive adjacent outcrops to the west could possibly be volcanic in origin. This possibility was suggested by earlier mapping of parts of the unit on the basis of the regular spacing and size

---
[a] U.S. Geological Survey.

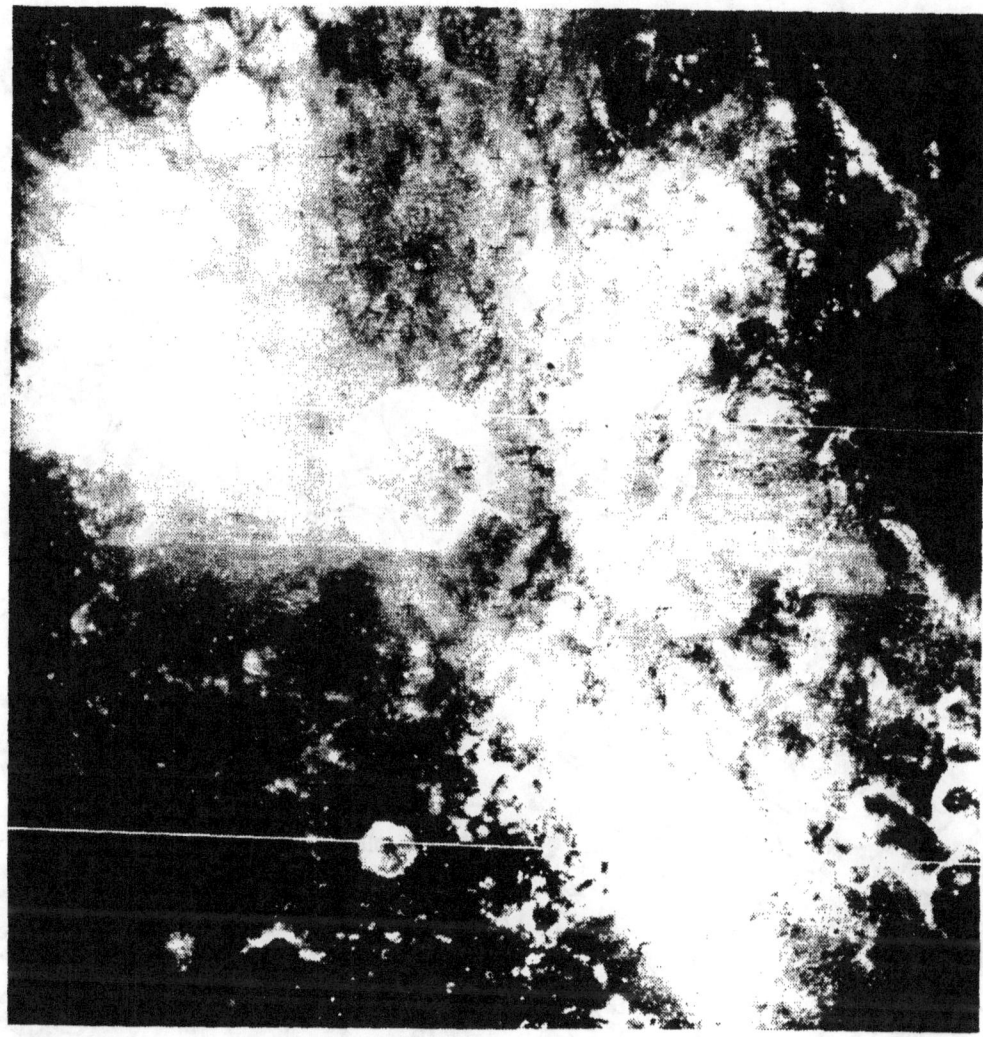

FIGURE 25-67.—Photograph of the region around Proclus Crater. This photograph serves as a base for the accompanying map (met AS15-M3-0958).

of the hills, an apparent result of crater mantling, and the presence of a few distinctive furrows and pits. The metric photographs, however, show that all these features can be explained geometrically by fracturing, so the issue remains unsolved.

In summary, preliminary study of the Proclus region shown in the Apollo 15 metric photographs suggests that an earlier predominantly volcanic interpretation of the terra in this area might be replaced by another interpretation in which mass wasting and fracturing play the major roles in producing different terrain types. Perhaps the entire Crisium basin rim consists of rock older than the basin, either uplifted or uplifted and ejected and then mantled to varying degrees by its own debris. The amount of debris at any locality depends on local topographic differences, which in turn depend on tectonics, mainly basin tectonics.

Because of this apparent lithologic homogeneity, the terra of the Proclus region is an undesirable objective for an extensive manned lunar-landing mission. Mare Crisium is an additional possible target, but the mare material might not add significantly to data already gathered at other mare sites.

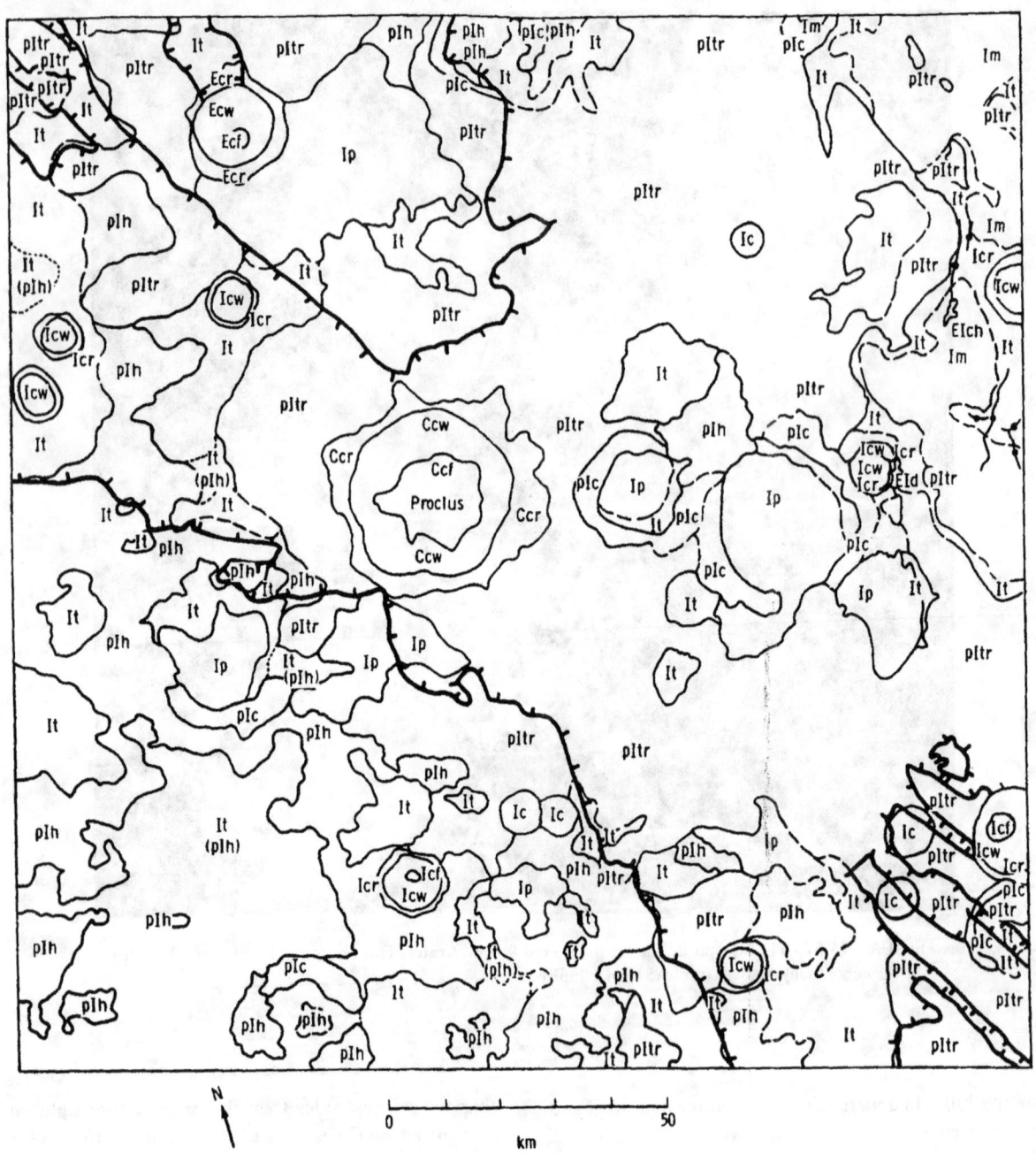

FIGURE 25-68.–Geologic map of the region around Proclus Crater.

Explanation

| Ccr | Ccw | Ccf |

Material of rayed craters

Ccr, rim material
Ccw, wall material
Ccf, floor material

} Copernican system

| Ecr | Ecw | Ecf |

Material of relatively fresh nonrayed craters

Subunits lettered as above

} Eratosthenian system

| EId |
Dark mantling material

| EIch |
Material of irregular, dark chain craters

| Im |
Mare material

| Ip |
Plains material

Flat relative to other terra units; more macrorelief and microrelief than mare; albedo intermediate between mare and rugged and hummocky terra

| It |
Terra material, undivided

Generally undulatory; various textural detail; similar in albedo to plains material

| Ic | Icr | Icw | Icf |

Material of moderately subdued craters

Ic, crater material, undivided. Subunits lettered as above

} Imbrian system

| pItr |
Material of rugged terra

Relatively sharp, bright, large, randomly oriented and unevenly spaced peaks; gently undulatory darker material in valleys

| pIh |
Hummocky material

Closely clustered, steep, rounded hills mostly less than 5 km across; mostly bright; varying amount of smoother darker material between hills

– – – – – – (pIh)

Contact

Dashed where approximate; dotted where buried (buried unit indicated by symbol in parentheses)

⊥⊥⊥⊥⊥⊥

Limit of principal Proclus rays

Hachures point toward bright region with faint radial lineations

———•———

Fissure

| pIc |
Material of subdued craters

} Pre-Imbrian

PART K

GEOLOGIC SKETCH MAP OF THE CANDIDATE PROCLUS APOLLO LANDING SITE

Baerbel Koesters Lucchitta[a]

A panoramic camera frame (fig. 25-69) was used as the base for a geologic sketch map (fig. 25-70) of an area near Proclus Crater. The map was prepared to investigate the usefulness of the Apollo 15 panoramic camera photography in large-scale geologic mapping and to assess the geologic value of this area as a potential Apollo landing site. The area is being considered as a landing site because of the availability of smooth plains terrain and because of the scientific value of investigating plains materials, dark halo craters, and ancient rocks that may be present in the Proclus ray material.

The map was prepared from a portion of an unrectified panoramic photograph, and an overlapping frame (pan AS15-9243) was used to obtain a stereographic image. The photographs, taken under high solar illumination, resulted in good definition of albedo features and poor definition of topographic relief under normal viewing. However, under stereographic viewing, the topographic relief is expressed with extreme clarity, and terrain units are easily delineated.

The area is located at latitude 19°30′ N, longitude 48°30′ E, approximately 50 km from the west edge of the Crisium basin and approximately 100 km north-northeast of the rayed Proclus Crater. The rim in this area is characterized by large, bright, randomly oriented peaks and closely clustered, steep, rounded hills (part J of this section). Plains material fills low, level ground between the mountains, and rays from Proclus Crater cover the region extensively. Numerous small, dark halo craters are visible on all older units.

The geology of the area is dominated by features of Copernican age. Older units are expressed morphologically but are probably not evidenced extensively. Because some units have gradational boundaries and cannot be sharply delineated, some of the contacts are only approximately located.

The lowlands shown on the map are bordered on the north by the pre-Imbrian hilly terrain (pIh). The boundary is marked by a distinct scarp. The domical, closely spaced hills are 1 to 2 km in diameter, and, where the hills are not covered by younger material, a smooth, even surface texture is evident. The origin of the hills could not be established without a more thorough regional investigation.

Imbrian plains material (Ip) is mapped in the area of low-lying, level, relatively smooth ground adjoining the mountains. The surface is gently undulating and very finely ridged, locally. The surface probably consists of some volcanic-flow material, mass-wasted debris from adjacent mountains, and a thin layer of Proclus ejecta material. These materials probably are mixed near the surface because of a rain of secondary debris from Proclus Crater. Buried Eratosthenian craters (Ecr) are present in the northeastern corner of the area.

The Copernican terra unit (Ct) has a sharp, hackly, rugged surface expression that extends in a wedge-shaped segment across the mountain scarp and ends in a broad lobe on the plains area. The surface expression is well delineated by a sharp scarp around the periphery. The configuration of this unit suggests that it originated as a debris or volcanic flow. However, the presence of secondary craters and subparallel ridges on the surface may indicate significant modification by secondary impacts from Proclus Crater.

Shallow craters and crater clusters of irregular shape (Csc) are thought to be secondary impact craters related to the impact event that formed Proclus Crater. These craters have narrow, low, locally discontinuous rims; however, those located in the mountainous terrain are generally deeper and have broader rims. The sharpness of the crater rims in clusters varies from highly subdued to very sharp and jagged. Only the sharpest secondary craters have abundant bright spots on the rims. Light halos are generally not present. Apparently, the surface has been so thoroughly reworked by secondary impact that the earlier secondary craters are noticeably more subdued and therefore are older in appearance. Very shallow craters or rimless depressions are either older

---

[a]U.S. Geological Survey.

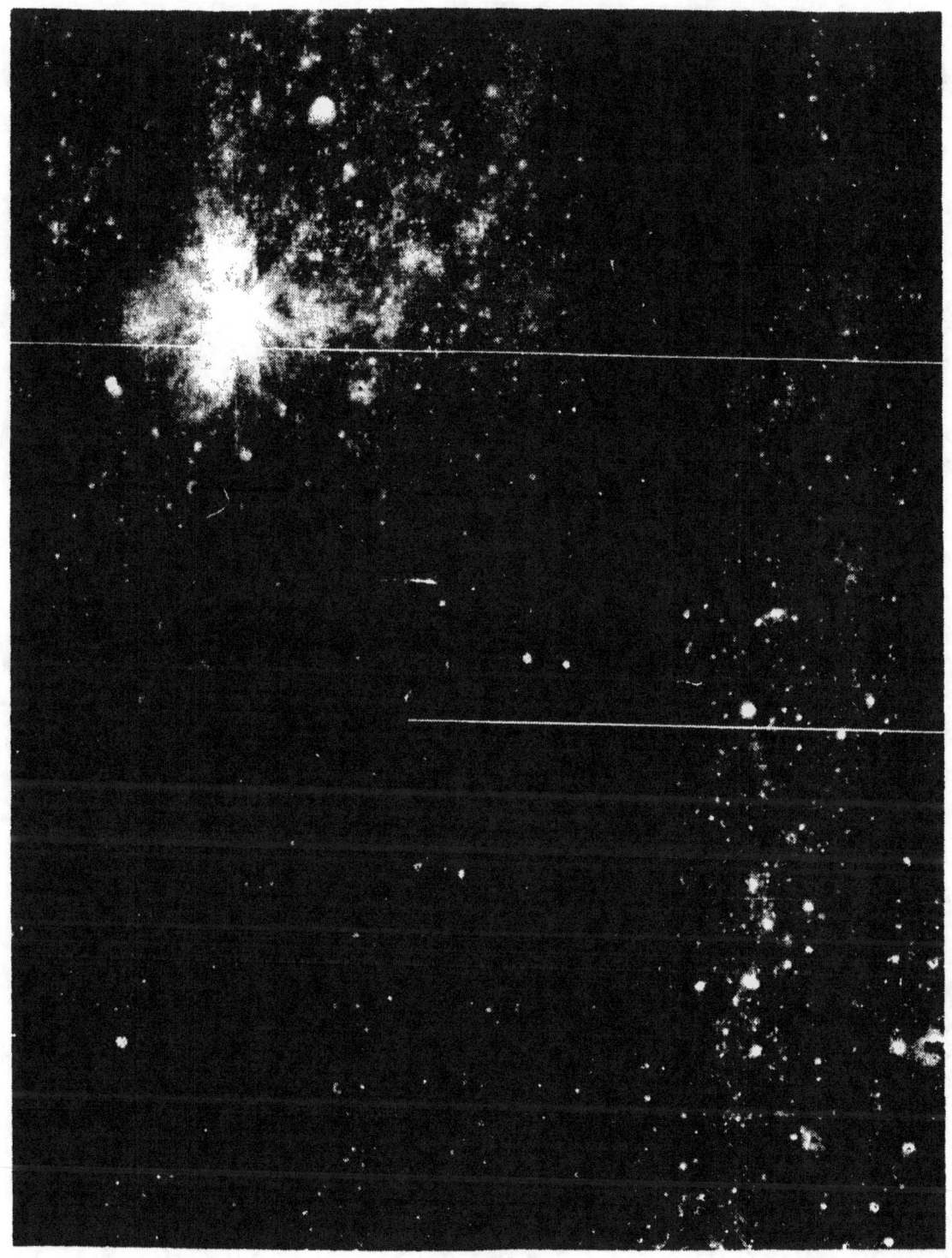

FIGURE 25-69.—Photographic base for accompanying map (pan AS15-9238).

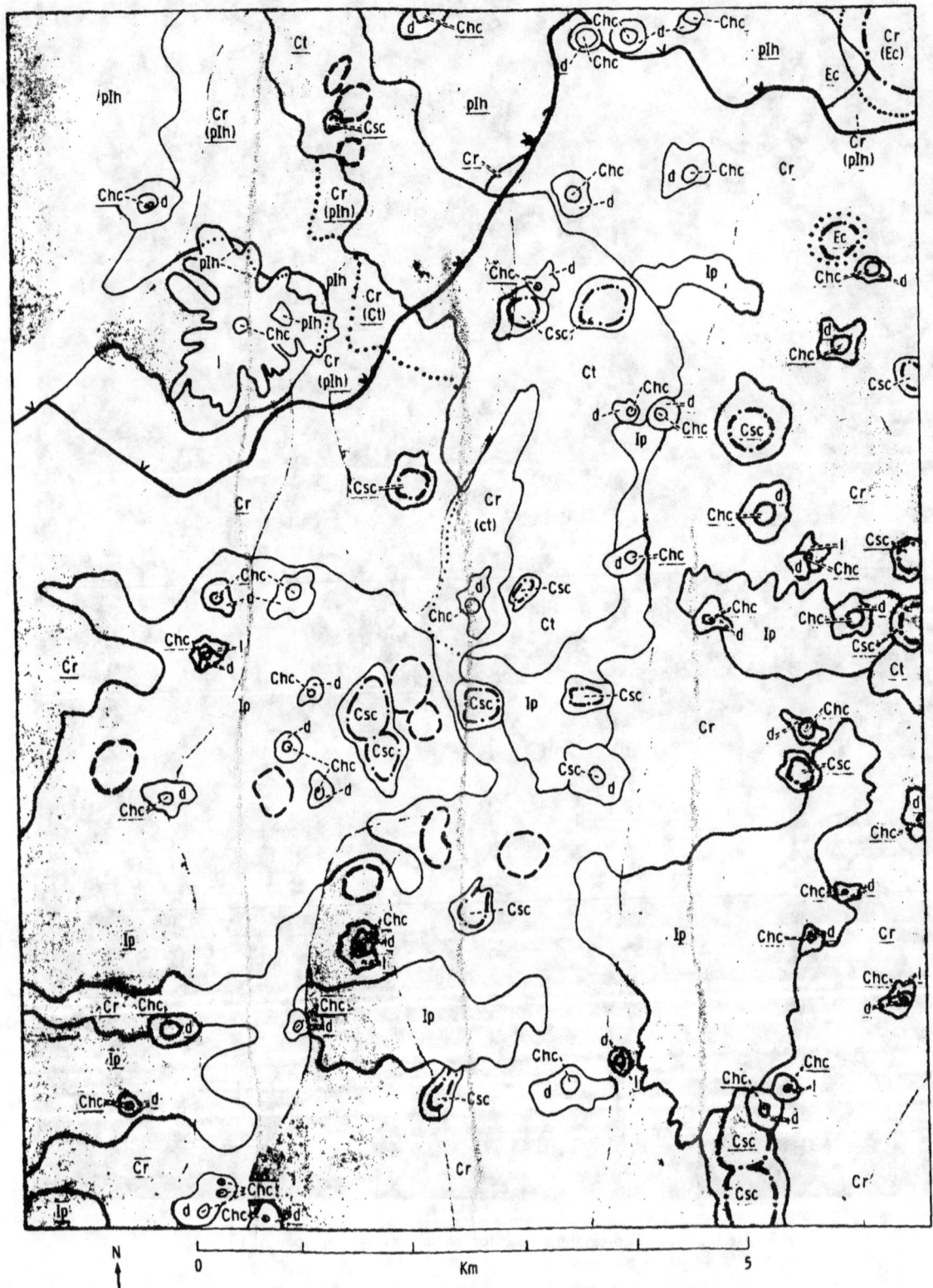

FIGURE 25-70.—Geologic sketch map of a candidate Proclus landing site.

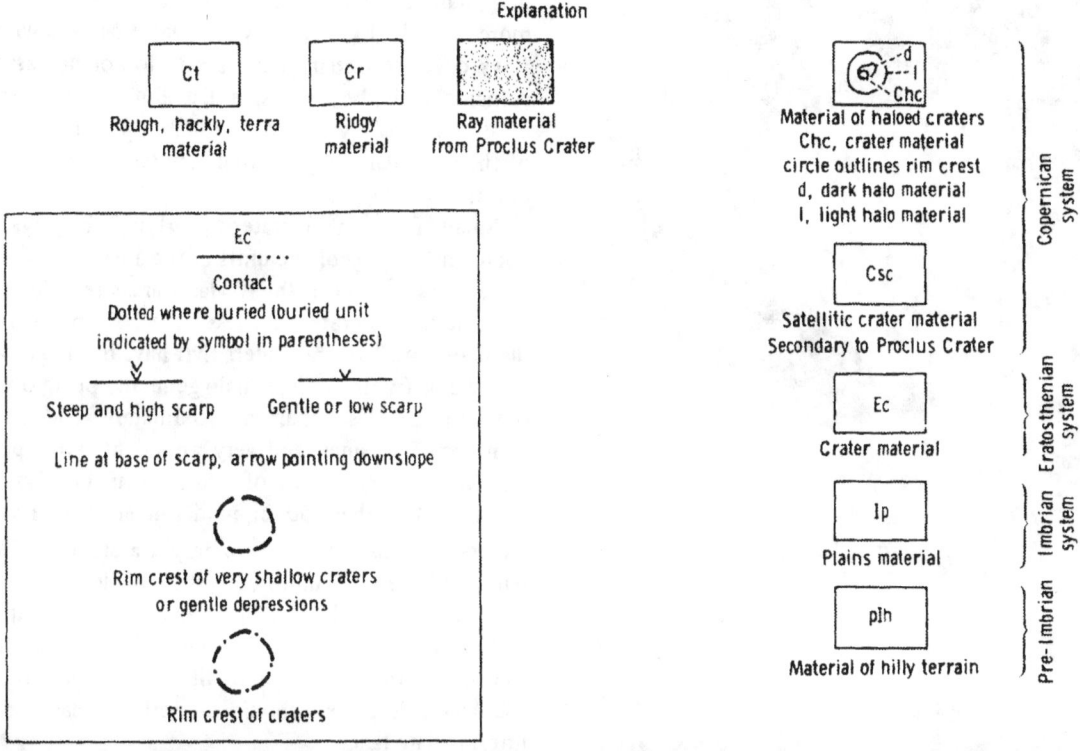

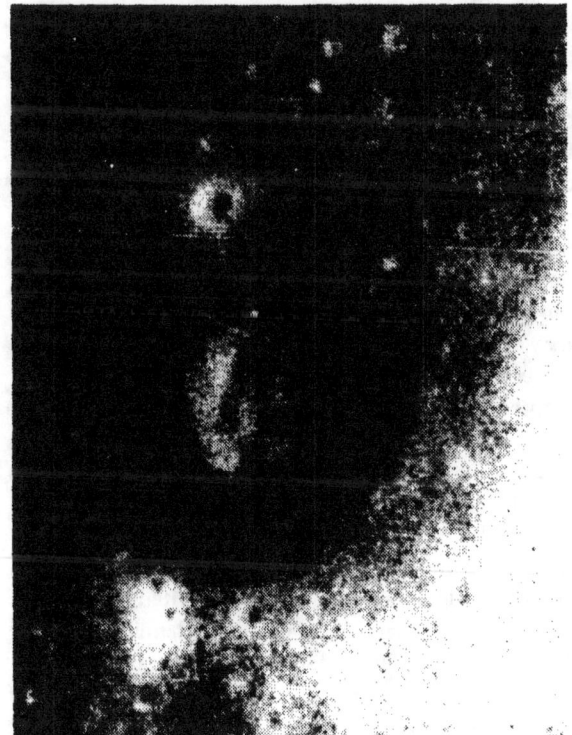

FIGURE 25-71.—Enlargement of dark halo crater, approximately 160 m in diameter (pan AS15-9243).

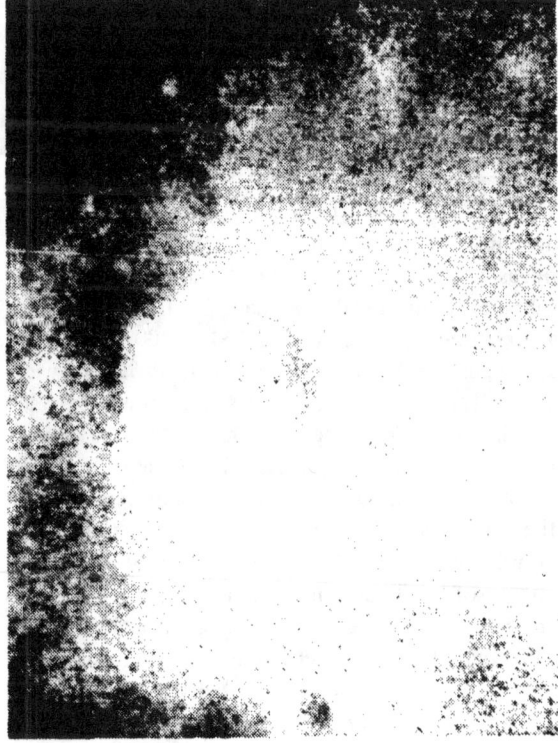

FIGURE 25-72.—Enlargement of light halo crater, approximately 150 m in diameter (pan AS15-9243).

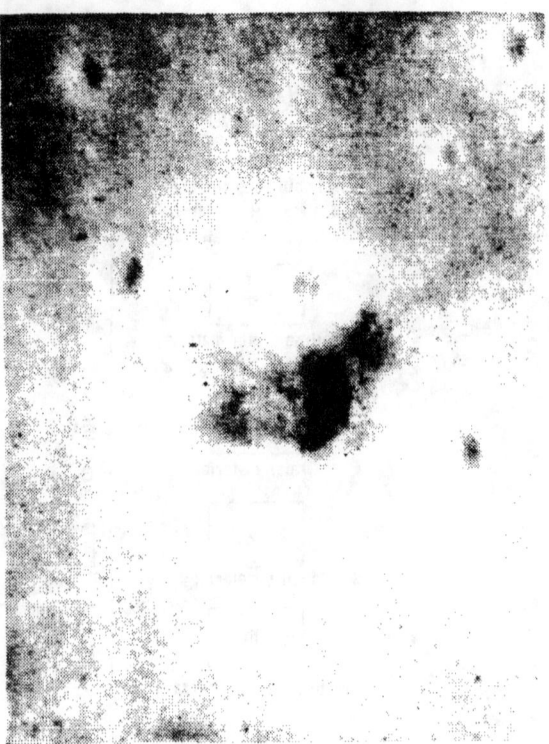

FIGURE 25-73.—Enlargement of mixed halo crater, approximately 70 m in diameter (pan AS15-9243).

craters buried by debris associated with the Proclus event or the earliest of the Proclus secondaries.

Copernican ridgy material (Cr), covering a large area of the map, occurs on plains material and overlaps mountains, craters, and terra material. The ridges, generally extending north-northeast, are linear, subparallel, and locally braided. The largest ridges are as much as 3.5 km long and 200 to 300 m wide and are separated by troughs as much as 500 m wide. Typically, however, the ridges and troughs are approximately 2 km long and 100 to 200 m wide. The ridges enter secondary craters locally and, in places, are continuations of rims of such crater clusters. On the high-resolution panoramic photographs, obviously, the ridges and grooves locally consist of chains of small craters; however, all the gradations from crater sequences to linear troughs can be observed. The ridgy terrain probably consists of a complex mixture of locally derived material and Proclus ejecta material.

Ray material extends over most of the area. Small, light halo craters and light specks are significantly more abundant in the rayed areas than elsewhere. Rays cover secondary craters but do not necessarily coincide with the craters in distribution. Deposition of the Proclus ray material probably postdated most of the secondary impacts that resulted from the same event.

Small, fresh, halo craters overlying the rays are abundant on all geologic units in the area. A survey of the craters located in the lowland areas revealed that most dark halo craters are less than 100 to 200 m in diameter. Most of the craters that have both light and dark halos (with the light halo generally being on the outside) are less than approximately 100 m in diameter. The small and very abundant, unmapped, light halo craters, most of which occur on rays, are generally less than 50 m in diameter. Larger fresh craters, such as Proclus secondary craters, either lack halos or have faint halos consisting of light and dark streaks. The dark halo craters that are mapped include all those with halos even slightly darker than the underlying ray material; therefore, a number of the dark halo craters should properly be classified as intermediate halo.

Dark halo craters are of interest because of an inferred volcanic origin based on analog studies with terrestrial cinder cones. However, the dark halo craters of the map area and vicinity may be impact craters that have excavated dark material for the following reasons: (1) The shape of the dark halo craters is not significantly different from the shape of the light halo craters that are considered to be of impact origin (figs. 25-71 and 25-72). (2) The various degrees of halo darkness and the occurrence of halos on brightly rayed terrain suggest that some dark halos may be more apparent than real. (3) The tendency for craters having halos of different brightness to fall into certain size ranges on the plains area may suggest that different layers are excavated on impact. (4) Because craters with mixed halos do occur (fig. 25-73), a mixed origin would have to be interpreted for consistency.

The panoramic camera photographs reveal a wealth of detail and are eminently suited for geologic mapping purposes. As a potential landing site, the Proclus area offers relatively rough plains terrain. Ancient rocks are probably present in abundance but are highly mixed with rocks of all ages. The presence of volcanic features has not been established.

PART L

SELECTED VOLCANIC FEATURES

Mareta N. West[a]

Preliminary examination of Apollo 15 orbital photographs indicates a large number of volcanic features. One area of exceptionally interesting volcanic activity is depicted in figure 25-74. Located approximately at latitude 25° S and longitude 123° E on the lunar far side, this region also is covered by panoramic camera photographs AS15-9954, 9956, 9958, and 9960 and by stereoscopically overlapping frames AS15-9959, 9961, 9963, and 9965.

On the southwest wall of the oblong crater, just west of the center of figure 25-74, is a distinct "strand line," marking the highest level reached by lava before cooling and withdrawal. This line is in shadow in the figure cited but is clearly visible in a panoramic camera frame (AS15-9960). A faint trace of this line exists in other parts of the wall. A prominent terrace (around all except the southern part of the outer edges of the floor) marks another state in the subsidence of the lava.

A system of fractures (fig. 25-74) may provide vents from which much of the smooth, dark lava has been extruded. The largest fractures are bounded by what appear to be broad lava levees, and some of the smaller ones contain strand lines.

Craters in various stages of burial are visible in the flooded area (beginning in the center of fig. 25-74 and continuing in a south-southeasterly direction). Two craters in the northern part of the area have breached walls and interiors that are partially filled. Two small craters near the southern edge of the photograph have been almost completely obliterated and are barely discernible as shallow pits; also in this

[a] U.S. Geological Survey.

FIGURE 25-74.—Far-side view of exceptional volcanic activity. Arrow 1, area in which strand line is most clearly visible on pan AS15-9960; arrow 2, fractures from which lava may have been extruded; arrow 3, breached and partially filled craters; arrow 4, barely discernible flooded craters; arrow 5, mantled remnants of uplands; and arrows 6 and 7, bulbous floor material (north toward top, met AS15-M3-2358).

FIGURE 25-75.—Southeast rim of Tsiolkovsky Crater, arrow showing location of plains-forming material (north toward top, met AS15-M3-1026).

FIGURE 25-76.—Northwest rim of Tsiolkovsky Crater, arrow showing location of surface flowage (north toward top, met AS15-M3-0763).

FIGURE 25-78.—Flow fronts west of Mt. Lahire (north toward top, met AS15-M3-1558).

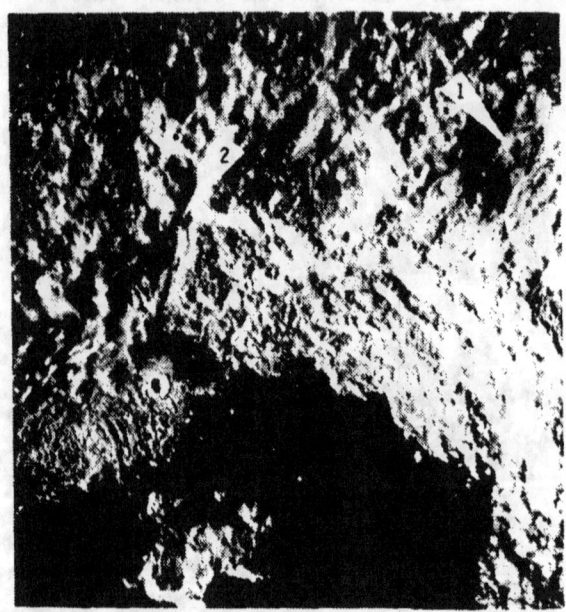

FIGURE 25-77.—Rim of Tsiolkovsky Crater. Arrow 1, lobate flow probably of volcanic origin; arrow 2, landslide (north toward top, met AS15-M3-0481).

area are remnants of uplands that have a smooth, mantled appearance.

Bulbous material on the floors of some craters also appears to be of volcanic origin. As seen on panoramic photograph AS15-9965, a few of the mounds in the larger of the two craters designated in figure 25-74 have summit pits. The bulbous texture of the material in the smaller crater appears to be less well developed than that in the larger crater, possibly indicating an earlier stage of volcanism.

Numerous interesting features were photographed in the vicinity of the far-side crater basin Tsiolkovsky. Smooth, plains-forming material (fig. 25-75) is present in parts of the ejecta blanket southeast of the basin rim. The smooth material is younger than the ejecta, as indicated by the absence of radial structure characteristic of the rim.

A debris flow on the northwestern part of the rim of Tsiolkovsky is depicted in figure 25-76; a topographic map of this same region is shown in figure 25-40. This area of the rim displays characteristics similar to those of some areas of the rim of Mare Orientale, especially to the northeast of the center of that basin. In the case of Tsiolkovsky Crater, the material has numerous parallel lineaments and may have decelerated rapidly, as indicated by a distinct scarp at the distal end.

What appears to be a volcanic flow in figure 25-77 occupies a depression in the northeast rim of Tsiolkovsky. This flow is visible in more detail on a panoramic camera photograph (AS15-8965) and on a stereoscopic frame (AS15-8970). The lobate form of the material in this flow suggests fluid emplacement and differs from a landslide in the wall of Tsiolkovsky. The slumped material also occupies a depression, but the outer boundary lacks lobes and roughly parallels a detachment scarp.

The usefulness of near-terminator photography for the study of small-scale detail of low relief is demonstrated in photographs such as that shown in figure 25-78. Numerous flow fronts not seen in photographs taken at higher angles of illumination are clearly visible. The presence of so many fronts confirms the belief that multiple flows are responsible for the filling of lunar mare basins. Prominent fronts are visible on Lunar Orbiter photographs, but the intricate, superposed relationships have not been observed previously. Mare materials northwest of the photographed area have been designated as of Copernican age. The sharp detail in the photograph and the presence of only small, fresh superposed craters confirm this age designation.

PART M

MARE IMBRIUM LAVA FLOWS AND THEIR RELATIONSHIP TO COLOR BOUNDARIES

Ewen A. Whitaker[a]

The Apollo 15 metric camera photography has provided a number of spectacular views of most of the southern and central portions of the unique system of major lava flows that stretch diagonally across Mare Imbrium from the general area of Euler Crater to a point east of Le Verrier Crater. The existence of this system has been known for some time through Earth-based photography and visual observations, but the low resolution of such photographs makes the mapping of the system a matter of some difficulty. Strom (ref. 25-34) has published a preliminary map of the northern and central portions of the system. Because of the greatly increased resolution, the Apollo 15 photography will permit the preparation of a far more detailed and definitive map.

That the lunar surface exhibits small but definite color differences has long been known. In 1929, Wright (ref. 25-35), for instance, published photographs of the Moon taken through infrared and ultraviolet filters, which show notable differences in certain mare regions. Positives and negatives of lunar infrared and ultraviolet photographs were combined so that albedo differences were removed as far as possible and only color differences were left. Two such experimental composites are reproduced in reference 25-34. Recent work on the same originals with much closer contrast control has produced results that are far more trustworthy than those composites.

Strom (ref. 25-34) showed that one prominent color boundary corresponds exactly with the eastern front of the northern portion of the flow system, the bluer material overlying the redder material. Flow fronts to the west of this boundary show no color variations because they represent subsequent flows of bluer material over the first and more extensive flow of similar material.

A preliminary examination of the metric camera photographs, viewed both monoscopically and stereoscopically, leaves little doubt that the source of the material comprising the major flows is situated to the west of Euler Crater and south of the limits of the photographic coverage. Some lava-flow channels located closely west-northwest of Euler are scarcely distinguishable from similar features found on Mauna Loa and other terrestrial volcanic fields. The lavas have flowed in a general northeasterly direction, indicating the direction of slope that existed in Mare Imbrium at the epoch of the eruption. The point of termination of the flow system, near Le Verrier Crater, is very near the center of Mare Imbrium. This fact implies that the older filling of redder lavas never completely obliterated the shape of the prelava basin and that the later, bluer lavas simply ran downhill to the lowest area at the center.

---

[a]Lunar and Planetary Laboratory, University of Arizona.

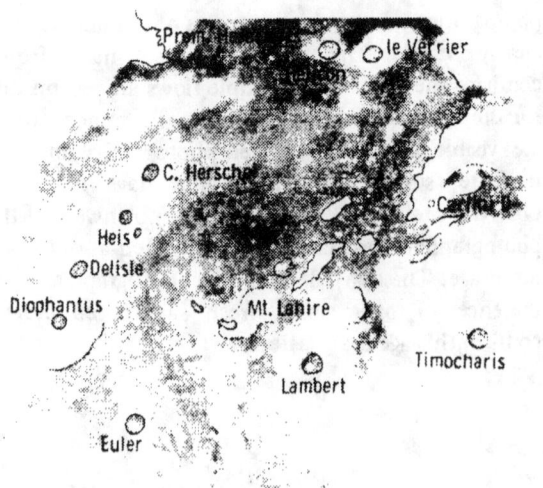

FIGURE 25-79.—Color-composite photograph (infrared minus ultraviolet) of Mare Imbrium. Color boundaries that correspond with visible flow fronts are outlined.

The photographs also show clearly that most of, if not all, the mare ridges included were formed after the solidification of the flows. Thus, several flows appear to cross ridges of substantial elevation without deviation, ponding, or change in thickness— an impossibility if the ridges had preceded the flows.

Comparison of the metric camera photography and the improved color-composite photography is difficult because of grossly different viewing directions and the general disappearance of recognizable landmarks in the composite. Furthermore, few color boundaries exist in the area of metric photography where the illumination angle is low enough to show the fronts of some of the very thin flows. Nevertheless, patches of redder material lying northeast of Mt. Lahire are found to be enclosed by low flow fronts (i.e., they are "kipukas"), confirming the theory that the redder material represents an earlier filling of the Imbrium basin. A portion of a color-composite photograph of Mare Imbrium in which redder areas appear lighter is shown in figure 25-79. Color boundaries that correspond with visible flow fronts are outlined in this figure.

Elsewhere in the lunar maria, the evidence is always that the bluer areas are more recent than the redder background. Thus, several instances are known where portions of ray systems lying on the older, redder surface have been obliterated by later, bluer deposits. Strom noted some time ago[1] that an apparent correlation exists between the degree of blueness and the titanium content of the regolith. Analyses of Apollo 12 and 14 materials also support this idea. Thus, it appears that, for any lunar mare where color differences are visible, the redder material represents an early filling of low-titanium-content basalt, while the bluer material represents a later (in some cases, apparently much later) inflow of titanium-rich basalt.

---

[1] Private communication.

PART N

AN UNUSUAL MARE FEATURE

Ewen A. Whitaker[a]

While the Apollo 15 panoramic camera photography was being rapidly scanned, a most peculiar feature was noted that presented a totally different appearance from anything seen in all the rest of the Lunar Orbiter and Apollo photography. The feature was missed on Lunar Orbiter IV frame H-102 because it is situated in a group of bimat marks. The feature is located at latitude 18°40' N, longitude 5°20' E in a small patch of mare material lying between the Haemus and the Apennine Mountains. This patch is abnormal in that it is an unbordered plateau; the surface appears to lie several hundred meters above adjacent mare patches.

The feature is D-shaped with a 3-km-long straight edge. Viewed stereoscopically, it is seen to lie perhaps a few tens of meters below the level of the surrounding mare, the latter presenting a convex meniscus at the line of contact. About half of the floor is covered with blobs of marelike material, reminiscent of dirty

---

[a] Lunar and Planetary Laboratory, University of Arizona.

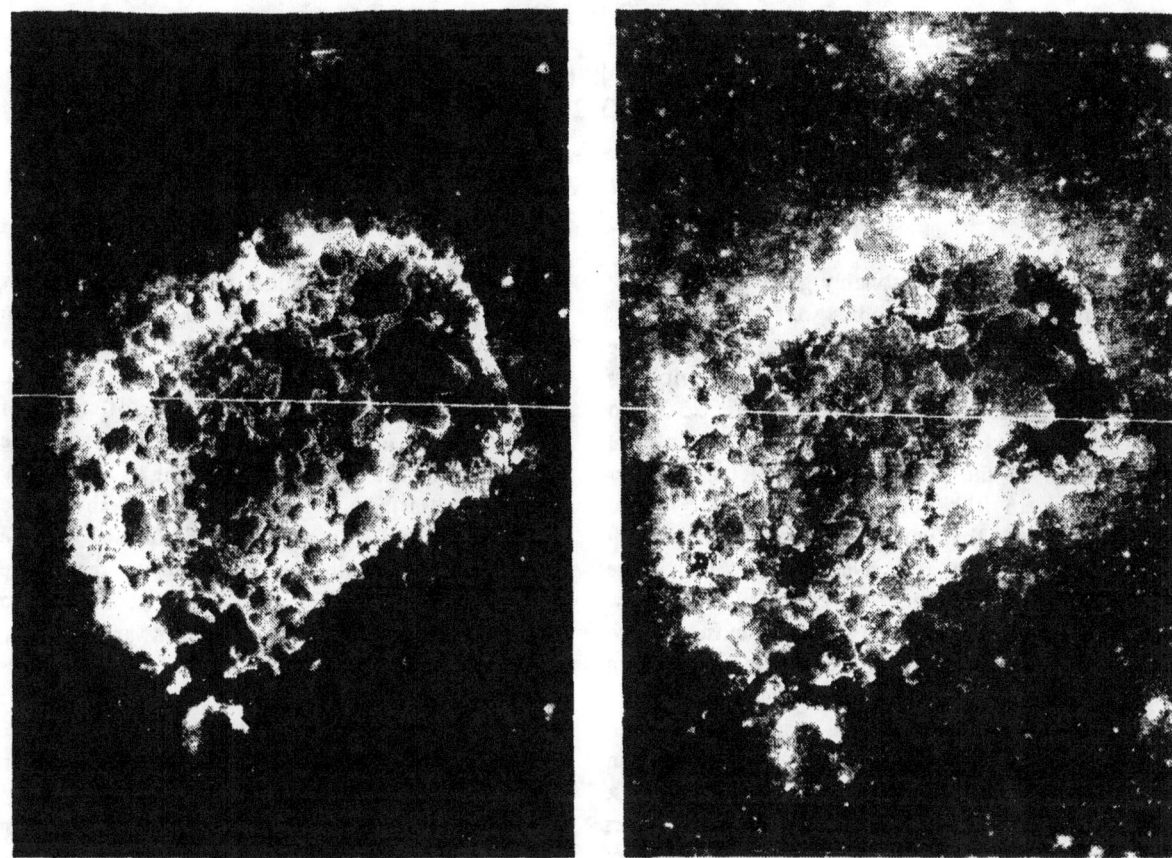

FIGURE 25-80.—Unusual mare feature of latitude 18°40′ N, longitude 5°20′ E. Feature may be viewed stereoscopically in these photographs (pan AS15-10176 and 10181).

mercury. Contacts between the floor and both the mare and blobs are frequently outlined with highly reflective material, perhaps sublimates. The floor also displays some darker areas that have noticeably different photometric properties from the mare surface and the blobs. The whole feature is seen to be almost devoid of small impact craters, thus differing from the surrounding mare. Two panoramic camera photographs of the feature are shown in figure 25-80, which may be viewed stereoscopically.

PART O

REGIONAL VARIATIONS IN THE MAGNITUDE OF HEILIGENSCHEIN AND CAUSAL CONNECTIONS

Robert L. Wildey[a]

Approximately 35 reasonably good candidates for specialized photometric studies were found during a thorough examination of the frames exposed by the Apollo 15 metric camera. Of these, the majority was of value in heiligenschein studies (refs. 25-36 to 25-38). A few were of value for limited-interval delineation of the photometric functions of crater walls, wherein it is now known from past Apollo Program studies that younger craters have walls much more Lambertian in reflective properties than those of the standard lunar surface (ref. 25-39). It has now become apparent that some difficulty exists in such crater-wall studies with regard to varying the domain of mathematical definition over which such photometric functions are delineated to encompass much more of phase-angle-brightness-longitude space than has already been carried out. Nearer to zero-phase angle, the brightness of the illuminated wall becomes so intense compared with the general luminance of the surrounding lunar terrain that it approaches the saturated region of the $D/\log E$ curve of the photograph where D is density and E is exposure. Conversely, for phase angles approaching 180°, most of the illuminated wall is seen so obliquely that the image size is small and reliable photometry is difficult. Also, difficulties inherent in the methodology exist—the descriptive geometry used to extract configurations begins to fail as craters become either too old or are imaged too near a vertical view. Nevertheless, some useful candidates for crater-wall studies have been found in the Apollo 15 data, although their analysis is too lengthy for incorporation in this preliminary report.

In the frames containing heiligenschein, a few things are visually obvious. First, no direct correlation exists with normal albedo, or even two distinct correlations within the highlands or within the maria. It was previously reported, on the basis of studies of the heiligenschein phenomenon on the Earth, that the magnitude of the phenomenon ought to be causally related to surface roughness. Thus, it should be possible to find a correlation of the size of the brightness surge with age, according to reasonable ideas on the modes of erosion and sedimentation under micrometeorite bombardment in vacuum.

A preliminary quantitative analysis of four frames (met AS15-M3-0139, 0149, 0161, and 0367) has been completed. Two of these frames show the zero-phase surge on maria and two show the surge on terrae. In each category of terrain, a rather clear age difference seems to be indicated by the abundance of small primary impact craters, although no attempt to compare ages across categories has been made. The curves resulting from brightness plotted as a function of phase angle have been found to differ markedly according to which terrain type (mare or terra) is involved. The magnitude of the heiligenschein is significantly greater for the maria than for the terrae (contrary to previously obtained results). Furthermore, the difference prevails not merely over the interval of phase angles from 0° to 1.5°, but from 0° to at least 5°. But, within each category, the curves agree rather well. In fact, if anything, a slightly greater surge (between phase angles of 0° to 1.5° only) is shown by the younger terrain, in comparisons of the two maria. Clearly, a multivariate analysis incorporating more complex geological properties is warranted.

---

[a]U.S. Geological Survey.

PART P

THE PROCESS OF CRATER REMOVAL IN THE LUNAR MARIA

L.A. Soderblom[a]

The processes by which craters disappear from the lunar surface have been of principal concern since the first high-resolution pictures of the lunar maria were returned by Ranger VII. Those pictures revealed that craters smaller than a few hundred meters on the lunar maria vary morphologically from sharp and pristine features to shallow, highly subdued depressions. The constancy of the population density of these smaller craters (with diameters of less than 100 m) was predicted by Moore (ref. 25-40) and Shoemaker (ref. 25-41), who concluded that these craters were in a steady-state condition.

The process under which these craters are being removed became the subject of some debate. Two basic models have been advanced for crater removal. The first model involves the catastrophic removal of craters by formations of new craters that overlap older ones (ref. 25-42). This model satisfies the observational criterion of constant crater density on surfaces of different ages—formation of a crater causes destruction of a crater. Such a model would also predict that craters should saturate the surface of the maria. Available photography, however, shows that these craters occupied only 10 or 20 percent of the surface. A second set of theories has been advanced according to which craters are gradually eroded by a rain of small impacts that individually produce only negligible changes in the crater form (refs. 25-43 and 25-44). These models satisfy the observed condition of steady state and explain the unsaturated nature of the population.

Specialized photography returned by the Apollo 14 and 15 missions demonstrates that both these processes are active and provides sound observational evidence as to their relative effects. The purpose of this paper is to evaluate the relative importance of these processes.

Observational Evidence

The Apollo 14 and 15 missions carried an orbital experiment designed to photograph the lunar surface very near the terminator at extremely low Sun-elevation angles (ref. 25-8). A high-speed film (Kodak 2485) was used to avoid long exposure and smear. The Apollo 14 photographs contained one terminator-crossing sequence, and the Apollo 15 photographs contained seven sequences. The sequence taken on the Apollo 14 mission is compared with a section of a Lunar Orbiter IV frame of the same area in figure 25-81. As presented, the photographs have similar resolutions. The striking difference in the appearance of the surface in the two photographs arises from the radically different illumination conditions. The Sun elevation in the Lunar Orbiter frame is approximately 19°, whereas it varies between 0° and 2.5° in the Apollo sequence. The important observation is that the crater population on this mare surface rapidly takes on a saturated appearance as the Sun-elevation angle drops below approximately 1°. Two enlarged sections of the Apollo 14 sequence, with solar-elevation angles of approximately 0.5° and 2°, are compared in figure 25-82. At the lower Sun angle, most of the craters are shadowed, and the population appears saturated. In the second framelet, the surface does not show this saturated condition. Hence, most of those craters have maximum interior slopes of approximately 1°. The observed fact is that the net accumulated impact flux required to saturate a surface with these craters (50 to 100 m diameter) is about the same flux required to erode a crater of that diameter to an interior slope of approximately 1°. For the first time, observational evidence exists that calibrates the gradual impact-erosive mechanisms directly in terms of accumulated impact flux.

In figure 25-83, an Apollo 15 near-terminator sequence is compared with a section of a Lunar Orbiter IV frame of the same area (solar elevation angle, 15°). These photographs have approximately four times higher resolution than the Apollo 14 sequence (fig. 25-81). Consequently, the fine structure of the surface is more evident. As the range of solar-elevation angle is only approximately 1°, the rapid onset of saturation at approximately 1° is not shown. Figure 25-84 is an enlarged section of figure

---
[a] U.S. Geological Survey.

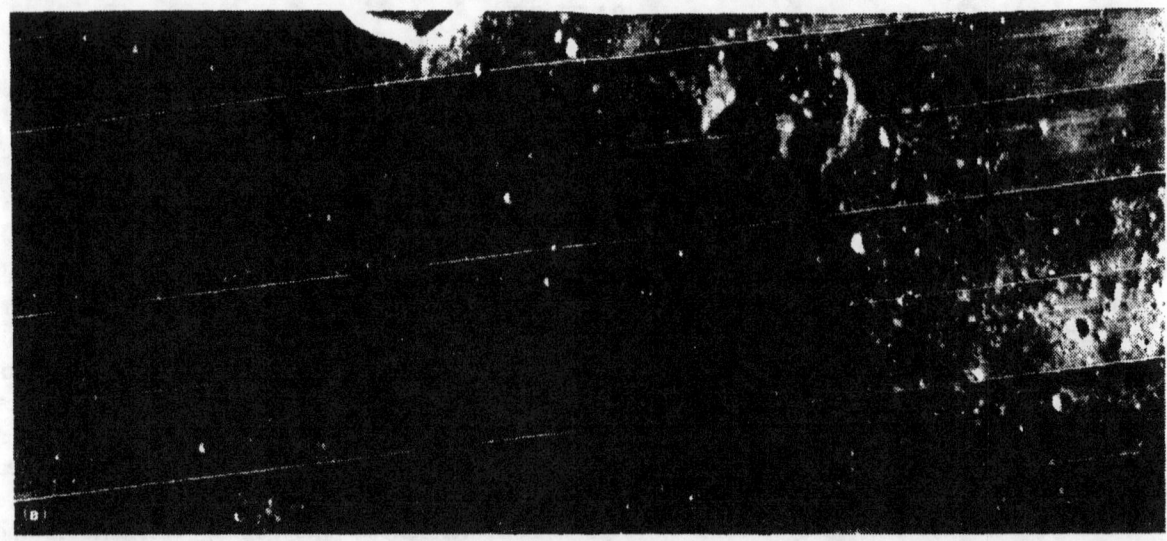

FIGURE 25-81.—Area south of Kunowsky Crater; the center of this 120- by 60-km area is at latitude 2° N, longitude 31° W. (a) Orbiter IV high-resolution photograph taken at a Sun angle of 19°. (b) Near-terminator photomosaic of same area at Sun angles of from 0° to 2.5° (AS14-78-10374 to 10377).

25-83(b). The shoulder-to-shoulder saturation of 100- to 300-m craters is evident. The saturation of the surface by craters a few hundred meters in diameter with slopes of approximately 1° is again apparent.

The information that is the most useful in understanding the surface process that removes lunar craters is the distribution of shapes of craters of a particular diameter. This kind of information is usually difficult to obtain photographically. Photometric techniques have failed because of large calibration errors in the available data combined with local variations in the photometric function and lack of knowledge about these variations. Photogrammetric techniques require extremely stable geometric and laborious reduction. However, the Apollo 14 and 15 near-terminator photography has provided, for the

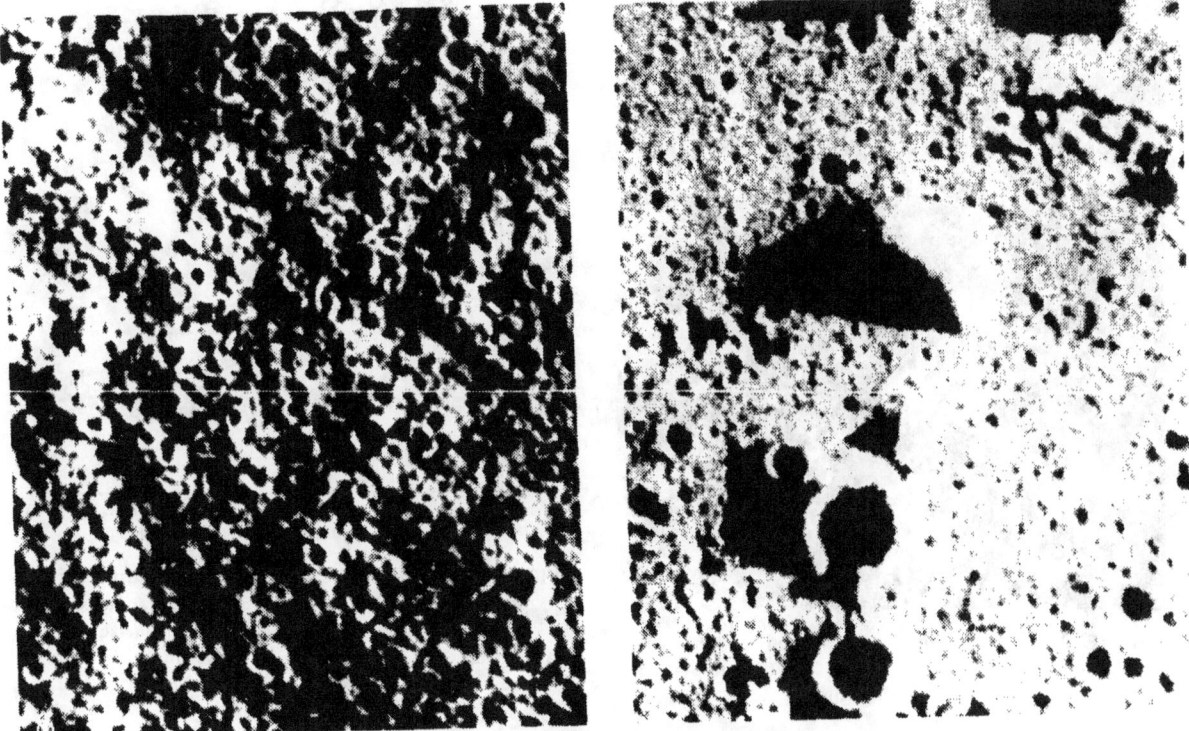

FIGURE 25-82.—Enlargements of portions of photographs in figure 25-81(b); the vertical dimension represents approximately 10 km. (a) Photograph AS14-78-10377 at a Sun angle of 0.5°. (b) Photograph AS14-78-10375 at a Sun angle of approximately 2°.

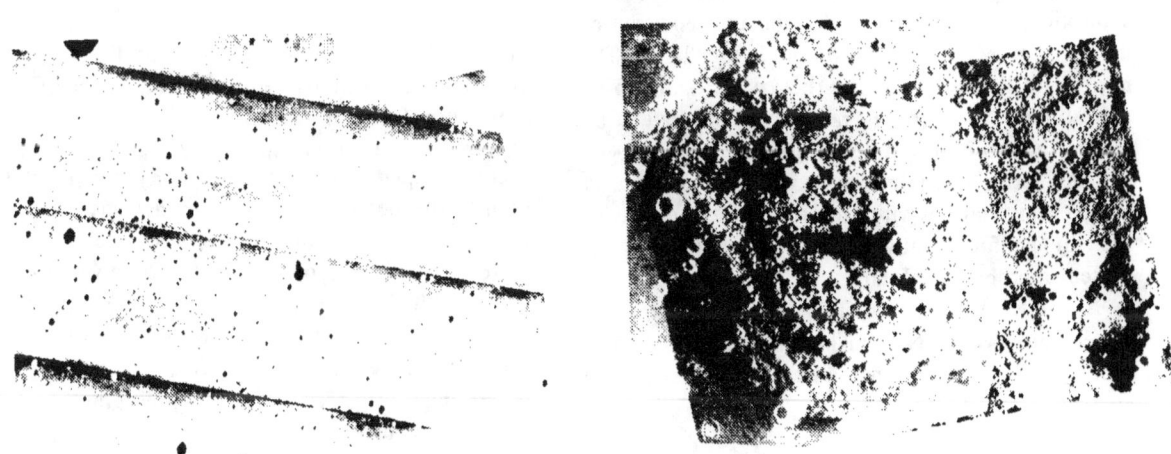

FIGURE 25-83.—Mare area at latitude 26° N, longitude 58°12' W; area shown is approximately 30 by 40 km. (a) Orbiter IV high-resolution photograph taken at a Sun angle of 15°. (b) Photomosaic of same area at Sun angles of from 0° to 1° (AS15-98-13347 to 13349).

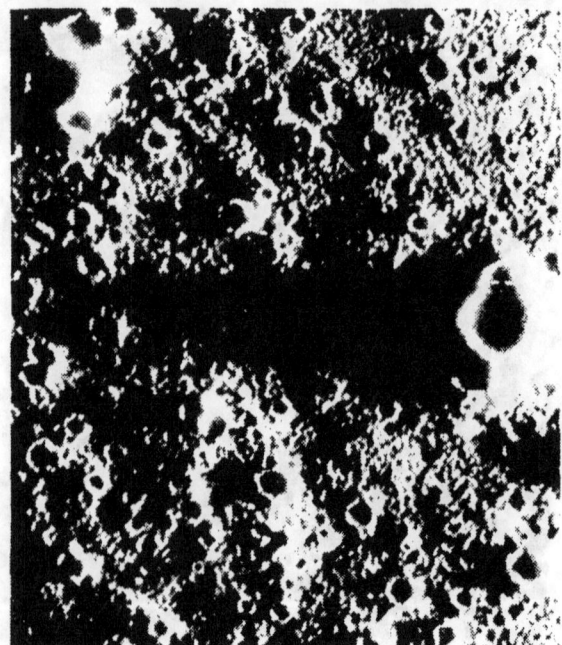

FIGURE 25-84.—Enlargement of portion of photograph AS15-98-13348 at Sun angles of from 0.1° to 0.4°; area shown is approximately 10 km square.

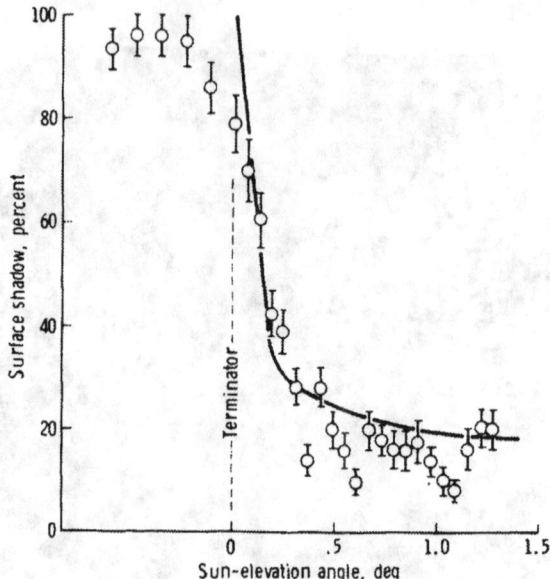

FIGURE 25-85.—Percentage of shadowed surface area in photographs AS14-78-10375 to 10377; data were obtained by point counts of shadows in the image. The error bars are based on counting statistics; the curve represents the best fit of the model discussed in text.

first time, unambiguous photographic documentation of the saturation conditions and slope distributions of craters in the lunar maria. This information is displayed by the distribution of shadows. For example, as may be seen in figures 25-82 and 25-84, these mare surfaces are saturated by depressions approximately 100 m in diameter with slopes of approximately 1°. A more sophisticated rendition of these kinds of data is shown in figure 25-85. The purpose of this part is to use the data of figure 25-85 to calibrate the erosion rate of lunar craters by comparing the observed distribution of morphologies in the saturated population with the integrated impact flux that has saturated the surface and has, at the same time, gently eroded the craters to the present morphological state.

A model for gradual impact erosion of lunar craters has been presented in reference 25-44. This model predicts that a lunar crater is eroded according to

$$S = S_i \exp \frac{-\alpha^2 F}{D^2} \qquad (25\text{-}2)$$

where S is the maximum interior slope (tangent of slope angle) at a time when the amplitude of the integrated flux is F, S_i is the initial slope of the crater, D is the crater diameter, and α is a constant. This was derived by assuming the craters were being produced according to

$$\Delta N = \Delta F D^{-3} \qquad (25\text{-}3)$$

where ΔN is the number of craters of diameter larger than D which accumulated per unit area per ΔF. This production function was used both to generate lunar craters and to erode them.

From equations (25-2) and (25-3), it is possible to predict the percentage of the surface shadowed as a function of Sun-elevation angle. By choosing a value of α such that the model fits the observed distribution of shadows in figure 25-85, it is possible to calibrate the rate of crater erosion as a function of the accumulated impact flux. This impact flux can be determined directly from the photography of the saturated surface.

The surface is saturated by craters larger than some diameter D_{min}; that is, the total area of craters larger than D_{min} is about twice the area of the surface. Because the number of craters increases so rapidly as a function of decreasing diameter, the surface is saturated principally by craters of diameters between D_{min} and a few times D_{min}. Larger craters are so scarce that they contribute only negligibly to saturation. Hence the saturated population can be represented by an average diameter $\bar{D}$.

Consider a crater of diameter $\bar{D}$ formed on the surface. Erosion of the crater immediately begins under a rain of impacts each of which produces only a minute change in crater form. Equation (25-2) describes this gradual smoothing process. While the crater is being worn down, it is also vulnerable to catastrophic destruction by craters of diameter equal to or larger than $\bar{D}$. If this happens, the crater is destroyed and a new crater of initial slope $S_i(30°)$ is formed that likewise suffers gradual erosion. The probability that a crater survives catastrophic annihilation after flux F has accumulated is

$$P = \exp\left(\frac{-2\pi F}{\bar{D}}\right) \qquad (25\text{-}4)$$

By combining equations (25-2) to (25-4), the density of craters as a function of slope is

$$dN = \frac{\exp\left\{-2\pi \ln^{1/2}\left[\frac{S_i}{\frac{S}{\bar{\alpha}}}\right]\right\}}{2\alpha S \ln^{1/2}\left(\frac{S_i}{S}\right)} ds \qquad (25\text{-}5)$$

Equation (25-5) provides an estimate of the number of craters per unit area dN which have slopes between S and $S + dS$.

To calculate the fraction of the area of a crater that is shadowed, the crater is assumed to be conical with an interior slope S. If the solar-elevation angle is S_s, the fraction shadowed is

$$R = \frac{S(S - S_s)^{1/2}}{(S + S_s)^{3/2}} \qquad (25\text{-}6)$$

By integrating the product of equations (25-5) and (25-6) over all slopes from the initial slope S_i to 0, it is possible to obtain an expression for the percentage of the surface that is shadowed for a given Sun angle, as shown by the curve in figure 25-85. The value of R was adjusted to obtain the best fit ($\alpha = 20$). Using the value of α in equation (25-2), it is possible to predict the morphological distribution of craters on surfaces of different age and different saturation levels. For instance, a surface saturated with craters 10 m and larger in diameter has received sufficient impact flux to erode a 250-m-diameter crater from an initial slope of 30° to a slope of 18°.

Conclusions

From the observations and analyses presented here, the following conclusions can be drawn. (1) The erosion of lunar craters in the size range of 10 m to 1 km is effected principally by small impacts producing minute changes in crater form up to the point where the crater is worn to an interior slope of approximately 1°. Below that slope, the annihilation process is dominated by the formation of younger overlapping craters. (2) The distribution of shapes of craters in this size range can be explained as resulting purely from impact-generated processes.

Near-terminator photography of the type described here has been found to be of great value in understanding the surface processes operative in the lunar maria. As indicated in reference 25-7, a profound need exists for a technique to establish reliable relative ages of surfaces from orbital photography. Only through such a technique will the complex and subtle stratigraphy of lunar maria be explicated. Imagery of the type discussed here is of incalculable value toward that end.

PART Q

CRATER SHADOWING EFFECTS AT LOW SUN ANGLES

H.J. Moore[a]

Comparison of Apollo 15 lunar-surface photographs taken at low Sun-elevation angles with photographs of an experimentally cratered surface at low lighting angles reveals some marked similarities. In both instances, smaller craters found between larger craters occur in such profusion that, locally, shadows of several small craters coalesce to form linear patterns and clusters. The fraction of area covered by shadows within these smaller lunar craters is so large that 30 to 40 percent of the total field of view is covered by shadow. This situation obtains when the frequency distribution of the smaller craters has reached a "steady state."

Experimental Surface

Personnel at the U.S. Geological Survey and the NASA Ames Research Center bombarded surfaces underlain by carborundum powder and sand with small projectiles and produced craters from 0.52 to 30 cm across. Details of the experiment have been described previously (refs. 25-24, 25-25, and 25-40). Bombardment of the surface was programed to produce a surface for which the cumulative frequency of craters produced per unit area per unit time is proportional to the reciprocal of the cube of the crater diameters, this being the same relationship which appears to describe the lunar situation. Initially, the frequency distribution of craters on the surface reflects, directly, the crater-production frequency distribution. With continued bombardment, the surface will no longer sustain the number of smaller craters produced on it, so that craters are destroyed by erosion and infilling. The larger craters still reflect the form of the crater-production frequency distribution, however. As bombardment continues, the sizes of craters that are eroded and filled extend also to the larger sizes. The end result is a steady-state surface in which craters are destroyed as rapidly as they are formed. This steady-state surface is characterized by craters of various sizes with various states of preservation up to a limiting size that is dependent on the age of the surface (ref. 25-45). For this steady-state surface, which is attained at the smaller sizes first and then proceeds to larger and larger craters with continued bombardment, the cumulative frequency of craters per unit area is proportional to the reciprocal of the square of crater diameter and is independent of time.

The experimental surface after a steady state has been achieved for all crater sizes is shown in figure 25-86. An earlier stage is shown in figure 25-87, where the steady state has not been achieved by all sizes. In that figure, the surface is heavily cratered by craters 0.52, 1.1, and 2.5 cm across. None of the surface has escaped modification by impact. For craters 2.5 cm and smaller, the surface has reached a steady state; but, for larger craters, the steady state has not been achieved. Close inspection of the surface in figure 25-87 reveals that dark shadows inside small craters commonly coalesce to form short beaded chains and clusters. This effect is particularly prominent on the left side where the illumination angle is smallest.

Lunar Surface

Studies of the lunar surface at various scales have shown that a steady-state surface is produced by impact cratering over a large range of crater sizes (refs. 25-44 to 25-46). Like the experimentally cratered surface, steady-state surfaces are achieved for the smaller sizes of craters first and then for progressively larger sizes as the surface becomes older and more cratered. If the size of the crater at which the steady-state crater frequency distribution joins the crater frequency distribution of larger craters is determined, then the relative age of the surface can be established (refs. 25-44 and 25-45).

Apollo 15 photographs of areas illuminated by low Sun-elevation angles are similar in appearance to the experimental surface. Such a surface can be seen in figure 25-88, where the smallest craters are also filled with shadows that form beadlike chains and clusters.

---

[a]Bellcomm, Inc.

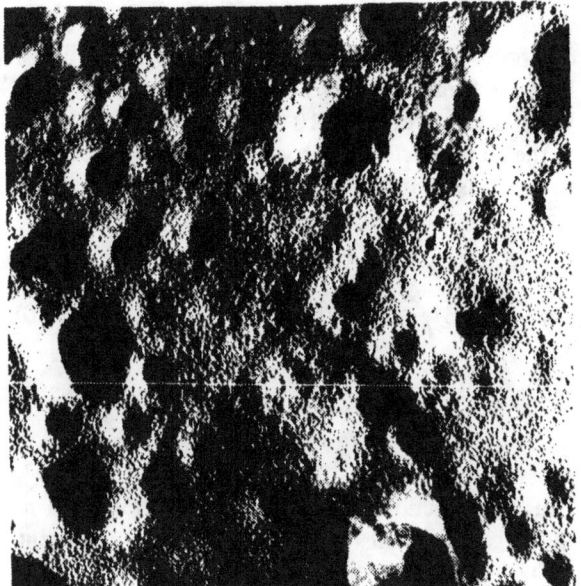

FIGURE 25-86.—Photograph showing experimentally cratered surface, in which the largest craters are approximately 30 cm across. The surface has reached steady state for all crater sizes. It should be noted that the morphologies of craters of a given size range from fresh and unmodified to subdued and eroded.

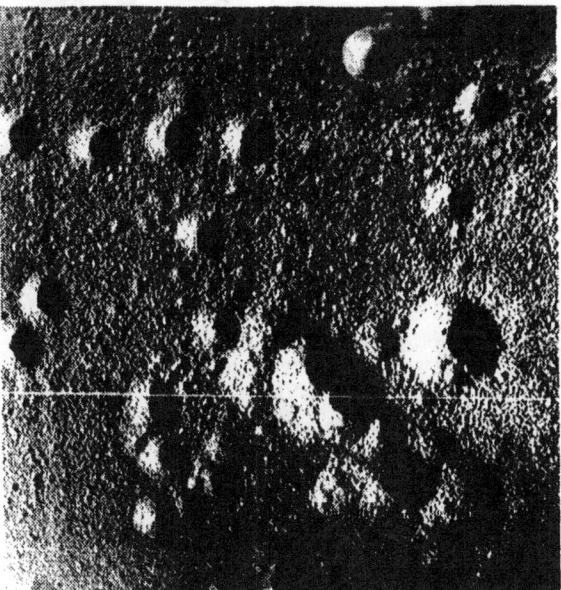

FIGURE 25-87.—Photograph showing experimentally cratered surface before steady state has been achieved for the largest sizes of craters. The largest craters are approximately 30 cm across. That the small shadowed craters form short beaded chains and clusters is noteworthy.

No visible part of the surface seems to have escaped cratering. Marked contrast in appearance exists between the surface when viewed at a very low Sun-elevation angle (fig. 25-88) and under a relatively high Sun elevation of approximately 17° (Lunar Orbiter IV frame H-157, framelet 963, 2.2 cm from data edge of standard print).

The largest crater in figure 25-88 is approximately 1400 m across. Small craters to a lower limit of approximately 20 m can be identified just to the east of this crater. Visual inspection of the photographs indicates that the steady-state condition for this surface begins for craters approximately 100 m across and extends downward to craters 20 m across and less. Preliminary estimates indicate approximately 30 to 40 percent of the surface is covered by shadow to the east of the 1400-m-diameter crater.

Discussion

The fraction of area covered by resolvable craters, which should be somewhat less than the fraction of area covered by shadow for photographs with very low Sun-elevation angles, may be calculated using the steady-state crater frequency distribution for craters

FIGURE 25-88.—Photograph of lunar surface in Oceanus Procellarum approximately 100 km west of the Aristarchus Plateau. The large crater just left of center is approximately 1400 m across. It should be noted that the small shadow-filled craters near the limit of resolution form beaded chains and clusters similar to those shown in figure 25-87 (AS15-98-13348).

from 20 to 100 m in diameter and then adding the area covered by larger craters for which the crater frequency distribution has the form of the crater-production frequency distribution.

For the steady-state crater frequency distribution (ref. 25-40)

$$N_s = 10^{-1.003} D^{-2} \qquad (25\text{-}7)$$

where N_s is the cumulative frequency of craters per m² smaller than diameter D in meters.

From equation (25-7), it can be shown that the fraction of area covered by craters A_s is

$$A_s = 0.336 \log_{10} \frac{D_2}{D_1} \qquad (25\text{-}8)$$

This yields a value of 0.235 for the fraction of area covered by craters with diameters between 100 and 50 m.

A crater-production curve of the form

$$N_p = KD^{-3} \qquad (25\text{-}9)$$

can be used to estimate the fraction of area covered by larger craters. Using $D = 100$ in equation (25-7) yields a value for N_s of $10^{-5.003}$ that must equal N_p at the same size. Thus, the value of K in the preceding equation is $10^{0.997}$. The fraction of area covered by craters larger than 100 m A_p is then:

$$A_p = 23.6 D^{-1} \Big|_{D=\infty}^{D=100} \qquad (25\text{-}10)$$

or

$$A_p = 0.236 \qquad (25\text{-}11)$$

Thus, the fraction of area covered by craters 20 m and larger is near 0.47. This observation is consistent with the estimate of 30 to 40 percent of the area covered by shadow, which should be a little less than the area covered by resolvable craters.

The correlation of shadowed area with crater distributions suggests the possibility that, in certain situations, automated determinations of the area covered by shadow, using film and densitometers, can be used to assess crater populations.

PART R

NEAR-TERMINATOR PHOTOGRAPHY

J.W. Head[a] and D.D. Lloyd[a]

For many years, it has been widely accepted that an examination of the lunar surface under near-terminator lighting conditions is extremely valuable to geologists and other scientists. Before the era when photography could be obtained from spacecraft in lunar orbit, a large percentage of the telescopic observations of the Moon (both direct and photographic) was conducted when the feature of interest was under near-terminator lighting conditions. When unmanned spacecraft were flown to obtain lunar photographs (e.g., Lunar Orbiter series), the mission parameters were selected so that photographs could be taken near the terminator. One reason is that, under near-terminator conditions (low Sun elevation), small changes in slope produce greater contrast changes than at high Sun-elevation angles. A related desirable phenomenon is that, at low Sun elevation, the shadow is longer than the object is high, thus increasing certain information about the object. For example, at $20°$, the exaggeration is 2.75 (cotan $20°$), and detectability and morphologic identification are consequently enhanced.

Historically, it has been difficult to obtain photography nearer the terminator than approximately $8°$ without severe underexposure. The Lunar Orbiter and the panoramic camera photographs were both optimized for photography at Sun elevations of $20°$ or above when photographing maria (although these cameras can be used to photograph lunar highlands at lower Sun elevations because of the higher albedo). Neither camera can produce the desired midrange exposures when operating nearer the terminator than $8°$.

Faster films could be used to obtain photography nearer the terminator than $8°$, but such fast films entail a resolution penalty, which, for most unattended camera systems, would affect all the photography obtained during the mission; that is, the film selected for photography at, for instance, $1°$ from the terminator would also have to be used for all the other photography.

The ability of an astronaut to change films in the Hasselblad cameras provided an opportunity for use of a very high speed film. Despite the fact that no image motion compensation is normally available for this camera (although such compensation can be obtained by rotation of the spacecraft), a preliminary set of four photographs from the Apollo 14 mission showed that photography within $0.5°$ of the terminator could provide lunar photography of special geological interest. The shadow length at $0.5°$ from the terminator is greater than the height of the corresponding object by a factor of 114.6 (cotan $0.5°$). Slight variations (less than $0.5°$) in nearly horizontal slopes produced significant variations in scene contrast (ref. 25-8).

Operation

During the Apollo 15 mission, 10 sequences of photographs were taken starting a few minutes before crossing of the terminator and continuing past the terminator (figs. 25-89 to 25-93).

The Hasselblad data camera was used with the 250-mm and 80-mm lenses at the maximum aperture. The film was a very high speed black-and-white film (Kodak 2485). Photographic operations were concluded by the command module pilot, as specified in the mission requirements document.

Results

Ten sequences (sets of approximately six photographs) were taken using magazine R. (This magazine also was used for other photography.)

The preliminary scientific results apply to three general areas:

(1) A sequence of photographs of the Mare Vaporum area showing maria structure in detail

(2) Four sequences of the Aristarchus Plateau area

---

[a]Bellcomm, Inc.

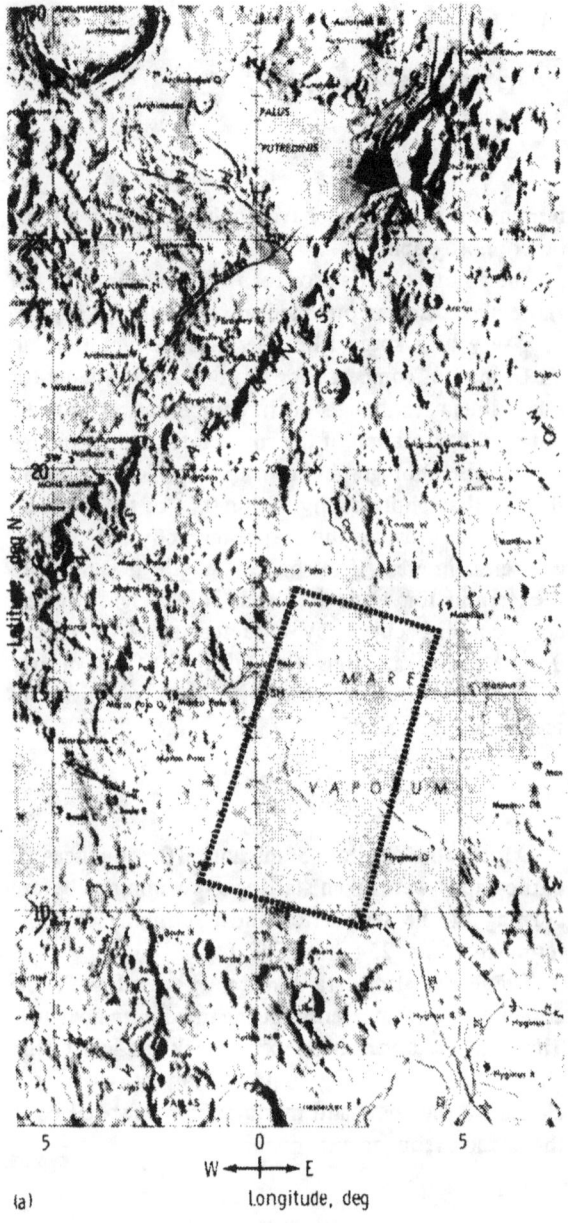

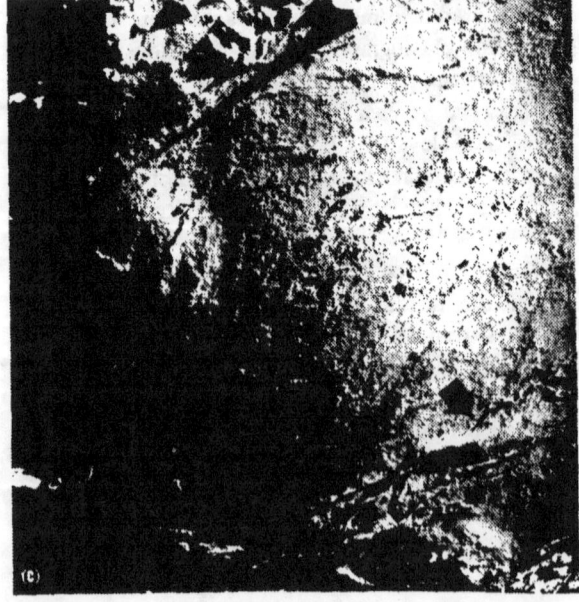

FIGURE 25-89.—Mare Vaporum area. (a) Map of area, with arrow indicating Apollo 15 landing site approximately 200 km north. (b) Lunar Orbiter IV photograph of 200- by 120-km area shown in part (c) of this figure. (c) Near-terminator photograph of central and western Mare Vaporum. Features made visible or enhanced in this low-Sun-angle view include a 25-km-diameter mare dome (upper arrow) and numerous mare ridges of varying widths and orientations (lower arrow). Many structures such as flow fronts, elongate craters, and depressions are visible in the near-terminator photography, which provides additional data regarding the surface and subsurface evolution of the maria (AS15-98-13302).

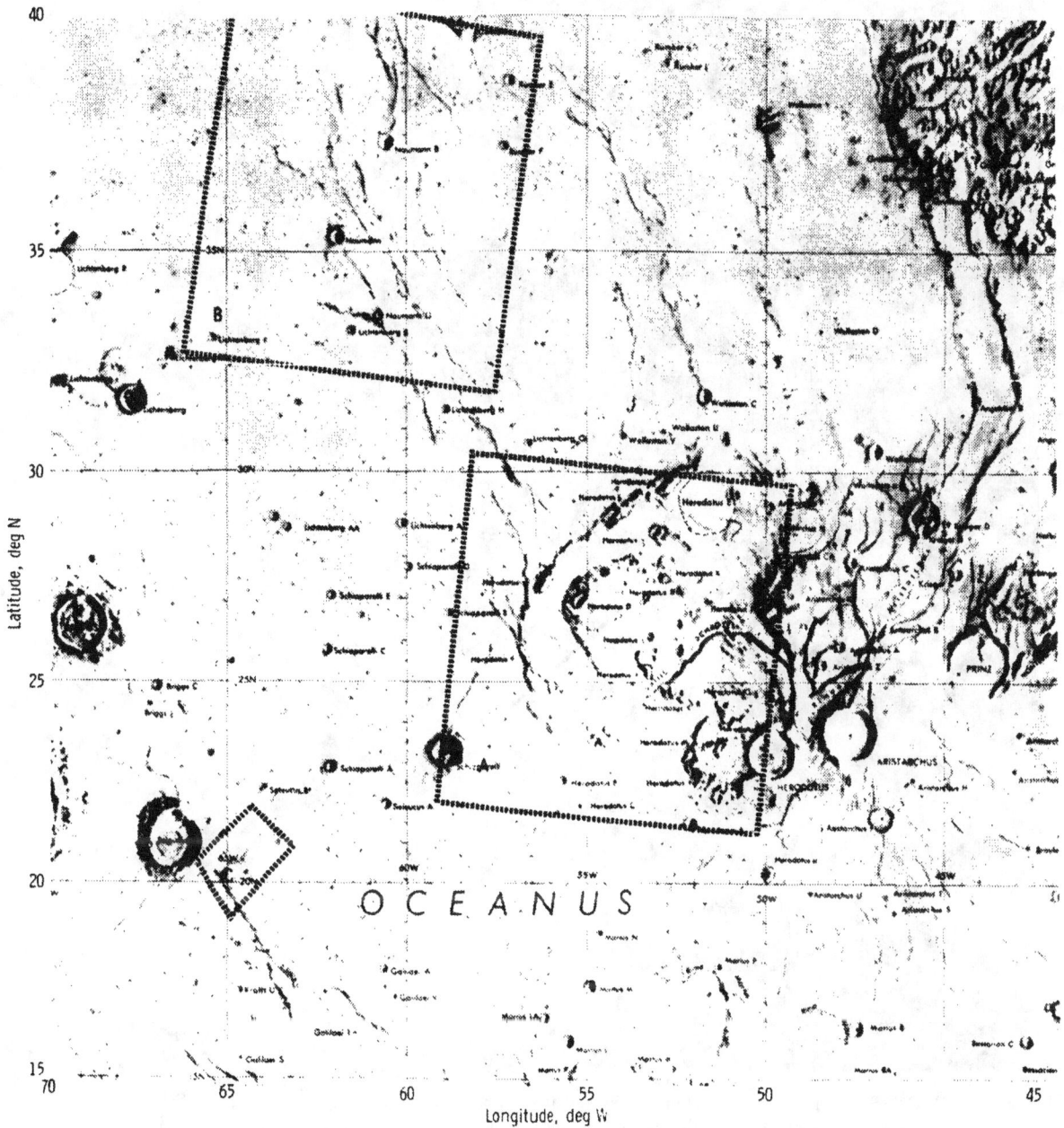

FIGURE 25-90.—Index map of northwest Oceanus Procellarum, indicating areas shown in photographs provided in figures 25-91 to 25-93 (A to C, respectively).

and the Oceanus Procellarum area west of the Aristarchus Plateau

(3) Five sequences of highland areas

Conclusions

Many photographs were obtained that show lunar-surface areas within a few degrees of the terminator; these photographs are of significant geologic interest. Many geological features stand out in a distinct manner not normal in conventional lunar photography, thus providing additional data on the surface morphology and the configuration of a large number of lunar-surface structures.

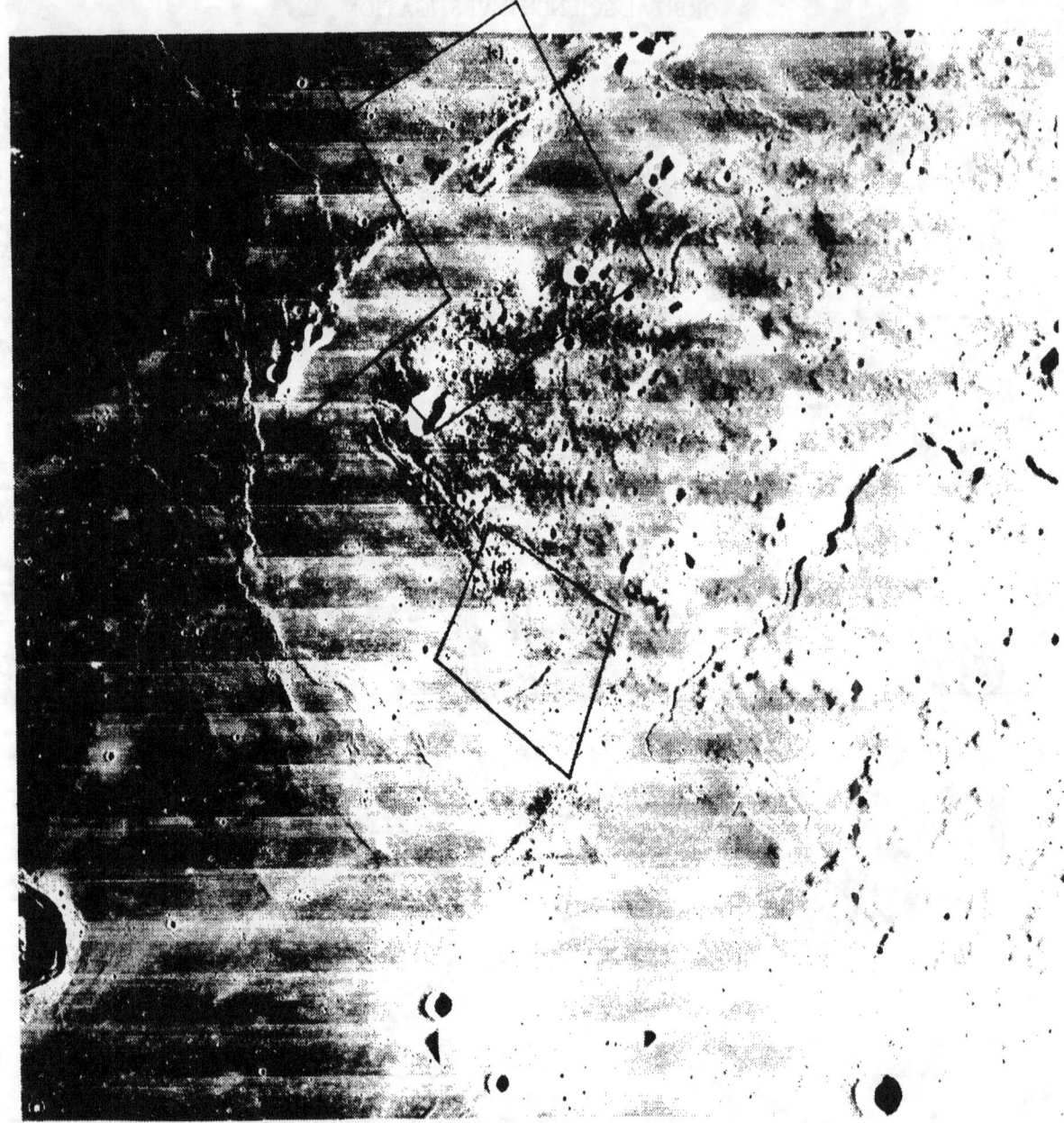

FIGURE 25-91.—Western portion of Aristarchus Plateau, including a significant part of Schroter's Valley. (a) Lunar Orbiter IV photograph, indicating areas shown in photographs provided in parts (b) to (d) of this figure. This area, approximately 250 km square, shows abundant evidence of volcanic activity. (b) Northwestern edge of Aristarchus Plateau. Higher crater density on the plateau and the highly sinuous rilles at the mare edge are evident. Mare ridges in the upper left continue northwest into Oceanus Procellarum, and numerous smaller sinuous rilles are seen that are barely visible in photography taken at higher Sun angles (AS15-98-13329). (c) Area east of that shown in part (b) of this figure. Features enhanced by the low Sun include the mare ridges and associated structures, areas of differing crater densities, and flow fronts (AS15-98-13331). (d) Southwestern edge of Aristarchus Plateau. Several straight and slightly sinuous rilles originate in the plateau and are obscured in the mare. Three units of varying crater density can be seen. The most heavily cratered includes the plateau; the intermediate unit is more marelike and appears to be the unit in which the rilles originate; and the least cratered area is in the mare and embays into the rilles (AS15-98-13345).

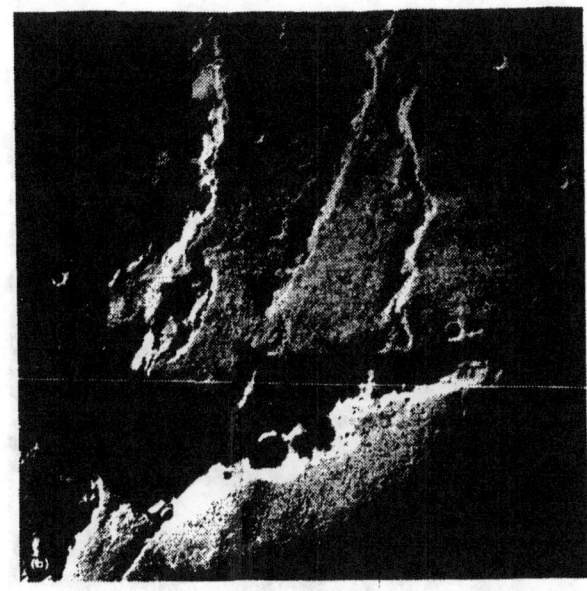

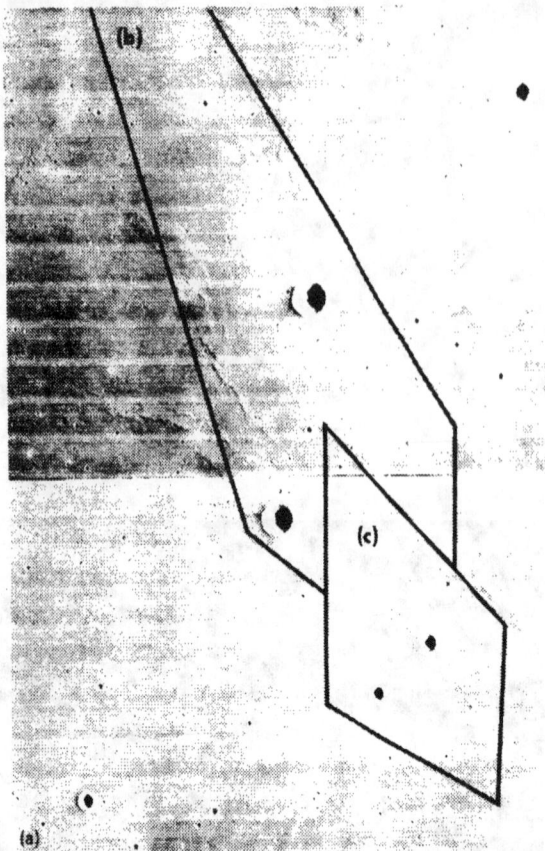

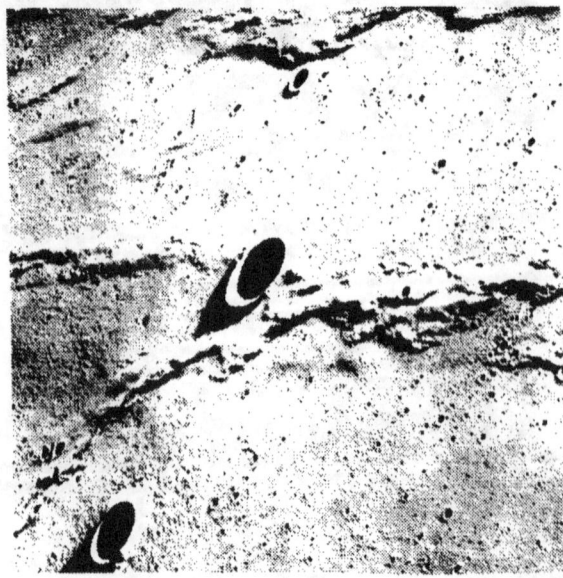

FIGURE 25-92.—Mare in Oceanus Procellarum between Aristarchus Plateau to the south and the Rumker Hills to the north. (a) Lunar Orbiter IV photograph, indicating areas shown in photographs provided in parts (b) and (c) of this figure. This photograph (showing an area approximately 240 by 320 km) indicates that the abundant mare ridges, which are characteristic of this area, trend in a northwest-southeast direction. (b) Mare ridges in the vicinity of Lichtenberg B Crater (lower left) and Naumann G Crater (center). Features of interest include subsidiary ridges on top of mare ridges, subdued isolated mare ridges, and en echelon structures perhaps related to regional structural patterns (AS15-98-13354). (c) Additional mare ridges to the northwest of Naumann G Crater (lower right). Naumann Crater (lower left) and Naumann B Crater (upper center) show hummocky ejecta blankets. Mare ridges are seen to follow angular structural patterns toward the horizon to the northwest (AS15-98-13355).

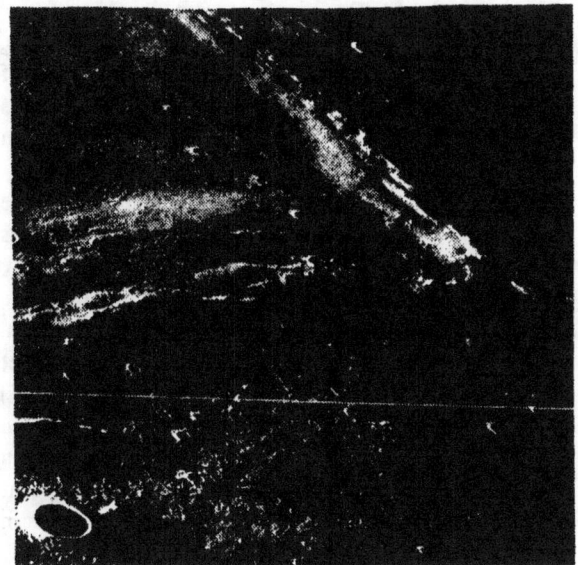

FIGURE 25-93.—Near-terminator photograph of intersecting mare ridges a few kilometers east of Seleucus Crater in western Oceanus Procellarum. In addition, the photograph shows a broad ridge of low slope that lies beneath the mare ridge but is several times wider than the mare ridge. The structures indicate that the material that produces the narrow ridge may bow up a considerably wider area underneath before coming through to the surface (AS15-98-13361).

PART S

FIRST EARTHSHINE PHOTOGRAPHY FROM LUNAR ORBIT

D.D. Lloyd[a] and J.W. Head[a]

During the Apollo 15 mission, 15 photographs of the Moon were taken with earthshine illumination. These were the first earthshine photographs taken from lunar orbit.

The photographs are of photometric interest, particularly because they involve double reflection of sunlight—by the Earth and then by the Moon—before photographic exposure. Certain published data on the mean illumination of the Moon by the crescent Earth predicted lower exposure values than were obtained for each measured area. The apparent albedo values obtained for the floor of Aristarchus Crater were anomalously higher than those obtained for the surrounding maria.

The first earthshine photographs of the Moon taken from lunar orbit were exposed during revolution 34. Two sets of photographs (a set of four and a set of 10) were taken at 1/16 sec and 1/8 sec, respectively. (One additional photograph showing portions of the Moon in earthshine was obtained as an unplanned result of the solar-corona photography.) The set of 10 photographs can be considered the basic set, and the preliminary set of four can be used for analytical comparison.

The 10-photograph set covered various types of lunar terrain including certain special lunar features. Six frames were of maria and were significantly underexposed; the locations of the other four photographs are shown in figure 25-94. The photographs themselves are shown in figure 25-95, which shows two views of Aristarchus Crater, one of Schroter's Valley, and one of Herodotus Crater.

Technical Discussion

Camera and Film Used

The earthshine photography was obtained with the

---
[a]Bellcomm, Inc.

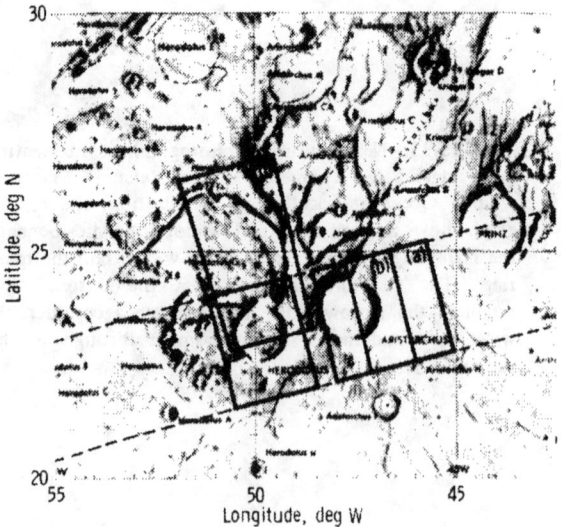

FIGURE 25-94.—Map showing location of earthshine photographs.

35-mm Nikon camera, which was required for the gegenschein experiment. This camera has a 55-mm-focal-length $f/1.2$ lens. This low f number provides a competence for low-light-level photography greater than that available on earlier missions.

The film selected was Ektachrome 2485, a high-speed black-and-white recording film. The response characteristic curve ($D/\log E$ curve, where D is density and E is exposure) for this film, as used for preflight exposure selection, is shown in figure 25-96. The derivation of the preflight-predicted film-exposure values for maria and Aristarchus is provided later.

The film has a lower resolution capability than is normally sought for aerospace photography and has historically been used for laboratory photographs of instruments, including cathode ray tubes. The resolving power of the film is 56 lines/mm for a target with a contrast of 1000:1 and 20 lines/mm for a target with a contrast of 1.6:1 (ref. 25-47). The potential quality of the film for photographing high-contrast targets was demonstrated on the Apollo 14 mission when it was used for near-terminator photography (ref. 25-48), but the predicted low-contrast resolution of 20 lines/mm was used in the preflight selection of the optimum shutter speed.

Operations

The camera was handheld by the command module pilot and was pointed out the hatch window. The spacecraft lights were dimmed, and the timing of the photographs at approximately 30-sec intervals was achieved by real-time command from the Mission Control Center.

All photographs were taken in accordance with the flight plan. The photograph of Schroter's Valley was oriented significantly to the north, a discretionary decision by the command module pilot involving the type of action encouraged during preflight briefings.

The earthshine photographs were taken on August 1, 1971, at 144:10:32 ground elapsed time (GET) (13:45 G.m.t.). At that time, the Moon/Earth/Sun relationship was as shown in figure 25-97. The eastern limb of the Moon was in sunlight, the subsolar point was at longitude 60.6° E, and the target was in earthshine. Earthshine comes from the portion of the Earth that is sunlit and visible from the Moon. Visualization of these conditions is aided by the photograph in figure 25-98, which was taken from Lunar Orbiter I under similar Moon/Earth/Sun conditions.

Lighting Conditions

The primary factor that determines the magnitude of the reflected light from the target is the phase angle g (fig. 25-97). This phase angle is defined as the angle between the vector from the source of illumination to the target and the viewing vector from the imaging system (camera in spacecraft) to the target. The phase angle is dependent on the longitude and latitude of the target. (Fig. 25-97 does not show the effect of latitude on g.)

To determine the direction of the source of illumination to the photographic target, the target position in lunar coordinates must be adjusted for the extent of lunar libration. In the east-west direction, the libration at the time of photography placed the Earth at longitude 6.0° W, as shown in figure 25-97. The photographic-target coordinate of longitude 47° W must be adjusted by this −6.0° to produce an angle of 41° of longitude between the source of illumination and the viewing vector for vertical photography. In the north-south direction, a lunar libration of 6.3° N existed. This value must be added to the 23° N latitude of the target if an exact value of g is to be determined. This latitude adjustment produces a value of g a few degrees greater than 41°. A value of g = 45° was used for preflight exposure-prediction purposes.

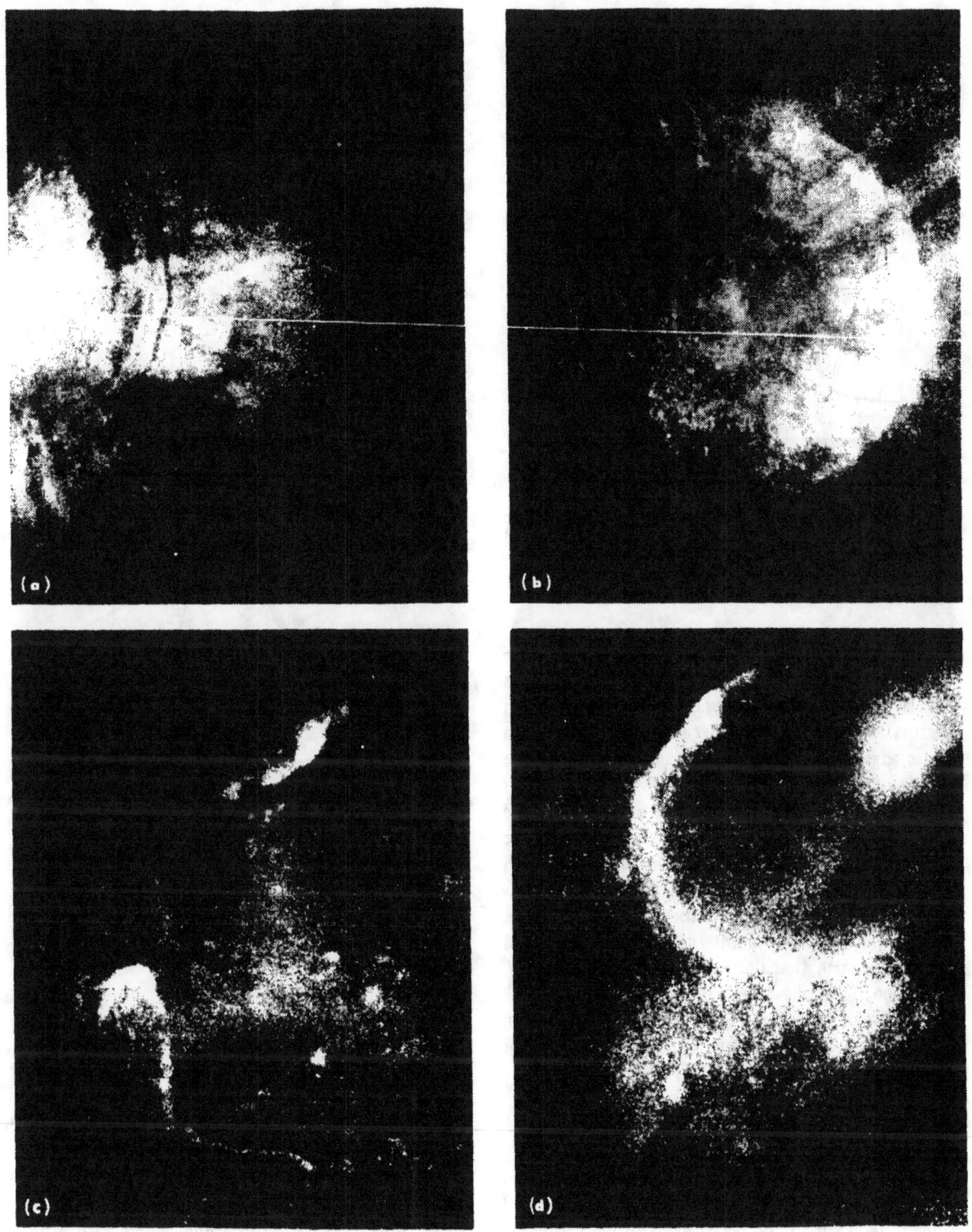

FIGURE 25-95.—Earthshine photography. (a) Aristarchus Crater (AS15-101-13591). (b) Aristarchus Crater (AS15-101-13592). (c) Schroter's Valley and Cobra Head (AS15-101-13593). (d) Herodotus Crater (AS15-101-13594).

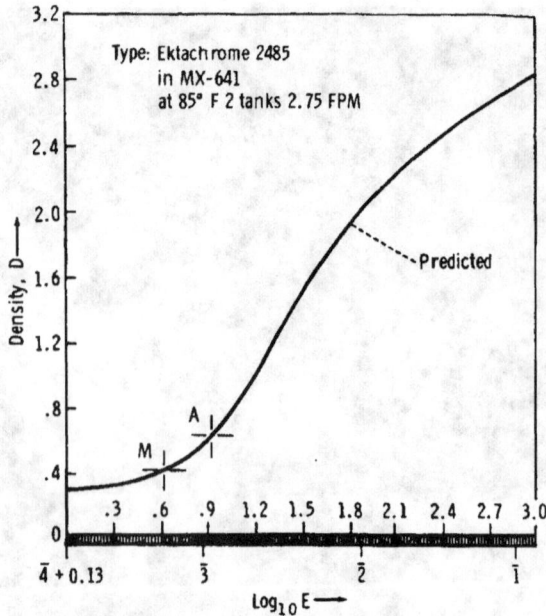

FIGURE 25-96.—D/log E curve showing predicted values of exposure for maria and Aristarchus Crater.

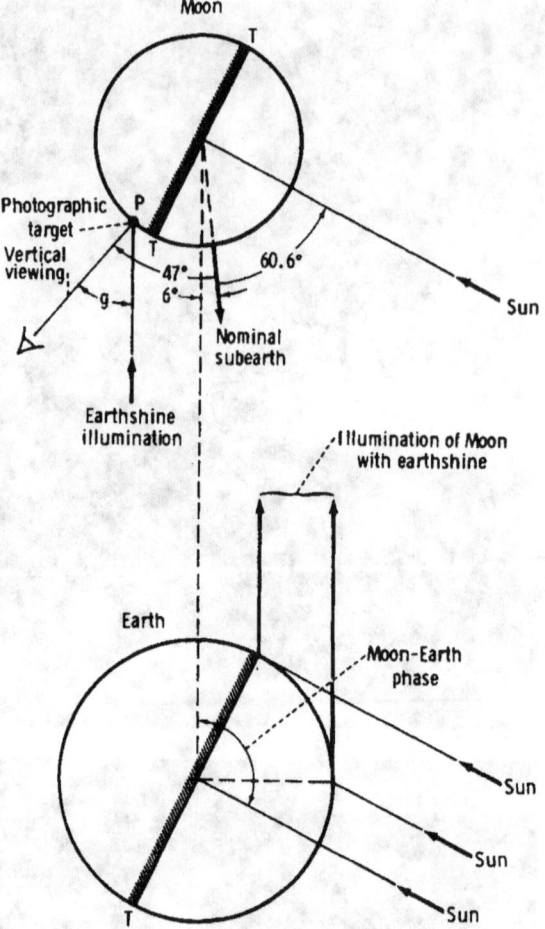

FIGURE 25-97.—Sun/Earth/Moon geometry to show earthshine illuminative conditions.

The primary factor that determines the magnitude of earthshine illumination incident on the Moon is the angle between the Sun/Earth and Moon/Earth vectors. This angle can be called the Moon/Earth phase angle. The Moon/Earth phase angle affects the extent of the sunlit Earth visible from the Moon. At large Moon/Earth phase angles, the magnitude of the illumination is significantly reduced. The magnitude of the illumination as a function of Moon/Earth phase angle is shown in figure 25-99, which includes data that were available before the mission (ref. 25-49). For a subsolar point at longitude 60.6° E and a libration such that the Earth is at longitude 6° W, the Moon/Earth phase angle is 180° − 66.6° = 113.4°.

As shown in figure 25-99, the mean illumination of the Moon by the crescent Earth is 1.35 lm/m² for a Moon/Earth phase angle of 113.4°.

Target Albedo

Albedo data based on photoelectric-photographic measurements from Earth are available. The preflight albedo values used for preflight exposure-prediction purposes were obtained from reference 25-27. A value of 0.09 was used for the maria area, and 0.18 was used for Aristarchus.

Predicted Exposure Calculation

The predicted exposure E at the film is given by

$$E = P_j u C t \qquad (25\text{-}12)$$

where P_j is illumination; $u = \rho\phi$ (where ρ is the target albedo and ϕ is the target photometric function); $C = L_t/4f_\#^2$ (where L_t is the transmission of the lens plus the window and $f_\#$ is the camera f stop); and t is the shutter speed.

When a crescent of the Earth is illuminating the Moon at a Moon/Earth phase angle of 113.4°, $P_j = 1.35$ lm/m². Taking this value, along with $\rho = 0.18$ for Aristarchus, $\phi = 0.33 \times 0.8$ for $g = 4.5°$ (where 0.33 is Fedorets data and 0.8 is a value used to adjust the Fedorets data to be consistent with the measure-

FIGURE 25-98.—Lunar Orbiter 1 photograph H-102 showing illuminative conditions similar to those in figure 25-97.

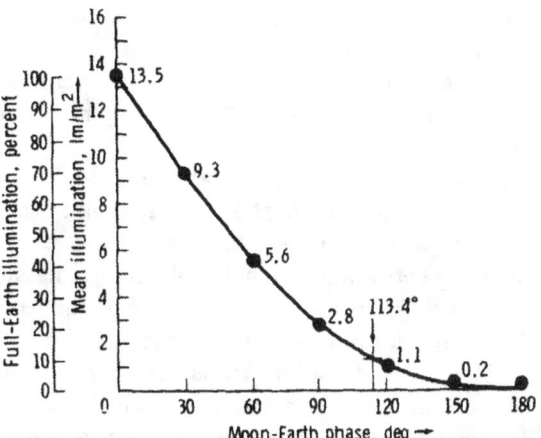

FIGURE 25-99.—Predicted values of mean illumination of Moon by crescent Earth.

ments of albedo data), $L_t = 0.8$ ($T_\# = 1.26 + L_w = 0.9$, where $T_\#$ is transmission number and L_w is loss through window), and $f_\# = 1.2$, equation (25-12) yields

$$E = 1.35 \times 0.18 \times 0.264 \times \frac{8}{4(1.2)^2} \times t \text{ m-cd-sec} \qquad (25\text{-}13)$$

If the shutter speed is selected as 1/8 sec, then $E = 1.11 \times 10^{-3}$ m-cd-sec. $\log_{10} E$ is $\bar{3} + 0.05$ for $\rho = 0.18$ (Aristarchus) and $\bar{4} + 0.75$ for $\rho = 0.09$ (maria).

In this calculation, a shutter speed of 1/8 sec is used and a slight underexposure of a target with an albedo of 0.18 is produced. The predicted density with this albedo is shown as A in figure 25-96. For an albedo of 0.09, such as that of the maria, the photograph can be expected to be significantly underexposed. The predicted density with an albedo of 0.09 is shown as M in figure 25-96.

Clearly, the shutter speed can be selected to produce any desired magnitude of exposure. If exposure were the only criterion, a shutter speed greater than 1/8 sec would have been selected to increase the predicted exposure of the maria. However, the 1/8-sec shutter speed was selected because of the effect of smear when no image-motion compensation is provided.

Selection of Shutter Speed

In the absence of smear, expected resolution on the film was 20 lines/mm (for low-contrast targets) or, on the ground, 110 km/55 mm × 50 = 100 m. The forward-motion smear at a shutter speed of 1/8 sec is approximately 210 m (for an estimated orbital velocity of 1.680 km/sec). Frequently, it might be argued that a smear of 40 percent of 100 m is acceptable and that 210 m is not. This degree of smear would require an exposure of 1/8 × 0.4 × 100 sec/210, but the resultant reduced film exposure would be so low as to underexpose the film completely and to invalidate the resolution data of 20 lines/mm, which is predicated on a reasonably exposed photograph. A faster film or a camera with a lower f stop or with image-motion compensation would be desirable.

Results

Measurements made from the primary set of 10 photographs provided density measurements from which the obtained exposure could be deduced. The density and exposure values obtained for general maria and for the floor of Aristarchus are shown in figure 25-100 (at M and A, respectively), which also shows the preflight predicted values.

If the measured exposure values are compared with the predicted values, the following results are obtained:

(1) For the maria area, reasonably close agreement exists; thus, future photographic results can be predicted with reasonable confidence. However, the difference between theory and results is such that it is highly desirable that further analysis be performed to clarify the cause or causes for the difference.

(2) For the floor of Aristarchus, the exposure value obtained is far greater than predicted.

If only the measured exposures are examined (no reference being made to the predicted value), a large ratio exists between the exposure for the floor of Aristarchus and the exposure for the maria. This ratio is approximately 7 (a difference in $\log_{10} E$ of approximately 0.85). This ratio implies an "apparent albedo" of $7 \times 0.09 = 0.63$—a very high albedo value, if it could be accepted as such. It should be noted that the measured ratio is independent of any possible errors in the estimates of factors used in predicting exposure.

The preliminary set of four photographs taken at a shutter speed of 1/16 sec produced grossly underexposed photographs. Clearly, such shutter speeds are not usable when the Earth illumination of the Moon is at large Earth/Moon phase angles (i.e., low levels of illumination).

It is desirable to obtain earthshine photographs when the Earth is near full (low Earth/Moon phase angles). Earthshine photography taken when the Earth is near full (early Apollo revolutions) could be performed with a shutter speed of 1/16 sec, thus reducing the smear to half that obtained at 1/8 sec yet producing reasonably exposed photographs. In highland areas, a shutter speed of 1/32 sec could be used, thus reducing the smear to acceptable levels.

Preliminary Geological Observations

The area of earthshine photography seen in figure 25-95 lies in the northwestern part of Oceanus Procellarum, a large, irregularly shaped mare area in the western near-side hemisphere of the Moon. The photography is centered on the Aristarchus Plateau, a complex structure characterized by relatively young structures interpreted to be of volcanic origin (refs. 25-50 and 25-51). Schroter's Valley is the most striking feature in the Aristarchus Plateau that might be related to surrounding volcanic structures. Cobra Head (figs. 25-95(c) and 25-101) is a deep crater marking the probable origin of Schroter's Valley. Both the main portion of Schroter's Valley and the smaller, internal, very meandering channel are associated with Cobra Head and continue through the Aristarchus Plateau for more than 160 km before terminating in Oceanus Procellarum. Earthshine photographs of this region show a bright area in the southern portion of the floor of Cobra Head. (This area is obscured by a shadow in fig. 25-101.) Other bright areas in the vicinity of Schroter's Valley and the Aristarchus Plateau correspond to bright-halo craters, older crater walls, steep-sloped domes and hills, and the walls of Schroter's Valley. The bright region in the floor of Cobra Head is anomalous by comparison with the rest of the floor of Schroter's Valley. A detailed analysis of this area is being undertaken because of its possible significance in relationship to questions on the origin and types of sources of sinuous rilles. The bright area may be related to bright walls in the initial crater or to light material associated with a deposit at the origin of Schroter's Valley. Bright ejecta from Aristarchus Crater seem unlikely because areas of the rille floor a few kilometers to the north, which are known

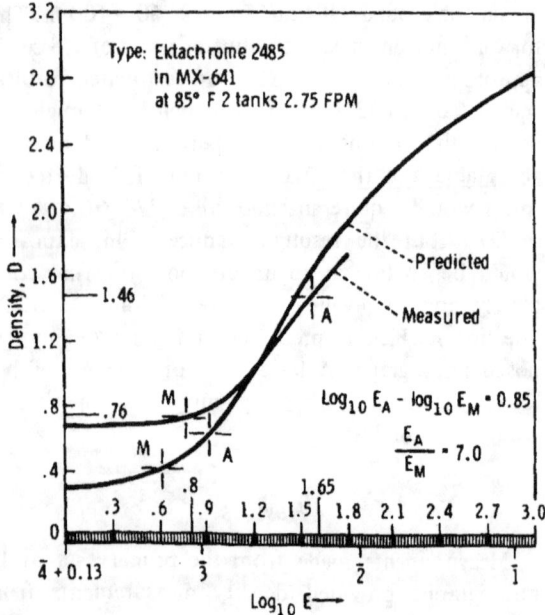

FIGURE 25-100.—Comparison of predicted and actual exposures of maria and Aristarchus Crater.

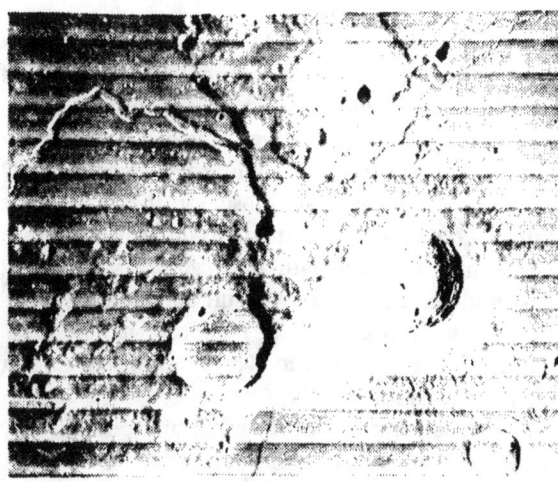

FIGURE 25-101.—Lunar Orbiter photograph of the Aristarchus Plateau region showing Aristarchus Crater (approximately 39 km in diameter), Schroter's Valley, and Cobra Head.

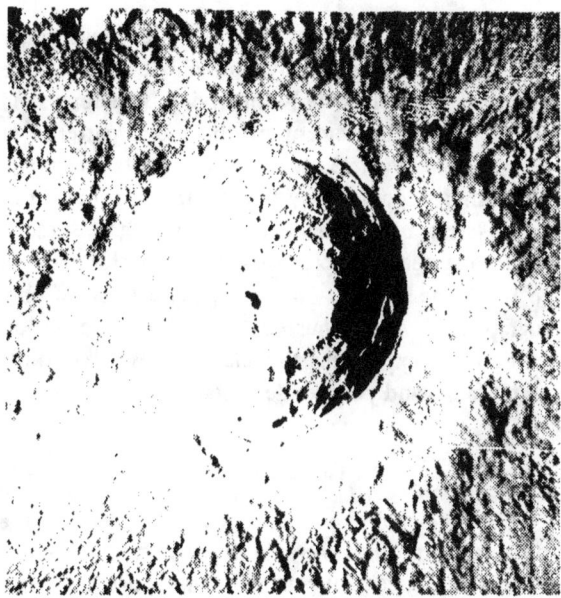

FIGURE 25-102.—Lunar Orbiter V photograph of Aristarchus Crater.

to contain Aristarchus Crater ejecta, show as dark areas in earthshine.

A second area of interest on the Aristarchus Plateau is Aristarchus Crater itself. Aristarchus Crater is approximately 39 km in diameter and is located at the southeastern edge of the plateau (figs. 25-94, 25-101, and 25-102). A series of slumped terraces is located between the crater rim and the edge of the crater floor. The crater floor is much less circular than the rim and contains an arcuate series of central peaks, which contrast with the ropy texture and series of low domes characteristic of the rest of the floor. Three sequentially concentric facies surrounding the crater have been defined (ref. 25-52). In sequence away from the crater edge, these facies are (1) a concentric or transverse dune facies immediately surrounding the crater rim (0 to 8 km) and characterized by a series of concentric dunelike structures 15 to 25 m wide, dark pools and flow structures, block fields, and a general lack of craters; (2) a radial or longitudinal dune facies (approximately 0 to 4 km) characterized by radially arrayed dunelike structures transitional between the concentric and rhomboidal facies; and (3) a rhomboidal facies characterized by a coarse and fine rhomboidal pattern formed by barchanlike crescent ridges and finer, chevron-shaped ridges, the apexes of which are radially arrayed around and point toward the center of Aristarchus Crater.

Aristarchus Crater is one of the youngest lunar craters in its size range and is of Copernican age. Based on crater densities within Aristarchus, Hartmann (ref. 25-53) has recently estimated its age as 1×10^9 yr, younger than Copernicus Crater and older than Tycho Crater. Coincident with the young age of Aristarchus Crater is a very high albedo, as previously discussed. Spectral-reflectivity data for Aristarchus Crater indicate that the crater has a curve characteristic of bright upland areas, in contrast to the surrounding mare regions, and this observation may indicate that the crater-forming event penetrated the volcanic deposits, excavating upland-type material and depositing it on the crater rim (ref. 25-54).

The most significant features of the Aristarchus Crater earthshine photography emerging from the preliminary analysis are the contrasting bright and dark bands located on the crater walls and rim and oriented radially to the crater (fig. 25-95). These contrasting areas are hardly apparent in terms of brightness differences in Lunar Orbiter V photography of the crater (fig. 25-102). The facies patterns outlined in reference 25-52 show correlations with the earthshine photography. Areas dominated by the inner concentric dune facies correspond to the bright bands on the rim of Aristarchus seen in earthshine photography. Darker areas are either (1) areas of concentric dune facies flooded between dunes by deposits of lower albedo or (2) areas on the rim and interior walls where postcrater or latest stage crater deposits, characterized by low-albedo flows, have obscured and blanketed underlying structures and facies. The enhancement of the radial patterns of distribution of these two facies suggests the possibility that structure may control the distribution of these latest blanketing deposits and that a radial fracture system may provide avenues for this material to reach the surface. Therefore, the earthshine photography of Aristarchus Crater may provide important data in explicating the relationships between the crater facies and younger deposits.

Conclusions

The predicted exposure of the maria areas and the obtained exposures were in reasonably close agreement. This agreement provides some rough confirmation of the assumptions and data used in the predictions. However, the difference between theory and results (a factor of approximately 2) is of scientific interest and merits further analysis.

The apparent-albedo values obtained for the floor of Aristarchus Crater were found to be approximately seven times greater than those for the maria. The ratio of 7 produces a computed apparent albedo of 0.63, which seems unnaturally high and which must be considered a preliminary result not yet analyzed in terms of appropriateness or inappropriateness as an indicator of normal albedo. This result merits further scientific analysis.

Preliminary analysis of the earthshine photography of the Aristarchus region shows the existence of new data relating to the evolution and modification of impact craters, to circular structures associated with sinuous rilles, and to the history of the plateau in general. Further analysis is being conducted on a continuing basis.

PART T

ASTRONOMICAL PHOTOGRAPHY

L. Dunkelman,[a][†] *R.D. Mercer,*[b][†] *C.L. Ross,*[c] *and A. Worden*[d]

In the Apollo 15 mission, the photographic observations of astronomical interest (other than lunar photography) conducted during the Apollo 14 mission (refs. 25-55 and 25-56) were continued and expanded. Included in the Apollo 15 mission were observations of solar corona, zodiacal light, the lunar libration region, the Milky Way near Delphinus, a total lunar eclipse, the optical environment in cislunar space, and lunar photography in earthshine (part S). A complete index of data frames of low-light-level exposures is in preparation and will be available from the author at the NASA Goddard Space Flight Center.

These investigations, although not formal experiments, were conducted under the guidance of the Apollo Orbital Science Photographic Team with the following goals in mind.

(1) To perform operationally needed tasks

(2) To obtain desired scientific data not practicably obtainable from other manned or unmanned spacecraft

(3) To take advantage of the unusual location, geometry, or ideal low-light-level optical environment extant in circumlunar missions

(4) To continue, under these highly favorable circumstances, the low-light-level astronomical observations begun in Project Mercury (ref. 25-57), expanded in the Gemini series (refs. 25-58 and 25-59) and previous Apollo flights (ref. 25-60), and conducted and reported as dim-light photography

(5) To use the capabilities of the $f/1.2$ 35-mm Nikon camera advantageously

As reported in reference 25-56, the Apollo 14 investigation in low-light-level astronomy was the first phase and was considered to be an operational test to

---

[a]NASA Goddard Space Flight Center.
[b]Dudley Observatory.
[c]High Altitude Observatory.
[d]NASA Manned Spacecraft Center.
[†]Investigator.

determine the feasibility of performing dim-light photography from the Apollo command module using the 16-mm data-acquisition camera equipped with a fast lens. Even though the format was generally too small, the camera provided an acceptable operational test. From this operational phase, it was learned that vehicle pointing and stability would permit long exposures, up to 4 min. The Apollo 15 mission time line, indeed, included many exposure requirements of 1 and 3 min and one at 4 min. The other significant difference between the Apollo 14 and 15 missions was that the Apollo 15 crew carried a Nikon camera, which operated entirely satisfactorily.

Solar Corona

Observations of the solar corona beyond 1.5 to 2.0 solar radii have been possible only during total eclipse when observations to approximately 5 solar radii are possible, or from balloons, rockets, or satellites that have extended the range to 8 to 10 solar radii. Observations were made from the lunar surface by Surveyor (refs. 25-61 and 25-62), and Bohlin of the Naval Research Laboratory infers coronal streamers to 15 solar radii in his analysis of the Surveyor VI and VII data. Coronal observations were therefore attempted on the Apollo 15 mission to provide further information on the two-dimensional form and radiance of the corona from 2 solar radii. The solar corona is the brightest of the several low-light-level phenomena photographed; typically, the radiance in terms of the mean solar disk falls off from 10^{-6} at the limb (1 radius) to 10^{-10} at 5 radii from Sun center.

The observing plan for the solar corona was designed to provide photographs starting at 70 sec before sunrise at which time the Sun was 3.5° below the limb of the Moon and extending to 10 sec before sunrise when the Sun is 0.5° below the lunar limb. All spacecraft control jets were inactivated except those required to maintain the lunar horizon in a fixed position as viewed from the spacecraft window. Exposures were then made every 10 sec during the period from 70 sec to 10 sec before sunrise. The lunar limb, therefore, occulted the solar corona at 14 solar radii when the photograph was taken 70 sec before sunrise, and the Sun effectively rose at 2 solar radii per picture through the sequence until the last picture with the Moon occulting at 2 solar radii. The field of view of the camera used was 37.9°; and, because approximately 2° of the field of view were blocked by the Moon, the total view of the corona extended to an elongation angle of 39° at 70 sec before sunrise and 36° at 10 sec before sunrise. To compensate for the wide dynamic range in the radiance of the corona, the exposure duration was varied from 1 sec at 70 sec before sunrise to 1/125 sec at 10 sec before sunrise. The sunrise sequence provided photographs of the eastern corona of the Sun, and the reverse procedure was accomplished starting at 10 sec after sunset to obtain photographs of the western corona of the Sun. Also, the sunrise sequence was repeated after 50° of solar rotation (4 days later) in an attempt to learn more of the three-dimensional structure of the corona.

Calibration for Solar Corona

Photometric calibration of the film was accomplished using a sensitometer box, developed by the High Altitude Observatory for the coronagraph experiment, on the Apollo telescope mount to be used on the Skylab missions. Reduction of the solar corona photographs will be accomplished by microdensitometer tracing and digitization to permit computer manipulation of the data. Ground-based observations of the corona were successful on both July 31 and August 4, 1971, from the High Altitude Observatory K-coronameter station at Mauna Loa, Hawaii. The K-coronameter observations provide radiance and polarization data from the solar limb to 1.5 to 2 radii from Sun center. By combining the Apollo 15 data and the K-coronameter data, it should be possible to map the form and radiance of the solar corona on July 31 from the solar limb to as much as 8° or 32 solar radii from Sun center. It should also be possible to map the interface of the solar corona and the zodiacal light.

Selected Preliminary Results

On July 31 at 17:30 G.m.t., the first sequence of sunrise photographs was started. The second picture in that sequence is shown in figure 25-103. The brightest star in the center of the frame is Regulus, approximately 20° from Sun center; just at the edge of the frame is Mercury at approximately 28° from Sun center; and the lesser stars form the head of the constellation Leo. It would appear that the zodiacal light is visible to some 15° to 20° elongation and the

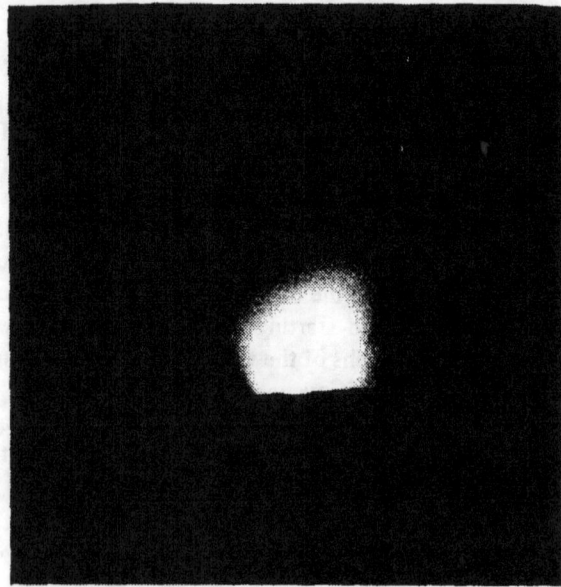

FIGURE 25-103.—The solar corona photographed approximately 1 min before sunrise on July 31 (AS15-98-13311).

FIGURE 25-104.—The solar corona photographed approximately 1 min after sunset on July 31 (AS15-98-13325).

solar corona goes in the other direction to possibly 8° or 10° from Sun center. The frame was taken 1 min before sunrise, so the Sun is approximately 3° below the lunar limb.

The final frame of the sunset sequence is shown in figure 25-104. The exposure has been estimated from the star streaking at approximately 9.2 sec. The brightest object at the end of the streak is Venus at approximately 3° from the lunar limb and 8° from Sun center. The bright star at the center of the frame is Pollux, and the other bright star near the edge of the frame is Castor. Quick-look analysis of this frame indicates a coronal streamer peaking out just below Venus or approximately 32 radii from Sun center, and the rest of the light continuing to the area of Pollux is K-corona or zodiacal light, depending on the desired frame of reference. The lunar surface is well illuminated by earthshine in this frame.

Some preliminary densitometry measurements of selected exposures and the calibration exposures reveal that the faintest surface luminance observable corresponded to one-tenth of the luminance of the night sky or approximately twenty 10th-magnitude stars per square degree, which corresponds to 2×10^{-19} cd/cm². This indicates that the f/1.2 Nikon camera with Ektachrome 2485 film is sufficiently sensitive for the performance of the scheduled Apollo 16 observations of the gegenschein/moulton region and other faint objects.

REFERENCES

25-1. Worden, A.M.; and El-Baz, Farouk: Apollo 15 in Lunar Orbit: Significance of Visual Observation and Photography. Abstracts with Programs. Geol. Soc. Am., vol. 3, no. 7, Oct. 1971, pp. 757-758.

25-2. Whitaker, E.A.: Sublimates. Section 2 of the Analysis of Apollo 8 Photography and Visual Observations. NASA SP-201, 1969, pp. 34-35.

25-3. Strom, R.G.; and Whitaker, E.A.: An Unusual Farside Crater. Analysis of Apollo 10 Photography and Visual Observations. NASA SP-232, 1971, p. 24.

25-4. El-Baz, Farouk: Light Colored Swirls in the Lunar Maria. Trans. Am. Geophys. Union, Fall Meeting 1971.

25-5. El-Baz, Farouk: Crater Characteristics. Section 2 of the Analysis of Apollo 8 Photography and Visual Observations. NASA SP-201, 1969, pp. 21-29.

25-6. Heacock, Raymond; Kuiper, Gerard P.; Shoemaker, Eugene; Urey, Harold; and Whitaker, E.A.: Interpretation of Ranger VII Records. Part II: Experimenters' Analyses and Interpretations. Calif. Inst. of Tech. JPL-TR-32-700, Feb. 10, 1965, pp. 149-154.

25-7. Wilhelms, D.E.; and McCauley, J.F.: Geologic Map of the Near Side of the Moon. U.S. Geol. Survey Misc. Geol. Inv. Map I-703, 1971.

25-8. Head, J.W.; and Lloyd, D.D.: Near-Terminator Photography. Section 18 of the Apollo 14 Preliminary Science Report, Part G. NASA SP-272, 1971.

25-9. Roosa, S.A.; and El-Baz, Farouk: Significant Results of Orbital Photography and Visual Observations on Apollo 14. Abstracts with Programs. Geol. Soc. Am., vol. 3, no. 7, Oct. 1971, pp. 687-688.

25-10. Anders, William A.; Lovell, James A.; and Borman, Frank: Visual Observations. Section 1 of the Analysis of Apollo 8 Photography and Visual Observations. NASA SP-201, 1969.

25-11. Stafford, T.P.; Cernan, E.A.; and Young, J.W.: Visual Observation. Analysis of Apollo 10 Photography and Visual Observations. NASA SP-232, 1971, pp. 1-4.

25-12. Shepard, Alan B.; Mitchell, Edgar D.; and Roosa, Stuart A.: Crew Observations. Section 2 of the Apollo 14 Preliminary Science Report. NASA SP-272, 1971.

25-13. Van Flandern, Thomas C.: Some Notes on the Use of the Watts Limb-Correction Charts. Astron. J., vol. 75, no. 6, Aug. 1970, pp. 744-746.

25-14. Baldwin, Ralph B.: Ancient Giant Craters and the Age of the Lunar Surface. Astron. J., vol. 74, no. 4, May 1969, pp. 570-571.

25-15. Carr, M.H.; Howard, K.A.; and El-Baz, Farouk: Geologic Maps of the Apennine-Hadley Region of the Moon: Apollo 15 Pre-Mission Maps. U.S. Geol. Survey Misc. Geol. Inv. Map I-723, 1971.

25-16. Swann, G.A.; Silver, L.T.; Sutton, R.L.; Schaber, G.G.; et al.: Preliminary Report on the Geology and Field Petrology at the Apollo 15 Landing Site. U.S. Geol. Survey Interagency Rept. 32, Aug. 5, 1971.

25-17. Hinners, N.W.: The New Moon: A View. Rev. of Geophys. and Space Phys., vol. 9, no. 3, Aug. 1971, pp. 447-522.

25-18. Hapke, Bruce W.: A Theoretical Photometric Function for the Lunar Surface. J. Geophys. Res., vol. 68, no. 15, Aug. 1, 1963, pp. 4571-4586.

25-19. Birkebak, R.C.; Cremers, C.J.; and Dawson, J.P.: Spectral Directional Reflectance of Lunar Fines as a Function of Bulk Density. Proceedings of the Second Lunar Science Conference, vol. 3, A.A. Levinson, ed., MIT Press (Cambridge, Mass.), 1971, pp. 2197-2202.

25-20. Hackman, R.J.: Geologic Map of the Montes Apenninus Region of the Moon. U.S. Geol. Survey Misc. Geol. Inv. Map I-463, 1966.

25-21. Greeley, Ronald: Lunar Hadley Rille: Considerations of Its Origin. Science, vol. 172, no. 3984, May 14, 1971, pp. 722-725.

25-22. Zisk, S.H.; Carr, M.H.; Masursky, H.; Shorthill, R.W.; and Thompson, T.W.: Lunar Apennine-Hadley Region: Geologic Implications of Earth-Based Radar and Infrared Measurements. Science, vol. 173, no. 3999, Aug. 27, 1971, pp. 808-812.

25-23. Wise, D.U.: Pseudo-Radar Topographic Shadowing for Detection of Sub-continental Sized Fracture Systems. Proceedings of the Sixth International Symposium on Remote Sensing of Environment, vol. 1, Univ. of Mich. Press (Ann Arbor, Mich.), 1969, pp. 603-615.

25-24. Gault, Donald E.: Saturation and Equilibrium Conditions for Impact Cratering on the Lunar Surface: Criteria and Implications. Radio Sci., vol. 5, no. 2, Feb. 1970, pp. 273-291.

25-25. Moore, H.J.: Geologic Interpretation of Lunar Data. Earth Sci. Rev., vol. 7, no. 1, Feb. 1971, pp. 5-33.

25-26. Muller, P.M.; and Sjogren, W.L.: Mascons: Lunar Mass Concentrations. Science, vol. 161, no. 3842, Aug. 16, 1968, pp. 680-684.

25-27. Pohn, Howard A.; and Wildey, Robert L.: A Photoelectric-Photographic Study of the Normal Albedo of the Moon. U.S. Geol. Survey Professional Paper 599-E, 1970.

25-28. Carr, M.H.: Geologic Map of the Mare Serenitatis of the Moon. U.S. Geol. Survey Misc. Geol. Inv. Map I-489, 1966.

25-29. El-Baz, Farouk: Geologic Characteristics of the Nine Lunar Landing Mission Sites Recommended by the Group for Lunar Exploration Planning. Bellcomm TR-68-340-1, May 31, 1968.

25-30. McCord, Thomas B.: Color Differences on the Lunar Surface. J. Geophys. Res., vol. 74, no. 12, June 15, 1969, pp. 3131-3142.

25-31. Zisk, S.H.: Zenith Azimuth Coordinates (ZAC) 4.04, Haystack Radar Map. MIT (Lincoln Laboratory), 1970.

25-32. Wilhelms, D.E.: Geologic Map of the Taruntius Quadrangle of the Moon. U.S. Geol. Survey Misc. Geol. Map I-722, 1971.

25-33. Cannon, Philip Jan: Lunar Landslides. Sky and Telescope, vol. 40, no. 4, Oct. 1970, pp. 215-218.

25-34. Strom, R.G.: Map, in Interpretation of Ranger VII Records. Part II: Experimenters' Analyses and Interpretations. Calif. Inst. of Tech. JPL-TR-32-700, Feb. 10, 1965, p. 32.

25-35. Wright, W.H.: The Moon as Photographed by Light of Different Colors. Pubs. Astron. Soc. Pacific, vol. 41, no. 241, 1929, p. 125.

25-36. Pohn, H.A.; Radin, H.W.; and Wildey, R.L.: The Moon's Photometric Function near Zero Phase Angle from Apollo 8 Photography. Astrophys. J., vol. 157, no. 3, Sept. 1969, pp. L193-L195.

25-37. Wildey, Robert L.; and Pohn, Howard A.: The Normal Albedo of the Apollo 11 Landing Site and Intrinsic Dispersion of the Lunar Heiligenschein. Astrophys. J., vol. 158, no. 2, Nov. 1969, pp. L129-L130.

25-38. Pohn, H.A.; Wildey, R.L.; and Offield, T.W.: Correlation of the Zero-Phase Brightness Surge (Heiligenschein) with Lunar Surface Roughness. Section 18 of the Apollo 14 Preliminary Science Report, Part F. NASA SP-272, 1971.

25-39. Wildey, Robert L.: Limited-Interval Definitions of the Photometric Functions of Lunar Crater Walls by Photography from Orbiting Apollo. Icarus, vol. 15, no. 1, Aug. 1971, pp. 93-99.

25-40. Moore, H.J.: Density of Small Craters on the Lunar Surface. Astrogeological Studies Annual Progress Report, Aug. 25, 1962–July 1, 1963, Part D: Studies for Space Flight Program. U.S. Geol. Survey Rept. 1964, pp. 34-51.

25-41. Shoemaker, Eugene M.: Preliminary Analysis of the Fine Structure of the Lunar Surface in Mare Cognitum. The Nature of the Lunar Surface, ch. 2, Wilmot N. Hess, Donald H. Menzel, and John A. O'Keefe, eds., Johns Hopkins Press, 1966, pp. 23-77.

25-42. Marcus, Allan: A Stochastic Model of the Formation and Survival of Lunar Craters. Icarus, vol. 3, no. 5 and 6, Dec. 1964, pp. 460-472.

25-43. Ross, Howard P.: A Simplified Mathematical Model for Lunar Crater Erosion. J. Geophys. Res., vol. 73, no. 4, Feb. 15, 1968, pp. 1343-1354.

25-44. Soderblom, Laurence A.: A Model for Small-Impact

Erosion Applied to the Lunar Surface. J. Geophys. Res., vol. 75, no. 14, May 10, 1970, pp. 2655-2661.

25-45. Shoemaker, E.M.; Morris, E.C.; Batson, R.M.; Holt, H.E.; et al.: Television Observations from Surveyor. Surveyor Program Results Final Report. NASA SP-184, 1969.

25-46. Trask, Newell J.: Size and Spatial Distribution of Craters Estimated from Ranger Photographs: Rangers VIII and IX. Part II: Experimenters' Analyses and Interpretations. Calif. Inst. of Tech. JPL-TR-32-800, March 15, 1966, pp. 252-263.

25-47. Anon.: Kodak 2485 High Speed Recording Film, Eastman Kodak Company Data Release.

25-48. Lloyd, D.D.: Near Terminator Photography (Within 1/2°) Obtained on Apollo 14. Bellcomm TM-71-2015-1, Apr. 20, 1971.

25-49. Anon.: Natural Environment and Physical Standards for the Apollo Program and the Apollo Applications Program. NASA M-D E 8020.008C, July 10, 1969, p. 5-2.

25-50. Moore, H.J.: Geological Map of the Aristarchus Region of the Moon. U.S. Geol. Survey Misc. Geol. Inv. Map I-465, 1965.

25-51. Moore, H.J.: Geological Map of the Seleucus Quadrangle of the Moon. U.S. Geol. Survey Misc. Geol. Inv. Map I-527, 1965.

25-52. Head, James W.: Distribution and Interpretation of Crater-Related Facies: The Lunar Crater Aristarchus. Trans. Am. Geophys. Union, vol. 50, no. 11, Nov. 1969, p. 637.

25-53. Hartmann, William K.: Lunar Crater Counts, VI, The Young Craters, Tycho, Aristarchus, and Copernicus. Commun. Lunar and Planet. Lab., vol. 7, part 3, 1968, pp. 145-156.

25-54. McCord, T.B.; Charette, M.P.; Johnson, T.V.; Lebofsky, L.A.; et al.: Lunar Spectral Types. J. Geophys. Res., vol. 77, 1972.

25-55. Dunkelman, L.; Wolff, C.L.; Mercer, R.D.; and Roosa, S.A.: Gegenschein-Moulton Region Photography From Lunar Orbit. Section 15 of the Apollo 14 Preliminary Science Report. NASA SP-272, 1971.

25-56. Mercer, R.D.; Dunkelman, L.; and Roosa, S.A.: Astronomical Photography. Section 18 of the Apollo 14 Preliminary Science Report, Part H. NASA SP-272, 1971.

25-57. Anon.: Mercury Project Summary Including Results of the Fourth Manned Orbital Flights, May 15 and 16, 1963. NASA SP-45, 1963.

25-58. Dunkelman, L.; Gill, J.R.; McDivitt, J.A.; Roach, F.E.; and White, E.H., II: Geo-Astronomical Observations. Washington Manned Space Flight Experiment Symposium, Gemini Missions III and IV, 1965, pp. 1-18.

25-59. Cameron, W.S.; Dunkelman, L.; Gill, J.R.; and Lowman, P.D., Jr.: Man in Space. Introduction to Space Science, second ed., ch. 14, Wilmot N. Hess and Gilbert D. Mead, eds., Gordon and Breach Sci. Pub., 1968.

25-60. Anon.: Analysis of Apollo 8 Photography and Visual Observations. NASA SP-201, 1969.

25-61. Norton, Robert H.; Gunn, James E.; Livingston, W.C.; Newkirk, G.A.; and Zirin, H.: Surveyor 1 Observations of the Solar Corona. J. Geophys. Res., vol. 72, no. 2, Jan. 15, 1967, pp. 815-817.

25-62. Bohlin, J. David: Photometry of the Outer Solar Corona from Lunar-Based Observations. Solar Phys., vol. 18, no. 3, July 1971, pp. 450-457.

APPENDIX A
Glossary

ablate—to carry away aerodynamically generated heat by arranging for absorption in a nonvital part that can fall away or vaporize

achondrite—a stony meteorite devoid of rounded granules

agglutinate—a deposit of originally molten ejecta

albedo—the percentage of the incoming radiation that is reflected by a natural surface

anhedral—pertaining to mineral grains that lack external crystals

anomaly—an area of a geophysical survey (e.g., a magnetic or gravitational survey) that is different in appearance from the survey in general

anorthite—a calcium-rich variety of plagioclase feldspar

anorthosite—a granular, plutonic, igneous rock composed almost exclusively of a soda-lime feldspar

apatite—any of a group of calcium-phosphate minerals that occur variously as hexagonal crystals, as granular masses, or in fine-grained mass as the chief constituent of phosphate rock

aphanite—a dark rock of such close texture that the individual grains are invisible to the unaided eye

aphyric—not having distinct crystals

apolune—the orbital point farthest from the Moon, when the Moon is the center of attraction

augite—one of a variety of pyroxene minerals that contain calcium, magnesium, and aluminum; usually black or dark green in color

bit—abbreviation of binary digit; a quantum of information

bleb—a small particle of distinctive material

bow shock—a shock wave in front of a body, such as an airfoil

breccia—a rock consisting of sharp fragments embedded in a fine-grained matrix

Brewster angle—the angle of incidence for which a wave polarized parallel to the plane of incidence is wholly transmitted

bytownite—a calcium-rich variety of plagioclase feldspar

chondrite—a meteoritic stone characterized by the presence of rounded granules

chromosphere—a thin layer of relatively transparent gases above the photosphere of the Sun

cislunar—pertaining to the space between the Earth and the orbit of the Moon

clast—a discrete particle or fragment of rock or mineral; commonly included in a larger rock

clinopyroxene—a mineral that occurs in monoclinic, short, thick, prismatic crystals and that varies in color from white to dark green or black (rarely blue)

coherent—a term used to describe two or more parts of the same series that are in contact more or less adhesively but are not fused

collimator—an optical device that renders rays of light parallel

comminution—the reduction of a substance to a fine powder

conchoidal—a term used to describe a shell-like surface shape that has been produced by the fracturing of a brittle material

cristobalite—an isometric variety of quartz that forms at high temperatures (SiO_2)

Curie temperature—the temperature in a ferromagnetic material above which the material becomes substantially nonmagnetic

dendrite—a crystallized arborescent form

devitrification—the change of a glassy rock from the glassy state to a crystalline state after solidification

diamagnetic—pertaining to substances having a permeability less than that of a vacuum

dielectric constant—a measure of the amount of electrical charge a given substance can withstand at a given electric field strength

dunite—a peridotite that consists almost entirely of olivine and that contains accessory chromite and pyroxene

ephemeris—a tabulation of the predicted positions of celestial bodies at regular intervals

epicenter—the point on a planetary surface directly above the focus of an earthquake

eucrite—a meteorite composed essentially of feldspar and augite

euhedral—pertaining to minerals the crystals of which have had no interference in growth

exsolution—unmixing; the separation of some mineral-pair solutions during slow cooling

fayalite—an iron-rich variety of olivine (Fe_2SiO_4)

feldspar—a group of abundant rock-forming minerals

flux—the rate of flow of some quantity, as energy or gas molecules

fractional process—separation of a substance from a mixture (e.g., one isotope from another of the same element)

Fresnel zone—any one of the array of concentric surfaces in space between transmitter and receiver over which the increase in distance over the straight line path is equal to some multiple of one-half wavelength

gabbro—a granular igneous rock of basaltic composition with a coarse-grained texture

Gaussian distribution—normal statistical distribution

gegenschein—a round or elongated spot of light in the sky at a point 180° from the Sun

holocrystalline—consisting wholly of crystals

hypabyssal—pertaining to minor intrusions, such as sills and dikes, and to the rocks that compose them

ilmenite—a mineral rich in titanium and iron; usually black with a submetallic luster

indurated—a term used to describe masses that have been hardened by heat; baked

intersertal—a term used to describe the texture of igneous rocks in which a base or mesostasis of glass and small crystals fills the interstices between unoriented feldspar laths

isostatic—subjected to equal pressure from all sides

lamella—a layer of a cell wall

lath—a long, thin mineral crystal

leucocratic—a term used to describe light-colored rock, especially igneous rocks that contain between 0 and 30 percent dark minerals

limb—the edge of the apparent disk of a celestial body

lithic—of, relating to, or made of stone

lithification—consolidation and hardening of fines into rock

lithology—the physical character of a rock, as determined with the unaided eye or with a low-power magnifier

magcon—a magnetized concentration

magma—molten rock material that is liquid or pasty

magnetopause—the outer boundary of the magnetosphere

magnetosheath—the transition region between the magnetopause and the solar-wind shock wave

magnetosphere—the region of the atmosphere where the geomagnetic field plays an important role; the magnetosphere extends to the boundary between the atmosphere and interplanetary plasma

magnetotail—a portion of the magnetic field of the Earth that is pulled back to form a tail by solar plasma

mascon—a large mass concentration beneath the surface of the Moon

maskelynite—a feldspar found in meteorites

massif—a mountainous mass

melanocratic—a term used to describe dark-colored rocks, especially igneous rocks that contain between 60 and 100 percent dark minerals

metamorphic—a term used to describe rocks that have formed in a solid state as a result of drastic changes in temperature, pressure, and chemical environment

microlite—small lath-shaped minerals, commonly plagioclase feldspar, occurring as minute phenocrysts in basalt

morphology—the study of the shape or form of geologic features

mosaic—a term used to describe the texture sometimes seen in dynamo-metamorphosed rocks that have angular and granular crystal fragments and that appear like a mosaic in polarized light

multiplex—to transmit two or more signals simultaneously within a single channel

norite—a type of gabbro in which orthopyroxene is dominant over clinopyroxene

olivine—an igneous mineral that consists of a silicate of magnesium and iron

ophitic—a rock texture characterized by lath-shaped plagioclase crystals enclosed in augite

peridotite—an essentially nonfeldspathic plutonic rock consisting of olivine, with or without other dark minerals

perilune—the orbital point nearest the Moon, when the Moon is the center of attraction

permeability—the ratio of the magnetic induction to the magnetic-field intensity in the same region

phenocryst—a large crystal of the earliest generation in a porphyritic igneous rock

photogrammetry—the science of obtaining reliable measurements by means of photography

photosphere—the intensely bright portion of the Sun visible to the unaided eye

pigeonite—a variety of pyroxene

plagioclase—a feldspar mineral composed of varying amounts of sodium and calcium with aluminum silicate

plasma—an electrically conductive gas; specifically a mass of ionized gas flowing out of the Sun

plutonic—pertaining to igneous rock that crystallizes at depth

poikilitic—a term used to describe the condition in which small granular crystals are irregularly scattered without common orientation in a larger crystal of another mineral

polarimetry—the measurement of the angle of rotation of linearly polarized light

poloidal—geometric shape of a dipole magnetic field

porphyritic—having larger crystals set in a finer groundmass

pyroxene—a mineral occurring in short, thick, prismatic crystals or in square cross section; often laminated; and varying in color from white to dark green or black (rarely blue)

rarefaction wave—a wave in a compressible fluid such that when a fluid particle crosses the wave in the direction of its motion, the density and pressure of the particle decrease

regolith—the layer of fragmental debris that overlies consolidated bedrock

remanent—pertaining to the residual induction when the magnetizing field is reduced to zero from a value sufficient to saturate the material

schlieren—tabular bodies that occur in pluton, generally several centimeters to several meters long

slickenside—a polished and striated surface that results from friction along a fault plane

solar wind—streams of plasma flowing outward from the Sun

spall—a relatively thin, sharp-edged piece of rock that has been produced by exfoliation

specular reflection—reflection in which the reflected radiation is not diffused

spinel—a mineral that is noted for great hardness ($MgAl_2O_4$)

sporadic—a meteor which is not associated with one of the regularly recurring meteor showers or streams

subhedral—pertaining to minerals that are intermediate between anhedral and euhedral

talus—a collection of fallen disintegrated material that has formed a slope at the foot of a steeper declivity

tectonism—crustal instability; the structural behavior of an element of the crust of the Earth during or between major cycles of sedimentation

tephra—a collective term for all clastic volcanic materials that are ejected from the volcano and transported through the air

terminator—the line separating illuminated and dark portions of a celestial body

translunar—outside the orbit of the Moon

transponder—a combined receiver and transmitter which transmits signals automatically when triggered by an interrogator

troilite—a mineral that is native ferrous sulfide

vermicular—a term used to describe a group of platy minerals that are closely related to the chlorites and montmorillonites

vesicle—a small cavity in a mineral or rock, ordinarily produced by expansion of vapor in a molten mass

vug—a small cavity in a rock

zircon—a mineral, $ZrSiO_4$; the main ore of zirconium

zodiacal light—a faint cone of light extending upward from the horizon in the direction of the ecliptic

APPENDIX B
Acronyms

AEI—aerial exposure index
ALSD—Apollo lunar-surface drill
ALSEP—Apollo lunar surface experiments package
amu—atomic mass unit
ASE—active seismic experiment
CCEM—continuous-channel electron multiplier
CCGE—cold cathode gage experiment
CDR—commander
CM—command module
CMP—command module pilot
CSM—command-service module
DAC—data-acquisition camera
dc—direct current
DU—digital units
ESRO—European Space Research Organization
e.s.t.—eastern standard time
EVA—extravehicular activity
FOV—field of view
FWHM—full width half maximum
GET—ground elapsed time
G.m.t.—Greenwich mean time
HEDC—Hasselblad electric data camera
HFE—heat-flow experiment
KREEP—potassium, rare-Earth elements, and phosphorus
LCP—left circular polarization
LCRU—lunar communications relay unit
LM—lunar module
LMP—lunar module pilot
LP—long-period
LRL—Lunar Receiving Laboratory, NASA Manned Spacecraft Center
LRRR—laser ranging retroreflector
LRV—lunar roving vehicle (Rover)
LSM—lunar-surface magnetometer
LSPET—Lunar Sample Preliminary Examination Team
LSS—lunar-soil simulant
MSC—Manned Spacecraft Center
PSD—pulse-shape discriminator
PSE—passive seismic experiment
RCP—right circular polarization
RCS—reaction control system
rms—root mean square
SCB—sample container bag
SEM—scanning electron microscope
SESC—surface environmental sample container
SEVA—standup extravehicular activity
SIDE—suprathermal ion detector experiment
SIM—scientific instrument module
SM—service module
SP—short-period
SRP—self-recording penetrometer
SWC—solar-wind composition
SWS—solar-wind spectrometer
TEC—transearth coast
TEI—transearth injection
TV—television
uv—ultraviolet
UHURU—Explorer 42
V/h—velocity/height ratio
VHF—very high frequency
WES—Waterways Experiment Station (U.S. Army Engineers)

APPENDIX C
Units and Unit-Conversion Factors

In this appendix are the names, abbreviations, and definitions of International Systems (SI) units used in this report and the numerical factors for converting from SI units to more familiar units.

Names of International Units Used in This Report

Physical quantity	Name of unit	Abbreviation	Definition of abbreviation
Basic Units			
Length	meter	m	
Mass	kilogram	kg	
Time	second	sec	
Electric current	ampere	A	
Temperature	kelvin	K	
Luminous intensity	candela	cd	
Derived Units			
Area	square meter	m^2	
Volume	cubic meter	m^3	
Frequency	hertz	Hz	sec^{-1}
Density	kilogram per cubic meter	kg/m^3	
Velocity	meter per second	m/sec	
Angular velocity	radian per second	rad/sec	
Acceleration	meter per second squared	m/sec^2	
Angular acceleration	radian per second squared	rad/sec^2	
Force	newton	N	$kg \cdot m/sec^2$
Pressure	newton per square meter	N/m^2	
Work, energy, quantity of heat	joule	J	$N \cdot m$
Power	watt	W	J/sec
Voltage, potential difference, electromotive force	volt	V	W/A
Electric field strength	volt per meter	V/m	
Electric resistance	ohm	Ω	V/A
Electric capacitance	farad	F	$A \cdot sec/V$
Magnetic flux	weber	Wb	$V \cdot sec$
Inductance	henry	H	$V \cdot sec/A$
Magnetic flux density	tesla	T	Wb/m^2
Magnetic field strength	ampere per meter	A/m	
Luminous flux	lumen	lm	$cd \cdot sr$
Luminance	candela per square meter	cd/m^2	
Illumination	lux	lx	lm/m^2
Specific heat	joule per kilogram kelvin	$J/kg \cdot K$	
Thermal conductivity	watt per meter kelvin	$W/m \cdot K$	

Supplementary Units

Plane angle	radian	rad
Solid angle	steradian	sr

Unit Prefixes

Prefix	Abbreviation	Factor by which unit is multiplied
giga	G	10^9
mega	M	10^6
kilo	k	10^3
centi	c	10^{-2}
milli	m	10^{-3}
micro	μ	10^{-6}
nano	n	10^{-9}

Unit-Conversion Factors

To convert from –	To –	Multiply by –
ampere/meter	oersted	1.257×10^{-2}
candela/meter^2	foot-lambert	2.919×10^{-1}
candela/meter^2	lambert	3.142×10^{-4}
joule	British thermal unit (International Steam Table)	9.479×10^{-4}
joule	Calorie (International Steam Table)	2.388×10^{-1}
joule	electron volt	6.242×10^{18}
joule	erg	1.000×10^{7} [a]
joule	foot-pound force	7.376×10^{-1}
joule	kilowatt-hour	2.778×10^{-7}
joule	watt-hour	2.778×10^{-4}
kelvin	degrees Celsius (temperature)	$t_C = t_K - 273.15$
kelvin	degrees Fahrenheit (temperature)	$t_F = 9/5\, t_K - 459.67$
kilogram	gram	1.000×10^{3} [a]
kilogram	kilogram mass	1.000×10^{3} [a]
kilogram	pound mass (pound mass avoirdupois)	2.205×10^{0}
kilogram	slug	6.852×10^{-2}
kilogram	ton (short, 2000 pound)	1.102×10^{-3}
lumen/meter^2	foot-candle	9.290×10^{-2}
lumen/meter^2	lux	1.000×10^{0} [a]
meter	angstrom	1.000×10^{10} [a]
meter	foot	3.281×10^{0}
meter	inch	3.937×10^{1}
meter	micron	1.000×10^{6} [a]
meter	mile (U.S. statute)	6.214×10^{-4}
meter	nautical mile (international)	5.400×10^{-4}
meter	nautical mile (U.S.)	5.400×10^{-4}
meter	yard	1.094×10^{0}
meter/second^2	foot/second^2	3.281×10^{0}
meter/second^2	inch/second^2	3.937×10^{1}
newton	dyne	1.000×10^{5} [a]
newton	kilogram force (kgf)	1.020×10^{-1}
newton	pound force (avoirdupois)	2.248×10^{-1}
newton/meter^2	atmosphere	9.870×10^{-6}
newton/meter^2	centimeter of mercury (0° C)	7.501×10^{-4}
newton/meter^2	inch of mercury (32° F)	2.953×10^{-4}

APPENDIX C

Unit-Conversion Factors—Concluded

To convert from –	To –	Multiply by –
newton/meter^2	inch of mercury (60° F)	2.961 × 10^{-4}
newton/meter^2	millimeter of mercury (0° C)	7.501 × 10^{-3}
newton/meter^2	torr (0° C)	7.501 × 10^{-3}
radian	degree (angle)	5.730 × 10^1
radian	minute (angle)	3.438 × 10^3
radian	second (angle)	2.063 × 10^5
tesla	gamma	1.000 × 10^9 [a]
tesla	gauss	1.000 × 10^4 [a]
watt	British thermal unit (thermochemical)/second	9.484 × 10^{-4}
watt	calorie (thermochemical)/second	2.390 × 10^{-1}
watt	foot-pound force/second	7.376 × 10^{-1}
watt	horsepower (550 foot-pound force/second)	1.341 × 10^{-3}
weber	maxwell	1.000 × 10^8 [a]

[a] An exact definition.

APPENDIX D

Lunar-Surface Panoramic Views

LUNAR-SURFACE PANORAMIC VIEWS

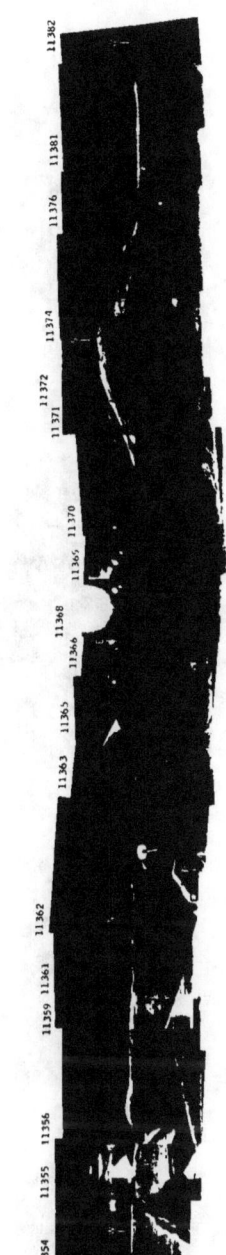

Figure D-1. Vertical stereophotographs taken during standup EVA. (a) Upper portion (AS15-85-11354 to 11382). (b) Lower portion (AS15-87-11730 to 11758).

LUNAR SURFACE PANORAMIC VIEWS

Figure D-2.— Panoramic views taken in vicinity of LM. (a) Northwest of LM (AS15-87-11785 to 11804). (b) Northeast of LM (AS15-87-11805 to 11821).

LUNAR SURFACE PANORAMIC VIEWS

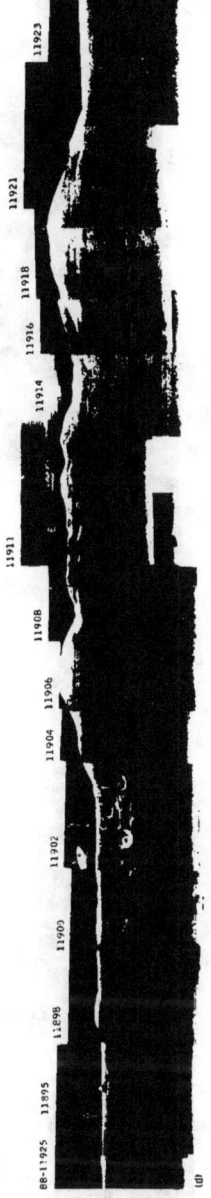

Figure D-2 (continued). - (c) Southeast of LM (AS15-87-11822 to 11840). (d) At point 100 m east of LM (AS15-88-11895 to 11925).

LUNAR-SURFACE PANORAMIC VIEWS

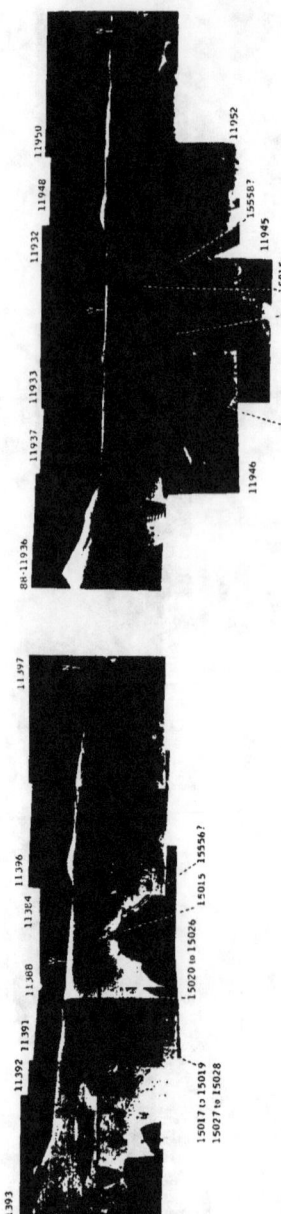

Figure D-3.- Panoramic views taken through LM windows. (a) View through left window before EVA-1 (AS15-85-11384 to 11397). (b) View through right window after EVA-3 (AS15-88-11932 to 11952).

Figure D-4.- Panoramic view of station 1 (AS15-85-11398 to 11415).

LUNAR-SURFACE PANORAMIC VIEWS

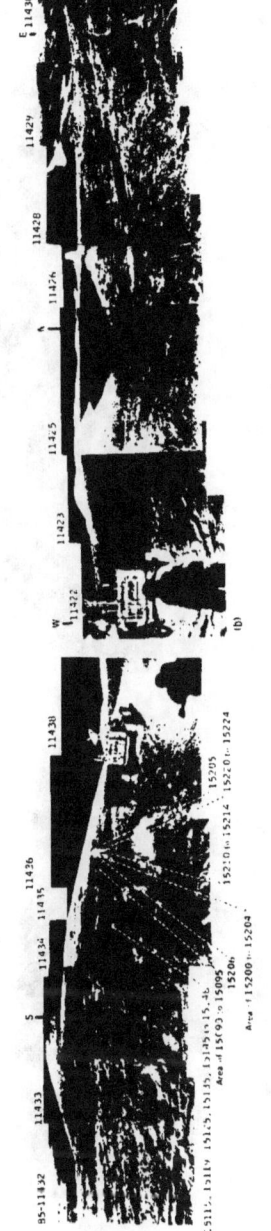
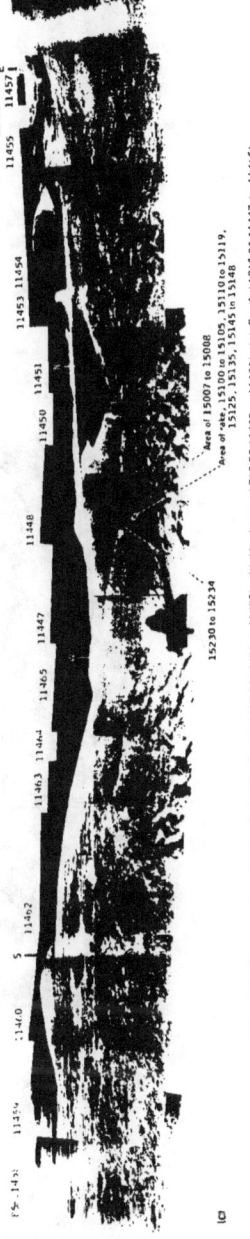

Figure D-5.—Panoramic views of station 2. (a) Northeast (AS15-85-11432 to 11438). (b) Northwest (AS15-85-11422 to 11431). (c) South (AS15-85-11447 to 11465).

LUNAR-SURFACE PANORAMIC VIEWS

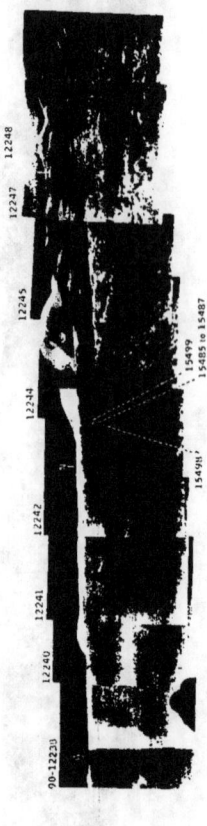

Figure D-6.— Panoramic view of station 4 (AS15-90-12238 to 12248).

Figure D-8.— Panoramic view from Rover of area between LM and station 6 (AS15-85-11473 to 11480).

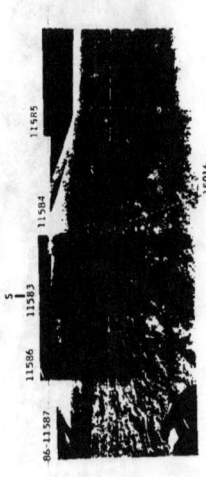

Figure D-7.— Panoramic view of station 3 (AS15-86-11583 to 11587).

LUNAR-SURFACE PANORAMIC VIEWS

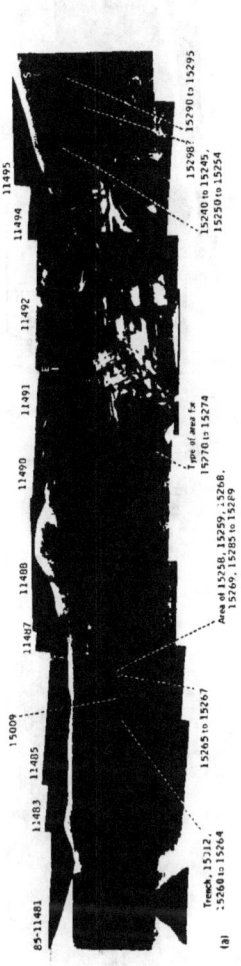

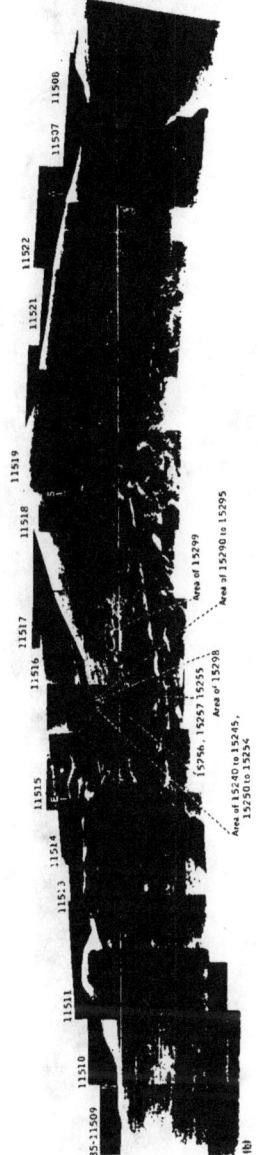

Figure D-9.- Panoramic views of Station 6. (a) East (AS15-85-11481 to 11495). (b) West (AS15-85-11507 to 11522).

LUNAR-SURFACE PANORAMIC VIEWS

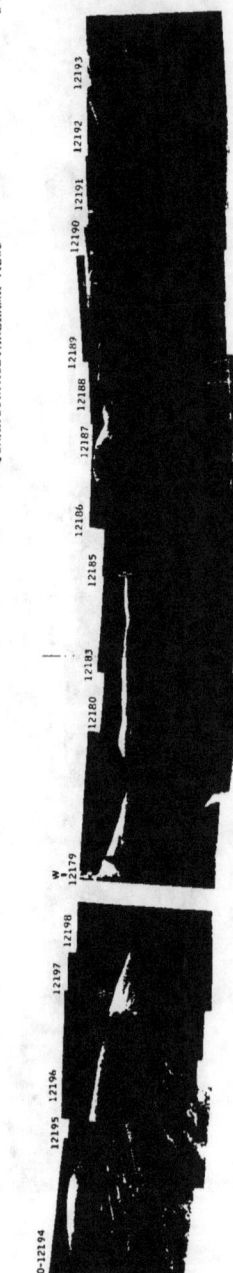

Figure D-10.—Panoramic view of station 6A (AS15-90-12179 to 12198).

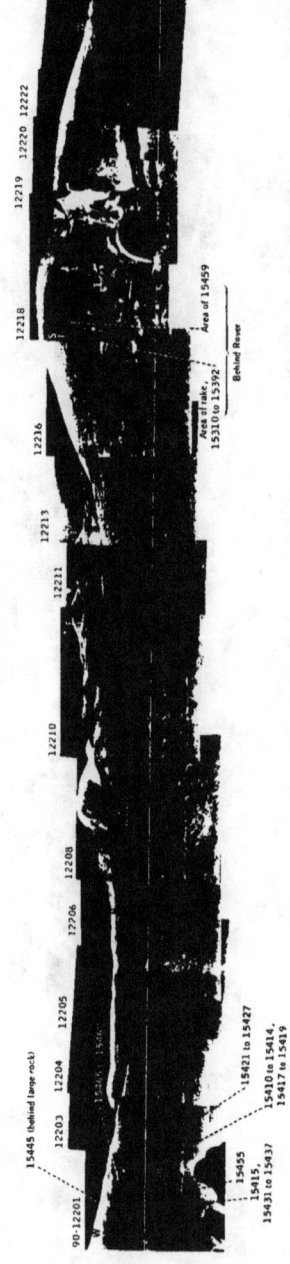

Figure D-11.—Panoramic view of station 7 (AS15-90-12201 to 12222).

LUNAR-SURFACE PANORAMIC VIEWS

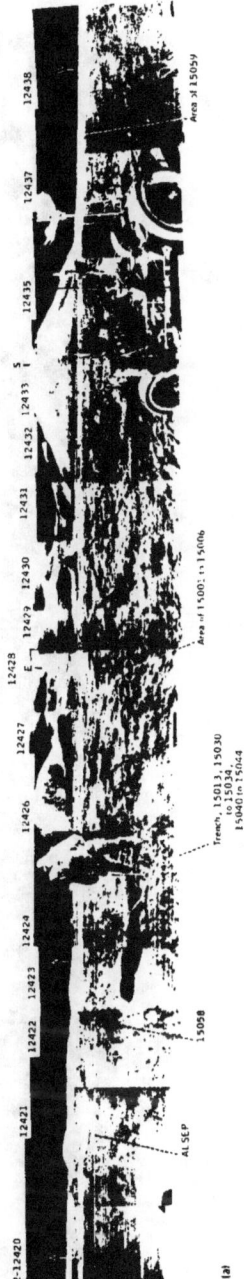

Figure D-12.- Panoramic views of station 8. (a) At end of EVA-2 (AS15-92-12420 to 12438). (b) At beginning of EVA-3 (AS15-82-11050 to 11064 and AS15-88-11879 to 11881).

LUNAR SURFACE PANORAMIC VIEWS

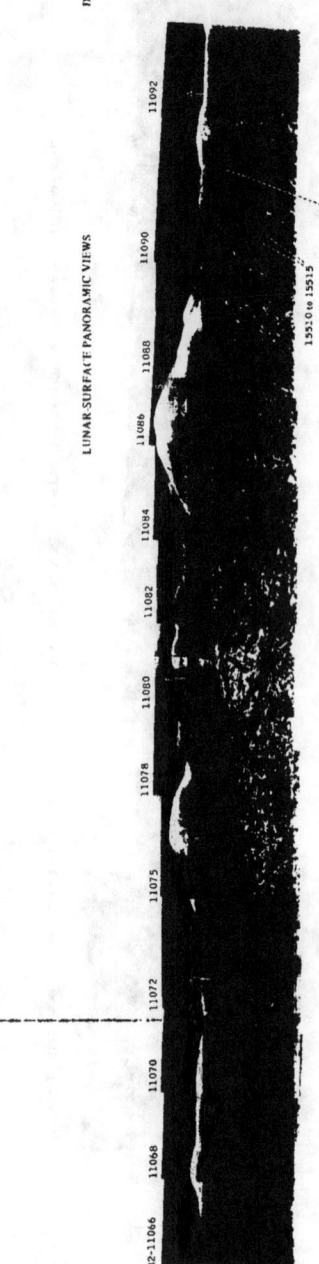

Figure D-13.- Panoramic view of station 9 (AS15-82-11066 to 11092).

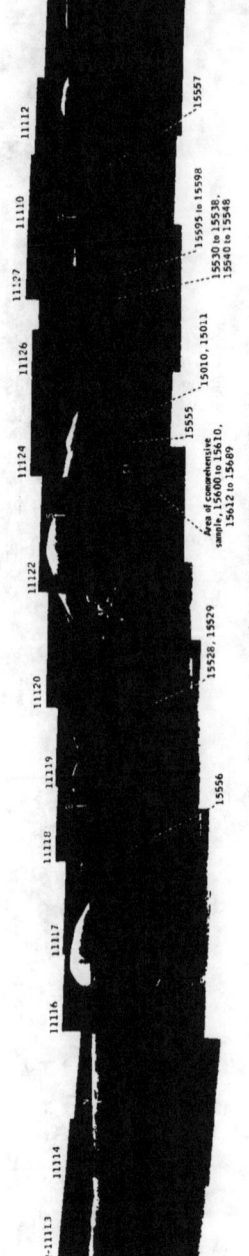

Figure D-14.- Panoramic view of station 9A (AS15-82-11110 to 11127).

LUNAR SURFACE PANORAMIC VIEWS

Figure D-15.—Panoramic view of station 10 (AS15-82-11166 to 11184).

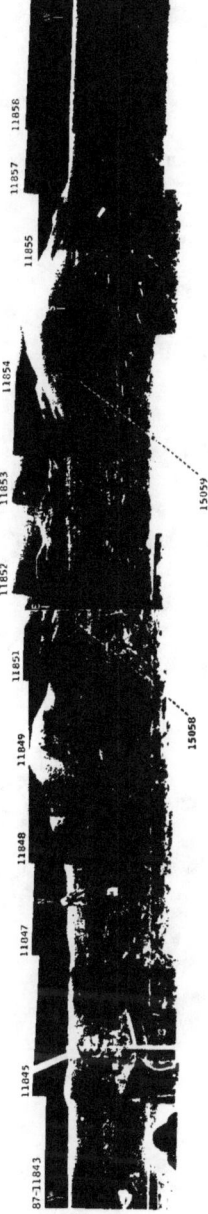

Figure D-16.—Panoramic view of ALSEP area (AS15-87-11843 to 11858).

www.ingramcontent.com/pod-product-compliance
Lightning Source LLC
Chambersburg PA
CBHW081713170526
45167CB00009B/3569